The DIGITAL HAND

JAMES W. CORTADA

The DIGITAL HAND

Volume II

How Computers Changed the Work of American Financial, Telecommunications, Media, and Entertainment Industries

OXFORD
UNIVERSITY PRESS
2006

OXFORD
UNIVERSITY PRESS

Oxford University Press, Inc., publishes works that further
Oxford University's objective of excellence
in research, scholarship, and education.

Oxford New York
Auckland Cape Town Dar es Salaam Hong Kong Karachi
Kuala Lumpur Madrid Melbourne Mexico City Nairobi
New Delhi Shanghai Taipei Toronto

With offices in
Argentina Austria Brazil Chile Czech Republic France Greece
Guatemala Hungary Italy Japan Poland Portugal Singapore
South Korea Switzerland Thailand Turkey Ukraine Vietnam

Published by Oxford University Press, Inc.
198 Madison Avenue, New York, New York 10016

www.oup.com

Library of Congress Cataloging-in-Publication Data

Cortada, James W.
The digital hand. How computers changed the work of American financial, telecommunications, media, and entertainment industries / James W. Cortada.
p. cm
Includes bibliographical references and index.
ISBN–10: 0–19–516587–X
ISBN–13: 978–0–19–516587–6
1. Information technology—Economic aspects—United States—Case studies. 2. Technological innovations—Economic aspects—United States—Case studies. 3. Business—Data processing—Case studies. I. Title: How computers changed the work of American financial, telecommunications, media, and entertainment industries. II. Title.
HC110.I55C67 2005
338'.064'0973—dc22 2004030363

9 8 7 6 5 4 3 2 1

Printed in the United States of America
on acid-free paper

To three great editors who taught me how to write books, then published them:

James Sabin
Paul Becker
Herb Addison

PREFACE

It is always of the utmost importance for us to be thoroughly masters of the economic history of the time, the country or the industry, sometimes even of the individual firm in question, before we draw any inference at all from the behavior of time series.

—Joseph A. Schumpeter, 1939

The great economist quoted above points the way for the search in understanding how modern economies work. He did more than most economists to call attention to the growing influence of technology on business and economic circumstances. This book takes one primary collection of technologies (computing) and a secondary set (telecommunications) and explores the path Schumpeter urged us to take because use of these methods changed profoundly how people did their daily work across the American economy in the second half of the twentieth century. Computers, in particular, influenced enormously how companies, and the industries in which they resided, functioned. Conversely, industries played a much greater role in the deployment of computers in the economy than was thought even just a few years ago. The implications of these findings are important both for scholars of the modern American economy and for managers working within it. This book describes how computers were used in a variety of industries, how they influenced companies and industries over time, and what, ultimately, were the technological "sweet spots."

This book is part of a larger project to document the role and effects of computing across the American economy over the latter part of the twentieth century. In the first volume, *The Digital Hand: How Computers Changed the Work of American Manufacturing, Transportation, and Retail Industries* (2004), I argue that information technology (IT) profoundly influenced the daily activities of a large collection of industries. I focus on the physical economy—that is to say, on the manufacture, movement, and sale of goods. In the course of looking at 16 industries, I describe how digital technologies so profoundly changed the nature of work that, in fact, a whole new style of doing business had emerged by the end of the century. Yet it is a story not complete, because it is still unfolding at the dawn of the new century. In this second book, I turn our attention to other sectors that have traditionally relied more on selling information rather than goods, and that also were early adopters of a variety of digital and telecommunication tools. As in the earlier book, I contend that the influence of the digital hand and its effects on economic activity were and are as important as management's actions on the economy and a natural extension of its role. Using the tools of the historian to write a narrative account, I demonstrate that process at work in the dozen industries described in this book. For the techniques used, and the rationale behind them, I encourage the reader to go to Appendix A of the earlier book.

My long-term objective is to study enough industries to be able to generalize, with confidence, about the role of computing in 80 percent or more of the economy. I intend to accomplish this task across three books. The first concerned the physical economy, this one more of the information-centered portion of the economy, and the third will look at the public sector (for example, government and education). I am learning through research across all three books that there are many common patterns of behavior that appear in scores of industries. But, just as the first book had to stand on its own, so too must this one for the industries it describes.

The industries selected for this current volume—from the financial, telecommunications, media and entertainment sectors of the economy—all share several common features that led to a natural grouping. First, the industries are extensive users of telephonic technologies. Second, they use a variety of computer technologies and have done so for the bulk of the period we are examining. Third, their central products are information, data, and experiences—not goods (such as cars or groceries)—and their primary inventory is information. Banks, for instance, keep more data about dollars than actual cash on hand; radio stations transmit information over the airwaves; while a phone company makes it possible for you and me to distribute both conversation and data. Fourth, they are increasingly being seen as core pillars of what has come to be called the Information Age or Information Society, so we need to understand in some detail the makeup of those pillars.

Fifth, and perhaps most subtly, these industries are not so much merging together, as many have argued, as they are increasing the interactions among them by using digital technologies. If that trend continues, one can easily expect that we will have moved to a postindustrial economy, one that also includes the manufacture, distribution, and sale of goods, not just the movement of information. As I demonstrated in the earlier volume and reinforce in this one, the emerging postindustrial economy is not one that is reducing its output of goods. Rather, the "new" economy continues to play out its historic role of providing the physical necessities and wants of life at the same time that it also expands human interactions considered vital to the functions of an economy and a society at large. The result is less a retreat from the role of the "old" economy than an evolution to a new style of accomplishing its historic purposes. To be sure, how work was done in 2000 differed in many ways from how it was done in 1950, but business agendas had not. Banks still had checking accounts, made loans, worried about bad credit risks, and served as the nation's channel of distribution for cash from the U.S. government to the public at large. Insurance companies still sold insurance and made profits. Radio and television stations still informed and entertained.

As in the earlier book, I look at how computers were used in specific industries over the last half of the twentieth century, again focusing on the American economy since it was the earliest adopter of the computer. I look at how management came to use computers and the influence the computers had on the industries. To demonstrate how profoundly computers affected the nature of work, I discuss how applications were extensively deployed within industries and subsequently brought about adoptions or changes at the industry and firm levels.

A subtext of this book is the lack of a computer revolution and indeed the digital evolution that occurred. In one industry after another, firms and their management teams incrementally adopted computers; over time, the accumulated uses and transformations brought about by these millions of individual adoptions constituted a metamorphosis so great that one could easily jump to the conclusion that a revolution had occurred. But the insight about how the change occurred has much to teach us about how to embrace a new technology and how, in fact, managers and their firms and industries change over time. Thus, we can begin to move from generalities and hype to a more solidly grounded set of operating principles and insights about change, innovation, and transformations in the modern economy at three levels: firm, industry, and national economy.

The case for looking at the adoption of the computer on an industry-by-industry basis was explained in the earlier book. To summarize: companies looked to their industry's publications, conferences, members, and competitors to determine how they should use computers and other resources. These other resources included management practices and changes in government regulations, because technology by itself was never all-pervasive, operating in some abstract way, but rather part of a larger mosaic of activities and influences that affected industries, individuals, and the technologies themselves. In turn, activities within industries did influence the attitudes of individuals and firms, who then influenced their colleagues and competitors. Some uses of computers were also unique to their industries, as happened with scanning in retail, ATMs in banking, and telecommunications in the telephone industry. To get down to the level of specificity needed in order to understand more precisely how computers affected the American economy and either helped or hurt its enormous success during the half century, we need to look at computing in specific industries. That is why the bulk of this book is a collection of industry overviews with descriptions of how they used computers. I consider this exercise an essential first step—but only a first step—toward fulfilling Schumpeter's dictate.

In the earlier volume, I used the prism of digital technologies, focusing on how these were used over time, to shed light on how the work of the U.S. economy evolved. Most uses of digital technologies—the computer chip for instance—were in the forms of computers and telecommunications. These uses were largely based on digital technologies, which is why I frequently use the word *digital* as a noun, instead as an adjective.

How This Book Is Organized

I have relied on the same approach used in the earlier volume. I look at the history of how information technology was used in a collection of industries, concentrating on the applications of computers and computing technologies at large. I tell the story of a group of industries from roughly 1950, when commercial applications of digital technology first appeared, to about 2005, although I ended the earlier volume with 9-11-2001. The industries included in this book met several criteria. First, and most important, they existed in 1950 or emerged as a direct result of the creation of computing and telecommunications technologies (for example, the cell phone business and video games). Second, I wanted industries that played very large roles within the American economy, such as the Banking Industry. Third, I chose industries that were innovative users of technology (such as the Banking Industry), early adopters (e.g., Insurance Industry), or whose influence on the rest of the economy was significant (such as the Telephone Industry). I favored industries that were either very large or that had so pervasive an influence that they forced or influenced other industries to adopt their ways of accommodating changes. They forced other industries to change, too, which often included deploying computing. The result of my selections is that we have a sufficiently large sampling to start generalizing about patterns of adoption and use of computing and drawing some conclusions about how people did their work within the U.S. economy. These industries vary enormously, as did their responses to the use of computers, thus illustrating the importance of diversity of application so necessary in a large, complex, modern capitalist economy.

The sectors chosen for this book both conform to and depart from traditional configurations of industries. The collection of banking, insurance, and brokerage industries fit nicely into what economists have long called the Financial Sector. These industries are closely related in many obvious ways. The same can be said, within reason, about the media industries selected for discussion in this book. We have print media (such as newspapers and book publishers) and electronic media (such as radio and TV). But we also are seeing borders between these industries blurring, which has been made possible by computing. For instance, a newspaper can be on the Internet, can feed data to radio and TV, and can still publish a physical paper or magazine. Telephone companies started out as part of "Ma Bell" (AT&T) but ended the century flirting with cable television, heavily involved and threatened by the Internet, and pondering what constituted the Telecommunications Industry. All these industries have their respective historic patterns of behavior, but they are also changing in fundamental ways as a consequence of the digital. In total, they account for nearly 40 percent of the economic activity of the nation, as measured by their contribution to the U.S. Gross Domestic Product (GDP).

I pay considerable attention to describing the structure and work of the industries in this book—particularly those in the less understood financial and telecommunications sectors—and to how they functioned in the second half of the twentieth century. This is done before discussing computing for two reasons. First, I assume that most readers are not familiar with what all these industries do for a living, so context is required to make sense of the digital story. Second, computing serves the business needs of a firm and industry, so understanding the business issues of an industry is essential to our story. What becomes crystal clear is that the history of computing is at least as much a business history as it is a history of the technology itself. I have argued in this and in at least a half dozen prior books that the most important story about computing is its applications, not its technological evolution. I find fascinating the inventors and engineers who created computers and they made them increasingly cost-justifiable, but the people who bought and used these creations are what make the history of computing important to understand.

I focus on the United States for three reasons. First, the United States was the initial user of information technology on a broad scale and, in time, proved to be the most extensive adopter in the world. So, it has the most experience with and has been influenced the most and the longest by this new class of technology. Second, the American experience has influenced profoundly the activities of other industries around the world and the economies of the United States and other economically advanced nations. Third, there is a huge body of material that we can rely on in support of observations made in this book.

There are some basic questions and issues to deal with in this book that contrast to what happened in manufacturing, transportation, and retailing. Because the primary inventory (goods) sold by the industries surveyed here is information (as opposed to physical products), we want to know if computing came earlier to these sectors of the economy and how it was used. How did that happen, and why? We know, for example, that banks spent more on computing than manufacturing for the entire period. Why? How? To what extent did industry-centric applications influence the types and rate of adoption of computing from one industry to another? How much influence did practices in one industry have on another, for example, computing in banking on insurance firms? In manufacturing we found that, over time, computing and telecommunications linked suppliers to large manufacturers in tight codependent relationships that today would be impossible to imagine breaking. Has that been happening in any of the sectors studied in this second book? A series of questions can also be asked regarding how computing affected economic activities in the service sector as compared to the manufacturing sector. The immediate answer is that the digital affected the service-sector industries in ways both similar to and different from manufacturing and retailing. Finally, what are the implications, particularly for management, in those industries that are growing fast within the American economy, such as media and entertainment—portions of the economy considered core to the Information Economy, to the look and feel of contemporary American society.

Two concurrent patterns of behavior emerge from the study of these industries. On the one hand, a reliance on computing increases for many of the same reasons we saw in industries concerned with the physical economy. This pattern involves more than just handling the purchase and deployment of desks and pencils in an insurance company, or performing normal accounting in banks and movie studios. The pattern concerns business justification, the role of industry-specific technologies, and applications as well. On the other hand, however, there also are activities and concerns specific to the industries reviewed in this book that we must account for, such as the effects of file sharing in the music and movie industries. I focus on patterns of behavior in these industries. By the end of the book I also provide comparisons with and contrasts to those industries surveyed in the earlier study on manufacturing, transportation, wholesaling, and retailing. In both sets of industries, however, computing has clearly provided more linkages among traditional and emerging industries, again demonstrating that technology has made borders porous between long-established industries. It is no accident, for example, that TV, movie, Internet, cable, and telecommunications seem to intermix a great deal, or that the monetary "coin of the realm" is not the only legal tender authorized by the U.S. government, but also includes credit cards marketed by banks through telemarketing, television, and traditional print media.

The endnotes are rich in the detail of what was required to document my sources. Where the literature on a subject is not well-known, as in the case of the Internet, I have added material to help those who wish to explore further the themes discussed in this book. The topic of this book is vast and there is a large and growing body of contemporary material on the subject that can be examined. I hope that others can flush out details I could not address in the trilogy.

As with the previous book, because I chose to write about a very broad set of issues, I have had to be almost arbitrary in selecting materials and industries and limiting the length of the discussion to keep it within the confines of one book. Thus, in many instances, I have had to generalize without fully developing explanations. For this I ask the reader to forgive me, or at least understand. The views I express in this book, and the weaknesses exposed, are not those of the good people who helped me, of my employer IBM, or my publisher.

This book would not have been possible to write without the help of many people: historians, business managers, archivists, and, of course, the team at Oxford University Press. At the IBM Corporate Archives, archivists over the past 20 years found thousands of pages of material for me to study. The same is true for those at the Charles Babbage Institute at the University of Minnesota, home of the world's largest collection of materials on the history of computing. Sharon Smith at LOMA, the Insurance Industry's leading association, and John D. Long at Indiana University helped me make sense of the Insurance Industry, while Steven Wheeler, archivist at the New York Stock Exchange, filled in valuable details on the early use of computers in his industry. Darryl C. Sterling at IBM has been watching telecommunications and media industries for years and carefully critiqued many chapters. James Lande at the Federal Communications Commission taught me how to navigate through his agency's mountain of material. Herb Addison, a retired book editor at Oxford University Press, who worked in the book publishing business for nearly the entire period covered by this book, as well as Ed Wakin, one of America's premier experts on the television and radio industries; both pointed out many opportunities to improve various chapters and taught me about their industries. I was privileged that both Kristin Thompson and David Bordwell could help with materials and critiques about media industries in general and about the movie industry, of which they know so much. I would not have been able to write on the photographic industry without the help of Kodak's Robert J. Gibbons and Gerard K. Meucher, both of whom were invaluable fonts of knowledge about their business and on sources of materials. Henry Lowood at Stanford teaches a course on the history of video games and has accumulated a vast amount of useful materials and insights that helped me make sense of this emerging industry. Phillip Meyer at the University of North Carolina exposed me to the importance of online newspapers, emphasizing the growing role this new use of the digital has had on the industry for some time. Always on hand to critique and advise me on each of my books—including this one—is one of the pioneer historians of computing, Professor William Aspry of Indiana University. Professor Arthur Norberg, most recently the Director of the Charles Babbage Institute at the University of Minnesota, provided continuous support and stimulating discussions about the business history of technology.

One can do important research, but if the results cannot be presented in a coherent manner and disseminated, it is wasted effort. I have dedicated this book to three editors who taught me how to write books and then turned around and published them. Jim Sabin, my first editor in the 1970s at Greenwood Press, got me started. Paul Becker talked me into writing about information technology and published the results at Prentice-Hall in the 1980s. Herb Addison at Oxford University Press took me to a whole new level of research and writing when I wanted to blend my dual interest in the history of IT with my ongoing analysis of how managers currently use this technology. Each spent an enormous amount of time shaping my views and action for which I am deeply grateful. If it takes a village to raise a child, it takes an industry to support an author.

The team at Oxford University Press that worked on this book was excellent. My editor, John Rauschenberg, was very supportive. The production staff, led by Lewis Parker, was very accommodating and professional. I owe a debt of gratitude to Kathleen Silloway, who copyedited the manuscript, and to Sandi Schroeder for indexing the book. Finally, I must acknowledge the role my wife, Dora, played because she made it possible for me to spend so much time working on the book to the dereliction of my duties as a member of the family and household.

James W. Cortada
Madison, Wisconsin

CONTENTS

1. Role of Financial Industries in the U.S. Economy 3

2. Uses of Computing in the Banking Industry 37

3. Deployment of Digital and Telecommunications Applications in the Banking Industry 80

4. Business Patterns and Digital Applications in the Insurance Industry 113

5. Business Patterns and Digital Applications in the Brokerage Industry 151

6. Role of Telecommunications Industries in the U.S. Economy 189

7. Business Patterns and Digital Applications in the Telecommunications Industry 227

8. Role of Media and Entertainment Industries in the U.S. Economy 263

9. Uses of Computing in Print Media Industries: Book Publishing, Newspapers, Magazines 293

10. Digital Applications in Electronic Media Industries: Radio and TV 336

11. Digital Applications in Entertainment Industries: Movies and Recorded Music 371

12. Digital Applications in Entertainment Industries: Video Games and Photography 412

13. Conclusions: Lessons Learned, Implications for Management 442

Appendix A: Role and Use of Industries, Sectors, and Economic Models as Concepts for Understanding Business and Economic Activities 479

Appendix B: How a Telephone Works and the Basics of Telecommunications and Networking 485

Notes 491

Bibliographic Essay 584

Index 601

The DIGITAL HAND

1

Role of Financial Industries in the U.S. Economy

On any given day, you can pick up any periodical and the chances are 10 to 1 that somewhere there will be an article or information dealing with the subject of electronics or automation. This flood of information has resulted in the subject's becoming the most popular since sex was invented.

—Harry E. Mertz, National Association of Bank Auditors and Comptrollers, 1955

For over two decades, economists and scholars have noted that the American economy was evolving into a services-based one and away from a predominantly manufacturing one, making that transformation the hallmark of the Information Age.[1] In fact, two things happened over the last half of the twentieth century that both support and contradict this observation. On the one hand, over time the service sectors of the U.S. economy indeed grew in dollar value and in percent of the total Gross Domestic Product (GDP). But on the other hand, the industrial component of the economy did not vanish. While the economic importance of new and more traditional services grew, such as those of lawyers, educators, and consultants, new products were invented, manufactured, and sold, such as computers, software, and medical imaging equipment. As I pointed out in *The Digital Hand* (2004), manufacturing contributed only slightly less of a percent to the total GDP over the past half century than it did in 1950, but it profoundly changed how it did its work. In this book, in which we look at important components of the services sector, I argue that a similar process of work transformation has occurred. Management across many industries within the services sector has deployed computing technology in such vast amounts and for such a broad array of uses that one can conclude that computers have profoundly changed the way work is done in America and what products and offerings have become available.

Indeed, one can demonstrate that computing played a more pervasive role in portions of the services sector than it did in manufacturing or retail. The Banking Industry is a case in point. At the start of the second half of the twentieth century, not a single bank used a computer; by the end of the century, what banks did, what services and products they provided, and how they competed and worked were deeply influenced by computing. Technology made it possible for not only new competitors but also more varied and novel offerings to emerge. As a percentage of their total budget (or sales), banks outspent manufacturing and retail companies by almost 200 percent over the past two decades. We cannot escape the conclusion that the role of computing in the Financial Sector expanded in important ways, even exercising spectacular effects on how various transformations occurred. However, the journey to a world filled with computers was not a simple one. The author of the outrageous quote at the start of this chapter made the same point with more sober language in the rest of the article, lamenting the confusion about technology rampant in his industry.[2]

One can even see physical results. In 1950, automated teller machines (ATMs) did not exist; today we count them in the hundreds of thousands. They nest in the lobbies of skyscrapers, in airports, and in shopping centers. The concept of banking from home was science fiction in 1950, yet today more than 20 percent of Americans conduct some financial transactions from personal computers.[3] In 1950, one could only do business with a bank in a branch office from 9 in the morning until 2 in the afternoon; after closing time, bankers needed the rest of the day to sort and post all the checks they had received in the morning. Today, one can conduct business around the clock, seven days a week. In 1950, government regulators restricted banks to operating only in the state that chartered them; today, we have national banking chains functioning all over the country. In 1950, money was cash—the paper bills and coins issued by the U.S. government. Today, we question what cash is: Is it what is credited in our debit cards or the available credit on our credit card? At the dawn of the twenty-first century, we talk about "e-money" as the next, natural evolution in cash that began with shells and clay objects thousands of years ago.[4] Bankers today even postulate that they are not in the business of handling money, but are instead managing information about money.[5] In short, much has changed. To be sure, much of the change was made possible by the combination of many processes, such as the elimination worldwide of the gold standard in 1971, the deregulation of the Financial Industry throughout the 1980s and 1990s, and, finally, the introduction of a large array of computing technologies, from the mainframe computer and PC to the Internet and software tools, which made possible new ways of working and competing.

But why did computing play such an important role in the world of banking? Banks moved information around—a core task of the industry—and the brokers' work was labor intensive. Computers facilitated the collection, sorting, management, and display of information in larger, more diverse ways than before. This made it possible for banks to offer new services and to compete in innovative ways while holding out the promise of lower operating costs. The latter, always a concern to executives in any industry, became more acute in banking as interest rates rose in the 1970s and 1980s and as competition from other banks and from outside the industry increased all through the second half of the century, particularly after 1980. Computers made it possible to break up one bank's virtual monopoly on banking while offering other banks the ability to increase scale and scope. The same was true in the Insurance Industry and with stock brokerage firms.[6]

The story of the role of computing across the Financial Sector is a complicated one. In 1950, the three pillars of this sector—banking, insurance, and brokerage—were independent industries operating essentially at arm's length from each other, as required by law. By the end of the century, the borders of these industries were quickly blurring as computing and deregulation made it possible for each to operate in the markets of the others. The forces merging these various activities were a blend of consequences derived from economic, demographic, regulatory, and technological changes experienced by the economy of the United States during the entire half century. One study, commissioned by firms in the Insurance Industry to understand why banks were encroaching on their markets in the 1970s and 1980s, summed it up this way: "The collective forces of change have prompted a blurring of distinctions among the traditional segments comprising the financial services industry."[7]

One of the technological consequences was the growing ability of customers, banks, and insurers to have interactive transactions directly between customers and financial institutions; this made it possible to easily sweep funds from one account to another, such as transferring excess cash out of one into another to optimize yields. That capability created a growing demand for the integration of various financial offerings. In particular, banks and brokerage firms went after the property and casualty lines of the Insurance Industry. Both had the capability to do this by the early 1980s because they had existing channels of distribution that included credit cards, networks of ATMs, and retail branch offices. Add to that their access to customers through loans, and checking and savings accounts, and one can begin to see how exposed insurance firms were in the market. Furthermore, both bankers and brokers knew how to handle money and financial instruments on behalf of customers. The same study cited above also pointed out that "savings and loans have a positive, consumer-oriented image; generally high level of online, real time data processing capabilities; and strong customer loyalty."[8] Insurance companies did not enjoy those advantages because their image was not always positive, their channels of distribution were more expensive (for example, cost 25 to 35 percent of premiums versus a bank's roughly 4–5 percent cost of assets, and for a broker closer to 1 percent). From the perspective of a banker, going after insurance business seemed logical since banks could compete on price of distribution cost alone. Since the Insurance Industry was highly fragmented, rivals were relatively small and weak, and bankers could differentiate their offerings more clearly.

Complicating things even more, non-financial institutions also became active in financial markets; for example, automotive and other large machinery manufacturers set up wholly owned subsidiaries to provide financing for their products to customers, bypassing banking and other financial industries. And it seems that everyone today is offering a credit card.[9] The history of computing in the Financial Sector illustrates clear patterns of behavior in each of the major industries within the sector. Recognizing these patterns clarifies how work changed, how technologies entered and exited, how they supported complex services industries and firms, hinting at managerial practices that could be expected to continue in the years to come. As occurred with manufacturing, process, transportation, wholesale, and retail industries, services industries had their own personalities and behavior that greatly affected how technologies of all types were used. The same held true for the services sector at large, and nowhere can this be more clearly demonstrated than in the Financial Sector. By telling its story, we can better understand how the American economy evolved away from the classic Industrial Age model to something that I believe is not yet fully defined but that may perhaps be the Information Age (although I am more inclined to characterize it as an as-yet-unnamed new age).

To understand the interactions among industries and computing in the American economy, it is important to look at sectors and industries. To start that process with the services sector, we need to understand what constituted the Financial Sector, placing it into the context of the larger American economy. Then we can discuss the role of these industries that comprise part of that sector, because they changed over the second half of the twentieth century. In subsequent chapters I review computers' places in the banking, insurance, and brokerage industries. I present evidence in those chapters of the extent to which firms in these industries deployed the technology and its effects over time. We will see that while common patterns existed, each industry also responded uniquely to the arrival and use of computing technologies. In the process, each industry refined and evolved its practices and industry-level personality.

Financial Sector in the American Economy

The Financial Sector of the economy is normally defined as consisting of the Banking Industry, the Insurance Industry, Stock Brokerage Industry, and the Real Estate Industry. The U.S. government, and economists in general, have long accepted this rough definition. For our purposes, however, we will discuss banking, insurance, and brokerage, leaving real estate out because its characteristics are more similar to such smaller services sector industries as the legal profession and accounting.[10] In addition, banks, insurance firms, and brokerage enterprises have far more complex and varied tasks and products and consist of larger enterprises encompassing more economic and managerial facets than real estate. To complicate things, however, there are various types of industries and firms within the

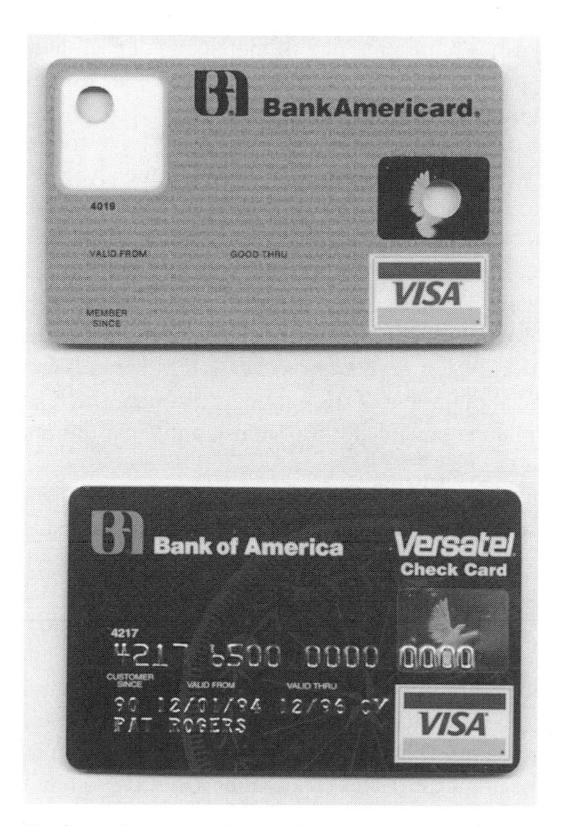

Credit cards were made possible by computing and telecommunications working together in an industry specific application. Some of the earliest credit cards issued in the United States. Courtesy Bank of America.

three major segments of the Financial Sector: commercial banking, international banking, and retail banking; there are life, property, and casualty segments in insurance; and brokers sell both stocks and bonds. There are also other companies participating in the financial segment, such as GMAC financing, which finances automotive loans for General Motors' customers. Some of the most visible institutions in the American economy function within the Financial Sector, such as Citicorp, J. P. Morgan, and the Federal Reserve (banking); Prudential Insurance or State Farm (insurance); and Charles Schwab, New York Stock Exchange, and NASDAQ (brokerage). Even the idea of "Wall Street" conjures up both stock selling and banking. Insurance and banking companies occupy some of the tallest buildings in dozens of American cities, and every town and city in America seems to have its own insurance offices and branch banking (or ATMs). In short, it was and is a sector that has more presence in more parts of the nation than manufacturing, surpassed only by the physical presence of retail establishments such as stores and restaurants.

Because over the past half century Americans have interacted with banks more than with the other two portions of the Financial Sector, we need some definitions. The most visible to people are banks. So, what is a bank? One industry guide defines it as "any organization engaged in any or all of the various functions of banking, i.e., receiving, collecting, transferring, paying, lending, investing, dealing, exchanging, and servicing (safe deposit, custodianship, agency, trusteeship) money and claims to money both domestically and internationally."[11] That is an impressive definition. What is clear, however, is that banks have historically been the primary financial plumbing of the economy through which flow a large variety of activities that make it possible for individuals and organizations to save, borrow, transfer, and spend money.

What is an insurance company? These are institutions that provide financial protection against adverse circumstances that cause economic loss by spreading the cost of risks across a large population of people and organizations. Life and property insurance are familiar examples, but insurance also provides protection against specific types of harm as well, such as the loss of a payload on a space ship or the costs of medical treatment. Insurance companies are also investment institutions where people can invest their funds in a variety of financial instruments. This means that insurance companies now play a similar role to that of banks, with their savings accounts and brokerage houses, in the selling and managing of stock portfolios. Insurance firms invest premiums in such volumes as to make them major investors in the U.S. economy.

Finally, what is stock brokerage? These are organizations that have historically done three things: sold a variety of investment products (such as bonds and stocks); managed portfolios of these instruments for individuals and organizations; and provided a variety of financial consulting and other services to corporations, government agencies, and individuals. We see this today, for example, in the sale of individual stocks to people and in the management of individuals' 401(k) retirement accounts. Through brokerage institutions, young companies needing capital from the financial markets can launch initial offerings of stocks. A more precise definition of a stock brokerage, and the stock exchanges within which they do important portions of their work, is an organized market and firms whose purpose it is to trade in financial securities.[12]

Over the years, these three elements plus real estate of the Financial Sector have made up growing portions of the U.S. GDP.[13] Looking at the total sector as defined by the government in table 1.1, we see that the entire Financial Sector doubled over the past half century. This meant it took "share" away from other sectors of the economy, such as agriculture and manufacturing. To put proportions into perspective, let's look at 1989 as a sample; this year reflected a decade's worth of deregulation but was before the advent of the Internet. Government economists reported that the Financial Sector constituted 17.4 percent of the GDP: the Insurance Industry made up 1.7 percent of the GDP, the Banking Industry 3.3 percent, and security and commodity brokers 0.8 percent; real estate accounted for 11.5 percent.[14] In 1997, the Financial Sector (insurance, banks, brokers, and real estate) accounted for 18.9 percent of the GDP; in 2000, it had grown to 19.6 percent.[15] The key point is that these industries increased their percentage of the total economic activity of the nation over time. Since the American economy also expanded enormously during the last half century, the actual raw dollar volumes for each industry did, too, and did so at a proportionately greater rate than the economy as a whole, as demonstrated by the growth in the share of the GDP. What is not reflected in these statistics are the financial activities of other industries, which also proved significant; for example, the previously cited case of GMAC that, in effect, was a lending bank owned by General Motors; the various credit card arms of retail establishments and petroleum companies. But most important to keep in mind is that every dollar that moved through the economy passed through the Financial Sector numerous times and, more frequently than not, did so as pieces of information in computers.

Employment data and the number of banks and branches also suggest the scale and presence of this sector in the economy. Banks in particular are illustrative because of the large numbers involved. Like the other industries in this sector, all through the half century the Banking Industry grew in size, number (usually employment), and volume of business conducted. In 1951, there were 14,132 banks in the United States and 19,396 branch offices. Just over twenty years later, there were 14,653 banks (1973) with 27,946 branches. In 1983, the combination of branches and banks totaled 55,960. In 1960, more than 672,000 people worked in this industry, and that number steadily increased to 1.5 million in 1980, then flattened out for the next fifteen years. During that fifteen-year period, however, the number of branches increased, and tens of thousands of ATMs were installed. Not until late in the century did employment levels shrink—a trend triggered by the combined effects of computing, competitive pressures, and, at the dawn of the new century, the economic recession.[16] Similar trends in employment in the other two financial industries demonstrate that the number of people and offices in this sector remained important components of a rapidly expanding services sector across the economy; this remained true during periods of peace and war, during prosperity and recession, and during an era when information and telecommuni-

Table 1.1
Finance, Insurance, and Real Estate as Percent of U.S. Gross Domestic Product, Selected Years (1949–1998)
(Percentage of total GDP)

1949	1959	1977	1987	1998
9.7	12.9	13.9	17.5	19.1

Source: Bureau of Economic Analysis, "National Income and Product Account Tables," available at http://www.bea.doc.gov/bea/dn/nipaw/Ta.

cations technologies did more to change the nature of work and who did it than possibly any other technologies had in the history of the nation.[17] In subsequent chapters I explore the interrelationships between the size and scope of the three financial industries and their use of information technology, because each affected the other. To characterize the situation, these interrelationships and their dynamics over time were intense and extensive.

It is essential to know what these various industries did, because computers and telecommunications were always adopted in the hope that they would improve the abilities of firms to carry out their core missions. Along the way, managers wanted to improve productivity by lowering operating costs or avoiding future expenses, and also by offering new services that would make a firm more competitive or secure from rivals. Firms approached technology first from the perspective of how it could help existing operations, then from how it could make it possible to do new things. These financial industries had a spill-over effect on all other industries. They financed investments and expansion, managed financial accounts, even physically moved money or the myriad of financial transactions that were the content of daily work, and insured it all.

What Banks Did

The tasks of banks are not as obvious to any individual outside the industry as are tasks of other industries, such as manufacturing or retailing. Much of the work of banks occurs behind the scenes, leaving the public to see only, for example, check writing, clerks handling retail transactions at a branch, or someone using an ATM machine. The public at large does not clearly understand many of the products offered by a bank, such as derivatives, secured distribution of loans, various types of loans (other than for home mortgages or cars), mutual funds, and so forth. Many would be hard pressed to describe the major roles of a bank, or even how money and checks move about the economy and how one's bank balance remains mostly accurate and up-to-date. Checks, debit cards and credit cards are ubiquitous and efficient—thanks overwhelmingly to computers and telecommunications—and mortgages and loans are managed effectively, crediting balances in a timely fashion. Corporations raise money to expand their operations while venture capitalists transfer millions of dollars daily all with the help of banks, to promising young firms eager to fuel the next information technology, innovation, or biological "revolution."

One brief guide for bank directors stated that the role of banks is to, "gather funds from depositors by providing services and paying interest and lend those funds to individuals and businesses."[18] In short, banks rent money and loan it out; they normally do not buy or sell it except when trading currencies from multiple countries. A bank makes money either by charging for services (for example, handling a "bounced" check) or through collection of interest on loans. A bank's expenses, from which revenues are deducted, consist of far more than the standard overhead costs of buildings, computers, and people; they also include interest paid on deposits and loans, leaving net interest and revenue income as the firm's profits.

The reason for explaining what might seem patently obvious is that in order to implement this business model banks have to deal with three kinds of risk: interest rates (which go up and down, sometimes rapidly and with no local control by a bank), credit (judging whether a person or firm will pay back a loan), and liquidity (availability of cash to fund loans). Over time, bankers acquired a formidable reputation for being not only conservative (that is, risk adverse) but also expert in managing risk. Another skill they honed was the ability to manage a large body of numerical data in ways approved by government regulators and reinforced by sound industry practices. These sets of skills made it possible for bankers to leverage capital (money) to generate profit for a bank.

To carry out their roles, banks perform essentially four basic tasks. They receive deposits (money, electronic funds, and checks) and pay customers' checks drawn against their accounts. This task is accomplished by mail, through customers' personal visits to a branch office, by ATMs, or via PCs and the Internet; since the mid 1990s, banks have also issued debit cards that debit a customer's checking account when used. Second, banks loan money to individuals for such things as the purchase of a home (mortgages) or automobile, or to businesses to handle routine cash flow and to invest in inventory and buildings. Third, since the 1960s they have issued credit cards, generating interest revenue from unpaid balances. Fourth, they offer a variety of financial services, ranging from providing billing for companies to acquiring and selling bonds to managing trusts. Table 1.2 lists a variety of bank services widely available over the course of the second half of the twentieth century. As the century progressed, the number of services increased, such as ATMs in the 1970s, debit cards in the 1980s, and Internet banking in the 1990s.

Table 1.2
Types of Bank Services, Late Twentieth Century

Receive demand deposits	Offer debit cards
Perform payroll	Supply credit to firms
Operate ATMs	Transfer funds
Receive time and savings deposits	Facilitate exchanges
Issue negotiable orders of withdrawal	Issue cashier checks, money orders
Perform automatic transfer services (ATS)	Rent safety deposit boxes
Offer discount notes, acceptances, and bills of exchange	Make loans, provide credit
Offer custodial services for securities	Offer credit cards
Manage trust funds	Underwrite securities
Conduct corporate trust services	Sell government bonds
Invest in government debt securities	Manage foundations
Transfer funds electronically	Exchange currencies

For a more complete list of services, see Charles J. Woelfel, *Encyclopedia of Banking and Finance* (Chicago, Ill.: Probus Publishing Company, 1994): 70.

Each of these offerings required banks to keep records of every transaction (inflows and outflows of dollars) and to tie those to the overall accounting records of the bank. The press of competition—and later legal requirements—to keep accounts current as of the day of a transaction increased pressure on banks to do faster bookkeeping. Banks' biggest, most time-consuming, and resource-draining activity of the second half of the twentieth century was cashing checks. While I say more about this in the next chapter, suffice it to acknowledge that throughout the period the number of checks the American banking system had to handle each year ran into the billions, and in most years tens of billions. Each check had to be recorded, sorted, then physically sent back through the banking system to the bank that originally issued the check to a customer. No single activity of a bank generated more discussion over more decades than check handling. No single function stimulated more speculation and wishful thinking about the "checkless society" or e-money. However, toward the end of the century, the number of checks cashed annually hovered at 60 billion, despite billions of transactions also conducted using credit and debit cards.[19]

Banks performed all these various activities entirely through headquarters and brick-and-mortar branches in the 1950s through the 1960s; after the introduction of various electronic tools they used electronic means as well. In the late 1960s, and more so in the 1970s, telephone-based services began (such as balance inquiries and electronic funds transfers); widespread use of ATMs started then, too. In the 1980s telephone-based and, later, PC-based home banking began; then, in the late 1990s, businesses and individuals began the historic shift now clearly underway to Internet-based banking.

What Insurance Companies Did

Like banks, insurance companies also manage financial assets and have developed a large body of skills in the area of risk management. Insurance is a mechanism by which an individual or organization can pay a small price (premium) to cover the cost of a large, always uncertain loss up to the amount of the insurance policy. Insurance companies sell insurance to many people or organizations to generate the money needed to cover the losses of a few individuals. While waiting for a loss to occur, an insurance company collects premiums and invests them to generate additional funds—funds that either will be needed to cover a loss or that will constitute part of the profit. The core competencies of an insurance company include the ability to predict how many and how big one's losses will be (the role of actuaries) and adjusting premiums (or selling more) to cover these risks under guidelines set by state regulators.

Insurance falls into three basic categories: life insurance (paid in the event of one's death), property and casualty insurance (to cover damage caused by a storm or an automobile accident), and medical insurance.[20] Risks are personal (such as loss due to death), property (such as loss of income due to weather damage), or liability (such as loss due to a doctor losing a malpractice law suits). In this book, we discuss the first two and leave medical until the next volume of the *Digital Hand.*

The heart of the insurance business is the art and science of managing risks.[21] Insurance companies rely on mathematical models, actuarial experience, increasing the number of insured people in a pool, and selling numerous other services besides insurance to raise funds, such as various whole-life insurance products, and managing financial investments for individuals. One thinks of the insurance industry's offerings and activities as falling into two categories: first, life and health; and second, property and liability. Some firms only specialize in one or the other; others work in both markets. Key activities involve creating and selling insurance to individuals and groups that minimize risk and optimize profits; managing large bodies of funds resulting from premiums and investments; paying out claims against insurance policies; and, like banks, managing a vast collection of data and paper. Over the past half century, just as with banks, insurance companies have broadened their offerings, made possible by the computer systems that accurately and cost-effectively handle large volumes of data and provide insights to facilitate the management of risk. Table 1.3 lists typical insurance offerings available during the second half of the twentieth century.

Insurance for property has always far exceeded in revenues that for life, with fire and marine first in volume, followed by an assortment of life and casualty offerings. It should be noted that not all insurance policies come from insurance companies. By the late twentieth century, one could acquire fire insurance from, for example, stock companies, even mutual fund companies. Additionally, until the last quarter of the century, insurance companies either sold just life or fire and casualty; today there are many firms that sell products in both categories.[22]

What Stock Brokerages Did

Brokers are salespeople and firms that sell stocks to individuals and organizations (such as to pension funds). They spent the second half of the century expanding their services to include the launch of stocks (initial public offerings, usually called IPOs) for new firms and industries to emerge in the period, such as the dot-coms, software, and computer hardware companies. They offered money market accounts in which individuals could maintain cash balances for buying stocks or writing

Table 1.3
Types of Insurance Services, Late Twentieth Century

Types of Insurance Services
Life insurance—term, whole, ordinary, limited payment, group, accident, health
Fire and marine—fire, marine, motor vehicle, tornado and other windstorms, earthquake, sprinkler leakage, water damage, riots, and other civil commotions
Casualty and surety—automobile and other forms of liability, workers compensation, theft and burglary, property, accident, and health
Financial investments—often investment offerings combined with whole-life insurance, can include financial counseling, management of 401(k), and other retirement instruments

Table 1.4
Stock Brokerage Services, Late Twentieth Century

Over-the-counter (OTC) securities, stocks, options, block trades, financial futures, money market funds
Automated stock transfers, stock loans, and cash management
Bringing new offerings to market, called initial public offerings (IPOs) on behalf of companies
Portfolio management advice and administrative support to client
Investment-banking services, underwritings, and bond swaps
Mutual funds, shareholder and investment accounting
Sale of commodities futures and commodities and options accounting
Accounting services: stock records, clearing and settlements, bookkeeping and statements, proxies and dividends
Clearing and settlement services

checks. Like banks with depositors, stock brokers provided their clients with periodic reports on balances—that is to say, about stocks bought and sold on their behalf—and on cash balances and margins. During the last decade of the century, stock brokerage firms made it possible for customers to buy stocks and directly manage their portfolios and accounts using the Internet. Over the entire period, and under the guidance of government regulators, brokers allowed individuals to buy stocks on margin—to expend cash at a level below the actual price of the stock—thereby, in effect, loaning people the difference. As occurred in banking and insurance, the variety of offerings and services provided by a stock brokerage firm expanded over the years. Table 1.4 lists commonly available stock brokerage services late in the century. What jumps out from the table is the variety of functions, many related to the administration of accounts, performed by stockbrokers in addition to the buying and selling of securities. Note also, the variety of financial instruments and types of buying and selling that occurred. While most of these were available throughout the second half of the century, permutations of each appeared over time, as both deregulation and re-regulation on the one hand, and the deployment of the digital on the other made possible new offerings.[23]

Part of the industry includes a collection of stock exchanges, which are organizations whose purpose is to provide centralized buying and selling of securities. We think of the American Stock Exchange, Chicago Board of Trade, and the New York Stock Exchange (NYSE) as obvious examples, but there were almost a dozen exchanges operating in the United States by the end of the century. Exchanges offered their members—stock brokerage firms—such services as displays of stock prices, administrative and technological infrastructures through which to accept and place orders for securities (usually stocks), rules and guidelines for corporations whose stocks were listed for sale by an exchange, and publicity for prices. In short, while exchanges did not buy and sell, they provided the physical and virtual marketplaces where such transactions occurred. Through a complex network of telecommunications and other digital technologies, firms all over the United States could conduct business through a physical stock exchange located elsewhere in the nation.[24]

The Special Role of Government Regulations

It would be difficult to overestimate the influence of state and federal regulatory bodies and laws on the character and practices of the Financial Sector. From the eighteenth century forward, banking in the United States has been subject to extensive regulation concerning who could issue money, where banks and insurance companies could be chartered, the extent to which they could operate in one or more states, and even what interest they could pay and charge for deposits and loans. By the end of the 1930s, there was a new body of laws and regulations governing the buying and selling of securities as well, crafted in response to practices that economists and public officials saw as having triggered the Great Depression and crises in both the banking and brokerage industries. Over the past two centuries a large body of regulations, and the audits that inevitably accompanied them profoundly dictated and influenced the behavior of management in all three industries. These ultimately affected their culture and practices. The same could be said for other industries studied in this book, such as all of telecommunications and electronic media (and even to a certain extent the Internet), and especially radio and TV.[25] Print media was immune from such pervasive controls due to specific provisions in the U.S. Constitution protecting the rights of the press, freedom of speech, and copyright privileges.

In order to appreciate the role of computing in these industries, it helps to have an understanding of the history of regulations in each industry. Each industry has its own set of federal regulatory practices which are watched over by federal regulatory agencies established for the purpose of enforcing national laws. States did the same with banking and insurance commissions.

In the Banking Industry, for nearly two centuries banks operated principally only within individual states, and federal regulations reinforced that pattern. Regulations also governed the movement of money, handling of checks, what interest rates could be paid for deposit accounts, and usurious practices. By the end of the 1930s, federal policymakers had established rules preventing banks, brokerages, and insurance companies from selling each others' products and services. The major modern regulatory law was passed in 1933, called the Glass-Steagall Act. It provided for deposit insurance, eliminated securities underwriting from commercial banks, and gave states permission to charter banks to operate across an entire state. The Securities Exchange Act enacted in the following year defined how the buying and selling of securities would take place for the next half century. In each decade since the 1930s, regulatory legislation appeared that defined practices concerning most aspects of banking and securities. Because of the nature of these regulations, banks only competed against other local banks until the 1979 Electronic Funds Transfer Act, which required firms involved in moving funds electronically to meet specific standards. This Act, applied all over the nation, continued to limit the flexibility of banks to compete.[26]

In the early 1980s, banks were given increasing authority to provide new services and offerings. In 1980, for example, Negotiable Order of Withdrawal (NOW) accounts were authorized, which allowed a customer to write checks against dollars held on deposit. In 1982, the Depository Institutions Act made it possible to offer money market deposit accounts and expanded the authority of thrifts to make loans. A series of laws passed in the 1980s responded to bank failures and further protected the deposits of consumers (a focus of regulators for decades). While it is a common impression that banks were deregulated in the 1980s, that, in fact, was not fully the case. For example, in 1991 Congress passed the Federal Deposit Insurance Corporation Improvement Act which, among other things, expanded the borrowing authority of the Bank Insurance Fund and the Federal Deposit Insurance Corporation (FDIC), constrained the use of brokerage insured deposits, expanded the power of regulators to deal with poorly capitalized banks, added additional regulatory responsibilities over foreign banks, and tightened existing capital requirements across the industry. In the late 1990s, additional regulations governed electronic banking and competition. Throughout the last three decades of the twentieth century, every major innovation in services and offerings grew out of laws that made it possible to do something new, or as a result of some regulatory body granting permission (for example, when Internet-based banking began in the mid-1990s). The effects of regulatory practices on competition and structure in the industry (such as bank mergers) are discussed in the next chapter; suffice it here simply to acknowledge the pervasive influence of government regulations on the actions of bankers.[27] Every new service or change in laws and regulations required new uses of information; each in turn led to new or different applications of computing in support of these activities.

Insurance Companies

Insurance companies were not immune from regulatory supervision and controls. Banking and securities regulations often included dictates to the Insurance Industry, such as laws and regulations governing the sale of securities and the protection of bank deposit accounts. State regulators focused on the pricing of products and services and chartering of firms, although national firms have been more prevalent in insurance than in banking over the past two centuries. In the Insurance Industry, the states have dominated the regulatory practices due to a nearly century-long legal debate about whether insurance is a commercial activity or not. In 1945, Congress passed the McCarran-Ferguson Act, which granted the states regulatory authority over insurance firms, providing it was done to the satisfaction of Congress. Over the last four decades of the twentieth century, however, federal regulatory bodies became increasingly involved, particularly in regard to the buying and selling of financial products and services that previously had been the preserve of the banking and brokerage industries. Therefore, such federal agencies as the Federal Trade Commission (FTC) and the Securities and Exchange Commission

(SEC) assumed more roles in the affairs of the Insurance Industry. The U.S. Congress itself also played an active role, primarily through a continuing series of investigations of various state regulatory practices over the entire half century. All this government attention resulted in insurance companies conforming to many practices dictated by both federal and state agencies and to asking for permission to offer novel products and services, all of which influenced the cadence of the industry's activities.[28] Insurance companies proved a lot quicker to embrace the digital than bankers, an important issue discussed in subsequent chapters; it sheds light on the nexus of innovation and regulatory practices in the modern American economy, a theme that surfaces as well in the telecommunications and media industries and reflects the effects of precomputer-era automation.

Brokers

Securities traders were not immune from the controlling hand of regulators. In the years immediately following World War II, this industry continued to be heavily regulated as a result of the Great Depression; like banks and insurance firms, state and federal regulators defined how closely one industry could tread on another's territory. There is a natural tendency in financial markets to integrate the services and offerings of all three—they deal with risk management, flow of money, and the buying and selling of financial securities. As state and federal regulations made it easier for all three to begin operating selectively in each other's domain, regulators who specialized in one industry became increasingly involved in the others. This proved to be the case with the securities portion of the Financial Sector, even though this industry also had its dedicated regulators who defined or approved the practices of the industry, always with an eye to protecting the nation's financial system and the individual consumer. With brokerages, regulatory practices focused on the activities of the individual firms and also of the stock exchanges themselves.[29] Beginning in 1934 with Securities Exchange Act, the latter saw the most regulatory activity because through it government agencies could control the actions of individual firms. Thus began, for example, the long-standing practice of registering brokers, stocks, and exchanges. Over the years a series of new federal laws and SEC rulings expanded the role of exchanges across a national market while defining the way electronic linkages could take place.[30]

The pattern of regulations almost reads like what we saw in banking. In 1933, Congress passed the Securities Act, which required the purveyors of new stocks to register an issue with full disclosure to protect potential buyers from false information. The Securities Exchange Act of 1934 and the Maloney Act of 1936 extended these guidelines to secondary markets and expanded the role of regulators. Other laws were passed in the 1930s and 1940s that continued to enforce making information available and protecting investors. In 1970, for example, Congress passed the Securities Investor Protection Act, which established the Securities Investor Protection Corporation (SIPCO) as an insurance company to protect investors from failed brokerages. Five years later, the Securities Act Amendments called on the SEC not only to create national markets for securities—a major transformation of the industry—but also to outlaw fixed brokerage commissions by which brokers could charge any fee for their services, resulting in fees declining over the next quarter century.[31]

Regulatory Development

In the twentieth century, the Great Depression of the 1930s influenced enormously the nature of regulatory practices, particularly from the 1930s through the late 1970s. In the '70s, deregulatory initiatives were launched that fundamentally changed many practices in these industries—transformations still underway and subject to controversial debates among economists and managers.[32] During the last two decades of the twentieth century, additional innovations in regulatory practices occurred. In banking, in the early 1980s, cross-state banking became permissible, and in 1987, the first bank holding companies could underwrite securities. The last deposit interest rate regulations disappeared in 1986. In the early 1990s, thrift banks were allowed to open branches wherever they wanted. Cross-market restrictions continue to be eliminated in the middle years of the decade, furthering cross-industry marketing of offerings, including insurance. In November 1999, Congress passed the Gramm-Leach-Billey Act, which eliminated many barriers faced by banks and bank holding companies to compete across the entire United States and over a variety of financial products.[33]

As time passed, and insurance companies became larger players in the securities markets, the federal authorities took on larger roles in the Insurance Industry. The same applied to the Brokerage Industry, particularly in the 1980s and 1990s. There is insufficient space in this chapter to plow through the details of regulatory practices late in the century; others have described these. However, one should keep in mind that new offerings, and the IT infrastructures and applications required to deliver and support them, were always subject to regulatory permission. Since state and federal regulations governed the activities of all three industries, that regulatory environment remained complex right into the new century despite a philosophical desire on the part of public officials to open up markets.[34]

Patterns and Practices in the Adoption of Computer Applications

The continuous influx of new uses of computers across all financial industries in the second half of the twentieth century ultimately played a significant role in how these industries operated; no company avoided being influenced by the digital. The three major components of the financial sector illustrate several patterns of behavior in integrating technology into this part of the service sector of the economy and fundamentally changing the nature of work.

In each financial industry, managers were a cautious lot, taking their time to learn about computers and their potential. They already had processes, dating back to the dawn of the century, that integrated use of adding machines, calculators, tabulators, and industry-specific devices of various kinds (such as the "ticker" in stock trading). Their prior commitment to and use of existing technologies served as both a braking mechanism and confidence builder while their managers looked at the potential of the computer. The effects varied, however. Banking's prior experience with precomputer IT served to slow down its adoption while insurers pushed ahead, becoming some of the most intense computer users in the U.S. economy in the early to mid-1950s.

Precomputer technologies were not ideal, giving managers reason to look at new forms of automation, including use of computers. For example, in the Banking Industry the surge in paperwork in the years immediately following World War II—especially check handling—strained the capabilities of existing technologies and processes for handling the growing volumes of documents and transactions. An official at the Mellon National Bank and Trust Company recalled the circumstances facing his industry:

> Since World War II, however, they [banks] have experienced such a growth in the demand for normal banking services that these capabilities have been seriously challenged. Not only has the growth in traditional bank operations served to increase the industry's volume of paper work and record keeping but it has been further intensified by the addition of many new banking services . . . all of which have added significantly to an already staggering work load.[35]

Similar thoughts could have been voiced by executives in insurance and brokerage firms as well. Built into the increased volume of business, of course, were rising costs. Even when these industries were heavily regulated, as they were in the 1950s through the 1970s, they were subject to the vagaries of changing interest rates and novel forms of competition that put pressure on their balance sheets, especially in the 1970s, creating incentives for constant improvements in productivity.

These industries were initially introduced to computer technology by their industry associations, magazines, surveys, and studies. The pattern was set early: editors of industry magazines introduced their readers to every new technology (hardware and software), followed up with testimonials about successes by early adopters, and subsequently published surveys on who was using the technology and for what. Information on computing was light in the 1950s, grew into a steady stream in the late 1960s and 1970s, and then appeared in large volumes in the 1980s. By the 1990s, almost every issue of every industry publication carried multiple articles on the use of computing. (Industry conferences followed a similar pattern.) There exists a strong correlation between the volume of publications data on deployment of the digital and the adoptions of computer-based uses of technology soon after an industry introduced it to its members. Each firm was afraid of being bitten by a competitive disadvantage if they did not move, or saw the potential for gains in productivity. However, data from surveys on deployment suggest that the process of embracing a new use of computers remained slow when compared to many other industries.

It is important here to acknowledge the role of computer vendors. They called on executives and technicians, made presentations at industry conferences, and published articles in their journals and magazines. The key selling strategy was to show industry decision makers how these technologies could be applied to streamline existing processes or to enhance a firm's capabilities and competitive posture. For most computer vendors, financial industries were major customers throughout the entire half century.[36]

One pattern of behavior within the Financial Sector more prevalent than in the manufacturing or transportation industries was the extensive role played by industry committees and organizations. Many of the activities in the Financial Sector required firms to transmit paper and information back and forth to each other, with the result that they needed shared standards for documents, information, and practices to facilitate smooth operations. For example, checks had to be of a common type, and a set of technical and operational standards had to be instituted so that billions of these pieces of paper could be physically moved around the country through millions of organizations, then through the tens of thousands of bank branches, and ultimately to the banks of origin for debiting accounts and storing. Initially, banks could take days to accomplish the task, but in time the Federal Reserve mandated faster clearing and crediting of transactions. The required synchronization could not occur just through automation and the use of computers by individual banks. Without a high degree of process coordination across the industry as well, check management could not occur. Examples of necessary synchronization could be cited in insurance and brokerage too.[37] The point is, coordination came early to these industries, indeed long before World War II. They also had high information-technology components from the earliest days, including extensive use of telecommunications. One did not see that level of coordination in large U.S. manufacturing industries, for example, until the 1980s.[38] Financial industry associations created the necessary study committees that made the various recommendations essential to facilitate migration from one standard or practice to another during the entire half century. These industry groups proved to be an important part of the history of how computers were adopted in the Financial Sector and, indeed, have continued to play a major role in influencing how the digital is used to the present.

Did the Financial Sector adopt applications from other industries? The evidence is mixed; the short answer is they did, but they did not do so to the extent evident in manufacturing, transportation, wholesale, and retailing industries. The Financial Sector was unique in its data-handling requirements and applications, and many uses in other industries just did not fit there. However, three general trends are evident. First, when a technology appeared of a general purpose type (such as the IBM S/360 in the 1960s or the personal computer in the 1970s and 1980s), industries across the entire economy studied and, then adopted them. Second, all industries looked at how to use specific technologies in ways unique to their own industries (such as database software in the 1970s and 1980s), often learning from early adopters in other industries. Third, industry associations across the spectrum focused on industry-specific problems (such as the MICR coding standards for checks in the 1950s and 1960s). Often, firms that did not have specific in-house skills reached out to other industries for management and programmers or to technology vendors and consultants for help.

Financial Sector managers, and by direct implication their industries, acted very conservatively, appearing even slow to adopt computing technologies. Why? These were three industries deeply experienced in the use of information technologies. I touched on the issue above—they had existing systems that worked—but there is more to the story than that. First, all three industries worked in a relatively stable economic environment; the combination of growing postwar demand for their services and a highly regulated world (read, where de facto franchises existed) favored status quo operations in the first three decades following World War II. To change operations posed more risk than not. However, as disequilibrious economic changes occurred in the 1970s (such as the oil crisis, inflation, and rising interest rates), competition and the squeeze on profits forced the hands of many managers, compelling them to take dramatic steps to improve their operational and marketing efficiencies. This happened at a time when dramatically improved digital technologies and telecommunications became available and more cost effective than earlier options. These developments stimulated the use of computing applications.[39] Furthermore, and perhaps ultimately more influential, deregulation in the 1980s led directly to the transformation of these industries by making it possible, for example, for firms to sell products and services previously restricted to one industry (insurance companies could sell investment instruments, banks could operate in multiple states) and international markets.

Three profound consequences resulted from the combination of changes in economic realities, use of digital technologies, and deregulation in all three industries. First, companies within these industries aggregated themselves into larger enterprises to gain economies of scale (productivity) and to expand their scope for economic growth, more frequently through mergers and acquisitions than by organic growth: companies became larger and their numbers fewer. Second, in all three industries over the half century, firms introduced a larger variety of services and offerings than they had before. For example, a customer walking into a bank in 2004 had access to dozens of services that simply were not available in 1950. This same customer also had more options on where to get those services: A customer in 1950 could only write checks drawn from a checking account at a bank; by the end of the 1980s, it was not uncommon to do so from a money market or brokerage account; and in the early years of the twenty-first century a customer could manage all financial affairs—checking, investments, and insurance—from a terminal at home, using widely available software products bought at an electronics store.

Third, and of enormous concern to managers in all three industries by the end of the 1970s, were the capabilities computers gave them (and their rivals) to pick off pieces of a competitor's business. These industries are full of examples: credit cards coming from multiple industries, large corporations bypassing the Banking Industry to fund debt, certain types of insurance acquired in the form of warranties, and so forth. When economists and industry experts talk about the desegregation of an industry, this is what they are referring to: the niche product and market attacks made on an industry. There is little question that the Financial Sector proved more susceptible to this kind of problem caused directly by the availability of information technology than any other industry in the second half of the twentieth century. Some firms were essentially neutered by the process, while aggregate profit levels began falling over time as practices changed. For example, the high brokerage fees charged customers to buy and sell stocks between the 1950s and the 1980s became a thing of the past when, in the 1990s, customers could conduct their own transactions online for a fraction of the cost of earlier times.[40]

The Emergence of a New Style of Business

All of these changes beg the question: Did the use of computing fundamentally alter the way firms operated? Did a new style of doing business emerge by the end of the century that did not exist in 1950? These are important questions to ask, because if the answer is yes, then we are entering a new way of doing business, perhaps the Information Age, or the Third Industrial Revolution, or even some other form for which we have yet to decide on a label.

The implications for management are even greater: the emergence of new business models, different sources of profit and loss, new sources of competition, sales, multiple channels of distribution, regulations, and value chains. Critical questions concern what to invest in, cost of admission to an industry, how to structure organizations, and even what new personnel practices are required. For industries that were historically risk adverse and are still emerging from the heavy hand of regulations, the prospects seem daunting, even frightful, made all the more disturbing by the fact that churn has been one of the hallmarks of this sector for over two decades, and there is no evidence to suggest a period of stability is coming.

Yet, much as occurred in other industries and the public sector, sufficient change occurred to make it possible to answer these questions in the affirmative; in short; yes, a new style emerged while providing industry management with a context in which to carve out new chapters in their future history.

We can borrow a useful model for defining the transformation from manufacturing. Historians of manufacturing industries have increasingly adopted the notion that there were styles of operation that changed over time, beginning with craftlike manufacturing (with its one-of-a-kind approach) evolving to the mass-production approach (known as the Fordist style)[41] and then to a digital style; other experts emphasize the coordination of suppliers and fabricators (the Toyota approach).[42] The use of information technology and reliance on more data proved a large change for manufacturing, transportation, wholesale, and retail industries.[43]

More than most scholars in recent years, economist Andrew Tylecote has argued the case for looking at manufacturing in terms of styles.[44] Taking our cue from him we can look at all the industries reviewed in this book and measure them against this framework. Briefly stated, one looks at how things were made or done, the way organizations were structured and evolved over time, and how people optimized work and generated profits. For instance, the introduction of electricity made mass-production assembly lines easier and more cost effective to run. The merging of telecommunications with digital tools (computers and software) produced the Internet, which later enabled home banking and the online purchase of securities. Such profound changes helped stimulate mergers of local and regional banks and stock brokerages into large national enterprises. The long-term move of information storage from paper to computers fundamentally changed the nature of how people worked, what skills they needed, and how many were required. Those kinds of changes made up the new look and feel of an industry, what Tylecote promoted as a new style.[45]

The financial industries described in this book experienced the same kind of transformation. Banking experienced a sea change as information technology altered services and products, how services were delivered, and how the industry's organization (and that of its member companies) evolved over the last half of the twentieth century. Insurance companies also underwent significant change as early adopters through their move to computers, while early and intense, proved initially less revolutionary.[46] The Brokerage Industry changed perhaps even more than did banking. Information technology, in hand with regulatory changes, knocked out fixed fees, facilitated the creation of large national chains, and led to an explosion in online trading by individuals. That last change was also accompanied by consumer access to enormous amounts of financial data that had been the province of brokers in prior decades. Many activities and circumstances continue to unfold during the early years of the new century.

Given all these transformations, we can begin to speak about the existence of a *digital style* by the end of the century. However, as of this writing (2005) all three industries are still finding their digital styles, because so many processes are being redesigned to exploit the Internet and other digital tools, and the consequences of mergers and the effects of foreign competition (also facilitated by the use of information technology and telecommunications) have yet to play out fully. In addition, the digital style is developing at various speeds and extent in all the industries looked at in future chapters. The key finding of this book, and that of its predecessor, is that over the past half century, fundamental changes have occurred in how the American economy works, driven by a combination of business, environmental (that is, regulatory) circumstances, and the extensive use of information technology.

Most important for today's managers, this process is far from over, because leveraging the Internet is just starting. There is the history of several significant prior information technologies to learn from, most specifically how extensively a technology (and its applications) were adopted before they were considered fully deployed and their derivative consequences were obvious.[47] Based on that historical record, one can expect to see further changes driven by existing information technologies until at least 2015 or 2020—in other words, during the remainder of the working lives of every middle and senior manager alive today. While vendors in the information processing world would argue that use of the Internet is occurring much faster than it is, surveys on deployment suggest that all the industries

ATMs changed how people received money and conducted financial transactions. This photo illustrates an ATM, circa 1980s. Courtesy Bank of America.

studied in this book are now (a) adopting this technology at a rapid pace but (b) in an evolutionary manner, integrating it into their operations when it makes sense and is cost justified.[48] Once a round of adoptions occurs, there are consequences that, in turn, lead to new applications, different forms of competition, and so forth. These, in turn, then cause more evolutionary changes. That is why we can speak of dramatic changes and about evolutionary transformations in the same breath and think of these as taking many years, even decades, to occur.[49]

The issue of style has also affected the technologies themselves. For example, the deployment of ATMs in the 1970s led to pressure by banks on manufacturers of such equipment to add new functions to those devices during the 1980s and 1990s. The same thing happened in the Grocery Industry, which made possible today's ubiquitous bar code.[50] Banks and brokerage houses pressured information technology firms to improve the quality of security software in the 1980s and 1990s, and even successfully lobbied the U.S. government to allow companies to use encrypting software (developed for the military community during the Cold War) to secure transactions conducted over the Internet in the late 1990s. Regulators, who were almost maniacal in demanding that financial firms maintain a high level of data security to protect the privacy of a customer's financial records, also proved to be a compelling force in the information technology world by demanding the use of more secure software.[51] In fact, how computer hardware, software, and telecommunications firms interacted with U.S. regulatory agencies is an important story that has yet to be fully told but that ultimately will probably demonstrate that technology's response to new business opportunities was influenced in significant ways by the activities of regulators and federal legislation.[52]

It is important to remember that, as the emergence of a digital style occurred, several other circumstances either stayed the same or evolved more slowly. First, and foremost, the banking, insurance, and securities industries may have evolved to the point where they were competing with each other, thus confusing their borders, but they did not go away. There is still a Banking Industry, for example, complete with its associations, magazines, annual conferences, and lobbyists. The same is true for the insurance and brokerage industries. Employees within these also identified with their industries in 2005 just as they did in 1950. The importance of this sense of identity cannot be overemphasized, because it was within the context of an industry's attitude toward and initiatives in computing that individuals and their firms embraced information technology and experienced the consequences of their actions.

The way companies were commanded and managed at all levels remained essentially unchanged over the entire period. That is to say, CEOs and CFOs had the same roles and missions in 2005 as they did in 1950; it was how they carried out their work that altered over time. That they engineered more mergers in the second half of the century than in the first, for example, did not change the fact that profits were made the same way. Life insurance salesmen still sold insurance in 2005 as they did in 1950, because their products remained so complicated that buying such offerings on the Internet remained illusive. Customers usually still had to sit across a table to talk to an insurance or financial expert to understand what it all meant. Banks were unable to get rid of brick-and-mortar branches all through the half century, though brokerages were more successful in eliminating them. Future chapters examine the "why," but the short answer lies in the nature of the products and services offered. One could easily acquire telephone service by calling a telephone company, which was much simpler than buying insurance over the telephone, despite heroic efforts by the Insurance Industry. On the other hand, handling claims over the phone, linked to large databases, was as possible in the Insurance Industry as it became for a telephone company to do. Brokerage firms still wanted to meet with clients but, in time, business could be done over the telephone as well.

It may seem a little odd that we have to acknowledge that the core role of a company and its industry remained essentially intact. After all, the basic laws of economics were not abrogated by information technology—or were they? In the 1980s and 1990s, a substantial debate occurred precisely over that issue. The debates included discussions of "productivity paradox" and the nature of technological influence on basic economics.[53] A distinguished duo of business professors, Carl Shapiro and Hal R. Varian, reduced the debate's intensity with a best-selling book that argued the obvious: that the laws of economics had not been retired because of the use of such technologies as the Internet.[54] However, people had confused the effects of technology on business strategies, which were strongly affected by innovations in IT.[55] And sociologists, doing good work, also wondered if in fact change was so profound that maybe economics *was* affected.[56] Even within the corridors of the Federal Reserve System, studies were conducted by economists to define the effects of technology, but ultimately most concluded that the laws of economics had not been abrogated.[57] I consider the issue of whether fundamental economics changed as a result of new information technologies. Furthermore, I do not anticipate that any changes in this technology in the foreseeable future will affect this fundamental conclusion. Firms still have to offer services and products that the public will buy and use; they have to make a profit doing so to stay in business; and competition will always be present, putting enormous and constantly innovative pressures on rivals.

To be sure, companies reorganized internally and within industries—but they also did not. If one were to look at the largest companies in each of the three industries in 1950 and compare this list to that of 2000, many of the names would be the same. What did change, however, were the smallest players in each industry. Seismic consolidations occurred as tiny firms in all three industries either went out of business or consolidated into regional then national enterprises, all made quite possible by the extensive combined use of computing and telecommunications in lockstep with regulatory permission. So absolute generalizations can be dangerous to make, but broad patterns were becoming evident.

Finally, what these three industries did, and their purposes in the economy, remained the same. Insurance companies sold insurance, banks cashed checks and took in deposits, brokerage houses sold stocks and bonds. They did that in 1950 and continue to do so more than a half century later. Computing and telecommunications only facilitated the execution of those activities. When information technology was subsumed to the purposes of the firm, deployment of IT worked well enough and created consequences both predictable and unintended that, in turn, caused new rounds of changes both welcomed and feared. That is why we can discuss how computing in banking, for instance, made it possible to offer new services faster and less expensively, but also helped insurance, brokerage, and even manufacturing and retail firms to pick off banking and brokerage products to sell. Sears's financial counseling services, brokerage firms' banking products, or AT&T's credit cards were opportunities and threats facilitated by the use of information technology. But the fundamental tasks of each remained the core of their businesses. The use of computing reinforced that reality not just in the financial industries but also, for example, in the telecommunications and media industries.

Ultimately, however, computing extended across many departments within firms and to all sizes of companies, made possible by the availability of new applications that could be carried out. While more is said about this point in subsequent chapters, a brief look here at some evidence helps reinforce the point that computing served the core missions of firms, and that the availability of applications (usually software but also knowledge about how to use computers for specific tasks) proved important. Figure 1.1 is a chart used by IBM to train its employees about the activities in the financial sector in the 1980s. It depicts both the normal functions of banks and also functions for which IBM either had hardware or software products. The figure is useful at two levels. First, it demonstrates that the time-honored work of banks had not changed profoundly from what might have

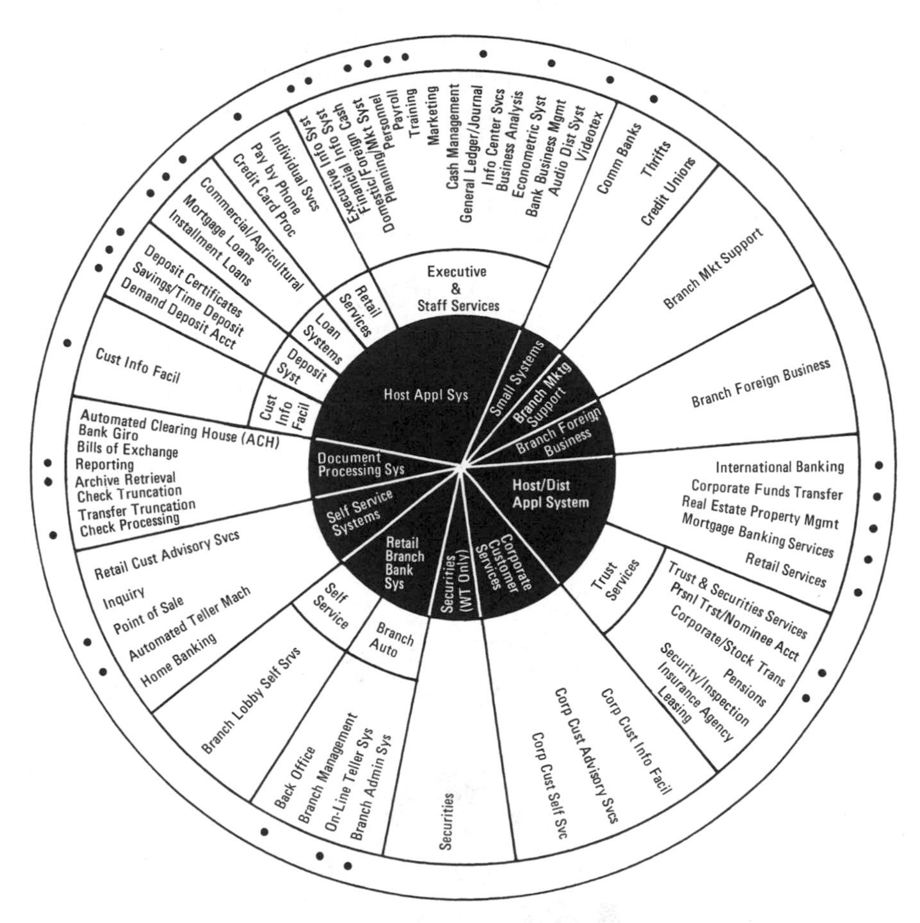

Figure 1.1
Financial applications, circa 1985. Courtesy IBM Corporation.

been done in 1950; second, it shows, that there were new offerings available and consequently new tasks to be performed. For example, we see there is a category called Payment Services—something banks did in 1910, 1950, and 2000—but also within that part of the chart one sees funds transfer which, by the mid-1970s, was available electronically (for example, one's paycheck could be deposited to a checking account electronically without use of paper). Document processing was another historic activity present over the past century, but note all the black dots suggesting that there were information technology tools available to facilitate a bank's ability to perform those tasks.[58]

The timing of this publication is informative because, by the late 1980s, the industry had been sufficiently de- and re-regulated to see the results, creating pressure on them to innovate and to become more competitive. The wheel chart demonstrates that novel use of computing was already well underway to improve the variety of services and to reduce operating costs. In the accompanying text for this figure, an anonymous writer noted that, "the retail customer is a major competitive target among financial institutions." Furthermore, "there is a growing recognition of the role of technology in increasing personnel productivity, increasing management control, and reducing the paper processing requirements of the industry."[59]

Figure 1.2, also from the same IBM publication, demonstrates a familiar pattern of already available applications for the core functions of the Insurance Industry. Similar statements existed about the role of IT for the purposes of making firms productive and competitive.[60] In the same publication, an IBM writer de-

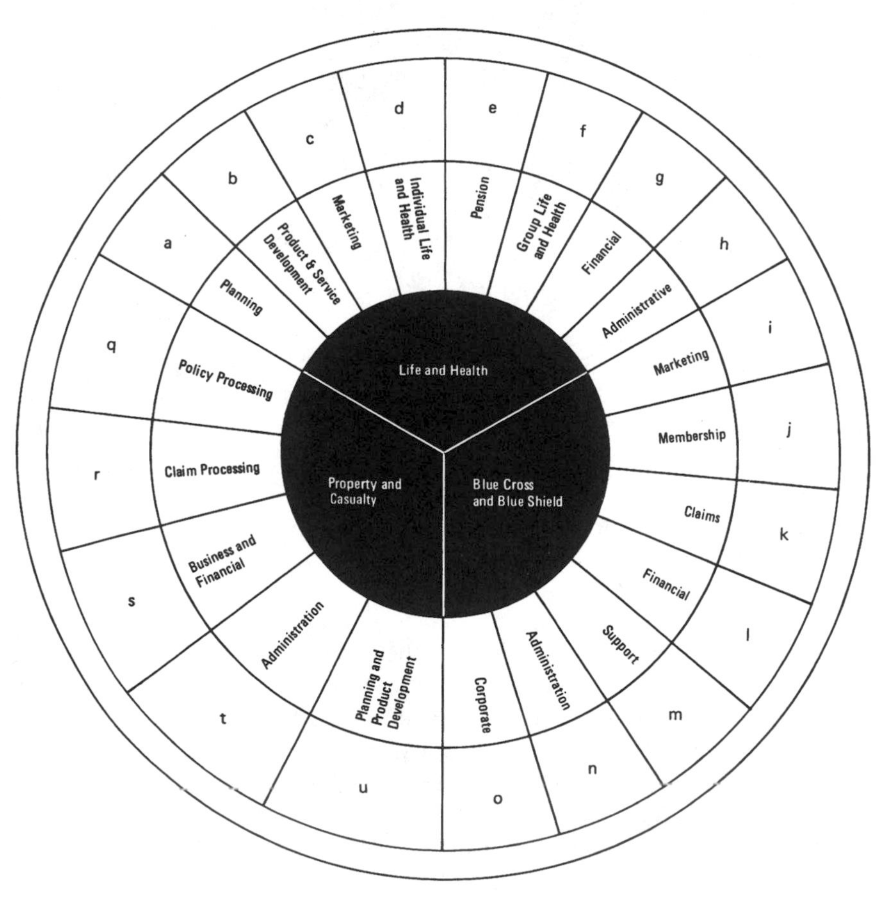

Figure 1.2
Insurance Industry Software Applications, 1988. Courtesy IBM Corporation.

scribed existing products that support back-office accounting, branch office information systems, trading, research and analytics, and investment management—cataloguing dozens of offerings, all in support of the core functions of the Securities Industry.[61] In an earlier version of the same publication, released in 1985, a similar wheel chart (figure 1.3) shows a slightly different pattern with regard to securities. Unlike banking and insurance, to whom IBM offered products in almost every functional area of the business, inside the securities wheel we see that the company focused on general security applications. In the late 1980s and early 1990s, IBM, along with numerous other firms, provided software products in support of many of the other functions noted on the chart, particularly for broker dealer systems.

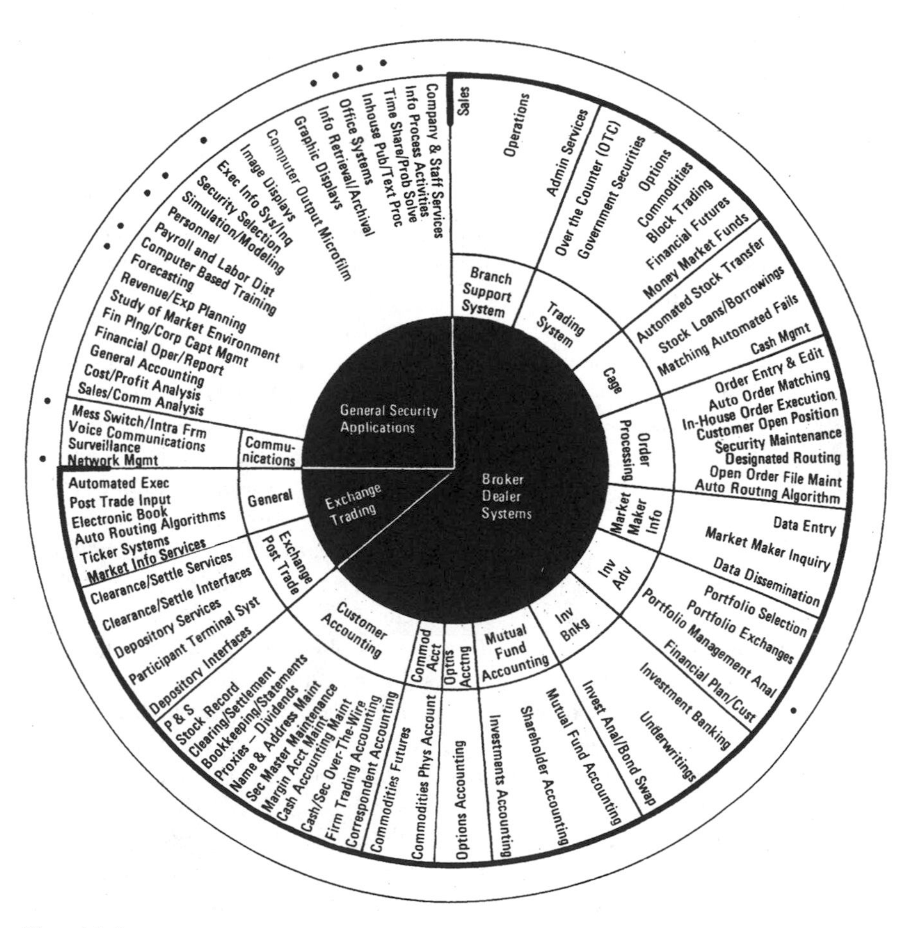

Figure 1.3
Security Industry Software Applications, 1985. Courtesy IBM Corporation.

In the late 1990s, many software firms created new products and services that began integrating the use of the Internet into these core functions.[62]

Changing Patterns of Consumer Uses of Financial Services

Finally, we should ask, did residents of the United States undergo some sort of change in their style of interacting with the three financial industries? There are three possible ways to answer this question that help us understand the role and effects of IT within the American economy. First, we can look at the services consumers embraced that were affected by technology. Second, we can identify how consumers interacted with these services through the use of IT. Finally, we can step back and ask what effect the wide availability of financial services had on how individuals managed their financial affairs—in other words, the influence of these industries on the mundane behavior of people. Answering these three questions makes it possible to determine if the notion of style can be extended to the individual and not simply be constrained within the borders of an enterprise or industry. The more that style differences have emerged across the economy, the more reasonable it would seem that fundamental changes did take place in the economy, thereby warranting a new view of how it is structured and works. The attributes of a new style also would allow future historians and students of business not only to name the era but also to document how to be successful in these new circumstances.

Let's begin with the first question, services consumers embraced. What will be shown in the next several chapters is that most new services and products offered by all three industries became popular with the American public. In banking, the various interest-bearing accounts, mutual funds, credit cards, and the ability to move funds from one account to another became ubiquitous by the end of the 1980s. Each of these offerings would not have been possible without IT applications hard at work. The instituting of telephone-based services, by which individuals could call their bank at any hour to check on balances or move funds from one account to another, proved equally popular. The same could be said for ATMs once the public came to regard them as reliable and safe, a situation achieved by the end of the 1980s. Although over one-third of all noncash transactions are now done via credit and debit cards, two-thirds are still conducted with the old paper-based checkbook or cash. This leads to another observation about consumers: they did not embrace all changes proposed by the industry, and the elimination of paper checks was clearly the most important example. By the end of the century, however, one can say with reasonable confidence that adult Americans who were either in the workforce or living in a home with a proverbial breadwinner not only used a variety of banking services that did not exist in 1950 but also had become comfortable with them. It was not uncommon for Americans of various socioeconomic classes to carry with them more than three credit cards, to use ATMs several times per week, and to manage multiple types of bank accounts.[63]

All through the second half of the twentieth century, the purchase of insurance products by both enterprises and individuals expanded, and nowhere more so than by the middle class. As that group in American society grew rapidly in number, beginning by the early 1950s, so too did their use of insurance. As they acquired mortgages they bought home-owner insurance along with life and automobile insurance. One could track the acquisition of homes and automobiles, for example, to see the phenomenon at work.[64] States even passed laws making it mandatory for individuals to carry automobile insurance, while banks insisted on fire insurance for homes they were backing with mortgages. As the vagaries of regulations evolved, along with interest rates and other offerings from within the three financial industries, people bought investment products from insurance companies, such as whole-life insurance. As the products evolved, American appetites for insurance products grew. Employers established a complex network of private/public health and life insurance programs for their employees, and by the start of the new century, 140.9 million Americans had some form of health insurance and 22 million were covered by "traditional" pension plans (excluding Social Security and 401(k)s).[65] In short, the social safety net first developed with the creation of Social Security in the 1930s, and expanded by other public programs to enhance medical, retirement, and unemployment protection, was further bolstered by the private efforts of the Insurance Industry to sell products that the expanding middle and upper classes of the American economy widely accepted. But even the poor were not left behind, as both pre–World War II government and the Great Society programs after the mid-1960s demonstrated. In short, just as Americans were acquiring a portfolio of banking services, so too did they accumulate a myriad of insurance products not available to earlier generations.

A similar pattern became evident in the world of securities. During the 1920s, Americans from all walks of life had "played the market," almost with the level of interest and intensity seen once again in the late 1990s. But in the 1930s and 1940s, American consumers backed away from bankers and brokers (many mistrusted them, on average having lost money in stocks). One can track the growing participation of consumers in the stock market as the last half of the century progressed. After the mid-1980s, the stock market experienced a huge positive surge in part thanks to the introduction and popularity of 401(k) retirement savings plans. Under these provisions, employees could set aside a certain pretax percentage of their salaries, matched to some extent by employer contributions, and invest in stocks and bonds. These accounts could be fed automatically through employee payroll deductions (all accomplished by payroll applications of computers). By the end of the century, approximately 55 million individuals had a 401(k) or similar type of retirement account.[66] Such huge numbers of individual investors entering the market in the last decade of the century were facilitated by consumer access to discount brokerage servives, the availability of technology to make trades (using a combination of PCs and telephone lines), and a booming economy.[67] While the recession at the dawn of the new century dampened investor enthusiasm for stocks, consumers continued to participate, maintaining or adding to their 401(k) and other investment portfolios. While the expansion of the middle class had more to do with Americans supporting an affordable appetite for stocks, regulatory changes, tax policies, and the availability of computer-based applications to deliver and manage these offerings, all combined to change fundamentally the mix and volume of people's net worth. In 1950 the largest investments most people would have had were their homes and/or automobiles. By the end of the century, while their homes were still their largest investment, it was not uncommon to see homes matched in dollar value by 401(k)s and other retirement/investment assets.

In short, over the half century, consumers acquired various banking, insurance, and brokerage-offered products. As time passed, they also embraced a wide variety of technologies to interact with these products. In 1950, a consumer had to walk into the offices of any of three industries to conduct business. By the end of the 1970s, consumers began having technically based options for interactions: electronic funds transfers of paychecks to checking accounts, telephone customer-support centers with banks of knowledgeable employees who could conduct business and provide information, ATMs with a growing list of applications, and use of PCs and the Internet to conduct transactions. At the dawn of the new century, over two-thirds of all homes had PCs, and over 25 percent were connected to the Internet.[68] Thus, the essential element for the use of computer-based financial applications was already in place. To be sure, there were failures and setbacks along the way. Consumers did not like home banking using telephones linked to TVs or special terminals, which were impractical and expensive. They took to credit cards quickly but were very slow to embrace debit cards[69] and they wrote more checks at the dawn of the new century than ever before, in fact, 36.7 billion just in 2003. In addition, the Insurance Industry has not yet found ways to interact with consumers through the use of the digital to the extent that bankers and brokers did.

What then, did all of these trends lead to? For one thing, we know that the public's appetite for financial products and services grew steadily all through the half century. Two students of corporate demographics, Glenn R. Carroll and Michael T. Hannan, identified the cause and effect and the balances involved in how the consuming public approached these industries, citing the case of banking in particular:

> As the economy expanded, more households and firms required banking services, ensuring a steady flow of new customers. Interest rates were low, so consumers—whether individuals, small businesses, or large corporations—became loyal customers.[70]

While their observations were based on the period before deregulation and heightened competition became prevalent in the 1980s, their perspective held for the entire half century because, from customers' viewpoints, individuals and firms simply treated all three industries as alternative sources for products and services for the reasons stated above: the growing economy, the availability of funds to borrow, social and tax policies that encouraged savings, and the financial necessities of a rapidly growing middle class.

Access to an expanding array of financial products and services altered how individuals behaved financially. To be sure, large corporations had long used every available financial service to advantage, but it was not until after World War II that small businesses and individuals began to do the same. But, begin they did and that change in behavior is a major source of evidence for the notion that people too acquired a new style of operating—one, in fact, more important than whether they used PCs to do their banking or walked into a brokerage firm to buy and sell stocks. The distinguished historian Daniel J. Boorstin pointed out that the ability to buy automobiles through installment payments in the 1930s started the whole movement of consumers considering the acquisition of large and expensive goods through loans, a strategy for purchasing automobiles well in place by the end of the 1950s, along with the practice of taking out mortgages for acquisition of homes.[71] Starting in the 1950s and 1960s, petroleum companies extended credit to consumers buying gasoline for their cars, thereby further reinforcing the emerging consumer practice of not paying with cash or checks for everything that one acquired. These practices of using debt and credit became important building blocks of the nascent consumerism that characterized so much of the American economy after the mid-1960s. This pattern of routinely leveraging debt and credit was further reinforced by a slough of credit cards, such as American Express, Visa, and MasterCard. BankAmeriCard alone had over two million customers in 1967, who charged $250 million. By the early twenty-first century, it was not uncommon for a household to charge 20 percent of their purchases to credit cards. Boorstin pointed out yet another change, a direct consequence of the credit card: "Credit, once closely tied to the character, honor, and reputation of a particular person, one of man's most precious possessions, was becoming a flimsy, plasticized, universal gadget."[72] He concluded in the early 1970s that "the cash customer had become the Vanishing American."[73] Though Boorstin overstated the demise of the cash customer, he was quite accurate in pointing out the rapid growth in credit loans and credit cards, both of which made it possible for consumers to acquire goods sooner rather than later—a shift away from prior practices.[74]

Conclusions

This history of financial industries demonstrates that computing became a major component of all the firms operating in this sector of the economy. As will be discussed in the chapters ahead, their annual expenditures of between roughly 7 and 15 percent of their budgets on IT was (and is) the least of management's worries. IT both liberates and constrains one's ability to succeed. With every new wave of important, transforming technology, the issues of success and failure, of efficiencies and productivity, exist. As Mary J. Cronin, an expert on business applications on the Internet, wrote, "The Internet has emerged as a key competitive arena for the future of financial services. Now the race for revenues, market share, and online advantage has begun in earnest."[75] She further commented:

> Moving financial services to the Internet creates a totally new competitive landscape. Instead of operating within clear-cut service boundaries, well-established financial organizations suddenly find themselves competing for customer loyalty and liquidity. What's more, the competition may come from a technology provider, a tiny start-up, or a telecommunications titan as easily as from within the industry. The advent of Web-based commerce has added new layers of complexity and unpredictability to the worlds of commercial and retail banking, mutual funds and brokers, back-end processors and front-end financial providers.[76]

Although Cronin published this comment in 1998, it could easily have been written in 2005, because the trend she identified was still unfolding. The central focus of this book is how this came to be the situation, and what effects prior uses of computing did either to enable or constrain managers as they attempted to run successful enterprises in the early decades of the twenty-first century. The history of these, and other industries, provides context for current events and suggests the extent of the paradigms these industries are either trapped in or may expect to move toward. But, just as Cronin responded to the circumstances of her day without significant benefit of historical insight, so too did managers operating in all the industries reviewed in this book.

The Financial Sector provides many sources of insight on the functioning of the American economy not available elsewhere. For a starter, as already noted, this sector has provided much of the financial plumbing for the economy, and that now makes it possible for other industries to enter the sector. So, while banking functions in the economy will continue to be needed, they will not necessarily be provided by the Financial Sector. Customers want insurance and stocks but many do not necessarily believe they have to acquire these from insurance salespeople or stockbrokers. Computing has clearly made real the new possibilities that are emerging in various nascent forms.[77]

The Financial Sector also serves as a large case study of how money and other instruments of negotiation became digitized over time, challenging even the wisdom of having actual coins and paper money issued by the U.S. Treasury Department. In a world still filled with barter, coins, and paper money, the notion of purely digital cash (actually, data that serve as cash) is an exception around the world, with only partial, advanced implementation in parts of Western Europe, North America, and in East Asia.

While pro-technologists charge forward with predictions of "cashless societies," they ignore the long-standing practices and predilections of consumers, individuals who accounted for two-thirds of the economic activity of the United States during most of the last half century. These consumers quickly embraced electronic securities trading but were reluctant to surrender their checkbooks. They loved their one dollar paper bills, but also made credit cards ubiquitous and, in the process, became the world's most extensive users of this digital tool. Understanding consumers' whims and catering to their desires made many financial institutions profitable and large; those who ignored them were relegated to a quick demise.

Historians hate to predict the future; being one, I share much of the same trepidation for the same good reasons, most of which boil down to the occurrence of unintended and unpredictable events. However, historians also admit that we all rise in the morning to work in a set of circumstances that existed the day before when we left work to go home for the evening. Economists are bolder in that they think the future is subject to certain principles that will apply in the future, an approach that applies here too.[78] I consulted much economic literature in preparing this book, not so much to steep the reader in theory as to understand trends.[79]

We know some rules of the road; perhaps the most important, to use the words of a distinguished historian, David S. Landes, is that "culture makes all the difference."[80] The same applies to the culture of industries, which should be evident with every industry case study presented in this book and its predecessor. Two observers of how enterprises have operated in American society, Thomas H. Davenport and Laurence Prusak, make the same point with regard to firms: "Culture trumps everything."[81] What firms have and do is so firmly influenced by their past that, following the lead of economists, we can safely expect existing realities to continue to influence future behavior. One quick example of heritage at work illustrates the normal course of events: when firms install new information technologies or applications, few dispose of all their old ones. Instead, they incrementally change their portfolio of equipment, software, and uses (applications) on a continuous basis. They fit new technologies into preexisting processes, only changing work steps after the fact. Management insists that newly refitted applications and processes work with, or side-by-side to, earlier ones. Banks in the 1960s and 1970s, for example, used state-of-the-art computers and later, databases, alongside adding machines and punched-card tabulating equipment. Stock brokerage firms used batch computer systems deep into the 1980s while simultaneously installing online account records. It made good sense to change this way because:

- It was cost effective.
- It minimized destroying careers and technical risks of failure.
- It proved least disruptive to ongoing operations.

The same held true for industries in the public sector, manufacturing, transportation, wholesale, and retail industries.

Consumers of all the these products and services changed as well. Just like the industries, their core missions did not change: people wanted careers, to raise families, to provide protection against financial and medical problems, and to enjoy an expanding standard of living. But they did not hesitate to embrace the new, frequently adopting computing and telecommunications as these technologies became easy to use and affordable. This resulted in their making financial industries important in their lives, leading to wide use of debt, credit, and different ways of accumulating wealth that would have been inconceivable to an American consumer coming out of World War II.[82]

The next chapter will focus on the role of information technology in the Banking Industry, describing how this industry changed, how it used IT, the extent of its deployment, and the effects of technology on the industry. To know what happened in banking is to understand what is occurring in other industries to one degree or another. While not a bellwether industry in adopting technology, it demonstrates many of the effects of computing in the services sector. It was easier for a services industry to pick up pieces of another services industry's markets than for manufacturing, retailing, or government. Nowhere was this possibility more evident than in the Banking Industry, and no single operational capability enabled that more than information technology.

2

Uses of Computing in the Banking Industry

Americans want action for their money. They are fascinated by its self-reproducing qualities if it's put to work.

—Paula Nelson, 1975

The history of the evolution of banking in America over the second half of the twentieth century cannot be written without discussing extensively the use of computing technology by members of the industry, by rivals in other industries, and by the customers. It is important to understand that no company, or industry, adopted computing simply for the sake of using computers. Computer applications were created to facilitate implementation of business strategies and to execute tasks efficiently and cost effectively. From the earliest days of the computer, bankers were aware that "the field of electronics" and "automation" held out the promise of improved efficiencies across their enterprises. They also knew that embracing these new technologies had potentially serious implications for their industry. In one of the earliest articles published on the promise of computing in banking, John A. Kley of The County Trust Company, of White Plains, New York, could not have been clearer:

> One can readily see that a realignment will have to be made in many instances in our present approach to problems. This will not be a mere substitution of equipment for existing manual or semimanual operation. It will represent a complete new philosophy in how to get things done.[1]

Leaders within the American Banking Association (ABA) also pointed out that industry coordination of technology adoptions would be required, thus ensuring that practical technical solutions would allow the greatest efficiencies to emerge that could then be leveraged across the entire industry. One contemporary member of the industry agreed that having the ABA lead the charge "seems to present one of the most complicated and urgent situations in the history of the organization."[2] Intra-industry cooperation remained an important hallmark of the history of computing in banking over the closing decades of the century.

By the end of the century, the world of banking had changed enormously. As one economist at the U.S. Federal Reserve noted in 1997, banking had become "the most IT-intensive industry in the U.S."[3] One of the most obvious consequences of this transformation was the decline in the number of banking companies, from over 14,000 in the 1950s to just over 6,500 by the end of the century. However, the number of bank branches rose from just over 14,600 in the 1950s to over 72,000. By the dawn of the twenty-first century, there were in excess of 324,000 ATMs installed across the United States. In 1950 nearly 100 percent of all banking was done through banks; by the end of the century, other financial markets had stolen share from the Banking Industry. Annual 3 percent growth rates for the industry over the last two decades of the century, for example, were normal, but so too were annual growth rates for the financial markets at over 13 percent.[4]

Industry watchers attribute the developments in banking to a variety of forces at work: regulatory changes, corporate consolidations, the implementation of a long list of technological innovations, changing appetites and practices of consumers and institutional customers, long-term expansion of the American economy, and introduction of effective competition from outside the industry.[5] Each had a role to play, sometimes greater, other times less so. For example, regulatory protection of the industry dampened enthusiasm for radical changes to existing processes such as those made possible by the arrival of digital computing in the 1950s. By the 1970s, however, with banks straining to implement new digital applications, deregulation and economic crises (such as spikes in interest rates and oil prices) jumpstarted a three-decades-long trend in bank consolidations and the birth of new sources of competition. The role of ATMs in the 1980s and 1990s was an important influence, as are the emerging effects of the Internet today. In short, the story of what happened to banking is complex, with IT written all over it. That story represents an important chapter in the history of the modern U.S. economy.

Evolution of the Banking Industry

In 1950, the banking system of the United States consisted of a network of state chartered and national institutions—a total of 14,676—and 12 Federal Reserve Banks, which served as a collection of bankers' banks, helping to move checks through the banking system, providing discount facilities, and injecting U.S. currency into the economy through private banks. By 1964, all commercial and mutual savings banks were members of the Federal Deposit Insurance Corporation (FDIC), which insured bank deposits. During the boom years from 1950 through the mid-1960s, combined bank assets saw a steady increase, rising from $191.3 billion to $401.2 billion in 1964. Most banks at the time were small, with deposits in the range of between $1 million and $5 million. In fact, in 1950 only 1,241 had branches (totaling 4,721), nearly half of which were situated in the same city as the headquarters office; national banking would came later. All banks operated within the warm cocoon of federal and state regulations that essentially made it possible for them to operate with a minimum of competition (even from within the industry) and a maximum of stability. The American Bankers Association played the role of useful lobbyist representing the industry's issues; it served as a coordinating point, looking at the effects of proposed regulatory changes, how to solve the growing problem of check volume, and other administrative and productivity issues. In short, banking was a highly regulated, very fragmented industry in these early years.[6]

Between 1964 and the end of 1973—a period of transition with global oil crises, growth in inflation, and an expanding American population—the industry grew slightly in size, with a total number of banks reaching 14,653 in the latter year. The more dramatic rise occurred in the number of branches (27,946), whose surge began in the second half of the 1960s. The industry was also a fair-sized employer with roughly 800,000 workers in 1960 and 1.2 million in 1973. Many of the employees were women, who made up 61 percent of banking's workforce in 1960 and 65 percent in 1973, though most were in lower level clerical and teller positions.[7] Work was predictable. As one executive in the period noted, "In banking's new world, the three traditional and basic functions—accepting and handling deposits, processing payments, and extending credit—remain essentially unchanged. It is the scope, type, and substructure of those functions that have been modified and expanded."[8] What changed occurred more in the second half of the 1970s, discussed below. Meanwhile, bank assets grew all through the 1970s, tripling over those of the 1960s.[9]

The massive changes that eventually came to banking started in the 1970s. They are important to understand because they represented a shift in strategy. Prior to that time, computing was designed to improve internal operational efficiencies, such as the handling of larger volumes of checks. By the mid-1970s, though, bank executives were beginning to worry more about competition, enormous variations in profits (caused by shifting interest rates and regulatory practices), and structural changes that came to fruition in the industry in the 1980s. All of those circumstances began influencing how the industry looked at computing technology. To be sure, bankers viewed technology as an aid to internal operations, much as they had in the 1950s and 1960s. But the introduction of new digital tools also made possible novel offerings and strategies, such as electronic funds transfer systems (EFTS) and ATMs, the latter holding out the promise that perhaps costly branch offices would go away (they never did).[10]

The financial landscape had been complicated for a long time; it was occupied by savings and loan associations, mutual savings banks, credit unions, life insurance companies, private pension funds, state and local retirement funds, "open-end" investment companies, security brokers, money market funds, and commercial banks. Several sets of events defined the 1970s and early 1980s such as the introduction of NOW accounts (demand deposits with interest) in the early 1970s; a growing number of bank failures; and the spread of NOW accounts across the nation in the late 1970s, which increased competition among all kinds of financial institutions for a consumer's deposit accounts. As competition heated up, bank managers became more aggressive in extending loans to individuals, businesses, even to whole countries, leading to a variety of innovations in consumer and commercial lending offerings. These included new ways of attracting funds, such as certificates of deposit, money market certificates, and even Eurodollars. The period also witnessed a large increase in the number of bank-holding companies that offered a wider variety of services across the nation than were provided by individual financial institutions. As noted in the early 1980s, besides deregulation and competition making this all possible, there "were technological gains in data processing, which were evolving toward giant capacities to hold deposits and clear transactions instantly."[11]

At the end of the 1970s, however, there still were many small banks; the great age of consolidation was yet to come. In 1979, for example, there were over 15,000 banks, with 75 percent holding aggregate assets worth less than $50 million. To put that volume of assets in perspective, in 1979 these 11,000 banks controlled less than 15 percent of all the assets held in American commercial banks. Just over 400 large banks controlled large amounts of assets (some 60 percent), which had been pretty much the story since the mid-1950s.[12] What is important to remember is that the largest banks were in the best position to exploit information technology all through the period, and also could afford to do it earlier than smaller institutions

IBM equipment being used to handle customer deposits and withdrawals, and check processing, 1963. Courtesy IBM Corporate Archives.

because of their advantages of scale and scope. While larger banks introduced computing all through the 1960s and 1970s to improve efficiencies, the great wave of productivity gains would come later. Blaming public policy that discouraged innovation by bankers in the 1960s and 1970s, some economists argued that, despite growing competition, new offerings came slowly into this industry and diffused equally lethargically across the economy: "Almost all the changes have been prepared externally and injected by energetic outside firms, often against the considerable reluctance of bankers."[13]

The number and variation of evolutions in the industry sped up in the 1980s, however, and continued at a rapid pace to the end of the century. For example, consolidations increased, occurring at rates of just over 3 percent annually, thanks to mergers and acquisitions (M&A). Results were cumulative and dramatic, with 11,894 banking institutions in 1984 shrinking to 6,578 in 2001, while the number of branches increased by nearly 50 percent from 50,000 to over 72,000. Gross total assets (GTA) grew by nearly 75 percent in the period ($3.44 trillion to $5.69 trillion).[14]

Bankers strived to become more competitive and thus worked to align technology, people, organizations, and investments in more innovative ways in the 1980s and 1990s. Increasingly, that meant relying more on IT. A few statistics suggest the kinds of alterations in the 1980s and early 1990s before the Internet began challenging the industry in additional fundamental ways. Megabanks doubled their assets in the period 1979 through 1994; small banks lost nearly half of theirs. Demand for loans increased all through the period, from $1.5 trillion to $2.36 trillion, and the percentage of loans to consumers rose from 9.9 percent in 1979 to 20.6 percent in 1994. Interestingly, one would have thought that with all the M&A activity, and the installation of so many ATMs and computers, the number of employees in the industry would have declined; in fact, that did not happen. In 1979 the industry employed nearly 1.4 million people; in 1994, 1.5 million. One could argue that since the volume of business conducted by the industry grew all through the period, the fact that employment only increased by a small amount was testimonial to the productivity gains achieved through M&As and use of technology.[15]

Regulators permitted more mergers; bank holding companies acquired more mutual fund enterprises, brokerage houses, and insurance firms, making it possible to offer a wide choice of financial products. Economists are not in agreement on whether M&A activity added or destroyed value, although the evidence suggests that the desire for economies of scale and scope were less important than were operational efficiencies, which remained high in the industry all through the period. M&A activities were driven more by the need to respond to competition from outside the Banking Industry and also from the ever-larger banks that expanded across the nation. Economists looking at the issue in the late 1990s concluded that "mergers and acquisitions . . . are a powerful force of change in the banking industry, impacting not only the geographical scope and product variety of the organization, but also affecting the underlying technological and managerial infrastructures of the banks."[16] They also noted that, in support of largeness and efficiency, banks were spending 20 percent of their budgets on IT by the mid-1990s.[17]

There was thus a high level of M&A activity at the dawn of the Internet's arrival—a level that was unparalleled in the U.S. industry's recent history. Keeping in mind how many little banks there were in the United States, in the 1950s, it is staggering to see that, by the end of 1995, the four largest banks in the United States held nearly 20 percent of the industry's deposits and some 23 percent of total assets. M&A activity had picked up momentum in the early 1990s, moving faster than at any earlier time. Between 1991 and the end of 1995, total bank deposits at the "big four" rose from 16 percent to 20 percent, and total assets from 17 percent to 23 percent. The numbers are made more dramatic by the fact that the sheer volume of deposits and assets also grew in this period. So too did the number of banks with assets of over $100 billion, increasing from four banks to 45.[18]

But perhaps the most important change that occurred in the industry during the late 1980s and early 1990s affected those who did the basic work of banking, such as collecting deposits and making loans. Two business school professors, Dwight B. Crane and Zvi Bodie, studied the shift, citing many examples: Sears Roebuck acquiring Allstate Insurance Group and Dean Witter, then spinning them off; NationsBank broadening its portfolio of products; State Street Bank & Trust Company choosing to narrow its scope of activities to those of investment managers; and so the list went on. These students of the industry concluded that "competition from nontraditional institutions, new information technologies and declining processing costs, the erosion of product and geographical boundaries, and less restrictive government regulations have all played a role."[19] They noted that the U.S. stock markets saw increasingly funded venture capital projects; that electronic transfer of funds played a profound role in the movement of money (such as via EFTS, credit and debit cards); that technology stimulated new ways of performing basic banking functions. Successful innovations were coming from various sources, not just banks, thereby fragmenting traditional roles of banks. Put in their words, the "old bundle of functions is fracturing."[20]

They cited the example of tight links between savings and mortgages breaking down as new mortgage-backed securities markets—which emerged in the late 1960s and grew in power all through the 1970s and 1980s—damaged bank earnings, compelling the U.S. government to authorize the Federal National Mortgage Association (Fannie Mae) and the Federal Home Loan Mortgage Corporation (Freddie Mac) to acquire bunches (often called pools) of mortgages. These agencies then converted them into collateral to back securities. Securitization thus made it possible for banks to sell off mortgages to other institutions around the country; this shifted assets away from banks to other enterprises, such as to pension funds, leading to further disaggregation of the mortgage business itself and reduction of the risk of holding bad loans. Thus, a loan could be negotiated by a bank with a consumer in one state, processed using another bank on behalf of the original bank, and sold to a third party. This chain of events proved to be a huge transformation for savings banks. Financial assets of consumers traditionally held by commercial banks also shifted, with deposits in these institutions declining from 49 percent in 1980 to 35 percent in 1994 as people moved funds to other locales, such as money market accounts. Companies also began acquiring loans from venture capitalists, not just from banks, and from other large enterprises (such as Ford Motor Credit Company).[21]

Finally, the general credit card business, initially dominated by traditional banks, over time consolidated into the hands of a few banks and, more importantly, into those of companies that specialized in credit cards. While local and national banks continued to issue credit cards in the mid-1990s, they issued them to customers nationwide, often through Visa and MasterCard, for example. By the mid-1990s, an individual or firm could conduct all banking activities without recourse to a bank.[22]

All the major studies of the Banking Industry of the 1970s through the 1990s discussed IT as an agent of change due to a variety of reasons: technical functionality, price/performance improvements over time, and so forth. But the business case can be stated in nontechnical terms. Computing in all its forms helped competitors in multiple industries gain access to information that had once been the total preserve of banks—thus enabling them to compete against the Banking Industry. Exploiting information, and hence IT, became possible only when regulators began allowing other industries to compete in banks' traditional turf—which occurred partly because IT made it possible. By the same token, banks invested most of their IT dollars in automating routine backroom processing, which contributed marginally to higher profits but also enabled some banks and many other institutions to leverage IT to enter new financial markets, taking share away from rivals.[23]

One unintended consequence of the injection of IT into the banking activities of the nation was the decline in cost of conducting financial transactions by individuals and firms not in the Banking Industry.[24] Put simply, it cost an employer less in processing fees to pay an individual in 2000, for example. In 1950 payroll was a labor-intensive process: the employer wrote an employee a check, supervised the process by hand, and mailed or delivered it by hand; then the employee took that check to the bank, where it was also processed by many levels of people. In 2002, software could calculate an employee's paycheck and in an instant deposit it to an account at a fraction of the cost of doing it in 1950. By the end of the century, many firms in scores of American industries billed and collected through EFTS.[25]

How Computers Were Used

To understand how information technology affected the structure and work of an industry, it is essential to look at how firms applied the technology in their strategic and operational activities. This process begins by looking at how IT was first used to improve existing core operations in the industry that were voluminous and expensive and to see those operations in non-IT terms. As that process of deployment evolved over time, it became possible for banks to use IT as a strategic tool, to help them expand business into more states (for example, issuing national credit cards, instituting ATMs) or achieve necessary scale to compete successfully. Only when that occurred did IT become so central to the functioning of the industry that it became the core processes, complete with their own names (such as EFTS, Internet banking). While the list of applications developed in the Banking Industry is large (see table 1.1, page 9), the essential ones can be boiled down to a half dozen, which are described in the remainder of this chapter and spill over into the next.

Crane and Bodie's reinforcing strategy for this approach can inform us about events dating back to the 1950s. My only significant departure from their approach is to insist that we answer more fully the questions about the extent of IT deployment by application, technology, and investment. This is less a criticism of their work than simply an enhancement, taking advantage of an historical record from which to learn. Their list includes six operational duties of a bank: tasks to make payments, pooling resources to fund projects, transferring economic resources, managing risk, collecting and using price information, and handling incentive problems. Their list of core functions is useful for the entire period, although I use more traditional banking terms with which to describe them.[26]

Demand Deposit Accounting and Check Processing

Most individuals and firms refer to bank accounts where they deposit cash, payroll checks, and so forth, as a checking account. Money goes in, and people write checks drawn on those accounts; banks pay those checks, reducing the amount in the account by the amount paid on the checks. Banks call these accounts demand deposit accounts. It is impossible to speak about checking accounts and check processing without discussing demand deposit processing. The lion's share of the paperwork and labor in banks in the twentieth century came from demand servicing of deposits and checks. In the mid-1960s, one banking vice president wrote that "40 to 50 percent of the nation's bank employees are allocated to the check-processing function," which at the time meant over 350,000 people processing over 15 billion checks.[27] The same executive described the check handling process, which essentially remained the same over the entire half century:

> Each of these checks must receive individual treatment. Some live exciting lives traveling halfway around the world; others have life spans lasting only a few hours and never venture out of the bank in which they were born. After being drawn by its maker, a check is eventually either deposited or cashed in a bank. After this, batches of checks are balanced and sorted to various geographical categories so that ultimately each check arrives at the bank upon which it is drawn. During the process, the batches of checks will be proven, that is, listed individually to arrive at a total amount for the batch, then sorted by each bank as they are routed through the check-clearing network.[28]

Aside from sheer volume, what made the process so onerous was the high content of labor involved in processing these checks. "It is estimated that each check is handled approximately ten times by an average of two banks before being returned to the maker in his statement. When these handlings are considered, it means that the 15 billion checks written in 1964 required 150 billion handlings!"[29] Let us impose on this banker one more time to describe the additional accounting required to complete the transaction:

> For every checking account there must be a signature card, ledgers and statements, or, in the case of automated accounts, balance journals and history records. Each customer who overdraws must receive an overdraft notice or a returned check notice. Changes of address must be noted appropriately . . . and reports for management and supervisory agencies must be produced.[30]

Years later an observer ensconced in the Harvard Business School wrote that "the early 1950s found the banking industry on the brink of a crisis. Check use in the United States had doubled between 1943 and 1952," and banks were predicting they would have to process a billion checks each year by 1955. "Banks were at a standstill, unable to expand, or, in some cases, even to keep pace with the increasing flow of paper."[31] Lamentations came from many places. For example, Paul Armor, working for the RAND Corporation in the 1960s, concluded in one of the earliest studies of computing's effects on the U.S. economy that it was a serious problem: "Next to government, the commercial banking system is the largest processor of paper. In 1964, for example, the system handled 15 billion checks in addition to its other financial transactions."[32]

The industry had long feared dire consequences from the rising flood tide of checks. From the early twentieth century on, banks sought ways to automate, that is to say mechanize, the basic tasks and processes for managing checks and checking accounts. At first, their challenge was to automate existing steps.[33] But by the start of the 1950s, bankers began dreaming of attacking the root cause of the problem—the check itself—by trying to get people to use alternative ways of moving money around; at first with newly designed checks, next with EFTS, then with credit cards, later with paying bills with home banking and debit cards in the 1980s, and finally, at the end of the century, with the Internet. As we have seen, they failed to kill the paper check, but they did manage to shift many payments to these alternative means. So, the history of deposit accounting and check handling is a story of incremental application of the digital to continuously reduce the cost of handling checks and to keep up with volume. Bankers did this while fighting competition that had alternative cash-handling services, such as credit cards.

All banks used noncomputerized office equipment, such as simple adding machines, in the 1950s to help in the processing and to do the necessary accounting. Most were electromechanical, some purely manual. Tellers also used these devices and shared procedures at the end of the day to "prove," that is to say, count and balance cash and checks received. Proof machines listed, sorted, and endorsed

Bank of America ERMA computer system, early 1960s. Courtesy Bank of America.

checks in one integrated process, and were ideal where large volumes existed. By the end of the 1950s, electronic check sorters could sort checks by customer, route, bank, and so forth.[34] The most famous of these early initiatives in automating check sorting and accounting was Electronic Recording Machine Accounting (ERMA), but more on that later.

Already mentioned in the first chapter was the work done in the 1950s to standardize checks by size and printing so that optical scanners could be used to "read" data on account numbers, routing information, and so forth, funneling data through punch card tabulators and later computers. This path was initiated by the American Banking Association (ABA), which began studying how to standardize checks in the early 1950s, issuing its first important report with recommendations on July 21, 1956. It called for the use of a standard Magnetic Ink Character Recognition (MICR) process as the basis for all bank-check processing, opting for a design already pioneered by the Bank of America and the Stanford Research Institute (SRI).[35] Figure 2.1 compares a check from early in the twentieth century and another from a period after the industry adopted the recommendation. The MICR coding on the newer check is at the bottom of the check. Bankers needed to be educated and they wanted to justify the cost of conversion. The ABA, along with U.S. Federal Reserve Banks, took the lead in convincing its members to adopt the technology by using industry committees, seminars, conferences, and articles published in its trade magazine, *Banking*.[36] In the first survey on the adoption of the technology, conducted in 1962 by the Federal Reserve Board, *Banking* reported "widespread use of MICR equipment and an increasing use in sorting."[37] Of the 974 Federal Reserve banks, 659 were preprinting checks using transit numbers and otherwise moving to the new format. Nearly half either had or were close to using electronic equipment to sort these new checks.[38] Another survey done that same year with 1,000 of the largest commercial banks demonstrated that these institutions were the first to deploy the technology. Of the 26 million checking accounts covered by this survey, 80 percent used MICR preencoded checks and deposit slips, while only 20 percent of the smallest banks used the technology only to process part of the paperwork involved in managing checking accounts.[39]

A major step forward in automating check processing took place when Bank of America began investigating how to solve the crisis posed by so many checks.

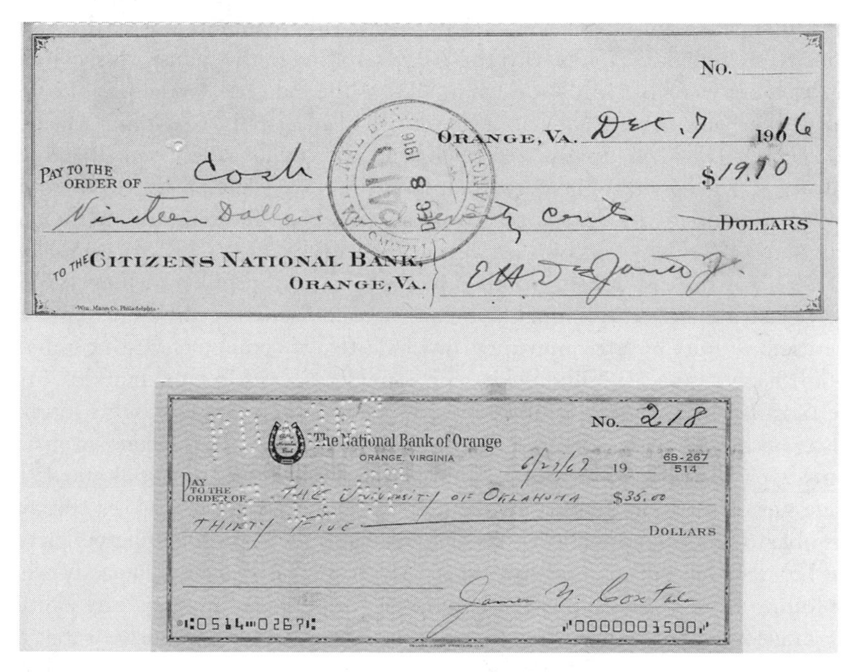

Figure 2.1
Checks over time. Note that the older check was larger than the newer one. They came in various sizes and fonts during the early decades of the century, while by the time of the second check, they were standardized in size, text, and format all over the United States. Also observe how the amount of writing on the check had been reduced, making optical scanning possible, with the technology capable of reading handwriting by the end of the century.

Working closely with the Stanford Research Institute (SRI) to identify a technical solution, the process spanned the entire decade of the 1950s. In 1950, Bank of America held 4.6 million accounts of various types, and its number of checking accounts was increasing at a rate of 23,000 per month. The bank's senior management recognized that the firm ran a significant risk of not being able to grow if it could not find new ways to process checks. With SRI, the bank explored alternative work flows, procedures and practices, and technical options throughout the early to mid-1950s, designing a computer system to perform basic bookkeeping functions then done manually for checks. These functions included credit and debit postings to accounts, maintenance of records of all transactions, maintaining current account balance records, responding to stop-payment and hold orders on checks, printing various standard reports on activities, and signaling to personnel when accounts were being overdrawn.[40]

Several lessons could be drawn from the experience, however. The bank spent the first half of the decade learning what it needed and how to apply IT and telecommunications, concluding it would have to change existing processes in dramatic ways in order to optimize technology most efficiently. By the mid-1950s, the bank realized it needed computing expertise in-house and so went out and hired it. The design of the new checks required the commitment of the entire industry, so it aligned closely with the ABA, resulting in the industry association's selection of this bank's MICR design in 1957 as the industry standard. In the mid-1950s, the bank made public its project, called Electronic Recording Machine Accounting (ERMA). In the second half of the decade, technical design and learning how to use computers with an IBM 702 moved the project to the point that by the end of the decade the bank and SRI had specifications for the construction of 36 systems, a contract won by General Electric.[41] ERMA went "live" in 1960, and the last system installed in the bank was installed in June 1961. It was one of the largest investments in computing made in the third quarter of the twentieth century by any company in any industry: 32 computers, costing millions of dollars, servicing 2.3 million checking accounts across 238 branch offices. Over the next year, the bank also moved nearly 2.4 million savings accounts into the galaxy of computers. To close out the Bank of America's story, the company added many applications to its original digital system throughout the 1960s and 1970s while supporting additional banks and more checking accounts. And yes, the bank was able to process more checks. Over time, it installed newer computers such as the IBM System 360 in the mid-1960s, which increased overall capacity while exploiting improvements in the cost performance of the technology.[42] In addition to using computers to drive down operating costs, the bank also insisted that the equipment itself be cost justified. As the cost of computer hardware went down, that became relatively easy to do. As one vice president at the bank recalled later of this period, "each subsequent hardware change has been predicated upon a reduction in unit operating costs."[43]

ERMA's story is important for several reasons. First and most obvious, this was a large, bold project that moved one bank into a new era of automation and electronics, driven by the need to scale up activities in a cost-effective manner.

Second, the project was so visible that it caught the attention of the entire Banking Industry and led other firms to begin implementing similar systems in the 1960s and 1970s. Third, as seen in so many other industries, those who moved first and most boldly to new digital technologies were always the largest enterprises, because they could afford the start-up expenses. In the case of Bank of America, the SRI study alone came close to costing $1 million; many more millions were spent before the bank even processed the first check or electronically updated an account. But over time, smaller banks were able to participate by using service bureaus and, as the cost of technology dropped, installing their own systems.

Vendors began introducing new systems and other products to exploit the check-processing applications, along with more integrated functions, mimicking the five accounting applications done by ERMA. For example, by 1966, IBM had introduced magnetic character readers that, combined with card read/punch equipment, worked with a 1441 computer in a service bureau; other banks began installing integrated systems from the same vendor.[44] IBM had bid on the ERMA project, but lost to GE, whose product was less expensive and perceived to be more viable. However, in the process IBM had worked out the details of how computers could be used in proof and transit applications and had started designing and manufacturing check-processing equipment. These were skills and investments that made the Banking Industry one of the most lucrative for the firm over the next forty years.[45]

By the mid-1960s, converting check processing to computers had become the single most important digital application in the industry. The ABA estimated in late 1964 that 79 percent of the largest banks in the industry had already converted, and that another 11 percent of the top tier enterprises were in the process of doing so. Among the smallest banks—those with assets of less than $10 million—40 percent had already converted, and another 40 percent were on their way. In total, about 66 percent of all banks used computers to process some or most of the check handling and checking account record keeping.[46] And nearly all banks in the United States were also encoding routing numbers on their checks with MICR ink.[47]

Debates about the "checkless society" did not stop, however. Bankers wanted it and lamented the fact that their customers did not. Survey data from the late 1960s offer evidence that the public felt the risk of forgeries remained too high, and that they liked writing checks; bankers responded by arguing that the public just did not know enough about the application and simply needed to be educated.[48] A survey conducted five years later (1973) showed nothing had changed: Consumers did not want any changes to the payment system, and bankers could not convince them of the benefits of even electronic funds transfer, for electronic posting of salary checks.[49] One writer, Peter J. Bannan, in 1980 uttered a now familiar lament in the industry: "Checks. Billions of checks. And every year more checks than the year before. The operational burden on the nation's banking system caused by this ever-increasing torrent of paper is a major cost for all banks, large and small."[50] This same writer posed a good question; "What of EFTS?" Following the classic model of the "turnpiking effect,"[51] EFTS did not ease the burden on banks but rather "increased it by making it possible to handle ever larger volumes of paper and by leading more of the public still further away from cash" and toward checks.[52] He accused banks of doing nothing to encourage firms and people to stop using so many checks. However, if computers had not already been in use, "it is conceivable that the entire system of payments and credit could by now have broken down, buried under an unmanageable and unceasing avalanche of paper."[53] This last quote calls out a fundamental consequence of computing in banking, namely, that as banks embraced computers, work changed so that more could be done but there could be no return to a precomputer approach. Digitally based processes now ruled the day. When that happened, one could then begin speaking about a digital style being extant in the Banking Industry. The phenomenon occurred in each of the major applications reviewed below, but any one of these by itself could have been sufficient to have caused a fundamental change in how the industry conducted its work and how costs and profits formed new value chains and business models. And so the transformation proceeded.

The ABA continued to report incremental progress in processing checks, citing case studies of innovations all through the 1970s and 1980s.[54] In the early 1980s, banks that could handle large volumes of checks were able to reduce processing costs the most, although smaller banks enjoyed similar benefits since the federal government had absorbed some of the costs of moving checks and data through the economy, a practice that changed in the 1980s when the Federal Reserve System started charging for services.[55] In the 1980s, a greater use of database software and other digital tools continued to streamline the process. The industry also began using bar code technology and scanners. In a testament to how this industry operated all through the period, Stanley Josephson of Republic Bank, an early user of bar codes, commented: "I'm a 30-year veteran of Operations. This is one of the few times, if not the only time, that I have seen a financial institution take technology developed in another industry and apply it to banking."[56]

Cost dynamics changed to a point where, by the mid-1980s, one could document the effect. In 1987, the ABA conducted a major survey on the use and cost of IT in the industry, noting that on average banks were spending about 7 percent of their total operating expenses on computing. Expenditures for processing checks hovered at 5.9 percent for large banks and 9.2 percent for smaller community banks. Costs of processing checks, as a percent of expenditures, declined in the mid-1980s by almost a percent a year as volume increased and costs of computing declined for the largest banks. For small banks, costs remained essentially the same as a percentage of total expenditures. Over the years, processing checks had been done in five ways: on the premises, by a holding company arrangement, with a correspondent bank, through some joint venture, or by way of a service bureau. Of the largest banks, 94.7 percent did it in-house, in 1987, while in that same year the second most widely used approach was a correspondent arrangement (19.1 percent). Small banks—those with less than $100 million in assets—were extensive users of all methods, but 60 percent did it in-house, nearly 20 percent with correspondent banks, and another 13 percent through a specialized service bureau.

Table 2.1
Number of Checks Processed in the United States, Selected Years (1970–1992)
(Billions of checks)

1970	1972	1974	1982	1985	1986	1987	1988	1989	1990	1991	1992
18.0	20.1	24.8	32.5	46.2	47.5	49.2	50.3	53.9	56.4	56.8	57.6

Sources: Booz-Allen & Hamilton Inc., "A Proposal to Assess the Banking Benefits of Escort Memory," January 16, 1985, proposal prepared for Burroughs Corporation, in possession of the author; Federal Reserve Board as reported in Patricia A. Murphy, "Electronic Check Clearing Alternatives Take Shape," *ABA Banking Journal* 85 (May 1993): 62.

So, different-sized banks had figured out their optimal economic options. This pattern of paths taken remained in force for the rest of the century.[57]

In the years immediately prior to the arrival of Internet banking (late 1980s–early 1990s), the volume of checks remained high (see table 2.1). The rate of growth in volume slowed in the 1980s and 1990s, when credit and debit cards made inroads as alternative payment options, and the trend continued into the next century. Banks hoped that customers would also start paying bills at their ATMs by the late 1990s, although that remained, like the notion of the "checkless society," wishful thinking.[58] It is telling that between 1998 and 2002, the ABA did not publish any major articles or studies on what to do about checks. Rather, the ABA shifted its focus to e-wallets, Internet banking, smart cards, and other forms of online banking that, if they took off, would reduce the demand for checks. It is as if the industry had stabilized on how to process checks and let the issue *per se* be displaced by other applications closer to the root cause of why checks had always existed—as an instrument for transferring money from one party to another. That shift to the transaction itself, and away from checking, suggested that perhaps sometime in the twenty-first century checks would finally disappear, to become instead quaint examples of twentieth-century ephemera.

As more often happens with a digital application, check handling evolved from a paper-based approach to a more electronic form by the dawn of the new century. The approach, called Accounts Receivable Conversion, or simply ARC, involves scanning a check when it is written, making the electronic scan the transaction record, and then destroying the paper document (i.e., the check). The electronic scan is then cleared in one to two days, as opposed to the four to five when a paper check is involved. In short, the scanned document is now an electronic payment. Most banks have not notified their customers of the new application, surprising many with the elimination of float and even the appearance of unfamiliar transactions in their monthly bank statements. During the first half of 2003, some 55 million checks were processed through ARC, a small number when compared to the 4 billion paper checks processed during the same period.[59]

The effects of using IT to manage check processing have been carefully studied by economists, and while data vary a bit from one study to another, the data from the 1980s and 1990s clearly demonstrate the value of IT to banks. In the period

1995 to 2000, the average annual rate of decline in use of checks hovered at 3 percent, while use of credit card payments rose by 7.3 percent and of debit cards by 35.6 percent. Combining these data tells us that the share of checks in financial transactions (checks, credit cards, and debit cards) fell from 80.8 percent in 1995 to 64.6 percent in 2000, a dramatic shift in just five years. This happened while the total number of checks cashed continued to grow, as did the economy at large. Use of Automated Clearing House (ACH) applications also dropped in cost. Data from the Federal Reserve have documented the savings to all concerned. Between 1990 and 2000, real unit costs for clearing transactions dropped by 83 percent.[60]

Savings and Investment Applications

It might seem logical to discuss EFTS applications next, since they are so intimately linked to the whole issue of check processing; but chronologically, EFTS came a bit later as an important component of our story, while savings and investment applications were some of the earliest to move to computers, a natural transaction from older punch-card uses. In prosperous, post–World War II America, the number of savings and investment accounts held by banks grew straight through the entire half century. For example, between 1945 and 1964, they doubled in number, from 40 million to 80 million, with a concurrent increase in deposits from $30 billion to more than $120 billion.[61] From the point of view of work for the banks, however, the more important statistic is the number of accounts, although the amount of deposits suggests an increase in activity per account, which translated into more work for bank personnel, hence more cost to service customers. For decades, banks had used a myriad of machines developed by such firms as IBM, NCR, and especially Burroughs to facilitate basic tasks like daily debiting and crediting of entries, performing interest calculations and then subsequently posting them, giving customers receipts for transactions performed, updating and reporting totals for a bank, and sending monthly statements to depositors.[62]

By the early 1960s, banks had started to use computers to handle some of the paperwork to log deposits and withdrawals, produce statements, open new accounts, post payroll deposits, close accounts, and print 1099 tax reports. In 1962, Congress passed the Revenue Act, which required all banks to report interest paid to depositors in excess of $10, using the IRS Information Return Form 1099. This form illustrates well the amount of data many bank transactions included and, thus, had to be tracked: name, address, social security number, and amount of interest paid. These had to be printed and mailed both to the IRS and to each customer—80 million just in 1965 alone! Another labor-intensive activity requiring tracking involved the one-year savings accounts called Christmas Clubs, popular in the United States dating back to before World War I and in wide use until late in the century. Year after year, consumers typically opened their Christmas Club accounts in January or February and closed them in December, when they withdrew the funds to pay for presents. Finally, lockbox deposit accounts for corporations were also widespread by mid-century and also had data-intensive work streams.[63] This service involved receiving checks at a bank on behalf of a customer, verifying the match of invoice amount and payment received, and depositing all checks, an application that became increasingly computerized.

It should be of no surprise, therefore, that some of the earliest banking functions that moved to computing involved, along with check processing, the servicing of a growing variety and number of savings accounts. From the earliest days of digital computing, bank managers recognized the need to have multiple applications increasingly integrated to provide comprehensive views of lines of business (such as savings) and, for customers, consolidated reports on all their own accounts within a bank. The earliest computer installations, dating from the late 1950s, involved posting transactions to checking and savings accounts. For example, at The First National Bank of Boston, which started using a DATAmatic 1000 general purpose computer system on June 2, 1958, staff posted 20,000 transactions on the first day. As bank officials noted at the time, this action represented merely a "first step in a long-range program of automating the whole scope of the Bank's massive record keeping and data processing, and the culmination of more than four years of intensive study and analysis."[64] The computer performed all the calculations and assignment of each item to accounts, sorting and presenting results in the manner required by bank personnel. The Worcester County National Bank in Worcester, Massachusetts, became the first bank to install an IBM Ramac 305 in the Banking Industry (in early 1958) to handle servicing of checking and other customer accounts. At this bank by the end of 1959, savings accounting had started to move over to the computer.[65] At the Philadelphia Savings Fund Society, an IBM 650 Ramac system began operating in 1960 following nearly a decade-long hunt for ways to improve the speed of processing while lowering operating costs. Because of the high cost of such systems, the bank also had to "run" additional services on the system to justify its costs and thus added functions now made possible by the new machines, such as producing more timely reports, handling a larger number of accounts, and integrating multiple account information.[66] Burroughs, long a leading provider of accounting equipment to the industry, also installed computers for similar applications.[67]

Bankers liked that computers could perform necessary calculations, sorting, posting, and reporting, but bankers also resisted altering the basic tasks that they themselves had historically performed. They just wanted these done faster, cheaper, and often more frequently or at more convenient times. Sacred amongst all was the production of the detailed customer ledger record, which documented transactions with the bank, because it was the source of data that tellers used for handling all customer-account inquiries and that bank employees used to understand the business on a customer-by-customer basis. So early digital systems had to reproduce these ledgers, a situation that did not begin changing until the availability of database software in the 1970s made it possible to take fragmented customer data and repackage them into more useful, integrated clusters that could, for example, be conveniently displayed on a teller's online terminal. Thus a customer came into a bank with a transaction to perform, and the output was still an update to the ledger, just as before, only the intermediate steps of calculation and data management transferred to the computer.[68]

Adoption of these applications proved to be a slow process for many of the reasons discussed in the prior chapter. Traditional bank and industry values and practices, the industry's natural conservatism, and a skepticism toward the new all contributed to the slowness. An ABA survey in the fall of 1962, for instance, reported that only 5.2 percent of 974 banks surveyed had moved online applications to computing. To put that trend in perspective, 6.4 percent had migrated regular checking accounts to computers, 12.9 percent loan processing, 14.4 percent customer payroll accounting, and 10.4 percent personal trust accounting.[69] As with checking accounting, however, the largest banks led the way. In late 1964, about 30 percent had transferred savings accounts to their computers.[70] By the end of the 1960s, however, the application had become nearly ubiquitous across the industry.

In the late 1960s, banks also began switching from batch to online systems, which gave managers, back office clerks, and branch tellers the ability to look up account information for customers and to gain nearly instantaneous updating of records. By moving check processing, EFTS, deposit accounting, and other functions to computers, by the early 1970s banks could expand their hours of service to customers as well as handle more transactions without necessarily adding staff.[71] In addition, use of automated teller machines in the 1970s and 1980s gave customers direct access to their savings and checking accounts for the first time, thereby offloading the task of many inquiries previously performed by bank clerks.[72] Shifting the task of conducting inquiries away from one's personnel to customers became a trend evident in almost every industry by the end of the century; in that regard banks were no different from firms elsewhere.

By the early 1980s, clerks, tellers, and bank managers were widely connected to a bank's records through online terminals, and customers could view similar data via ATMs and telephone response systems. In the 1990s, customers had account access through their PCs with Internet banking. Savings and trust account management had automated this application as much as any industry operating in the American economy. Savings account processing was no longer labor intensive, though check processing continued to be. As early as the dawn of the 1980s, customer inquiries were being handled much faster. A U.S. Bureau of Labor Statistics (BLS) economist, while noting that as early as the 1960s online terminals were used to handle inquiries, commented that by the early 1980s, "tellers respond[ed] to requests for bank balances in a few seconds with online terminals, compared to several minutes by former methods."[73] Industry observers began reporting increases in teller productivity of 20 percent from using terminals and integrated account files to handle inquiries. Furthermore, as inquiries were brought online, other messages regarding accounts could be transmitted immediately, such as information about lost or stolen credit cards. An additional bump in productivity also occurred of some 4 to 5 percent because online electronic journaling could be performed, by which all updating, calculations, and posting were done by computers responding to the immediate request of tellers made through an online terminal.[74]

However, since the number of transactions a customer performed with their savings accounts were few when compared to the number of checks they wrote or loans they paid, the level of interest expressed by consumers in using IT for participating in savings account applications proved small in comparison to the interest expressed by banks. Consumers were not hostile to the notion of adding or moving money in and out of their savings accounts; they were just more focused on other higher volume transactions, such as bill paying. In a survey done in mid-1995, for example, consumers did not even report accessing their savings accounts online as a home-banking application for them. It was a given that they could access balances in all accounts. They saw such access as an integrated view of multiple accounts at the same bank, followed by the ability to stop payment on checks, request copies of old checks, order new ones, pay bills, and apply for various types of loans and other forms of credit. In short, savings accounts had sublimated into a portfolio of activities consumers and banks now interacted with; no single function stood out alone.[75]

Loan Processing Applications

Loans are one of the fundamental building blocks of banking. Over the past half century, banks have made essentially three types of loans: consumer installment, mortgage, and commercial. There have been other types as well, such as credit cards (discussed below), but these three form the core. Consumer installment loans grew in importance to banks in the second half of the twentieth century, moving from having been unattractive prior to the late 1940s to a point where for some banks they were their major source of income. Most automobiles were financed by consumer loans, as well as various other goods (such as refrigerators and other large home appliances)—all in an era of rising personal income. Mortgage loans, a unique feature of the American banking industry, were provided too because so many residents of the United States were able to acquire homes due to a variety of circumstances that need not detain us here. The salient observation is that the vast majority of Americans bought homes with the help of a mortgage. In fact, throughout the entire period, housing costs and mortgage interest rates were topics of great interest to public officials, economists, and individuals. And commercial loans were often the largest dollar category of loans issued by banks. These loans were used all through the period to fund capital improvements, such as construction of stores and factories, or to finance inventory or cash flow.

All three types of loans involved numerous processes for handling data collection and decision making; of course, volumes of paperwork were integral to the processes, both in the 1950s and still at the dawn of the new century. Each type of loan required bankers to understand the financial condition of the applicant for a loan, perform credit checks, determine the risk the bank faced in the event that a customer might not pay back the loan, establish a competitive price for the loan (closing costs, interest, and points), and then manage the process for the life of the loan, including collections. All three tasks involved the use of a great deal of information over a long period of time. Consumer loans, in particular, not only created their own mountain of paperwork, but also represented a fundamental departure from pre–World War II practices in America, away from an earlier time when banks avoided making loans to people. With rising incomes and boom times in later decades, however, it made sense for banks to seek out their business. But, for many banks, pursuit meant developing new processes for handling loans; part of that set of new processes involved computing. Not until late 1963, for example, was there enough activity to impel the ABA to conduct one of the earliest surveys to determine how many consumer loans had been made by the industry. Their finding: 552 banks reported over 22.3 million loans.[76]

For loans of all types, there are numerous documents and reports: applications, credit reports, the loan notes themselves, descriptions of collateral, account records, payment books (usually with payment coupons), account summary reports, bank loan portfolio reports, late notices, bad debt reports, and much more.[77] Early on, data handling was a strong candidate for partial automation. As early as 1958, at least one bank used an IBM Ramac 305 to help manage the paperwork.[78] But in those early years, the industry focused more on finding equipment, for example, to sort and tabulate checks. There were no loan management software packages; banks had to write their own, yet these tools did not require unique equipment, just normal input/output devices (for example, punch cards and printers), general purpose digital computers, and programming languages.[79] Electronics in loan data management became, however, one of a series of applications that helped cost justify the otherwise very expensive computing technology of the day. The initial problem, of course, was that there was such a low volume of loan transactions, when compared to check processing, that computing did not provide economic advantages over other business machines. As volumes of loans increased in the 1960s and 1970s, however, that circumstance changed. However, as one banker noted, since early transaction volumes were light (perhaps one payment per customer per month to process), the application offered bankers an opportunity to learn how to use computers. As bankers tightened their record keeping and tracking for past due accounts, they discovered that computers flagged overdue accounts better than clerks, leading to earlier collection practices, collecting both amounts owed and overdue charges. John N. Raleigh, assistant vice president at the United States National Bank in Portland, Oregon, for instance, reported that as a result his bank doubled its late-charge income.[80]

In its fall 1962 survey, of the ABA noted that nearly 13 percent of its respondents performed some consumer loan processing with computers, while for mortgages the number came closer to 8 percent, and for commercial loans 2.9 percent. This set of data suggests that the application was still in a highly embryonic state.[81] Bankers could not cost justify automation with low volume applications. In the case of mortgages a level of complexity also proved problematic, as one industry observer noted in 1963:

> Mortgage accounting is traditionally complex. The number of taxing authorities, insurance provisions, and mortgage types and bases result in expensive programming costs. Returns from computerizing mortgage accounting can be realized only in high volumes.[82]

Table 2.2
Mortgage Reports at the First National Bank in St. Louis, circa 1967

Accumulative Transaction Journal	Exceptions Report	Name & Address Index Cards
Auditor's Confirmation Request	File Maintenance Register	New Account/Payoff Report
Billing Register	Insurance Expiration Report	Posting Journal
Billing Slips	Investor Report	Property Tax Bill Request
Card Coupons	Late Notices	Slow Loan Report
Concentrated Loan Report	Ledger Recap	Teller Journal
Delinquent Notices	Loan Profile Report	Trial Balance—Long
Escrow Analysis	Mortgage Statements	Trial Balance—Short
Escrow Maintenance Report	Name & Address Change Report	Trial Balance—Statistical

Source: IBM, *First National Bank in St. Louis Online Savings and Mortgage Services* (White Plains, N.Y.: IBM Corporation, undated [1967]): unpaginated, last page, Box B-116-3, DP Application Briefs, IBM Corporate Archive. This document also reproduces sample reports from the bank, and includes illustrations of the installed IBM S/360 systems.

By the mid-1960s, banks were putting various loan record keeping and reporting functions on computers originally acquired to handle checks and savings accounts.[83] Table 2.2 lists the various mortgage reports that one bank produced using a computer, suggesting how the application grew in complexity even in the early years.

About the same time, bankers began speculating about the possible uses of computers to perform credit-scoring, or determining how credit worthy an applicant is for a loan. Credit-scoring is based on factoring a number of data points, among them past loan payment history, income levels, and length of time in the same home or job. Some in the industry began viewing this application as a perfect use of operations research (OR). Using actuarial techniques already understood as far back as the late 1930s, and which had spread to many industries (retail, petroleum, and mail order houses) and to banks by the 1960s, OR and computing seemed a perfect match. Both needed to handle large volumes of data and navigate a complex of mathematical algorithms. Forecasting the future performance of a consumer is difficult even after a half century of credit-scoring, but made more possible if a computer could help assess various types of information and provide a score (that is, points) that suggest that "if someone has *x* number or more of points, they are credit worthy and should be given the loan."[84] Yet much lay in the future because of the lack of software. Arrival of third-generation computers raised questions about how best to use these new machines, although consensus developed around the need for less data and more refined information.[85]

All through the 1970s and 1980s, much of the paperwork associated with collecting, scoring, and managing loans evolved into major applications for computers at banks of all sizes. Software products also became available. However, computers failed to resolve the problem of failed savings and loan banks, bad loans to countries and companies, and the regulatory inhibitors discussed in the previous chapter.[86] Yet the management of loans with computers proved important in holding down operations costs as the number of loans increased in volume and importance. One set of statistics suggests loans' importance: In 1960 about half the assets of a bank were loans; in 1995, they made up four-fifths. Over the period, commercial loans as a proportion of the total shrank, making it possible for others than banks to fund these kinds of loans. In the same time frame, consumer loans increased, both in number and dollar value. Companies could raise their own capital and manage the paperwork with computers, due to the increase in mortgage loans. This was manageable because banks could aggregate loans and sell them to other financial institutions (called securitization), thereby spreading the risk of bad loans among many institutions (secondary markets for mortgages and later credit card receivables), thereby creating new products. One other consequence emerged in the mid- to late-1990s: the availability to consumers over the Internet of large amounts of data on loans from multiple institutions, thereby increasing competition for loans among banks.[87]

Loan processing and management changed incrementally. By the end of the century, these changes had accumulated in sufficient quantity to have altered practices substantially from what they had been in 1950. For example, as late as the second half of the 1970s, a potential home owner and the bank's loan officer would have filled out a variety of documents, including the loan request, using both a typewriter and a pen; they would have collected prior tax reports on the property and income; the bank would have ordered up a credit report on the individual and prepared other requests for title searches, surveys, and a number of disclosure documents (truth in lending requirements of the U.S. government). A closing would have taken place (if all went well) about six to eight weeks later, and would have taken an hour or longer to complete with a bank official and a clerk, lawyers, and the home owner signing dozens of papers in multiple copies. Now picture a loan process at the end of the century. The consumer would have already used the Internet to check on terms and conditions at a variety of banks, would have selected the most attractive, and would have filled out the application online using a personal computer linked to the bank's computer system. Credit-scoring would be done quickly, often automatically, linking the bank's system to a preferred credit bureau that could also perform other checks, while an e-mail notice to a title search company and to a surveyor could be generated. Closing could be done within ten to fourteen days and involve signing perhaps only two copies of a handful of documents—still more than the bank would prefer, and still due to the same government regulations as before. Records of the entire transaction, and subsequent management of the loan, would be processed via computer. In the late 1970s, consumers would have received a book with coupons to include with their monthly mortgage payment; at the end of the century, coupon books were almost a thing of the past, as banks simply automatically deducted the amount of the mortgage from one's checking account.

Hibernia National Bank branch office using IBM CRTs attached to a central computer to provide in-branch processing, such as loans, 1989. Courtesy IBM Corporate Archives.

As the century moved forward, expert systems were in place with all kinds of loans, not just mortgages, making decision making easier and less risky, which sped up loan processing as well. As with other applications, however, adoption of expert systems came slowly. As one industry observer noted, "Bankers seem reluctant to embrace it fully." Why? The reason cited was the same one seen for so many other applications: "Bankers are by nature conservative."[88] In reality, the truth was more complicated: insufficient knowledge about the technology, lack of common definitions needed to use expert systems, and the dearth of effective systems lingered all through the 1980s and 1990s, despite continuous improvements of these software tools.

Credit and Debit Cards

Credit cards represent an important form of credit—loans in effect—that were unique to the second half of the twentieth century because they had not existed in earlier centuries. To be sure, there had been predecessor paper cards and other forms of retail credit processes in the first half of the century, but the credit card as we know it today is clearly an icon of the second half of the century in the United States. Of the types of loans handled by banks and other financial institutions, credit cards became one of the largest as defined by the number of credit cards (that is, loans) and by the amount of dollars that funneled through this application. Debit cards proved to be far less significant, although by the end of the century they too were becoming widespread in the United States. The integration of computing and cards illustrates how drastically technology affected the structure and practices of the Banking Industry.

Let us begin by understanding how a credit card works. It is a plastic document that has embossed on it an individual's account number, the name of the bank or credit card company that issued it (such as Visa or MasterCard), an expiration date, and, on the back, a magnetic stripe that contains similar data to that on the front. When a consumer presents the card to pay for merchandise or service, the retailer swipes the card through a reader that combines the data from the stripe with purchase information (store name, amount being spent, what for, and when). The card-reading device automatically dials up the merchant's acquirer (the bank handling the processing of transactions for the merchant) to notify it of the proposed transaction. The computer is programmed to check with the credit card's computer-based system to make sure the consumer has enough of a credit line to be allowed to charge the purchase, and the system approves or "declines" (disapproves) the transaction. If approved, the cardholder's credit-card record is updated to reflect the amount and particulars of the transaction and is billed at the end of the billing period for that and all other transactions, including interest on the unpaid balance. A variety of reports on transactions, accounts, and so forth are generated for the issuing card bank, for the consumer, and for any other entities involved in the transaction. The merchant is credited with the sale and is paid at the end of the month. If declined, the merchant either retains the card (for example, if it is reported stolen) or denies the consumer the opportunity to make a purchase using that card for payment. This is a very simple description of the process, but it essentially outlines how all of such purchases have worked over the past half century.[89]

Before delving into the history of the credit card, it should be put in a context relevant to banking. The year 1996 was the first in which measurable volumes of sales over the Internet occurred—sales that overwhelmingly used credit cards for payment. This development demonstrated the importance of the card to the industry and, even more so, to the U.S. national and global economies. In that year, consumers in the United States spent a total of $4.2 trillion (both on the Internet for goods and services and on non-Internet expenditures). They paid for 57 percent of their expenditures with checks, another 21 percent with cash, and 22 percent with payment cards (both debit and credit). In 1984, the year for which we have the earliest reliable data on consumption, consumers used their credit cards for only 6 percent of their payments, checks for 58 percent, and cash for the final 36 percent. As one student of the credit card pointed out, "the increase in card usage came at the expense of cash, not checks."[90] Computers made all this possible because, as the functions of computers improved and their costs went down, electronic monetary transactions became more attractive to many industries, and thus the credit card quickly became an important tool in the economy with which to conduct business. Then came the Internet, where nearly 100 percent of all transactions occurred with credit cards as the medium of payment. In 2003, Americans had spent $51.7 billion over the Internet, up from $31 billion just two years earlier.[91]

Though retail petroleum and department stores had experimented with various types of charge cards as early as the 1920s, banks did not get into the game until the second half of the century. The modern history of credit cards began with the Diners Club, established in 1950, to provide travelers with a way to pay for meals and entertainment anywhere in the country. On the heels of the success of the Diners Club, some 100 banks introduced credit cards in 1953 and 1954 for use in their own geographical markets. They also wanted to capitalize on the growing number of department stores in suburban areas, but these ventures failed because managing such programs proved complicated. Of the original group of initiatives, none survived as independent offerings. In 1958, American Express, Bank of America, Carte Blanche, and Chase Manhattan Bank each introduced universal credit cards.[92] Banks recognized the opportunity, realizing it was a way to service small merchants who could not afford their own store-branded credit cards. By the start of 1961, 40 banks had launched national credit cards, although those from the largest two banks (Bank of America and Chase Manhattan) proved most successful. The Banking Industry spent the 1960s and 1970s learning about this kind of consumer credit. As late as 1977, most households that had credit cards placed only 3.4 percent of their expenditures on these; by the end of 1997, the volume amounted to 20 percent of a household's expenses. In short, the great boom came in the 1980s and 1990s.[93]

The key innovation that helped recruit retailers to honor cards, and to link them to the issuers of cards, was not technology so much as the emergence of card associations. In 1966, Bank of America established such an organization called BankAmerica Service Corporation through which banks could issue a credit card usable nationwide. The Corporation cleared accounts between banks while creating a nationally branded card, what eventually would be called Visa. A second association, called Interbank Card, also started up in 1966, offering a nationally branded card that in time became MasterCard (1980). It is important here to recognize credit cards' popularity in the years that banks learned how to market and use them. In 1969–70, what would become Visa had 243 issuing banks, that is, organizations that issued credit cards to consumers; that number grew to 6,000 in 1998. By 1998, MasterCard had over 6,500 issuers.[94] Their ease of use, combined with a long economic boom, made the credit card one of the most popular financial tools introduced into the U.S. economy in the twentieth century. While only 16 percent of all households had a credit card in 1970, by the early 1980s, almost all adults did, including many low-income people. By the end of the century, most adults had several.[95]

How to make a profit from this additional consumer loan offering was a mystery for all bankers in the beginning, but bankers recognized the business opportunity that lay before them. As one industry commentator declared in early 1968, "Charge account banking has suddenly mushroomed from a participation by a relatively few banks to several hundred in the space of a year."[96] Fearing loss of opportunity and possibly other banking services to rivals, bankers increasingly wanted to issue cards. The key was to issue enough cards and to have a sufficient volume of transactions to make the program profitable. It was not exactly clear to bankers in 1968 how much was enough or how best to be profitable, but they knew that the key to success was to control credit.[97] Taking part in credit-card associations (and their computer systems) at least controlled some of the costs of bad debts for issuers. As volumes increased over time, however, the paths to profit became clearer. Some banks created their own computer systems to process transactions to help minimize risk of unpaid debts and how best to authorize transactions. Texas Bank & Trust Company, for example, installed an IBM System 360 computer system in 1967 to handle the anticipated volume of authorizations that would be needed for its BankAmericard. As the bank's vice president, Thomas N. Overton, described the process used by tellers and clerks, "using typewriter-like IBM 2740 communications terminals in their office, they can ask for and receive information about any active account at any time without perceptively interrupting other work the computer is handling."[98] Merchants telephoned a teller to get approval for any transaction beyond a minimum level. A cashier would type the request on a 2740 keyboard and the computer would check the account to determine if the card had sufficient credit ("open to buy"). If yes, the merchant was authorized to conduct the transaction; if not, then the teller's supervisor had to decide whether to authorize, comparing the amount the consumer wanted to spend against how much open balance still remained. Having access to balance information in order to decide whether to authorize a transaction thus reduced the risk to the bank of a consumer exceeding his or her credit limit.[99]

The challenge, of course, was how to expand the use of cards and at the same time control the problem of consumers' exceeding credit limits and thus the potential for unpaid balances. Credit card associations and banks increasingly relied on computers to provide centralized, online credit information and national exchanges of authorization systems. All during the 1970s, banks and credit card associations worked out the details of the needed authorization and clearance applications.[100] But there still was much impatience and frustration in the industry. As one frustrated banker put it in 1974, "The time for division, petty squabbles, and parochialism is over. We are either in the card business together or we will fail as bankers."[101] The big fear of delinquencies did not materialize as a problem in the 1970s; authorization processes worked to control the perceived risk. By the mid-1970s, 48 percent of all banks were involved in credit cards in some way or another, with the largest banks most fully committed (81 percent) while the smallest lagged, at 39 percent.[102] At the end of the decade Professor Donald P. Jacobs, teaching at the Kellogg Graduate School of Management, wrote in *Banking*, "Thus far, the major innovations center around the credit card. A small number of banks have demonstrated credit cards can be issued and customers serviced that are domiciled at great distances from the banks' physical facilities."[103] By the 1980s, credit cards had become the most profitable product in a retail bank's portfolio of offerings.

By the early 1980s, banks were also discovering that the credit card associations were now becoming competitors as well as providers of new sources of consumer loans. Although not rivals in the early 1970s, a decade later Visa U.S.A. and MasterCard International were beginning to enter markets for money-market funds and national ATM networks, as well as those for travelers' checks and debit cards.[104] Additionally disquieting in the 1980s was the trend that, as the volume of credit card transactions grew, their profitability finally fell. This became evident beginning late in the second half of the decade as competition within the credit card business increased. Profits declined because the cost of maintaining funds were rising in an era of inflation along with the price of fighting competition. Some bankers questioned if the market was saturated, and the suspicion arose that computers had done just about as much as they could to contain risk.[105]

Incremental changes to credit card applications helped, such as the introduction of improved scoring models, integrating cards with other financial and service offerings, and even embedding microchips in cards. All of these innovations kept the necessary computer applications in flux during the late 1980s and early 1990s. Meanwhile, card usage kept growing, reaching 128 million accounts in 1988 and generating 10 percent of the entire Banking Industry's profits that year. Cards like Visa and MasterCard nearly displaced retail store-brand cards as well, thereby increasing volumes of transactions pumped through the associated telecommunications and digital systems. Observers of the industry noted at the dawn of the 1990s, credit card systems had always been the most automated in the industry. The technology made it possible to add new services that took advantage of the databases, software, and telecommunications infrastructure that connected banks, customers, and merchants. Databases with customer information on purchasing habits began to inform bankers and retailers on marketing issues, which made it possible for credit card issuers to pioneer credit-scoring models and, later, use of credit bureaus. Computing made it possible to offer tiered rates for different types of customers, to handle efficiently large volumes of customer calls, and to streamline collection services.

By the end of the 1980s, as one study of the industry pointed out, "the requirements for processing credit cards are so systems-intensive that processing is itself considerably more concentrated than is card issuance."[106] By 1987, credit card processors—firms whose business it was just to handle such transactions—did half of all processing. This included authorization services, collecting data on transactions, posting transactions to accounts of cardholders, processing remittances, and producing statements and reports for all parties concerned. By the late 1980s, the business became extremely concentrated into fewer, larger firms because of the economies of scale made possible by high volumes of automated activities. Two firms in the business, FDR and Total Systems, combined controlled 61 percent of the market for such services.[107]

The credit card business is perhaps the most visible example of a technology-enabled desegregation of banking in the American economy. When Visa and MasterCard began siphoning business away from banks in the early- to mid-1980s, one could claim that the process of computer-enabled desegregation had started. When the impetus came from outside the industry and the rest of the Financial Sector, however, then one could argue here was a second, more potentially significant case of something new happening. That second phase in the transformation of the credit business began at the dawn of the new decade. All through the 1970s and 1980s, banks had generally enjoyed high profit margins on cards, but by the early 1990s one industry reporter noted:

> Banks clearly are no longer the only credit card player in the business. Taken for granted for years as a stable, personal financing tool, credit cards today are the center of a feeding frenzy, and banks are having to fight harder to ward off nonbank organizations determined to share the feast.[108]

Rivals competed for customers with high credit ratings, hence low risk, to them. Larger banks merged, creating fewer but more powerful rivals within the industry, such as Chemical Banking Corporation and Manufacturers Hanover Trust. In 1990, 5,900 organizations issued credit cards, but by 1995 57 percent of all accounts were in the hands of only 16 institutions. M&As and financial problems at some firms made it easy for others to pick up entire portfolios of accounts and simply transport them to their own computer systems, often continuing with the same card-processing vendors. In addition, new entrants appeared, such as the Ford Motor Company's Financial Services Group and the AT&T Universal Card, the latter debuted in March 1990 and quickly amassed over 10 million accounts. In 1990 Discover Card and AT&T together controlled over 15 percent of the market, and Citicorp was the single largest card provider, with 10.5 percent of the market. By now, American Express had 8.7 percent of the market and BankAmerica 3.5 percent. The industry started to offer different terms such as low or no interest charges, or no subscription fees—all of which were easy to manage with software. Consequently, consumers became quite savvy about the various existing terms and conditions for cards and in the early 1990s began switching accounts more frequently than before, feeling no compulsion about keeping their credit card with the same bank that had either their checking or savings accounts.

In fact, at the same time, consumers were spreading their financial accounts across multiple organizations, perhaps placing mutual-fund accounts with a brokerage firm while keeping their checking at a local bank and their credit card account with some firm not even in their own state.[109] The Banking Industry was particularly concerned with AT&T with its credit card business and the financial promiscuity of their customers, who did not hesitate to change the providers of various financial services. As competition increased, profit margins were squeezed.

By mid-decade, software tools had come along that made it possible for customers and banks to integrate services, creating new opportunities to re-aggregate offerings.[110]

The next twist in technology came with the introduction of smart cards. These cards look like normal credit cards but have a microchip embedded in them, permitting cards to hold a great deal more data than the magnetic stripe alone. Industry watchers accused banks once again of being slow to exploit the new technology in the early 1990s. Already by 1993, some industry experts were claiming that transaction costs for a smart card were lower than for a traditional "magstripe" card. Universities started issuing them as identification cards in combination with debit cards to students. Retail and manufacturing firms began providing them to employees to manage the process of buying and selling goods and supplies. Essentially a debit card, the first major introduction of this technology in the United States came in 1996 when Visa introduced a smart card for use at the Summer Olympics games in Atlanta, Georgia. The launch did not prove successful and, as of the end of the century, the Banking Industry had yet to fully embrace the new technology. The biggest problem with the cards involved their technical infrastructure; they could not use the more primitive existing networks already in place to support ATMs and stripe cards. Just to refit an ATM was estimated at $2,000 to $3,000 each in 1995. In short, the business case for smart cards remained elusive in the 1990s for merchants and bankers, both of whom would have to shoulder the cost of conversion.[111] In addition to application issues, concerns regarding data security, and industry attention to the explosive success of the Internet made the smart card a potential technology whose future had not yet arrived or at best, remained uncertain.

Banks preferred to focus on credit cards, which by now had become the currency of the Internet. Since back-office infrastructure to support card transactions was expensive to build and maintain, and there were companies like First Data Corporation that focused on just providing credit card transaction services, bankers proved reluctant to move to a new technology. They already knew what one commentator in the industry publicly stated: "It was no longer a banking but rather a technology business."[112] Consolidating volumes of transactions meant sticking with what already existed and worked. Hence, the hunt for the economies of scale justified the mergers of so many large banks, all of whom were extensive marketers of credit cards in the 1990s.[113]

Paralleling the expansion in credit cards was another type of card, the debit card. During the mid-1980s, this concept led to a major drive by banks to attract customers and merchants. Unlike the smart card, debit cards by the mid-1990s had finally begun to be accepted in ways that mimicked the experience with credit cards in the 1960s and early 1970s. So what is a debit card? It is a tool quite different from a credit card, "platforms that different providers of debit services employ to distribute those services to consumers and merchants."[114] A bank could offer a debit card through Visa, for example, or through an ATM network. In short, multiple brands could sit on the same card: one's bank, MasterCard (for example), and possibly even the name of an ATM network.

Begun in 1969, debit cards started as ATM cards issued by one's bank for withdrawing cash from one's checking account. At first customers could access only their bank's ATMs, but soon could connect to networks of ATMs from any number of banks. The debit card evolved to a card that had on it a consumer's bank balance, that could trigger a withdrawal from a checking account, and that could be used in a store (beginning mainly in the 1990s) instead of a check to make a purchase. Merchants were credited with the amount of the transaction (less any fees charged by the credit card company) via the servicer, such as Visa. MasterMoney and Visa appeared as initial universal debit cards in the early 1970s. Later PIN (personal identification number) pads appeared. A customer swiped the card through these pads attached to a cash register and, as with an ATM machine, entered a PIN code to activate the transaction. ATM cards could provide the same function as Visa debit cards, for example, while one could also have an ATM card that could not act as a debit card, or use one's credit card (possibly also from Visa or MasterCard) at the same retail outlet. The point is, cards served various purposes.

Though this all sounded wonderful, execution proved faulty. All of these permutations of the basic payment transaction created confusion among consumers and merchants during the final three decades of the century.[115] At the dawn of the millennium, new services were still being introduced by major credit card providers. Some network systems such as Visa, were online and providing instant updating of files, but others, such as MasterCard, remained off-line and updated later, which meant that one could use PIN pads with the former but not with the latter. These were problems that could not be fixed universally since that would have required simultaneous upgrades of all payment networks. That never happened, and so the confusion remained in the U.S. market.

Despite all this, the public finally began embracing debit cards in the 1990s. Most consumers' ATM cards were converted to debit cards by their banks. Late in the decade, merchants began installing PIN pads, which linked to existing ATM communications systems; some industries, such as the Grocery Industry, finally made it possible for customers to pay with debit cards; while banks encoded their cards so as to function in a variety of industries. In the same decade, Visa and MasterCard increased their advertising to promote debit cards, and finally the confluence of all this with improved technology and software made transactions easier and faster to perform, driving down processing costs for merchants and card-issuing banks. The surge in the number of transactions with debit cards came after the mid-1990s, rising from 400 million purchases in 1994 to over 2 billion by 1998.[116]

Two experts on debit cards asked a wonderful question, the kind that intrigues students of technology: Which came first, the ATM or the debit card?[117] It is a chicken-and-egg problem. While we will have more to say below about ATM technology (which includes a debit card), let's deal with the question now. If merchants aren't on board, one cannot attract customers, and if customers aren't on board, what merchant wants to participate in a debit program? The ATM came first, of course, and the problem was to get merchants to install PIN pads (which were needed to use a debit card), which they finally did in large numbers in the

1990s when the pads themselves had dropped sufficiently in cost. To put things in perspective, in 1984, there were 2,200 PIN-pad–equipped terminals in the United States. By 1990, the number was at 53,000, and it reached 1.6 million by 1998.[118] Bankers were reluctant to promote debit cards until merchants would accept them and had POS technology (that is, the PIN pads) available at their cash registers. That began to happen when the competition between Visa and MasterCard heated up and they both began promoting debit cards to merchants. Bankers, however, proved slower to embrace the cards than merchants or customers, as the old question of costs and profits never disappeared. An industry commentator noted that, "for banks, the hardest thing to accept about POS debit is that there is no clear profit stream immediately associated with the service."[119] It took years to sort out who to charge and how much; technological issues had nothing to do with it, with the exception of getting enough PIN pads installed in stores, and even there the issue turned on where lay the economic gain, not on the issue of could the technology do the job.[120]

By reviewing the evolution of the credit, smart, and debit cards, one can discern a pattern of evolution common to many types of technology. In each instance, the cards' functions, technologies, and processes developed over time. The credit card has the longest history; debit cards came next and gained momentum by the end of the century; while the smart card remains a potential third wave. In each case, new functionality, supported by an increasingly sophisticated network of customers, merchants, service providers, and bankers, along with increased automation and always changing economic dynamics, has made for a continuously complicated process. With the advent of the new century, cards of all types had displaced some checks and money. For e-commerce (that is to say, transactions between individuals and companies or government agencies),[121] cards had become the coin of the realm, whose implications for the sovereign power of nations to mint and distribute money are just now being seriously discussed.[122] However, already certain is that almost every new banking product of the last two decades of the century has relied on cards of some sort—cards that are coupled to existing networks of computing and processing and that involve telecommunications, continuously enhanced ATMs, and, now, the Internet. Perhaps all these are evolving into the ultimate digital form, what Elinor Harris Solomon has so perceptively called "money messages."[123]

Automated Teller Machine Applications

ATMs are machines connected to one or more telecommunications networks that can update transactions in one's checking or savings accounts. They are the most visible form of self-service banking. In 1950, they did not exist; all interactions by a customer with a bank had to be done by mail, telephone, or visit to a branch, and only during the limited hours a bank was open to the public. By the end of the century, one could interact with a bank's computers using the Internet, telephone systems, and, of course, ATMs, 24 hours a day, seven days a week, almost anywhere in the world, regardless of where one's account was physically located.

The first step in this journey toward round-the clock self-service banking occurred with ATMs, beginning on September 9, 1969, when the first ATM opened at a Chemical Bank branch on Long Island, New York. By 1989, there were over 72,300 ATMs, and by 2000 over 300,000. As of the early 1990s, almost 90 percent of all routine functions done with customers had been automated, largely using ATMs.[124]

The original intent of an ATM was to automate many of the functions of a branch teller—an intent that remains to this day. This included allowing customers to use a machine to withdraw cash from their checking or savings accounts and to inquire about account balances and bank services. Later, customers could make deposits and transfer funds from one account to another. A customer gained access to an ATM system by inserting a plastic debit card, the first debit cards in fact in the industry, which told the computer who the user was, then displayed a prompt on the screen for that individual to enter his or her PIN on the keyboard. The system compared the two and if it got a match, the individual was allowed to perform a transaction. The device communicated with the bank's computer to make sure that person's account had sufficient funds for a withdrawal; if authorized, the computer instructed the ATM to dispense cash to the customer, at the same time updating account records. Updates to records were made usually in batches overnight during the 1970s.

By the end of the 1980s, some ATM systems were completely online, which meant instant updating. This made it possible to add more functions, such as making account inquiries or transferring funds. By the end of the decade, ATMs could also accept deposits. The ATM equipment was specialized, while mainframes to which they were connected were general-purpose computers. Also by the end of the 1980s, companies like IBM and NCR sold specialized software products for handling transactions.[125] At the beginning of the 1980s, ATMs were attached to private networks owned by banks.[126] In the 1980s, regional, then national, networks appeared, making it possible for banks to share ATMs. A customer could use an ATM at any bank to perform transactions with their own bank. Information went to the right account, thanks to software written for that purpose. Network owners charged a handling fee, which came out of the bank transaction fee a customer paid for using the ATM.

Despite the convenience of anytime-anyplace banking, the public took slowly to the systems, partly because of mistrust and unfamiliarity, partly because banks charged a fee to do what was free if customers walked into a branch, and partly because there were not many ATMs installed in the early 1970s so it wasn't always convenient. In the period from the 1970s to the end of the century, the most popular transaction was withdrawing cash, followed by making account inquiries and, by the 1990s, making deposits and transfers.[127] Despite all this, as late as 1989, half the ATM cards issued by banks were never used by customers.

Early experience with ATMs did not resolve the basic question faced with all new technologies: was the technology cost justified? Conventional wisdom held that as the expense of having tellers rose, ATMs could reduce transaction costs by replacing the more expensive human beings. On the other side of the equation, however, was the cost of the machinery, development and maintenance of applications and telecommunications software, and the expense of the network. Just to complicate matters, by the end of the century those people who used ATMs conducted more, often smaller, transactions with their banks than they did when they had to physically enter a branch to conduct business. Once again we see the turnpiking effect at work. So, the Banking Industry engaged in a three-decade debate with itself about the merits of ATMs.[128]

Adding to the debate were the technical difficulties to overcome, particularly in the early 1970s. Standards presented a host of issues involving establishing magstripe standards and off-line versus online functions. Tied always to these discussions were concerns over costs. In 1973, for example, a bank could purchase a typical ATM for $25,000 or more, but that price did not include costs of installation, security devices (such as cameras), or promotion of the application through advertising. In addition, cards were not free; banks had to pay to print and mail them to customers.[129] As with so many other applications, only the largest banks could afford to implement first. By roughly late 1973, the 25 percent of all banks with ATM networks were larger.[130] But smaller banks began adopting the technology as a way of meeting increased competition banging at their doors; rivals who provided Saturday hours or access to regional ATM networks. One banker at the San Diego Trust & Savings Bank, Lowell G. Hallock, reported that to do the job right and make the initiative cost effective and productive required installing many ATMs, not just a few experimental devices. He observed that as his volumes of transactions increased, banking by mail declined by 15 to 20 percent. The initiative was launched, he said, "to meet the competition of Saturday banking and extended banking hours, which were very heavy in our market."[131]

But nagging questions about cost justification remained. An article in the *ABA Banking Journal*, published over a decade after the introduction of this technology, asked if ATMs were "worthwhile for bankers." Competition was a source of justification, of course, even after it became apparent that it would come from both banks and credit card providers such as American Express, which had 85 machines dispensing travelers checks in 1981. Those who felt ATMs were justified reached that conclusion when the volume of transactions exceeded prior levels; but all the evidence was anecdotal in the 1970s.

In 1981, the ABA conducted a formal survey about ATMs and reported several findings. First, ATMs had become "the most visibly successful of the various automated retail services offered by banks."[132] (The others were POS banking and the capability of paying bills by telephone.) Second, over half of the largest banks now had ATMs, hence wide acceptance of the technology existed in the industry. The data in this report, however, made it abundantly clear that small banks—those with assets of less than $50 million—had a way to go, as only 10 percent had ATMs. A third finding pointed out that the larger the bank, the more transactions there were at each ATM, illustrating a path to cost justification. When asked how many transactions were needed to break even, respondents answered with over 5,300 per month per machine, with some answers as high as 6,597 or as low as 3,875. Interestingly, the larger the bank, the more transactions were needed to justify a network of ATMs.[133] Fourth, fraud losses through ATMs—a major concern from the earliest days of the application both to customers and bankers—appeared moderate, 12.5 percent of all small banks, and 82.8 percent of all the largest banks experienced some fraud, with annual dollar losses ranging from an average of $709 for small banks to just over $26,000 for the largest banks. All reported taking actions to reduce these amounts by using better software tools and security cameras.[134] Despite any concerns though, installations increased, and national ATM networks did too. In fact, in the early 1980s, eight companies were building national networks, including Visa, MasterCard, and Cirrus, the latter of which became one of the largest networks by the end of the century.[135] As ATMs became widely available, any competitive differentiation they may have provided earlier began dissipating since the proverbial "every bank" now offered this service.[136]

Despite nagging concerns about cost justification, ATMs became so popular in the 1980s that they became the price banks had to pay to be in the retail banking business. Once that happened, there now existed yet another data point to suggest that the Banking Industry had moved to a new digital style of doing business. By the early 1980s, a third of all customers used their ATM cards and nearly half of all banking clients had one or more ATM cards. As an ABA study reported in 1986, "next to credit cards, automated teller machines may be electronic banking's best success story."[137] Customers liked the 24-hour, seven-days-a-week convenience. ATMs were also faster than tellers and easy to use. However, customers did not like the fact that ATMs were out of order as much as a third of the time (or else did not accept their ATM cards), or the consequent record keeping. Furthermore, banks began installing ATMs off their premises such as in shopping malls, grocery stores and in large office buildings. The report commented that a third of all customers resented the transaction fees, a complaint that remained throughout the last three decades of the century. Over 85 percent of ATM users preferred to use machines installed in branches, as opposed to those in shopping centers and elsewhere. Part of that preference, of course, was due to the fact that in the early years of ATMs most were installed in banks.[138]

The debate on the economics of ATMs went on. One industry observer in 1987 stated that "while the consensus is that ATMs are cheaper than tellers, that they are 'profitable,' it is difficult to obtain figures."[139] The ABA published a detailed analysis of the issue, yet concluded that getting hard data remained problematic.[140] A couple of years later consultants working in the industry reported that use of ATMs had not reduced the number of bank branch costs:

> In as recent a year as 1987, for example, only 8.7 percent of all branch transactions were conducted through an ATM. This low penetration is one reason the number of banking locations and tellers (600,000) has not decreased substantially in the past 15 years. The costs of the ATM technology itself are too high relative to the cost of the teller it is displacing.[141]

Yet this team also reported a very tantalizing potential: "An average ATM transaction costs from 40 to 60 cents versus 90 cents to $1.20 for a teller transaction."[142]

Having ATMs installed outside of banks, having many of them, and encouraging customers to use them had become the high road to justification—exactly what happened in the 1990s.

Equally of concern to bankers was competition. In the 1980s, ATMs proliferated across the entire industry; most banks either had their own networks or participated in regional ones, making ATMs a basic requirement to do business, with functions virtually the same from one machine to another. Sharing ATMs in particular proved vexing for the larger banks, which had their own expensive networks, just as regional networks became an opportunity for those who did not. In 1982, some 11,000 ATMs operated in shared networks; by 1988, that number had grown to 65,000, or 90 percent of all ATMs then installed in the United States. At this point, the networks began to consolidate. In the late 1980s, there were hundreds of ATM networks. MasterCard, however, acquired Cirrus, and Visa picked up PLUS, demonstrating that the ATM market would follow the path of consolidation that had already become evident in the credit card business.[143]

In short, the Banking Industry transformed to a digital style. As the consultants above observed, "Banks strongly believe that to attract checking account customers they must make the associated debit card attractive. A wide choice of machine locations is now mandatory in order to do so."[144] By the end of the 1980s, bankers had resolved to push forward with ATMs to deliver more services to customers while, at the same time, reducing the number of tellers. The major banks also launched another attempt to reduce the number of branches in their ongoing crusade to lower operating costs. ATM volumes increased such that, by the early 1990s, some 50 percent of all teller transactions were going through this digital technology.[145] But the debate about the death or survival of branches continued right up to the point when the Internet made its presence felt in the industry. Those advocating the demise of the branch likened these brick-and-mortar assets to dinosaurs on the verge of extinction. Others (who, in the hindsight of the early twenty-first century, turned out to be correct), argued the case that despite the use of many types of digital technology (which they still called "electronic"), branches were not going away. Rather, some banks would have to close (due to overbranching), while others might specialize, or have limited services. Even the definition of what was a branch was changing along with how customers used bank channels for services.

No bank was able to eliminate its competition by integrating electronic weapons into its operations. In 1995, John P. Hickey, the chairman and CEO of Bank One Texas, observed, "Just as QVC (the shoppers channel) and the mail order business have not killed retail stores and the introduction of TV didn't kill the movie industry, the growing use of telephone, ATM, and computer banking transactions will not kill the branch, at least not in my lifetime."[146] He pointed out that banks were also streamlining their networks of branches and ATMs, centralizing back-office operations (an important reason for M&As that continued all through the 1990s), and offering an increasing number of electronically based bank services, such as PC- and Internet-based banking.[147]

Hickey was observing what had happened with so many technology-based applications, involving digital and other types; namely, that concurrent uses of multiple ways of doing work would and did coexist. Banking was not exempt from that broad pattern. Banks, just like enterprises in other industries, adapted existing processes—in this case branches—to leverage various available technologies. One survey of bank branches in 1995 pointed out that:

> Some of these are all-electronic facilities, some are manned, "full-service" branches, and some are a little of both. Many are "in-store" branches, set up inside supermarkets and mega-marts, but many others are on city streets, in country towns, and even in office buildings.[148]

Blending technologies into prior practices became an acknowledged feature of banking in the 1990s. The strategy received its most public recognition when the ABA's new president for 1996, James M. Culberson, Jr., chairman of the First National Bank & Trust, advocated using technology where it made sense as, for example, in shopping malls—which had become, in his words, his own bank's new "downtown."[149]

As the Internet became an increasingly important tool during the second half of the decade, branches and ATMs continued to grow in number and popularity across the entire United States. Between mid-1995 and the end of 1997, banks installed over 50,000 additional ATMs, bringing the total to more than 175,000, while in the same period transaction volumes climbed from 9.7 billion to over 12 billion per year. BankAmerica had the most ATMs in 1997—roughly 7,500—although an archrival in the retail banking business, BancOne, had 7,350, with both planning to add more ATMs. In short, ATMs dominated retail banking. Depending on which surveys one cited, by the end of 1997, between 50 and 65 percent of all households used ATMs. Smaller ATM machines were appearing in retail locations and office buildings, as newer, less expensive models came onto the market along with lower telecommunication costs. Banks were permitted by regulators to charge fees for transactions and did not hesitate to do so, despite consumer and Congressional objections.[150] The fees, however, helped cost justify these systems.

Because of the variety of ways one could bank in the United States, and by extension globally, it is useful to take a snapshot of consumer practices at the dawn of the new century. Banks clearly showed all through the twentieth century that they would bend to the will of their customers, even if individual bank executives resisted for a while. A 2001 ABA survey of consumer habits (see table 2.3) reported an increase in online banking in the late 1990s, with use of ATMs at a record high. As the ABA's survey team concluded, "what bankers increasingly believe is that retail banking is all about choice, convenience, and service. No one delivery channel covers the waterfront for all customers."[151]

With ATMs, everything that could be done with this technology in the United States had gone global by the early 1990s. An American consumer could use credit and debit cards in almost any country in the world by the early 1980s, and in ATMs by the early 1990s. Add in the ability to bank over the Internet in

Table 2.3
Banking Practices of the American Consumer, Circa 2001

% Who Used	Type of Service	% Who Used Most Often
90	Traditional branch	51
64	ATM	29
49	Telephone to live teller	4
47	Automated telephone service	6
22	Computer or online banking	5
26	Nontraditional branch	4

Source: Adapted from "What We Have Here Is Channel Evolution," ABA *Banking Journal* 93 (September 2001): S14.

the late 1990s, and one could bank from anywhere in the world where there was a telephone line and at least a charged battery for a laptop computer. In short, digital retail banking had become ubiquitous and global.

Electronic Funds Transfer Systems (EFTS)

In understanding banking applications, it is important to distinguish check-writing automation from electronic funds transfer systems; though related, they are different. Yet discussing EFTS at this juncture allows us to bring together a series of digital applications that, while starting independently of each other and remaining technologically apart for many years, slowly converged toward the end of the century. That process would be lost in our discussion if we moved directly to the Internet because the truth of the matter is that the Banking Industry used telecommunications extensively and in a variety of ways beginning in the early decades of the twentieth century. A discussion of EFTS also demonstrates how crucial the complex electronic "plumbing" of the industry had become very early in the second half of the twentieth century.

"Electronic funds transfer systems" refers to the telecommunications networks that move information about in the Banking Industry in order to perform a financial transaction. The task of the first EFTS was clearing checks from one bank to another, using the Federal Reserve's network. In the second half of the twentieth century, other EFTS applications emerged, such as the movement of data across networks to conduct a credit card transaction or to enable a point-of-sale terminal in a retail establishment with a debit card.[152] The Internet is yet another illustration of financial transactions conducted with telecommunications. Even in the 1960s, banks were beginning to offer services to businesses that relied on telecommunications. These services included paying bills for customers by electronically debiting their accounts and transferring money to creditors; and allowing companies to pay their employees electronically, transferring the amount of a paycheck directly into an employee's checking account without recourse to paper. In fact, this last application developed in two stages: first employees could elect to be paid electronically but still receive a paper pay stub; then some enterprises would pay employees electronically, giving them electronic pay stubs accessible through e-mail. A similar pattern developed with corporations vis-á-vis their bill paying either electronically, by phone, or using the Internet.

Over the course of the half century, a variety of networks emerged to handle the various types of electronic movement of money: the Federal Reserve's system; dedicated networks to clear credit card transactions; private banking networks in support of ATMs; public networks for use with ATMs; telephone banking systems; electionic data interchange (EDI) and, most recently, the Internet. Throughout this time frame, banks acquired a considerable body of experience with telecommunications, weaving the above technologies into the fabric of their operations early in the twentieth century, and adding to their portfolio of telecommunications as new technologies became available. The reasons for this interest remained relatively constant throughout the entire period. As two experts on the industry noted in the early 1970s, "Financial institutions throughout the country are being swamped by the ever-mounting flow of paper that accompanies their activities."[153] The governors of the Federal Reserve, also "swamped" by the growing volume of checks, frequently encouraged their own internal organization and the industry at large to use electronics as a way of lowering costs and handling the volume of transactions.[154] The emphasis of most bankers in the 1950s and 1960s centered on just handling the growing volume of checks; later their interest extended also to lowering operating costs. A similar pattern of interest and deployment existed in the brokerage business which too was being inundated with paper as the sale of stocks and bonds increased in post-World War II America. In fact, all institutions in the financial sector were burdened with mounds of paper; eliminating these was a prime justification for IT.

The Federal Reserve had pioneered transfer applications using the telegraph early in the century, and by 1950 the first modern EFTS application came into being with Bank Wire, which allowed banks to transfer money to other banks. In the mid-1960s, a few banks experimented with point-of-sale applications involving consumers. The Federal Reserve once again led the way with the opening of its automated Bank Wire facility in Culpepper, Virginia, in 1970. Banks could transfer funds among themselves without recourse to paper. This was the second modern EFTS application in the United States.[155] Initially using teletype equipment, banks cleared interbank payments through their local Regional Reserve Banks over a network that allowed all 12 Federal Reserve Banks to communicate among themselves and individually with member banks. Eventually the technology grew in capacity to handle more transactions; it also increased its transmission speeds and deployed new hardware and software tools as they developed over the course of the century.[156]

Experiments with payments made electronically, instead of with checks, progressed in various parts of the United States in the early 1970s. These included processing business-to-business transactions through a bank and a network, providing the ability of businesses to pay employees electronically, and even setting up early trials of individuals paying personal bills through their banks.[157] As noted earlier in this chapter, all through the 1970s and 1980s the largest push to use EFTS applications came from the deployment of credit cards across the petroleum, retail, travel, entertainment, and banking industries. By the early 1970s, banks were looking at how to implement what by then bankers called automated clearing houses (ACH). As one early proponent of the application described it:

> In its most simple form the ACH represents the electronic counterpart of a standard check-clearing facility which is established and operated on behalf of local or regional associations of commercial banks. In this case the data that is normally captured and processed from checks is replaced by a system in which magnetic tapes are used for the computer-based capture and transfer of the necessary administrative and payments information between participating financial institutions.[158]

First conceived to address the volume of checks, ACH systems quickly acquired other responsibilities, such as doing the same data exchange for credit and debit cards. The most important systems in the country were the Federal Reserve Wire System, Clearing House Interbank Payments System (CHIPS)—used by New York banks essentially to clear Eurodollar transfers—the Society for Worldwide Interbank Financial Telecommunications (SWIFT) to connect large foreign and domestic banks; and finally various credit card networks. By the late 1970s, these various networks were transferring tens of billions of dollars each day, and every month more banks enrolled in them.[159]

The extent of activity in the Banking Industry in the various EFTS applications and their prerequisite telecommunications is, quite frankly, remarkable. One thinks first of the telephone companies or of manufacturing firms as being early users of telecommunications, but clearly banks (and stock brokerage firms to a similar extent) pioneered an enormous variety of applications conducted over networks. Table 2.4 gives a simple chronology of events and milestones just from the late 1960s (when POS technology began diffusing across the banking and retailing industries) to the mid-1970s (when all the basic elements of what constituted pre-Internet EFTS in the United States were in place). A clear pattern of industrywide behavior emerged. First, the ABA and the Federal Reserve took steps to encourage the industry to embrace the applications—the former through studies, the latter through actual implementation of networks. Second, industry associations and groups formed either to study or build EFTS networks. Third, all these initiatives were discussed publicly in the trade press: the events, applications, and their significance.

In the early 1970s, however, a debate was taking place about how to cost justify these systems. Obviously, keeping up with the paper flow provided a powerful incentive, but not necessarily a profitable one. One survey of individuals working on a system reported that "an electronic payment system, including payment of regular bills, payroll deposit, and on-line point-of-sale terminals, can make only a dent in the flow of paper checks."[160] At best, these techniques could displace only 30 percent of all checks (circa 1972), so other sources of justification were

Table 2.4
Major Events in the Emergence of EFTS in the U.S. Banking Industry, 1968–1980

Year	Event
1968	Formation of Special Committee on Paperless Entries (SCOPE) in California
1969	Georgia Tech studied role of EFTS for the industry; published results in 1971
1970	Formation of Clearing House Interbank Payments Systems (CHIPS)
	ABA/Arthur D. Little offered the industry considerable data on the need to use EFTS
1971	Publication of Monetary and Payments Systems (MAPS) report
	Creation of Committee on Paperless Entries (COPE) by 5 banks and Federal Reserve, launching the Atlanta Payments Project
	City National Bank, Columbus, Ohio, launched POS test called POST1; Hempstead Bank, Long Island, N.Y., does the same
1972	Incorporation of Mutual Institutions National Transfer System (MINTS)
	Wilmington (Del.) Savings Funds Society launched WSFS Plan, first EFTS application for consumers integrated into pre-existing banking services (e.g., checking)
	ACHs open in Los Angeles and San Francisco, Calif., by SCOPE, followed by formation of California Automated Clearing House Association (CACHA)
1973	Incorporation of Society for Worldwide Interbank Financial Telecommunications (SWIFT)
	Second ACH implemented in Georgia
	Bank card association launched national credit card authorization networks
	In-Touch home phone to computer billing service launched, first phone banking application
	ABA created Payments System Policy Committee
	First National City Bank, New York, installed thousands of branch and merchant terminals
	Introduction of Transmatic Money Service by First Federal Savings and Loan, Lincoln, Neb.
1974	Bellevue Exchange electronic facility became operational as joint venture of 15 banks to link ATMs, telephones, and other services
	Formation of National Automated Clearing House Association (NACHA)
	U.S. Congress authorized creation of National Commission on EFTS
	U.S. Treasury Departments initiated direct deposit of Social Security payments
1975	ABA launched large-scale EFTS strategy study
1978	Electronic Funds Transfer Act became law, protected consumer data
1980	First videotext projects launched as part of home banking

Source: "Major Events in Evolution of EFTS Are Occurring with Rising Frequency," *Banking* 67 (May 1975): 79–80, 82, 84, 114; ABA, "125th Anniversary Timeline, 1970–1979" and Ibid., "1980–1989," undated, http://www.aba.com/aba/125/timeline1970–1979.htm and Ibid. "1980–1989."

needed. At the time, the best thinking seemed to be that bill payment for individuals and businesses could be the ticket because it could be fully automated and enough volume existed. Furthermore, surveys of potential users suggested that 60 percent would consider using the service.[161] Bankers expressed growing concern that rivals in other industries, (insurance firms, thrift institutions, even service companies) could use such a system. In 1972, one banker warned that to win in this market would require significant marketing and technological investments in order to gain the benefits of lower operating costs, warning that insurance companies could gain enormously for the same reasons.[162] Yet it was not clear if there should be one, a few, or many networks—nor which types of services should be offered through EFTS.[163]

By the end of the 1970s, multiple networks had emerged that survived into the next century. Besides the Federal Reserve networks for clearing checks, those of Visa and MasterCard for credit transactions, and hundreds of private ATM and POS networks, some branded networks emerged. One of these was TYME, which was founded in 1976 as an ATM network and which, over time added more services: POS, debit and credit card transactions, transfer of money from one customer account to another, loan payments, and so forth.[164] In August 1981, federal regulators gave federal savings and loans permission to cross state lines with their ATM networks, increasing pressure on traditional banks while making possible further expansion of EFTS applications, into national EFTS networks. While bankers in the late 1970s and early 1980s still questioned and debated the economic benefits of EFTS, nonetheless they installed them. Simultaneously, industry press and banking consultants nearly unanimously touted their benefits. The fanfare continued all through the 1980s.[165] Over time, additional services were added, such as direct deposit services through an ACH in the late 1980s. The potential in this area was enormous; the Federal Reserve had noted that only 8 percent of the 2.16 billion paychecks issued in 1988 had been deposited electronically. While that percentage represented a large number of transactions, the other 92 percent seemed an even grander prize.[166]

As use of electronic payment systems expanded in the 1990s, the applications evolved and new firms entered the business. One industry consultant sounded an alarm in 1994:

> The pace of change in the payments systems is quickening. The banking industry's dominant role is threatened. Individual banks and the industry as a whole must begin now to make strategic resource allocation decisions if the industry is not to lose its preeminent position.[167]

The systems had become very reliable and efficient, and banks were lowering their operating costs, in part due to better technology; but consumers took all of that for granted and were now looking for more services, such as information. Rivals outside the industry began offering such new information-based services, such as credit cards, home banking, and ATMs, all integrated. GEISCO, Merrill Lynch, and AT&T were early movers into this new consumer market and EDI-based services for companies (such as check sorting and processing, and lockbox services), while some enterprises created their own clearing services, such as General Electric.[168]

Over the prior four decades, electronic payment systems had become ubiquitous across the industry. Even as early as the 1970s, government economists were documenting the spread of EFTS, although characterizing most of the implementations as "being tested" by large banks solely for the processing of checks. A decade later, ATMs had come into their own as the second largest EFTS application. In addition, ACH transactions had climbed from 50 million in 1976 to more than 557 million in 1984 (a service offered by 56 percent of all banks), with no end in sight. By 1984, wire transfers were in wide use among medium and large banks, and 39 percent of all banks had electronic payroll deposits; yet POS terminals were still in limited use. During the second half of the 1980s and all through the 1990s each type of EFTS spread across the entire industry to all sizes of banks, to stock brokerage houses and exchanges, and to large portions of the Insurance Industry.[169]

The arrival of the Internet brought with it new issues and old questions. The form of payment most widely used was the credit card, which Internet sites processed by using existing EFTS networks. Banks pondered the implications of this practice and how EFTS could and should link to the Internet and banking. The old questions about data security and privacy and the security of transactions surfaced again in force and was not calmed until encryption software was widely adopted by Internet merchants in the late 1990s. This time, however, while bankers asked the same kinds of questions as with prior systems, they seemed more comfortable with the inevitability of the Internet playing an important role. In a survey conducted in 1995—really the first full year of measurable e-business deployment in the United States—bankers expected a huge surge to occur in the use of credit cards, digital purses, and even electronic checks.[170] EFTS networks that existed before the arrival of the Internet did not go away; they coexisted with the expanding Internet of the late 1990s and continue to do so today. While there will be more on the Internet in the next chapter, suffice it to point out here that by the year 2000, bankers were puzzling over how to blend old and new networks together. Consumer business pushed the case for more Internet-based transactions (such as use of credit and debit cards as opposed to checks) while commercial bankers were still more conventional in their use of EFTS (such as to manage POS networks). Meanwhile, both sets of customers were supported by the older networks (for processing checks, payroll deposits, and so forth). So at least as of 2000, the Internet did not save banks money; rather, it became a new channel for delivery services already existing in the industry's cost structure—a channel, however, that held out the potential of bringing new business for existing portfolios of transactions at banks of all sizes.[171]

Conclusions

Slow to embrace the use of computers and telecommunications in the beginning, as each new application worked on computers and was affordable, first the large, then ever smaller banks embraced these. It was a dazzling array of applications that, in time, became emblems of the economy—credit and debit cards, ATMs, even Internet banking. They were visible in the economy: in the form of several credit cards in every adult's wallet, in the hundreds of thousands of ATMs located in every type of public space in America and in all industrialized countries. They all linked together through a patchwork of telecommunications. It all seemed so marvelous to the promoters of modern banking; and all contributed to the view held by sociologists that society had become "networked."[172]

But to understand the degree to which this was so, we need to look at patterns of deployment, to understand general consequences to the industry, and to examine more closely the role and effect of the Internet. How technology came into this industry suggests the paths of learning taken by other services industries. The first step is to understand what the applications were—the objective of the current chapter. The second is to weigh their importance in affecting the nature of work in the American economy.

3

Deployment of Digital and Telecommunications Applications in the Banking Industry

There is evidence of mounting interest and speculation in our industry about the possibilities the field of electronics may have in store for us.

—John A. Kley, 1954

Suddenly, in less than five years, a gigantic new market has emerged. Consider its profile: 30 million people; up-scale income; highly intelligent; average age about 26; reachable anywhere in the world at a small fraction of the cost of mail or telephone.

—Bill Orr, 1994

It is the nature of historians to catalog events by periods, to give those eras names, and to define their characteristics, all for good reason. These acts of cataloguing, naming, and definition help the scholar to make sense of patterns of behavior and to accent features of a particular time. Since every era is the child of a prior one, we expect the present to carry within it the heritage and influences of an earlier time; in this, historians are never disappointed. Cultural anthropologists call these "handed-down values," political scientists refer to "political realities" or "political preconditions," while historians may simply call them "historical precedence." But the point is the same: practices of a prior time help define those of the present.

As should be obvious from the discussions in the prior two chapters, the Banking Industry has a distinct personality, one that at its core is influenced by regulatory practices, competitive pressures, and information technologies. By describing what might artificially appear to be distinct periods in the deployment of IT, we can clarify how this industry embraced technologies, used them, and experienced the consequences. The interaction of events involving adoption, deployment, and consequences enrich our understanding of how the modern industry operates.

To be sure, describing phases is largely an arbitrary exercise, but there are some practical criteria that we can use to do this in a sensible way. In examining the role of technology throughout American industries, I have found that exploring three issues sheds light on the behavior of companies and industries.[1] First, it is important to bring back together all the various digital applications to understand how important and related each was to the other. In the prior chapter I simply described these as if they were independent developments; but reality was quite different, because multiple applications were used simultaneously and existed always in varying stages of deployment. Tied to this latter circumstance were management's concerns at any time, which always influenced the rate of adoption of the digital and, in the Banking Industry, also of telecommunications.

This leads to the second element: presenting historical research as phases or periods in the industry. No period is distinct from any other, that is to say, one era does not end on a Friday and a new one start on Monday morning. In reality, one period blends gradually into another such that at some point we can speak of a new style (in our case, the digital) as being more the case than not. Complicating this process of identifying periods is the fact that the "future" comes sooner to one company or corner of an industry than elsewhere. For example, the establishment of an all Internet–based banking company in the 1990s, Security First Network Branch (sfnb.com), represented a new era in banking for the employees and customers of that firm, but at the same time all other banks still lived within a prior existing paradigm that did not include use of the Internet to deliver services to customers. If the time should ever come that most banks operate like SFNB, some future historian will be tempted to describe that post-1995 period as different from what had existed before and will probably give it a name, possibly the Age of Internet Banking or hopefully something more insightful.[2] However, it is possible to sense when a new era is starting or is upon us, as is implied in the quote at the head of this chapter; note it was made by a banker who spent the bulk of his career in a pre-computer age.[3]

The third influence on the selection of periods involves consequences. When a new technology becomes relatively pervasive in an industry, there is a shift in how firms do their work and interact among themselves. In other words, the transformation will have already occurred before we recognize it. The results can be positive, such as the ease with which records of financial transactions could be updated (as with credit card processing using computers and telephone lines) or negative (as occurred when nonbank financial institutions competed effectively against bankers in offering financial services with PCs). In any period of transformation, some people gain advantages, others lose them.

The periods described below are arbitrary; that is to say, they can easily be changed, renamed, and re-dated by future historians armed with more information than I have in 2005. The value of calling out periods, however, is that it gives us a starting point for examining the industry's technological history in some practical manner; at the same time, it helps to highlight the importance of IT to this industry and the economy at large—a central theme of this book. For a manager or employee in any industry, but particularly in the Banking Industry, periods suggest how decisions were and are made to acquire and, later, replace various technologies and applications, enhancing decision making in the process. In short, it is all about context.

First Period: Dawn of Modern Automation (1950s–1970)

Banking had long been "high tech" by any standard. In any branch office in 1950 one could view a scene of IT clutter, with boxy adding machines at every teller's window and larger metal calculating "engines" nearby that updated account balances; and in the back offices more noisy adding and calculating machinery competed for space with other devices to handle checks and account balances. Loan officers sat behind dark wooden desks in the lobby area with black metal typewriters for processing loans and new accounts; staff used more typewriters in the backrooms. A banker in 1950 would have thought he was very much "state-of-the-art" and staying current. If the ABA's *Banking* magazine reflected the patterns of life in banking—and I contend that it did (and does)—then clearly interest existed in the 1950s and 1960s in innovation with technology. Prior positive experiences with information technology inclined the Banking Industry at large to keep an open mind and gave bankers sufficient experience with information-handling machines to be realistic about the limits of mechanical aids to data processing. This community of business managers demonstrated that they exercised this choice by taking their time in embracing the digital until they understood how it could be used profitably.

The process of embracing the new technology is not a story of computers ignored for a long time. Choice implies appreciating available options. From the early days of computing, bankers learned about automation and computing. In fact, advocates of the digital were active in the industry, not the least of which was the ABA. Getting ready for the new "electronics" became a mantra splashed across the pages of *Banking*. C. M. Weaver, for example, wrote many articles for this publication with a common theme: "Now is the time to study the means by which banking can benefit to the utmost from developments in electronics."[4] Banking automation became a major focal point for articles, conferences, and industry-sponsored committees in the late 1950s and 1960s. These provided explanations, justification, and ultimately the tipping point to action. Hardly any monthly issue of any banking magazine appeared without at least one article focusing on the general theme of automation and, later, computing. These were years when "firsts" got attention, such as when one of the first computers was installed in the industry, a 1.5 ton Univac 120, delivered to the Western Savings Fund Society of Philadelphia in the spring of 1957 through an upper floor window using a construction crane.[5]

This first period in the era of the computer spanned a good 20 years and was marked by several patterns of behavior. Bankers started to learn about computers and to use them for the first time, either in-house or through some service bureau arrangement. Early applications involved account management, loan and savings processing, and, of course, handling checks. As in other industries, what to use this technology for and how to build the business case for it dominated much discussion. How all these new tools squared with those already cluttering the offices of bank headquarters, and especially branches, remained of real concern. All of these issues swirled around, compounded at the same time by the technology, which itself was rapidly evolving into more usable forms. Computers in this period went from being one-of-a-kind, experimental, costly devices, to commercially available, mass-produced, reliable systems offering increasing amounts of speed, capacity, and software tools with which to write new programs. Even software application products aimed at the industry became available.[6] During these years, the computer and telecommunications industries figured out how to connect their technologies and make computers and telephones "talk" to each other. Simultaneously, interaction and work with computers evolved from just batching with cards and tape to also real-time, online dialogues using terminals attached to local and remote computers.[7]

Paper checks being read by an automated check processor using MICR, 1958. Courtesy IBM Corporate Archives.

Finally, the period evolved into one during which the industry began enjoying its first successes with computers, most notably a firm's ability to keep up with the processing of checks and updating transactions in checking and savings accounts.[8] Yet not all bankers were satisfied with the progress made so far. For example, C. Russell Cooper, an assistant vice president at the Harris Trust and Savings Bank, in 1965 commented on the state of things in his industry:

> Our data processing accomplishments to date have been primarily a piecemeal approach to information processing primarily for the purpose of clerical savings. In addition most banks are producing large detail "machine run-out" reports to management.
>
> As a result, one of the disturbing facts related to computer usage is the realization that contribution to executive decision-making and control is not happening.[9]

And always, the question was asked, "Does It Pay?"[10]

By the early 1960s, however, survey data from the industry demonstrated that automation was moving ahead across a broad front, reflecting a pattern of adoption that continued all through the technologically innovative period of the 1960s. MICR coding of checks represented one of the most important initiatives of the late 1950s and early 1960s, when almost half of a sample of 974 banks reported already moving to the new format for checks. One survey, done in 1962, turned up the fact that 46 percent of the banks questioned responded that they either had a computer or were about to install one. Table 3.1 lists the most widely used digital applications in this early period. Of course, an installed application could mean many things, such as incidental use of a computer (such as loan management) or extensive use (as in check clearing). We have no way to be more precise in understanding how computers were used based on currently available data, although anecdotal evidence strongly suggests that data input remained manually intensive in the mid-1960s, with the exception of check processing. Returning to the survey, two observations jump out from the data. First, bankers had already identified and installed a wide variety of applications, all of which remained core uses of computers for the rest of the century. Second, between 6 and 10 percent of all banks used these applications, suggesting that adoption was spreading across the industry.

Bankers looked over each other's shoulders to see what was being implemented and frequently moved to do the same. Savings applications were moved to computers at a quickened pace during the mid- to late-1960s, and all the applications listed in table 3.1 continued to appear in an ever-increasing number of banks.[11] In one ABA survey conducted in 1966, the association asked bankers to articulate their reasons for embracing computers. The top two reasons bankers gave were to improve internal operations (76 percent) and to freeze or reduce operating costs (71 percent). About half reported wanting to improve management reporting (54 percent) and increase operational flexibility (47 percent). Only 44 percent thought they could add new services by using computers, but this percentage grew steadily all through the century. Only one-third thought to use com-

Table 3.1
Computer Applications in the American Banking Industry, 1962
(Expressed as percentage of banks)

Application	Banks Using	Banks Anticipating Using	Total
Regular checking accounts	6.4	21.6	28.0
Special checking accounts	9.4	19.3	28.7
Savings	5.2	12.5	17.7
Other time deposits	2.6	4.1	6.7
Consumer loans	12.9	16.6	29.5
Mortgage loans	8.0	11.8	19.8
Commercial loans	2.9	7.0	9.9
Transit	0.7	8.2	8.9
Payroll	14.4	11.8	26.2
Personal trust accounts	10.4	8.8	19.2
Transfer agency, registrar	6.3	6.6	12.9
Other trust accounts	7.6	6.7	14.3
Other accounts	13.6	7.7	21.3

Source: Adapted from Herbert Bratter, "Progress of Bank Automation," *Banking* 55 (September 1962): 48.

puters to meet competition (34 percent), an indication that the industry had yet to enter the great competitive phase of its history, which began in the 1970s and continued right into the twenty-first century.[12]

Looking at the initial use of computers in the Banking Industry, say from the perspective of the late 1960s or early 1970s, what could one conclude were the effects of this new technology on banks? Fortunately, we have some comments from the period made by observers interested in this question. Installation of computers required creation of electronic data processing (EDP) departments staffed with programmers, operators, and data-entry clerks, especially in the largest banks. These were new departments that sat side-by-side with traditional ones (personnel, general operations, comptroller) and usually reported to an executive one level below the president. With the emergence of the EDP department, there now existed a group motivated to promote the further use of computing.[13] One perceptive bank executive posited in 1968 that bankers' initial experience with computers influenced how they thought about the way to operate their banks. He bristled at the notion that executives were reluctant to embrace the new technology, but also understood the broader managerial issues facing them:

> Most modern managers are fully aware of what computer technology can mean for increasing management efficiency. While no manager will deprecate the importance of quantitative skills and their contribution to decision-making, these skills alone do not provide management with the ingredients to solve some of the most perplexing problems of our time.[14]

These comments from Willis J. Wheat were written late in his professional career while vice president of The Liberty National Bank and Trust Company of Oklahoma. Wheat, not prone to naive views of how best to run a firm, believed that bankers had not yet integrated the technology into the overall operations of their firms, and urged them to understand better the potential economic benefits of the computer.[15]

Finally, what can we say about labor productivity in the 1950s and 1960s? Because of the low usage of computers in the 1950s we can quickly dismiss the first decade of computing by simply stating the obvious: computers had no appreciable effect on labor productivity until at least the early to mid-1960s. Beginning in the second half of the 1960s, however, and extending through the 1970s, computing had a positive effect on the productivity of internal operations. One U.S. government economic study reported that this productivity increased at a pace of 1.3 percent per year between 1967 and 1980. But computing never functioned alone; people always operated this technology in tandem with other circumstances at work. Two economists described this positive increase in productivity within commercial banks:

> The rise in banking productivity was associated with strongly expanding customer services and with advances in computer technology and their rapid diffusion throughout the industry. However, the spread of branch banking, while enhancing access to banking services, somewhat retarded productivity improvement, partly because scale economies became less favorable.[16]

They also pointed out, however, that it was about the same when compared to the productivity of the rest of other nonfarm business sectors. It would be pure conjecture at this point to try and answer the question why, because until we understand more fully than we do now the effects of computing across many dozens of industries in this period, we will just not know. As one would expect during the early stages of a new technology's implementation, however, output should increase faster earlier than later and, in fact, that is what happened here. Output from banks, defined as number of demand-deposit transactions, lending activities, and fiduciary tasks, rose at twice the pace as the rest of the business sector in the U.S. economy, with an annual average rise of 6 percent from 1967 through 1980. Between 1967 and 1973, output rose at 7.8 percent a year. Then, as the initial benefits of automation wore off, and the effects of growing competition and other economic influences (such as the recession of the early 1970s) were felt in the mid- to late 1970s, annual productivity dropped to 4.8 percent.[17]

Gathering together all the evidence presented so far, one can conclude that banks had made a good start in using the technology as best they could, given computing's state of evolution and costs. Nearly one-third of all banks ended the 1960s using computers in some fashion or another; almost all involved some form of telecommunications, if for no other reason than to clear checks. The Federal Reserve pushed bankers faster into the world of telecommunications than they might otherwise have gone, the ABA provided leadership and the compelling case to drive forward computerization of check processing, and IT vendors focused enormous attention on this industry.[18] Bankers bought in because of their concern that they would be unable to handle increasing volumes of checks and out of a growing awareness that, once installed, computers could be used for an assortment of other applications. This is exactly what happened in the 1970s. Actions in the 1950s, however, extended practices of the 1940s, which one Federal Reserve study argued were characterized by the use of more people rather than incorporating better equipment to improve operations.[19] The introduction of MICR in 1958, however, became the event that expanded general use of computing by bankers, providing a basis for improvements in operations that began appearing in statistics in the 1960s and reflected in the deployment of computers by nearly all commercial banks by 1980 (97 percent).[20]

Second Period: When Computerized Banking Became Normal (1970–1990s)

In the quarter century spanning from 1970 to the mid-1990s, the Banking Industry was transformed from a heavily regulated into a national and international industry experiencing intense competition from within and outside of its traditional borders. During this period, banks left behind the operational models that had so characterized their industry over the previous centuries and arrived rapidly in a new world, that of the digital style. If ever there was an example of the concept of technical or economic momentum at work, the integration of computing in this industry is it. Surveys on deployment conducted over the entire period showed not only increasing numbers of banks implementing various applications, but also higher percentages of budgets allocated to IT and higher rates of expenditure and usage as time passed. In short, it was in this period that banks were second only to manufacturing the most intensive commercial users of computing in the U.S. economy.[21] Furthermore, if not for the massive introduction of the Internet into the business side of the economy, which forces historians to acknowledge its arrival by giving it its own "era," I would have labeled the period from 1970 to the present as one, because they shared common trends and practices. However, because the Internet did profoundly and rapidly begin altering again bank practices, I chose to close this second period with 1995, recognizing all the while that patterns evident as late as 1994 continued into the next century.

Let us begin analyzing this period by looking at patterns of adoption. Figure 3.1 shows the percentage of banks that either had one or more computers in-house or used a service bureau or some other bank's computer to run their applications over a twelve-year period. The chart makes two trends very clear: that the percentage of banks acquiring computing kept increasing, and second, that the rate of adoption quickened as time passed. A U.S. Bureau of Labor Statistics study provides more data behind these trends. In the early 1970s, banks added facilities for processing as well as new services. The major collections of applications were electronic data processing for handling demand deposits and savings accounts; ATMs and cash dispensers; regional check-processing centers, which were first

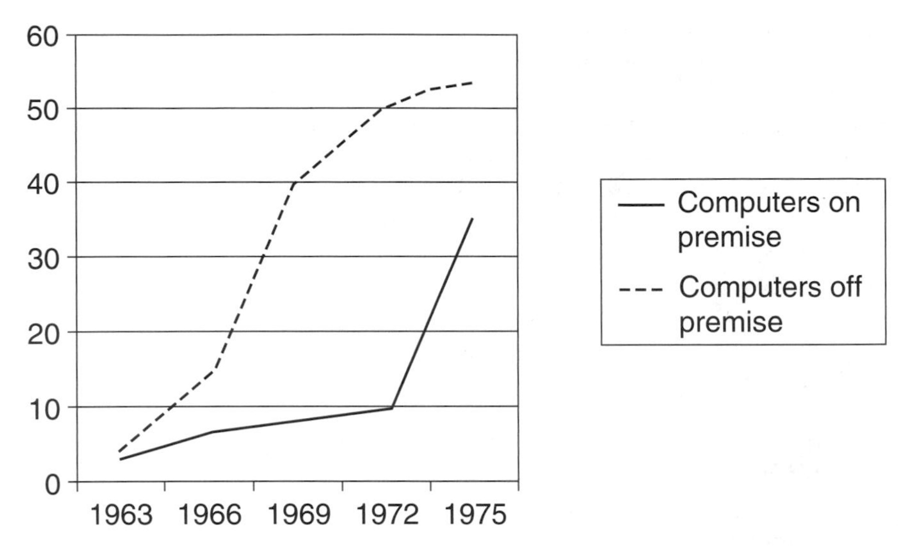

Figure 3.1
Percentage of U.S. banks with computers, 1963–1975.

started in the late 1960s; a series of automated customer services, such as billing, payroll handling, and correspondent bank services all using excess capacity of in-house computers; and a variety of EFTS applications, with major projects in early test stages by the end of 1974.[22]

In support of these applications, telecommunications expanded across the industry throughout the 1970s, with a third of the associated expenses involving contacts with other financial institutions. In 1978, for example, the largest banks proved the most extensive users of both voice and data communications, but small and medium-sized banks all relied on telephones to conduct some of their business transactions; 80 percent of mid-sized banks also used data networks, while only 24 percent of very small banks.[23]

As noted in the previous chapter, the number of ATMs increased steadily in the 1970s. As with other technologies, large banks did more sooner than did smaller enterprises. In 1980, only 8.4 percent of the smallest class of banks—those with assets of under $50 million—had ATMs, while at the other extreme 80.7 percent of banks with assets of between $1 billion and 5 billion were extensive users of ATMs. To put the machines into further context, the smallest class of banks had on average just over 1 ATM each, while banks with $1 to 5 billion in each had nearly 30 on average. The very largest banks, only 65 percent of which were involved in retail banking had just over 100 machines per participating firm.[24] The data also suggest that the clutter of adding machines, calculators, and bank-accounting machines were finally disappearing, replaced by online teller terminals connected to computers. By the end of the 1970s, 14.1 percent of the smallest banks had installed these terminals; 63.2 percent of the largest banks; and 37.3 to

Table 3.2
Data Processing Costs as Percentage of Total Operating Expenses for Banks, Selected Years, 1972–1985

Year	Under $100M in Assets	Assets of $100–500M	Assets of $500 and Over
1972	4.0	6.0	5.0
1974	6.0	7.0	7.0
1976	6.5	7.5	7.4
1978	7.4	7.9	8.5
1980	6.1	6.6	9.1
1985	6.9	8.1	12.8

Source: Derived from data in "How Do You Compare on Prime Pass Rejects?" ABA *Banking Journal* 79 (May 1987): 62.

48.4 percent of the mid-sized banks.[25] It would not be long before all tellers had them.

We also have some data on the cost of IT as a percentage of total operating expenses for the 1970s. Since this is a measure long used by all industries, and the IT Industry itself, it is worth examining this data. Table 3.2 shows the percentages for three sizes of banks for the period 1972–1985. Several patterns become evident. First, small and medium-sized banks spent essentially the same percentage on IT, between 6.1 and 8.1 percent, which is higher than across the entire U.S. economy but far less than for the largest banks. The largest firms began spending proportionately the same as smaller institutions, but as time passed that changed as they began operating nationally, offering more services and processing larger volumes of transactions, which drove up the amount banks had to spend on computing. By the late 1980s, expenses had drifted up for the entire industry.[26]

The relentless push to install new applications and to upgrade, expand, or modernize old ones continued in the 1980s. With the rapid injection of IT into the burgeoning industry it was no wonder that the U.S. Bureau of Labor Statistics came back into the commercial banking sector to gauge trends. (Interestingly, employee productivity in the period 1978–1981 did not increase; in fact it worsened, then picked up over the next several years.) A combination of continuously improving productivity of computer technology, increased demand for online services by customers and employees, and expansion across larger geographic markets contributed to the further spread of IT. It seemed everyone went online, which meant a large increase in the number of networks as well. The BLS reported that by 1984 all large commercial banks used computers for the kinds of applications discussed in the previous chapter. The same held true for the use of online terminals by employees, especially in the largest banks. EFTS deployment, which had been at about 2,000 installations in 1972, rose steadily to almost 60,000 largely in 1984, done to reduce the rate of growth in the population of tellers while providing corporate payroll departments and customers a useful service. ACH transactions jumped from 50 million in 1976 to over 557 million in 1984. The only slowing growth in usage occurred with point-of-sale terminals, which were only in their early stages of introduction.[27]

I could continue presenting more data covering the late 1980s and early 1990s to demonstrate the obvious: that all banks were extensively using computers in all manner of ways.[28] Rather, I would prefer to offer only one additional statistic that suggests clearly the symbiotic relationship between IT and the consolidation of banking firms. In 1992, 13 percent of all banks—the largest—spent 78 percent of the entire Banking Industry's expenditures for IT. And what did those expenditures go for? A third went for salaries of IT personnel, another third for hardware, 13 percent for software, and 10 percent for telecommunications; the rest went for miscellaneous expenses. M&As caused expenditures to decline because managers exploited technologies as opportunities to consolidate large data centers and IT operations in the late 1980s and all through the 1990s.[29]

What issues did the bankers themselves focus on in this second period? As in earlier times, they spent time and money incorporating IT into the basic operations of their firms, initially to lower operating costs, later to constrain competition, and all-in-all to deliver new services. Even a casual reading of the industry's contemporary literature would show these basic activities along with bankers' more sophisticated concern about the best way to use IT and the consequences of management's actions. Bankers debated the pros and cons of in-house computing,[30] the potential size of a market for a particular application or use of a technology (for example, ATMs),[31] POS,[32] EFTS,[33] ACH,[34] other hardware and software,[35] and how best to cut operational costs with IT.[36] Bank cards of all types and check processing commanded continuous attention as well.[37]

Nearly a half century after the arrival of the digital computer, some bankers were still asking if the technology improved or hurt customer service.[38] To be sure, it was still a good question to ask, since so many different uses of the digital had emerged over those decades. Contemporary literature suggests that the costs and benefits of using any particular technology were never settled. Many debated problems of check processing, POS, ATMs, and so forth, questioning constantly their benefits and their costs. No other industry that I have studied seemed so unsettled about its use of the digital as this one. The same observation can be made about important issues and consequences that derived from the use of IT and telecommunications, such as the continual angst over security of data and privacy of a customer's transactions.[39]

In the 1970s the extensive infusion of the digital into the daily operations of banks led to a growing discussion of the effects of technology on managerial practices. This occurred in parallel with other debates we have reviewed in earlier chapters, such as its effects on the role of branch banks, relations with customers, and, of course, on the size of a bank and its competitive posture. Like managers in all industries, bankers argued the pros and cons of centralized and decentralized management as made possible by computing. As one banker put it, "When the technology of computer-to-computer communications was still primitive and computer processing of banking transactions was all batch, there was little choice but to allow each regional center to develop its own computer system in its own manner. And each did."[40] Nothing was compatible. To exploit technology in more effective ways as his bank grew, "we had to establish central control to attain standard systems."[41] "Since computers helped to impose standards across many bank practices, bank managment personnel always remained connected to computing. We already saw the expectations bank executives had that computers should reduce the number of tellers. The industry at large anticipated computing to help train those that remained.

But the industry as a whole recognized that it lived in a period of constant change, resulting from use of the digital. There was general acceptance of the notion that a combination of factors, however, worked together to cause change, emanating from economics, regulatory practices of state and federal governments, and, of course, technology. Inflation, for example, squeezed profits as banks tried not to raise costs to consumers (actions largely affected by federal regulations); inflation also put pressure on wages in a labor-intensive industry; and computing made possible new forms of competition. Bankers understood the existing links between regulatory practices and competition, and also involving technology.[42] For example, deregulatory activities in the early 1980s (when AT&T was broken up) caused bankers to ponder the implications of lower telecommunications costs as the long-distance telephone business began to open to competition. With extensive networks for ATMs, EFTS, and so forth already in the industry, telecommunications proved important. As one observer at the time noted, "Banking is communications." While bankers were hesitant to engage in national debates over telecommunications policies, the same person argued that "they've got to treat communications as importantly as AT&T does."[43] To be sure, AT&T competed with the Banking Industry since it served as a third-party provider of networks for financial transactions offered by others.[44]

If we leave aside the bankers and their industry's spokesmen and examine the changes brought about in the behavior of the industry by technology, we can leverage some hindsight. Three economists recently asked a very good question: "How do banks innovate?" Frances X. Frei, Patrick T. Harker, and Larry W. Hunter thought of the question in both technological and competitive terms. For one thing, they discovered that bankers reacted to possibilities of losing market share and in response to consumer demands by embracing PC banking in the 1990s. But they had always done this: PC dial-up services in the early 1980s, then, use of off-the-shelf home financial software tools in the mid-1990s, and, of course, Internet-based banking later on. Merely reacting never seemed enough, though; there always were problems with implementation and in persuading customers to accept new approaches. So, getting positive reactions to new uses of technology often posed the biggest problems for bankers. As the economists concluded, "the fundamental challenges to innovation in PC banking are not technological per se but arise from the complex set of organizational choices to implement such a service for the consumer."[45] Integrating various technologies and marketing programs presented the true challenge to change, let alone innovation. In observing specific banks in the 1980s and 1990s, they noticed that the innovative initiatives were facilitated by vendors of IT products working closely with bankers to develop new ways of using the digital. IT involvement also included suggesting changes in the way bankers organized and ran operations in a bank. The authors concluded that "the bottom line . . . is that services industries such as banking must develop a new generation of management talent to play the role of systems architect—one who can blend technical knowledge with complex organizational design issues to drive innovation through their firms."[46] Their observations were published in 1999,[47] but the process continued. For example, IBM and J. P. Morgan came to an agreement at the end of 2002 for IBM to take over the majority of the bank's computing operations and manage them in ways that would allow the bank to continue innovating its daily work procedures and services under a long-term contract.[48]

The structural changes that came to this industry as a result of economic, regulatory, and technological factors caught the attention of many observers, mostly economists and consultants. The majority of their interest concerned developments that occurred in the 1970s and 1980s, but these remained important issues through the 1990s as well.

Central issues for any industry concern how to use technology to create value, increase profits, destroy competition, and grow. As noted in the first chapter of this book, economists and consultants who have observed the Banking Industry have accepted that computing altered the structure of the industry, against whom it competed, and what services it offered. As a reminder, IT changed banking in four fundamental ways: it created new services that were subject to more scale economies and fewer diseconomies; it enhanced existing banking services that lent themselves to greater scale economies than earlier technologies; it made it possible for banks, particularly larger ones, to expand their risk-expected return frontiers; and it lowered managerial diseconomies of scale by improving control and monitoring functions.[49] Give the final word to a student of information technology's effects on the U.S. economy, Harvard University's Anthony G. Oettinger, made in 1964: "Automation affects not the mere mechanics of banking, but the very foundations of banking; not the individual bank, but banking systems and the national and international economics in which they are imbedded."[50]

What we still need is research on the degree to which value and profits were created and destroyed in the Banking Industry. Many statistics demonstrate that banks made money in good economic times, but when interest rates were low banks faltered and they blamed the regulators and competition from outside their industry. In one of the first thoughtful discussions of the issue, Thomas D. Steiner and Diogo B. Teixeira argued that the industry was stagnant in the 1970s and changed to survive. Today, we know that much of the blame for productivity lags in the 1970s and 1980s can be placed at the feet of the transition to a more competitive environment caused by deregulation, not computing. Technology played a profound role in that transformation, making it possible for the industry to become more cost effective over time. However, stagnation continued despite investments in computing until the 1980s and 1990s, at which time it was clear how IT had inherently changed the economics of the industry. Shifts to digital delivery systems and the proliferation of software tools to handle derivatives and credit checks and commitments alone increased economies of scale in transaction-processing to such an extent that some economists have concluded it was by more than 50 percent just in the 1990s.[51] Large banks responded better than smaller institutions to changing circumstances, particularly in the 1980s, resulting in continuously improving net income for the largest players all through the 1980s and stagnation or slower growth for smaller banks.[52] Yet even the latter could leverage IT for positive gains.

The biggest change occurred in their balance sheets. Steiner and Teixeira demonstrated that large banks de-emphasized less profitable services, such as large corporate and international loans, and pursued more profitable ones, such as consumer loans and credit cards. Income from fee-based services also increased, all made possible thanks to computing and telecommunications. Citing the case of Citicorp, one of the most aggressive large banks of the period, consumer loans went from 22 percent of the bank's loan portfolio in 1980 to 55 percent in 1988.[53] Expenditures on technology rose faster than any other form of expense, all in support of computer intensive consumer services that represented its most profitable line of business. Did technological leadership affect the changes in profitability and portfolios? Steiner and Terxeira argued yes. The essential technical infrastructure of a bank—mainframes, networks, databases, and so forth—did not offer specific competitive advantages; yet they were vital as supports to other technological activities that did, such as digital applications that delivered new or less expensive services. Base technologies were needed for defensive reasons because everyone had them; they had become a requirement to participate in the industry.

Though bankers worried about whether they needed the most current or novel technologies, experience suggests the answer depended on why a technology was used and when it was implemented. Economies of scale, time, and execution determined advantages. The basic requirement to have infrastructure alone points to the conclusion that the industry had moved to a digital style by the end of this second period. However, it was a second level of computerization—the applications, and not just information technology—that made it possible to automate work flows, largely to control operating costs. Competitive advantages came from the use of skills and the creation of proprietary offerings. The industry invested enormously in both levels of technology in the 1970s and 1980s. Steiner and Teixeira argue, however, that there was yet a third set of technological uses that supported business strategy; marketing, that is, sales that would lead to new offerings attractive to the market. By the end of the 1980s bankers were using computing (applications) to sell more or different services, not merely to hold down costs.[54]

The evidence presented in the last two chapters suggest that the second and third levels were intimately tied together with no differentiation made by bankers, who understood the difference between the technical plumbing of the industry and applications. The latter were always defined as services, such as loan processing or credit and debit cards. Competition usually occurred in delivering these services, so speed to market with new applications was always the most important source of potential revenue and profits. As we have seen, banks did not always move as fast to exploit technology in this second period as they might have. So, many applications quickly became commodities, such as ATMs. Every bank had them, and few attempted to differentiate their characteristics. If the industry moved to a new offering, then everyone did. Applications that promised the highest return on investment came to market first, which is why wholesale banking (cash management, funds transfer, trading, and so forth) did not get as much attention as retail banking (such as ATMs, POS, credit cards, checking, and personal trust offerings).[55] But because everyone offered these at the same time,[56] one would have to conclude that the vast majority of expenditures on IT supported "routine tasks," as our two consultants called these, and that only 10 percent of these services, in their opinion, led to truly differentiated offerings. Yet these "routine" services were the bread-and-butter offerings that generated enormous quantities of revenues in the industry.[57] Today we know that improving back-office operations by using computing did yield economic benefits to banks, particularly the largest. In fact, these benefits were already evident during the second half of the 1980s.[58]

We have increasing evidence that consolidations in the Banking Industry in the 1980s and 1990s improved cost efficiencies overall, that is, across all operations and not simply in those subject to automation. One still needed differentiation in products and services, and competition was capable of scaling and competing too.[59] Hence, simply focusing on the effects of one application or offering does not tell the whole story, even though economists and industry watchers constantly do this, particularly with EFTS.[60] How broad a sweep of events is necessary in order to understand the role of computing in this industry is still open to some discussion. Historians are inclined to look for a broader variety of issues interacting with each other, which in this industry were the confluence of regulatory, competitive, technological, and economic dynamics, all functioning within an industry that had its distinct handed-down values and practices.

Allen N. Berger and his colleagues have studied the industry in a broad fashion over many years and have persuasively demonstrated that, in this second period (1970s–1990s), a confluence of deregulation, changes in the nature of competition, and the infusion of large quantities of the digital affected the performance of the Banking Industry. Commenting on the decade of the 1990s, when many IT-based offerings became commodities, Berger concluded that "cost productivity worsened while profit productivity improved substantially,"[61] because banks raised revenues and drove down operating costs to prop up profits. As for computing, "banks provided additional services or higher service quality, which may have raised costs but also raised revenues by more than the cost increases."[62] He also collected evidence that by using computers, banks could control the quality of loans they made so that they were able to expand operations to national markets, relying on technology to contain the risks (credit-scoring applications) that bankers of old could do simply by knowing local markets and individual loan applicants.[63] This evidence is an antidote to those arguments that deregulation or competition alone could have caused the massive restructuring of the industry that occurred between the 1970s and the end of the century.

By the early 1980s, online systems to collect, analyze, and present business and financial data were widespread across many industries. Courtesy IBM Corporate Archives.

All that said, we are left with one inescapable conclusion: during this second period, banking became very high-tech, so much so that even the history of the Information Processing Industry cannot be written without taking into account the practices of the Banking Industry. It was also the period when consumers—now the darlings of the industry—also began shopping in other industries for their financial services. One industry observer, Robert E. Litan, in his book-length analysis of the industry of the 1980s, minced no words:

> Americans are putting lower proportions of their wealth into traditional depository institutions such as commercial banks and savings and loans. Instead, they are depositing funds in other financial intermediaries, principally pension and mutual funds. Similarly, business borrowers are increasingly turning away from banks and toward finance companies and the securities markets.[64]

Third Period: Banking IT in the 1990s

In the 1990s, computers contributed mightily to the wave of loan securitization that took place. That trend made it possible for other industries to buy mortgages and access the banking market, especially through credit cards. In short, computing in this third period had become a double-edged sword of profit and risk for bankers.

There was a cumulative effect on the style of operations of a bank in this period. Bertrand Olivier and Thierry Noyelle, two European economists, observed

> a tendency by firms to try to use distributed data processing to move away from Taylorism—a fragmented mode of work introduced earlier in large service firms and often reinforced by the introduction of the first generation of centralized data processing—and replace it with a new division of labor emphasizing the recombination of tasks within a work process, in other words, emphasizing totality over fragmentation.[65]

Thus, as banks decentralized, they shifted authority, responsibility, and control to remote offices or to different groups of employees from before. Decentralized computing in the 1970s had come about because that form of technology started to prove less expensive than first-generation centralized processing. American banks in particular had aggressively decentralized in their hunt to lower labor costs and, in the process, had started tampering with the predigital style of doing their work.

No sooner had bankers and economists begun coming to terms with the role of computing in the industry when along came the Internet. It arrived quietly in this sector of the economy that had long been familiar with telecommunications and had made the historic shift to consumer banking. Thus its leaders and opinion makers thought they understood their markets and how best to leverage technology. They were in for another round of surprises.

Fourth Period: Banking in the Networked World (1995 to the Present)

From 1997 on, it became almost impossible to write about trends in American business without giving enormous attention to the role of the Internet. Commentators routinely reported an ever-growing number of Internet users in the society at large and highlighted specific uses within industries.[66] No other computer-based technology was as closely monitored as it unfolded. Even the digital computer in the 1950s and 1960s and the PC in the 1980s failed to receive the same level of attention. The emphasis imposed by economists, industry watchers, business school professors, consultants, and policymakers on discussions about the role of the Internet proved so great that this fixation overshadowed many other pre-existing trends and activities.[67] The Banking Industry was not exempt from this focus on the Internet.[68]

As a result, an observer of the modern Banking Industry faces several challenges in looking at the post-1995 era. For one thing, nothing really altered between late 1994 and early 1995; people still went about their business of writing checks, spending money, borrowing more (indeed a great deal more); banks and brokerages launched IPOs at a record pace; the U.S. economy prospered; and the Banking Industry continued to increase its reliance on the broad range of digital applications. But during the years from roughly 1995 forward, the Internet became a body of technologies embraced across the American economy by consumers and businesses; again, as in other industries, banking participated.[69] Because of both its importance and the attention it received, it is easy to focus on just the Internet's role and forget that it incorporated a host of prior applications of computing. To understand the post-1995 period in this and in so many other industries, one has to keep in mind that coexistence of multiple uses of computing and telecommunications was, in fact, the way of life.

Prior experience with networked-based applications had created in this industry an enormous body of knowledge about computing and communications at many levels of a bank's organization. By the time the Internet became commercially attractive to use, this body of knowledge helped bankers determine when best and how to exploit the Internet as yet another channel for distributing their services. That prior experience, as we have seen, when coupled to regulatory practices and the way things were generally done in the industry, made it fairly predictable that bankers would not rush out and embrace the Internet with wild enthusiasm. Concerns about regulatory permissions and serious issues related to data security and customer privacy held back the industry—as did the perennial questions of costs and benefits. From the point of view of timing, two cumulative effects became evident. First, though slow at first, banks did embrace the Internet; interestingly, their pace of adoption proved essentially not so different from what occurred in manufacturing and retail industries. Second, at the time this chapter was completed (2005), the Internet was continuing to influence the structure, offerings, and competitive landscape of the industry, much as had earlier technologies—and creating opportunities for growth and threats to profits and market shares.

Before discussing either the general use of various technologies, their effects on the industry in this last period of its history, or even the specific role of the Internet, let's take two brief metaphorical snapshots of Internet usage in the United States. The Internet grew in importance as the industry went about its business in the late 1990s, and the two views should highlight what happened between 1995 and 2000. I propose that the first one be of events from 1995 to 2000 to provide context for discussing the overall use of computing in this industry. This period is also useful because the importance of consumers to the Banking Industry in the late twentieth century meant that their access to the Internet, attitudes toward it and experience with the technology served as a significant gating factor in its deployment in banking, just as in so many other industries at the same time. It would seem that consumer attitudes were even more influential than the regulators. The Clinton Administration recognized that deployment of PCs was key to economic growth and so focused on promoting use of IT. Officals strongly encouraged people to get on the "Information Highway." In other words, bankers could not use the Internet as another channel of interaction with customers unless the customers used the technology. This symbiotic relationship paralleled the earlier experiences of television networks and TV manufacturers when expanding their business, made possible largely by the availability of both programming and affordable receivers that the public would buy. There was a similar issue in the early days of minicomputers and PCs, when sales of the hardware needed software applications to grow beyond a certain point. Later, sales of software packages could increase once a large number of PCs were in place in homes and offices. And even in the Banking Industry, consumer influence was seemingly greater than that of the regulators and the bankers themselves. We have only to think about the effects consumers had on the industry with regard to paper checks, PC banking, and debit cards to recognize their influence over the industry. Whenever bankers butted heads with consumers, the former gave in and neither side enjoyed the experience.

One can view the consumer/supplier readiness synchronization in banking as a classic example of the chicken-and-the-egg conundrum, but that would be an incorrect analogy. The crucial point is that both had to arrive almost simultaneously. Furthermore, the act of arrival was not a single event, rather a multitude of incremental steps taken—a continuum so to speak—that affected each side of the supply/demand equation. Therefore, it would be more useful to use a different analogy, say that of a child's seesaw, where the activity's pleasure occurs as a result of the back-and-forth feedback caused by the actions of both participants. This pattern of interaction is consistent with the findings of others looking at the developmental dynamics of a technology's adoption and evolution by a society.[70]

Building on our analogy of the seesaw, in 1995 just over a third of all households in the United States had a PC. The year before, Netscape had introduced its browser, which nearly instantly revolutionized the Internet, because its software made it possible for people to interface easily with the Internet by English-language commands and thus increasing the desire of people to acquire PCs. The largest online service provider of the period, America Online (AOL, established in 1989), saw explosive growth, claiming 10 million subscribers in 1997; as a whole, 19 million people used online services from all vendors that year.[71] Meanwhile, what was happening with online banking? Checks continued to dominate as a preferred vehicle for conducting small dollar, noncash payments, followed by credit cards. Table 3.3 presents results from a study conducted in 1995 by the Bank of International Settlements (BIS) regarding usage of various cashless payment instruments in the early to mid-1990s. The data demonstrate two things: first, non-Internet transactions were extensive; and second, the Internet was not yet a player in the industry because the study makes no comment about Internet-based banking. With only about a quarter of all homes equipped with a PC in 1995—the instrument needed to access the Internet—one should not be surprised, since web-based banking played such a diminutive role at that moment. All the surveys conducted in the mid-1990s regarding electronic banking concerned telephone usage, ATMs, debit cards, and so forth, but not Internet banking.[72] It was not yet a factor worthy of attention.

Now let's fast forward to 2002 to see how many people were using the Internet.[73] Some of the most useful studies done at the time concerning usage of the Internet in the United States came from the Pew Internet and American Life Project, which continuously surveyed a broad spectrum of the American public on various issues related to use, producing many studies each year. The Pew reported that in January 2002, approximately 55 million people used the Internet at work, up from 43 million in 2000. Over half of all American homes and businesses had

Table 3.3
Usage of Cashless Payment Instruments in the United States, 1990–1994
(Transactions in millions)

Instruments	1990	1991	1992	1993	1994
Checks	55,400	57,470	58,400	60,297	61,670
Debit cards	278	301	505	672	1,046
Credit cards*	10,478	11,241	11,700	12,516	13,682
Federal Reserve ACH**	941	1,059	1,190	1,346	1,526
Federal Reserve ACH Direct debits	487	573	654	739	847

†Totals are not precise due to rounding of data by type of instrument.

*Includes cards from all sources, not just banks (e.g., oil companies, retail stores, telephone, etc.).

**Paperless credit transfers.

Source: Drawn from various studies presented in Arthur B. Kennickell and Myron L. Kwast, "Who Uses Electronic Banking? Results from the 1995 Survey of Consumer Finances," unpublished paper, July 1997, available from the Division of Research and Statistics, Board of Governors of the Federal Reserve System, p. 4.

access to the Internet. In 2002, over 58 million people had made at least one purchase over the Internet, a jump from the 40 million who did in 2000. The Pew reported that 14 million people had conducted some form of Internet-based banking in 2000; that number jumped to 25 million the following year, suggesting an inflection point for the Banking Industry, a time arriving when enough people were online to make their kind of banking potentially a major business opportunity. To put this observation in perspective, the jump from year to year was 79 percent, and over the same period the buying and selling of stocks by individuals online jumped by 30 percent, use of online auctions grew by 83 percent (largely due to the e-Bay phenomenon), buying travel services rose by 59 percent, and purchase of products increased by 45 percent.[74] In short, by 2001, a large percentage of the population had used the Internet for a variety of business transactions. This is not the place to discuss who used the Internet; suffice it to point out that in the early 1990s, upper-income families did so more than other economic groups, but by the year 2000 people in all economic brackets were involved. One could prudently conclude that many people had integrated the Internet into their routine financial activities, among these Internet-based banking.

Yet, banking over the Internet was only one small part of the larger mosaic of digital applications used by bankers and their customers in the late 1990s. Bankers, as we know, focused on a variety of managerial and technological issues related to the operations of their firms. For example, right up to the end of the century, bankers still struggled with the issue of bank branches. On the one hand, customers liked them and, in fact, more had to be opened. On the other hand, these new branches often were smaller and contained more technology to reduce the labor needed to run them. The president of Commerce Bank in Cherry Hill, N.J., Vernon W. Hill II, observed in late 1994: "We don't see any indication that customers are leaving their branches," because "customers want that face-to-face interaction with their banker."[75] Yet bankers kept trying to find ways to have fewer branches. In the spring of 1995, a senior vice president at First Union National Bank, Charlotte, N.C., Jude W. Fowler, pontificated that "the banking industry will suffer the same fate as the dinosaur within the next five years unless the brick and mortar branch banking system is cast off in favor of more nimble delivery alternatives."[76] He thought that future competitors to banking would be IBM, Microsoft, AT&T, EDS, General Motors, USAA, and others. As late as the end of 1996—after Internet banking had appeared—customer survey data suggested that Fowler may have overstated his case for implementing more technology. As two consultants at the time noted, "Most consumers grudgingly accept the inevitability of technology, but they don't necessarily like it."[77] But, customers did want instant access to up-to-date information about their accounts around the clock.[78] Small banks saw the Internet as a way to compete effectively with larger banks. Some banks (both large and small) also wanted to set up microbranches in supermarkets, yet even this was a problem for consumers who still preferred a separate building for their branches. Most bankers, however, spent the rest of the decade trying to balance their need for branches with various forms of cost-reducing technologies in their mix of offerings and ways for customers to access them.[79] This strategy was evident in dozens of other industries and proved consistent with how people had adopted technologies for centuries.[80]

Debit and smart cards remained part of the technology discussion right into the new century. As in the 1980s and 1990s, bankers debated when the market for such products would take off and how they could exploit opportunities yet drive down operating costs. Although debit cards were enjoying considerable success in Europe—a trend noticed by the United States industry—bankers remained uncertain about the U.S. public's acceptance of these up to the end of the century.[81] Bankers also continued to closely link discussions about the evolving role of ATM cards as consumers and bank officials tried to work out what new services made sense with this technology.[82] Two issues continued to influence the debate: concerns about how secure one could make information flowing electronically—it became a major one with regard to the Internet—and how best either to block competition from outside the industry or to launch attacks against rivals as, for example, in the sale of insurance and other financial instruments. Physical security also captured increased attention after the terrorist attacks on September 11, 2001.[83]

I take it as a sign of managerial maturity regarding technology when an industry moves beyond just discussions on how to use a specific technology, and begins to understand and exploit the implications of its use. While one can debate when the Banking Industry reached this level of maturity, it is clear that by the late 1990s members of the industry were treating technological improvements and IT strategy increasingly as strategic business issues. By the mid-1990s, CEOs were in office who had personally experienced implementing various IT applications. For them, IT had become strategic and so they were involved.

There were circumstances that simultaneously pulled at their sleeves; none more so than the Y2K issue. The widely held opinion in IT circles was that when January 1, 2000, came around, calendars in computers might not reflect the new date and in fact might revert to 1900, with the result that many calculations dependent on dates (such as mortgage payments) would be fouled. That problem existed because programmers in the 1960s and 1970s, and even later, never thought to design their systems to operate past 1999. No sector of the economy felt the pang of this threat more so than the Financial Sector. Government regulators weighed in, demanding that key financial applications be rewritten or replaced so that they would not suffer from the Y2K problem. As a result, the banking and other financial industries spent billions of dollars—historians do not yet know how much—revamping most of the major application software. Two consequences became apparent as a result of this initiative in the several years following this forced march. First, many old systems were replaced with newer, state-of-the-art versions, which meant banks did not have the desire or financial wherewithal to add new applications in the several years following Y2K. Second, it diverted management's attention away from adding new applications in the period 1997–2000 as banks wrestled with this problem. Clearly, vendors selling IT products and services to the industry did very well in those years but far less so in those that followed.[84]

Since we are so close to those events, other unintended consequences will probably become clearer at a future time that we cannot recognize today. However, since repair of older systems often represented a drain on expenses, bankers still fretted over how to make profits—even though the nation was experiencing an economic boom in those years. Outsourcing of IT became an attractive strategy for controlling rising expenses for IT; banks also attempted to use the Y2K crisis as a way to integrate disparate applications, thereby reducing maintenance costs while simultaneously responding to customer requests for more integrated pictures of their dealings with banks. As Patrick Ruckh, a senior vice president for IT at the UMB Banks, put it, "There is no silver bullet in banking . . . so we're putting islands of information together."[85] Late in the century bankers increasingly came to realize that information useful for themselves and their customers pointed to a possible new path to profits, at least theoretically.[86]

Uses of the Internet by banks and customers represented a mixture of older applications transferred to the novel medium and the addition of new issues. Initially, existing electronic applications and information sharing appeared on the Web, along with information about offerings, terms and conditions, and so forth. Next, Internet banking became increasingly an interactive medium for transactions between banks and clients. That in turn led—at least it has so far—to reconfiguring old and new services into packaged offerings, and simultaneously to the desegregation of services (such as people paying bills through a service provider, bypassing their banks' checking accounts and even bank-issued credit and debit cards, all resulting in loss of fees for banks). The possibility of providing new services and creating different fields of competition made the Internet seem like a huge break from the past for so many industries, banking included. It is why so many people routinely use the word revolution to explain it all. But Internet banking is only one of many important sets of technologically based forms of financial transactions in play, albeit an important one. Because we are in the very early stages in the use of the Internet in all manner of financial transactions, use of vitriolic or extreme language, like the word revolution, seems be premature in the first decade of the twenty-first century, despite the technology's implicit promise of dramatic changes. As experience with other technologies adopted by bankers and their clients has demonstrated, acceptance and wide deployment takes time. What little data exist about the use of Internet-based banking suggests that this pattern of moderately paced adoption is again occurring. As if we needed a reminder, people are still using their paper checks.

Bankers "discovered" the Internet at the same time as colleagues in other industries. For example, the ABA published its first article on the subject in November 1994. Quotes in this chapter appeared as the first comments published in the *ABA Banking Journal* about the Internet. The article went on to note that the reason banks have not rushed to the Internet grew out of concerns for "security of the funds they handle." As the author of this article commented, "the free-for-all Internet culture seems to invite electronic eavesdropping and larceny."[87] While this argument was legitimate, until 1993 one could not responsibly conduct commerce over the Internet anyway; it was not until several years later that the IT community (and its vendors) had worked out many of the issues related to security by introducing various encryption tools. Meanwhile, bankers took to using e-mail as a way to communicate as quickly as other people across the economy. Conventional wisdom should hold that bankers would have been slow to embrace the Internet; however, given their extensive prior experience with IT and communications technologies, they got into it at a reasonably fast pace, asked good questions about the technology (such as about data security and cost justification, like managers in many other industries), watched the technology unfold, and began using it as it made new functions possible.

This was a technology, however, where consumers turned the tables and pushed it on businesses, including banks. Barely six months after issuing the first publications in the industry about the Internet, the ABA reported that some 15 million people in the United States were managing their financial affairs using Microsoft's package *Money*, Intuit's *Quicken* and *Quickbooks*, and Block Financial's *Managing Your Money*. At the time, roughly 40 percent of all banks offered some form of home banking, so the question now was what role the Internet would play in this market?[88] Table 3.4 presents a list of home-banking activities circa mid-1995. It is important to call out that while Internet-only banks soon came into existence and caught the attention of many in the business community at large, (not just other bankers and regulators), the fact remains that the Banking Industry was already extensively involved in IT-based banking, having learned a great deal in the previous three decades about networked financial activities. Keeping that prior experience in mind, we can make two observations about the data in this table. First, it reminds us that the largest banks in the industry were extensively involved in networked banking projects. Second, the variety of different forms

Table 3.4
Home-Banking Initiatives by the U.S. Banking Industry, 1995

Banks offering personal financial management software (e.g., Microsoft's *Money*)
BankAmerica, Chase Manhattan, Citicorp, First Chicago, Michigan National, NationsBank, U.S. Bancorp

Banks offering PC home banking with proprietary software
Bank of America, Chemical, Citibank, NationsBank, United Missouri Bank, Wells Fargo

Banks offering consumers screen telephone services
Citibank, NationsBank, State Bank of Fenton, Michigan

Banks experimenting with cable TV offerings
Barnett Banks, Chemical, Meridian, NatWest, Shawmut

Banks testing screen-phone offerings
Bank of Boston, Chase Manhattan, Chemical, Crestar, Mechanics, Wilber National (Oneonta, N.Y.)

Banks offering 24-hour inbound/outbound telephone banking services
Bank of America, BayBanks, Chase Manhattan, Chemical, Citibank, First Union, Huntington Bancshares, Meridian, National City, NationsBank, NatWest, Wells Fargo

Banks (and thrifts) offering services through commercial online enterprises
Through Prodigy: Barnett Banks, Banc One, Boatmen's, Chemical, Chevy Chase, Coamerica, First Interstate Bank of Denver, Hamilton Bank (Miami, Fla.), Meridian, Midlantic Bank, NBD Chicago, NBD Detroit, Pennsylvania Savings Association, PNC, Wells Fargo, Wilmington Trust Co.

Through CompuServe and America Online: None, but mortgage companies, insurance firms, and credit card networks use these services

Through Internet: Cardinal Bankshares, Lexington, Ky., became first bank to gain regulatory approval (May 1995) to offer banking services over the Internet; over 400 banking institutions had World Wide Web pages

Source: Modified from lists in Penny Lunt, "What Will Dominate the Home?" *ABA Banking Journal* 87 (June 1995): 38.

that networked-based banking was taking in the 1990s demonstrated that bankers were leveraging all available technological options, with the largest banks building a large body of in-house knowledge of such systems that they applied later to Internet banking.

The industry had learned some lessons that stood it well in taking on the Internet. The largest body of experience brought to bear involved PC banking of the 1980s which, quite candidly, had failed, largely because banks had attempted to make customers use proprietary software. In the open architecture culture of the Internet, banks and consumers had more options available to them. For banks, the Internet offered the possibility of presenting more services and connecting to a larger body of software already used by consumers and other organizations. For consumers, the Internet provided the chance to link together multiple services and providers of financial services. Thus, for example, a bank could offer to communicate to *Quicken* and *Money*. Bankers and software vendors continued to look for common platforms, each searching for the new holy grail of the "virtual bank," while banks continued to worry about security over the Net.[89] Bankers first set up web pages where they could place information about their existing offerings and how to contact them. In the period from 1994 to 1995 there was little interactive banking going on over the Internet, but by late 1996 banks were beginning to offer services over the Internet and, over the next several years, added to their collection of such services. These included bill paying, fund transfers from one account to another, mortgage transactions, bank-to-bank payments, and Internet-based EDI, to mention a few.[90] Bankers realized very quickly that consumer commerce on the Internet was being paid for with credit cards, posing threats (that some other organization would handle the transactions and collect fees for these) and opportunities (to do the same).

As that realization unfolded in front of the leadership of the industry, a crucial event became the proverbial "wake-up call" for the industry. It was the establishment of Security First Network Bank (SFNB), considered the first Internet bank. Rarely can an historian point to one event as decisive in germinating so many others, without referring to assassinations, military attacks on nations, and so forth. This event was not a negative one, nobody died, but bankers quickly realized that their world was changing again. On October 18, 1995, the Security First Network Bank allowed customers to conduct banking transactions over the Internet—a first in the Banking Industry; by the end of the year, bankers were reading about it in their trade journals, having heard about it at conferences and through conversations among themselves. Within two weeks of its start-up, 750 individuals had opened accounts with SFNB. The bank was part of a traditional brick-and-mortar institution established in Louisville, Kentucky, the prior year that, in its first year had opened only 187 accounts. Equally startling, the first accounts on the Internet were noninterest bearing, that is, the services offered through the Internet were sufficiently attractive to customers that the bank did not have to entice them with the traditional offering of interest payments on balances or ice chests or sets of glasses and dishes (conventional methods used to attract new customers). When the bank opened, some 12 million people were using the Internet with dial-up services in the United States. This small bank could compete with large banks and other financial institutions around the country because the telecommunications infrastructure was, in effect, free, driving down its overhead; customers already had the necessary hardware and telephone lines with which to interface with the bank. SFNB used encryption software originally designed for the U.S. military to protect data in transactions. The bank spent a year cultivating regulators, ultimately receiving permission to offer Internet services by the U.S. Office of Thrift Supervision after demonstrating the quality of the data security in place. SFNB's creation caught the attention of bankers all over the world as hundreds of consumers quickly responded by opening accounts. SFNB recognized that its future depended on expanding the market for online financial services and the market's acceptance of its services. It worked; by the end of the year—after introducing Internet banking—it had over a thousand customers.[91]

Soon after, banking on the Internet began to spread, involving traditional banking services offered around the clock, seven days a week, with the ability to link multiple services together (for example, credit and debit card processing, along with checking and savings accounts, and so forth) for consumers and businesses; and for commercial enterprises, transfer of financial instruments, payroll, EDI, and others.[92] Payment systems also moved to the Internet, and as before, not all were operated by banks. Others included Amazon.com (online retailer) and Yahoo! (search engine and portal). To be sure, all the large banks offered such services as well.[93]

The pace picked up in 1997 and 1998. One industry study estimated that 200 banks offered Internet services by late spring 1998; Wells Fargo, then the leading provider of web-based services, claimed at that time to have 450,000 customers, up 50 percent over the prior year. The number of "pure" virtual banks, that is to say with no brick-and-mortar branches, remained tiny. Instead, existing traditional banks added Internet services to their portfolio, just as companies did in so many other industries. Early providers of Internet services did it because they either thought it was a profitable business to enter or because they saw it as "an unavoidable necessity."[94] Many banks moved forward to at least establish a website; industry survey data show that some 30 percent had one in 1997, 34 percent the following year, and nearly 42 percent in 1998. With this technology, a bank's size did not matter, perhaps because the relative cost of creating a site was not so great when compared to embracing earlier digital applications. In the period 1997–1998 almost all banks, large and small, offered information about their offerings on the web; half included information about the geographic community in which they operated and offered two-way e-mail, while a third provided online financial calculators. Only about 15 to 16 percent made it possible in this period for someone to open an account or apply for a loan over the Internet, and slightly less allowed their customers to view account balances. However, bill paying grew sharply in 1998–1999. Interestingly, in a 1999 ABA study, 47 percent of those surveyed reported they did not plan to offer Internet-based services. Reasons ranged from customers not having requested the service to other priorities attracting the attention and investments of a bank. Much like others in the United States, bankers reported personally using the Internet, so ignorance of the Net was not the issue. At the time, branches still proved to be the most effective way to reach individual and institutional customers. ATMs, automated clearing house direct-debit services, and phone banking were still overwhelmingly popular channels as well.[95] But the industry, as with earlier technologies, continued to debate the costs and benefits of using the Internet and how best to leverage it.[96]

It appears, however, that between the second half of 1998 through the first half of 1999, Internet banking began to take off as an important component of the industry's offerings. By the start of 1999, about 7 percent of all households were involved in some form of Internet banking—viewing account balances, transfering funds, electronically paying some bills, and even doing limited business banking. Opening accounts and online loan applications were also gaining steam, and the market seemed huge since nearly half of all homes had PCs and over 98 percent telephones; in short, the technical infrastructure on the demand side of the equation was in place.[97] Key players in this market included SFNB, CompuBank, First Internet Bank of Indiana, and NetB@nk (all Internet-only institutions), while among the traditional banks highly ranked by consumers for their e-banking services were First Tennessee, Salem Five Cents Savings, Citibank, and Wells Fargo.[98]

Part of the demand for online financial services came from the brokerage market, where several million Americans were trading online in a boom period for the stock market. Day traders had become the latest class of go-it-alone entrepreneurs, driving demand for discount brokerage services. This trend created a problem for bankers, however. As one commentator noted, "the personal online trading account is fast becoming the online personal finance account." The first went to the discount broker; the latter was the historic preserve of the banks. As the same observer noted, "to banks, the biggest retail threat comes not from BankAmerica or First Union, but from Charles Schwab & Co. and E-Trade, or maybe a web portal or wireless giant."[99] All the forecasts had households—the banks' most important customers for over two decades—increasing their use of brokerage firms for an ever-expanding list of financial services that had been the preserve of banks—bill paying, even checking accounts.[100]

But bankers did not rush out to get on the Internet willy-nilly. We know, for example, that in early 2000, about 53 percent of all banks had a website; but that meant that 47 percent did not, although a little over 7 percent of that latter group intended to have one. Meanwhile the industry was learning about web traffic patterns, what services to offer, who was making money on the Net and how, and so forth. Survey data varied from one source to another in 1999–2002, but the pattern was clear, and Internet-based banking increased over time.[101] Even at the end of 2000, when bankers were surveyed on the top managerial issues they would face in the following year, technology and the Internet ranked only fourth and fifth, respectively, after such other issues as growth, privacy and regulatory issues, asset quality, and the state of the U.S. economy.[102] Yet by spring 2001, over 65 percent of all banks had a website, and many were already using second generation websites with services. In fact, nearly half now offered Internet-banking services. When asked how they felt about the Internet, the banking community was split roughly equally among three views: the Internet was a defensive weapon, an offensive tool, and merely another channel for business. Of those that offered services over the Net, nearly half made it possible for customers to check account balances and transfer funds, and nearly as many enabled paying bills. Less than a third offered such other services as stop payments, check ordering, online loan applications, check images, or bill presentment and online securities purchases.[103]

Web aggregation had also begun, albeit slowly. Web aggregation is the process by which an individual's financial account information can electronically be collected so that a person can see it all at one site, making possible one-stop shopping for electronic services or to plan or review portfolios, and so forth. Privacy and other electronic issues had been resolved by 1999, making it possible for this new application to work in a practical manner. Demand for such services proved quite strong. For example, one survey done in 2000 reported 43 percent of the respondents believed that within five years they would be doing most of their financial transactions online; only 7 percent said none. Just over half the respondents already had had experience with electronic transactions with financial institutions, creating the mental infrastructure necessary to demand more services or at least understand the trend toward the digital.[104]

The pace of adoption aside, what is very clear is that the demand for online banking was coming from several sources. These included individuals, online shopping (e-commerce), online stock purchasing, and the growing use of the Internet by companies and public agencies to conduct business (e-business), all of which were shifting the financial activities of the nation away from just brick-and-mortar, paper-based transactions to increasingly electronically based forms. Competition also emerged in intense forms in electronic markets by the early 2000s. The battlefields were more than simply the Internet: they included getting the attention of customers with front-end interfaces who had the ability to access more financial services, which called for integrated standards for seamless access; payment and transaction systems, which called for secure and common standards so that multiple institutions could share financial data of transactions; easy and integrated information use and access, calling for back-office systems that could support these functions; a variety of trust and security brokerage services, which had to overcome regulators' reluctance to let multiple financial industries sell each other's products and services; and data management and information mining, which raised issues regarding privacy laws and practices both in the United States and in other countries (especially in Europe).[105]

Finally, we should discuss the most businesslike issues of all: profitability and costs. Since the Banking Industry's experience with the Internet is still in its early stages as of this writing, with less than a decade of experience to call on, generalizing on patterns of economic behavior is at best a premature exercise. But as the industry's history suggests, without at least some historic perspective to work with, bankers will be inclined to put off investments in any technology that clearly does not have a track record of demonstrated profitability or practical use behind it. So what do we know so far? For one thing, the credit card business has remained highly profitable. Furthermore, during the last two decades of the twentieth century, the number of credit card transactions continued to increase; in the second half of the 1990s, the Internet can logically be seen as responsible for stimulating a sizable amount of that increase. The Federal Reserve and the Banking Industry continued to use private networks for clearing checks, and that was not going to go away. Use of debit cards promised to lower those costs, but these cards were not yet part of the Internet story. In fact, internet banking was such a new phenomenon in the mid- to late 1990s that the volumes of transactions probably were too small to spin off profits. Having a website that just makes information available is not a profit generator; it is either a sunk expense or the cost of doing business. Not until 1999 did survey data suggest that banks were beginning to ramp up transaction volumes on the Internet. So any discussion about profits should be constrained to a period after early 1999. By outsourcing some network-handling functions associated with the Internet, as the industry had learned to do with credit cards, banks were able to control some costs. But most banks failed to reinforce their brand over the Internet as of 2004. Instead, they were too closely linked to branded credit cards, such as Visa and MasterCard, which left the industry vulnerable to competition from commodity-based rivals. We know from survey data presented earlier that consumers were promiscuous when it came to whom they went to for financial services, a classic example of the absence of branding benefits and commodity-based competition at work. As one banker noted as early as 1999, "American consumers came to learn that access to loans, payment systems, and even cash were not necessarily related to any bank they happened to patronize."[106]

A major economic study on the impact of the Internet on the U.S. economy conducted in 2000 concluded that much of the value that could come from new technologies, such as the Internet, "will likely come through intangible improvements in products and the creation of new types of products, distributed in new ways." It argued further that "it is similarly difficult to estimate the gains from structural changes when in so many industries one observes only a limited diffusion of new practices." What the researchers did observe, and felt strongly about, was that "price transparency, differential pricing, and disintermediation" would play important roles in how the Internet would affect an industry; they had the financial industries in mind when they reached that conclusion.[107] Price transparency, usually the result of offering basic services, for example, could become the cost of doing business. This would support the argument that the playing field has moved from atoms to electrons, and that to participate in banking, one had to offer services over the Net.[108]

In short, the discussion needs to move away from simple return-on-investment (ROI) calculations and instead focus on the effects the Internet was having on the structure of the Banking Industry and its rivals in other financial industries. There is the prospect that the Internet was such an altering force that, unlike computers, which led banks into the digital, this technology might lead to different configurations of industries. Simple calculations of ROI then become increasingly trivial discussions. Instead, investing in new services—such as the integration of multiple offerings and sources of financial information—would give banks the chance to charge more. In this situation, better service would divert a customer away from solely judging the value of a service by its price. Bypassing a bank or other financial institution—disintermediation—is a risky game. One can try to lure customers away from the competition with better services, but customers can move just as quickly with the Internet to another provider. We just do not have enough data to know how much of that is going on or what the costs of that behavior are to the industry. However, bankers began sensing that the game was rapidly being played for higher stakes. One simply put it this way in 1999; "banking has become an information business."[109] Further, "the real issue for banks . . . is to understand that just as the boundaries between banking and other financial services are eroding, so, too, is the difference between 'technology' and 'content.' "[110]

For those who would insist on a more quantifiable discussion and less attention to issues of grand strategy, the one major study on banking's profitability on the Internet, conducted at the end of the century, provides some insights. Done by economists, it modeled the behavior of banks that used a service provider's Internet services (Digital Insight [DI]) rather than analyze broadly the activities of many banks using a variety of Internet services. The base of data proved sufficient to report results, because its categories of benefit were consistent with the findings of economists regarding benefits of technology; they also match patterns the Banking Industry experienced with earlier networked-based applications. Three sources of profitability in using the Internet were (and continue to be) available to banks: increased customer retention, lowered costs of transactions, and increased cross-selling opportunities. Other than for very small banks or those banks with a high percentage of their customers not using the Internet, a "typical" bank using DI's services enjoyed a net present value of over $5 per customer. That, in turn, translated into a return on investment of 60 percent. Internet banking made it possible for banks to collect additional revenues from various transaction and user fees for such services as online bill payment, lending, cash management, and services on portals. Despite the investments needed to use the Internet, cost savings in combination with operational efficiencies reduced unit costs, particularly for large banks that could enjoy economies of scale. Cross-selling additional services to existing customers became an opening banks pursued and contributed to some higher rates of retention.[111]

But, in the final analysis, while all this data makes sense to a manager and probably will be borne out by experience, that process has yet to play out; we are still too close to the events. A study conducted by the Federal Reserve in late 2002, however, reaffirms that the Banking Industry's expectations aligned with this study. Bankers interviewed for this study, however, felt the need to find ways to determine the "business case" for new offerings. These bankers saw the benefits of receiving electronic payments and associated data straight through their systems with no human intervention, and in speeding up settlement of transactions.[112] But let's give the bankers the last word. In mid-2002, the ABA asked a number of community bankers to assess their experience with the Internet. Nearly 3 percent said their websites were already profitable, while another third said theirs would be soon. Nearly 40 percent reported their websites were not profitable and probably never would be, but they were being used as a defensive strategy to ward off competition from other banks and financial rivals. The bankers said that benefits of cross-selling and transactional revenues lay in the future, not in the present, and that they expected portals with services to be a major source of revenue sometime, along with Internet-based bank-to-bank transactions by way of financial aggregation. The ABA study examined the question of how many customers a bank needed in order to be profitable. When the number of people in the United States using the Internet was compared to the size banks involved, the evidence suggested that in mid-2002, banks would need about 17 percent of their customers online; at that time, some 8 percent were using the Internet for banking. In short, the industry had a way to go, but not so far to reach critical mass.[113]

Recent Trends and Current Issues

By 2000–2002, it had become quite clear to banking executives, industry watchers, government regulators, and economists that the electronic delivery of services by banks was far less expensive than were earlier methods. Furthermore, they also knew that additional electronic services were both cost effective and attractive to customers. Issues concerning the Internet, cashless checks, and so forth all centered on the key issue of how best to make electronic payments. Customers already understood the advantages of the digital, as well. For example, as banking fees rose for handling payroll "checks," companies began renewing their insistence that employees accept direct deposit of payroll "checks." It normally cost 75 percent less to pay an employee by some electronic means. An emerging method for paying digitally that came into early use by 2002 was the use of payroll cards that functioned very much like a credit or debit card; a bank would post an employee's paycheck to the account, often issued by Visa or MasterCard. By late 2003, some 14 million households used such instruments, in effect making them "unbanked," as the practice came to be known.[114]

As the number and type of electronic financial transactions increased, the more pressure there was to accept electronic signatures in ways legally acceptable as binding and as secure as signatures on paper. The Federal Reserve Board advocated means to reduce the number of paper checks because of their enormous cost. In October 2003, Congress passed, and the president signed into law, the Check Truncation Act, which went into effect in October 2004. This legislation made it legal for banks to clear checks electronically by approving digital images, thus no longer having to transfer paper checks back to the banks of origin or make banks establish agreements with other banks and organizations to do that work. The Federal Reserve, having noted the rapid increase in electronic transactions—fivefold in the previous two decades, to be precise—expected that in 2004 these might exceed the number of paper checks and thus wanted to continue pressing for more efficiencies in electronic transactions. But in addition to the normal economic reasons for encouraging electronic transfers of check data, there was the desire to stop dependence on air transportation of checks since the attacks of September 11, 2001, which had led to a total shutdown of air traffic for nearly a week and disrupted the normal clearance of checks.[115]

By early 2003 banks were interested in expanding use of all manner of electronic transfer of data, not just checks or use of credit cards. Even something as mundane as mail presented enormous opportunities for cost savings. The Social Security Administration and the U.S. Treasury Department managed to get the press's attention regarding the cost of mailing or transmitting checks. In 1999, for example, the Treasury issued 880 million checks, of which 68 percent were electronically distributed. The Treasury—and thus American taxpayers—could have saved $180 million if the checks had all gone out electronically. Bankers, however, had already started to shift to electronic means. Between 1996 and 1999, mail-based billings and payments had dropped by 18 percent for the simple reason that digital was cheaper than "snail mail." The trend to convert to the electronic continued slowly in the early years of the new century.[116]

The effects of banking promoting use of the Internet in the early years of the new century began turning up in the data. Growth in use continued such that by mid-2004, some 22 million customers of the ten largest banks in the nation had used their online accounts that year, representing an increase of nearly 30 percent in activity over the previous year. Over the longer period from 1996 to 2003, growth in online use had grown tenfold and appeared to be on a faster growth rate in the middle of the new decade. The industry had turned an historic corner. In 2003, Americans made more payments using electronic means (credit cards and debit cards, for example) than paper checks (44.5 billion electronic payments versus 36.7 billion checks). Ultimately, the milestone was not reached because of Internet banking. Rather, electronic payments had become more convenient to consumers than paper checks.[117]

Conclusions

The Banking Industry served as the central financial infrastructure of the nation throughout the second half of the twentieth century. While that pre-eminent position is being challenged by other financial industries in the new millennium, the Banking Industry has remained one of the most important within the economy. It is also the one industry that most dramatically illustrates what might be an emerging set of practices and patterns in the services sectors of a highly digitized economy of the future. The experience of this industry teaches us at least three lessons. First, much of its work lends itself to computerization, since its "inventory" is information, not just coins, and all its paper ephemera lends itself to digitization. For other industries with similar characteristics, computerization becomes a nearly irresistible force to be reckoned with, because computing is quick, cheap, and can handle large volumes of transactions. Second, once an offering is digitized, and once regulators are comfortable with alternative business models and players, one can pick apart a rival's offerings, to recombine with those already in one's portfolio to create new products, new competitive advantages, even new firms and industries. The Banking Industry was one of the first in the American economy to experience this. That phenomenon alone presents interesting questions about how the economy of the United States might evolve over the next half century.

Third, the case for digitizing proved so compelling that this industry was able to move from a precomputer style of operation to one that we can now call the digital style, in effect within the working careers of one generation of bankers.[118] Yet we run the danger of overstating the case because much of what banks did, and now do, is the same: they make loans, hold deposits, and cash checks. The difference is that they use computers so extensively that to run a bank without them is now impossible. What are we to conclude from such a statement? As is occurring in the manufacturing and retail industries, the Banking Industry is still in the early stages of transformation to an as yet unknown paradigm. We see glimpses of it, but that is all at the moment. The only solid fact is that this industry relies heavily now on computers and telecommunications, and that we can reasonably expect its practices, culture, and offerings to continue changing over time. Because we are still in a transitory phase, the best any historian can do is to remark that bankers have not been shy about using technology. Yet they continued to worry extensively about doing so, and experienced new forms of competition and structural changes unanticipated a half century earlier.

But theirs was not the only industry to experience change brought about by computers and telecommunications. In the next chapter we will see that the Insurance Industry did too, and, like banking, its experience remained a source of important lessons about how the economy of the United States was transforming away from an earlier, purely industrial model.

4

Business Patterns and Digital Applications in the Insurance Industry

The insurance business is unique among businesses in the quantity of statistical data that are compiled and in the degree to which such data are utilized in everyday practice.

—Norton E. Masterson, 1970

The Insurance Industry presents historians, economists, and business leaders with its own case study of how information technology affects companies and industries. In the Banking Industry, we see an industry whose products and structure were inherently changed as its firms and rivals used computers and telecommunications during a period of extensive revisions in regulatory practices. In the Insurance Industry, we also see shifting regulations, but the wide use of the digital and communications over the entire half century, with new offerings made possible, and rivalry from other industries, fundamentally influenced the issues the industry watcher presented above.[1] Yet the industry raises many different questions, perhaps the most important of which concerns whether it changed fundamentally because of the use of computers in our time frame. The Insurance Industry has always been one of the most extensive users of data processing in the private sector. Its record of aggressive utilization of information technologies dates back to the 1890s; at least one major firm, Prudential, built its own data processing equipment before World War I.[2] Because the industry was such an extensive user of computing before the arrival of the computer, the question arises: Had this industry, in effect, made sufficient operational and structural changes such that its adoption of the computer was merely a continuation of a style of operations that mimicked the digital before modern IT was even invented? Put another way, was the Insurance Industry the first services industry to have transformed into some Information Age sector long before any student of the economy thought to notice? If so, what lessons can we absorb about the adoption and consequences of technology?

Answering these questions is partially out of the scope of this book because it would require comparing operations prior to the arrival of the computer with those that took place afterwards. We do not yet have a clear picture of IT-influenced operations in the industry pre-computer, although, we are not without some information about this earlier time.[3] In this chapter, however, our purpose is to fill in details about the post-1950 period, reviewing digital applications, the extent of their deployment, and so forth, so that at some future date a student of the industry can answer all those questions and begin to differentiate more precisely the role of information—as opposed to information technology—on the services sector of the economy.

We cannot, however, fully put off the pre-1950 experience because if, as I posit, this industry had already made its initial transition to an information/technologically based business decades before, it did not make another shift until changes in regulatory rules, competition, and IT began concurrently in the 1980s. That prior transition would have created circumstances that explain why it embraced the computer faster than other industries in the 1950s. Unlike so many others, for whom the 1950s were a start-up period for computing, the Insurance Industry was very active in the 1950s installing computers. As soon as the technologists had figured out how to connect telephone lines to computers and to create online applications in the 1960s, the Insurance Industry again quickly adopted those new ways of using IT. In fact, this industry collectively embraced innovations in computing technology earlier than many other industries throughout the second half of the century.

Before delving into a description of the industry and its use of computing, what can we say about banking compared to insurance that would explain adoption of computing or the speed at which it happened? Both industries were subject to extensive government regulation of their practices and offerings. Regulators opened up the industry to more competition at about the same time in each, beginning in the 1970s. Like banks, insurance companies enjoyed relative calm in their markets in the 1950s and 1960s and then experienced many changes. As with banking, insurance has a few central activities, in this case selling and managing insurance policies, collecting and investing fees, and controlling risks. Both had to handle massive quantities of information—bankers to collect it (transactions), insurers to process and calculate it. To be sure, these statements are gross generalizations, but intended only to emphasize the tendencies that affected the kinds of IT applications each adopted.[4]

There were differences, of course. While both managed money in massive amounts, insurers were more involved in inventing funds than bankers because they needed access to large volumes of cash to pay out in cases of disasters, deaths, and the like, as well as to supplement income received from premiums. Bankers tended to move money through the economy by way of checks, credit and debit cards, and loans. The Insurance Industry also occupied a different physical space in the economy. Banks were almost entirely highly decentralized operations, since the bulk of their work was done in tens of thousands of branch offices; they had few large "headquarter" facilities staffed with thousands of employees. In fact, all through the past century, the ratio of back-office employees (also working in branches) supporting front-office employees had evened out from several in the back for every individual in the front to nearly an even split, all made possible by technology. In the Insurance Industry, there existed a dichotomy. There were large centralized facilities, employing tens of thousands of people to manipulate data to calculate risks, determine to whom to offer a policy, constantly update policy information, and create new financial offerings while managing large portfolios of investments. Often insurance companies owned and occupied some of the largest buildings in an American city. At the same time, the industry also had thousands of local agency offices. At the risk of being too judgmental, one could argue that their work was mathematically more intense than that of bankers, and they needed to manipulate more data, if for no other reason than to perform their underwriting and financial management responsibilities.[5]

In this industry, thousands of people operated out of very large or very small offices—always smaller than a bank's branches, many out of their homes. Some worked for an insurance company while others were independent agents—that is, they sold products of multiple firms. The information needs of an agent were different than those of a bank teller, although both sold products and services and interacted with the public. A brief example of how technology was used suggests the impact of physical dispersion and applications. Those who worked in large office buildings tended to need massive amounts of computing capability because of the digital applications they used, hence the industry's propensity to rely on large centralized mainframe technologies all through the half century. On the other hand, insurance agents, operating in small offices, normally needed smaller, less sophisticated computing, and thus had a propensity to use minicomputers and laptops (in the 1990s), and to rely extensively on telecommunications to tap into the power of regional or national data centers.[6] While both groups used telecommunications to move information, bankers had more experience with it, having used communications technology since before World War I to handle check clearing. One did not need to use telecommunications in insurance until the 1960s, when its cost and convenience justified the use of online systems to begin handling various transactions. Again, these are gross generalizations intended just to suggest degrees of difference in emphasis in the use of digital and telecommunications technologies.

A Prudential Insurance Company of America agent uses an IBM PC Convertible—predecessor of the laptop—to describe his products. About 20,000 agents used this technology beginning in 1986. Courtesy IBM Corporate Archives.

Evolution of the Insurance Industry's Structure and Role

So what did the Insurance Industry look like in the 1950s and 1960s? In 1950, there were 649 life and health insurance companies in the United States; that number grew to 1,550 in 1964, reflecting an enormous growth in the standard of living of Americans accompanied by an appropriate increased interest in and affordability for their products. By 1964, many of these 649 firms were small, with assets totaling less than $5 million, and roughly 10 percent were mutual insurance firms, that is, were owned by their policyholders. Bear in mind that these were important because, by 1964, they accounted for 60 percent of all policies in force and roughly 70 percent of all assets of all American insurance companies. In 1950, twenty-two firms had been in business for over a century. Assets had grown from $64 billion in 1950 to nearly $150 billion in 1964, far exceeding inflation, solid proof that it was golden era of growth for this industry. Mortgages accounted for a third of these assets, another 24 percent were in commercial and industrial bonds, the rest in government financial instruments and miscellaneous issues. One final statistic rounds out the picture of the life insurance industry: in 1950 life insurance policies had a combined value of $234 million, but that value climbed to $800 billion in 1964. This last set of statistics is important because life insurance companies make money in two ways: from premiums collected on policies and from earnings generated by investments. So the bigger the pool of investments, the more a firm could make in earnings. The combination of more policies sold and earnings generated meant that the industry was profitable, growing, and constantly needing to handle more data in more sophisticated ways.[7]

What was the situation on the property and casualty side of the Insurance Industry in 1950 and in the years immediately following? There were many more firms providing fire, casualty, burglary and other protections in the 1950s than life insurance, and by 1962 some 3,500 firms existed in the United States. Cash turned over extensively, unlike at life insurance companies where cash was more stable and held in more types of investments. Property and casualty always held more liquid quantities of cash to have available for immediate disbursement when claims were filed by clients. In the property insurance business (using 1962 as an example) companies brought in $24 billion in premiums but paid out $15.5 billion in losses—a highly profitable business but nonetheless one with a high cash flow. In the 1950s and 1960s, the demand for house and automobile insurance caused an enormous expansion in this business, and with it came the requirement to process tens of thousands of policies with frequent transactions per policy. To put size in relative perspective, however, life insurance companies were the larger of the two segments of the industry, having more assets, larger companies, and more fees ($30.7 billion in 1964, compared to roughly half that for property insurers).[8]

One could also ask: What about health insurance, what did that look like? For the purposes of this book, we will not break this group out as a separate line first, because firms in this segment of the life insurance business, were seen to be part of the life side and second, because they had IT needs that were a mixture of what the life and property and casualty enterprises used. Furthermore, it is my intention in a subsequent book to deal with health insurance firms and issues as part of the Health Industry since they were and are so intricately woven into the health and medical sectors of the economy and how each used computers. To discuss it here would distract us from the larger Insurance Industry.

What did the combined life/casualty industry look like late in the twentieth century, just as the Internet was beginning to affect insurers all over the world? Using 1997 as a benchmark, we know that there were 40,717 insurance carriers in total, and over 131,000 insurance agencies, brokers, and services. Collectively, they earned nearly $200 billion just in fees and employed over 2.3 million people. Employment data suggest how intensively they worked with information. Insurance carriers employed 1.6 million workers, with a payroll cost to the industry of nearly $68 billion, while agents and brokers, and their employees—those who most interacted with customers and sold policies—employed just over 700,000 people, with a payroll of $25.5 billion. As a percent of the Gross Domestic Product (GDP) it was a small industry, representing only 2.3 percent of the total, but because of its role in providing an economic safety net, its impact on the nation far exceeded that statistic.[9] The industry maintained that proportion of GDP pretty much

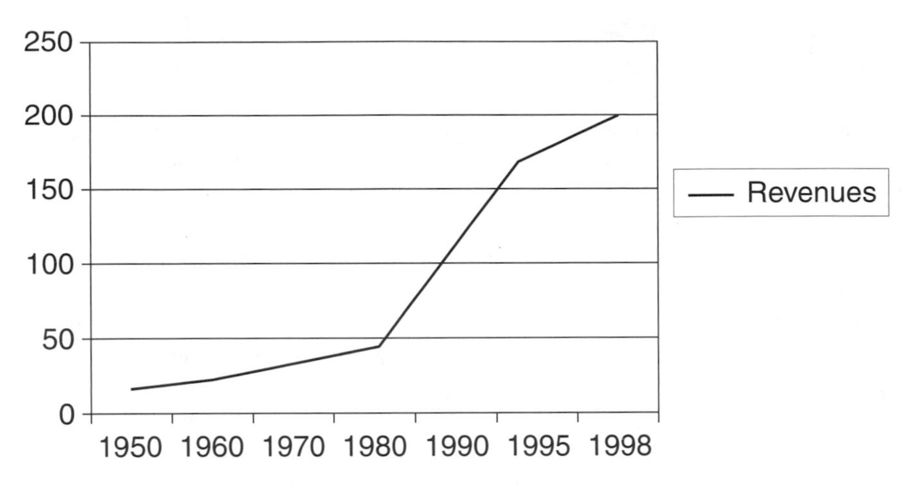

Figure 4.1
Revenues of U.S. insurance industry, selected years, 1950–1998.

throughout the entire half century. Yet, as figure 4.1 illustrates, the industry did well over many years.

In 2000, the American Insurance Industry was well underway with another round of technological changes, this time stimulated by the availability of the Internet, at the same time that the nation had entered a period of national economic recession. It remained a very large, complex industry. One-third of all life insurance premiums in the world came out of the U.S. economy, for example, with $465.6 billion in premiums, and Americans carried the highest per capita insurance in the world. When combined with the property and liability half of the industry, total assets had grown all through the 1990s, hovering at $4 trillion in 2000. There were over 1,100 life insurance companies and nearly 2,500 serving the property/casualty side of the industry. Life insurers were massive financial institutions, with 55 percent of their assets invested in corporate bonds and another 13 percent in mortgages. The property/casualty side of the industry generated revenues from a wide variety of sources: personal automobile insurance provided 39 percent, commercial 36 percent, homeowners insurance another 12 percent, workman's compensation 8 percent, and miscellaneous other policies the remaining 5 percent. This side of the industry kept 10 percent of its assets in cash and short-term notes, about another 35 percent in government and corporate bonds, and only 22 percent in stocks. Important for our story were the levels of concentration in the industry: in life the top ten firms accounted for over 40 percent of the ordinary life market and over 65 percent of group annuity market. We could offer up similar types of statistics to demonstrate the property/casualty side of the business had also been concentrating.

Across the industry, however, offerings had become rapidly commoditized, which meant firms had to compete on price. That is why computers at the end of the century were again being used to drive down costs while adding customer-friendly capabilities. In the early 2000s, for example, they exported data entry and file management to India and other low-wage countries, using communications and digital technology to move, process, and use data. One expert on the industry noted that personal lines of business were the most concentrated because of the opportunities to leverage economies of scale, exploit name recognition, and command enormous market power. Commercial lines experienced even greater commoditization, but had no brand recognition beyond the normal financial ratings provided by Best, Standard & Poor's, and others, yet they could reach efficient scales at smaller sizes than the life side, and ultimately their customers were more sophisticated in how they procured insurance.[10]

All through the half century, business volumes grew. We saw above the growth in the 1950s and 1960s; this trend continued in the 1970s, despite turbulence in the U.S. economy as the nation experienced inflation, recession, and went through both the Civil Rights Era and the Vietnam War. The period from the 1980s to the end of the century saw new levels of computing and telecommunications spread across all industries as they embraced the personal computer and later the Internet.

Much like their financial counterparts in banking and stock brokerage firms, the two types of insurance industries—life/health and property/casualty—operated relatively independently of each other at mid-century. Regulations, laws, customs, and an expanding U.S. economy essentially kept that structure in place throughout the 1950s and 1960s. As in banking, however, and for many of the same reasons, the ground began shifting in the late 1970s, causing a number of fundamental changes. First, what started out in the 1950s as a patchwork of state regulators slowly began changing, with federal regulators increasingly influencing practices in the industry, primarily because this industry was such a large participant in the world of financial investments. Recall that this industry's income was made not just from premiums; it also earned from investments in stocks, bonds, mortgages, and other financial instruments, especially on the life side. Second, throughout the last three decades of the century, insurance firms started to compete with banks and brokerage firms for customer share, offering services that overlapped each other, thereby beginning to blur some boundaries during the early years of the new century among the three major financial industries. Because we know that use of computing reflected operating concerns of an industry, understanding the changes underway by the early 1980s is crucial to our appreciation of what happened with IT in insurance.

Two industry watchers described what happened just before and during the 1980s:

> Until the eighties, the insurance industry exemplified the stability of the economy. Everyone complained about competition, but most firms had increasing sales and profits each year. Interest rates were slowly rising. That made the investment people look smart. The pressure for more cash to invest became more intense. In the eighties everything turned around. The vaunted stability

> of financial markets evaporated and, at the end of the decade, interest rates hit what would have been contemptuously called "banana republic" highs. Policy holders started to view their policies as financial assets rather than as evidence of mortality, and borrowed on, or surrendered policies until the companies had severe liquidity problems.[11]

The industry experienced growing competition and inflation that really started in the late 1970s and continued especially through the 1990s. Consumers began playing a greater role in the 1980s and 1990s when, for example, they increasingly concluded that yields from investing in whole life insurance policies could be improved by switching their investments to other financial products, such as stocks. What had once been a highly profitable product now had to be replaced quickly with new offerings that could successfully compete with banks and brokerage houses. Profits declined in the 1980s, and on the property/casualty side of the business, some observers labeled the situation an "insurance crisis" because rates increased and available insurance decreased, possibly driven by the rise in tort costs, although that remains subject to some debate.[12] Thus, during the 1980s, property and casualty insurance firms transformed rapidly into commodity businesses. The wider availability of the Internet in the second half of the 1990s simply reinforced that reality, because consumers could now compare prices and choose their providers on that basis.[13]

While below I deal with the special role the Internet played in this industry, suffice it now to point out that the two types of insurance companies did not attain any sort of financial or marketing equilibrium by the dawn of the new century. Consolidations continued, following a pattern reminiscent of the Banking Industry, although not to the same extent of concentration. Banking companies continued to move into the insurance market.[14] Simultaneously, by the end of the century it appeared that Internet-based financial services were also depressing profit margins for the industry at large, although it is too early to be definitive about the features and implications of such activity.[15]

Insurance Applications

The Insurance Industry took an early interest in computing. In 1947, just two years after the introduction of the ENIAC—the first operational digital computer in the United States—the Life Office Management Association (LOMA) established a committee to explore possible uses of the new technology.[16] In 1954 the Metropolitan Life Insurance Company installed the first digital computer in the industry.[17] GE had installed the first computer in a commercial setting earlier in the decade at its home appliance manufacturing plant in Lexington, Kentucky, heralding the arrival of computers in a commercial setting and was a milestone event that caught the attention of all industries. Other systems were already in use within government and academic settings. Back in insurance, by the end of 1955, 17 systems had been installed by various companies in the industry, driven by the need for better ways to handle the growth in information and paper. In 1954—the year the first computer system came into the industry—insurance companies employed more than 350,000 people, of which nearly 75 percent worked in offices, at a time when the demand for insurance was increasing all over the nation. With so many people involved in manual work, systems that could automate their activities were special. So how did insurance companies use these early systems? Principally, companies used these systems to process traditional accounting applications, such as generating bills and tracking budgets, and to handle large volumes of simple transactions, such as calculating and reporting policy owners' dividends and commissions for agents. The reasons for using computers with these applications were straightforward; they were easy to process; cost benefits of reduced labor could be calculated, since management already knew how much manpower it took to perform these functions; and they had already been partially automated, using punch-card tabulating equipment.[18] The most enthusiastic early users of computing for these applications were large insurance companies that, unlike many of their property and casualty counterparts, often had larger numbers of accounts to process on which they performed fairly routine calculations with minimal variation; they also had prior experience with information processing equipment, such as tabulators.[19]

In the mid-1950s additional insurance companies followed the same pattern, moving basic accounting and transaction processing to computers, one application after another. As their technical and line staffs became more familiar with the capabilities of computers, additional applications were cost justified and moved to these systems. Popular among these were actuarial analysis and statistical forecasting, both of which were numerically intensive applications and, in the case of life insurance, fairly stable. Policy records remained in various forms, ranging from punch cards to billing records similar to those that banks had used in the 1930s and 1940s. Consolidation of account records did not begin until the late 1950s, when larger tape storage systems became available and, soon after, disk drives.[20] In the 1960s, as the latter technology dropped in cost and could handle larger volumes of data, more records and information per record moved to computers. With that transition underway, additional computer-based applications involving customer records and sales became cost effective, fast to perform, and convenient. By the end of the 1960s, online access to data made it possible to update files very quickly, while providing agents and home office personnel with near-instant (real-time) access to information.[21]

In understanding how computing came into the industry, it is important to keep in mind that differences between life and property/casualty highly influenced adoptions throughout the half century, demonstrating that a firm's business reality largely dictated rates of adoption and use of computing. The various products of a life insurance company tended to be very similar to each other within a product line; therefore, a computerized system could handle many, if not most, offerings for a long enough period of time to cost justify a computer system. The same applied to health insurance products. Changes, such as extensions of coverage, could easily be programmed into existing applications. In the property/casualty environment, however, products varied far more than in life insurance, with the result that any digital application written in support of an offering could not necessarily be leveraged to conduct the transactions of other products, which made computing more complicated and difficult to cost justify. In the property/casualty world, agents also combined multiple offerings to sell comprehensive policies, which meant that using computers to perform the calculations necessary in support of this kind of offering was used more extensively in support of agents. Finally, life insurance policies did not change as frequently as property/casualty products, again reducing for the latter the time needed to recapture investments in what was expensive technology in the 1950s and 1960s. In the 1950s, life insurance companies gravitated toward large computer systems, while the property/casualty firms continued to add punch-card tabulating applications which were less expensive to implement. In 1956, a few of these firms, located in New England, mitigated the cost issue by acquiring collectively a digital computer system that they shared. Larger firms could justify their own systems as the decade progressed.[22]

In the spring of 1953, IBM announced the availability of the 650 computer, which it began shipping to customers in 1954 and in volume in 1955. It became the company's most successful computer product of the decade. Its introduction made available a small, workable system that made it possible for the first time to move many applications from old tabulating and accounting equipment to computers. For example, in February 1956, the Pan-American Life Insurance Company in New Orleans installed a 650 system to perform premium billing, commission accounting, and dividend accounting; to track issued policies; to calculate collection of premiums, dividends, reserve valuations, and policy loan accounting; and to determine annuity and settlement option payments.[23] An automotive insurance company, Nationwide Insurance, used a 650 system with its prior punch cards to do inquiries on records, to calculate costs of premiums for proposed insurance policies, and to bill, thereby relying on a large library of pre-existing machine-readable data. Premium billing quickly became the major application because of the high volume of paper involved and the fact that every policy holder had to be billed twice a year. Thus Nationwide, which also sold other types of insurance, quickly became a major user of computers. In fact, by the fall of 1957, it had installed 31 applications worth listing (see table 4.1) ultimately, they were ported over to computers during the next couple of decades. At the time, Nationwide was also programming two other applications, a monthly claims summary report and a guaranteed premium-payment accounting system. Note the mixture of traditional accounting applications (such as payroll) and industry-specific uses (like property fire reinsurance calculations).[24] That mixture characterized the collection of applications in all insurance companies during the rest of the century. Other firms were quick to install computer systems and all had used tabulating equipment before, some dating back to the early years of the century (e.g., Mutual Benefit Life Insurance Company since 1914, Equitable Life Insurance Company of Iowa, 1920, Farmers Insurance Group, 1920s, Metropolitan Life Insurance Company, 1910s).[25]

By the late 1950s, insurance companies were beginning to report on using

Table 4.1
List of Sample Applications at Nationwide Insurance, 1957

Agents' commission calculation	Job time analysis
Agents' experience analysis (run twice annually)	Mortgage loan-investment accounting
Auto new business premium verification	Paid and issued policies
Auto renewal premium calculations	Payroll
Budget comparisons	Payroll expense distribution
Credit union accounting	Payroll group insurance dividend
District managers retirement (payroll)	Payroll personnel analysis
Duplicating cost analysis	Policy six months analysis
Electronic actuarial processing (internal reports)	Property fire insurance calculations
Expense distributions	Unearned premium reserve
Expense reallocation of agents' commission (a payroll application)	Unearned premium reserve (auto)
Incurred but not reported report	

Source: "Transceiver Network Facilities Premium Billing," undated [1958?], *Office Automation Applications*, III, D2, p. 9, Market Reports 55, Box 70, Folder 1, Archive of the Charles Babbage Institute, University of Minnesota.

computers at industry conferences, resulting in continuous sharing of information on which applications were beneficial. Management teams had learned in the early 1950s that computing made sense, but they spent the rest of the decade figuring out how to do this through trial and error, by learning from others who had gone before them, and with the assistance of such computing vendors as IBM and Sperry Univac. A cursory look at the agendas of various industry conferences reveals that many presentations were made by both employees of the Insurance Industry and by vendors about the kinds of applications listed in table 4.1. At the LOMA Automation Forum in 1959, for example, representatives from Boston Mutual Life, State Farm, New York Life, Massachusetts Mutual Life Insurance, Equitable, and Northwestern Mutual Life gave talks on their experience with computers. Participating vendors included National Cash Register, Electronic Services, Inc., Moore Business Forms, Remington Rand, IBM, and AT&T. For years the roster at these annual conventions read like a *Who's Who* of the insurance and computer industries.[26]

Companies paid very close attention to the rapidly changing features of computing in the late 1950s and early 1960s, seizing upon new forms as they made sense to them.[27] Management in these years focused on pure cost reductions with little understanding that implementing computers could also lead to changes in the way their firms functioned; that realization took until the 1970s.[28] As one manager said in 1961, "Complete integration of data transmission and processing on a nationwide scale will reduce our operating costs more than $4,000,000 annually within the next two years."[29] Similar quotes could have been heard at any of dozens of presentations made at the LOMA automation conferences of the 1950s and 1960s. As they were touting the rationale of cost displacement, however, managers were also coming to realize that computers could also make information become more conveniently available, particularly after the adoption of disk drives, which provided direct access to data. However, even in the days of tape files (1950s), access to information proved faster than with earlier punch-card systems.

It was this ability to do things quicker that enabled the long process of transforming work. One data processing vice president in 1960, Robert A. Greenig of Mutual of New York, described what he could do with his newly installed IBM 1401 system:

> The daily cycle means that transactions can be processed the same day they are received. . . . This results in more up-to-date and accurate information which can be passed on in improved service to all . . . policyholders. This daily schedule is made possible only by the tremendous calculating and data-handling speeds and capacity of this new electronic system.[30]

He could post a premium, make a loan, and update a customer's master record in one day (instead of the several that it took before) thereby speeding up his business, customer service, and cash flow to the firm. At the tactical level, users began learning that savings could be had, as one executive noted in 1959, "where we use electronic equipment to bring about the integration of related functions and to reduce the number of files to be maintained separately." A compelling objective became reducing the number of people to "touch" a file to complete a transaction. In addition, users realized that manual maintenance files "get out of balance with one another and added effort becomes necessary to reconcile the discrepancies which result." [31] Moving to one master file, and later to database management systems (DBMS), could drive down costs.

The decade of the 1960s saw wide adoption of the many applications. A comprehensive survey of the industry conducted in 1963 cataloged 22 different applications and the number of firms using them in the property insurance side of the business alone. The top five, based on how many were implemented, were unearned premium reserves (80), agency statistics (79), premium accounting (78), various agent statistics (65), and premium billing (64), which also tied with reserve for unpaid claims management and commissions accounting. After the arrival of online systems in the second half of the decade, other applications that had only just started to be integrated increased in popularity, such as claims processing, which was easier to enter into a system online than was batching, issuing policies, and various general accounting updates.[32]

It is customary in the historical literature to speak about the introduction of IBM's System 360 family of 5 computers and some 150 related products in 1964 as revolutionary, and to dwell on how these systems "changed everything." To be sure, the introduction of the first family of computers, in which one could move from one size system to a bigger one with minimal rewriting of applications, was a compelling innovation welcomed by the data-processing community, because it meant that migrations from one system to another would not be as expensive or onerous to perform as in the 1950s and early 1960s. The orthodox interpretation holds that, with the arrival of the S/360, and its copycat competitors in the 1960s, use of computing expanded across the U.S. economy. To be sure, innovations in technology—and clearly the S/360 was a major one—and the consequent reduction in the cost of computing that it promised did stimulate double-digit growth rates in the revenues of the Computer Industry in the 1960s—in IBM's case, doubling revenues in less than a half dozen years. The Insurance Industry exploited technological innovations as they came to market, and the S/360 was no exception. This products and its rivals' proved to be very popular across the industry for many insurance firms, their first opportunity to embrace computing, much as large insurance companies had done with Univac computers in the early to mid-1950s, and with IBM's mid-size 650 and larger 1401s in the mid-1950s to early 1960s.[33]

However, the Insurance Industry's experience demonstrates that the major turn to computers began well before the arrival of the S/360. It is vital to understand, as we begin discussing the 1960s and beyond, that this industry had already started to exploit a series of technologies that in fact became crucial elements of that new generation of computers. While I do not want to minimize the historical significance of the introduction of the S/360, the history of the Insurance Industry suggests that prior experience created some circumstances that made the S/360 inherently attractive and that some users of computing were already exploiting components of the new generation of technology.

Briefly stated, when IBM introduced disk drives in the late 1950s, the Insurance Industry, because of its massive files, adopted the new technology, which it used right alongside tapes and cards. By the time the S/360 was announced, the largest firms in the Insurance Industry, and particularly in life insurance, already had a half-decade's worth of experience with disk files and had worked out many of the balances among uses of disk, tape, and punch card files. This industry also embraced second generation computing languages, such as COBOL, which IBM itself resisted in the early 1960s. (COBOL made it possible for companies to write larger, more complex programs in support of ever-more sophisticated applications by the time the new generation of computing became available). Thus, by the arrival of the third-generation computers (such as the S/360), the Insurance Industry was pushing the limits of second-generation computer ability to receive data from such input/output equipment as tape and disk drives (due to slow channels, or paths, to the mainframe), as well as the limited memory in computers in which transactions were performed (think of it as the size of the playing field for a football game—more players means one needs a bigger field—and conversely, four-year-olds playing soccer use smaller fields). In addition, insurance companies were beginning to experiment with online systems in which terminals were connected to computers via telephone lines or other wiring schema to access information directly.[34] So the third-generation of computers arrived at a time when the Insurance Industry was well-positioned to exploit their features.[35]

Recall that the period of the 1950s and early 1960s was still a golden age for insurance. Demand for all its products was rising, profits were good, regulators kept things quiet, and the U.S. economy was expanding. Both individual and organizational purchasing power resulted in increased demand for affordable insurance. Utilization of computers steadily expanded through the introduction of third-generation computers, which exploited the continuing transformations of digital technology.[36] Life insurers led the way, but as we saw earlier, property and casualty firms were not far behind.[37] By exploring how each segment of the industry embraced computing, from the mid-1960s into the 1990s, we can appreciate how the digital changed the nature of work in the industry at large.

Life Insurance Industry's Adoption of Computing (1965–1998)

All through the 1960s, old tabulating applications for accounting and policy management were ported over to computers. Consensus never developed, however, about the best uses of data processing in the decade. However, actuarial calculations and pricing, because they were relatively stable in life firms, were continuously moved to computers.

In the second half of the decade the concept of databases attracted the attention of data-processing personnel. In a nutshell, the notion of DBMS involved separating records (such as customer name and address) from programs so that each price record could be accessed by multiple programs; previously, every program had to have a complete copy of all pieces of the records it processed. In the late 1960s, DBMS software became available to help manage records; in an industry such as life insurance, with its massive files, this was an important innovation. By the early 1970s almost all the major firms were using DBMS, such as Aetna Life and Casualty, Employers of Wausau, Federated Mutual, John Hancock, Liberty Mutual, Northwestern Mutual Life, and Travelers.[38] In the 1960s, and extending into the 1970s, older applications that housed large quantities of data, such as those with master customer records of accounts and policies, were rewritten to be accessible via telephone lines and terminals for real-time lookup, which exploited DBMS, online computing, and larger disk files.[39] All through the 1970s, insurance companies were doing what occurred in banking. "Wherever possible, companies are turning from the typing and filing of paper documents to the electronic recording, storage, and retrieval of data," even e-mailing information as opposed to delivering data to its remote offices through the U.S. Post Office.[40]

What kind of data lent itself to inclusion in DBMS applications? Typical were data for policy applications, policy master records, policy claims, alpha indices, investments, general accounting, personnel, marketing, and banking.[41] In the 1970s additional applications were added, much along the lines illustrated in figure 1.2 (page 28). Also convenient, particularly for smaller firms, was the availability of industry-specific software products—in fact, there were over a hundred vendors with several hundred packages by the start of the 1980s. Table 4.2 lists many of the applications for which there existed commercially available software packages, demonstrating that there readily existed a richly textured demand for such uses.

As the 1970s and 1980s progressed, more and more life insurance companies linked their offices through telecommunication networks that comprised long-distance telephone services, telephone banks for customers, and terminal access

Table 4.2
Life Insurance Industry Software Products, 1980

Agency accounting	Group pension administration
Agency sales aids	Group pension proposals
Agent status	Group pension valuation
Credit life insurance	Individual consolidated life and health insurance
Debit life insurance	Individual issue and new business
Flexible premium annuity administration	Individual pension administration
Group claims	Individual policy update and inquiry
Group consolidated processing	Variable annuity administration
Group new business	

Source: Drawn from data in Charles H. Cissley, *Systems and Data Processing in Insurance Companies* (Atlanta, Ga.: FLMI Insurance Education Program/Life Management Institute LOMA, 1982): 83–84. For a description of vendors providing these, see Ibid., 84–90.

to home office mainframe computers via telephone lines (direct and dial-up). Much of the operations involved data transfer and lookup functions. Recall that at this time in the Banking Industry, efforts were underway to use EFT to move money; in the Insurance Industry less than 15 percent of all life insurance firms used this application, due to a combination of the related high cost, resistance to new approaches, a desire to keep cash float, and concerns regarding data security. The most common from of EFT (employee payroll) remained little used even as late as 1980, with one LOMA survey putting usage at 13 percent.[42] One of the few intra-industry cooperative efforts occurred in the area of telecommunications with the establishment of the Insurance Value Added Network Services, better known simply as IVANS. Formed as a nonprofit organization in 1983, IVANS promoted the use of low-cost telecommunications on the property/casualty side of the business. In time, it became a service that served both life and property and included agency systems vendors, managed care providers, and others. Originally established to benefit independent agents, IVANS eventually became an important intra-industry network, supporting agents with applications and access to networking. By 1992, though still aimed at the independent property/casualty agent, it had over 45,000 users, many relying on IVANS' supply of online applications.[43] It was not until nearly the end of the 1970s that independent life insurance agencies began installing minicomputers to help control the large volume of data and paperwork, handle accounting functions, provide sales illustrations, and track follow up with clients.[44]

The introduction of word processing into the industry in the late 1970s enabled "the use of sophisticated machinery to facilitate the production of typed correspondence and reports."[45] Early machines were stand-alone, later ones attached to mainframes and minicomputers; in the 1980s agents also used personal computers. Survey data from 1980 suggested that almost all major life insurance firms used the application, but deployment appeared spotty, particularly in field offices. Companies were also began using microfilming, optical character recognition tools, and facsimile.[46]

By the early 1980s, traditional functions of an insurance company were being augmented or rapidly automated by the use of computers. The largest firms already had massive systems that could handle thousands of transactions daily. Some applications were housed in centralized systems, while others were distributed via terminals and minicomputers. Firms increasingly focused on increasing productivity by putting computing into the hands of field organizations and individual insurance agents as the cost performance of computing continued to improve. Arrival of the personal computer made other applications possible, such as desk-top financial planning and proposal preparation.[47] Underwriters—those who do risk analysis to determine whether to insure someone and what to charge—had long been users of computers, but in the 1980s they obtained new hardware and software with which to model applicants and to conduct analysis.[48]

The overall interest in applications from the late 1960s deep into the 1980s was set on an unwavering projectory of reducing costs of operation in combination with speeding up access to information. As new technologies became available, they were adopted. Agents received more tools in the 1980s than at any earlier time, largely due to the ability of home offices to give agents access to databases and because of the widespread availability of terminals and PCs.[49] In the early 1990s, while industry focus remained on using technology to hold onto agents and to provide more accurate customer service, concerns about containing operating costs dominated much industry thinking about IT. Due largely to IT, work shifted from home offices to field offices in the 1980s and 1990s, although survey data suggest participants in the industry were split on whether that was a good trend or not.[50] Part of the ambivalence arose out of the promulgation of numerous applications and the use of multiple generations of software and hardware. An executive at one large firm described the problem agents faced:

> Infoglut built up over a period of years to the point where the field force received information and software in a chaotic fashion. As a result, those in the home office wondered why they received a lesser response than expected without realizing that the field force, bombarded from all sides, was unlikely to respond well to any of the material.[51]

Life companies paid less attention to these problems than did firms operating on the property/casualty side of the business. Large firms debated whether to bypass agents using technology, while others wanted to continue to use their agents as their primary channels, using IT as a tool. It was a debate that intensified after the arrival of the Internet, as technology made it possible for customers to start comparing services and offerings, forcing insurance companies to pay more attention to consumer needs than they had in earlier decades.[52]

Before turning to the property/casualty insurance side of the industry, summarizing the inventory of applications in life companies on the eve of the Internet's influence on the industry helps to clarify the status of IT in this sector of the American economy. A study published by LOMA in 1995 conveniently collected credible details. All major providers used computing to process transactions for individual life insurance products. Subsystems included processing new business, issuing policies, administering claims, and handling various customer services and inquiries. Administration of policies increasingly occurred with IT, involving such mundane activities as billing premiums, accounting for collections, processing policy loans, calculating and generating policy dividends, producing lapse warnings, and changing a customer's address, beneficiaries, and mode in which premiums were paid. These systems by now were highly automated, with human intervention only when a customer called or wrote in to change a piece of data or request a transaction. Another widely used application involved claims processing, which appeared in the late 1980s and involved knowledge-based systems to help evaluate claims, control fraud, and generate payments. In addition to the applications listed above, a large collection of others in support of processing for group life and health policies were in wide use by the early 1990s, covering the spectrum from selling a policy through to processing of claims and reporting financial results. As with individual claims, group claims became highly automated. Finally, there were all the traditional accounting transactions applications that had been in place in the 1950s and that had been continuously upgraded and modernized over the decades.

Because insurance companies invest premiums and dividends, all large firms had IT systems supporting this. IT handled such mundane activities as buying and selling of investments (such as stocks, bonds, mortgages) and calculating returns and costs. Companies that managed pensions also had tailored a suit of applications to manage such programs as establishing eligibility, making payments, and so forth. Finally, the buying and selling of insurance by one insurance firm to another, called reinsurance and widely practiced in the industry, involved a variety of applications, particularly after the mid-1980s when firms sought ways to limit their liabilities.[53] In short, by the end of the 1980s, the industry still did what it had in the 1950s, but now it relied on computers, particularly online systems, to help perform part or all of the steps.

The picture just painted of IT usage in 1995 remained essentially the same to the end of the century with one exception, the Internet (discussed below). A subtext of the Internet, however, that became evident in the industry by the start of the new century was the growing need to be, in the parlance of the day, more "customer focused." Large firms began altering their applications to support that new emphasis.[54]

Property/Casualty Insurance Industry's Adoption of Computing (1965–1998)

Companies in this sector of the industry were slower to embrace computers in the 1950s largely because they either were smaller, and hence could not afford the expensive technology, or did not believe that the digital could help them with their far more complex, variable products than those of life insurance. Nonetheless, as with life companies, in the 1950s some firms did experiment with computing, and by the early 1960s companies and independent agents were relying on service bureaus for some services, because this approach put the onus to maintain technical skills (and their cost) outside the organization while lowering prices for transactions or "machine time."[55]

Property/casualty IT requirements were similar to life insurance requirements. Internally, insurance companies needed help in performing underwriting functions, processing claims, supporting marketing and its analysis, planning, and maintaining financial and accounting controls. Externally, they had to manage policyholder records, conduct transactions with independent insurance agents, provide information to regulators in those states in which they sold products, and share different sets of data with industry associations. Because the variety of products proved far more extensive than in life insurance, one can imagine the permutations needed by the industry in each of these areas. But as with life insurance firms, companies in this sector of the industry added to their collection of applications all through the second half of the twentieth century such that, by the late 1990s, no major function in the industry occurred without some use of computing while, again as with life insurers, some applications had become almost completely automated.[56]

Most of the documented case studies on the use of computing in the Insurance Industry in the 1950s come from the life insurance side of the business. We do know, however, that a great deal of processing in property/casualty firms occurred using tabulating equipment and punch cards in larger enterprises in home offices, though independent agents remained paper-bound and manual in large part during the decade.[57] We also know that, as in the life side of the business, the arrival of disk drives provided new opportunities to access information. In the case of the Norfolk and Dedham Mutual Fire Insurance Company, for example, the company installed an IBM Ramac 305 system at the end of the decade to help with premium distribution tasks and to collect data on premiums and accounts. Another application tracked losses, insured accuracy of calculations, and so forth. Finally, routine accounting applications, such as general ledger and payroll, were moved from older tabulating equipment to the new system.[58] Casualty insurance applications were some of the earliest, however, to move to computers. Allstate Insurance, General Insurance Company, and Argonaut Underwriters, for example, introduced Burroughs 205 systems in the late 1950s extending into the early 1960s, when these firms began another round of technology upgrades. State Farm Mutual Insurance Company became an early and extensive user of IBM's 650 systems for all its lines of business, while Univac 60 and 120 systems were also deployed in such firms as the American Automobile Insurance Company and at the Life and Casualty Insurance Company in Nashville, Tennessee. However, over 80 percent of all the mainframe systems installed in the industry in the second half of the 1950s went into life insurance firms.[59]

Thus, by the late 1950s/early 1960s, companies were beginning to embrace the digital, often beginning with policy accounting applications, many of which had been done before using tabulating equipment from IBM and accounting machines from Burroughs and NCR, to name a few. With computers, more data could be collected per policy. Data were intensively transaction oriented, that is, information accumulated as part of premium transactions. Automatic rating of policies proved to be an attractive early candidate for automation, especially for private individual automotive insurance, since firms usually had high volumes of policies and thus mimicked some of the same characteristics of large life insurance offerings. There were many of them, and losses and premiums were relatively predictable. Direct billing to customers by computers became another early application, although it was fraught with problems if a customer received a bill with an error. Insurance companies also hoped that computer underwriting would become possible and all through the 1950s and 1960s attempted to do this using digital means, but by the 1970s it became clear that human underwriters were vital, with computers providing assistance in the modeling of options. Early attempts at automating underwriting involved using a scorecard of several variables (such as age of a driver, accident history) that could be used to score an applicant and in turn dictate the premium to be charged or to determine if a policy should even be issued. The industry tried again in the 1980s and 1990s with expert systems which proved more effective but never led to full automation of the tasks involved.[60]

All through the 1960s, technology came down in price while functions and capacities expanded, with the result that on the property and casualty side of the industry more firms could afford newer systems. They also wanted to use computers whenever possible. With the large volume of data needed to run a firm, this was no surprise. For example, in the automotive insurance business, it was not uncommon in the 1960s to pay 55 to 60 percent of a premium on losses. Understanding that phenomenon and tracking what happened to the other 40 to 45 percent of the premium provided powerful incentives to automate data collection and analysis.[61] In the mid- to late 1960s, large firms began paying considerable attention to finding ways to communicate over telephone lines to information housed in computers with their agents. Managers began noticing that they could get information sooner if it were in computers, both for their use and in response to requests for data from regulators, all of which happened in a decade when the volume of policies rose.[62] A survey conducted at the start of the 1970s suggested that property/casualty firms had identified and implemented more than 40 different types of applications. To put that in some perspective, the same survey found that life insurance firms had implemented twice as many different uses for computers in the 1950s and 1960s.[63]

Lurking as an issue all through the 1960s and to the end of the century was the relationships between independent agents and insurance companies, which over time involved computing. The property/casualty side of the industry had more independent agents than life, and both sides had agents who worked as well for national companies. The agents' paperwork grew all through the period, along with business volumes. The independents also had to deal with multiple firms, which meant various collections of forms and systems. National companies wanted to have these agents sell more of their products rather than someone else's and sought ways to promote this with a variety of incentives, both financial and functional. Independent agents wanted to reduce their volume of paperwork and to retain the flexibility to sell products from multiple firms. Insurance companies wanted to drive down the costs of agents and increase sales volumes. In short, tension existed all through the period.[64] A casual reading of industry magazines of the entire era demonstrate that few months went by without discussions of the issue and, about the role of computing in driving down costs and paperwork, holding onto good agents, and the like.[65] In the 1950s and 1960s, however, computing played a relatively minor role in company-agent relations.

By the early 1970s, however, the subject began drawing more attention. Companies had been pushing systems out to agents, and the latter were not happy with the results. One agent argued that the situation was getting out of control; note that the word "application" here is not used to describe a computer-based task but rather a form filled out by a customer:

> New revised applications arrive by mail daily, requiring more and more information to feed the insatiable appetite of the computer. We stand on the brink of a serious disaster in the American agency system. The competent agent who takes the time to underwrite risks, check policies for accuracy, keep up-to-date on coverages [*sic*] and other industry changes, and perform all of the other service functions of an agent, finds that it is becoming increasingly more difficult to make a profit.[66]

All through the period others complained that computers were not as helpful as advertised. As one reporter commented after attending a conference of insurance executives, "there is a lot of evidence that the computer is still far from doing what it is capable of doing for insurers."[67]

Because branch and independent agents became the targets of renewed interest in applying the digital, the subject warrants more attention. As much as 80 percent of an insurance company's work—tasks related to underwriting, rating, coding, typing, filing, and reporting results—was done in remote offices. Most employees were low-level clericals who turned over frequently and had to do many things simultaneously. So it was also not uncommon for up to 40 percent of transactions to contain errors. By the early 1970s, companies were automating data collection for premiums and losses, essentially replacing keypunching with terminals and word processing to correct errors before they entered a system. DP (data processing) departments converted old batch systems to online versions, Most of these activities involved personal lines of business, much of which was straightforward and standard work. Beginning in mid-decade, however, firms began trying to automate portions of the commercial lines, which were the least automated in the industry. This collection of offerings often represented 65 percent of a company's book of business, proved consistently more profitable, and yet most labor intensive. It was also complex, requiring many iterations to sell and much dialogue between agent and home office. The fundamental problem, however, was that to sell commercial business required use of vast amounts of data and rules; in some cases over 3,000 tables might be consulted. The industry spent the rest of the 1970s and all of the 1980s working on the problem of increasing productivity and accuracy in this segment of the business.[68]

In the period 1968 to roughly 1975, when adoption of third-generation computing and online systems (including early use of DBMS) took place across many industries, property/casualty firms still lagged. Observers complained about the lack of good software applications written by programmers who did not understand the insurance business. Yet others took the industry to task for focusing more on cost reduction and less on leveraging technology for new uses.[69] As more information became available in computers, more agents and managers began relying on new bodies of data, often accessible through telephone lines and terminals connected to remote computers. Modeling options for proposals and investments with "what if" analysis also made their way into the business by the early 1970s. Telecommunications, however, became the most popular new technology of the early 1970s, and expanded all through the decade to deliver information and speed processing of applications for insurance and claims settlement. By mid-decade, reports of productivity increases through the use of online systems and databases began appearing, often the result of faster processing that reduced errors in performance and the number of employees needed.[70]

Because of higher losses than normal in the mid-1970s, caused by a combination of economics and weather-related factors, industry managers and data-processing experts renewed their interest in using computers to improve underwriting and marketing. In addition to a growing market orientation to improved customer-service support such as through call centers, managers continued to look for ways to improve productivity. In short, the agenda for IT had not changed fundamentally from what it had been in the 1950s, 1960s, and early to mid-1970s. But incremental increases in the use of IT did result in a greater role being played by the digital.[71] Improved word-processing applications, for example, made it possible to standardize certain types of correspondence, contracts, and other documents, along with the work necessary to prepare them.[72] By the end of the decade, however, observers pointed out that just automating was not enough, that making information available was more important because the insurance business was not as labor intensive as it was information intensive. As one commentator noted in 1980, "the proper choice may be better information rather than simply a few less people."[73] It was a perspective already evident in both the banking and brokerage industries, among credit reporters, and even in the life insurance end of the industry. Members of the industry were becoming quite savvy, however, about the implications of IT. One observer noted in 1981 that IT might actually destroy, for example, the independent agency system, because direct providers could bypass inefficient agents. In fact, if agents did not improve their efficiency and effectiveness, insurance companies would be able to do what banks and airlines had already done—namely, solve some of their product-delivery productivity issues.[74]

Thus, perhaps the single most contentious area of IT focus in the 1980s involved agents: how to control them, how to get them to sell more, and how to improve everyone's productivity. The pages of *Best's Review*, for example, were crowded with articles on this all through the decade and beyond. The same issue concerning how to manage agents was true for underwriting. At times there would be discussion of an issue related to agents that one would have thought to have

Services representative at USAA (United Services Automobile Association) using an image processing system to scan incoming paper documents. Most processing at this insurance company was paperless, 1989. Courtesy IBM Corporate Archives.

been resolved years earlier. For example, *Best's Review* published an article in April 1982 entitled "Some Thoughts for Agents Considering Computerization."[75] It is the sort of article one might have expected to read in the 1960s, not in the 1980s. On the other hand, the industry was facing serious issues linked to IT. In the 1980s, banks began selling insurance; this new competition did not escape the attention of insurance agents or their firms. Unlike the bankers who wanted regulators to be their primary protectors from insurance and brokerage rivals, the Insurance Industry, while also eager for help from regulators in blocking out bankers and brokers, took on the competitive challenge as something it had to address through better products, services, and productivity.[76] Meanwhile, the mix of applications being installed and used remained essentially the same in this decade.[77]

During the second half of the 1980s, however, two new applications appeared in force which dominated the industry through the 1990s, driven by improved technologies that were becoming effective and cost justifiable: expert systems (a primitive form of artificial intelligence) and imaging (data scanning). Although very primitive, early expert systems provided some of the largest firms with assistance, such as help desks, and helped agents by doing automated policy writing and policy issuance. A near-expert system at Aetna made it possible to invoice a policy while simultaneously updating various general ledger files. I call this a near-expert system because it did not require use of artificial intelligence, although in the 1980s expert systems could make a decision based on a formal set of rules preloaded into a computer. Other firms began experimenting with underwriting systems that could suggest options that agents then either accepted or modified. Such systems could scan thousands of submitted applications for policies, quickly determine which could be approved automatically (and do so), and indicate which ones still required human intervention, thereby saving time and manpower to process.[78] By the end of the century, numerous variations of this application were in use in the industry.

More of a bread-and-butter application, optical scanning of paper documents came into its own in the property/casualty area of the Insurance Industry in the 1980s and 1990s. The technology had been around in various forms since the early 1960s. In fact, Mutual of New York, an aggressive user of IT, had used OCR (optical character recognition) technology since the early 1960s to collect premium-payment data from premium stubs, automatically preparing punch cards that could be used to enter this information into a computer.[79] Two decades later, the issues remained the same: reducing the quantity of paper, getting data into a computer quickly, both accurately and automatically, and all the while reducing the labor content of work. The United Services Automobile Association (USAA) became the most publicized user within the industry of OCR in the 1980s. In 1981 this company decided to start moving to a paperless environment by entering as much data into computer systems over the next few years as possible. Online access to information helped reduce the need to consult customer paper files. The firm next began relying increasingly on online data entry to eliminate millions of manually produced worksheets. The next step was to scan documents which would convert data into digital formats which in turn could be accessed by IT systems and through online terminals. Over the years, USAA applied OCR to a variety of files and applications—initially policy service and underwriting files, because of their enormity—but later other records. Thanks to dropping costs of computing and disk storage, the firm was able to convert the majority of its paper records to digital files by the mid-1990s, and in the process became a role model.[80]

Filing transactional data slowly became one of the most widely adopted OCR-based applications in the 1980s and early 1990s, reducing the volume of paper records that had to be maintained, although never totally eliminating them. Some of the applications were massive. For example, the Automobile Club of Southern California began the 1990s by generating 33,000 pieces of paper each day and had an inventory of 118 million others on file, which required 140 people to take care of them. Impressed with the early experiences of USAA, this organization embraced the new technology, implementing it all through the 1990s. At the same time, other companies were installing their initial scanning systems.[81]

In the years immediately before and after the wide incorporation of the Internet into insurance operations, twin issues also attracted much attention: the continuing role of agents, and how to increase customer-centered services. The most intense interest seemed focused on agents, and for the same reasons as in prior years: cost, operational control, and productivity and sales. Agents by the late 1980s were either embracing automation for the first time or upgrading earlier systems, the latter typically in the larger independent firms. *Best's Review* dedicated nearly a whole issue in 1991 to a discussion of such topics as how insurance companies should deal with agents, the pros and cons of regional carriers, the effects of mergers, and so forth.[82] At the heart of the matter was how insurance carriers wanted to distribute their products and, for the agents, how to maintain control over their own operations and access to customers.[83] A similar debate was underway in the Automotive Industry regarding dealers.

In the case of insurance agents, IT also influenced events. As one observer recalled:

> During the 1960s and 1970s . . . explosive growth in information technology triggered major changes in the insurance distribution system. In particular, the advent of large-scale mainframe processing systems and sophisticated data communications capabilities resulted in centralized operations and made possible innovations like company rating and issuance of policies and direct billing. As a result, many insurers, who were satisfied that they had significantly reduced the administrative burden of agents and brokers, came to view intermediaries almost exclusively in a marketing and sales context.[84]

Insurance firms remained hierarchical, however, with the home office at the apex, regional offices next, and branches or agents at the periphery of a value chain or at the bottom of a pyramid organization chart. Home-office systems housed many applications, but at the other end of the distribution channel (agencies) were "dumb" terminals or PCs. Minis and PCs were equipped with software designed for agencies in the 1980s, but the hierarchical status quo remained essentially intact. By the early 1990s, however, insurers were attempting to use IT to reach new markets with new offerings, often accessing customers directly through telecommunications linked to PCs, posing a direct threat to the traditional agency system. The existence of networks, such as those provided by IVANS, also created the possibility of upsets in the industry. Meanwhile, independent agencies were leveraging technology; 85 percent, according to one survey done in late 1992, did this to improve their customer service, internal productivity, and ability to interact with multiple insurance firms.[85] By the mid-1990s, it looked like a power play was beginning for access to customers and control of sales channels, with IT the weapon, although we are too close to the events to be certain. The trail of literature from the period suggests that as early as the fall of 1994 agents were using the Internet to get together to discuss issues, already bringing them closer together as a community and hinting at further changes in the traditional relationships between insurance firms and agents.[86]

The second major use and effect of IT in the late 1980s and all through the 1990s involved sales and marketing, as competition within and from outside the industry caused all involved to upgrade their customer service capabilities. All of this occurred as customers increasingly employed their PCs, telephones, and later the Internet to pick and choose providers. Insurers were warned: "If the industry as a whole proves unable to satisfy customers, the result will be socialization, in whole or in part, of the insurance market. Consumer satisfaction is all-important in our world."[87] IT was becoming a strategic weapon in the world of marketing by analyzing and grouping customers, targeting them for direct advertisements, and quicker underwriting. It became common practice to take laptops to accident sites to help resolve estimating for reimbursements. It seemed the proverbial "everyone" was re-engineering their "distribution channels."[88] Old systems were seen as inadequate; one consultant accused the industry of simply increasing capacity without fundamentally redesigning old applications. Incremental improvements and changes seemed the order of the day.[89]

Finally, in these years before the full blossoming of the Internet across many American industries, firms continued to upgrade and refine their underwriting applications, which increasingly were able to take work off agents and firms and to automate decision making. One of the important advances took place in the 1980s, with the development of underwriting databases. Operating similarly to a credit database, multiple firms contributed claims data to a central pool that insurance companies could access to determine how attractive a potential customer might be. For example, such a database could (and did) house a person's entire lifetime history of automotive accidents anywhere in the United States; when that individual moved to a new city, prior records were available to an agent in the new state. Two claims database organizations were already in business by 1989, and by the mid-1990s loss control and claims management had become highly automated; in short, no agent did any underwriting without interacting with an underwriting IT application.[90] Before discussing the effects of the Internet on the industry, however, it is important to look at the extent of deployment of all manner of IT to gain the necessary context with which to appreciate the possibilities that the new technology presented to the industry.

Information Technology Deployment (1950–1990s)

The largest firms in the Insurance Industry were the most extensive users of information technology throughout the twentieth century. They had the greatest number of policies, therefore the most transactions to perform. Consequently, they had the greatest opportunity to leverage incremental improvements in productivity. Given the fact that when compared to other existing options, tabulating equipment in the 1910s and computers in the 1950s were expensive to rent and operate it should be no surprise that the largest firms were the enterprises that could most afford the newer technologies of their day. The dynamic of new technologies costing a great deal softened as the second half of the twentieth century progressed, but overall IT expenses remained high since so much hardware, software, telecommunications, and DP staff were needed. Declining costs of hardware and software, however, played an important role in the justification of new IT applications since minicomputers and personal computers, for example, simply cost less when they first became available. More organizations, such as smaller insurance companies, could afford to begin automating their work. When high-level programming languages (such as COBOL) became widely available in the 1960s, even large enterprises saw the cost of writing new applications begin to decline slowly. As software products aimed at the industry became readily available, particularly in the 1970s and beyond, the cost of maintaining applications declined as did the cost of storing information in computers. It would be difficult to underestimate the significance of these long-term trends in the evolution of IT on the industry's decision to continue automating work.[91] In addition, the way technology was put together proved influential on the rates of adoption. Professor JoAnne Yates has clearly demonstrated how the IBM 650 was so similar in function, even in how it looked, to prior systems that insurance companies could replace older IBM calculators with this one, thereby rapidly increasing their computing capacity and minimizing disruptions to existing work.[92]

While she did not intend to demonstrate that this high degree of compatibility may have contributed to the industry's lack of fundamental structural changes in the 1950s, her evidence nonetheless supports that contention. Existing applications were now run on faster, larger machines, and for decades IBM was the preferred supplier of computers in this industry. Compatibility not only made it possible for the industry to embrace computers but also reinforced existing norms of behavior and organization that were not altered until market conditions (such as competition) and changes in regulatory guidelines began forcing substantive transformations in the industry in the 1980s.

Thus what we have for the 1950s and deep into the 1960s is an industry in which all the largest firms had moved quite rapidly to install digital computers. Indeed, by the end of 1955, there were over 20 mainframe systems already installed in this industry, more than in any other.[93] The largest firms also were quick to embrace every new technological development that came along, including the use of telecommunications. In the 1960s, those firms that employed 80 percent of all insurance workers used computers. By the end of the 1960s, use of online systems began to fundamentally alter how this industry worked. First, the ability to do inquiries in real time not only reduced paper in an office but also made it easier to find a fact or answer a question. Data in computers could also be collected in ways to give management, underwriters, agents, and others information about how the business was functioning and to inform pending decisions. Agents in field offices were the least equipped with IT in the 1950s through the 1970s, but in the following decade began a slow, steady increase in adoption that continued into the early years of the new century.

Through the 1950s, 1960s and 1970s, an ongoing battle was waged in many industries between management trying to lower the labor content of work through automation, and the staffs and unions striving to protect jobs. U.S. Bureau of Labor Statistics documented many instances of industries reducing their populations of workers because of their adoption of various types of technologies, not just computers.[94] However, the Insurance Industry stood out as an exception to this pattern because, as Professor Yates observed for the 1950s, business was expanding so rapidly and to such a large degree that any savings achievable through automation simply led companies not to hire as many people as they would otherwise have done. In short, computers did not cause layoffs; hence, there was less resistance to its introduction by staffs most affected by them. The number of new jobs in IT increased all during the period, as did expenditures on them.

It is still not clear if the same comments could be made of the 1980s, a period of enormous transition in the industry as it began dealing with economic strains, competition, and changing patterns of regulatory rules. The one group that began sensing it was under siege were agents in both types of insurance companies, beginning in the 1980s, particularly on the property/casualty side of the industry. Increased merger activity in the 1980s and 1990s also threatened job cuts, but only in part directly attributable to the digital. (We know for the period after 1995, for instance, that direct customer access to insurance offerings over the Internet raised the specter of job losses in the industry.) IT organizations increased in size, and in the 1980s it was not uncommon for insurance companies to spend some 2 percent of their gross revenues on computing and more on telecommunications.[95] The rationale remained the same: to avoid or contain the continually rising costs of labor-intensive work while improving the speed, quality, and accuracy of services provided to both customers and employees in the industry.

The adoption of computing was not completely without other forms of controversy, however. A study conducted by Diebold in 1971 demonstrated that confusion had existed for some time in the industry about how best to incorporate computers. While the Diebold study focused only on the life side of the industry, its findings held true also for property/casualty firms to a far greater extent than in 23 other industries examined in the survey. At the time of the study, the most extensive users of IT were life insurance companies who had turned to computers in the 1950s and 1960s to reduce costs. Other considerations were very secondary at the time, such as the ability to access new types of information. By the time of the Diebold study, however, there existed a growing perception within the industry (both life and property/casualty) that the digital's potential was barely being exploited. As one industry observer noted in 1972, "The early concentration on cost savings, logical as it seemed at the time, built certain assumptions into the development of computer systems that were at odds with the later shift to computers as instruments of corporate policy, writers of insurance contracts, and providers of information."[96] Nonetheless, deployment of IT to handle accounting, policy management, and even sales and marketing was widespread in the life insurance end of the industry by the late 1970s.[97] Another survey by LOMA late in the decade reinforced these findings, namely that computer systems were benefiting the industry, but could provide more.[98]

Economists at the BLS observed that by the end of the 1970s almost every major function in the industry relied extensively on computers. This technology had also made it easier for insurers to offer various products that required extensive underwriting calculations, particularly for life insurance. As they noted, advances in IT in the 1960s and 1970s had

> led to the merger of a vast electronic data base and computation capability with on-line communications networks, output devices, and office operations equipment. Utilizing this technology, many of the major operating functions for all insurance lines may be performed by a central computer on demand from a terminal in the home office or in any field office . . . with the results made available either through visual display devices or hard copy.[99]

In this survey, the BLS observed computers in wide use for billing, collections, actuarial research, underwriting, and claims processing. BLS also noted some minicomputers in departments within headquarter organizations, though not in the field. The number of clericals required declined while the population of IT experts increased.[100] BLS also noted that databases and telecommunications were now spread across the industry. OCR technology was utilized, primarily for premium billing and collection operations, and for promotional campaigns. PCs were virtually nonexistent.[101]

The 1980s saw new technologies of all types installed to replace earlier generations of systems, and smaller computers (such as PCs) permeating into the corners of all mid to large insurance companies. Various surveys from the period demonstrated that the classic applications of the past were diffused even further, such as premium billing and collection applications. The largest growth in usage, however, was in field operations. These surveys also indicated that satisfaction with the technology's performance had improved in the 1980s over prior decades. Expenses were being managed at satisfactory levels, new offerings were made possible thanks to computing, and more varied IT tools facilitated speed and ease of work. Those that had the greatest amount of experience with computing found their ability to control operations one of the most important sources of satisfaction with the technology. Satisfaction levels exceeded 85 percent in some surveys. Newer users continued to implement traditional applications, such as accounting, billing, and word processing, and were also highly satisfied with their new tools.[102]

One empirical study demonstrated that those firms which invested the most in IT in the decade enjoyed the highest levels of profitability in the industry because they were able to perform at lower operating costs than their rivals.[103] The findings also showed a correlation between growth in sales and containment of expenses, with the greatest effect on controlling costs. My own studies of now some 30 industries confirms the same pattern, that IT had a greater influence on controlling an enterprise's internal costs than it did on expanding sales in the period prior to the use of the Internet in marketing offerings. I emphasize "before the Internet" because, while we are still in the very early stages of the Internet's deployment and hence in understanding its actual effects, it appears initially that the equation might be changing; but more on that below.

As use of IT was expanding, users in the industry were becoming aware of the kinds of findings cited above as they were occurring, though it is still difficult to determine from extant records if reports of high levels of satisfaction were accurate.

This awareness grew out of both personal experience and from reports of what other firms and industries had experienced. One can argue, however, that since this industry had a great deal of experience with data-processing technologies, it was normally in a better position than other industries to exploit new tools and to maintain a realistic set of expectations for what these could do. Circumstantial evidence bears this out because, while quite pleased with the results obtained, members of the industry did not hesitate to complain about poor technical service and support from vendors, problems with software packages not working as advertised, and chronic complexity of implementation—issues evident in all industries.

The BLS conducted a major study of the life and health insurance portions of the industry in the late 1980s, looking at the role of PCs, PC networks, OCR applications, and e-mail. Mainframes were not reviewed *per se* because they had been in wide use for decades. Some brief statistics suggest rates of IT integration. Computers had allowed firms both to decentralize and centralize work, shifting processing either to remote offices or to concentrate work in large data centers. One result was that work forces remained stable in the 1960s through 1980s. The industry segment surveyed had spent $902 million in capital expenditures in 1967, and $21 billion in 1988. In the 1960s less than a quarter of all capital expenditures went for equipment of any kind, not just computing; 75 percent went into construction of buildings. In 1988, only 20 percent went for offices, the rest for equipment. As the BLS report noted, "the primary reason for both the rapid growth in capital investment and the change in emphasis from structures to equipment has been the industry's investment in computers and related equipment."[104] Computer equipment had only accounted for 10 percent of physical assets acquired in 1967 but had grown to 79 percent in 1988, and amounted to $7 billion in outlays. It noted that from 1967 to 1988, the industry had increased its expenditures annually by 30.6 percent. The rest of U.S. services industries had collectively only increased their expenditures by an average of 24.8 percent. Either statistic on rate of expenditures is important evidence of how the economy as a whole had embraced computing in the second half of the twentieth century, but the fact that the Insurance Industry had exceeded that rate is even more impressive proof of the importance it had placed on use of the digital.[105]

The BLS survey showed that almost every insurer used a mainframe and online systems, and that over 90 percent had PCs; nearly half had networked their PCs. About one-fourth used optical scanning, while half used e-mail. BLS characterized the use of word processors as "widespread." The one major exception to extensive deployment occurred in the area of electronic funds transfer (EFT), where BLS found only limited use because customers were not yet allowing insurance companies to deduct premium payments from their checking accounts. The major observation involved PCs; with so many installed, workers could independently look up and analyze data necessary for their work. Another important change in the 1980s came with the shift from keypunch operations to direct data entry at the point of origin of information.[106]

Taking our analysis from an industry wide view down to the level of the agency, we can see that, by the late 1980s and all through the 1990s, computing

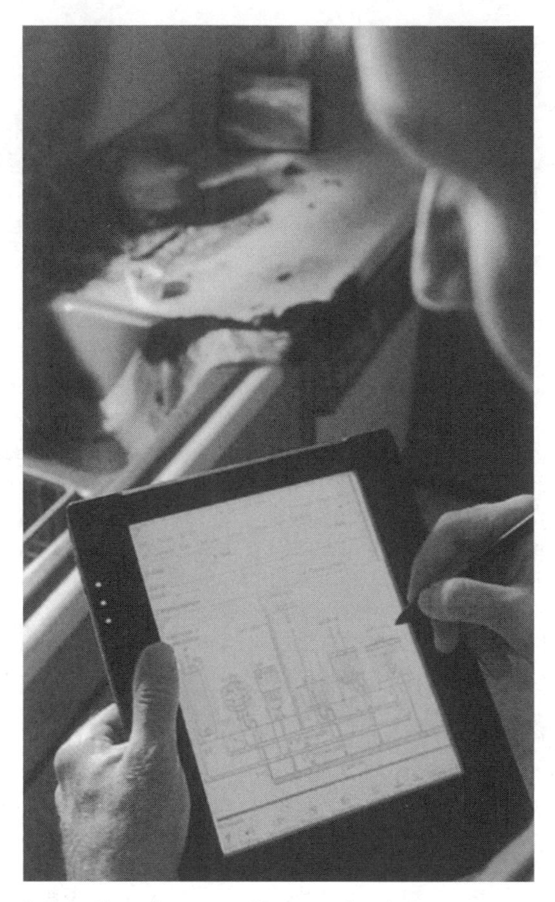

A State Farm Insurance Company fire claims representative using an IBM thinkpad 730T in 1994 to estimate damages, no longer using manuals, paper, and pencils. Courtesy IBM Corporate Archives.

continued to affect the work of field personnel. Some independent agents were writing (i.e., selling) less expensive business than agency companies because they could switch from one provider to another to obtain the best deal; as time passed, electronics made the act of switching and communicating easier. Of the 50,000 to 60,000 independent agents in the late 1980s, only 20 to 25 percent were equipped with IT, however, so we must be careful about sweeping generalizations derived from computing as the agents were the last to automate in any decade.[107] For example, in the mid-1990s agency companies in property and casualty were extensive users of IT for underwriting activities, but only about 5 percent of their field organizations were; the agencies preferred to retain that function in headquarter locations.[108] Various applications continued to be set up at the agency level through the 1980s and 1990s, and by the early 1990s, surveys demonstrated that about 85 percent of all agency offices in the United States used some form of computing—including independent agents who made their own capital investments. The reasons for continued investments were as before: to provide better service, make staffs more productive, and to handle larger volumes of business with existing employees. Applications that provided the greatest number of benefits of automation included, in descending order of value to users, proposal support, data on historical losses and claims, information on applications for insurance coverage, customer profile information, and scheduling. Agents reported that the greatest contributions made by computers to their productivity in descending order were customer profiling, collecting customer application data, proposal support (producing declaration pages), and then underwriting. And as we saw in earlier decades, the largest independents were the most computerized (88 percent) while the smallest were the least (28 percent), with the majority hovering in the 68 to 70 percent range.[109] To avoid any confusion, note that all of the agency surveys were of property/casualty agents. Life insurance agents were more automated and had been for a longer period of time, thanks to the investments of the insurance companies for whom they worked. But on both sides, the primary compelling motivation was increasing competition from within and outside the industry.[110] As in earlier decades, in the 1990s, life insurance firms continued to outspend property/casualty firms on IT just in new amounts of dollars. In fact, by 1995, over one-third of life insurance companies were spending over 13 percent of their operating budgets just on IT, while casualty firms spent closer to 10 percent. Surveys of the industry dating back to the early to mid-1990s considered computing ubiquitous across the industry.[111]

The Role of the Internet in the Insurance Industry

The issues faced by the industry in the 1980s and 1990s (such as competition, cost containment, productivity) remained active right into the new century, as did the patterns of applications adopted and used by both sides of the industry. One-third of U.S. respondents in a 1995 survey thought the Internet and other electronic means (interactive television, kiosks, commercial online services) would profoundly affect their industry. A growing, driving force in the 1990s was a population of consumers in the industrialized world who were comfortable with using technologies (such as ATMs and the Internet), who sought out their own sources of information on products, and who increasingly used that data and access to the Internet to find the best offerings possible.[112] Thus the Internet became a force to deal with as the decade of the 1990s progressed. As this chapter was being written (2005), it appeared that the Internet had the potential to alter some fundamentals in the industry such as organizational structures, who sells insurance, and how, though that had not as yet materialized.

As occurred in banking, brokerage, and many other industries, the volume of discussion about the role of the Internet became so intense by the late 1990s that understanding what actually had been its role is difficult to define. If one reads the industry literature, such as *Best's Review,* to gauge the pulse of interest in the Internet, then we could conclude that the insurers only began paying attention to the new IT beginning in 1994. These first articles on the topic touted the virtues of the Internet, next described its possible uses, published articles arguing that the industry at large had been slow in embracing the new technology (publishing these admonitions as early as 1995), and documented the emergence of a potential new channel of distribution, not the least of which was the role of agents.[113] Some executives resisted the notion of using the Internet in the mid-1990s, going public with their skepticism about the value of selling insurance over the Internet for a variety of reasons: concerns regarding security, the complexity of their products, and angst about conflicting with agents. Additionally, some limited market survey data from the period suggested that customers hardly saw the industry on the Internet, and when insurers were present they had poorly designed and informed websites.[114]

One could thus reasonably expect managers and executives in the industry to have formed some expectations about the future of the Internet. In a survey done in 1995 regarding the role of IT, conducted by the Economist Intelligence Unit, respondents saw the single most important change coming from intense competition from banks and other financial institutions and the second largest change coming from the Internet. In fact, 72 percent of the American respondents felt this way. While these respondents reported very low demand by consumers to buy or interact with insurance companies over the Internet, expectations were high that all that would change in the future. Consumers would want to comparison shop—which is exactly what happened by the end of the century—and would want to query claims status, obtain receipts and illustrations, buy coverage, and pay premiums. By the early 2000s, each of those applications had become possible and implemented by some of the largest carriers in the industry. But the industry was slow to get going. In the 1995 survey, only 9 percent of respondents said customers could even get information about offerings off a company's website. The majority of those surveyed, however, said they were planning on moving to the Internet such applications as inquiries, payments, policy updates, viewing of bills and status of accounts, claims submission, and the ability to purchase policies.[115] In short, insurers expected the Internet to play a major role in their future.

Initial applications mirrored what was happening in other industries. First, it seemed "everyone" needed to have a web page; in fact, most insurance companies did that in the mid- to late 1990s, providing information about their firms, offerings, and how to contact them. These were static sites, however, and customers could not interact online with a firm. Most customers were not attempting to buy insurance online in the mid- to late 1990s anyway, so the form of the websites matched the fact that both insurers and customers were learning what the Internet was all about.[116] But some customers did try—with little success—to comparison shop on the Internet as early as 1995; the products were difficult to compare, even term life insurance. Over time, though, websites appeared dedicated to providing comparative data for term life and automotive insurance. By the start of the new

Table 4.3
Sample U.S. Insurance Industry Websites, 1995–1996

Insurance Companies and Resources on the Net http://lattanze.loyola.edu/users/cwebb/hotlist.html
Insurance Companies in the World Wide Web http://www.tne.com/glossary/glossary.htm
Insurance Information Institute Gopher://infx.infor.com:4200/
Insurance News Network http://www.insure.com/
Insurance Products and Services Listing from Galaxy Search Engine http://galaxy.einet.net/galaxy/Business-and_Commerce/General-Products-and-Services/Insurance.html
Ins Web http://www.insweb.com/
National Association of Insurance Commissioners http://www.naic.org/

Source: Suzanne E. Stipe, "Tap the Internet's Endless Supply of Information," *Best's Review* 96, No. 9 (January 1996): 77; also includes data on search engines and other sites relevant to insurance.

century, hundreds of insurance websites existed, including some that were industrywide (see table 4.3) and some interactive ones that allowed customers to do comparison shopping, submit applications, and engage in other communications with insurance firms. A number of services began to appear on the Internet to facilitate the process, with forecasts calling for 5 to 10 percent of all sales of term life and auto insurance to come over the Internet in a matter of just a few years.[117] Early evidence suggested that price was the primary motive for shoppers to compare offerings online. Nowhere was this more so than in the sale of automotive insurance by 2002, although slightly less than 30 percent of Internet users had even sought information about automotive insurance, and slightly less than 10 percent had searched for health, life, or home owner insurance data.[118] Sales of insurance online at this time was still in its formative stages, though online vendors were starting to appear as they had in banking and brokerage—for example, Progressive for automotive insurance.[119] There were few pure players, as most e-vendors were established insurance companies that used the Internet as yet another channel for distributing their offerings.

More so than any other information technology to appear since 1950, the Internet received more attention earlier within the industry, both by its publications and the variety of authors who were managers working in the industry. Much of their discussion centered on the potential of the Internet, what it was all about, and how poorly insurers were using it. "Insurance malls" (virtual sites where one could comparison shop) were making their appearance in the late 1990s, alarming insurers who rightly understood that such a capability would cause prices for insurance to drop, as they began occurring by the end of the century, especially

Table 4.4
Top Ten Applications in Agencies with Computers, 1997

Application	Percent Using
Rating	94
Word processing	91
Accounting	74
Billing	72
Application data	70
Diary/suspense	64
Customer profile	63
Dec pages	52
Electronic mail	51
Historical losses/claims	46

Source: Derived from survey conducted by Acord Automation & Interface Survey, reported in Bill Coffin, "Agents' Favorite Software," *Best's Review* 98, No. 1 (May 1997): 64.

with term life and automotive insurance. Other types of insurance were not immune from the trend either.[120]

Management quickly came to a census about how the Internet affected the problem of independent agents vs. insurance carriers. Agents continued to acquire digital applications, some of which they began funneling through the Internet. By the late 1990s, the breadth of applications used by agents either on the Internet or within their own internal systems was impressive, indeed comprehensive, demonstrating that computing had become both ubiquitous and broad in scope at the agency level. Table 4.4 lists the applications a survey found agents used in 1997, though the survey is not completely credible because some applications appear too low, most specifically accounting. It is hard to imagine that any agency did not use computers to do accounting; perhaps those who reported not using a system had outsourced that work to an accounting firm which, of course by this time would have been an extensive user of IT. So, if anything, the results are understated.[121]

Another long-standing concern was the role of banks in the industry. In the late 1990s and early 2000s, banks and other firms expanded into the Insurance Industry, and technology continued to play a crucial role in their invasion. Intuit Inc., which made financial software, and Essex Corporation announced a partnership to sell insurance over the Internet in early 1998. The major event, however, was the merger of Travelers Group from the Insurance Industry with Citicorp from the Banking Industry in 1998, creating Citigroup Inc., one of the largest companies in the world. As the reporters at *Best's Review* told the story, "the ability to organize huge amounts of information about customers is transforming the selling process."[122] The merger leveraged the capabilities of IT. The next major step, of course, was to obtain the approval of a myriad of regulators. But other banks also saw the opportunity that IT, and more specifically the Internet, provided them to sell insurance. The market was huge—$735 billion in premiums alone in 2000 were sold—and banks only participated in 6.1 percent of those sales. Banks were already far more capable than insurance companies in cultivating customers via the Internet, and it made sense to add insurance products into their pre-existing offerings. As of 2004, however, most of that lay in the future, though by 2002 over 150 banks had started down that path.[123]

Economists working for the Brookings Institution looking at the role of the Internet across many industries at the dawn of the new century were sanguine about results so far with the technology:

> Penetration of even simple commodity products, such as term life, has seen rapid growth but remains insignificant in relation to overall volume in the industry. For example, life insurance originations over the Internet represented less than 1 percent of total originations in 1999 and are not expected to exceed 15 percent by 2003. Moreover the majority of these originations are in term life, the simplest insurance product and one of the least profitable.[124]

They noted similar trends in property and casualty insurance, such as in auto and home owners' insurance. Price comparison dominated insurance activity on the Internet, and thus, if anything, agents were becoming targets for disintermediation by the insurance providers themselves and even somewhat by banks. In 2000, agent commissions began dropping, as they had started to in the brokerage business, directly because of the Internet's ability to displace their tasks. If that trend continued deep into the new century, then we would have a direct case of profound structural changes caused by the Internet. Economists looking at this industry observed, however, that whatever services in general existed on the Internet were simply "an extension of capabilities that may have already existed and have been made less costly and more prevalent by the Internet."[125] Slow adoption of the Internet in this industry could be attributed to four causes: insurance products were complex and thus not easy for consumers to comparison shop and "buy off the shelf"; manual underwriting was maintained to prevent instant quotes and issuance from occurring despite the fact that applying for insurance online was becoming widely available; the industry fear that making insurance too widely available could result in the acquisition of unprofitable customers; and agents still played a central role in finding customers and working with them, so the industry wasn't ready to disrupt the distribution mechanism already in place. Agents saw the Internet as a threat to their livelihood, especially independent agents who could easily direct business to firms that proved less aggressive in using the Internet to stimulate new business. As the economists quoted above concluded, "the insurance sector has been only mildly affected by the Internet, due to many structural factors of the industry and the nature of its products."[126] Despite reservations, in 2001 the industry as a whole expended $16.8 billion on all manner of IT, not just on Internet-based projects, and in the following year raised that amount to $18 billion.[127] As of 2005, it appeared the industry was on track to spend roughly $20 billion that year.

Conclusions

I began this chapter by asking if the Insurance Industry had been profoundly changed by its extensive use of IT during the second half of the twentieth century. The question links to the broader one asked of all industries, namely: did it move to a digital style and thus, by inference, would it never return to an earlier mode of operation? Let's deal with the second question first. To qualify as having moved to a digital style of doing work requires that an industry's key work activities and processes be so intricately based on IT that it could not realistically revert to an earlier mode of operation. What this industry's experience indicates are three patterns of behavior. First, from its earliest days, its largest companies used computers to increase their ability to process already semi-automated functions, such as accounting and file management. Over the next half century, files, forms, and analysis of data became activities continuously performed with the help of computers, software, and telecommunications. Second, computing spread slowly across the footprint of the industry, first appearing in large headquarters and regional offices in the 1950s and 1960s, then in field offices in the 1970s and 1980s, and, by the end of the century, in nearly every agent's office, even in their automobiles in the form of laptops. That pattern injected the digital into the work of nearly every employee, especially after the introduction of the PC in the 1980s. Third, tasks dependent on activities across the industry and various enterprises became increasingly integrated, thanks to software and telecommunications that made it possible to share information through databases.

This chapter has documented the spread of technology and applications. The life insurance end of the industry moved quickly to computers because of their large files and relatively standard offerings and practices. Thus one could reasonably conclude that the life side of the industry had adopted a digital style by the end of the 1980s, perhaps even as early as the mid-1980s. The property/casualty end of the industry took longer to arrive there, but we can conclude it had by the mid-1990s, which is quite late when compared to many other industries. I cannot conceive of either side of the industry being able to function today in a non–IT-intensive environment.

Now as to the structure of the industry, we have a different situation. Professor JoAnne Yates first raised the question in a dialogue about theories of human behavior, which hold that social structures (in this case companies or industries, but also how things are done) influence what members do.[128] Technology plays both a reified and recursive role in any organization, which means it influences people and people influence it. Based on her study of computing in the Insurance Industry in the 1950s, Yates concluded that computers did not change the work of clerks because tasks continued to be done as before, and thus there were no cultural or institutional changes. Evidence provided in the foregoing pages allows us to take her conclusion further to argue that computing did not fundamentally change the industry's structure during the second half of the twentieth century but it did have an enormous effect on how work was done.

Several circumstances, however, remained fairly stable. The life and property/casualty businesses continued to operate almost independently of each other, with notable exceptions when a large firm offered both types of products, which started to happen following the 1970s, though companies still often kept independent divisions. Channels of distribution remained the same, that is to say, the use of agents. To be sure computing made it possible to centralize underwriting, for example, but underwriting also continued essentially the same, with only slight improvements thanks to simulation and modeling software late in the century. Regulators played a more important role in altering the structure of the industry by allowing banks to sell insurance, insurance companies to sell financial services, and holding companies to operate in more than one financial industry or region. Computers and telecommunications simply provided the infrastructure to make those changes operational in a cost-effective manner. To be sure, had telecommunications and computers not existed, such structural changes could have been caused by regulators, as those kinds of changes had occurred in pre-computer times, as well. Finally, the nature of the products offered remained more subject to the whims of regulators and some market forces than to the capabilities of computers. Even after the arrival of the Internet, that statement held true for most insurance products.

In short, the old business model of centralized control of insurance activities by providers remained intact throughout the century. Mainframe computers, which housed master records, simply reinforced the authority of the home office. The incredibly large expense of refitting many application software tools because of Y2K did not alter that reality, despite the trend developing at the end of the century of once again distributing applications to regions, districts, and agents, much as had been attempted to a limited extent in the 1970s and 1980s with minicomputers and later PCs. As one advocate of the use of the Internet in financial industries had to conclude, "the mainframe is still an industry workhorse" despite its inflexibility in some instances in a changing market.[129]

The Internet became the first form of IT to appear since the introduction of punch-card equipment nearly a century earlier that held out the potential to cause structural changes in how the industry organized itself. Once the regulators made it possible for multiple financial industries to participate in each other's markets, the Internet facilitated the necessary marketing and operational tasks. Customers who had learned to use the technology for banking and stock transactions could, in theory, next be expected to click on insurance. While that has yet to be borne out, all three financial industries are forging ahead by aggregating financial offerings. Should the market reach a point where customers intermix their buying of financial products such that banking, investments, and insurance were merely variants of a common pool of actions, then one might expect insurance companies to reorganize themselves, the old notion of form following function. Mergers of banks and insurance companies portend of possible things to come. The fundamental conclusion we can reach, based on the historical record so far, is that the Internet has only just started to have an effect on the Insurance Industry. As occurred in other industries, insurance executives at first froze action in the proverbial headlights of the new technology, learned more about it, then began morphing along with the technology, just as had occurred with so many prior innovations in IT. But because the process has just barely begun in the Insurance Industry, the question of structural change remains unanswered. The experiences of other industries, however, suggest that the industry will look different by the middle of the twenty-first century.

Given what we have learned about the banking and insurance industries, one can conclude that the introduction of computing does affect how individual tasks are done, but also that the scope of change can simultaneously be constrained by circumstances of a nontechnical nature, such as government regulations. We have also learned that some technologies have the capacity to stimulate changes in organizations, such as have already occurred in banking and to a lesser extent in insurance. For a more extreme case of the effects of computing, we now turn to the Brokerage Industry. Because of the extent of influence of the digital on this industry, and the importance to the national economy of the financial offerings of its firms, we cannot complete our understanding of the role of the digital in the financial sector without examining what stockbrokers were doing.

5

Business Patterns and Digital Applications in the Brokerage Industry

People warned him he'd have all those weird technical people working for him. But as the low-cost broker, he felt we had to have control of the technology.

—Dawn Lepore, Chief Information Officer, Charles Schwab & Co., 1979

Nowhere did computing and telecommunications seem to play a deeper role in transforming companies than in the Brokerage Industry. The Brokerage Industry comprised a collection of firms reluctant to change long-established practices, and often had to be driven to transform by government regulators, most notably the Securities and Exchange Commission (SEC) and the U.S. Congress, even by clients and by its own experts on computers. Yet this industry's experience demonstrated how IT could be used extensively for decades without fundamentally changing many of its basic processes. This may seem like an odd finding since it has become *de rigueur* for most commentators on IT usage in the late twentieth and early twenty-first centuries to proclaim loudly and vociferously how computers have changed everything. Even thoughtful commentators do not hesitate to use such terms as *revolution* and *blown to bits* to describe the passing of the Old Order.[1] Indeed, in many of the industries described in this book, and in the previous volume of *The Digital Hand*, that did happen. But in the instance of the Brokerage Industry, change of that dramatic a nature did not start to take place until late in the century. To be sure, those who worked in the industry would argue differently, but when compared to so many other industries, one cannot escape the conclusion that if looking for a revolution in industry operations, due to computing, one would have to wait a bit longer to see the results. This is not to say that computers were not used; indeed this industry, more than many others, spent vast sums on IT, particularly in the last two decades of the century. From PDAs and cell phones to PCs and CRTs, the digital's various artifacts seemed to have appeared on every industry member's desk.

Thus, while many firms in this industry were limited users of predigital information technologies in 1950, by the end of the century some of the most pervasive images one could conjure up of such companies were of terminals on the floor of the New York Stock Exchange (NYSE) or of batteries of screens in brokerage offices displaying dense quantities of numerical data in various colors. Technology facilitated an extraordinary expansion of stock trading in the United States, and improved the efficiencies with which capital was raised, owned, and deployed.[2] This was also an industry, however, that in retrospect had many incentives not to install computers as the use of IT contributed to the collapse of the high fees brokers charged their clients and to the displacement of some workers.

The importance of this industry in the functioning of the American economy can hardly be overstated, which is why understanding the role of IT and telecommunications in its firms is so important. The Brokerage Industry has two fundamental responsibilities: to raise capital to fund growth and transformation of companies across the economy, and to facilitate creation of wealth for individuals and other firms. Without capital, modern companies cannot exist let alone invest in equipment, distribution, and products. Brokerage firms and banks raise money for companies by offering stocks and bonds for sale through IPOs and other financial instruments. Individuals and other firms invest in companies through the medium of stocks, which are sold in the secondary market (that is, when you and I buy stock). The American brokerage industry has long been considered one of the most efficient in the world and one of the most attractive in which to conduct investment activities. It is situated in the largest advanced economy on earth, has the greatest number of regulatory and legal safeguards for investors, and often operates in good economic times. Thus, this industry has served globally as a major engine for economic development.

Each industry in this book allows us to explore various facets of the effects of technology on the operations of the economy at the firm and individual levels of work. The Banking Industry demonstrated how the digital facilitated the rapid and safe movement of cash and credit across the U.S. economy. Today, nearly 70 percent of all business transactions are conducted by consumers; it is impossible to imagine how that could be so without the efficient operation of a sound banking system. Looking at the Brokerage Industry in 1950, however, we see that 100 percent of all transactions were *not* done directly by individual stockholders but rather by stockbrokers. As demonstrated in Chapter 2, technology contributed profoundly to efficiencies that increased over time and to the ability of individuals to obtain information and later place orders. The Insurance Industry relied on technology to reduce risks of loss in investments in buildings, possessions, lives, and future earnings of people and organizations. Put in insurance terms, managing risks was a crucial role played by all firms in the industry. Computing helped, for example, to securitize investments and insurance policies while improving risk analysis. So, while banking made it possible for people to have purchasing power, insurers protected individuals and organizations from losing the assets they had accumulated. The combined efforts of these two industries contributed to the increased aggregate wealth of the nation. The Brokerage Industry made it possible for people and organizations to invest easily anywhere it made sense in the United States' and global economies, most notably in the ubiquitous economic institution of the century, the corporation, and in a myriad of government bonds. When the efforts of the Brokerage Industry are set alongside those of the other two financial industries, we begin to comprehend how essential technology was in the performance of the American economy from the 1950s forward.

Finally, this industry allows us to explore further how the digital improved speed of operations, influenced the nature of structural changes in the industry, and how millions of individuals participated in its activities who did not before. The latter may be the most important reason to explore the digital in brokerage because, as a general statement, in 1950 only large organizations and the wealthy owned stocks and bonds in significant quantities; very few members of the middle class did. By the dawn of the new century, nearly two-thirds of American households owned stocks and bonds, and millions of individuals bought and sold stocks almost with the same ease as they bought books from Amazon.com; by logging onto the Internet. The dispersal of stocks throughout all levels of society is a remarkable story in itself, one facilitated by the implementation of IT and telecommunications across all sectors of the economy. That development alone would justify our examination of the role of the digital in this industry.

But there also are limits to the story of the digital. At mid-century, this was a regulated industry, one in which brokers and exchanges operated where existing circumstances led to a highly profitable business, one they did not want to disrupt. It was an industry plagued by periodic bouts of scandal, forcing the SEC and the U.S. Congress to intervene. This industry also faced severe problems in handling growing volumes of transactions as the second half of the century unfolded. Many of these issues had less to do with technology and more with the structure of the industry and how it functioned. So the tale about the role of computing and telecommunications is not the whole story, but it is nonetheless an important part of its modern history.

Evolution of the Brokerage Industry

The quote starting this chapter came from the Chief Information Officer of one of the most distinguished firms in the Brokerage Industry and refers to the strategic importance of IT not only for the head of the firm but also for all senior executives at Charles Schwab. By the early 1990s this firm had become one of the most "high-tech" in the industry, so profoundly influenced by technology that it could serve as the industry's poster child for the digital style.[3] However, we would not have found anything particularly unusual about most brokerage firms in 1950 when it came to their use of IT. In fact, most used an assortment of adding machines, teletypes, telephones, and a vast array of paper forms and records. Thus, the transformation of many firms, along with the industry's institutional infrastructures, such as the New York Stock Exchange (NYSE) and a handful of regional exchanges, exemplified how things were affected because of technology.

There are essentially two major components of this industry which are both unique in their roles and yet inextricably tied to each other. One consists of all the investment banks and brokerage firms that launch new stock offerings and buy and sell securities on behalf of their clients. The second component comprises the various exchanges themselves, such as the NYSE, NASDAQ, and AMEX (American Stock Exchange). The first generates activities in the form of new offerings, and the second runs the transactions. That pattern has not changed fundamentally since the late eighteenth century, when the industry first came into existence. One must conclude that the history of IT in this industry is about uses made by both communities within it, and equally important, how technology has aided interactions between them. Put another way, IT has helped brokers run more efficient operations, understand their markets and competitors, and manage accounts with clients. The exchanges have used IT to handle accurately and quickly the ever-growing volumes of transactions, all the while increasing the availability of information for those interested in stocks and bonds, which has grown to include individual investors and the public at large, not just members of the industry. A third story to tell concerns the interactions among brokers and exchanges made more diverse and sophisticated thanks to the use of various technologies.

Before discussing the effects of computers and telecommunications, we need to understand the industry and its history since 1950. In 1950, small brokerage firms dotted the landscape, primarily in large American cities. A few national brokerage firms existed, while those major banks dealing with primary securities (such as IPOs) were clustered in a few cities, most notably in New York, Chicago, Boston, and San Francisco. There were a number of regional stock exchanges, most importantly the NYSE (which had evolved out of earlier exchanges dating back to May 1792)[4] and others in such cities as Boston, Detroit, Philadelphia and Cincinnati. The industry was regulated by the SEC, which came into being in the early 1930s when federal regulators separated brokerage activities from those of insurance and banking institutions while also imposing regulations that increased the safety in trading in securities, a mission undiminished at the dawn of the twenty-first century.

In 1950 the broker was king. Usually a man, he put together financial deals to provide corporations with capital through issuance of stocks and corporate bonds, which he then sold to institutions and wealthy individuals. All transactions came through his hands, and he charged a fixed fee per share that was not discounted, even for large volume orders. Only those who were members of an exchange could trade, thus they were the only ones who could charge fees. As the

Early use of Burroughs equipment by a stockbroker, 1963. Courtesy Charles Babbage Institute.

American economy expanded in the 1950s and 1960s, the volume of transactions handled by members of this industry grew, as did the number of firms, employees, and customers. Table 5.1 documents the volume sold on the NYSE, which throughout the entire twentieth century was the largest exchange in the nation, reflecting the massive growth of the entire industry. The numbers of stocks bought and sold on the American Stock Exchange were also high: 107 million shares in 1950, which grew to over 1.6 billion shares in 1980, then to over 3.3 billion in 1990.[5] Similar stories could be told of other exchanges as well.

Contrast 1950 with 1997, at which time the Internet was beginning to affect the industry. In that year the industry consisted of 45,029 firms, employing over 100,000 people. In 1997, the industry generated 1.5 percent of the nation's gross domestic product (GDP), serviced roughly 25 percentage of the American workforce, and supplied the vast majority of investments of all insurance and pension funds in the United States.[6] We do not need to document in detail expansion of business volumes over the half century; that has been done effectively elsewhere.[7] It is enough to recognize that the industry had expanded enormously over the period.

However, the consequences of such expansion are important to note. First, large institutional buyers began balking at the high fixed fees brokers charged for

Table 5.1
Volume of Transactions on the New York Stock Exchange, 1940–2004

Year	Shares Stock (Millions)
1940	207.6
1950	524.8
1960	766.7
1970	2,937.4
1980	11,352.3
1990	39,836.4
2004	367,099.0

Source: New York Stock Exchange, http://www.nyse.com (last accessed 3/12/2005)

each share of stock bought and sold. These buyers put enormous pressure on the industry, regulators, and Congress to break that practice, which happened in the 1970s and led to a significant increase in competition and sharp declines in brokerage fees. Second, since this industry had not modernized its IT infrastructure in the 1950s and 1960s to keep pace with the growing volumes of transactions, it collectively found itself in a position of constantly playing catch-up similar to the experience of the Banking Industry in trying to keep up with the ever-growing volume of checks. Like the bankers, the brokers muddled forward, just managing to keep up. Third, technology made it possible to expand the number of participants in the industry (individuals buying stocks directly), while making it possible to add banking-like functions (such as the ability to write checks) to brokerage offerings.

Between 1950 and 1997, however, many important events occurred in the industry that cannot be ignored. Economists studying the industry recognized the role of technology more directly than for any other financial industry. "Beginning with the trans-Atlantic cable that linked the London and New York markets for foreign exchanges, a series of technological innovations has transformed the securities industry, changing not only the pattern of financing but also the regulation and governance of institutions that provide the infrastructure for financial flows." In the same paragraph, they noted that "trading floors have been abandoned in favor of an electronic interface in some stock exchanges," although acknowledging that not all yet had done so (for example, the NYSE still had humans trading on its floor).[8]

The evolution of the securities markets from 1950 to 2000 is a remarkable story because of how much they reshaped themselves. A large variety of offerings became widely available, including market funds, credit market equities, corporate equities, mutual funds, and stocks, to mention just a few. Institutional investors, such as pension and mutual funds, invested heavily in stocks, in 1999 controlling

37 percent of financial wealth in the private sector. In 1950, only 17 percent of all consumers owned stock, but from 1960 on this rate was elevated to about 25 percent.[9] But consumers were also invested in other financial products such as pension funds, which accounted for slightly more than stocks directly; roughly 10 percent were in mutual funds, all of which were offered or managed by brokerage, insurance, and banking firms.[10] Institutional investors played an extraordinary role over the period in launching, selling, and managing investments for other firms and consumers. Many new stocks were issued first through the primary market (to institutions buying IPOs) and then through the secondary market (where consumers and other institutions bought stock), particularly after 1980. In that year, 149 IPOs were launched, a number that grew each year almost to the end of the century, reaching 873 issues in the peak year of 1996. The costs of launching new offerings also declined throughout the period, while their size increased.[11]

The industry managed to find many new ways to break up investments into different forms and offerings, which lowered risks to investors while reaching new markets in novel ways. Technology made it possible, as one economist put it, "to carve up and quantify various bits of risk throughout the world to the party that is most willing and able to absorb them."[12] In turn, that capability made it possible to lower the cost of raising funds by companies within the American economy. Brokerage firms participated in the securitization of investments much like banks and insurance firms. Just like computers could calculate bits of risk, telecommunications could transmit transactions around the world. But lest anyone think the digital hand alone made all this possible, we must remember that regulatory bodies governing the work of banks, insurance companies, and brokerages encouraged these trends.[13]

Another nearly invisible transformation occurred over time, made directly possible by technological innovation.[14] Starting about 1970, trading electronically drove down costs of transactions while increasing the speed with which they were conducted. In turn these developments increased the number of transactions (called *turnovers* in the industry). Two costs declined: the expense of physically taking and executing orders, and the fees charged. The first was a function of normal efficiencies one could gain from using software to handle transactions and manage risk, the second the result of a combination of increased competition, regulatory changes, and the ability of individual consumers to submit orders to brokerage firms after the arrival of the Internet to this market in the 1990s. While economists are still debating how much the savings were, we know that the cost of performing a transaction dropped by roughly a third in the last several decades of the century, while execution fees declined by some 70 percent.[15] It is a trend that brokers resisted throughout the period, taking the position that computers and regulators were directly engaged in attacking their incomes. But the trend continued right into the new century and as of 2005 had not ended. Furthermore, as the cost of performing transactions declined, the number increased—the turnpiking effect also evident with frequent withdrawals of cash from banks after wide deployment of ATMs—because it became easier, quicker, and cheaper to do them than in the 1950s or 1960s. Economists have calculated that each decline in cost of a transaction by 10 percent stimulated an increase of 8 percent in turnovers in the 1980s and 1990s.[16]

As occurred in banking and insurance, mergers and acquisitions made it possible for larger investment banks (those that issued IPOs and other initial investment instruments) and brokerage firms to merge. As in the other two industries, technology facilitated these corporate mergers by improving economies of scale, managing risks, and delivering new, often integrated, offerings. The 1980s became the greatest period of mergers for brokerages, as well as so many other industries. By the start of the 1990s, there existed a number of very large firms capable of building global IT infrastructures that in turn allowed them to reach more customers. Ranked in order of number of offices and employees alone in 1991, those included Merrill Lynch (510 offices with nearly 40,000 employees), Shearson Lehman (427 offices and 31,000 employees), Dean Witter Reynolds (499 offices and nearly 16,000 employees), and a firm intimately connected to insurance, Prudential Securities (336 offices with 17,000 workers). Just those four firms alone controlled over $17.5 billion of capital.[17] All were extensive users of IT.

The role of the SEC proved important, often decisive in determining policies and practices in this industry. As with banking and insurance, over the decades brokerage firms were given piecemeal permission to sell a larger variety of products and services, while regulators implemented controls to ensure the interests of consumers were protected and that competition thrived. The SEC played an active role in promoting the use of IT and in influencing its use, having encouraged the industry to automate operations as early as 1962. The SEC's most important push for technology actually occurred early in the 1970s when it launched an initiative encouraging development of a national market as well as maintaining existing regional trading markets because they stimulated competition. The SEC's early actions culminated in the U.S. Congress passing amendments to the Securities Exchange Act of 1934—the base legislation governing the work of the SEC—in 1975 that called for the creation of a national market system. This directive essentially motivated much of the work of the SEC for the rest of the century. One of the basic reasons this law passed was that data processing and telecommunications had finally made it possible to create a national trading market where, in theory, everyone could see what the prices were for securities anywhere in the world and could trade regardless of geographic location. The SEC wanted to use this approach for trading to improve efficiencies, increase competition, and make more information available to all parties involved in a transaction.[18]

It was an extraordinary piece of legislation because it was one of the first regulatory laws to emerge because of the capabilities of the digital and, in this case, to alter how markets worked. While the ideal, friction-free trading market has yet to appear, enormous progress has been made. Today, thanks to the wide use of portals and data streaming services on the Internet investors have access to vast quantities of information—indeed almost as much as a broker—complete with real-time trading quotes. The historic process of consolidating markets that had started in 1900 sped up after 1975. In 2000, two markets dominated: the NASDAQ, with over 4,700 stock issues valued at $3.6 trillion, and the NYSE, with over 2,860 issues worth $12.3 trillion. Combined, they handled $15.9 trillion in securities out of the $17.1 trillion total traded in the United States. In 1999, 78.7 million individuals owned stock, most of which they had acquired through these two markets, which were both linked electronically to brokerages. In the prior year, only about 11 percent of all individuals had used the Internet to buy and sell stocks, so the two exchanges remained the *de facto* national market.[19]

A second initiative by the SEC also had its roots in technology. Like other regulatory bodies in the United States, this one fostered competition. One of its central collections of regulations are called the Manning Rules, which require the NASDAQ to execute a customer's order before executing its own so that broker members do not take advantage of price differentials to make quick profits. Beginning in the 1960s, another related set of initiatives involve the continuous implementation of rules reducing the amount of time it takes to execute and clear transactions, moving from weeks and days earlier in the century to three and now on its way to one. The latter of the two targets—from a few to one day—are increasingly being made possible by computers, which can expeditiously collect orders, update records, receive and pay cash, and do whatever else is needed to complete a transaction. In 1996, the SEC issued a series of order display rules which required market makers, electronic communications networks (ECNs),[20] and exchanges to take bids they received that better the "national best bid and offer" (NBBO) and to execute the order immediately against its own inventory, to display the better price as part of its quotations, or to send limit orders to other market makers who would then have the same choices to make. In short, quotes from ECNs had to be integrated in the NBBO. That could only occur if using computers.[21]

Four major events in this industry are profoundly linked to the story of IT. The first, already mentioned, was the combination of the SEC breaking the back on fixed fees and demanding that the industry create a national market, both of which started in the mid-1970s. The second involved the market crash of October 1987, when the stocks dropped by just over 500 points on the NYSE, which translated into a 21 percent decline in the total value of stocks. The third was Y2K, which led to major overhauls in IT applications across the industry in preparation for the year 2000. The fourth was the terrorist attack on the Trade Towers on September 11, 2001, which pointed out both the durability and vulnerability of the market. In each instance, the SEC, firms, and exchanges took IT-centric actions to reduce risks, repair damages, and exploit new opportunities. By the end of the century, of all the financial industries, brokerage was the one that had most integrated IT into its daily tasks, business strategies, and organizational forms. This is not to minimize the role of the SEC or the U.S. Congress—they were and continue to be important—nor the resistance of brokers and exchanges to changes that limited either their fees or their ability to make profits by differences in asking and selling prices during the process of selling and buying of stocks. Yet the IT side of the story is as crucial to understand, because many regulatory actions were made possible by the capabilities of computing and telecommunications.

IBM 1620 installed at Associated Press to tabulate and communicate stock quotations to newspapers, 1963. Courtesy IBM Corporate Archives.

Early Brokerage Digital Applications (1950s–1975)

To understand the evolution of applications in this industry, we need to discuss how both firms and exchanges used IT. It is impossible to tell the two stories separately because they are so intertwined. This approach clarifies and supports the basic research theme of this book, how the work of industries changed as a result of their use of computing. But first, how much technology was in use in the early 1950s? We have a description from 1950:

> On entering, he finds himself in a large room filled with the noise of ringing telephones and clicking telegraph instruments. Small groups of men are watching a large blackboard covering a whole side of the room. A ticker in the center near the blackboard is printing symbols on a narrow tape that is eagerly watched by one or two young men who write quotations upon the board, rushing quickly from one column of figures to another. Occasionally, one of the men watching the board calls a clerk and gives an order which is quickly put into the hands of a telegraph operator.
>
> Every large brokerage house, if it is a member of the New York Stock Exchange, will have a quotation board. On this board sales of the different

> stocks traded on the floor of the Exchange are marked up, sale after sale, as they come in on the ticker tape.[22]

What about the NYSE itself, what did that look like? We have a description from the late 1950s which makes no hint of any computing at work. Along the edge of the trading floor were telephone booths to which brokers at the brokerage firms called in their buy and sell orders to their colleagues physically situated on the floor. Stockbrokers took their orders to 30 trading posts, each specializing in specific stocks among the portfolio of the 1,519 stocks sold at this exchange, still a physical activity in a physical setting. The exchange had "some 35 miles of aluminum tubing" under the floor to move paper from the telephones booths to the posts. As one observer noted at the time, "through these tubes flow the reports of sales for the ticker tape."[23] And, of course, the floor also used blackboards. But the industry at large had problems:

> The failure of several major brokerage firms in the late 1960s and early 1970s can be largely laid to their inability to process the volume of transactions their sales forces were generating. Often the problem was caused by neglect of the critical functions . . . little was spent on automation and trade-processing. Instead, these firms attempted to handle the increased processing needs by simply hiring more clerks, usually inexperienced, to handle problems in the same inefficient ways that had already created the problems. This simply compounded existing problems and left the brokers with even more fails than previously, until ultimately they had to pay the piper.
>
> Fortunately, the financial community learned its lesson well.[24]

Alex Benn, an advertising agent working in the industry in the 1960s and 1970s, was not so kind when recollecting the events of the late 1960s and early 1970s:

> Of the many reforms of 1969–1975, the opposition to nearly every reform that would benefit investors was formidable. Even proposed measures that would benefit members of the New York Stock Exchange were also sometimes vigorously opposed by many members of the Exchange as well—sometimes by a majority.[25]

The former report went on to describe the reforms implemented, such as the automated systems that reduced the need for printing and delivering physical stock certificates to complete a sale; the second described the many sins of the industry.

In reading their comments, one can argue that brokers were some of the most backward when it came to use of IT in the 1950s and 1960s, because volumes of transactions were low enough to handle with manual and semi-manual approaches, and there were few or no incentives to improve efficiencies or to share more data with clients.[26] That situation, however, did not mean the total absence of computing and other forms of IT, although the pickings were slim at best. Brokerage firms studied the new technology at the same time as did banks and insurance companies and continued to use adding and calculating machines of the past and, of course, the industry's emblematic ticker tape system which had been in use since

1867.[27] They also used a new class of electronic calculators that had come into existence in the years immediately following World War II.

In the mid-1950s, IBM obtained orders from brokerage firms for ten of its new system, the 650. By early 1957 these systems were being used to handle orders and allocation of new stock offerings at the very largest firms, the ones that had sufficient volume of business to cost justify their expense.[28] But interest had been building for a couple of years, finally resulting in what probably was the first installation of a computer in the industry by the Francis I. Dupont & Company, which installed an "electronic brain" in November 1955 to speed up calculations as to what stocks a client had, margin balances, and other account records. Later, brokerage firms and regional exchanges installed successive generations of computers, including the widely adopted IBM S/360 systems of the 1960s.[29] As one manager at Merrill Lynch (sometimes also credited as the "first" to use a computer in the industry)[30] reported, with the three machines the firm ordered, the branch offices had "available, at market opening each day, computations done electronically overnight in the firm's office in the heart of New York City's financial district," calculating market values of all marginable securities, a customer's equity by account, and any additional margin required.[31] Meanwhile, exchanges continued to use tabulating equipment to handle volumes. For example, both the NYSE and AMEX used high-speed calculators and punch-card tabulating systems to perform routine bookkeeping. Despite the introduction of computers, one commentator in 1955 noted that "the typewriter, adding machine, cash register and accounting machine do most of the exacting bank and brokerage work."[32]

Recognizing the need of brokerage firms to have computing capabilities to process transactions and client accounting in the late 1950s, IBM announced that it would establish data centers which would sell time on its products, with the first established in the heart of New York City's financial district.[33] IBM's records of the 1950s, however, suggest that the level of commitment to computers remained quite low. Yet rumblings of change were in the air because the volume of stock transactions began rising. There were early signs that the increased volumes of post–World War II sales were beginning to strain existing work processes by the mid-1950s, a problem that would grow to crisis proportions by the early 1960s and that nearly brought operations to a halt in 1968 and 1969. One broker at W. E. Hutton & Company complained in January 1955 that "we're swamped." A partner at Haupt & Co., reported that "our people are working overtime—some as late as midnight." High-speed calculators had already proven the value of electronic assistance. One broker at Kidder, Peabody & Co., noted that such equipment "simply gobbles up statistics so that we can clear out by 5:30 p.m. even on the busiest days. Overtime is a thing of the past."[34] These quotes came at the end of a week in 1955 when volumes of shares sold had climbed some 25 percent above average (from over 3 million shares per day to over 4 million per day) and nearly 50 percent per day over averages the year before. There was talk of curtailing the number of trading hours at the NYSE to help manage the avalanche of paperwork associated with trading and clearing. Brokers feared employees would quit.[35] The problems would worsen before the industry addressed them in an aggressive, comprehensive manner.

Yet as early as 1956 some firms began nibbling away at the problem. That year Bache & Company installed one of the first IBM 650 systems in the industry; in 1959 it upgraded to an IBM 705 Model III to handle stock transactions, the first brokerage firm to install this machine; Bache & Company also used this machine, which handled 780 trades per minute, to do accounting for purchases and sales, maintain stock records, perform margin and bookkeeping, and prepare monthly customer statements. Merrill Lynch, Pierce, Fenner & Smith, Inc., decided to do the same in 1960, but chose to use the larger IBM 7080 system. Other large firms followed suit, including Bear, Stearns and Company and Dean Witter & Company.[36] Regional exchanges, eager to compete against the giant NYSE, often were also early adopters of IT. A notable example in this early period was the Detroit Stock Exchange, beginning in 1961, which relied on computing to provide daily accounting records on stock trades.[37]

Although the NYSE lagged in bringing on new technologies in these early years, its management saw the need for increased capabilities to handle the volume of transactions and thus began following the lead of some of the smaller regional exchanges. For example, in the early 1960s they began the long process of automating some of the collection and dissemination of trading data, but they were very careful not to eliminate the jobs of the exchange's floor personnel. Management focused on upgrading its ticker-tape system to keep up with the new stocks and the need for faster printing. Internal marketing correspondence from IBM documented the NYSE's desire, however, also to improve back-office accounting work by late 1964. The NYSE issued a series of press releases announcing its intention to automate the transmission of market information to provide its brokerage members current stock prices, bid-asked quotations, and volumes by telephones linked to computers.[38]

But why did a surge in interest in computing develop near the mid-1960s, long after that phenomenon had become apparent in insurance, banking, and manufacturing industries? In simple terms, until the mid-1960s the industry had not yet reached the point where many of its members believed they would not be able to handle the growing volume of transactions. The processes had become some of the most paper-intensive in the American economy. The problem was becoming particularly acute at the NYSE, the nation's leading exchange, where surges began reoccurring in the early 1960s. A particularly severe one happened in 1961 when daily volumes reached over 1 million, as compared to 767,000 the year before. Meanwhile, average daily volumes also increased; in 1963, for example, volumes averaged 1.1 million, up from 962,000 the year before.[39] Daily peaks increasingly drove up daily volumes on individual days. That pattern became quite severe in 1963 but on December 6, 1965, 11.4 million shares were traded, up from the average 6.2 million shares sold normally. At the time, the exchange was also forecasting that average daily volumes would reach between 9 and 11 million shares by 1975; in that year they actually reached 18 million. Brokers, clearing houses, and exchanges were reaching the point where they could not handle the paperwork: orders, stock certificates, transaction-clearing documents, account records, and more. Floor traders were still writing orders on little pieces of paper that often were as undecipherable as a doctor's prescription. Between 25 and 40 percent of these orders literally could not be processed because of the awful handwriting. But so long as the firms were making money, sloppy processes were affordable.

Another angle of the delayed timing to ask is: Why did it take so long? A number of answers have been offered over the years. One widely touted argument was that too many people felt computing would lower their incomes or eliminate their jobs.[40] A second explanation held that the culture of the industry ensured that its leaders (both at firms and exchanges) did not care enough about backroom processes and therefore ignored them.[41] A third argument played out the logic that so much money was being made that quick fixes, such as hiring more clerks or paying for overtime, although not optimal, were sufficient and affordable.[42] Each holds some truth. Floor personnel stiffly and successfully resisted computing and fundamental changes to existing ways of doing work, and no more so than those employed at the NYSE. It is also true that many senior executives in large brokerage houses paid insufficient attention to internal operations; there are just too many documented cases of this situation to ignore. However, regional exchanges clearly demonstrated a willingness to implement new processes.[43] We have too little data to generalize about the attitude and practices of small brokerage firms, but the practice of throwing more labor at the problem of growing volumes did continue while some managers at the NYSE and at many other firms were starting what turned out to be a long, tedious effort of reforming basic processes and obtaining the regulatory approvals to do so.[44]

The extant evidence leaves us with the conclusion that the industry was slow to change until it had serious economic incentives to do so. When the crunch of growing volumes began eating into profits and created other unacceptable operational and sales problems, the industry and its regulators responded, taking about as long as other industries to transform basic work processes, finally injecting computing into core processes. The industry as a whole did not transform any more than it needed to, however; the problems faced were so complex and severe that detailing some of these is essential if we are to understand how management reacted to the issues and to the practicalities of the digital.

As volumes increased in the late 1950s and throughout the 1960s, clearing a transaction became extremely challenging to complete. Firms were required to physically transfer a certificate from a seller to a buyer; failure to do so meant that other sales could not be completed. In turn, that meant that the physical transfer of related documents (such as floor reports, reconciliation reports, and confirmations) could not occur, further slowing the process of buying and selling at a time when volumes were sharply rising. Essentially, cash would not transfer from one brokerage firm to another until the process was completely executed, because brokerage houses paid other firms and institutions upon receipt of stock certificates they bought for clients. If brokers bought shares for customers but could not collect instantly from those who had purchased them, brokers were often left without

Financial "backroom operations" were cluttered centers of activity in all financial institutions. This is a Bank of America site in California, 1970s. Courtesy Bank of America.

sufficient working capital.[45] Major firms feared going out of business. By the late 1960s, there were over 700 brokerage firms; at the end of the decade the paper crisis had killed off over 200 of them.[46]

The importance of the paper crisis of 1968 to 1969, therefore, cannot be minimized; it was of epic proportions. To be sure, firms had failed to make the kinds of investments needed to handle growing volumes. The exchanges came in for their share of criticisms as well. The NYSE always maintained a public position that it had the necessary computing power to handle growing volumes and blamed its problems on the brokerage firms. In one report on the crisis it argued that "the jam-up began when the transaction left the floor and moved into the offices of the brokerage firms, which have to perform scores of separate operations in order to complete just a single transaction."[47] NYSE's apologists argued that the firms had not anticipated the growth in volumes. In fact, officials at the exchange argued that the NYSE took the lead in guiding the industry to its first industry-wide attack on the problem.[48] Outside the industry, however, the exchanges and brokerage firms were all treated as one dysfunctional sector of the economy.

In 1971, the SEC issued its analysis of the situation, criticizing the entire industry for its poor operational practices: paper-intensive processes, discrepancies in records, misuse of funds, and even theft. By the late 1960s, theft alone had grown to over $100 million in "lost" securities. The SEC accused the industry of having few good managers (most executives had grown up in sales, not operations) and of not having sufficient capital to cover their needs. Sales-heritage executives had invested more in opening sales branch offices than in improving operations. As investors began complaining to the Congress, the SEC and NYSE finally began taking action.[49]

First, since the clearing problem could not be solved overnight, in January 1968 the NYSE and over-the-counter markets (OTC) began closing offices at 2:00 P.M. rather than at 3:30 to allow back-office staff more time to clear that day's transactions; the tactic failed in the face of continuously increasing volumes of transactions, and increasingly complex problems. Second, the Banking and Securities Industry Committee (BASIC) came into existence in March 1970 to start addressing problems over the longer term. One of these problems involved the stock certificates, without which buyers and sellers were reluctant to conduct transactions.

The industry created an organization called the Central Certificate Service (CCS), renamed in 1975 the Depository Trust Company, to handle the necessary logistics, yet even it proved to be only a small improvement. Meanwhile each exchange had been independently implementing various computer-based applications to reduce the flow of paper and speed up reconciliation of accounts and transactions. In July 1972, the NYSE and the American Stock Exchange had started running a jointly owned subsidiary, called the Securities Industry Automation Corporation (SIAC), for the purpose of operating computer applications for both exchanges. It took over the bulk of the IT equipment, data-processing personnel, and communications infrastructures of the two exchanges, and assumed responsibility for performing various clearing functions, beginning with the processing of work for the Stock Clearing Corporation and that of the American Stock Clearing Corporation. In time, it did improve efficiencies, as well as expose a whole new generation of managers in the industry to computers and to their benefits. After setting up the Continuous Net Settlement System (CNS), it began working on a more automated order-processing system, called the Common Message Switch (CMS), which went "live" in March 1976. Next, it added the Market Data Systems for the two exchanges.[50] As several students of the crisis concluded, "the great back-office crisis of 1968–69 forced the previously complacent securities industry to turn broadly to automation for the first time. And this new technology paved the way for the first crude interfacing between various exchanges around the country."[51]

Conventional accounts of the industry almost uniformly hold that there had been little or no use of IT in the 1960s.[52] Yet as described earlier in this chapter, beginning in the late 1950s a few incremental attempts to use computers had begun, a process that expanded throughout the 1960s. Regional exchanges proved more aggressive than the NYSE in deploying computers because they wanted to drive down costs and speed up execution of orders to remain competitive.[53] There developed growing recognition across the industry, however, that just installing computers to speed up existing processes would not be enough, even though the risk of using automation could cost people their jobs. Two commentators on the industry in the late 1960s observed:

> The industry has developed to the point where it can no longer properly cope with growth by merely adding a few people or taking more space or adding another piece of office machinery. It must develop and adopt new ways of doing things. Securities firms are not oblivious to the need. They are now undergoing a quiet revolution in their operations. Most conspicuous has been the wide-scale application of electronic data processing equipment to the whole gamut of paper work operations. But automation offers only a partial solution to the securities industry's problems.[54]

Because brokerage firms interacted with each other (such as in the transfer of stock certificates), much like banks did with checks, improvements in one firm or brokerage were simply not enough; reforms had to occur across the entire industry, which had yet to be addressed by the community. In fact, in the late 1960s, vendors and experts on computing were still explaining to management across the industry how computers could help operations[55] at the same time that the industry was launching a public relations campaign to demonstrate how it was automating.[56]

Bear in mind, however, that this industry had been using telegraphic and telephonic systems for over a half century and continued to upgrade these, which often led a firm to use computers linked to telecommunications by the time third-generation general purpose computers became available in the mid-1960s. Telegraphic and telephonic systems had long provided brokerage firms and markets with reasonably up-to-date price quotes on stocks using ticker displays, and tape printouts. Blackboards were replaced by electronic versions (such as, those from Teleregister Corporation, later Bunker-Ramo Corporation). In 1960, Scantlin Electronic Corporation introduced Quotron, through which brokers could look up current trading data on demand for a specific stock; they used a desktop device that printed the information on a strip of paper. Ultronic Systems introduced the Stockmaster the following year. Attached to a computer, it provided a nationwide fast service offering similar data to Quotron except on a display screen; in 1962 Teleregister did the same. In 1964, brokers could start using small cathode-ray screens (CRTs) to start looking at quote data (using a system called Telequote III), bringing about the birth of the modern image of brokerages having cluttered collections of CRTs. Among the three services, the industry had installed over 30,000 desktop devices in over 3,000 brokerage offices by the late 1960s.[57]

Matching orders became an important application of the digital in the 1960s, as it increasingly merged computing with telecommunications at both firms and exchanges. Teletypewriter systems had historically displayed data as visual symbols on paper or screens. Computers enhanced teletypewriter systems, used throughout the century, by making possible automated message-switching, providing, for example, rudimentary checks for order accuracy, and rejecting obvious error situations without involving clerks or brokers. Large firms in particular began using computers linked to telephone systems in the 1960s. In 1969, Goodbody & Company became the first brokerage firm to admit that its system, which provided its brokers with CRTs to observe transactions, worked satisfactorily. Service bureau applications of a similar nature also became available to smaller firms, using, for example, Western Union's SICOM service. Telephone systems came into wide use by the 1960s, providing similar data by voice. Large capacity telephone networks were adopted across many parts of the industry, often in tandem with teletypewriter applications. By the late 1960s, these telephone systems were transmitting data in digital form and recording data on magnetic tape and later on disk drives.[58]

The pace of transformation of work processes could only speed up, however, if the NYSE picked up the pace. It had been under siege for its inability to handle growing volumes since the early 1960s. In fact, as early as 1963, the NYSE issued a report defending its operations on days when volumes had surged in May 1962. But it had to admit that it just barely handled the workload and announced it would install a computer to speed up data on transactions going to ticker tape, while storing trading data in the order in which they were received, making it possible to produce a variety of end-of-day reports. With typical press release hyperbole, the exchange announced that "this new equipment will mechanize all present manual operations in the Exchange's stock ticker and quotation services. It will be able to handle volume in excess of 16 million shares a day."[59] Over the next three decades, the NYSE installed a variety of computerized applications, but as in 1963, it seemed frequently behind in what the industry or public wanted or needed.

The story of computing at the NYSE in the 1960s and 1970s can be told in one of two ways: either as a tale of continuing introduction of new digital applications over the entire period or as an ugly story of resisting regulatory initiatives by the SEC to create both a national trading market and to reduce the economic (some would argue near monopolistic) power of the exchange. Both versions, however, are intertwined. Together they demonstrate how IT influenced the course of events and at the same time was part of the broader movement across the entire U.S. economy of improving the speed and access to information.

In the 1950s and early 1960s, the SEC essentially allowed the Brokerage Industry to police itself. However, a series of complaints about quality of service and unethical behavior, followed by a scandal at the American Stock Exchange in the early 1960s regarding the sale of unregistered stock, awakened the SEC to the need to play a more active role. Over the next decade and a half it implemented a number of reforms, one of which occurred when it broke the system whereby brokers and exchanges established fixed fees. Mayday 1975, as it was called in the industry, saw fixed brokerage commission rates eliminated, leading to increased competition and, over the next few years, a sharp decline in commissions from approximately 39 cents per share ($1.45 in 1990 dollars) to 3.5 cents per share by 1991. Resistance to the change was so intense that the SEC became, in the process, highly motivated to consider other, more comprehensive changes, some made possible by the simultaneous improvement in technology.

The SEC started to look at the notion that all those interested in buying and selling stocks should have access to information on current prices and be able to place orders through any exchange. This approach stood in contrast to the practices of the 1960s, whereby one had to go through a broker, and a broker could only trade in one exchange at the prices quoted in that exchange. SEC officials began to promote the idea of a system combining computers and telecommunications that would make it possible for the NYSE, the American Stock Exchange, regional exchanges, third and fourth markets, and others to compete among themselves, thereby opening up competition and reducing the need for further regulation. During the early 1970s, when the SEC pushed forward its ideas, the NYSE resisted this as an attack on its sovereignty in the market.[60] The industry gave way partially in 1972 when it agreed to implement a consolidated tape system,[61] and by April 1976, the system became fully operational, operated by SIAC on behalf of the industry. It initially only reported trades that had been executed in the various exchanges, not current offering prices, which meant that brokers still could not compare prices across exchanges. But it was a start. The SEC also wanted a consolidated quote system which the NYSE and AMEX also resisted, since a better price quote at another exchange would cause brokers to take that business transaction there. But the SEC won its point and in January 1978, after a decade of resistance, the system went live.[62]

All through the 1970s, the SEC promoted its idea of a national market system, one made possible by using computers; in theory, one would be able to match and execute orders automatically when prices matched (agreed), preserving, in effect, an auction market structure. While one could debate if the technology of the 1970s would have been up to the task, in the 1980s it did develop the necessary capacity and sophistication. The SEC worked on the project in pieces. For example, it sought a national consolidated limit-order book by which all orders of any given stock at a given price could be entered into one system. This met heavy resistance, though, as it could have cost floor personnel (floor brokers and specialists) at the exchanges their jobs, while the NYSE as a whole would have suffered because its role would have become essentially irrelevant. The exchanges responded with a counteroffer in 1978 (approved by the SEC in 1983) to establish the Intermarket Trading System, which allowed traders to trade in any market, seeking out the best prices for buying and selling, and also brokers to trade in their home exchanges. In short, it was a stop-gap measure made possible by technology that slowed the SEC's desire to create a truly national market in the 1970s.

Meanwhile, the push to provide brokers with real-time data on prices for stocks not listed by the major exchanges had progressed. For example, as early as 1971, the National Association of Securities Dealers (NASD) installated two Univac computers, which displayed instantly prices for many stocks; this created the exchange's offering called the the National Association of Securities Dealers Automated Quotation System, or more familiarly, NASDAQ.[63] For the first time, one could see current quotes for a large number of stocks, regardless of where they were traded, thereby reducing the spreads on prices and hence earnings opportunities of brokers. In the 1980s this system began executing orders in an automated manner.[64] The creation of NASDAQ can be characterized as a major transformative event in the industry because it put it on a path toward a national, integrated system, one whose base infrastructure was digital.

Creation of NASDAQ was well received both by the industry and the press at large. *Forbes* magazine headlined an article on it, "NASDAQ We Love You."

Key journals described its operations and significance, ranging from the highly influential data-processing magazine, *Datamation*, to *Fortune*, a rival of *Forbes*. Even outside the industry a realization had been growing that it needed to automate many of its basic functions, just as was occurring in so many other industries. NASDAQ represented the first example of that change at work visible to the public at large.[65]

Another aspect of the story—what work actually transferred to computing—is a busy one, although one with mixed results. A quick look at the NYSE's website shows a long list of technologies and applications installed over the years; table 5.2 lists those put in from 1957 through 1978. The milestones appear rather tame, however, and in some cases nearly trivial. However, a slightly different chronology of the industry as a whole, in which the NYSE was a central player, offers a different perspective, one demonstrating more substantive IT-related activities at the exchange (see table 5.3).

The NYSE, like the industry it served, had flirted with computing in the 1950s, indeed first used digital technology in 1959, but not until the 1960s did it begin implementing computers to attack back-room processing problems that kept growing. Along with other exchanges and firms, in the 1960s it, began using computers to tabulate results of sales and to perform other accounting functions. Computers made it possible to record transactions faster than older methods that used office calculators, as well as to collect more data. Optical scanning equipment worked side-by-side with punch-card machines to collect trading data. In late 1966, the NYSE's president, Keith Funston, who had led the charge to introduce

Table 5.2
Technological Milestones as Reported by the NYSE, 1957–1978

Year	Event
1957	Ebasco Service reported—first report suggesting possible applications
1964	900 Ticker installed—replaces black box ticker, at twice the speed
1965	Radio paging system installed—eventually replaced annunciator boards
1968	Central Certificate Service—CCS established to transfer securities electronically, later done by Depository Trust Company
1971	Securities Industry Automation Corp. organized—SIAC provided computing services to NYSE and Amex
1972	Depository Trust Company established—centralized storage of paper securities and electronically transferred stock ownership
1976	Designated Order Turnaround initiated—DOT electronically routed small stock orders
1977	Intermarket Trading System began—ITS linked electronically to NYSE and other exchanges, making it possible for brokers to access all markets for best prices before buying or selling a security

Source: "Historical Perspective, Timeline at a Glance," "Technology," http://www.nyse.com (last accessed 3/7/2005)

Table 5.3
Major IT-Related Events in the U.S. Brokerage Industry, 1959–1978

Year	Event
1959	IBM 650 installed to handle stock clearing and other services
1963	SEC published scathing report on the industry's operations, called for automation
1964	Quotation Service introduced as the first application of the Market Data System
1965	Market Data System began reporting regularly on trades
1968	NYSE established Central Certificate Service (1972 called Depository Trust Company)
1968–69	Paperwork crises in most brokerage backroom operations
1970	NYSE member Donaldson, Lufkin & Jenrette became first brokerage firm to offer its own inventory of stock to buyers, challenging NYSE exclusive control of member-selling activities
1971	SEC announced it wanted consolidated tape and quote systems and national market system
1971	NASDAQ system for over-the-counter quote displays launched
1972	Chicago Mercantile Exchange given government permission to trade futures contracts on its International Monetary Market NYSE and Amex established Securities Industry Automation Corporation to develop computer-based support systems First money-market mutual funds established
1973	Chicago Board of Trade started trading options
1975	Mayday, when NYSE fixed brokerage commission rates practices terminated Consolidated tape system implemented Congress passed amendments to 1934 Securities Act, calling for "linking all markets"
1976	NYSE introduced Designated Order Turnaround (DOT) electronic transmission system
1977	Merrill Lynch launched Cash Management Account, using computers
1978	Composite quote application launched Cincinnati Stock Exchange first to launch a fully electronic execution system

Sources: Drawn from data in Marshall E. Blume, Jeremy J. Siegel, and Dan Rottenberg, *Revolution on Wall Street: The Rise and Decline of the New York Stock Exchange* (New York: W. W. Norton, 1993): 270–272; NYSE, *Automation in the Nation's Marketplace* (New York: NYSE, undated [circa 1965/66?]): 11.

computing to the exchange, proudly announced that his Market Data System had captured all trading data which was then displayed on the ticker, ensuring this application received considerable press coverage.[66] The NYSE next began working on trying to partially automate clearing and settlement steps, along with such other tasks as accounting and odd-lot switching. In February 1969, the exchange began tracking certificates for brokers as an early attempt to reduce, then eliminate, the need to transfer physical stock certificates as part of a sale, providing an electronic processing for clearing. By the end of 1972, about 75 percent of all transactions were being done without having to create and transfer certificates.[67]

By the end of the decade, the exchange had begun providing an automated electronic book to replace the manual system of tracking prices of stocks performed by the floor "specialist," signaling the beginning of the end of a two-decade long internal battle by floor personnel to protect their jobs.[68] Other initiatives included the Automated Trading System (ATS), launched at the start of the 1970s, to collect orders at the edge of the trading floor so as to eliminate the need for brokers to be on the floor itself when they had orders of 100 or more shares. The system was automated unless a broker asked a specialist to "stop" a stock, which was a request to guarantee a current price but delay execution, done in hopes of getting a better price. ATS looked much like the one just implemented by NASDAQ. The NYSE's first attempt to implement this application led someone to come in over the weekend before the pilot was to go live and damage the equipment. While that event did not stop its launch, the system was hardly used.[69] Floor personnel simply continued to fight and block attempts to automate their work.

A second application, called the Block Automation System, was intended to facilitate brokers selling large blocks of stock to each other, much the way a new system in 1969 did that had been introduced by Instinet, a private system competing with the NYSE. This second initiative failed because specialists flooded it with successive buy and sell orders, rendering it useless. Furthermore, it only covered stocks traded on the NYSE, whereas Instinet handled all publicly traded stocks. Students of the exchange later noted that the system failed because employees of the exchange did not want it to succeed, specifically floor traders, specialists, and even some brokers, as it would have allowed any brokerage firm to bypass exchange members to trade directly.[70]

One major system, called DOT (Designated Order Turnaround), did prove highly successful. It was born out of concern over ComEx, an automated system for handling small orders developed by the Pacific Stock Exchange, and a similar system under development at the Philadelphia Stock Exchange. Until DOT went "live" in 1976, floor brokers walked all orders across the floor of the exchange to the specialists' booths, where the floor brokers bid against each other and the specialists to get the best possible price for their clients. DOT allowed a broker to send in an order electronically from his or her firm's computer directly to the specialist, who then either executed it against a quoted price or obtained a better price for it. After the order was executed, DOT transmitted a confirmation back to the originating broker, who then in turn telephoned his or her client with the good news. In time, the application further "automated out" human intervention. (In 1990 a successor system called SuperDOT began handling pre-opening market orders and post-opening orders as well.[71]) By 1983, DOT had made it possible for investors to post orders of fewer than 100,000 shares to the NYSE's electronic order book. In the 1970s and early 1980s, other exchanges had already sped up their trading processes, making that of the NYSE seem slow and clumsy, even costing brokers since prices could change in the midst of executing an order. In time, DOT had become a core application of the exchange, in effect moving the NYSE from being an auction market to being a dealer's market by eliminating the crowd of brokers competing on the floor at any particular post. With DOT, the specialists could execute orders on the electronic book, or against some other order they had pending, or even against their own personal inventory. DOT also made program trading possible for the first time. That meant an investor could trade dozens or hundreds of stocks simultaneously using computers because DOT and SuperDOT could send large numbers of orders to the floor for immediate execution.[72] Program trading became a major brokerage process in the 1980s, when all brokerage firms, and, in the 1990s, day traders, routinely used computers to buy and sell stocks.

Brokerage Digital Applications (1976–1998)

The history of computing in the Brokerage Industry over the next two decades is a tale of widespread use of technology which, in turn, stimulated major changes in the work of the industry. In 1976—to pick an arbitrary date half way through our period of study—the industry still functioned essentially as it had over the previous two decades, despite the introduction of new IT applications and the addition of more speed and function to traditional telecommunications systems. As automation and IT-based access to data expanded across the industry, and as individual investors started using PCs and then the Internet, the brokerage business began changing again. It also continued to experience enormous growth as increasing volumes of shares were traded and a larger proportion of the American public became stockholders. At the NYSE alone—still the nation's largest exchange—volumes grew at a compound rate of 15.3 percent between 1975 and 1990. Figure 5.1 shows stock market volumes for the NYSE and NASDAQ for the period 1985 through 2002, again documenting the expansion underway. These various sets of data suggest how the closed club of the industry came under enormous pressure to open up from both the public and from regulators and elected government officials, all during a period when technological innovations could facilitate more brokerage tasks than ever before.

Computing increased as functions were added and technology improvements exploited. DOT, for instance, which began operation in March 1976, expanded to the point where, late in the 1980s, it handled on average over 35,000 orders per day, accounting for roughly half the orders executed daily at the NYSE. As DOT became more vital to the exchange's operations, it had to be upgraded, if for no other reason than to handle growing volumes. Its sequel, SuperDOT, introduced in 1984, initially could handle orders of up to 5,099 shares (as opposed to 100 in the beginning with DOT) and more multiple orders than DOT. The exchange integrated into SuperDOT another system, called Opening Order Automated Report Service (OARS) to collect and store opening orders sent daily to the floor so that specialists could establish that day's opening prices. Frequently

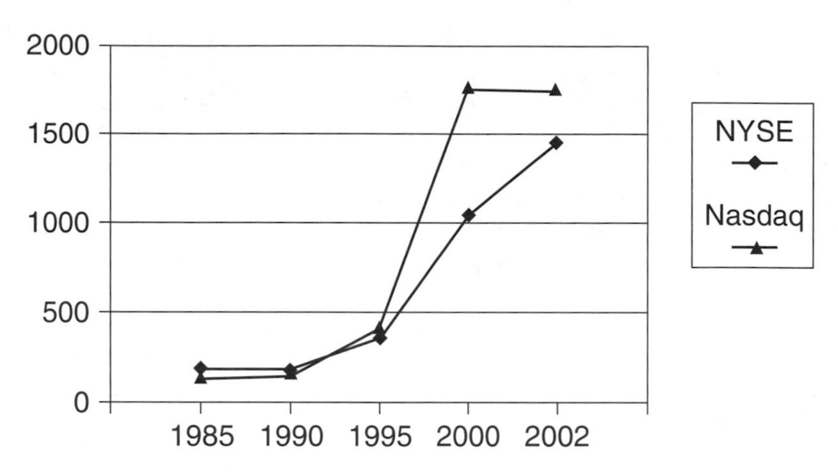

Figure 5.1
Stock market volumes, 1985–2002 (daily average millions of shares).

20 percent of a day's orders in the 1980s were placed at the start of trading in the morning, so the system clearly facilitated the work of the specialists. Despite this additional capacity to handle orders, once again the exchange was rudely shocked when, on August 1, 1982, over 132.6 million shares were traded, straining people and systems to the near breaking point. To alleviate capacity issues the exchange also expanded its trading hours in response to competition from other exchanges in both the United States and around the world. In 1985, for example, it started trading a half-hour earlier than before, and in 1991 it added a 45-minute "after hours" trading session.[73]

Brokerage firms continued to expand their internal IT capabilities as well to maintain current customer accounts and gain online access to that data, along with information from the major exchanges. Key applications by the end of the 1970s had already involved order handling (purchases and sales), cashiering, managing margins, updating stock records, issuing dividends, managing proxy duties, and opening new accounts.[74] The introduction of personal computers into the Brokerage Industry in the 1980s brought computing capability to the desktop, and by the early 1980s, the largest brokerage firms were using them to perform trades and order processing and, by the end of the decade, to offer cash management services to clients. As mutual funds became popular in the 1980s, additional IT applications were added to manage these and to perform necessary back-room accounting. Trading and clearing functions remained the centerpiece of IT developments across the industry throughout the 1970s and 1980s.

In the mid-1980s, brokerage firms were spending over $400 billion a year on IT. In the early 1980s, only the largest firms were investing heavily in computing. But that practice spread across the industry and to smaller firms to the extent that annual expenditures grew threefold by the early 1990s over the those of the 1980s. Online access to information had become the one feature of IT in the 1970s and

1980s that caught the attention of every level of the workforce in the industry, which drove growing amounts of expenditure, especially as the years went by:

> Online trading capability was management born and bred. It was designed to quench the thirst of traders, institutional salespeople, and account executives to have at their fingertips an accurate, up-dated trading inventory of any security. Functionally, in an online system, trades are recorded immediately right through to order processing, and the inventory is updated right away. In some online trading systems, securities firm branches are capable of accessing the trading inventory.[75]

The PC helped move along the process by placing on a broker's desk a variety of data drawn from exchanges, in-house systems, and vendors supplying financial and other economic information. One commentator noted in 1988 that the PC "is perhaps the most significant communications and information dissemination development that is likely to hit an industry in so short a span of years."[76] During the 1980s, a variety of software products became available that helped firms and individual investors. One survey of 25 of these software products, written in 1986, described them as applications with which to manage individual portfolios of investments, to access online databases, and, of course, to price data on stocks that had previously been the virtual monopoly of the industry's professionals as late as the start of the 1980s.[77]

By the early 1980s, a whole new set of applications began appearing in the form of information products from firms specializing in collecting financial and economic data useful to the entire financial sector. For example, Standard & Poor's established Compustat Services, Inc., to sell information in machine-readable form that by 1986 covered 6,400 companies; its initial offering was a magnetic tape, but later it went online. Stock data also became available in machine-readable form from the Center for Research and Security Prices at the University of Chicago, and the *Media General Financial Weekly* built a database which, by the late 1980s, had information on over 3,000 companies. Interactive Data Corporation (IDC) became one of the more popular providers of data, both in batch form (that is, in a non-online form) and later online. It collected daily price and volume information from both the NYSE and AMEX, and later from OTC transactions. IDC's service was a time-sharing offering to firms and individual investors that helped continue the historic opening up of the Brokerage Industry to ever-larger databases of financial and economic information.[78] There were dozens of such offerings aimed at two audiences: the professional portfolio manager at a brokerage firm, bank, or financial advisory company; and the individual investor working from his or her home PC.

Two snapshots of available applications suggest how computing had expanded across the industry. IBM catalogued for its employees its products in 1988. For branch-office functions (where retail customers buy and sell securities through their brokers), IBM used PCs and had relations with over a dozen software business partners providing brokerage trading tools.[79] The 1999 NYSE *Fact Book*, which highlighted its previous year's activities and transaction volumes also catalogued

Table 5.4
Major IT Applications at the New York Stock Exchange, Circa 1990

SuperDOT—Electronic order-routing system used by member firms to transmit market and limit orders to NYSE and report back executed orders. In 1990 it processed an average of 138,000 orders per day for 182 subscribers. Considered the umbrella application for others listed below.

Opening Automated Report Service (OARS)—Accepts member firm's pre-opening orders, comparing buyers and sellers, presenting imbalances to specialists to assist in setting opening bell prices.

Market Order Processing—Designed to execute and report market orders from member firms. In 1990, 98.5 percent of all orders placed through the system were completed within two minutes.

Limit Order Processing—Designed to collect and execute orders when specific price limits are reached.

Electronic Book—Used by the specialists to handle large volumes of transactions. Uses a database to record, report, research market orders; reduces paperwork and errors. In 1990, contained roughly 2,000 stocks.

Post Trade Processing—Compares orders received with those processed, providing audit trail prior to settlement and delivery of stock securities.

Source: NYSE, *NYSE, Fact Book 1991* (New York: NYSE, 1991), also available at http://www.nyse.com, 21.

its major IT applications (see table 5.4). These applications had become uniformly available at other exchanges in the 1980s, some even in the 1970s.

However, the NYSE did not always lead in the use of technology. One observer described the problem in the 1980s from a broker's perspective:

> Thanks to the new automated execution systems introduced at most of the regional exchanges, a single telephone call gave a broker confirmation of an order within a minute. But on the New York Stock Exchange, where the order had to pass through a mobocracy of floor brokers and specialists, the process still took five minutes—a long interval to a retail broker. And the wait was just part of the problem: Typically, brokers trading on the Exchange would have to make three phone calls to complete a market order: first to obtain an order from their client, next to confirm a selling price and obtain a purchase order to use the proceeds of the sale, and finally to confirm the purchase with their client.[80]

Yet the management of the exchange was not usually trying to block modernization. The normally implacable resistance of floor personnel and even sometimes brokers slowed the deployment of fully automated systems in the 1970s and 1980s. To address this, senior managers of the exchange launched an initiative called R4 (Registered Representative Rapid Response) to execute small orders at the best price available on any U.S. market without having to send it to the floor of the exchange. This time, however, the SEC feared such a system might severely limit

Merrill Lynch & Company installed PCs in 1988 to process securities transactions and collect and use other data. Picture is of a brokerage office in New York City. Courtesy IBM Corporate Archives.

the competitive position of the smaller exchanges, as well as change the nature of trading at the NYSE away from an auction-oriented market, which R4 did not provide for, and possibly damage too severely the exchange's liquidity. In 1982 the SEC gave the exchange permission to run a six-month test, but it proved too cumbersome and confusing for many brokers, so it quietly died.[81]

The trading process remained in stasis throughout the 1980s, although the exchange invested some $600 million during the decade to upgrade all its computers, software, and telecommunications. Only in the 1990s did the exchange start converting its ways of doing business, as opposed to simply automating existing practices. In fairness to the NYSE, we should acknowledge that the same problems existed to one degree or another in all the other regional exchanges. As late as 1990, for example, the Chicago Board of Trade called for upgrades in its IT infrastructure but warned against making any fundamental changes to how trading was conducted within the exchange.[82]

What was happening with the NASDAQ? Recall that in the early 1970s the SEC began advocating for the creation of a national market for trading securities. Ezra Solomon, a professor of finance at Stanford University, noted at the time:

> [B]rokers will have to be freed from the constraints with respect to where they execute a trade for a customer, including constraints on the size and source of their own commissions. In short, rules will have to be devised which promote the idea that any customer can and will get the best trade available on his broker's home exchange nor just the best trade for which his broker can earn a commission.[83]

When NASDAQ came into existence in February 1971 to address the needs of the over-the-counter market, it provided an initial step in the direction of a national market structure. Over the years it built a series of digital applications in support of the OTC market, starting with its first system in 1971, which was a computerized trading system whereby market makers could transmit quotations electronically to those interested in participating in the market. Next, NASDAQ added what it called the Small Order Execution System (SOES), with features similar to SuperDOT, in which all registered NASDAQ market makers had to participate and through which all stocks listed by NASDAQ could be bought and sold. Upgrades included a limit-order matching feature, whose primary audience was retail customers.

By 1990 the NASDAQ had already become the third largest exchange in the world, after the NYSE and the Tokyo Stock Exchange. One of the key reasons for its rapid growth had been its relatively successful deployment of IT over the years. Such a computer-based trading system was creating a de facto national market along the lines originally described by Professor Solomon.[84] But not all NASDAQ stocks were traded this way, even as late as 1990. At the end of that year, 4,132 stocks were traded on the NASDAQ, along with another 2,576 one could get to through its system, excluding some 46,000 stocks that were frequently of companies that had gone bankrupt, were shells for corporations, or even were fraudulent operations. So conversion to digital ways was not fully completed but nonetheless, the NASDAQ made available the majority of stocks traded in the United States.

The Brokerage Industry came into the 1990s having finally implemented a variety of core digital applications. In the years immediately preceding the entry of the Internet into the industry, business volumes had exploded, extensive automation had occurred, and use of computing and telecommunications had become relatively ubiquitous. Large brokerage firms had equipped their branch offices with online systems to look up stocks and manage accounts. Refinements in the early 1990s involved filling in missing applications or augmenting capacity and functions. In short, by now the industry had extensively embraced the digital, even if many of its basic processes remained essentially the same as ten, even twenty, years earlier. Pressure to conduct around-the-clock trading and to link to exchanges in other countries added more incentives for the industry to rely on technology to meet the demand of brokers and clients. Various technologies were having new effects on the industry, however:

> What has changed in recent years is not the existence of intraday volatility, but the speed with which it is communicated to large numbers of people. Computers, television, and other methods of information transfer have heightened the awareness of the public to stock market volatility, even if the behavior has not changed dramatically.[85]

The stock market crash of October 19, 1987, when the Dow dropped by over 500 points, compelled both the SEC and the industry at large to implement safeguards (called "circuit breakers"), including procedural and software brakes that could slow or stop programmed buying and selling of stocks following large changes in the Dow index.[86]

Leaving aside for now the role and effect of the Internet on the industry, other issues capturing the attention of the industry need acknowledgment. For instance, an extensive debate occurred in the late 1980s and early 1990s regarding the effect of computerized trading that linked stocks, options, and futures markets. But it remained an IT-centric discussion since it did not radically change behavior in the industry.[87] However, since the NASDAQ also received considerable criticism for how the markets performed in October 1987, it reformed its electronic trading system for small orders in the early 1990s and, along with the rest of the industry, implemented breakers. Thanks to the Internet and media coverage of the industry, the public became more aware of the detailed movements of stocks. As a secondary by-product of so many individual Americans owning stocks, pressure increased for more access to information, often delivered through inexpensive American telecommunications, the Internet, and mass media, including TV. The crash in 1987 energized the SEC to assess the industry's overall handling of IT, setting agendas for the watchdog agency for the early 1990s. This time other branches and agencies of the federal government began assessing the industry, particularly the U.S. Congress and the U.S. General Accounting Office, a further signal that once again there was growing concern about the effectiveness of this industry's use of IT.[88]

One final event needs to be mentioned: Y2K. Like the banking and insurance industries, the Brokerage Industry perceived a crisis if its systems did not function after the end of 1999, and for the same reasons: old software that could not handle the date change, hardware with internal clocks that also might not manage the data change, and the extensive use of COBOL over the previous 35 years. The SEC proved as aggressive in demanding that this industry modify or replace all its software as any other government agency and, in fact, went further by demanding that the industry test all its key systems and have them pass a Y2K compliance test long before January 2000. The media exacerbated pressure on the industry, which paid more attention to the Y2K issue's effect on the stock market than on any other industry, including banking and insurance. The Brokerage Industry in the United States dutifully assessed its software, remediated many systems, and hired outside programming contractors to help meet its deadlines.

Fixing Y2K took several years to complete at a cost of billions of dollars—this at a time when the industry needed to invest in modernizing and expanding existing systems to handle a more globalized brokerage industry, and when transaction volumes were rising faster than the exchanges and brokerages houses could easily handle. In 1997, the industry spent $12.7 billion on IT, and its members knew that expenditures would rise to at least $16 billion by 2000, making this the second largest noninterest-related expense after employee salaries in their budgets. Important projects competed with Y2K for resources and management attention in the late 1990s, however. These included converting systems to handle the anticipated adoption of the European Euro; switching quotations from fractions to decimals (decimatization); enhancing the National Securities Clearing Corporation's Automated Customer Account Transfer System (ACATS), which processed intercompany customer account transfers; introducing new accounting standards for handling commodity derivatives; and implementing the Order Audit Trail System (OATS) to transmit NASDAQ equity order, allocation, modification, cancellation, and execution data to the SEC. Y2K had the distinction of being the one project that absolutely could not be late. Thus, the industry faced a variety of complicated changes in technology (both hardware and software), applications, processing capacity, and customer demands all at the same time. As a whole, the industry devoted over half its application development and maintenance resources to Y2K by 1997, making it the single largest IT initiative in over a quarter of a century. Work had begun in the mid-1990s, peaked in 1998, and was field tested and finalized in 1999.[89]

Converting old programs to be "Y2K compliant" was not complicated; it just took resources to grind through the work. As a result, for the first time in its history the industry used contract labor in massive quantities. One study estimated 42 percent of all project development resources were devoted to the initiative.[90] Some 5,400 brokerage firms were involved in the effort. Since they all linked electronically either to other firms or to exchanges, they were required to fix any Y2K problems they had and to test their applications by exchanging specified amounts of data with other members of the industry. It would be difficult to underestimate the extent of the effort involved because of how electronically interconnected the industry was (and is) to banks and insurance companies, all of which also had to be compliant. To further complicate matters, "Y2K compliant" included all the government agencies that worked with these industries, such as the Federal Reserve Bank and the SEC itself. All three industries met their deadlines and on January 1, 2000, there were virtually no problems in the American financial sector.[91]

The consequences of this enormous effort only became evident in the early 2000s. First, since many old software programs and machines had been eliminated, replaced, or upgraded by 2000, there was very little "tired iron" or "old code" left in the Brokerage Industry. That proved so much the case that when the industry suffered through another recession in the early years of the new century, it at least did not face as many capital or excessive operating expenditures for maintenance of existing applications as it had half a decade earlier. Second, since Y2K expenditures came at a time when the market was roaring with success, one could point out that the industry was not burdened with a massive financial hangover from its Y2K experience. Third, along the way incremental improvements were made to major applications, many of which were intended to enhance connectivity to other global financial markets so as to provide more integration of various products and accounts, much as happened at the same time in both the banking and insurance industries. Account and transaction settlement processes remained a major problem at a global level, although largely fixed within the U.S. market. In summary, the industry entered the new century with a relatively refurbished, up-to-date infrastructure of IT technology, applications, and telecommunications in all segments, from regulators to exchanges, from brokerages to individual customers.

Role of the Internet in the Brokerage Industry

If one technology held out the promise of fundamentally disrupting long-standing patterns of behavior and practices in the Brokerage Industry, it was the Internet

Table 5.5
Online Financial Services, Circa Early 1980s

Charles Schwab & Co., Inc.	Modio	Telescan
CompuServe	Newsnet	Vickers On-Line
Dow Jones News Retrieval	Nite-Line	Warner Computer Systems
FCI-Invest/Net	Radio Exchange	
Fidelity Investor's Express	The Source	

Source: Thomas A. Meyers, *The Dow Jones–Irwin Guide to On-Line Investing: Sources, Services and Strategies* (Homewood, Ill.: Dow Jones-Irwin, 1986): v–ix, xi–xiii.

with its potential access to ever-growing amounts of information and to interactive applications hitherto only used by members of a brokerage firm. While it is widely accepted that the Internet began changing the nature of American work beginning in the mid-1990s with the arrival of tools making it easy to use this telecommunications network, significant transformations driven by the marriage of telecommunications and computing had been underway since the 1980s.

Prior to the wide availability of the Internet, services existed that one could subscribe to by dialing up a telephone line either maintained by a vendor or that relied on an 800-number. Users needed communications software that made it possible for PCs to interact with the online service. These were offerings that one subscribed to and that gave an individual access to a variety of services and sources of information. Table 5.5 lists some of the most widely used ones in the mid-1980s in the United States. [These existed in other industries as well, allowing one, for example, to purchase computer equipment online (e.g., Comp-U-Store) or to keep current with sports news in the 1980s (e.g., Sports Report).] But the number of users remained small and early attempts to start online brokerage services were just stirring and had not yet become widespread.[92]

One of the earliest online brokerages to succeed was DLJdirect, which connected over telephone lines and through investors' own PCs, beginning in 1988.[93] But it was not the first; earlier in the decade Charles Schwab and Fidelity began offering similar services. The case of Charles Schwab is instructive because it illustrated how IT could, in combination with nontechnical circumstances in an industry, bring about important changes. Table 5.6 lists major events in the company's history which led the firm to become so reliant on digital applications. When fixed commissions for stock trades were eliminated on May 1, 1975, this firm decided to provide discounted services. In 1976, customers could telephone a call center, ask brokers for quotes on various stocks (the brokers used Quotron), and place orders—all at fees lower than a traditional full-service broker would charge. By 1978, the firm was computerizing its basic operations to drive down costs, making it possible for the firm to charge even lower discounted fees and still generate profits. It began operating a back-office settlement system as its first major application in 1979, running on an IBM 360 system. Next, it installed an online

Table 5.6
IT-Based Milestones at Charles Schwab & Co.

1976	Uses Bunker Ramo System 7 to provide stock quotes to clients
1977	Establishes telephone call center
1979	Acquires back-office settlement system running on IBM 360
1980	Introduces 24-hour weekday stock quote service
1982	Introduces Schwab One Account, a cash management service
1985	Introduces three online services: The Equalizer, SchwabLink, SchwabQuotes
1988	Launches Advisor Source as a referral services
1989	Introduces Touch-Tone trading and quote applications
1993	Replaces Equalizer with StreetSmart online trading system
1996	Launches Internet-based trading (eSchwab)
1997	Opens one millionth online account; makes SchwabLink available on Web
1998	Schwab.com launches; reaches 1.8 million online accounts
1999	Introduces online After-Hours trading, Retirement Planner, MySchwab
2000	Launches additional foreign Internet-based services
2002	Launches Schwab Advisor Network

Source: John Kador, *Charles Schwab: How One Company Beat Wall Street and Reinvented the Brokerage Industry* (New York: John Wiley & Sons, 2002): 273–277.

order entry system, considered the first such application in the industry. Customers were subsequently given direct access to the online order system, thereby bypassing the need even to talk to a broker. The system collected data on customers, such as their buying habits, which informed management about potential opportunities, trends, and so forth. In the early 1980s, all major brokerage firms in the United States were implementing similar systems. Schwab grew all through the 1980s; computing made it possible for it to offer various products and make large amounts of information available to customers. Schwab also had a network of branch offices, much like banks. A handful existed in the late 1970s, over a hundred at the end of the 1980s, nearly 400 at the end of the 1990s and over 430 by 2003.[94] All had access to Schwab's growing suite of applications.

The story of this brokerage firm is important because it demonstrated to the rest of the industry how technology could be used to attract and service larger numbers of customers while driving down operating costs. The latter became urgently necessary once the system of fixed fees went away. The progression from a telephone to PCs, then to a combination of telephones and PCs, and finally in the 1990s to Internet access to brokers and information sources traces the historic transformation of the Brokerage Industry from a relatively closed society into a more porous industry.

The 1990s saw online investment activities shift to individual investors who could now could dial into a brokerage service and place orders, buy various financial products and services, and manage their own accounts. In 1995, the first online

Internet-based service was introduced, and by the end of 1997 there were over 9 million online accounts (although not all were active) conducting over 500,000 trades daily. A large portion of that business consisted of day traders—people who would buy and sell stocks, frequently in one day, and make and lose money based on the variations in prices of a stock. Day traders relied on Internet and PC access and information at thousands of sites on the Web.[95] The numbers involved became quite astonishing. In 1995, there were approximately 300,000 online accounts; the following year that number jumped to 1.2 million, and the next year to 3.2 million. In 2000, the SEC reported that 7.8 million people traded online for an average of 807,000 shares each day. In that year, over 200 stock brokerage firms offered online trading services to its clients, though only five collectively dominated over half of the Internet stock market: Charles Schwab (27.5 percent), E*Trade (12.9 percent), TD Waterhouse (11.6 percent), Datek (10 percent), and Fidelity (9.3 percent).[96] Collectively they supported several million clients doing over 300,000 daily trades.[97] Thus, the Internet, in combination with PCs and the willingness of some brokerage firms to exploit the technology, had fundamentally initiated the restructuring of the Brokerage Industry.

Regulators and industry insiders were aware of what was happening. One New York state official reported in 1999 that

> online trading is revolutionizing the securities industry in several critical ways. The advent of the online brokerage industry is fundamentally changing the relationship between broker-dealers and their customers by allowing individuals to manage their own investments in a manner never before possible. Members of the public can now readily access a range of sophisticated research materials and financial data—once available only to market analysts—directly from the websites of most online trading firms. Equipped with this information, investors can, if they choose, make an independent evaluation of stock performance. Investors can also place their trades without the assistance of a registered securities representative by entering an order and transmitting it to an online firm for execution by traditional mechanisms.[98]

As in so many other industries, brokerage firms and exchanges did not start with such interactive systems. Rather, they established websites on the Internet which provided basic information about offerings and how to contact them. As late as 1997, only half had websites, although just over 40 percent of those who did not stated in one survey that they planned to have one by the end of the year, and another 21 percent by the end of 1998. The same survey reported that every brokerage firm with 4,000 or more employees had a website, as did 84 percent of those with between 500 and 4,000 employees, indicating a ubiquitous technology.[99] The sites evolved rapidly, however, from simple sources of information with no interactive capability in 1995–1996 to order-routing systems by 1998, though they still required the intervention of a broker to actually execute an order coming in over the Net. As late as 1997, most investors were still using these sites primarily to conduct research and gather market data, but that practice began changing by mid-1998 as online order systems became more direct and interactive.[100]

An old nemesis once again made its appearance, though, despite efforts by this industry to build digital infrastructures: it had problems with capacity. Online trading volumes reached a record high of 499,476 trades per day in the first quarter of 1999, an increase of 47 percent in volume over the previous quarter. During the same period, general market volumes only grew by 4 to 5 percent, yet in that first quarter of 1999, about 16 percent of transactions took place online, resulting in what one New York official characterized as "a rash of outages." He went on to report that "most of the large online brokers had some sort of technological difficulties during the first few months of 1999."[101] A list of examples read like a *Who's Who* of the industry. Crashes were common at all the brokers' sites at the time, leading New York state officials to investigate and then to pressure the industry to fix its problems, which also included lost orders and the inability of customers to sell at prices they wanted.[102]

Yet the historic shift to online services continued. In 2000, online brokers managed in excess of $800 billion of their customers' assets, which represented about a third of all retail stock trades that year. They handled more then 10 million accounts that year. Full-service brokerage firms also launched online brokerage services, starting with Morgan Stanley Dean Witter and Merrill Lynch in 1999. At the dawn of the new century, online brokerage still remained highly concentrated with less than ten firms accounting for roughly 95 percent of online brokerage assets.

Why had the Internet become such an attractive technology? Economists Eric K. Clemons and Lorin M. Hitt argue that this technology represented a good fit for the industry:

> Customers require a great deal of timely, text-based, and numeric information that can be easily delivered through a website. From a retail investor's perspective, the trading process is relatively standardized, with the actual transaction typically requiring no intervention by a market professional. The previous generation of technology—direct phone calls to order taking brokers—was fraught with inefficiencies, such as errors in the communication of orders, limited ability to authenticate the customer, access problems . . . , and overall high costs, both to the customer and to the firm.[103]

They noted that the ability to unbundle services from transactions gave firms the flexibility to charge less for the latter. When increased volumes were stimulated by the lower fees, it became possible for firms to offset their costs of the technology and its applications that otherwise management would have perceived as too expensive. Thus the firms finally were able to achieve important improvements in productivity in their front offices as a nice complement to the increases they achieved in back-office operations in the 1980s and early 1990s. Back-office productivity also benefited from the volume of scale-intensive operations during the great boom in the industry at the end of the century. Acquiring new customers, however, remained very expensive, even in this new environment.[104]

The implications were already evident. Perhaps the most obvious was the shift in information balance-of-power to individual investors. While commenting on the role of the SEC, William Lyons, of American Century Companies, noted that was changing as early as 1998: "Most of our federal regulation . . . is built on the premise that there should be a minimum amount of information available to investors to guide decisions. Now, if anything, we have an excess of information and surfeit of information."[105] The dark side, of course, as another of his colleagues noted, was that "on the Internet any firm can look enormous, research can sound credible, individuals can seem qualified, all without any real foundation."[106]

Both trends were becoming realities in the marketplace. The Internet had brought increased access and convenience to all players in the investment market, lowered the cost of access to information, and pushed further globalization of trading. Online brokers were in the ascendancy, transforming their Internet-based services into financial hub sites to provide their clients with even more information, forcing old-line brokerage firms to expand their advisory services to add value. And they were winning. A 1999 study of the trend provided credible evidence that online brokers were spending less than the old-line firms to acquire and retain customers.[107]

Recent Trends

In the late 1990s, if we had wanted to identify recent trends in the industry involving the use of IT, we would have noticed that firms were focused on managing existing large enterprise-wide systems used to handle trading, risk management, clearing, customer support, and accounting. By the end of the century the list would have additionally also included several items: delivery of products and services directly to customers and organizations over the Internet, and, of course, Y2K. In 2000, however the Bull market became a Bear, and projects slowed.[108]

Then a catastrophe occurred. On September 11, 2001, many brokerage firms in New York—the financial capitol of the United States—lost their large operations centers, the back offices of their companies, in the World Trade Center attacks. The telecommunications infrastructure used by the industry and the local telephone exchange in New York also ran through the World Trade Center and was destroyed. The New York Stock Exchange was located several blocks away and so its IT operations were not harmed, but the entire market was shut down by the U.S. government. One of the early concerns was the extent of the effect of the terrorist attacks on the nation's financial system. The SEC's priorities were to get the industry up and running and then to make it more secure from future attacks. It achieved the first goal within a week as telecommunications and computer vendors, along with IT employees across the industry, quickly improvised, setting up data centers in New Jersey, switching to backup sites, and the like. Harvey L. Pitt, chairman of the SEC, reported to the U.S. Congress just 15 days after the attack that the U.S. capital markets were the strongest in the world and "most resilient" and that the biggest problem was reconnecting telecommunications.[109]

The SEC, brokerage firms, and exchanges, spent the next two years working on issues related to backup and recovery of IT and telecommunications infrastructures.[110] A key initiative on the part of the SEC was to ensure the industry had sound management of IT, indeed, prior to 9/11 the SEC had expressed concern about various practices of online brokers and had begun examining what new regulations it needed to implement in response to the growing use of the Internet.[111] These initiatives were subsumed in the overall actions taken in the wake of 9/11. All of this took place against a backdrop of an economy in recession, then anemic recovery, and another round of industry scandals. The NYSE, however, continued to experience growth in volumes, which necessitated investing in additional IT equipment, permitting as much as a 50 percent increase in transactions while providing offsite backup and redundancy for existing applications. In 2002 the NYSE launched OpenBook and the NYSE Broker Volume services, and in 2003 Liquidity Quote, all of which provided customers additional access to information and options for executing orders.[112]

As the industry entered 2004, most of the Brokerage Industry's recovery issues related to 9/11 had been resolved. Application development returned to integrating various systems to provide enhanced customer services.[113] As usual, the largest firms in the industry led the way. For example, Merrill Lynch and Smith Barney each committed $250 million to develop a desktop system for its brokers, while Prudential and PaineWebber each allocated $150 million and Dean Witter just slightly less. These five firms alone invested $800 million in just one application—a stark contrast to the early 1970s.[114] Furthermore, all these innovations had to be available on a global basis.[115]

The NYSE once again became embroiled in discussions about the extent to which it should automate floor trading. Complaints about slow orders processing were resurrected. Reflecting the tactics of a former chairman in the 1960s, the NYSE's chief in 2004, John Thain, championed the elimination of the 30-second rule, caps on volumes, and other practices that inhibited speedy transactions. However, as in prior decades, no radical recommendations were brought to the board of directors, even though NASDAQ and ECNs had been trading electronically for years. As in prior times, the SEC began revisiting its guidelines, particularly in light of a series of scandals in the industry in 2003 and 2004 involving mutual fund trading across the industry and a highly overpaid chairman of the NYSE.[116]

On April 21, 2005, the NYSE announced that it was merging with Archipelago, a leading electronic stock trading company, although the details about when that would occur were not described. This turn of events was of historic proportions, as it would bring the NYSE into the mainstream of how exchanges were now using computers and communications. The transformation would allow the NYSE to battle against electronic exchanges, particularly the NASDAQ. In short, the announcement potentially signaled the end of a 213-year practice of conducting transactions through its specialist auction process, involving people on the floor of the exchange.

Conclusions

The industry had ended the century as a major user of IT and telecommunications, finally reaching a point where it was beginning to acquire a digital style. It was only a beginning, however, as it still had brokers on the floors of the NYSE and brokerage firms were still trying to find ways to control orders and markets. The major change in play as of 2005 was the expanding, yet still limited role, of the Internet on a global basis. While I have argued that the industry was collectively slow to embrace the digital, it nonetheless did what other industries had done: adopted it in extensive quantities once it made economic sense to do so. Slowness should not be seen, therefore, as an indictment of the industry, as a criticism that either the NYSE or some brokerage firms were slow to understand the potential usefulness of computing. Various individuals resisted its adoption, but that was because they knew how it could harm them personally. Managers in various exchanges, including the NYSE, however, pushed forward its implementation, with large brokerage firms leading the way (contrary to conventional wisdom, which holds that small firms are normally the most innovative and nimble in any industry). Large firms could afford the costs of equipment, people, and software, and had the most to gain because of the potential economies of scale and scope such applications offered, both in reaching markets and later in facilitating mergers and acquisitions. Thus by the end of the century, every fundamental task performed by all firms in this industry had a significant IT component integrated into it. Starting in the 1970s, small brokers could use existing applications through service bureaus and networks (such as the Internet or an exchange's IT infrastructure). To be sure, the industry remained highly complex, still burdened with paper, and subject to the dictates of various regulatory commissions. Global settlement processes were not yet in place, and the seamless national market the SEC envisioned had yet to be fully created.[117]

This industry demonstrated a pattern evident in so many other parts of the economy that proved essential to the creation of industrywide applications, namely, the use of associations, special bodies, and task forces. Working independently of each other in the 1950s and 1960s, exchanges and brokerage did a poor job in leveraging technology and ultimately realized that, for practical managerial, economic, and political reasons, they had to coordinate their initiatives across the industry. That realization made possible such organizations as BASIC, CCS, and SIAC. Thus, like the Banking Industry, once these organizations were operating, momentum built for the introduction of extensive and important uses of IT and telecommunications across the industry. Halfway through the first decade of the 21st century, no evidence exists to suggest that this situation is likely to change.

There are several other issues that we need to consider before turning to the experience of the Telecommunications Industry. First, we should remember that two fundamental events occurred with regard to technology. Automation lowered the cost of trading in the United States, primarily by facilitating the execution of large numbers of transactions and back-room processes. It also facilitated an equally grand achievement, the speeding up of transactions. The combination of lowered costs and speed contributed to an increase in transactions, which in turn contributed to the decline in the cost of equity.

Second, technology led directly to the creation of computerized market auctions, which achieved wide acceptance in the 1990s. However, of all the brokerage industries in the world, that of the United States still remained essentially the same at the end of the century as it was in earlier years. Trades still had to be completed on a trading floor through the intermediation of human beings, unlike in many exchanges around the world where computers performed the entire transaction, while economic optimization of computing in the United States had not yet been fully realized. In fact, when compared to European exchanges, two economists concluded that the cost of capital could decline an additional 4 percent if fully automated trades could occur. Extant evidence suggests that in the case of exchanges it was cheaper to build a fully automated one than the kind normally used in the United States.[118] The NYSE has rightfully proclaimed over the years how it has invested enormously in technology and had cajoled its members to do the same. Indeed, the NYSE spent more on computers and telecommunications than any other exchange in the world, but because it did not fundamentally change its operating processes concerning floor trading, this investment did not result in the same level of efficiencies evident in fully automated markets. So there is room for more economic benefit.

How did the NYSE and its member firms embrace computing? The NYSE, and the industry at large did what so many other American industries did: they took an evolutionary approach to the adoption of computing. It was a cautious approach. That approach often meant that the cost of investments in IT were higher than those experienced in other countries and therefore begs the question whether opportunities for increases in productivity were lost in the process that could have benefited the industry itself. Is this an industry where supplanting old processes rather than incrementally altering them would have (and still might) make more sense? It is not a moot question, because as global integration of trading occurs, along with more globalized competition for making investments and acquiring capital increases, what other brokerage industries have done will have a direct effect on the cost performance and competitiveness of the American industry. The largest brokerage firms are already wrestling with how to perform in a global market. Thus there is the potential to reverse the roles of exchanges and regulators by being the constituencies that push for maximizing the use of digital applications.

We next look at the Telecommunications Industry, which was crucial to the evolution and success of the brokerage and banking industries over the past century. That would not have occurred if the Telecommunications Industry had not progressed in its delivery of cost-effective, highly reliable services. To do that required the use of computers in massive and imaginative ways. And because so many other industries became increasingly dependent on telecommunications and, therefore, on this industry, we cannot fully appreciate the transformation to the digital underway across the American economy without understanding what happened in the world of the "telcos."

6

Role of Telecommunications Industries in the U.S. Economy

The packet-switching network was so counter-culture that a lot of people thought it was really stupid. The AT&T guys thought we were all beside ourselves; they didn't think that interactive computing was a move forward at all.

—Vinton Cerf, 1996

The Telecommunications Industry is a vital part of the modern communications infrastructure that facilitates the flow of information in support of economic activity and the private affairs of people. Its primary role is to move voice, data, images, and even sound across all industries and to and from all people. It would be difficult to exaggerate the importance of telecommunications in the functioning of contemporary life. Any parent of a teenager instinctively knows this just by watching the volume of telephone conversations their child has using this technology. No industry in America operates without constant use of the telephone, and over 95 percent of all households have at least one. By late in the twentieth century the Telecommunications Industry accounted for about 15 percent of the U.S. GNP, actually doubling its share in the economy, between 1992 and 2001 creating half the new jobs of the decade, and absorbing nearly a third of all new capital investments. By any measure, this industry is an important player in the economy.

The Internet, now used by over two-thirds of all people living in North America, is simply the latest variant of telecommunications added to the existing inventory of earlier communications infrastructures. Distinguished sociologist

Manuel Castells has argued its importance quite simply: "The Internet is the fabric of our lives."[1] Conventional wisdom holds that the Internet gave U.S. society a high-tech identity, a nearly 100 percent original, "made in America" phenomenon. Those who argue that Information Age is modern society's label have drawn upon this nation's involvement with computers. However, until people widely began linking to computers and telecommunications in the late 1960s and early 1970s, dispersion of computing across the economy had been invisible. Once the process opened and became public, however, the public saw the start of over three decades of hyperbole about the arrival of the Information Age. Without telecommunications, it would be difficult to imagine the digital hand playing an important role in so many industries. Therefore, we can safely conclude that the significance of the Telecommunications Industry is so great that it should make anyone's short list of critical components of today's world.

The Telecommunications Industry, however, is not just about the Internet; in fact, most commentators treat the Net as if it were separate and apart from the industry and, until the early 1990s, so too did the industry. By World War I, the industry consisted of telephony and had become an important part of the information-handling mosaic of many industries; by 1939, 35.6 percent of all U.S. households had a telephone, and by 1950 it was 61.8 percent. Businesses were even more connected with over 40 percent in 1939 and over 75 percent in 1950.[2] Early in the twentieth century, the American "Bell System" was recognized as the finest telephone system in the world, and the nation had quickly integrated it into its economic affairs. For nearly a century before the creation of the Internet there was POTS, "plain old telephone service."

The story of digital technology in the Telecommunications Industry is complicated by the fact that there are two aspects to it. The one that seems to get the most attention is about how users of telecommunications embraced the digital—be it the Internet or the telephone. The other aspect is about how the industry itself used digital technologies, for example to transport ("switch") conversations across the nation or to produce a monthly bill. This experience is less known and parallels our focus on the role of the digital in industries.

It is important to link the two themes together to help address several basic issues posed by the behavior and circumstance of this industry. The primary and most urgent issue is to help explain why the Telecommunications Industry remained in such a state of disarray in the twenty-first century at the exact moment in history when its role in the economy of the world was the greatest it had ever been. It is an important industry affecting the economy in many ways. This industry serves as a superb case study on how an industry's culture, economic circumstances, and a nation's public policies influence the adoption of a technology.[3] A great number of digital innovations came out of this industry, such as the invention of the transistor, so one would think that such an intimate understanding of the technology would naturally lead telecommunication companies to all manner of business activities and to faster and more thorough leveraging than any other industry.

Yet nothing could be farther from the truth, as hinted at in the quote launching this chapter. Vinton Cerf and his colleagues developed telecommunications based on digital technology which became the basis of the Internet yet found the Telecommunications Industry resistant. This resistance occurred even though many engineers and scientists in the industry (such as those at Bell Labs, where three scientists had been awarded Nobel prizes for inventing the transistor) understood what they were talking about.[4] The blunt reality remains that the place of digital technology in this industry was often far more complex than in other industries, largely because of the effects of this technology on both consumer behavior (which normally cannot be controlled or anticipated) and on the business model of the industry itself. So, we are left with a number of interesting questions to consider: Why and how did it use the digital? What is this industry's relationship to the Internet? What influence has the digital had on its role and structure? Why is this industry so unstable at the moment? Of all the industries studied in this book, this one is perhaps in the greatest flux and yet so crucial for policymakers and management in all industries to understand.[5]

The ensuing analysis of the industry is complicated by the fact that it is still churning, that is, has yet to achieve some equilibrium. In fact, the industry may be so early on its journey that it is difficult, if not impossible, to predict a likely

AT&T's AMA system, installed in the Philadelphia area and used in the late 1940s and early 1950s. Reprinted with permission of AT&T.

outcome, although many commentators have expressed their opinions.[6] (That is not the case with most other industries, since their use of the digital often settled into more or less predictable patterns.) Thus it is essential to examine carefully this industry's experience over many decades. In addition, there are relevant implications for all the other industries discussed in subsequent chapters, most notably those comprising the media and entertainment sectors. They are becoming so dependent on telecommunications that one has to ask, whether they are becoming the new Telecommunications Industry. Is the role of the traditional Telecommunications Industry—largely the transportation of conversations and data by telephone—even remaking itself such that these functions are now moving to some other industry? Will the American icon, AT&T, disappear, gone the way of Pan American Airlines or ITT? In 2005 SBC announced it was acquiring AT&T, leaving Ma Bell's future a tantilizingly open question. It is parallel to pondering if someday the U.S. Postal Service might disappear as other industries, companies, and technologies take over with a new communications infrastructure.[7] By any measure, these are emotionally very disruptive possibilities, but also economically quite important.

Defining the Telecommunications Sector

Today's confusion regarding the identity of the industry has many causes: disruptive technologies, complex regulatory practices, managerial attitudes within firms, changing requirements of customers, massive media coverage, to name a few. But there is also an essential confusion caused by the multiple definitions of "Telecommunications Industry." As information technologies (such as computer chips and software) merge with other technologies (such as telecommunications, the Internet, even print and graphic media platforms), what is "in" or "out" of the telco industry remains unsettled. Since industries are like clubs—in that firms within them work with, compete against, or copy each others' practices—definitions are important. This is particularly so given that the industries covered in the rest of this book have become so heavily dependent on telecommunications that they either have frequently been the targets of acquisition by a traditional telephone company (such as AT&T going after cable TV) or have begun to acquire the characteristics of a telco, such as newspaper companies owning TV and radio stations (subject to regulation by the same agency that governs telcos).

Industries in the United States are first and foremost defined by the U.S. government for purposes of tracking, regulating, and taxing. For decades the master list of industries was the U.S. Standard Industrial Classification (SIC) system. Industries and subindustries were assigned an identification number, called a SIC code, and that number was used to organize and store data from all manner of sources, both electronic and printed, about a predefined set of firms within industries. It was all very useful and convenient for firms, industries, government agencies, economists, and historians. Between the end of World War I and the 1990s, the telecommunications sector consisted largely of telegraph and telephone service providers and other firms that made components for them.[8] Simultaneously, the main regulatory body overseeing this industry, the U.S. Federal Communications Commission (FCC), had its own definition of what constituted the communications sector, based strictly on its charter of responsibilities as defined by the U.S. Congress. That scope meant the FCC would regulate "wire and radio communications service," which included telegraph, telephone, and all manner of radio communications.[9] After the emergence of television and satellite communications, these too were added to the scope of the FCC's responsibilities. In short, the FCC viewed telecommunications as consisting of wired communications (as in POTS) and wireless (as in radio, TV, and satellite and cell phones). Squarely affecting these segments was the FCC's concern about what role it should play regarding the Internet.

The U.S. Census Bureau tracks a great deal of economic data by industry. In response to the emergence of new industries over time, and also in reaction to the enormous growth in the services sector and in many industries involved in the creation, sale, and movement of information, it developed with Mexico and Canada a new classification system in the 1990s to replace the Industrial-Age SIC system. The immediate impetus was the establishment of the North American Fair Trade Agreement (NAFTA), which called for the gradual integration of all three national economies and obviously required a tracking mechanism. It all culminated in the North American Industry Classification System (NAICS), which redefined industries and placed industries within sectors. In 1997 the U.S. Census Bureau began publishing current economic data using NAICS coding and restating some prior years' economic data the same way, a process still underway. However, in addition to creating a sector called Information, (about which more will be said in subsequent chapters), it also defined one called Telecommunications, beginning with data from 1997. It described companies in this industry group as "establishments primarily engaged in operating, maintaining or providing access to facilities for the transmission of voice, data, text, and full motion picture video between network termination points and telecommunications reselling. Transmission facilities may be based on a single technology or a combination of technologies."[10]

This definition is different from the SIC model and the FCC's because it clearly and more broadly recognizes the variety of contemporary communications and includes, for example, one form of movies. Table 6.1 lists the component industries included in this definition, along with additional data to demonstrate that this new taxonomy of telecommunications represented a large slice of the U.S. economy.

NAICS also has other categories of players in the industry that we should not lose sight of, such as telecommunications equipment rental and leasing; telecommunications line construction, which is part of the Power and Communication Transmission Line Construction category in NAICS; telecommunications management consulting services, which is part of Other Management Consulting Services; telecommunications networks (wired), which is listed under Wired Telecommunications Carriers; telecommunications resellers, which is listed under Telecommunications Resellers; and telecommunications wiring installation con-

Table 6.1
NAICS Hierarchy for Telecommunications (NAICS code 5133), 1997

Title	Establishments	Sales ($B)	Paid Employees
Telecommunication sector	30,012	$260.5	1,010,389
Wired telecommunication carriers	20,815	208.8	815,427
Wireless telco carriers (no satellite)	6,386	37.9	146,302
Telecommunication resellers	1,656	7.6	30,028
Satellite telecommunications	521	5.1	11,931
Other telecommunications	634	1.1	6,701

Source: U.S. Census Bureau, "1997 Economic Census: NAICS 5133 Telecommunications," http://www.census.gov/epcd/ec97/industry/E5133.HTM.

tractors (who normally install high-speed lines in homes on behalf of an Internet provider), which is listed under Electrical Contractors. To complicate matters, NAICS also reports some combined industries, such as the increasingly important "cellular and other wireless telecommunications" which, in 1997 (latest year for which we have NAICS data), consisted of 2,959 companies, with sales of nearly $21 billion and a paid staff of 75,857 people. This classification consisted primarily of those engaged in operating cellular telecommunications, but not paging.[11] Even the latter, however, was also regulated by the FCC.

Over time, commentators on the industry broadened their definition of what was in or out of the sector, usually based on either the FCC's scope or what the technology logically suggested, with influence from data organized by SIC codes. Telegraph and telephone services dominated definitions until the early 1960s, when satellites became important; later, portable and cellular phones, and finally the Internet. Paging and microwave systems also came along in the 1960s and remained important components of the industry until the 1990s. Telegraphy began to shrink in importance, in the 1960s and 1970s, and long-distance and local telephone services were by-and-large combined prior to the breakup of AT&T in 1984. Most observers thought of telephones as the centerpiece of the sector, though, and clustered radio and television apart from phone companies for the entire decade, despite the fact that the FCC regulated all three groups. By the mid-1990s, some commentators and the FCC were defining the telecommunications sector as comprising those firms which were carriers or public network operators (like AT&T or MCI), equipment suppliers or vendors (that is, who made or sold such things as optical fiber, microwave equipment, satellites, wireless networks, switches and routers—Motorola, Cisco, and Lucent are examples), and value-added services (VAS), often also called value-added networks (VANs), which provided high-speed data networks for corporations and individual access to the Internet (including AT&T and Sprint, along with many smaller enterprises). Over time, definitions also came to include more of the information-processing world and those industries that provided data or, to use a late 1990's phrase, "content."[12]

The situation became even more complex as the end of the century approached. Communications and information technologies were being discussed together as a matter of course, including by the same government agencies that earlier had catalogued telecommunications as consisting essentially of telephone, telegraph, wireless, radio, and broadcasting. For example, the U.S. Bureau of the Census began publishing reports on media usage by consumers that included not only telecommunications and cellular phone usage but also TV, cable, recorded music, and newspapers.[13] These shifting views reflected the merging of different distinct technologies, such as those made possible by the Internet. In addition, companies traditionally thought to be in one industry or sector moved into another, as happened when AT&T flirted with cable television in the late 1990s, and large media companies with their distribution of their music, TV, and movie products over cable and phone systems into homes, and by the early 2000s to cell phones. From the perspective of consumers, these industries' music, television programs, movie products, and cable services comprised cable TV.

The Internet reflects the problem of industries tripping over each other. The Internet came into existence outside the old Telecommunications Industry as a result of military, government, and academic sources exploiting digital technology in the 1970s and 1980s. Then it moved into the mainstream of the economy in the 1990s via telephone dial-up and later high-speed service providers from both the old telecommunications sector (local and long-distance providers) and specialized carriers permitted by the FCC to deliver services over existing telephone lines or by cable. Access to the Internet via cellular phones came next, and by the early 2000s, there was speculation that perhaps the Internet would absorb all communications as consumers moved rapidly toward wireless telecommunications.[14]

For the purposes of our discussion about the role of the digital in these various telecommunications industries, I shall consider the Telecommunications Industry to comprise wired telephone, wireless telephone, satellite, and the Internet, with a tip of the hat to microwave transmission, PDAs, and pagers. I discuss radio and television as part of the media industries later in this book. This taxonomy makes sense because telephone companies tended to identify with other telephone providers, not radio or TV, even though on occasion a phone company considered entering the radio or TV business (conversely, radio and TV companies never wanted to get into the telephone business). Applying the adage of "birds of a feather stick together" as the operative principal supports this breakdown; otherwise, we would have to think of the Telecommunications Industry as a vast chunk of the American economy—an unwieldy and misrepresentative grouping.

The Changing Telco Sector in the American Economy

The telecommunications sector of the U.S. economy has always been large. Prior to the 1950s and through the mid-1980s, the dominant provider of local telephone service was a collection of 22 companies that made up the Bell System, and included AT&T, the fabled Bell Laboratories, and Western Electric. By the start of the twenty-first century, over a thousand small regional telephone companies also provided services through AT&T's telephone network. Because of its size, and the fact that many regulatory agencies took an interest in the Bell System, there is much statistical data available on it. In 1950, Bell generated $3.3 billion in operating revenues; 20 years later, it enjoyed $17.3 billion in revenues, clear evidence of a continued growth during a remarkable period of economic prosperity for the nation. The Bell firms were also very large employers, growing from 534,751 people on their payrolls in 1950 to almost 793,200 employees in 1970. This nearly 50 percent increase occurred despite a variety of steps taken in that intervening 20-year period to automate various functions to lower the need for more workers.[15] The Independents, as the smaller non-Bell firms were called, also enjoyed growth and a substantial position in the economy. In 1950, there were 379 such companies, most serving single communities or a few counties; in 1970 that number had grown to 684 providers—even after some consolidations had already occurred.[16] In 1950, independents' operating revenues totaled just over $270 million; in 1970, they were nearly $2.8 billion. They collectively went from employing 63,000 people to 142,000 in 1970.[17] If we add together the two sets of providers of telephone service, in 1950 they collectively generated $3.57 billion in revenues, and in 1970 $20.1 billion. In 1950, the combined telephone businesses employed 600,000 people, and in 1970 that number had grown to nearly a million.

While the telegraph component of the industry shrank in size and importance all through the half century, it nonetheless remained active in the years 1950 through 1970, despite dropping from 179 million telegrams transmitted within the United States in 1952; to under 70,000 in 1970, all at a time when the volume of phone calls and the size of the population continued their steady growth. In 1950, there were nearly 24,000 employees in the domestic telegraph business, in 1970, just over 24,000.[18] In short, telegraphy was a technology whose time had passed and was slowly disappearing, while its productivity obviously had not improved.

The centerpiece of the telephone business was the Bell System, a regulated monopoly that dominated the domestic phone industry, a galaxy of interconnected commercial enterprises. In 1950 the system enjoyed revenues of nearly $3.3 billion in revenues that spun off $472 million in net income. In 1980—just three years before the breakup of the Bell System—revenues had climbed to $50.7 billion, generating just over $10 billion in net income. If we add all the AT&T and Bell operating companies' employees together, include Western Electric (the manufacturing arm of the Bell System) and Bell Labs, the entire system grew from just over 600,000 employees in 1950 to 1,044,000 in 1980.[19]

Before examining the size of the telephone business in the post-1980 period, we should look at a few statistics on the demand side of the story. In 1950, nearly 281 persons out of a thousand had access to a telephone; in 1970 that number had climbed to 583. Looked at another way, in 1950, 61.8 percent of all households had at least one telephone; in 1970, the number reached 90.5 percent. The volume

For most Americans, this is the image of their telephone company: a female switchboard operator. In this case, it is a traffic operator at Southern Bell, Atlanta, Georgia, in 1955. Today, almost all her functions are automated. Reprinted with permission of AT&T.

of daily conversations increased more than fourfold in the period to over 4.3 billion each day.[20]

More people used more telephones per capita in the United States by the mid-1950s than in any other country in the world. The level of usage continued to climb right to the end of the century, although it started to slow since penetration rates were already so high. After the breakup of the Bell System in the early 1980s, the demand for telephones continued to grow, and after the arrival of the Internet demand for additional lines increased. Demand grew as the cost of phones (purchase instead of leasing) and charges for long-distance calls both declined. Following the break-up of AT&T, the number of independent telephone companies remained roughly the same to the end of the century—about 1,300 firms in the 1980s and 1990s—yet the number of access lines kept increasing, from roughly 29 million in 1988 to nearly 169 million in 1995, the first year in which we can argue that Americans used the Internet on a near massive scale. That latter year, on average people made 11.6 billion telephone connections each day.

Productivity increased in the industry late in the century, thanks in large part

Table 6.2
U.S. Telephone Industry Features, Selected Years (1985–1996)

Feature	1985	1990	1995	1996
Access lines (millions)	112	130	166	178
Local calls (billions)	365	402	484	504
Toll calls (billions)	(NA)	63	94	95
Average monthly residential local telephone rates (in dollars)	14.54	17.79	19.49	19.58
Average monthly single-line business telephone rate (in dollars)	38.39	41.21	41.77	41.83

Source: U.S. Census Bureau, *Statistical Abstract of the United States, 1998* (Washington, D.C.: U.S. Census Bureau, October 29, 1998): 575, Table 919; http://www.fcc.gov posts semiannual reports on wireless and wireline adoption in the U.S.

to competition and the implementation of digital technologies, which we can see reflected in employment and revenue figures. In 1988, all telephone companies combined in the United States employed 639,000 people; in 1995 that number had dropped to 477,000, even though the volume of transactions they supported had increased.[21] A surge in demand for telephone services continued all through the 1980s and 1990s. Revenues grew as well. In 1990, for example, the industry generated $160,482 billion in revenues; six years later (1996), they reached $238,069 billion.[22] Table 6.2 displays a variety of data to demonstrate that growth in revenues over a longer period of time. This industry's resistance to redesigning its processes to optimize its use of the digital parallels the resistance evident in many other industries. However, when the industry has completed that process—one currently underway—we could expect a further increase in productivity on the order of one magnitude, if the experiences of other industries is an indicator of things to come. That development would dwarf the improvements experienced by the Telecommunications Industry in the 1980s and 1990s.

While more is said below about the Internet, data from 1998 already document the injection of this new class of telephony and telecommunications into the economy. In that year, thousands of companies in scores of industries moved aggressively forward in their use of the Internet to inform the public of their offerings, to integrate the technology into their internal operations, and to sell and conduct other forms of business with customers. In 1998, nearly 200 million people had access to the Internet, and roughly 10 percent used it within any 30-day period. While these government statistics may be subject to some errors because of the difficulty of collecting such data, they are a start in helping us understand the nascent telecommunications media beginning to work its way into American life. For example, these same data show that usage of the Internet split evenly between home and work.[23] Subsequent data collected by the U.S. Census Bureau report that over 75 million people had accessed the Internet within a 30-day period in 2000, up by over 150 percent from just three years earlier.[24] In the same year, the

U.S. Census Bureau reported that over 94 percent of all households had telephones, 42.1 percent a computer, and 26.2 percent access to the Internet and over 25 percent (86 million people) had access to cell phones.[25] In short, the nation had an extensive telecommunications infrastructure.

Cellular telephones came into their own in the 1990s. While wireless phone systems had existed all through the half century, the cell phone as we know it today is a phenomenon of this time. Usage expanded rapidly, from 5.3 million subscribers in 1990 to over 55 million by the end of 1997. This subsector of the Telecommunications Industry employed only 21,382 people in 1990; in 1998, there were over 109,380 workers. Due to economies of scale, competition, and improved digital and analog technologies, average monthly bills for subscribers declined from approximately $80.90 in 1990 to $42.78 in 1997 (table 6.3 translates this pattern into number of users).[26] All calls (local, long distance, wired, wireless or cellular) dropped in aggregate cost. By 2004, for example, one could make long-distance calls for between 3 and 5 cents per minute on both cells and regular phones; in many instances, they were "free" in that they had become so inexpensive that service providers found it easier just to charge a flat fee per month rather than to try documenting and billing individual transactions. In the beginning, cell phones were analog; later they became available in digital form, and by the end of the century, analog formats were in full retreat. In the 1990s, subscribers moved to digital phones, which proved convenient, had far more functions (such as caller ID) and were more reliable, at about 16.7 percent year-over-year, such that by mid-2003 92 percent of all subscribers only used digital phones.[27] The move to digital technologies was an important reason for the expanded use of cell phones in the 1990s.

Leaving aside statistical data, a quick overview of some critical events in the industry's history explains in part how the nation became so wired. The regulatory

Table 6.3
Estimated Number of Wireless Subscribers in the U.S., selected years, 1984–2004
(as of end of year)

Year	Number of Subscribers
1984	91,600
1985	340,213
1990	5,283,055
1995	33,758,661
2000	109,478,031
2004	169,457,535

Source: Various reports, Cellular Telecommunications and Internet Association, http://www.files.ctia.org/img/survey/2004_midyear/752×571/Subscribers_slide4.jpg.

activities, most specifically of the Federal Communications Commission, provided the central influence on the activities of this industry, though on occasion the Antitrust Division of the U.S. Department of Justice also played a crucial role in the industry. Federal and state regulators determined what services the Bell System and other carriers could offer the public, and at what prices. During the entire period, the FCC wanted to encourage competition within the industry while simultaneously imposing on it regulations that forced companies to provide telephone service to all areas of the country, such as to unprofitable rural regions—the concept known as "universal service." The thinking behind this policy was that everyone in the nation should have access to telephone service, a view carried over into the era of the Internet when the Clinton Administration, for example, made it a critical initiative to "wire every classroom" and give to everyone access to the "information highway."[28]

In 1956, as part of the settlement between the Antitrust Division and AT&T, the firm agreed to focus on doing business only in the regulated phone market, to license out its various patented technologies, and not to compete in the computer business. As the FCC opened up the telephone business to new entrants, tensions over antitrust matters continued to the time of the historic breakup of the Bell System in the early 1980s. As table 6.4 catalogs, a variety of events occurred that

Table 6.4
Chronology of Key Competitive Milestones in the U.S. Telecommunications Industry, 1950–1984

Year	Events
1950	Hundreds of independent local phone companies existed
1955	Hush-a-Phone allowed to be sold to AT&T Customers
1956	AT&T and Antitrust Division signed Consent Decree
1960	FCC issued Above 890 Decision, opening microwave transmission market
1962	Congress established Comsat Corporation to provide satellite communications
1963	MCI filed for permission to provide long-distance service
1968	FCC approved sales of Carterphone
1969	16 telephone companies listed in *Fortune* 1,000 MCI granted permission to build microwave network for long-distance telephone service
1971	FCC opened up long-distance market to other firms
1972	MCI began offering its service in competition with AT&T
1974	Antitrust Division launched investigation of AT&T FCC allowed customer-owned equipment to connect to the Bell System
1978	Non-AT&T manufactured phones sold to Bell System customers MCI allowed to sell local service through its Execunet offering
1984	57 telephone companies listed in *Fortune* 1,000 Breakup of AT&T implemented

increased competition in the industry. It is an important point to make because conventional wisdom holds that competition did not begin until after the breakup of AT&T. Long before that event, however, there were hundreds of independent phone companies in the United States, most of which were profitable, and the FCC had made a long list of decisions that fostered competition. However, we need to keep in mind that the pace of dismantling the regulated industry picked up in the 1970s as part of a broader national initiative by various administrations to open up the entire American economy to greater competition. Many reasons have been suggested for why the deregulatory movement occurred in the United States, but the most widely embraced argument is that increased global competition threatened the vitality of the American economy and thus companies had to be given the freedom to compete and, if necessary, be pushed into becoming more competitive for the good of the nation's economy. For that reason, the Telecommunications Industry was opened to other firms to provide less expensive or functionally richer services, as documented in table 6.4.[29] In short, deregulation did not mean allowing companies to do more of what they wanted without regulatory approval so much as it was a process of allowing more competitors to operate in a regulated industry. Thus no account of any important activity in this industry can be considered complete without acknowledging the important role of government, particularly the FCC, in computing and other technologies.[30]

The biggest single event in the history of the Telecommunications Industry during the twentieth century, and perhaps one of the most important of the last quarter of the century for the U.S. economy at large, was the breakup of the Bell System and, more specifically, of AT&T. As time passes, however, the old Bell System seems to be piecing itself back together *de facto* fashion through mergers of old Bell companies; when viewed alongside the ascendancy of both the Internet and cellular phone services, the event that seemed so thunderous in 1982, and that went into effect on January 1, 1984, may come to be seen only as part of a larger historic process in which the industry reconfigured itself in response to both national policies regarding competition and access, and the emergence of new technologies (mostly digital). Each result created economic incentives for redefining the players in the industry and the services they offered.

The two leading students of the breakup of AT&T, Peter Temin and Louis Galambos, argued convincingly just a few years after the breakup that it was due largely to the acceptance of the notion in regulatory and other governmental circles that vertical integration of businesses and industries made more sense than horizontal integrations "under the aegis of public authority."[31] The negative reaction by the public to the Vietnam War, political fallout of Watergate, rapidly increasing global competition, and spiraling prices for oil products compromised the public's confidence in the regulatory practices of its national government in the 1970s and early 1980s. The process of deregulation, which had started before 1970, picked up momentum through the 1970s. What role did technological innovation play in this? Temin and Galambos claim not much—a point I partly challenge in this chapter. They argued that innovations did not matter because AT&T had already been barred from entry into, for example, the computer business. More relevant was the fact that the demand for telecommunications in the United States was expanding too quickly, both in volume and complexity, for one company to satisfy it all. One major point they make is that AT&T's management also proved far too inflexible in its negotiations with the FCC and other government officials. They blamed AT&T's "institutional rigidity," which made it difficult for management to change its historic position that the Bell System should be kept intact and be run by AT&T's managers. Federal regulators were not burdened with such a rigid institutional culture and thus were more able to think of new ways to constitute the industry. Aspiring entrants into the industry aligned with various government agencies, ranging from the FCC to the Department of Justice (DOJ) to Congress and even to the White House, to gain access and consensus within the industry. In fact, these two historians argue that the government may have moved too quickly to deregulate AT&T.[32]

So what exactly happened? After years of negotiations with the Department of Justice, a few court appearances, and a lot of posturing in public and behind the scenes, in January 1982, AT&T and the DOJ agreed to the following:

- AT&T would divest itself of all the Bell operating companies.
- The companies would be reconstituted into seven regional yet independent Bell operating companies (the RBOCs—NYNEX, Bell Atlantic, BellSouth, Ameritech, Pacific Telsis, Southwestern Bell, and US West).
- Operating companies would only provide local and intrastate telephone services in their respective geographies.
- AT&T would offer long-distance phone service but not local call service.
- AT&T retained ownership of Western Electric and Bell Laboratories.
- AT&T would pay the RBOCs for access to local lines and would engage in a complex scheme of billing to ensure that customers in profitable service areas helped to fund universal access to all regions of the country.
- All of these actions would be completed and in place on January 1, 1984.

There were other terms considered of lesser importance at the time. The RBOCs could not manufacture telecommunications products and customer premise equipment or offer long-distance services provided by AT&T's rivals such as MCI and Sprint; nor could they offer information services, let alone any other products or services that were not part of the "natural monopoly" regulated by the FCC. Not until January 1991 did the government lift the ban on RBOCs providing information services. The timing proved excellent since in the 1990s data transmission over telephone lines expanded dramatically.[33]

The historic agreement went into effect on schedule. It is a testament to all the key participants in the industry that telephones worked on January 1 the same way as they had on December 31, 1983. While the public found their bills more complex (local phone charges and those of a long-distance carrier, which later subscribers had to select from a group already in the industry), the nation did not experience any disruption in telephone services.[34] During the remainder of the 1980s, the industry wrestled with the consequences of the breakup of the world's largest corporation—consequences of which were still being felt two decades later.

Between 1984 and 1996 a plethora of activity occurred within the industry which led to the passage of the Telecommunications Act of 1996. Since the 1950s, an accumulated momentum of competition had been growing in the industry as other firms saw the handsome profits being made by AT&T, which had to fight off attempts to open its markets. With the introduction of microwave technology in the 1950s for transmitting long-distance telephone calls, AT&T began to lose its stranglehold.[35] Then came the request of a firm to add a device onto phones, called the Hush-a-Phone, which AT&T resisted. The device was minor, with no electronics—a product which fit over a phone's mouthpiece to direct sound into the handset while blocking outside noises from being picked up. At first the FCC sided with AT&T, but then a court overturned the decision in 1955, for the first time making it possible for a non-Bell enterprise to attach its device (or component) to the network. Then came the fight over the Carterphone, which connected mobile radio-telephone systems to the existing wired telephone network. In 1968, the FCC allowed Carterphone to access the Bell System. Another big battle occurred over the introduction of long-distance competition into the industry, which began in the 1950s thanks to the introduction of microwave transmission. To make a long story short, the FCC allowed Microwave Communications, Inc. (MCI), to provide service in 1969 after a decade of legal battles. MCI did well all through the 1970s,[36] then the breakup of AT&T occurred.[37]

The years immediately following the breakup saw competition for the long distance market heating up, with already established firms doing well, such as MCI and Sprint. As new entrants made their way into the industry, pressure on regulators increased to rework the large body of regulations and practices governing the behavior of the industry. Entrants seemed to come from everywhere. Long Distance Discount Service, founded by Bernie Ebbers in 1983 in Mississippi, for example, emerged; the public later came to know the firm as WorldCom. Resellers, as these firms were called, resold AT&T long-distance calls at a discount. In 1995, there were more than 300 resellers, although most were small firms. In the period 1984 to 1989, AT&T went from being the industry giant to controlling only 80 percent of the market, with MCI and Sprint enjoying 8 and 6 percent market shares, respectively.[38] In 1995, when AT&T's market share had shrunk to 55.1 percent, the FCC declared that AT&T was no longer a monopolistic threat and thus not subject to as many regulatory restrictions as before. Meanwhile, the combination of competition and the introduction of digital technologies made it necessary for all competitors to lower costs of long-distance calls all through the 1980s and into the 1990s. In the mid-1990s, AT&T split itself up into three firms: the long-distance and cellular phone business; Lucent Technologies to absorb the manufacture of network products and those items made by Western Electric; and its acquisition of NCR, which allowed the company to participate in the computer business and was expected to begin changing the complexion of the industry. Meanwhile MCI invested in a fiber-optic network which went live in 1987, sporting over 150,000 miles of the new cable across the United States. Its market share grew in the early 1990s to 17 percent.[39]

Battles over control of the local telephone business also heated up. In that market nearly 1,400 companies vied for the local call business, mostly battling the RBOCs. Wireless business also expanded with new applications such as paging and eventually cell phones as regulators continued to open, and as digital technology began affecting the industry significantly. But before discussing those developments, we need to understand the highly controversial Telecommunications Act of 1996.

In the 1990s the FCC still regulated the activities of TV and radio broadcasters, as well as cable, wire, wireless, and satellite communications. As providers emerged, they were generally kept apart from each other and treated as individual economic sectors; but over time technological innovations challenged that model since, for example, digital technologies made it possible to deliver similar services in different ways (TV programming by cable and satellite companies, telephone service through RBOCs or cell phone, and the like). The law of 1934 establishing the FCC had called on the agency to keep these various businesses separate, a task that became more difficult to do, resulting in the creation of a complex network of regulations. Business problems developed as well. One common example concerned the ban on local television stations merging with local newspapers; however, newspapers could buy television stations in cities in which they did not publish. The Internet only complicated the issues facing the entire telecommunications and media sectors in the United States and therefore the FCC. Various telecommunications and media companies lobbied Congress and the FCC for numerous changes in laws and regulations in the 1980s and 1990s. Finally, the FCC decided that a massive redesign of rules was necessary. Its chairman during the Clinton Administration, Reed E. Hundt, explained, "We planned to undo 60 years of rules and reshape the structure, rights, and responsibilities of the cable, satellite, television, and wireless industries. But, as we told everyone, our plan was to do so with modest steps phased in over time, perfectly predictable, in no way unsettling to the markets."[40]

In February 1996, President Clinton signed into law a broad, complex set of regulations. It removed various legal restraints on the Telephone Industry to compete in the local exchange market, established the ground rules by which RBOCs could compete in the long-distance phone business, and recognized the technological conversion now made possible by the digital across all forms of telecommunications and broadcasting—allowing, for example, local carriers of telephone service to offer competitive local phone services for the first time in over a century. The law established ground rules for resale of long-distance calls, portability of telephone numbers, dialing parity, access to telephone numbers and other services (such as operators), how directory listings would be made, and access to rights-of-way. He also set up guidelines for how new entrants into the market could access existing RBOCs' lines and for pricing that allowed both the FCC and state regulators to govern rates. Sweeping sets of rules for satellite, wireless, radio and television broadcasting were also enacted.[41]

The consequences of this legislation were immediate. The revision in assigning telephone numbers to phone companies made more numbers available for households that wanted, for example, second lines to hook up fax machines and

Testing AT&T telephones (handsets) at Western Electric's plant in Hawthorne, Illinois, in 1946. The production system for these highly reliable devices remained essentially the same through the 1960s. Reprinted with permission of AT&T.

PCs. This new supply of telephone numbers in turn stimulated increases in the number of new area codes in the late 1990s. By the early 1990s, even cable television companies began considering local telephone service, while long-distance carriers AT&T, MCI, WorldCom, and Sprint had convinced the FCC to allow them to enter that market. They spent the years after enactment of the law in February 1996 expanding their beachheads. By the end of 2000, new entrants into the industry doubled their market share of local phone lines from 4.4 to 8.5 percent of the total 194 million lines nationwide.[42]

A round of mergers began as the old Bells merged. First to start the process were Bell Atlantic and NYNEX, which the FCC approved in 1997. The second major effort involved MCI and WorldCom, which the FCC allowed in 1998. Next SBC joined up with PacTel and Ameritech. AT&T, eager to enter local and new markets, merged with Teleport Communications Group (1998) and also with Tele-Communications, Inc. (TCI), while Qwest bought US West so as to expand into local exchange services. At the end of the century the FCC approved the merger of GTE (a large independent local telephone company) with Bell Atlantic. All these mergers and acquisitions enhanced the portfolio of services that the new law now allowed firms to offer, many of which resulted from the improvements in telephony and communications technology in the 1980s and 1990s.[43]

Those involved in the development of President Clinton's telecommunication policies were not in full agreement on the results. Hundt, the FCC chairman in the late 1990s, has argued that the law made it possible for vast new investments to be made in the Telecommunications Industry. He believed that the installation of new telephone lines in companies and homes grew directly out of his agency's actions and the new law. However, he was candid enough not to take full credit: "The data explosion was of particularly unforeseen dimension. In 1999 data traffic doubled every 90 days," which, in hindsight, was driving demand for more lines and services than the law of 1996.[44] President Clinton's chief economic advisor at the White House, Nobel Prize–winning economist Joseph E. Stiglitz, proved less sanguine, arguing that deregulatory features of the law contributed to bursting the great bubble and created the recession of 2001: "deregulation in telecom unleashed a Gold Rush," with various firms striving for dominance in their sector of the industry. He explained how the Gold Rush hurt the industry and the economy:

> Those who argued for deregulation said it would produce more competition as different companies vied for market share. But there also was a strong belief in the idea of "first mover advantage," the possibility that the first firm in a particular market might dominate. Companies believed they faced a game of winner take all and so spent furiously to make sure they would dominate. In the end, this frenzied overinvestment helped create the excess capacity that overhung the U.S. economy and brought on the downturn that began in 2001 and lasted for more than two years.[45]

Perhaps miscues or imperfect deregulatory policies allowed an industry to behave in ways that, in hindsight, proved harmful to the economy. But it is also a case of an industry behaving the way it did in large part because of what the digital had made possible. Overinvesting in satellite communications and fiber optics, for example, was driven by the thinking this economist pointed out. But regardless of why things happened the way they did, in 2002 to 2004 the nation had excess network capacity, markets that were somewhat more constrained than before 1996 (yet far more open than in 1984), and an industry that was having difficulty sustaining its investments in companies and networks.

The Job of Telcos

Before we can appreciate the significance of the events described above, we need to understand what telephone companies do. One has to live in many other countries to marvel that Americans always seem to have a dial tone when they pick up the receiver; that was true in 1950 and remains the case over a half century later. Most Americans have never been inside a telephone company, let alone understand its vast technological base. In the parlance of business professors, the industry's "value chain" consists of several sets of activities:

- basic and applied research on such topics as telephony, physics, and electronics

- development of such products as switches, telephones themselves, and other paraphernalia needed to transmit voices and data
- manufacturing, of everything from telephones to switches, to cables
- creation and maintenance of the global networks of wires, satellites, and other technologies that make it possible to transmit telephone conversations and data
- sales and customer support, which range from taking an order to installing a telephone, to billing customers for services, to repairing equipment and restoring services after bad weather
- dealing with regulators, which varies from making sure local and national regulatory agencies receive data they need, to ensuring firms are allowed to raise prices and make profits, to supporting the extensive lobbying that underlies that process.

Over time, these roles remained the same despite the change in players. For example, while Bell Labs did the lion's share of research on the technologies underlying telephony deep into the 1980s, by the end of the century others came to dominate the field.[46] How and what was done within these roles also shifted over the years due to regulatory rulings and technological innovations. While AT&T had historically invested higher percentages of its sales revenue in R&D in the pre-breakup era, that percentage declined over time such that by 1999 it hovered at 0.9 percent. By the end of the century telecom firms in general spent far less than many other industries, as demonstrated by table 6.5.

The story was complicated by what a great student of telecommunications pointed out. MIT's recognized expert on telephony, the late Ithiel de Sola Pool, explained in the early 1980s that "a revolution in communications technology is taking place today, a revolution as profound as the invention of printing. Com-

Table 6.5
Investments in R&D by Select Global Industries, 1999
(R&D as percentage of sales revenues)

Telecommunications	2.6
Automotive	4.2
IT Hardware	7.9
Media and Photography	4.2
Pharmaceutical	12.8
Software and IT Services	12.4

Sources: Data in Martin Fransman, *Telecoms in the Internet Age: From Boom to Bust to . . . ?* (Oxford: Oxford University Press, 2002): 218; "Financial Times R&D Scorecard," *FT Director*, September 19, 2000; *InformationWeek* 500 annual survey.

munications is becoming electronic."[47] Even at that time (early 1980s), he already saw a metaphorical digital hand changing this industry. He explained why this technological base was becoming practical:

> The recognition and reproduction of a digital signal is more reliable. If a signal has weakened substantially in a long transmission, it may be hard to recognize its exact frequency or amplitude, but whether the bit has sounded at all or not is more certain. A repeater, therefore, can restore the original on-off pattern more exactly. If a message consists of unit pulses instead of continuously varying ups and downs, there are many ways in which one can manipulate the pulses electronically.[48]

Beginning in the 1920s, Bell Laboratories had the major responsibility for conducting research on telephony in the United States. Bell Laboratories, a storied national treasure, normally presented as the source of massive quantities of valued technology, had its scientists and engineers focus on three fundamental types of research:

- issues related to general and applied physics and chemistry
- electrical properties of science and materials useful to telephony, such as the work done with silicon that led to development of the transistor in the 1940s
- engineering required to create better telephonic technologies, such as improving transmission of messages, switching, and signaling.

The complex and detailed history of Bell Labs has been told by others;[49] however, it is important to note that in one fashion or another Bell Labs engaged in the study and development of almost every major telephonic technology and innovation through the 1980s, with the notable exception of packet switching (for more on this, see Appendix A and the next chapter).

Research on digital means for collecting and transmitting data and voice had been underway at Bell Labs as far back the 1930s. Since the mid-1940s, research topics had included digital transmission, solid-state devices (what we would later call transistors and computer chips), digital switching systems, computers and computer networks, and by the 1970s, new functions one could perform using telephones thanks to the availability of digital technology (such as call forwarding and conferencing). Bell Labs and also AT&T at large often took well over a decade to transfer digital developments from laboratory to market, which created opportunities for other firms operating outside the Bell System to move quickly to agitate for regulatory approval to enter the market.[50] Table 6.6 lists some of the many impressive research initiatives in digital technology undertaken by Bell Labs before the breakup of the Bell System. Yet, while Bell Labs and AT&T experimented with digital switching devices in the 1970s, Northern Telecom in Canada introduced its own digital switch, which provided digital telephone systems for the first time in the more traditional analog devices of the day.[51]

Conventional wisdom holds that Bell Labs invented all the major innovations in telephony in the twentieth century. However, that view does not reflect the whole truth. Though scientists at Bell Labs did various types of R&D on digital

Table 6.6
Key Research Events in Digital Technologies by Bell Laboratories, 1926–Early 1980s

1926	AT&T filed for patent on pulse-code modulation (PCM)—the basis of much future research in digital communications by Bell Labs
1937	Concept emerged of relay digital computer (Complex Number Computer)
1939	First digital relay computer built at Bell Labs (Model 1)
1943	Model II computer put in service
1944	Model III computer put in service Bell Labs launched major research initiative on PCM
1945	Model IV put in service
1946	Model V built for National Advisory Committee for Aeronautics
1947	Invention of the transistor
1950	Model VI relay computer put in service
1950s	Bipolar digital radio signaling emerged
1962	Installation of first Bell digital transmission systems (T-1 carrier system)
1960s	Research on interactive graphics began Digital network research on nodes connected to closed transmission loops
1970s	Development of Bell Labs Interlocation Computing Network Research and development of Digital Data System (DDS) and other digital network devices and protocols
1980s	Research focused on protocols for data packet switching

Source: *A History of Engineering and Science in the Bell System*, 4 vols. (Murray Hills, N.J.: Bell Laboratories, 1978–1985).

themes over the years, for example, the development of packet switching and digital transmission of data and telephone calls, which contributed to the establishment and wide use of what we eventually called the Internet, occurred outside of Bell Labs.[52] Others who were not part of the Bell System could exploit technological innovations or take innovative products to market quicker than AT&T or some of the RBOCs. For example, in the 1960s, creation of a microwave telephone transmission system led to the establishment of MCI. The Ethernet came from Digital Equipment Corporation (DEC) and Xerox in the 1970s, which also rivaled the Bell System's offerings for networking.[53]

Until the breakup in the 1980s, Western Electric did the vast majority of manufacturing of nearly all the various components of the Bell telephone network in the United States. It was not a small operation; in 1971, for example, Western Electric employed 206,000 workers.[54] For decades, the manufacturing arm of Western Electric converted Bell Labs' innovations into the products and devices that could be used by the Bell System. Like so many other manufacturing companies in other manufacturing industries, it implemented within its own internal operations many of the digital applications described in the first volume of the *Digital Hand*.[55]

The company began manufacturing its first digital switches in 1976, called the No. 4 ESS. However, as the historians of Western Electric pointed out:

> Western Electric had spent nearly five years and some $400 million in developing the new switch. Meanwhile, AT&T held off on the decision to implement development of a local digital switch, preferring to rely on incrementally improved analog electronic switching systems that Western had been installing since 1965. Not until the mid-1970s did Bell Labs begin developing a local digital switch that could be utilized by the operating companies.[56]

Bell operating companies put pressure on AT&T to introduce new products to block antitrust pressures and stave off competition. Northern Telecom, for example, rolled out its DMS-10 digital switch and continued expanding its new line of products aimed at businesses in North America right into the 1980s. By 1989, Western Electric was focused on the manufacture of switching equipment, transmission systems, media products, operations systems, and cellular devices and systems, all of which now had many digital components and software as part of them. In 1991, Western Electric folded into Lucent Technologies.[57]

Just as the operating companies were experiencing competitive pressures from outside the Bell System, so did Western Electric, and not just from Northern Telecom. Cisco became a competitive threat as it grew from its founding in the early 1980s in Silicon Valley. Its founders developed an efficient router, which is a device that "sends" telephone calls (voice or data) from one network to another (later it could also select optimal paths through which to send messages). Routers in the Internet world were the essential links from one network to another, often likened to bridges that connected communities on opposite shores of a river. All through the 1980s, Cisco's products were sold to firms building local and national networks, many of which in time became part of the Internet. In the 1990s, it competed against Nortel (the new name for Northern Telecom), Alcatel, Siemens, Ericsson, and Lucent in the rapidly growing market for digital telecom equipment.[58] The purpose of presenting this brief review of Cisco is to demonstrate that as the century moved toward the new millennium, the old Bell System proved less able than a growing list of rivals in competing for market share in a continuously expanding telecommunications market where increasingly digital technologies came to dominate its infrastructure.

The final piece of the telephone value chain concerned the installation, maintenance, sale, and service of phone services, which remained the preserve of all the telephone companies discussed above. However, there now was one new set of products that relied on digital technology that flowered after the breakup of AT&T: cell phones. Far more important than satellite or microwave technologies and offerings, the cell phone came into its own in the 1990s. Because it presents the real possibility of displacing traditional wired POTS in the near future, we need to understand its role in the Telephone Industry. Even before the arrival of the cell phone, portable communications had existed via radio, either from a radio common carrier or even a Bell company, but these proved expensive and not very useful. Cell phones began making their appearance in the American market as early as 1981, and by 1990 over 5.4 million users had, in effect, created an important new telecommunications market in the United States. By 1996, over 44 million users subscribed to this service, and over 100 million at the end of the century. Initial subscribers used these in automobiles, but quickly the phone became attractive to individuals in all walks of life and for every conceivable reason. As with other technologies, debates and marketing wars took place over technical standards, and rival firms competed for the rapidly growing market. Cellular technology took time to reach the market, with delays caused by regulatory intransigence and the normal evolution of the technology to a usable cost-effective stage of development. Scientists at Bell Labs had worked out the basics of the process in the late 1940s, but it was not until the 1970s that the phone's economics made it attractive enough to sell as an offering. In 1975 the FCC began preparing the market for this technology by allocating a portion of the radio spectrum for its use. In 1977, it allowed the Illinois Bell Telephone Company to launch a pilot service, and in 1981 this firm offered the first commercial cellular service in the United States. The FCC subsequently licensed two providers per market area.[59]

Standards wars in the cell phone business were at times geographical in scope. Three major technical standards emerged in the 1980s and 1990s, each developed to create self-sustaining markets for telecom vendors in different region. Thus, for example, the European Union rallied behind one standard which supported Nokia and Ericsson as vendors of choice. Likewise, Japan had its own standard with its own satellite of vendors (such as Samsung, Sony, Matsushita), while the American market was supported predominantly by Nortel, Qualcomm, Lucent, and Motorola, but also by international vendors trying to gain a hold in the United States.

By the late 1990s, key cell phone applications had expanded beyond telephone calls to include access to data, the Internet, and other devices to locate information. By the early 2000s, one could use a cell phone to download and transmit messages and e-mail and to take and send photographs. The functions of wireless PDAs and pagers were folded into the handsets as well. Pagers and mobile telecommunications contributed to the emergence of yet another application called text messaging and the necessary underpinning networking. Each line of product and application ported back and forth across all manner of telecommunications. The market consisted of the manufacture and sale of handsets and of the transmission services. By the end of the century, the three largest manufacturers of handsets were Nokia, Motorola, and Ericsson, with Nokia the dominant player. The largest providers of cell phone service in the United States included AT&T Wireless, SBC/Bell South, Sprint PCS, Verizon, and VoiceStream.[60]

One other important wireless telecommunications product often ignored in discussions about digital telecommunications needs to be acknowledged: the nearly ubiquitous PDA, or personal digital assistant. It is a handheld device that stores data, such as telephone numbers and addresses, and that can communicate with the Internet and one's own company or agency network and computer system. The first widely used device was introduced in 1996 by Palm, which was the dominant player through to the end of the century. Other brands included Handspring, Motorola, Qualcomm, Nokia, and Sony, which were joined in the game by traditional computer manufacturers, such as Compaq and Hewlett Packard, and consumer electronics firms, such as Casio, Sharp, Sony, and Research in Motion. By late 1999, annual shipments of PDAs exceeded 2 million, doubling by the early years of the new century to the point where PDAs were now generating revenues in excess of $1 billion a year.[61] PDAs operated off software, including specially designed Microsoft Windows operating systems and applications, Java, and Linux. Application software was built to run on top of these various operating systems, much like application software products are built to run on top of mainframe and PC operating systems. In the wireless world, these new sets of software products were called mobile client applications. Examples from the late 1990s included WAP, Java 2 Micro Edition, and BREW. Dominance of operating systems and the applications they supported became another technology battlefield as the industry entered the new century; the outcome remains yet to be determined, and future studies of the PDA will have to address many of the same issues faced by today's historians and economists in trying to understand the software and PC markets and various digital technologies of the twentieth century.

From when PDAs were first developed in the mid-1990s until as late as 1999, not all had wireless functions. In the early years of the new century they acquired wireless capability, which was driven by their intrinsic advantages over cell phones. These included larger sized screens, local processing and storage of information, and a broader array of methods for data input. Adding the wireless function proved a natural next step in the deployment of yet another digital application. Some of the earliest providers of wireless PDA services included AT&T Wireless, GoAmerica, OmniSky, Palm.net, and Verizon, again exhibiting a combination of pre-existing teleco players and new entrants.[62]

Finally, we need to ask, what happened to pagers? They were first introduced in the early 1990s as a one-way communication device to alert users that they had a telephone message; later they acquired the ability to provide two-way communications. By the late 1990s, annual shipments hovered in the 25 million unit range. Sales flattened into the new century as wireless PDAs and cell phones cut into the pager market, but even then this market was not trivial. In 2000, Motorola, which dominated 70 percent of the market, generated approximately $1 billion in sales just for these little devices. The total market for pagers ran at $1.3 billion in 2000.[63]

In summary, one can conclude that the "Ma Bell" telephone system Americans enjoyed in 2000 had been substantially altered from what they used in 1950. Dominated in the beginning by a nearly monopolistic firm, and based on analog technology, the Telecommunications Industry ended the century heavily committed to digital technologies, highly competitive, and with many new players functioning in sectors of the industry nearly devoid of any heritage Bell firms. Injected into the process were brand new goods, services, and applications—the Internet, communicating PCs, cell phones, PDAs, and pagers. Satellites, which first went into commercial use in the early 1960s and contained many digital components, remained an important part of the U.S. telecommunications infrastructure,[64] while microwave communications and telegraphy began declining.[65]

Largely driving the evolution of the industry were the twin influences of regulatory policy—the FCC and the U.S. Congress—and the adoption of ever-changing digital technologies.

Patterns and Practices in the Adoption of Computer Applications

One of the untold stories of modern American history is that of the transformation of information, both printed and verbal, into digital formats during the second half of the twentieth century.[66] When we combine telecommunications with the amount of information available in digital formats by the end of the century, we can project a fundamental shift in how humans receive, use, transmit, and store information vital to social and economic activities—in short, what sociologists and futurists were predicting would develop.[67] At the heart of that transformation is the role played by the Telecommunications Industry, which is a much broader story than merely the activities of one industry. It includes the Internet, whose origins came from outside the industry, and the diversity of telecommunications available to users, ranging from radio to cell phones, from POTS to PDAs, from telegraphy to satellites, from microwave transmissions to the emerging radio-tagging systems for merchandise. Consider first, however, the transition of non-digital data to the digital from voice, phonograph records, radio transmissions, and printed media. While we do not know how much information was in digital formats in 1950, given the number of computers in existence in the world, and the paucity of storage available to those machines, we can reasonably conclude that it was miniscule.[68] Over time, that circumstance changed dramatically.

The increased reliance on digitized information moving from point-to-point via communications can be described in a fairly straightforward manner. As companies and government agencies installed computers, they also stockpiled information in digital formats for use by their computers. As the number of computers installed increased over the years, so too did the inventory of data in digital form. That story is very well-understood, including the capacity of all installed storage devices.[69] From the first transmission of information between computers over telephone lines—normally attributed to 1940,[70] but with practical experiments launched at universities, such as MIT, in the 1950s,[71]—the movement of data from computer to computer, and then from people to computer picked up momentum in the 1960s with the introduction of online applications for such work as airline reservations and order processing.[72] By the early 1970s, companies were transmitting information (data) to each other from one computer system to another via telephone lines, what came to be called electronic data interchange, or simply EDI. EDI became a widely deployed application of telecommunications by the early 1980s, one that continued to expand right to the end of the century.[73] The U.S. government, working with universities, in effect invented e-mail with the wide deployment among federal, university, and defense contractors in the early 1970s of what eventually came to be called the Internet. Usage of that family of technologies—computers and telephone lines—expanded quietly all through the 1970s and 1980s. In addition to EDI and the future Internet, companies also maintained private telephone lines that customers used to send information or that employees leveraged to transmit data (such as their internal e-mails, like the widely used PROFS system from IBM).

One other use of telecommunications for the transmission of data had emerged: fax machines. These large, clumsy, and slow devices had been installed in corporations and government agencies in the late 1970s, but during the 1980s, fax machines became smaller, transmitted information faster, and dropped in cost.[74] By the early 1990s, units became available for under $500, thereby extending use of data transmission to the home and small businesses. That development paralleled the growth in use of PCs equipped with the capability of communicating with each other, first over dial-up telephone lines then via the Internet. By 1998, people had high-speed dedicated telephone line access through telephone companies and TV cable providers.[75]

We do not yet have a good grasp of what percentage of telecommunications traffic in the United States over time was pure voice versus pure data. However, we know that as time passed two events occurred. First, the total number of minutes of transmissions kept growing. Part of the reason can be simply attributed to the growth in the nation's population. However, the volume of telephone minutes grew faster than the population, leading to the conclusion that per capita use of phones had gone up. The second is that the mix of voice and data changed as the percentage of total transmissions resulting from data grew. The growth in the movement of data through telephones lines is part of a much larger national transformation. In discussing Phister's early inventory of equipment, note that the amount of data in digital form in 1950 was small. However, we are still left with the intriguing question of how much of the mix of telephone traffic was voice versus data. The technology itself has not helped provide answers. As transmissions have increasingly become digital, the difference between voice and data has shrunk because a digital network perceives it all as data; thus the FCC made no serious effort to collect related information. More important were issues surrounding how long calls lasted, and their changing costs, about which a great deal of information exists.

In the 1990s, when much voice and data traffic was digital, the FCC tracked long-distance calls (both internal and international) and could report that annual growth in traffic grew by about 15 percent, amounting to billions of minutes. Using the same international call traffic to illustrate costs, the FCC reported that between 1980 and 1999, the cost per minute for a call declined from $1.34 to $0.53. In that same period U.S. billed minutes increased from 1.6 billion to 28.4 billion, and growth rates in volumes actually increased right into the next century. The FCC's data are as endlessly telling as they are fascinating. For example, in 1950, the telegraph sector of the business accounted for 80 percent of all international service revenues, but at the end of the century telephone service accounted for 95 percent of all international services. We can conclude from these various pieces of numerical data that there was more traffic and that some of it was data.[76]

We can approach the topic in a very different way, although still indirectly, by looking at information stocks from around the world In the late 1990s and again in the early 2000s, Hal Varian and Peter Lyman, both at the University of California–Berkeley, calculated how much information from around the world was in digital form. Their findings provide early evidence of a massive, worldwide conversion of information into digital formats. While not complete, and by their admission still a rough cut, their findings are as interesting as they are startling. As of 2000, only about 3 percent of all the information in the world was being produced in printed form; the rest was in either digital or other formats. Data stored on magnetic storage devices (diskettes, hard drives) constituted the fastest growing format, largely because the cost of magnetic storage was dropping (indeed had been dropping steadily since the 1950s); indeed, individuals owned about half of all digital data in the world (as kept on their PCs). Varian and Lyman concluded that the United States produced 25 percent of the world's text, about 30 percent of all the photographs, and just over 50 percent of the stored magnetic data in digital format. One other interesting finding: in 2000 people wrote an estimated 610 billion e-mails, and if we simply extrapolate from the fact that over 50 percent of the users of the Internet were in the United States, then one could conclude that half those e-mails traveled through the American telecommunications network.[77]

I have described the results of their research to call out that how telecommunications was used in the United States was driven by a much larger transformation that was building a head of steam. Data transmissions and e-mail were rapidly becoming "killer apps" in the telephone business, and by the end of the century both voice and data were traveling through digital technologies, either through the more traditional telephone networks or the Internet. As this chapter was being written in 2005, the number of telephone calls funneled through the Internet was increasing sharply. The digital had become the transformative technology of the Telecommunications Industry by the end of the century.

Finally, since we know that the Internet was playing a growing role in telecommunications and that until the dawn of the new century all traffic on the Internet was data transmissions (text, pictures, music, video) and not telephone messages, we can at least look at some information on numbers of users to get a sense of volumes and rates of deployment. We can begin by asking the fundamental question: What is the extent of use of the Internet in American society? The Pew Foundation has conducted dozens of studies on American use of the Internet, and a quote from a Pew study from late 2002 offers a take on the effect of this new form of telecommunications:

> With over 60 percent of Americans now having Internet access and 40 percent of Americans having been online for more than three years, the Internet has become a mainstream information tool. Its popularity and dependability have raised all Americans' expectations about the information and services available online. When they are thinking about health care information, services from government agencies, news, and commerce, about two-thirds of all Americans say that they expect to be able to find such information on the

> Web. Internet users are more likely than non-users to have high expectations of what will be available online, and yet even 40% of people who are not Internet users say they expect the Web to have information and services in these essential online arenas.[78]

All of that happened essentially since 1994, both on the home front and at work. Recall, however, that Americans had used online systems since at least 1984. So the Internet arrived in the 1990s into a society that already had some experience with telephone-based data communications.[79]

The Internet became such a widely used tool that from its earliest days government agencies and academics began collecting data on how many users existed. While the data are voluminous, they are not always as precise as one would like, they are good enough to suggest strongly how important the Internet had become in the telecommunications mosaic of the United States. U.S. government surveys of the late 1990s documented different pictures in the mosaic. First, in 1985 there were about 8.5 million PCs installed in homes, a number that continued growing right through the end of the century: 15 million in 1990, over 30 million in 1995, and in excess of 60 million in late 2001. As a percent of households, in 1985 it was just under 10 percent, in 1995, over 30 percent, and in 2001 56.5 percent. The government began tracking Internet subscriptions in the home and found that, of those homes that had a PC in late 2001, just over 88 percent also had a subscription to an Internet service.[80] What is most remarkable is how fast over half the homes had acquired Internet service—in less than a decade—and how it happened at a time when over 95 percent had at least one phone installed. Rapid adoption of Internet service in businesses paralleled that in homes, according to a growing body of evidence demonstrating similar rates of adoption of this new telecommunication tool.[81]

Recent Developments

So far our discussion of the changes during the 1990s has centered on the arrival of the Internet and the meteoric rise of wireless phones. But the changes experienced by this industry extended far beyond those two developments. In July 2002, Michael Powell, Chairman of the FCC, described this industry as in a state of "utter crisis." In reaction to Powell's assessment, another commentator, Randolph J. May, noted several months later that "since then, the situation has only worsened."[82] By the end of 2002, the negative score was ugly indeed:

- a half million people out of work
- $2 trillion loss of market capitalization
- 23 telecom companies bankrupt

Between 1997 and 2001, $880 billion in investments went into hundreds of telecommunications companies, and $10 billion went into wireless alone in 2002.[83] In 2003, WorldCom declared bankruptcy and the industry experienced a variety of scandals which led to criminal charges against senior executives. In October

2003, AT&T announced it had made errors in stating its financials for the prior two years. In short, the news seemed all bad, and the public had not heard good reports from this industry since at least early 2000. No sector of the industry was spared negative press; even the high-growth wireless business seemed under siege. For example, CNNMoney's website carried a story in March 2003 headlined "Cleaning Up the Wireless Mess," in which it argued there were too many suppliers and too much competition.[84] In short, it was a large industry that had severe problems. It is ironic because these issues existed at the exact moment in history when demand for telecommunications was the highest, when society as a whole had become dependent on it. While we are too close to the events to know all the details, enough is understood to develop useful context. The problems were global, and they varied by continent; but the focus below is on the American situation.

While it seems that every commentator on the industry has a different point of view about the problems, they can be summarized briefly enough. The first problem developed during the course of the 1980s and 1990s when regulators around the world worked to make the industry more competitive, to make it more subject to market forces, and to eliminate state protection, as we saw with the breakup of the Bell System in the United States. As competition increased, the inefficiencies of specific firms became obvious and had to be addressed. Under state protection, there had been less incentive to modernize processes and technologies, but after deregulation firms had to become competitive and thrive, all of which was expensive. Part of the transformation involved moving from just having simple voice and leased-line services to offering all the functions available on phones today (call forwarding, conferencing, and so on) to handling larger volumes, such as the data streaming off the Internet. The introduction of wireless technology also put a strain on traditional telco firms.

A second problem involved the enormous volume of investments made in the 1990s in several areas. First, the major telecommunications firms in the United States spent nearly $90 billion laying nearly 40 million miles of fiber optics networks. Others invested in satellite communications systems in support of wireless communications, which cost additional billions of dollars. The FCC auctioned off spectra for wireless communication, and, as happened in Europe, firms bid high for the access only to find themselves burdened with billions of dollars in long-term debt that could not be readily paid down by revenues soon enough to service these financial commitments. The anticipated market for wireless services did not take off as quickly or as sufficiently as needed. As debt mounted, stock markets became impatient, since they measure performance in terms of revenue growth and profits. When the U.S. economy slipped into recession in early 2000, investors retreated from telecommunications stocks, thereby diminishing the volume of liquid assets available to these companies. All of this happened as competition was driving down costs for the newly increased bandwidth so much as to make it a commodity. In short, prices for phone calls and data transmission dropped faster than companies could acquire more customers and when the transition from earlier technologies of the pre-1990s had not yet been completed. As two IBM telecom-

Table 6.7
Revenue of Select U.S. Carriers, 1999–2001
(Rates as percentages)

Carrier	1999–2000	2000–2001
AT&T	1.0	−5.4
Bell South	3.7	−7.7
SBC	3.7	−10.6
Sprint	16.5	10.4
Verizon	11.2	3.8

Sources: Annual reports for each carrier, 2001; IBM Institute for Business Value, annual reports on telecom revenues.

munications experts explained, by late 2002, "confronting the bleakest market in the history of the industry, telecom service providers face the daunting task of balancing short-term earnings expectations with building a longer-term growth platform." Furthermore, "the challenge is compounded by the inability of carriers to access the new capital required to address these problems effectively." They concluded their assessment as a situation consisting of "negative cash flows, rising operating expenses, crushing debt burdens and battered financial markets," making it nearly impossible to transform.[85] Their explanations were widely shared by many observers of the industry. Table 6.7 suggests the severity of the problem in declining revenue facing the American industry at the dawn of the new century.

However, one can argue that the dilemma of either wired or wireless companies is not so impossible to solve. For example, the wireless part of the industry went through such an intense period of high growth in the late 1990s that its managers focused on getting any customer rather than on acquiring the most profitable customers. When that business slowed at the end of the century, management began cleaning up the problems created by unbridled growth. In the process, they had to reconcile their accounts, identify and drop unprofitable customers, move them to lower-cost service alternatives, and provide new offerings to the more profitable ones. Since organic growth in this market did not vanish in the early 2000s, vendors were working through figuring out how to extract additional revenues from existing customers in order to support growth strategies.

Bankruptcy protection became an inevitable step that many new entrants in the industry had to take in the 1990s since they had not had enough time to become profitable before the downturn of the economy. The list of bankruptcies is impressive: WorldCom, Global Crossing, 360Networks, Winstar, PSINet, Exodus, Global, and Arch Wireless. This group of firms had a cumulative asset size of over $156 billion.[86]

By the end of the century, a further compounding problem was the oversupply of long-haul carriers, many of whom had invested in the optical fiber networks that were helping drive down costs of broadband. The logical action one would

Table 6.8
Debt Load of Selected U.S. Telecommunications Firms, 2001
($ billions)

Carrier	Total Debt (Billions)	Debt/Cash Flow Ratios	Debt Ratings (S&P) (October 2002)
Bell South	$20.1	2.52	A+
Qwest	24.9	6.20	B−
SBC	26.2	1.77	AA−
Sprint	5.3	1.16	N/A
Verizon	64.3	3.25	A+
Total	194.3		

Sources: IBM Institute for Business Value, annual reports.

have expected was further consolidations in the industry, but with too much capital debt (see table 6.8), this action could not be taken with the same frequency as we saw in the 1990s. Debt was for overvalued network assets (what economists call "asset overhang")—for infrastructures that technologically were becoming outdated but to which the firms had long-term financial commitments. The discrepancy between the depreciated value of existing networks and the expected or real value of these assets was great and the result inevitable: a decline in corporate earnings. The rapid growth in wireless markets exacerbated the situation.

Adoption of wireless telephony in the American market also stressed existing providers, since not all of them had a wireless offering sufficient to offset the loss of wired revenues. By the fall of 2003, 7.5 million residents in the United States had only a wireless phone, not the traditional telephone from their local ex-Bell service provider. Cell phones now accounted for 43 percent of all U.S. phones. To put into context the speed of conversion from old wired technology to the new, in 2000 cell phones accounted for 37 percent of all phones.[87] So the transition first evident in the late 1990s showed no signs of ending. By the early 2000s, the industry seemed to be facing too many wireless carriers such that prices dropped rapidly, putting pressure on cash flows and profit streams. Six providers dominated this segment of the market, competing intensely, yet were unable to consolidate due to pre-existing debt loads that were too great to make mergers attractive, let alone feasible. Thus their first priority was to fix their balance sheets, particularly for the former Baby Bells: Verizon, BellSouth, and SBC. Both AT&T Wireless and Sprint were potential takeover candidates, due to the depressed price of their stocks, but potential buyers could not afford to make such a move in 2003. As one observer noted, "the wireless sector faces an almost impossible-to-solve dilemma. Consolidation is needed in order for the sector to get better. But the sector needs to get better before consolidation is likely to happen."[88]

These problems were not just the industry's burden; investors across the United States had invested deeply into this industry, injecting $880 billion between 1997 and the summer of 2002: $326 billion into stocks and bonds, and another $554 billion through bank loans. As an industry watcher at IBM noted at the time, this amount was almost as much as the combined market capitalization of 14 out of the 17 largest firms in the United States.[89] Unfortunately, between June 2000 and June 2002, when the U.S. stock market experienced a severe drop during a recession, investors lost nearly $2 trillion in value from investments in this industry. In the process, approximately 500,000 workers in the industry were laid off; in 2002 alone, this industry accounted for 20 percent of all layoffs in the United States.[90] Every observer who attempted to quantify the extent of the problem used massive numbers, as in this example from 2003 by Om Malik, a writer for *Business 2.0*:

> Poof—$750 billion gone! With over 100 companies bankrupt and an equal number that have shut shop, as many as 600,000 telecom workers are now without a paycheck. WorldCom is bankrupt, Global Crossing is decimated, PSINet has been sold for peanuts, and Genuity, a company as old as the Internet, sold its assets for a mere $250 million, a fraction of its one-time worth. These are staggering numbers for an industry that accounts for a sixth of the U.S. economy.[91]

Before moving to other topics, we should look at what happened with the technological convergence of the Internet and telephony. That process proved disruptive to the industry. The Internet had already created a surge in demand for data transmissions, crowding out some of the requirements for voice transmissions. It is a dilemma because some 80 percent of all revenue for this industry continued to come from voice services. The integration of voice and data over the Internet called for new technologies to bridge the two and thereby capture revenue for the traditional voice providers. It is too early to know how that will play out, but it is not too soon to understand the relative merits of Internet and telephone technologies as they now appear. Table 6.9 briefly summarizes the salient strengths and weaknesses of each in 1998; and by the early years of the new century, telephony was becoming increasingly available over the Internet from ISPs able to compete competitively with traditional providers of POTs. Reliability rose, costs dropped, and wireless came to the Internet too. Meanwhile, in 2003, the FCC allowed long-distance phone companies to start providing local service, while permitting the ex-Bells once again to offer long distance in conjunction with local service.

The consequences of increased demand for wireless telephony and the further deployment of digital technologies, when coupled to the industry's improving balance sheet, led to another round of consolidations and restructuring of the industry in the early 2000s. Because a day does not go by without the media discussing these changes, it is important to have a clear understanding of what was happening as this chapter was being written. Three major consolidations occurred between 2002 and 2004: Verizon and MCI (a RBOC and a wireless firm), SBC and AT&T (also a RBOC with AT&T), and BellSouth with some minor cell phone acquisitions. The three combined accounted for 68.5 million customers for local phone service and just over 140 million wireless telephone accounts. In 2005, two other

Table 6.9
Comparison of Internet and Telephone Technology, circa 1998

Internet Technology	
Strengths	**Weaknesses**
Intelligence resident at the terminal	Provides no differential services since all packets are treated the same
Uses packet switching and store-and-forward routing	No easy way to differentiate between more or less important traffic
Can reroute messaging, therefore not dependent on a specific switch	Lacks control mechanisms for managing bottlenecks
Resistant to failure	Difficult to predict performance
Can operate over extremely heterogeneous collections of access technologies	Difficult to introduce new protocols or functions into the network
	Many security problems
Telephone Technology	
Strengths	**Weaknesses**
End nodes require little or no intelligence	Overallocates bandwidth to achieve high levels of availability
Can support diverse collection of low-cost devices	Switching infrastructure is determined by statistics of voice call traffic
Highly optimized to provide voice service	Difficult to introduce new services
Serves as a true utility, with extensive resilience to failures	

Source: Extracted from discussion in Randy H. Katz, "Beyond Third Generation Telecommunications Architectures: The Convergence of Internet Technology and Cellular Telephony," *Mobile Computing and Communications Review* 2, no. 2 (April 1998): 1–5.

vendors were in the market: Sprint and Nextel on the one hand and Qwest on the other, adding an additional 13 million local phone customers and just over 32 million cell phone accounts. So, it appears that the industry is once again being reconfigured and consolidated as firms attempt to shore up their market shares as new rivals with new digital products continue entering the market.

Did IT Lead to a New Style of Business in the Telecommunications Industry?

As of 2005, one would be hard pressed to find a commentator about the Telecommunications Industry willing to deny the deep-seated importance of both digital technology and the Internet on the current and future lives of this industry. One would also find it difficult to find thoughtful observers willing to deny the growing importance of digital communications on the work and play of the American public.[92] Thus, we are in the situation where providers and users of both this industry's services and those of potential rival industries (such as cable firms) are all affected by the digital. The digital hand has made it possible for companies to transform how they operate, made new applications available, and allowed new entrants into the industry who have redrawn the business and economic landscape. In 1950, long-distance telephone calls were major and expensive events in the life of an American; today they cost a user and provider almost the same as a local call. In 1950, almost no nonvoice data were transmitted by the industry; today we believe that more than half of all its traffic is data, not conversations.

The Internet was one of the great technological surprises of the twentieth century. Few, it seemed, predicted it. Certainly, every forecast generated by the industry that I read written before 1985 never mentioned it, even though the industry expected transmission of data to grow. Was the industry blindsided on digital technology? The opening quote to this chapter would lead one to believe that the answer is a resounding "yes," although we now know that the industry had extensively studied digital technologies as early as the 1930s. What industry leaders always believed was that for both voice and data, the volume of telephone traffic would grow all through the period. What is surprising is how early the industry recognized that data would be a growing part of its business. For example, an industry-sponsored study in 1970 led by Paul Baran (an early developer of digital telecommunications) and his colleague Andrew J. Lipinski noted that industry observers forecast "data terminal growth will be large, increasing from some 600,000 terminals at present to about 8 million by 1985."[93] Their study also documented that many of those whom they interviewed expected "stored-program electronic control" would be widespread by 1985.[94] The phrase is technical code meaning computing (either digital or analog, but mostly digital). Their survey was remarkably bullish on the role of the technology: "A rising percentage of all switches will be controlled by stored-program electronic control, increasing from the 1970 estimated value of 4 percent to 50 percent by 1985, and 86 percent by the year 2000."[95] The question remains, however, of why the digital did not make more headway in the industry in the 1970s, and not until the late 1980s.

The answer lies less in the realm of technology and more in the culture of the industry. Temin and Galambo, in reporting on why the Bell System broke up in the early 1980s, argued that AT&T management was reluctant to change many of its practices, and certainly faced a difficult problem in adopting new technologies.[96] Others in the industry, however, were impatient, either because they were champing at the bit to enter the industry or they were Bell operating companies facing competition from emerging rivals such as MCI. Add in the enormous churn stirred up by the breakup of the AT&T empire in the 1980s, which distracted everyone in the industry from logically moving forward with new technologies, and one begins to see how painful, even slow, the transformation to the digital had been. Inject into the mix the Internet in the 1990s, along with such new devices and supporting technologies as wireless telephones, pagers, and so forth, and we again can see that the move to the digital was resisted and slow for the greater part of the half century. Once it picked up momentum, it became nothing less than a technological tranformation of profound proportions.

An AT&T ESS No. 4, used for long distance telephone calls, installed in Chicago and at three other locations in 1978. This one was in Atlanta, Georgia. Courtesy AT&T. Reprinted with permission of AT&T.

As extensive as the use of digital technologies had become in the industry by the end of the century, it is not fully clear that a digital *style* of operating had been adopted. If we think of changes in style as being on the order of magnitude of, for example, moving from craft manufacturing to mass-production techniques, or from steam to electrical motors, it appears that the Telecommunications Industry had only transformed partially its style by the early years of the new century. It manufactured telecommunications equipment with the same modern practices evident in other manufacturing industries, adopting the techniques at roughly the same time (the 1960s to the 1980s). So for that portion of the industry, we can clearly state that it reflected the digital style of production. Its products were largely based on digital technologies, making it possible to offer new goods and services such as call forwarding or caller identification and functions now routinely provided in cell phones and PCs. The great beauty of digital technology is its inexpensive flexibility which can be molded into so many forms and uses. Slowly, often under duress of regulatory mandates and competitive pressures, the industry integrated the digital into its technical infrastructure, and eventually in its goods and service offerings. In the early years of the new century, one could see that the digital existed in every corner of the industry.

It seemed to be less influential, however, in the industry's business structure. The old Bells were merging or buying each other out, showing signs of being aggressive in thriving; in the process they were *de facto* partially restoring the Bell System of pre-1984. We still had regional telephone companies interacting with their customers nearly the same way as in 1950. The FCC still had its twin objectives of encouraging competition and insisting on universal access to telephone service. To be sure, much had changed, yet much had not. As late as 1997, over 80 percent of AT&T's revenues still came from long-distance calls. In fact, it was not until the end of the century that Michael Armstrong, AT&T's CEO and a one-time IBM executive, could restructure the firm to reflect the new realities influencing his company and industry. As he told the public in February 2000, "you have to understand that" the long-distance business "was going to go away" because of increased competition and the decline in the price of calls due to technological innovations: "technology and deregulation are taking the middle of the [long distance] phone call away."[97] The availability of digital technology made it possible for him to change AT&T's strategy. In his words:

> The strategy was about bundling services that travel over the same network. For example, on our wireless network, we bundled a local wireless call, a roaming charge, and a long distance call—and charged a flat rate for it. That redefined the whole industry.[98]

Consequently, because of the digital, he decided to break AT&T into four companies to provide wireless, broadband (the cable part of his business), business services, and consumer services. Similar transformations were underway across the industry. The overcapacity of networks, overcommitment of capital, national recession, and accounting scandals all complicated the process of transforming the managerial and institutional practices to a digital style. As this book went to press (2005), however, the process was well under way. Until it has progressed further, we can only, yet ironically, conclude that of all the industries studied, one of the last to transform to a digital style was perhaps the Telecommunications Industry.

Conclusions

Perhaps we expect too much too soon of this industry. A universal theme evident in all American industries is that enterprises did not embrace digital technology until it made economic sense to do so. The Telecommunications Industry proved no exception. In 1990, two students of the industry reported the issue of affordability was a primary determiner for timing the installation of electronic switching systems, including digital computing and use of computer chips.[99] As demonstrated in the next chapter, applications played a decisive role in the pace of adoption. That explanation goes far to explain why certain digital applications, understood in some corners of the industry decades before they were widely adopted, were not embraced sooner than later. In short, the heavy hand of regulatory practices did not excuse management in the industry from making sound business judgments. That they did not always do so does not change the basic economic realities of what constitutes sound management practices. But let us understand that there are always multiple sides to a story. Regarding what many considered over-expenditures on spectrum, one could argue that the fees paid for the licenses could never be paid through existing revenue streams drawn from the wireless business and any possible increases potentially available from wireless. On the other hand, spectrum was a finite resource given the way governments distributed it, and thus ultimately a telecommunications company had to have it in order to participate in the market. Therefore, although clearly overvalued at the time of purchase for the term of the licenses, management bet that these investments would generate sufficient return on investments to have warranted these expenditures. The proverbial jury is still out on what the final results will be. Thus, while the industry came to adopt the digital, it did so on its own terms and in consideration of its perceived best interests.

Before moving to another observation about the role of technology in this industry, there is a second contextual conclusion we can make and it concerns the "bubble" economists and observers talked about in the late 1990s and early 2000s. Economist Joseph E. Stiglitz, perhaps more rationally than some other commentators, labeled the entire period of the 1990s as "roaring," with all sectors of the economy, managers, public officials, investors, and commentators, swept up in what Alan Greenspan, Chairman of the Federal Reserve Board, called the "irrational exuberance." That exuberance involved banking, cable television, and other industries as well, not just telecommunications. While Stiglitz emphasized the role of regulators in determining the activities of a modern economy, he only offered a tip of the hat to technology. In the case of so many industries that had become extensive users of the digital, however, the potential uses of this technology informed, indeed provided reasons for, some of the behavior that either enhanced the capabilities of a company or industry to make money or that created opportunities and crisis for the economy at large. While it would be convenient to place the blame for all of the problems of an industry at the foot of one collective source, such as the regulators or some ill-executed business mantra (such as first mover advantages), reality is always proven by historians to be far more complex and subtle.

Finally, the third observation is that the United States came into the second half of the twentieth century with an excellent telecommunications infrastructure that contributed mightily to the welfare of the country at large, making it possible for it to embrace so many new forms of the digital. This observation is not simply about reliable POTS (plain old telephone service), although that was extraordinarily important. Rather, the nation did not use much digital technology in 1950, so good phone service was peripheral to our story; it would become crucial as industries began relying on data transmission, later new telecommunications made possible by digital, and the nation at large when digital technologies led to declining costs of various telephone services that affected economic and social behaviors in the 1960s and 1970s. It was the whole industry, with all its various components, that affected the nation—the very existence of Bell Laboratories, for example, with its hundreds, later thousands, of employees and ex-employees working in the U.S. economy. They developed the transistor and many other forms of digital technology; through their interactions with Bell System employees and with experts in academia and in other research laboratories, they spread a vast body of knowledge about telephony across the American economy. As the industry became more open and competitive after 1984, the ready flow of information about the technology increased. This process awaits its historian, but we can note that it mimicked the patterns in the Computer Industry and in the increased stockpile of the nation's knowledge about programming and software. Those two more digitally intense cases have been studied sufficiently to provide a model of how others can examine the spread of telephonic knowledge through the economy.[100]

But let us be clear about AT&T and its breakup. Many of the innovations developed within Bell Labs were locked up by AT&T's senior line management, and not until the breakup of the firm did knowledge about many leading telecommunications technologies transfer across the entire industry. In fact, one can argue that even in the early years of the new century, the industry was still reacting and absorbing the flood of this knowledge. What we are learning in the early 2000s is that part of that flood of new knowledge, and hence new uses of technology, included the effects on organizations, their structures and processes, which are still in transition to new forms. What the historical record demonstrates is that the consequences of fundamental change are subtle and take a long time to work their way through an industry and an economy.

In the meantime, there were droves of people working in various Bell Labs, hundreds of thousands of employees in the Bell System and its competitors, operating in every state, in nearly every county, in all cities, and in many towns. They sold telephones, installed and maintained them, and, in the process, created an enormous demand for telephone services. Regulators at both the state and federal levels facilitated promotion and adoption of telephones by businesses and individuals, much as the U.S. government had done, and even funded installation of electricity across the country as early as the Great Depression of the 1930s. Revenue figures for the Bell companies and its competitors for the entire past half century, and deployment data collected by various government agencies, are numerical scores documenting results. For the bulk of the twentieth century, Americans became the most extensive users of telephony in the world, and often the earliest to adopt new applications, such as data transmission. To be sure, as the nation reached the new millennium it did not maintain leadership in all submarkets in the industry; we have the case of the cell phone as the operative example, with Japan and Korea ahead early on. One could argue that Europe was behind the United States because Americans use their cell phones more extensively than Europeans—almost double in terms of minutes—even though more Europeans have cell phones. But the broad conclusion is clear: the United States used telephones and related technologies extensively, while knowledge of the technology, and thus of its potential uses, were broadly available. If there was a surprise in the story, it is how quickly users of telecommunications embraced digital technologies, almost always wanting more and wanting it earlier than the industry could provide.

The next chapter explores how the digital was used, telling the story in roughly chronological order, since knowledge and use of the digital in one period influenced what happened in subsequent ones. With that story told, we then can move to a discussion of how media and entertainment industries were able to exploit both digital technologies and telecommunications so effectively that in the early years of the new century one could note that American culture was once again undergoing relatively important changes, this time driven at its core by digital technology.

7

Business Patterns and Digital Applications in the Telecommunications Industry

In general, telephone systems are wonders of the world.

—Martin Mayer, 1977

Digital technologies infiltrated the Telecommunications Industry and its customers slowly, incrementally working their way into the nation's telecommunications infrastructure over a long period of time. There were no "big bang" events that one could point to that were as public and dramatic as, for example, the breakup of AT&T. While the industry conducted research on digital technologies, even built its own digital computers in the 1950s, it didn't incorporate the digital technologies until the same time as such low-tech industries as insurance and banking and even later than such high-tech industries as automotive and aerospace. Individuals at home were not offered the opportunity to use digitally enabled telecommunication services until the 1980s, and even then, they did not usually know that they were "going digital," although they were already aware of the existence of computers and were buying PCs. As demonstrated in the previous chapter, the industry had a large stock of telephonic technologies of a nondigital nature that worked well. The industry functioned in an economic and regulatory environment that hardly motivated its members to transform existing infrastructures to digital formats. During the period from 1950 through most of the 1980s such was generally the case; only afterwards did circumstances change rapidly.

Behind the scenes, the breakup did have one effect that became increasingly evident as the industry passed through the 1980s, namely, it forced the telcos to assess not only their cost structures for the first time but also their capital efficiency in preparation for a newly competitive world. As a result, firms began redesigning their accounting processes and metrics so as to understand better their costs and various sources of revenues, as well as where best to lower costs in the 1980s and beyond.

Because the digital entered into this industry so slowly, it is difficult to imagine how much had changed by the early years of the twenty-first century as a result of all the change in the industry and, therefore, assess the social and economic effects of the telephone. For the telephone we have a model of what issues to address through the work of social historian Claude S. Fischer, who studied the role of the telephone in American society in the decades prior to 1940. He discussed many of the same issues covered in chapter 6, acknowledging the importance of technological innovation continuing right into the second half of the century. The results were mixed however. On the one hand, for example, the cost of making telephone calls dropped in half between 1950 and 1980, but on the other hand, people made telephone calls much for the same reasons as they had done decades earlier.[1] He pointed out that the telephone was essentially an "anonymous object," because it was everywhere. Fischer suggested that the telephone enabled private citizens and people at work to perform contradictory tasks: become more "modern" through their ability to conquer space and time, yet preserve traditional functions such as staying connected with relatives and home communities. Use of the telephone in the second half of the century played a supportive role in making it possible for people to perform easier, quicker, or less expensively the social and economic functions of American society.[2]

The past half century of this industry's experience with the digital can conveniently be parceled into three periods. The first, covering the birth and introduction of digital technology in this industry, covered the period 1950 to 1984, the year when the breakup of AT&T took place. As in other industries, this was an era of early discovery, experimentation, and initial implementation of basic administrative record-keeping applications. The second, surveying the years 1984 to 1996, saw the build-out of implementations started in the earlier period and the arrival of the Internet into the mainstream of American life. The third, 1996 to 2005, was subject to the influence of the Telecommunications Act of 1996 and the wide adoption of both the Internet and wireless telephony across the entire American economy, displacing a significant portion of the business that the local telephone companies had held as a monopoly for decades. These eras are partially arbitrary, since events do not radically change from one day to another, but they offer us the opportunity to make some sense of a great number of events and trends central to the modern history of the United States.

First Period: When the Digital Was Born (1950–1984)

In 1950 neither the Telephone Industry nor its customers used digital technologies to communicate; it was the stuff of R&D buried in the laboratories of AT&T,

universities, the military, a small number of companies, and a few federal agencies. But the decade of the 1950s was also filled with activity as telephone services continued spreading across the nation. For example, independent phone provider TDS saw a business opportunity to exploit existing technologies by providing better customer service than that offered by others. Many of their competitors still provided difficult and expensive connections between communities or the still extensive use of party lines, giving TDS the opportunity to seize upon those circumstances to grow. TDS provided private lines, access to customer service representatives, and prompt service.[3]

Nothing explains better the state of telecommunications in 1950 than looking at what telephone users did. Those living in a rural community, for example, shared a telephone line with others in their town or part of the county on what was called a party line. To use the phone, an individual picked up the receiver and listened to determine whether someone sharing their line was already in conversation; if so, good manners dictated the individual hang up the phone and wait until the line was free. (Or, one could be nosy and rude and silently listen to the conversation in progress!) To make a call from the United States to Europe, one had to call the telephone operator, request an appointment for a time when a line to Europe could be made available, and then wait for the operator to dial the call at the appointed time. One might normally wait a half a day or even to the next to place the overseas call. These calls were also quite expensive. Often only business managers and high government officials would make such calls, or relatives to announce deaths, births, or other major family events. Walter Wriston, a former chairman of the board at Citibank, noted that in the 1950s and 1960s "it could take a day or more to get a circuit. Once a connection was made, people in the branch would stay on the phone reading books and newspapers all day just to keep the line open until it was needed."[4] Americans who had grown up with the telephone prior to 1950, when service was limited and expensive, used their phones sparingly. Even as late as the early 1990s, it was not uncommon for members of that generation to cringe slightly when the phone rang in anticipation of bad news, which was always delivered as quickly as any telecommunications technology could bring it. They also used telephones less frequently than their Baby Boomer children.[5]

Until the 1960s, long-distance telephone calls were placed by live operators on behalf of telephone users. There was no speed dialing, call forwarding, touch-tone dialing, conferencing capabilities, speaker phones, or answering-machine services. By the end of the century, however, the digital had made all that possible, normal, and ubiquitous. While long-distance calls were special events in this nation at mid-century, they were part of the fabric of every day life by the end of the millennium. Henry M. Boettinger, an executive at AT&T, commented in 1977 that "the telephone is so much a part of our lives that use of it is habitual rather than conscious."[6]

When examining deployment of a technology it is customary to begin by studying the supply side of the story, that is to say, the role of the innovation providers. Once done, historians turn next to users of a technology to understand how they embraced it, to what extent, and to what consequence. Chapter 6 was an exercise in looking primarily at the supply side of the story, but we need to look at it further. Customers did not introduce the digital into the industry, the industry did that to itself, and in the 1950s, Bell Labs played the key role. Scientists and engineers at the laboratory built a series of relay computers between the late 1930s and the mid-1940s for research purposes and also for the U.S. military, called the Model I through VI.[7]

However, in 1950 Bell Labs installed a Model VI for its own use, primarily to do complex research-related calculations. As a result of their positive experiences with these machines, managers at the labs decided to build a digital system to handle AT&T's complex calculations and data-handling in billing customers for telephone calls. Called the Automatic Message Accounting (AMA) computer, it collected information on telephone numbers that made and received calls and the number of minutes of each call, the type of call, and other relevant data. The heart of the application was the AMA performing the arithmetic to determine what to bill a customer. This application very quickly became one of the most extensive early uses of digital computers in the United States. In the early 1950s the Bell System had some 25 million customers who made nearly 200 million telephone calls each month. While the number of calls per capita was statistically small by today's standards, that number still represented 200 million sets of calculations per month. A veteran of Bell Labs from the 1950s, M. M. Irvine, recalled that the system "worked well." He noted that "these calls had to be billed to the customer accurately and on time. At the same time, the billing system had to be a low cost operation to add only a negligible amount to the billable phone call."[8] The Bell System installed some 100 of these systems; many remained in operation until the mid-1960s, by which time most had been replaced with commercially available digital computers.[9] The success of the AMA encouraged management at Bell Labs to build and acquire other digital equipment for research and other forms of telephonic simulation studies throughout the 1950s and 1960s.[10]

Given its heritage as the center for almost all major telephonic research in the United States, it should be no surprise that when management at Bell Labs wanted computing, it instructed its staff to build such machines, as we saw with the Model computers. In the 1950s, management decided to design a digital computer system to handle the business operations of the Bell companies. In 1956, it established Bell System Data Processing (BSDP) to do this. Two years later, however, management recognized that the Computer Industry had come far enough to provide regional Bell companies with adequate computing to handle business operations and so shut down the BSDP project. That action was quite a departure from the culture of the Bell System, which had historically made everything it needed at Western Electric. For commercial vendors of computers the new (computer) industry's products opened an important market, one that in a few years resulted in the sale of some 100 computer systems to Bell to handle normal business applications, such as billing, payroll, account management, inventory control, and finance. Various computers went into the Bell System that were the major commercial products of the day, primarily from Univac and IBM, including the IBM Card Programmed Calculator (CPC) (first installed at Bell Labs in 1952); IBM 650 (in Bell Labs in 1955); and the IBM 704 (1958); these started at the laboratories and then expanded into the operating companies.[11]

In the 1950s, switching technology advanced substantially, and although based on electromechanical and other nondigital technologies, they are worth noting because these transformed how telephony functioned and, in the 1960s and 1970s, were applied to digital computers. The mix of the two technologies improved automatic dialing and so spread, encouraging the Bell System to continue improving its switching equipment and, in the 1960s, to start converting to digital machines. The entry of microwave and satellite communications into the industry in the mid-1950s through the mid-1960s improved long-distance calls and lowered costs. In time these technologies relied increasingly on digital components, particularly satellites, because telephone networks needed small electronic parts and programmable subassemblies.[12] By the mid-1960s the digital had made it possible to start transmitting calls in digital formats, further improving the audible clarity of those conversations.

Installation of No. 1 Electronic Switch Systems (ESS) in the 1960s brought some digital benefits to earthbound wired systems, although not fully until the late 1970s. The No. 1 ESS proved to be a milestone in the history of the industry because it enabled new services that in time became commonplace. These included call waiting, call forwarding, and three-way calling. The ability to transmit data became perhaps the most important development of the 1960s, as digital switches worked with new cables to move information. These cables, often known in the 1960s through the 1990s by such names as T-1s (also called T1) or D-1s,[13] initially were laid to carry long-distance telephone calls. They were largely displaced by optical fiber systems in the late 1990s.

Other developments made possible by use of digital technology also entered into the American telephone system in the 1960s and were widespread by the 1970s. These included standard three-digit area codes, single-slot pay phones (in which any American coin could now go into one slot), touch-tone dialing, and the introduction of 800 services, whereby the receiving party paid for a call.[14] In short, during the last 30 years of the century almost all the functions we now expect of our phone systems were made possible by the digital. However, these infrastructures, which dated from the 1960s, also served as evidence for critics of the Bell System that little new had come from the Bell companies since that decade, an argument used by MCI and others to request that the FCC allow new firms to compete against AT&T and its regional companies. But we need to put things in more precise context, particularly in regard to the extent of deployment of the digital in Bell's network. In a short memoir published in 2000, Ian M. Ross, the president of Bell Labs from 1979 to 1991, reflected back to the digital, which he noted had really entered the network in 1962. However, he candidly commented that "I think it is fair to say that the major implementation of digital networking took place in the 1980s." While planning for such applications had

An AT&T ESS No. 1 switch at Succasunna, New Jersey, which shows the early use of computing in switching systems, 1965. Reprinted with permission of AT&T.

begun in the early 1980s, he admitted it was a story of the period following the breakup of AT&T.[15] In short, if there is a criticism to levy against the Bell System, it is that its transition to the digital was painfully slow.

Internal publications of the Bell System demonstrated that the Bell companies were grappling with how to use computer systems to improve operations, much like companies in any other industry. In addition to the research and development work at Bell Labs, there was the much larger community of hundreds of thousands of Bell employees running a business with all the same needs as other companies: accounting, finance, billing, market analysis, inventory control, and other forms of record keeping. By the late 1950s, Bell companies were using computers to process payrolls, rating tolls (pricing calls) and customer billing, among other applications. One internal publication from 1965 acknowledged, however, that all was not right within the Bell System concerning computers: "For all their good work, though, computers have caused problems. Managers often complain that their organizations are upset, that customers become dissatisfied, or that they don't get the information they need from the computers."[16] In the 1960s, billing problems occasionally resulted in a random customer being charged thousands, even hundreds of thousands, of dollars because of a computer malfunction, and the event invariably became the subject of much, often humorous, press coverage. However, all firms experienced these embarrassing, although rare events, including the U.S. Internal Revenue Service, when it would notify incorrectly (and also inadvertently) a taxpayer that he or she owed millions of dollars, often because of a decimal point being in the wrong place in a calculation. As of 1965, however, accounting and customer-oriented applications using digital computers remained limited in the Bell System;[17] even internal literature spoke in the future tense about how they would be used.[18] Management had plans under development to implement common applications across the system to do accounting, support customer services, and provide internal operations with data and record keeping applications.

Where computers already operated in the industry, their use paralleled that in other parts of the economy. For example, in the fall of 1965, the Iowa-Illinois Telephone Company installed one of the first IBM System 360 in the industry to handle accounting, inventory control, cost accounting, payroll, and personnel records. The S/360 initially handled 120,000 subscriber bills per month, processing the work of 200 exchanges across nine states.[19] As an IBM press release of the period noted, "in revenue accounting, the computer will process telephone message toll rating, inter-company toll settlements and commercial accounting."[20] General Telephone of Indiana followed suit in 1966. Some applications were highly specific to the industry. For instance, the Wisconsin Telephone Company used a digital computer to provide spoken answers to over 30,000 questions each day regarding costs of long-distance calls posed by telephone operators. In short, it was a primitive price lookup system that responded with voice answers. As the company's assistant vice president of operations, John D. Fitzpatrick, described it, "the voice is very life-like."[21]

In a collection of press releases from the Telephone Industry from the period of the late 1960s through the 1980s preserved at the IBM Corporate Archives, we gain glimpses of the patterns of adoption of the digital. For example, in June 1969, AT&T announced that it would soon install an IBM System 360 Model 50—a large machine by the standards of the day—to perform circuit layout engineering, which is work done to design and produce long-distance telephone lines. We learn that in 1970 Ohio Bell Telephone Company and Southwestern Bell had computerized a process whereby employees using optical scanning equipment analyzed 225,000 hand-printed trouble repair tickets that recorded customer service problems and the actions taken. Meanwhile, Bell Labs announced it was using a number of IBM systems to allow engineers to communicate with each other on development projects by late 1971. The same year Illinois Bell announced it was adopting one of IBM's newest products, the very large System 370, Model 155, to produce a raft of operating reports "more easily and quicker than in the past." The purpose of this system was to "permit our managers at all levels to make better decisions on subjects ranging from finance to construction to daily business operations. The result will be better services to subscribers."[22] These records show that through the late 1960s and early 1970s, telephone companies all over the United States had installed digital computers for normal back-room processing in support of routine industry operations and exuberantly announced the news to the public.

In the 1970s the industry also began injecting the digital across all parts of its networks.[23] This was a silent transformation not evident to the public, but the subject of much attention within the Bell System.[24] The common carriers launched a major initiative between 1972 and 1974 to upgrade their data communication lines, moving from digital transmission capabilities of only 4,800 bits per second to 7,200 per second. While by today's standards those transmission rates were painfully slow, it was nonetheless a near doubling in speed. By 1972, ESS switches were being installed in the United States at the rate of one per day. Despite what would appear to be a rapid pace of transformation, one report from the period projected that because of the size of the network it would take until the end of the century to complete the task.[25] Data-handling services became important to the Bell System as so many industries now relied extensively on this technology to transport information. In a report to the U.S. Army in 1972, consultants at Auerbach Associates wrote that "the securities, travel services, and time-sharing services industries are currently the primary users of data communications systems. Health services, educating [*sic*], banking, insurance, manufacturing, retail sales, and real estate account for the remaining systems now in operation" (see table 7.1).[26] The same report noted that AT&T managed 217 million voice channel miles, and that it planned to increase the number to 600 million by the end of 1975. If one added in the lines managed by the independent telephone companies, the grand total would then have been 800 million by mid-decade.[27] The network needed to be upgraded and expanded before the arrival of the digital, but Bell was slow to invest in new equipment. In fact, as measured by the invested costs for switches in 1970, when decisions to expand were being made, the total value of digital switches was a mere 4 percent and could possibly rise to 50 percent by 1985. By the end of 1980, the Bell operating companies were already handling, on average, over 500 million calls each day.[28] The historical record demonstrates once again that AT&T knew what had to be done, but obviously was not going to move very quickly to accomplish the task.[29]

Others in the industry also were converting to the digital, beginning in the

Table 7.1
Telecommunications Applications across American Industries, 1972

Airline reservations	Computer-aided instruction	Multiple-warehouse order processing
Branch banking	Credit checking	
Brokerage ticker services	Hotel reservations	Truck or rail system management
Brokerage transactions	Message switching	

Source: "Final Report: Survey of Current Communications Technology (Present–1976), Subtask II-A," p. 9, Auerbach Associates, 1972, CBI 30, Auerbach Papers, Box 130, folder 10, Archives Charles Babbage Institute, University of Minnesota.

1970s. In 1977, TDS phone company installed its first electronic switching system controlled by software in a subsidiary phone company (Waunakee Telephone Company) located in Wisconsin. The event proved important for the firm, because it was its first step in converting from electromechanical to all-electronic switching, for the first time exposing many customers to computer-based telephony. That exposure translated into such new services as call forwarding and speed calling, both of which had been impossible to provide with prior technology. That same year TDS installed other digital switches in the Midway Telephone Company and at the Mt. Vernon Telephone Company, completing the cutover in 1978. To be sure, these early systems were more electronic than pure digital, but pure, 100 percent digital switches were installed at the Somerset Telephone Company in Maine and a local office switch at the Kearsarge Telephone Company in New Hampshire in 1979–1980.[30]

Figure 7.1 allows us to look at a snapshot of what applications were widely deployed across the industry in the late 1970s and early 1980s. This figure, published in 1985, had been in use for a number of years within IBM to instruct field personnel on digital applications in this industry. As with other industries, those who developed the wheel chart indicated that the Telecommunications Industry required traditional business applications (such as finance and customer service), as indicated by the dark inner circle, along with a myriad of applications in support of the unique functions of the industry, as detailed in the outer circle. Applications with dots just beyond them were those that one computer vendor (IBM) could support at the time with software and hardware.[31] It is quite clear from the wheel that cross-industry applications dominated—what IBM was selling and the Bells were buying (software and hardware) and were in support of uses evident in most industries at the time. Contemporary literature demonstrates that interest in these applications and the management of telecommunications in general had spread widely across the whole economy, not just within the Telecommunications Industry.[32]

The billing application proved so crucial to the functioning of the business operations of the system that it warrants additional focus. As mentioned before, AMA computers processed billing calculations, tracking the number of calls and customers into the millions. All through the 1970s and 1980s the percent of calls that companies tracked and billed individually kept growing at roughly 3 percent a year, creating greater dependence on computing to accomplish the work accurately, cost effectively, and in a timely fashion. It was a complicated affair; even the output, which one would think should be a simple bill, was complex. As an internal Bell publication documented, however, Bell companies produced three types of bills:

> The full-detail format provides all details on each call, including date, answer time, calling number, called number, duration, and dollar changes.
> The bulk-format provides three entries on the bill, showing (in dollars) the usage amounts, the allowance amount, and the total usage charges over the allowance amount. All applicable calls in the billing period are totaled in one

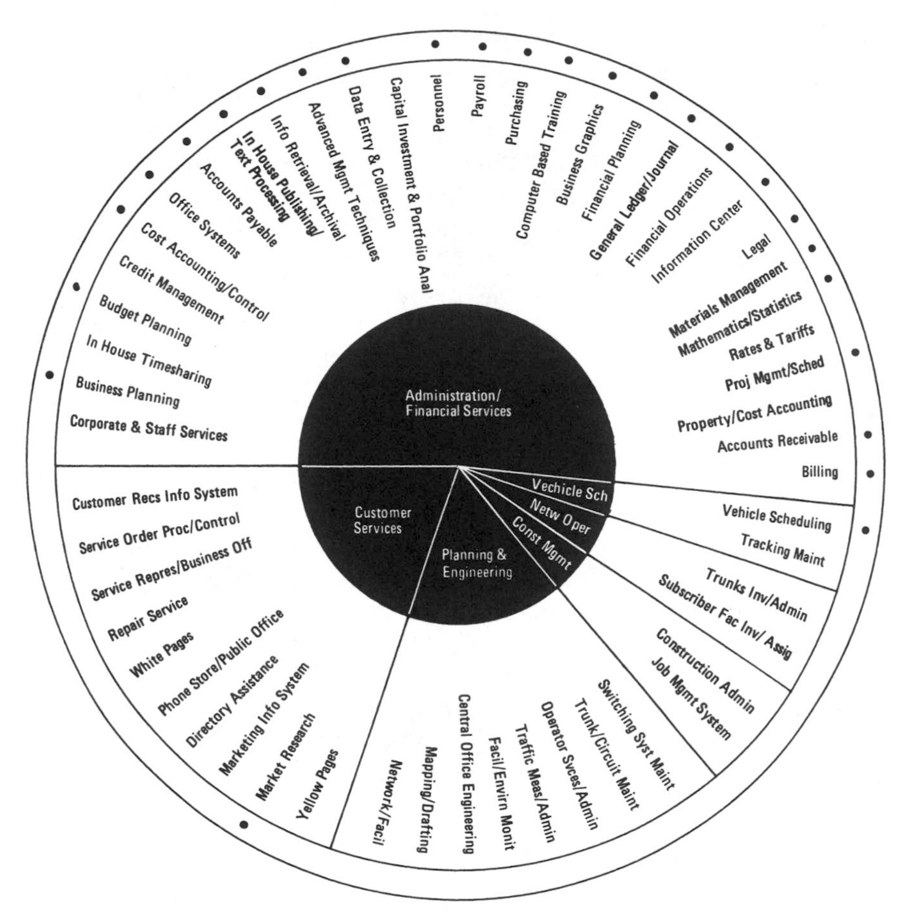

Figure 7.1
Telecommunications applications, circa 1985. Courtesy IBM Corporation.

> line on the bill. The summary-bill format provides the bulk-bill information and a summarization by zone called and rate period (date, evening, night), showing total calls, overtime minutes, and the associated dollars.[33]

Earlier in the decade, only toll calls were tracked in such detail; but as the years passed, companies were required to expand the scope of their tracking and billing. In pre-computer times, toll calls were initially manually logged on tickets then later by electromagnetic counters associated with each line; then the data (or paper) went to a regional telephone company center called Revenue Accounting Offices (RAOs), where clerks prepared customer bills. In the 1960s and 1970s, new methods for trapping this kind of data emerged that were both more automated and continuous, eliminating the need for operator-prepared billing by the end of the 1970s, beginning with a pilot operation in 1972.[34]

Computer vendors had little or no presence in the engineering part of the business, which was still dominated by tools developed internally by either Bell Labs or Western Electric.[35] Beginning in the 1950s, and extending through the 1970s, Bell's various plants specialized in increasingly more sophisticated products. By the end of the 1970s, craft workers had largely been replaced by machines specializing in the manufacture of complex circuit boards, for example, while computerized devices performed soldering functions; clean rooms came to be common in the company's factories. Workers were retrained or replaced to ensure that Western Electric had employees skilled in the use of modern automated equipment.

There were various results of the transformation in manufacturing processes. For one thing, the workforce shrank from 215,000 in 1970 to 153,000 in 1975. For another, a plethora of new products was produced, such as digital-switching hardware. Even in mid-decade, however, use of digital transmission equipment was still limited to short-distance networks; that is to say, from local switches to AT&T's toll-switching systems. However, in 1976, Western Electric brought out its first fully digital product line, the No. 4 ESS, after a half-decade of product development. Soon after, competition from Northern Telecom put more pressure on the company to introduce other competitive products and to further improve its internal efficiencies.[36] By the end of 1982, Western Electric had shipped 91 No. 4 ESS systems to the Bell System, and had orders for an additional 22 for delivery by the end of 1985.[37] To put these numbers in perspective, by the end of 1983, the installed No. four ESS switching systems (84 out of the total 718 switches of all types then in the system) handled just over half of all traffic.[38]

To better understand these applications in the context of what was happening in the wider world of customers and the economy, the new element in the 1970s was the massive increase in demand for data transmission services. In fact, demand for such capabilities exceeded that for traditional telephone services. The widespread integration of computers into so many industries drove up demand because their owners wanted connections over networks. Increased speed, performance, and declining costs of modems and terminal equipment simply fed this demand. Packet switching made it possible to consider deploying massive data transmission systems in manageable ways.[39]

Thus by the mid-1980s, the Bell System was well on its way to embracing the digital. By the time of the AT&T breakup, the RBOCs already had over 400 major computer systems installed. Large mainframes came from IBM and Univac, while applications running on smaller systems were designed either to operate with DEC or Hewlett-Packard equipment. In addition to needing equipment to handle complex applications, the entire system also wanted to slow the growth in the number of employees. An internal AT&T study on computing completed in 1983 and reported out the following year illustrated the expectation of the effects of the digital over the next several years (see table 7.2). Internal operating manuals of the period clearly demonstrated that there was hardly any important operation

Table 7.2
Anticipated Automation of Bell Operating Company Operations

Components of Automation	1981 Actual	1985 Anticipated	Growth (%)
Employees (thousands)	841	886	5
Terminals (thousands)	114	180	60
Minicomputers	4,200	5,800	40
Large mainframe computers	315	500	60
Terminals per employee	0.14	0.20	50

Source: Adapted from table in R. F. Rey (ed.), *Engineering and Operations in the Bell System* (Murray Hill, N.J.: AT&T Bell Laboratories, 1984): 604.

performed in the Bell System that did not involve some use of digital computing, with most major processes heavily dependent on this technology.[40]

All through this first period, AT&T and the RBOCs were aware of the importance of telecommunications to the nation and its growth. Studies on demand for communications and public pronouncements on the subject reflected the growing sophistication of information handling, its merger with communications technology, and the need to provide sufficient capacity and capabilities, with and without the help of regulatory permission. For example, Norman R. Holzer, a district marketing manager for AT&T in the 1970s and early 1980s, wrote in an Insurance Industry journal in 1981 that

> We in the Bell System and you, the users and managers of information, have much in common: we're all in the information handling business; we must operate under both state and federal regulations; we are all challenged by changing consumer attitudes and expectations; we all are subject to intense competition; inflation raises havoc with our ability to manage our capital resources, and continued growth is absolutely essential to our survival.[41]

He described research done within AT&T on the demand and costs of communications at the dawn of the new decade that illustrated the magnitude of AT&T's interest in the issue. Holzer reported that all businesses in the United States spent some $660 billion on all manner of office communications, representing between 20 and 80 percent of their operating costs in the 1980–1981 timeframe. Face-to-face conversations between people consisted of an estimated 30 percent of the total costs; meetings added another 42 percent and documents a further 25 percent. He reported that about 3 percent went for data communications and video and image applications. Holzer estimated that about 80–85 percent of the total reflected salaries of all the participants in communications, and only 10 percent was for hardware and services such as those provided by the Telecommunications Industry. Holzer concluded that the opportunity to improve the productivity of the nation's businesses, therefore, existed just by optimizing telecommunications. He also postulated that the future required a highly digital telecommunications infrastructure:

> [This future] could only be met with the expanded use of digital computing: The majority of the expenditures today are on voice-based communications with data communications representing a smaller but rapidly growing segment at an 18 to 22% increase per year. Even in 1990, well over half of these dollars will still be spent on voice communications. Data will still be growing rapidly, and by the year 2000, this relationship will continue to hold.[42]

Holzer noted how fast the cost of computing and communications were dropping while salaries were rising, arguing that "technology will continue to improve basic and enhanced voice communications so that they will remain the most natural and preferred mode of interpersonal communications."[43] His was both a challenge to his readers and industry yet an optimistic view of the future of telecommunications.

Lest intent be confused with reality, at the time of the divestiture of AT&T in 1984 the core of the business—switching—had only 4,000 digital local and toll switches of the roughly 21,000 total in the nation. Some 14,000 other switches were still old electromechanical systems, the rest analog. One survey estimated that of the electromechanical switches, between 50 and 65 percent were based on technology first introduced before 1900.[44] So, while digital systems were being introduced rapidly, there was much yet to replace. The Bell System was truly a network in transition.

Meanwhile, throughout the late 1970s and early 1980s, a variety of digitally based applications of telecommunications were embraced across the American economy. They can be conveniently catalogued as voice, image, text, and even hybrid applications (see table 7.3). What is striking is how varied the applications had become by the mid- to late-1980s. In addition to the ubiquitous accounting applications, those listed in table 7.3 were all made possible by the digital.

Some options gave companies new ways to deal with customers. One was the

Table 7.3
Communications Applications in the United States, circa 1975–1990

Voice	Data	Image	Text	Hybrid
Cellular radio	Electronic banking	Cable TV	Electronic mail	Videotex
Inward WATS	Database access	Voice teleconferencing		
800 number service	Computer conferencing	Graphics		
900 number service				
Dial-in service				
Voice mail				

Source: Thomas J. Housel and William E. Farden III, *Introduction to Telecommunications: The Business Perspective* (Cincinnati, Ohio: South-Western Publishing Co., 1988): 173–196.

800 numbers that made possible highly computerized customer-support centers. Indeed, call centers represented some of the earliest users to integrate computers and telephones, some as far back as the 1960s, using automatic call distributors (ACD).[45] This application continued growing in popularity and use with companies all through the rest of the century because it provided them with dramatic increases in the productivity of personnel.[46] Another option, Dial-it services (such as dial-a-prayer and dial-a-temperature—gave individuals information over the telephone. Then there was voice mail which, by the end of the century, had become *de rigueur* for companies and many individuals to have. Like so many other uses of the telephone, once the digital became available, so too did this function. First available in 1980 when VMX Incorporated went live, voice mail now so well known that we hardly need describe it, except to point out that software for such services resided in digital PBX systems and, by 2001, as a set of instructions in one's personal wired or cell phone.[47]

Other applications also went into effect. For example, ATM machines came into their own in this period. Image applications ranged from cable television to video conferencing. The latter made its first appearance in 1964 at the World's Fair complements of AT&T's "picturephones." For three decades there was insufficient bandwidth to support this application; not until the 1990s did it finally begin taking hold. And by the mid-1980s, graphics, such as bar graphs produced using PC software, could be transmitted as digital data as held true earlier with such text applications as electronic mail over dial-up lines and, later, as software offerings over the Internet.[48] In short, the industry kept adding new functions to telephone systems.

Second Period: The Digital Comes Into Its Own (1984–1994)

January 1, 1984, is a useful date to artificially demarcate when the rate of change sped up. One could postulate that the use of the digital would have grown without the breakup of AT&T: the only significant issue to debate might have been how fast. But the Bell System broke up and companies moved forward with their own plans to modernize, compete, and offer new services. The period 1984 to 1994 is bound by two clear events: first, the breakup of the Bell System, and second, the wide deployment of the Internet.

In 1984, expectations that the digital would expand throughout all the applications and most processes and offerings in the industry were high. Packet switching and other new approaches to handling telephone messages indeed continued all through the 1980s into the 1990s, building out and transforming the nation's telecommunications infrastructure into today's modern form. Internal networks also went digital. Looking back on the late 1980s, one observer reported that "local networking providing such digital communication services to government agencies, large businesses, and educational institutions were common in the United States by the end of the 1980s."[49] The reason, of course, was that the use of integrated circuits made these functions both practical and economically viable when compared to older alternatives; both conditions were essential to have simultaneously.

Customers expected continued improvements in services, from such mundane things as single bills for multiple services to additional functions on their phones and, for organizations, PBX capabilities. Flexibility, now made possible by the digital, began affecting demand. Two observers of the industry wrote in the mid-1980s: "The network should be capable of fast call set-up and disconnect times, a wide range of calling rates and holding times, and variable bit rates for user transmissions. There is increasing demand for cryptographic security for business and industrial transmissions."[50] The same commentators reported that AT&T and the Bell operating companies were responding to these demands: "Digital transmission systems are being implemented rapidly by satellite, microwave, coaxial cable, and optical fiber,"[51] with the Bell system having already laid over 130 million miles of digital communications. They accurately predicted that by the end of the 1980s, over 80 percent of the network in major metropolitan areas would be digital.

As others recalled of the mid-1980s, "all new systems had to have stored-program control and digital switching networks."[52] Telephone companies were no different from their customers when attracted to the use of the digital. Witness the experience of TDS, a small independent phone company:

> [It] was surprising, or maybe not so surprising, how rapidly the use of computers caught on throughout TDS. A network of computers was installed to communicate data between the Regions and Madison and Chicago. This network was gradually extended to include data communications between local companies and their Regional Headquarters. Computer-driven word processors rapidly replaced typewriters. As time went on, the MIS group integrated computers into the systems they developed.[53]

Like so many companies, TDS adopted more computer-based applications than its management might have thought earlier on were necessary, and in the 1980s, these often required telecommunications as well.

The initial cost of changing over whole networks to the digital proved high for the entire Telecommunications Industry, which explains why AT&T and a few large providers were the first to convert, followed by others as costs dropped. One quick example demonstrates that fact. In just two short years between 1987 and 1989, AT&T alone spent approximately $8.7 billion in modernizing its network, focusing on installing digital switches, transmission equipment, and writing software in support of new digital capabilities. By the 1990s, it had over 1.8 billion circuit miles of digitally controlled communications, most of it for local transmission. The economics of the digital made long-distance digital communications attractive as optical fiber came into play, hence the emphasis in the 1990s on laying new fiber across the nation. AT&T laid the first coast-to-coast fiber network, in this case for General Motors in 1986.[54] Software became a major component of AT&T's new applications, displacing expenditures for just hardware; by the early 1990s, software expenses comprised 50 percent of Bell Labs' products, up from 10 percent a decade earlier.[55]

Similarly, Sprint's "pin drop" long-distance digital network served as a direct challenge to AT&T, causing the latter to quicken its initiative to modernize its network. Sprint's was the first commercial network built entirely out of new technologies in modern times as a pure "green field" project, that is to say, built from scratch. The only others that possibly would qualify as brand new from the ground up were, of course, the satellite networks of the 1960s. But Sprint clearly represented a major milestone in the nation's transition to digital networks.

But the industry had also come to expect that overall the cost of the technology would smooth out. As installations increased in volume in the 1970s and 1980s, costs for equipment declined, thanks to economies of scale, while the capacity of the new equipment to handle more calls increased, providing another cost benefit.[56] In addition, improved reliability lowered maintenance and manpower expense. Software made it possible to increase efficiencies and to lower costs of managing networks; the equipment even used less electricity. These developments continued to keep down costs of transmissions over the rest of the century. The effects were cumulative and should not be underestimated. As several students of the industry's economics observed in the 1990s:

> Technological innovations and associated changes in relative costs are also altering the structure of networks. Digitalization has virtually eliminated the traditional boundary between switching and transmission, reduced the interface costs, and enabled more efficient and flexible use of equipment. In the day of all analog-equipment, transmission and switching equipment required hybrid connections. As digital transmission technology was introduced analog-to-digital conversion required interface relay sets between switches and transmission equipment. Now that digital switching is becoming the norm, the system is becoming much more transparent.[57]

In short, while capital expenditures were always high, the potential offsets were too. It is an important point to keep in mind when trying to understand why this industry was willing to invest so heavily in technology in the 1990s, to the point of driving it into debt at levels beyond those seen in other industries. Since the industry's prior experience had been positive, however, it counted on similar advantages in the 1990s when bidding on wireless licenses or in installing millions of miles of fiber optic cabling.

Transformation and deployment picked up speed in the early 1990s. John S. Mayo, the Bell Labs director during the first half of the decade, recalled later that "the biggest innovation was the move away from a large, centralized processor controlling a local environment to networks of microcomputers controlling a local environment, but also interconnecting regionally, nationally, and globally."[58] Technology itself drove this fundamental change in how networks were configured. As he noted, "During this period, microcomputer systems were doubling their processing power every year, compared to every three years for mainframe systems."[59] But, as we saw with every other industry implementing the digital, the desire to increase labor productivity and to drive down unit costs did not go away. As Mayo explained bluntly:

> One of the challenges in network evolution was to reduce the substantial labor and time needed to provision network services. Distributed network intelligence allowed a single technician using a console and software to do what formerly took many technicians and a fleet of trucks visiting various sites in the field—thereby eliminating the bottlenecks of geography, labor, and time. The capability to provide a broad spectrum of services was set up in advance so that the user or service provider could activate it on demand, much as electrical service to homes and businesses is preprovisioned and then activated when needed.[60]

As the cost of the technology dropped, though, other vendors found it attractive to enter the market and compete against the Bells. For example, Sprint marketed its network as 100 percent fiber optic by the early 1990s—"You can hear a pin drop." Four major fiber optic networks were built in the 1990s by AT&T, MCI, Sprint, and the National Telecommunications Network. Already by year-end 1991, they had laid about 2.4 million miles of fiber, and local companies laid an additional 4.4 million miles. By the early 1990s, other functions, such as paging and cellular phone services (made possible by new compression technologies that allowed more conversations to be transmitted over existing wireless networks and through cells),[61] were common. It should be no surprise that the Internet drove a wide expansion in demand for digitally based services. Surveys done in the mid-1990s showed that demand for additional telephone lines into homes, for example, was driven by the need to support computers, online services, the interests of children, and more personal and business calls, and to fax documents. Home offices increased in number in the mid-1990s, driving up the need for more lines and functions. For the second half of the 1990s, demand for second lines in American homes grew at double digit rates (up 10 percent in 1995, 13.4 percent in 1996, 16.5 percent in 1997, and so forth).[62] In short, increased affordability and use of more online services and the Internet were by-products of available digital applications.

Thus, advances in computing and telecommunications for this period on made it possible for the nation's workforce to become dispersed and yet still do the work of so many industries. That capability required the much more sophisticated set of communications managed services described in this chapter. The number of branch offices expanded, as in both the banking and insurance industries; workers spread out, as in the Transportation Industry; and white collar and call center employees could work out of their homes. That fundamental socio-economic change in the physical location of people required more broadband coverage, multiple networks, portability of applications over multiple networks, improved levels of service and security from employers and telecommunications providers, and more investments by firms in many industries in the products and services of the Telecommunications Industry. All of these requirements became even clearer and important in the post-1994 period.

Third Period: Time for the Internet and the New World of Wireless (1994–2005)

This last period saw the largest number of changes in telecommunications in the United States since the introduction of telephones in the late nineteenth century, making this one of the most confusing industries to understand. The question here is: What did people do with the digital in the 1990s that they had not done earlier? The answer: the Internet and wireless telephony. To be sure, prior uses of telephony continued—wired telephones at home and at work, private PBXs in office buildings, some wireless communications that relied on phones—but the major change was the addition of these two new technologies as major components of American life. Usage grew so rapidly that by the end of the century, one could hardly imagine a time when Americans did not use the Internet or didn't have a cell phone. The churn witnessed in the Telecommunications Industry in the late 1990s to early 2000s may someday be attributed to the rapid adoption of these two technologies and not just to overextension of debt, development of wireless communications networks, and deployment of fiber optic networks.

How People Used the Internet

In the early 1990s, many in the United States saw that the Internet was going to be the new "hot" digital application of the period. A vast body of literature began appearing on its history,[63] technology,[64] management,[65] economics,[66] politics and public policy,[67] effects on society,[68] and use. For our purposes, it is important to understand its use because the Telecommunications Industry as a whole had an important role in each of these sets of issues. As of mid-2005, it was not yet clear whether the old Telephone Industry would end up the leading provider of the Internet or if it would be some other industry such as cable TV. Since the industry's major firms had decided in the 1990s to become providers of the Internet's telecommunications, the Net is so far an important emerging component of the industry and we already know has had a profound influence on its behavior and economics. Thus, if for no other reason, the topic of how people used the Internet is part of the larger story of how the digital came to be used in the Telecommunications Industry.

To be practical about the matter, one can divide the history of the Internet into two periods: that prior to the availability of browsers (early 1990s), and that after. In the years before browsers, that is to say, from the early 1970s to the mid-1990s, the primary users of the Internet in the United States were associated with the defense establishment, the academic community, and a smaller group of computer-savvy individuals. E-mail was the primary application of the time, a form of communications that evolved to provide a means for rapid, short, interactions among individuals collaborating on projects and issues. As use of the Internet slowly extended outside these realms in the 1980s, e-mail continued to be the most important application, though other new ones began emerging, such as chat rooms. But the use of the Internet was not massive; in fact, it remained largely limited to possibly as few as tens of thousands of users. Extant data is not clear on the matter, but we know it was far less than in the post-browser period, when the number of users quickly climbed into the millions.[69] As discussed in chapter 6, since one needed a computer with which to communicate via the Internet, access remained limited until the PC (with modem) became ubiquitous across the economy. In the second half of the 1980s, however, the necessary infrastructure was created for individuals to access the Internet via telephone lines and later DSL and cable connections.

Use expanded rapidly in the years following the availability of browsers, which made access to the Internet more convenient and even more important and provided the ability to search for materials over the Net. E-mail at first continued to be the primary "killer app" of the Net but, by 1997 whole industries were making available information about themselves, their products, and services over the Internet via their websites. By the late 1990s, one could place orders and have other interactions with companies. Also by the late 1990s, individuals could establish their own websites, populating them with family photographs, newsletters, and memoirs such that by the dawn of the new century, billions of pages of material existed on the Internet.

David H. Brandin and Daniel C. Lynch have looked at applications on the Internet from a techno-centric perspective, which is a useful way to begin understanding applications of the Internet. Brandin and Lynch identified e-mail as the earliest application, which they contended amounted to over 80 percent of all traffic over the Internet in its early days, with the transfer of files also as a major application. In the formal language of engineers, this was referred to as using file transfer protocols, which made it possible to transfer data from one computer to another over a telephone line, an application dating back to the early days of the Internet. A third application, called Telnet, made it possible for a user to access a remote computer over a TCP/IP line.[70] Uses included opening and reading e-mail stored on a remote computer, unlike today where one simply downloads mail and then opens it. Usenet became the application by which one could access and use online bulletin board systems to post and read material, and did much to stimulate use of the Internet. People used this application quite extensively to find information on any subject, with whole websites devoted to providing data. Usenet began in 1979 as an application shared by Duke University and the University of North Carolina; in 1986 news was first broadcast over the Net by ClariNet Communications.[71]

The World Wide Web, or the Web, acts as a massive integrator for the Internet, through browser tools that retrieve information anywhere on the Internet. Browsers made it possible for people to read material on their screens and to receive and send graphics. It is difficult to underestimate the importance of this technical development, as without it one would be hard pressed to believe that the Internet would have become as widely used a tool. Two additional vital developments were search engines (software tools that sought out specific types of information) for cooking a turkey and finally software to provide security for data (so crucial for commercial transactions). Intranets were also created (networks for restricted use, such as only within a company by employees) and extranets (which the public at large could access).[72] All of these applications emerged and were widely adopted in the 1990s, and their histories have been already prolifically documented, so they need not detain us here.[73]

To be sure, though technical functions are important prerequisites for any discussion about how end users applied these developments, from the perspective of either business or society, once they are available and implemented, the focus shifts to other matters. In our case, the real story is what kinds of information and activities leveraged the Internet. As time passed, the list of users and applications grew in volume and diversity. In the period roughly 1994 through 1997, e-mail and bulletin boards dominated traffic over the Internet. Between 1996 and 1998, a massive, uncoordinated effort took place across the entire economy to establish websites and populate these with information such as product brochures, contact data, and the like—a process that continues to the present. As more information became available and more PC users with access to the Internet, traffic grew. One reliable estimate placed the number of users in the United States and Canada at 90 million by the start of 1999, with another 35 million in Europe.[74] Businesses took note of these numbers and began providing customers with access to their firms, first by making available access information, then by enabling orders. By the end of the century, roughly one-third of working Americans had access to the Internet at work and had begun using it to communicate, find information, and integrate the technology into their work processes.

Volumes grew rapidly. Between 1992 and 1998, corporate e-mail worldwide went from an estimated 1 million to 167 million, with over half in the United States. E-mail from offices exploded, with one study suggesting that some 3 billion were sent each day around the world by 1998.[75] That number grew to over 30 billion per day from all sources worldwide by 2004. Commentators began talking about the existence of a "Digital Economy," an emerging new robust channel for commerce.[76] While it is difficult to fully validate these numbers, there is enough to suggest that they were not far off the mark. Electronic markets were becoming the new way to do business. Three respected experts on the Internet gave us a sense of what this new world might look like, citing a live example in September 1999:

> A basement computer room at Buy.com headquarters in Aliso Viejo, California holds what some believe is the heart of the new digital economy. Banks of modems dial out over separate ISP accounts, gathering millions of prices for consumer products: books, CDs, videos, computer hardware and software. Specially programmed computers then sift through these prices, identifying the best prices online and helping Buy.com deliver on its promise of having "the lowest price on earth."[77]

By 2000, some 80 percent of all Internet traffic was being driven by electronic commerce within business enterprises, that commerce over the Internet, rather than by EDI for example, and had reached some $100 billion.[78]

By 2000, the Pew Internet and American Life Project was reporting that the key applications on the Internet of veteran users were, in order of priority, e-mail (97 percent), searches for information (87 percent), obtaining news online (78 percent), seeking job-related data (65 percent), purchasing products (64 percent versus 31 percent for new users), finding health-related information (63 percent), accessing financial information (54 percent), and listening to music online (already 45 percent and 33 percent for new users). Since downloading music became a major issue for the Music Industry (discussed in chapter 11), suffice it to note here that already by 2001 one-third of all veteran users, and nearly as many new ones (27 percent), were reported downloading music off the Internet. Only 18 percent of veteran users, however, were buying and selling stocks over the Internet.[79] These uses made up an extraordinarily broad set of applications for a technology that had only become widely available less than a decade earlier.

So who were these users who so rapidly appeared across the American landscape? They included students, senior citizens, the religious, small businesses buying and selling, people who wanted access to government services and information, shoppers, others doing the business of Congress, those buying airplane tickets, e-mailers, political campaigns, and individuals booking hotel rooms—the list is almost endless.[80] All ages participated; in fact, one government survey noted that in 2001 over half the nation had gone online (143 million, or 54 percent of the population). To be sure, the largest demographic groups using the Internet were children and teenagers, but no group was absent. The Telecommunications Industry scrambled to provide additional telephone lines and broadband services, and by September 2001, 20 percent of all Internet users accessed broadband services with the most widely deployed applications still the familiar ones (45 percent for e-mail, one third to search out information) at home and involving one-third of the American workforce.[81]

Major events in the life of the nation now involved the Internet. Much as people had turned to the telephone before to discuss major national events, or to television to see them, they folded the Internet into their suitcase of information-handling tools. Following the September 11, 2001, terrorist attacks, one study reported "that [Americans] were aggressively using all the means at their disposal to get information about the unfolding crisis and make [*sic*] contact with their networks of loved ones and friends."[82] They were still anxious consumers of TV news and restless users of the telephone, but they had also reached out to the Internet.[83] Within a month of September 2001, use of the Internet to discuss terrorism, the crisis, and other related matters skyrocketed.[84] Users also went to the Internet for training and development of new skills, to help loved ones deal with illnesses, to select schools and colleges for children to attend, to make financial decisions, to find jobs, to locate places to live, and to keep up with the war in Iraq in 2003.[85] As table 7.4 demonstrates, the variety of uses of the Internet kept growing.

While there are many statistics on the extent of use of the Internet, they are not perfect and, in fact, sometimes vary. But they are all consistent in one respect: they all bear out that the volume of users grew rapidly in the period 1994 to 2004, as did the variety of uses.[86] The data also suggest that online purchasing of goods

Table 7.4
Daily Internet Uses in the United States, 2002

Activity	% Performing	Activity	% Performing
Access Internet	57	Find political news and information	11
Send e-mail	50	Send instant messages	11
Find information	29	Watch video clip or listen to audio clip	10
Find news	26	Obtain financial information	10
Surf for fun	22	Obtain information on schools and training	10
Search for answer to specific question	19	Access a government web site	9
Obtain information on a hobby	19	Play games	7
Check on weather	17	Bank online	7
Conduct job-related research	17	Find health and medical information	6
Learn about a product or service before buying it	14	Participate in chat group	6
Find information about movies, books, other leisure activities	13	Listen to or download music	6
Obtain scores of sporting events	12	Obtain travel data	6
		Download music to one's PC	5

Source: Derived from a study done by the Pew Internet & American Life Project, January 21, 2003, http://www.pewinternet.org/reports/chart.asp?imge=Daily_A6.htm. The study involved the daily activities of some 64 million U.S. residents, and included 60 different kinds of activities. Only the most widely used are listed above.

and services had started in the late 1990s, when vendors added reliable encryption and other security software to protect the privacy of transactions. The public trusted most sites of small businesses, newspaper and television news, banks, insurance, and stockbrokers. Over half the population in 2001 also trusted nonprofit organizations and most U.S. government websites. So these institutions had made progress in creating environments that would motivate people further to use the Internet. There were also less trusted sites, such as those dedicated to e-commerce or providing advice to consumers.[87]

By mid-2003, some two-thirds of the population used the Internet, although to varying degrees. Interestingly, the rest still did not. Between 2000 and 2002, the percent of men and women using the Internet had increased from less than 50 percent, along with percents of all ethnic groups, ages, and across all income levels. The same held true for all levels of education, and rural, suburban and urban centers. Some 80 percent of all users found e-mail useful, despite the growing problem of spam beginning to clog e-mail by the end of 2002. A similar percentage believed the Internet was an effective source of information, but also that it could be dangerous (56 percent). Slightly less than half thought it expensive (43 percent), and half that it proved entertaining, even though they still complained that the Internet was difficult to navigate (40 percent).[88] In short, Americans had become comfortable with this new form of telecommunications.

The Internet was contributing to the transformation of American society into a far more information-intensive culture than in the past. The role of telephony in that process can be understood through the activities of telephone companies in providing access such as dial-up, digital subscriber line (DSL) service, and so forth. In the 1980s and early to mid-1990s, most people accessed the Internet dialing an 800 number. Dedicated, high-speed lines, however, became available to both businesses and individuals during the second half of the decade from local telephone companies and, in some places, local cable companies.[89] Because about 90 percent of American homes subscribed to cable, cable companies were well-positioned to provide DSL. As of early 2003, customers were subscribing to DSL services more frequently from cable companies than from telephone service providers. The number of users also grew substantially in the early years of the new century. For example, between the springs of 2002 and 2003, the number of users of various broadband services from all sources went from 13 to 21 million; DSL alone went from 7 to 9 million home users. To put this sharp growth into context, in mid-2000, there were only 5 million home subscribers to broadband services of all types. By mid-2003, roughly one-third of all homes had some form of high-speed communications provided either by telephone or cable providers, creating a highly competitive environment at a time when demand for such services was expanding rapidly, and reaching over 40 percent of homes and businesses by late 2003.[90] As of this writing, users in the United States had a plethora of options for acquiring this new application of telecommunications, creating opportunity and angst for the traditional Telecommunications Industry. Table 7.5 illustrates a pattern typical across many parts of the United States; it catalogues various providers of Internet access in Madison, Wisconsin, a community of nearly 400,000 people that includes the University of Wisconsin, large insurance firms, some 600 biotechnology companies and the state capital.

We should also point out that in the early 2000s, a new service came online that allowed people to send voice messages over the Internet, known as voice over IP (VoIP), which posed a major threat to the revenue streams of telephone companies.

There are some survey data to indicate what users did with their high-speed and dial-up subscriptions. Table 7.6 suggests patterns of use as of early 2003. By that time, Americans were online an average of 25 hours and 25 minutes per month from home and 74 hours and 26 minutes at work—over twice as long as any other nationality.[91] E-mail remained the most used application, although estimates and surveys varied from 35 percent of all Internet activity to as high as 84 percent.[92] A growing application on the Internet was peer-to-peer (P2P) file-sharing, which is the creation and exchange of data and media over the Internet (or earlier networks). This activity expanded dramatically in volume after tools were developed to facilitate such sharing, for example, Napster for downloading

Table 7.5
Internet Access Providers in Madison, Wisconsin, Metropolitan Area, circa 2003

Provider	DSL price/month	Type
America Online (AOL)	$54.95	ISP
AT&T Worldnet	None	Old Bell
Badger Internet	N/A	ISP
Berbee	$150	ISP
CenturyTel	$29.95–$49.95	ISP
Charter Pipeline	$29.95–$39.99	Cable
CiNet-Customer Internetworking	None	ISP
CoreComm	None	ISP
EarthLink	None	ISP
Efanz	None	ISP
Global Dialog Internet	None	ISP
Jefferson Internet Services	None	ISP
JVNet Internet Services	None	ISP
MHTC.net	$39.95–$59.95	ISP
McLeodUSA	$31.95	ISP
Merrimac Communications	None	ISP
Microsoft Network	$34.95–$49.95	Software
Powercom	None	ISP
SBC Yahoo!	$26.95–$159.95	Old Bell
TDS Metrocom	$24.95–$50	Independent
TDS Telecom	$19.95–$54.95	Independent
TerraCom	None	ISP
Ticon.net	None	ISP
United Online	None	ISP
Verizon Online	$35	Old Bell

ISP = independent service provider. The reason some firms did not charge is because they provided this service as part of some other offering for which they charged.

Source: Data collected by Judy Newman, "The Trend for Internet Users Is Speed," *Wisconsin State Journal*, October 19, 2003, p. A5.

and transferring music files in the late 1990s (see chapter 11). By 2003, KaZaA software had become another widespread application, with millions using this software to download music, for example. Most downloaded P2P files were music, often MP3 files, accounting for approximately 60 percent of all the files on people's PCs in 2003.[93] P2P files also included, however, play lists, images, video, software, and MP3 files, with audio gaining the lion's share (over 60 percent).[94]

If we step back and look at the broad patterns over a longer period, say from

Table 7.6
How Americans Used Their Internet Connections on a Daily Basis, 2003

Application	Using Broadband (%)	Using Dial-Up (%)
Reading news	41	23
Conducting work-related research	30	15
Participating in groups (chat rooms)	12	15
Streaming multimedia	21	7
Creating content	11	3
Downloading music	13	3

Source: Pew Internet & American Life Project, http:/www.pewintlnet.org (last accessed 3/12/2005).

about 1997 to the end of 2004, several trends become quite clear. The FCC reported the most important one in mid-2004 when it called out that Americans had reduced their use of residential phone lines by 50 percent since 1997. Consumers had other options, such as cell phones and the Internet, with which to transport conversations and data. The second trend, growing out of the first, was that information technology providers, such as software firms and other telecommunications companies, were displacing the old Telephone Industry. As Daniel Berninger, a student of the industry, observed:

> The data shows what we all know from day to day experience—the information technology industry is annexing communication as an application—email, IM, VoIP, e-commerce, etc. We have fewer and fewer reasons to use plain-old-telephone-service. The numbers reflect a world where RBOC's focus their energies on regulatory battles allowing them to raise prices, and the information technology industry works to deliver even more powerful communication platforms and applications of the Internet.[95]

He likened the process to a "hostile takeover of communication by the information sector." This mega-trend became the epicenter of the industry's crisis.

Before moving to a discussion of wireless communications, it is valuable to note how much of the world's information flows through the Telecommunications Industry. Today the vast majority of the world's information is electronic, moving through radio, television, telephone, and over the Internet. Estimates prepared by researchers at the University of California—Berkeley suggest that in 2002, nearly 97 percent of information still moved over the telephone (dial-up or private networks) and an additional 2 percent over the Internet, leaving radio and television to share only a fraction. In short, the Internet was the second largest channel for the movement of electronic data and communications on a global basis. We know that across the world there existed 1.1 billion main telephone lines, of which 190 million were in the United States, which housed one-third of all Internet users. Above we noted that U.S. users of the Internet spent twice as much time using this tool as populations in other countries. Putting it all together leaves us with the conclusion that Americans funneled a vast majority of their electronically based information over both the Internet and telephone. By any measure, the Internet had become an essential element of the nation's style of working and playing.[96]

The Age of the Cell Phone—Finally

While it is too early to describe the long-term consequences of wireless communications, it is easy to see how the wide use of this technology could have fundamental influences on the way Americans work and play, and equally important, the configuration of the Telecommunications Industry in this country. During the first years of the new century, the rate of adoption of wireless phones in the United States far exceeded that of the 1990s and came closer to patterns already evident in Western Europe. Before 2000, adoption had been affected by the lack of national technical standards, spotty coverage, confusion over offerings, and high prices. Combined, these factors go along way to explaining the way demand evolved over the decade for wireless services from the mid-1990s to the mid-2000s.

Wireless, or cell phone, applications are not limited simply to telephones unattached to copper lines. There is a growing list of devices (hence uses) that go beyond telephones, pagers, PDAs, and wireless Internet and included sensors on machinery and electronic tags on vehicles and large inventory. While most wireless in the 1990s consisted of telephone and pager services, that limited list began broadening by the end of the century and continued expanding during the early years of the new millennium. It is important to keep in mind that it was not always obvious in the 1990s that wireless held out any real potential for displacing wired POTS—hard to believe in hindsight, since now one would be hard pressed to find any commentator saying anything but that the future of wireless will be massive and at the expense of wired telephony.[97] Only one held a contrary view—none other than George Gilder, the unabashed proponent of the growth in bandwidth. He displayed some reservations as late as 2000: "The key error is the assumption of widespread cellphone [sic] coverage. Even in the U.S., cellphones reach less than 20 percent of the territory. At least 50 percent of the country will never be economically served by cellular."[98] Since his forecasts about technology were often too optimistic and subject to the risks of any prediction, we can conclude from this quote and from what actually occurred that he remained consistent in his understanding of what was happening.[99]

Expanding beyond the cell phone market by focusing on wireless, computing and consumer electronics vendors armed themselves with a strong compunction to optimize connections to each other and were not denied. In wireless, it became so complicated a process that one needed intelligence in a product or device to optimize any connection. So service providers had to open their business to other vendors whether they liked it or not, just to create business.

It was the proverbial foot in the door to the wireless business that brought others from both industries, such as Nokia and Samsung from the radio transmission business. Quickly the wireless business equipment vendors segmented into two camps: those who were good at transmission equipment (both devices and tranceivers) and those specializing in central office applications (for example, switching centers, registers, and controllers). Therefore, IT companies were able to rush into territory that should have belonged completely to the Telecommunications Industry, and the digital provided the excuse and ability to do so. The interplay between telecommunications and IT companies has yet to be studied by economists and historians, but remains a subtle but important example of the shift of work and opportunities among industries.

The list of companies attempting to enter the market for wireless is a long one, extending beyond the Telecommunications Industry to IT providers (such as IBM, HP, and Microsoft) and to major consumer electronics firms around the world.[100] In the early to mid-1990s, however, their devices were still expensive. Yet interactions among industries were intense; one example of corporate genealogy illustrates the point: HP had acquired Compaq, which had acquired DEC, which had made specialized systems for telecommunications networks. The lines between IT and telco professionals were blurring, making it harder to distinguish a network engineer from an IT professional. Digital technologies were blurring the lines between both industries and skills of their workers.

It turned out that a little pessimism about the cellular situation in the United States was not without cause. The FCC's chairman during the Clinton Administration, Reed E. Hundt, proved quite candid:

> With the Commission's [FCC] connivance, the local telephone company meanwhile charged huge prices for connecting wireless calls to the existing network. Lack of competition and these high interconnection charges raised prices for mobile service to about ten times higher per minute than for wire-based service. Cellular phones therefore were primarily for the well off. The United States wireless industry then marched down a path of high, anti-competitive prices that gave America by the early 1990s the worst cellular system of any developed country in the world.[101]

He resolved to fix that problem and in 1993 obtained from Congress authority to auction bandwidth to service providers, making available to them the necessary spectra (radio waves) to transmit calls and data. In 1994, the FCC auctioned raised $7.7 billion for the U.S. Government and resulted in smaller, lighter weight handsets. The spectrum that was released operated at a higher frequency than previously issued, and in order to optimize its use, companies had to develop network technology to be more efficient. That requirement meant the manufacture of smaller range cell site transceivers. The tradeoff in design was initially higher priced devices, but each device was to be closer to a cell site when it made a connection, so it would not have to be as big as before and would not require as much battery power. We care about such technical issues because smaller, lightweight devices proved attractive in the growing consumer market for wireless devices. With new and old providers, competition now heated up, prices fell through the 1990s, and the number of cell phone users increased. Growing volumes began taxing the industry's ability to handle the increased number of calls, compelling further investments in digital networks in the second half of the 1990s. In turn, that led to another wave of even smaller, more lightweight handsets, longer battery life, and overall cost reductions. Vendors began adding new services to their cell phone networks, such as an assortment of PDA functions and, by the end of the century, the ability to receive and send e-mail via cell phone.[102]

By 1999, the cell phone market was one of the fastest growing in the U.S. economy. It seemed every business professional had to have one, creating a 14 percent growth rate in business demand just between 1999 and 2000. As the recession set in during early 2000, demand for expensive cell phone services slowed; by the end of the year only 2 percent of all the 108 million cell phone subscribers in the United States had signed up for mobile data services. The economy was part of the problem, but so too was the cost of these services, leaving a dearth of web-enabled cell phones at the end of the decade.[103]

Meanwhile, technological innovations continued, driven "by the fundamental technologies of silicon and software," according to Daniel C. Stanzione, the president of Bell Labs from 1995 to 1999.[104] Without diverting our attention to a discussion of the evolution of the technologies, suffice it to say that standards for national communications were proposed, discussed, and partially implemented, while a third generation of technology made its way through development, emerging at the new century. 3G, as it increasingly came to be known, could handle the large data transfer capacities required for wireless Internet, driving down costs of transmitting information.[105]

The result was economic. As one industry observer noted at the turn of the century, "wireless communications has achieved the price, utility, and service levels necessary to attract both the business market and the mass consumer markets."[106] Despite a recession, as of late fall 2002, over 105 million subscribers were fueling a fundamental shift in American telephony, having expanded the demand for services through subscribership by more than 36 percent between 1990 and 2000, and revenues in excess of 27 percent each year. It reflected the same kinds of dramatic growth rates experienced by the nation with the Internet.[107]

So how were people using wireless technology in the year 2000? The age groups most likely to use the technology were from 25 to 55, although over time all ages increasingly participated. Retail customers made up between 65 and 75 percent of the market. About one-third had acquired phones for reasons of security and safety—they could call for help with a flat tire or if they felt in danger; in the earliest days of wireless phones, this was the selling argument used most frequently by vendors. Another one-third acquired these phones for convenience; and just over a quarter got them to conduct business. Of all the subscribers in the nation, one survey suggested 57 percent were personal users, another 27 percent were business, and the rest were both. Declining costs often determined one's service provider. Business users conducted conversations and created the greatest demand for data transmission. Increasing productivity beyond the boundaries of an office proved essential for e-mail, movement of data and reports back and forth, and in the activities of sales personnel. One study done in 2000 found that 30 percent of business users engaged in e-mail, faxing, and voice mail over their cell phones; another 15 percent of people from all walks of life used wireless for banking, and just slightly fewer (13 percent) to access the Internet. Other uses included mobile communications, telemetry, playing games, and telematics.[108]

As users became more comfortable with the technology, they began relying on it at the expense of traditional land lines. By 2003, about 15 percent of the 134.6 million U.S. users were reporting that cell phones were displacing their land lines, although only 3 percent had actually canceled their traditional wired phone service. Clearly, a transition was in progress, one not yet completed. As with broadband and DSL, a large number of options kept appearing for customers, which helped intensify competition. Using the same city of Madison, Wisconsin, to compare offerings as we did with Internet providers, table 7.7 lists cell phone providers as of late summer 2003.[109]

Despite the positive report about the expanded use of cellular telephones in the United States, progress was slowed by regulatory practices. In March 2004, Walter Mossberg, who is widely listened to regarding technology, reflected the views of many in the Telecommunications Industry and within IT circles when he reviewed a new generation of cell phones introduced in Europe: "Many of these advances are unavailable in the United States, where we have crippled our wireless phone system by failing to adopt a single transmission standard and by handing too much power to slow-moving wireless carriers." He was miffed that the Europeans were about to launch broadband cell phone services, which were not yet available in the United States. However, he also acknowledged that the Americans

Table 7.7
Cell Phone Providers, Madison, Wisconsin, Metropolitan Area, circa 2003

Provider	Monthly Charge
AT&T Wireless	$29.99
Cingular	$29.99
Einstein PCS	$39.99
Nextel	$39.99
Sprint	$40.00
U.S. Cellular	$35.00
Verizon	$39.99

Note that while the price ranges are very close, they represent various plans, number of minutes, different charges for long distance, minutes, roaming, and activation fees.

Source: Judy Newman, "Landlines Not Needed, Some Phone Users Decide," *Wisconsin State Journal*, August 10, 2003, pp. A1, A5.

did a better job in "melding of a phone, an organizer and serious e-mail capabilities in a small, portable device."[110] His comment about e-mail represented only the tip of a new iceberg as the wireless Internet emerged.

The success of wireless in the United States spurred a variety of interesting investments in alternative wireless technologies and products, each of which carried its own economic and business advantages and disadvantages: LMDS, MMDS, Terabeam, mesh wireless, UWB, and WiMax, to mention a few. Accessing the Internet with a PC, PDA or some other device without benefit of wired communications, such as using radio communications just as with a cell phone, increased. More people in the United States can now access the Internet in a wireless fashion allowing people to use their cell phones as PCs, for example, doing research, sending e-mail, playing games, or placing orders for goods and services. This merger of telecommunications and the Internet is simply the latest iteration of the process of merging of computing and communications underway for some 40 years.

This new use of multiple technologies is such a recent development that it does not yet have a history, and accurate numbers on how many users are involved remain sketchy. [We know that by late 2003 more cell phones were accessing the Internet than PCs.] Forecasts of future use speak of the application becoming nearly ubiquitous.[111] In a lead story on wireless for a November 2003 issue, *Time* magazine noted its rapid expansion in business, home life, and even among those Americans living in recreational vehicles (16 percent).[112] Starbucks, the coffee retail chain, had completed enabling over 2,400 locations with wireless capability by late 2003, making it one of the largest, if not most extensive, wireless fidelity (wi-fi) standards networks in the United States. Many other companies across numerous industries were enabling the technology at their facilities, leading the way for home owners. The pattern was historic in form: businesses adopted new technologies before individuals since they had the resources to invest in them first, though as of late 2003, only about 5 percent of all employees had started using this application at work.[113] Table 7.8 lists some of the early corporate adopters at the turn of the century.

The implication for telephone companies was obvious, as they struggled to figure out what role they would play in this emerging new application. As people increasingly moved from wired to wireless communication—not only costing tra-

Table 7.8
Early Major Corporate Users of Mobile Internet Applications, Circa 2001–2003

BEA Systems	IBM	Starbucks
Expedia	Oracle	Ticketmaster
Ford Motor Company		

Sources: Jouni Paavilainen, *Mobile Business Strategies: Understanding the Technologies and Opportunities* (London: Wireless Press/Addison-Wesley in partnership with IT Press, 2001): xi; "Wireless Society," *Time*, November 3, 2003, A23–A24.

Table 7.9
Leading U.S. Cellular Phone Providers, June 2003

Provider	Number of Customers (millions)
Verizon Wireless	34.6
Cingular Wireless	22.6
AT&T Wireless	21.5
Sprint PCS	15.3
Nextel Communications	11.7
T-Mobile USA	11.4
Total	117.1

Source: Adapted from data in "Calling the Winners," *Wisconsin State Journal*, November 16, 2003, p. D1.

ditional phone companies market share but also pitting them against numerous and different competitors—the industry faced this nagging, yet critical issue. The story does not yet have an ending, but the key contenders could well be those firms that have already established a significant presence in the wireless marketplace (see table 7.9).

Other Uses of the Digital

The years spanning 1994 to 2005 saw the proverbial "business as usual." Fiber optic cables were laid under many streets in America, more telephone subscribers signed up (often for second lines for Internet hookup or for growing businesses), and all switching and transmissions were going completely digital. Annual expenditures by all households grew steadily in the period, from $690 in 1994 to $849 in 1999, continuing into the new century, driven by rising fees for local phone service and second lines. Put in other terms, monthly bills went from just under $50 in 1994 to roughly $70 in 1999. This growth in expenditures was part of a much larger historic process. In 1984—the year of the breakup of the Bell System—annual average household expenditures for telephone services were $435, or just over $36 a month; these grew steadily over the years, despite the fact that the cost of individual phone calls declined, primarily for long-distance conversations. What is interesting, and a reflection of the growing income of Americans in this period, is that these sums represented the same proportion of household expenditures over the period 1980 to the end of the century, about 2 percent.[114]

Business volumes remained high. In 2000, for example, the U.S. economy expended $200 billion on telecommunications services: $100 billion for local telephone services, another $90 billion for long-distance expenditures, and the rest for other services. Expenditures for local calls reflected growth rates year over year equal to that of the GDP, but for long-distance, growth rates were double digit.

Lines dedicated to data transfer were increasing at nearly 200 percent each year by the end of the century, while voice traffic grew at single-digit rates.[115] At the end of the century, Americans used more telephones per capita than did the population of any other nation in the world.

Converging technologies made it possible in the last years of the decade for people to send facsimile transmissions at home and the office in ever-increasing numbers, linking this activity to e-mail, even to voice mail because now one had a suite of ways to transmit data, and often people used a combination of these approaches. Teleconferencing became ubiquitous in businesses and relatively common with home telephone systems. Automated attendant systems using touch-tone telephones and smart cards also spread widely across the economy. The ability to access large information databases continued expanding so that people could look up their bank balances, check on their company 401(k) funds, and order tickets to events through automated voice systems, paying for these using credit cards. These applications seeped into our daily lives, making it difficult to remember when they were not widespread. As one commentator of the time explained:

> Although these applications require many times the bandwidth of a telephone call, digitization offers the ability to compress the signal in order to reduce bandwidth requirements, thereby increasing even more the capacity of telecommunications networks to transmit information. The result is that video signals may use only a fraction of a broadband channel, and voice signals may also be compressed.[116]

Demand did not diminish. In his letter to stockholders in 2001, Charles R. Lee, the CEO of Verizon, captured the essence of what was happening: "Verizon's success will depend on our ability to bring to our customers the next generation of broadband, wireless data and many other products and services that technology innovation makes possible."[117]

Conclusions

Over the last half of the twentieth century, Americans used every new form of digitally based technology that became available, attracted by devices' functionality, cost, and reliability. The traditional perspective of the consumer, or the firms and agencies they worked for, is that cost performance determined the extent of demand. While cost was and remains a very important consideration, there was more to the story than dollars. Households did not hesitate to spend more on telecommunications, to acquire more telephone lines, augment them with wireless phones, and add PCs with which to access the Internet. The same held true for businesses. A teleconference was always cheaper than having a staff travel to other cities for monthly meetings; online applications were just plain less expensive than manual or old "batch" uses of computing. My suspicion is that the Telecommunications Industry could have charged a great deal more for its services over the past half century and have collected if the FCC would have allowed that.

Justification for all this new technology, from the perspective of its users, was not based solely on what it cost to use a new form of communications versus an existing or prior one, as is normally the case when comparing one new tool or technology to another. Rather, benefits and costs were of a new communications offering versus either another way of doing things (such as snail mail versus e-mail, land lines versus portable phones) or to gain access to new capabilities previously unavailable (e.g., use of the Internet to find information or place an order).

In an earlier study I conducted on how Americans had used information in their private and work lives over the past three centuries, I concluded that one common feature was their simultaneous use of multiple types of information technologies. In the course of a day, a resident in North America in the late twentieth century would as a matter of course listen to the radio, watch a little television, use both a land line and a cell phone at work, look up a telephone number on a PDA, spend five minutes reading a book, and, of course, glance at the newspaper and at what the mail man had delivered. Book sales were never higher, magazine subscriptions were extensive, and more people watched more television and cable than in earlier decades.[118] We have seen the same pattern at work with their use of telephony. With the one major exception of picturephones, Americans used every new function introduced by the Telecommunications Industry and its growing array of competitors from other industries.

The industry and the FCC understood instinctively this near-insatiable appetite Americans have for telecommunications. In fact, the FCC's multi-decade strategy of insisting on universal access reflected that national will. Insisting on a competitive environment and subsidized universal access was simply a strategy to meet that demand in a way consistent with the nation's attitude toward capitalism. The severe financial difficulties faced by the industry at the end of the century, caused by significant and rapid changes in technologies on the one hand and competition on the other, can be traced to the great hunger Americans had for new forms of telecommunications, which cost the industry a great deal to keep up with and that, if not met, would have resulted in certain corporate death. That reality had been understood only in highly theoretical terms prior to the breakup of the Bell System in the mid-1980s. With the ensuing competitive field, however, new technologies that had been percolating for years exploded onto the market, requiring fundamental transformations in existing business activities of not only the Telecommunications Industry (to become more efficient and productive) but also their customers (to enjoy and exploit the new tools). The enormous investments made quickly in infrastructures created large debt overhangs that beset the industry as it entered the twenty-first century. Thus, if customers had not desired new forms of telecommunications, the industry would not have had to worry about fiber optics, packet switching, the Internet, cell phones, wi-fi Internet, cable, PDAs, and so forth; POTS might have been sufficient with its land lines.

A major component of any history of this industry across the second half of the nineteenth and the entire twentieth centuries and now in the new millennium was, of course, the role of technology. In the previous chapter we looked at how the industry responded to new technologies, resisting their adoption in some cases

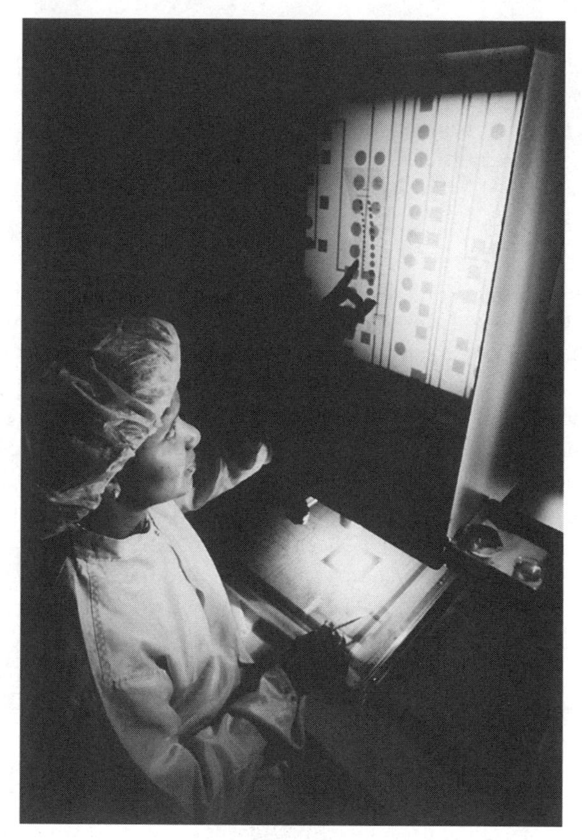

Advanced basic research continued in telecommunications companies right through the twentieth century. Bell Labs researcher at work, early 1980s. Reprinted with permission of AT&T.

but, in aggregate and in the end, embracing every one of them. Customers reacted to technological innovations, responding to the functions and costs involved. Telephony was perhaps the most extreme example of a widely used set of technologies that Americans and managers in organizations knew little about. Every neighborhood and company had its PC jockey, its radio ham, and valued expert on automobiles. However, very few people outside of the Telecommunications Industry understood communications technologies, and most of those either worked in the IT community or in universities. Even the proverbial teenage hacker knew little about telecommunications, despite knowing a great deal about software programming.

Ignorance of how a telephone works did not stop people from using the technology. In that sense, they were both practical and trusting. They were practical in that they cared less about technical standards and technologies and more about what their phone systems could do. In fact, they simply did not know, for example, that the industry was moving from analog to digital switches in the 1980s and 1990s; instead, they just knew that at some point they could get caller ID or voice-mail options. They were trusting for two reasons. First, over the course of a century, AT&T and the regional Bell companies had provided the nation a level of service unequaled anywhere in the world. That sense of quality and reliability remained a feature of the industry deep into the 1990s, and it was not questioned until a competitive field made choices complex (in selection of cell phone providers, and for a short time in the 1980s, long distance providers)—and until the accounting scandals in the early 2000s. Thus users believed the technologies would work until proven otherwise and, for the most part, they did. The one exception arose in the quality of cell phone services, which remained a severe problem in the second half of the 1990s, presenting performance issues not resolved at the time I was completing this chapter (2005).[119] Second, Americans who had experience with European cell phones knew what quality should be and did not hesitate to pressure regulators and legislators to demand improvements. The FCC and the Congress played crucial roles in addressing this issue, hence their initiation of new rules in November 2003 making it possible for Americans to keep their cell phone numbers but transfer them to other vendors, thus motivating vendors to improve service.[120]

Today any observer of the industry has a point of view about the effects of the Telecommunications Act of 1996. My files are fat with articles from academic journals and newspapers, and several books on the subject clutter my bookshelves.[121]

This chapter demonstrates the main consequences of the Act. The law did not slow or stop the flow of new innovations and offerings from reaching the market or from influencing the efforts of telco management to improve its internal operations. Three flood gates opened the industry to that development: First, the breakup of AT&T in the 1980s with the subsequent many rulings and actions taken by the FCC. Second, technological innovations and circumstances could still enter the picture with absolutely no say on the part of the industry or even of the FCC; in this case I am referring to the creation, expansion, and wide use of the Internet, a development that could eventually become part of the Telecommunications Industry. Third, consumers of telecommunications services were always eager to embrace every new development and offering, often faster than the industry itself. These patterns suggest that, regardless of the law, economic incentives to continue developing and expanding use of telecommunications would have existed anyway.[122] That left commentators with only one fundamental issue to discuss: to what extent did the new law speed up, slow down, or make more or less efficient the nearly 160-year-old process of injecting telecommunications into the American economy?[123] The original law covered so many aspects of telecommunications, from telephones to cable, from broadband to broadcasting, from pay phones to digital TV, that one could debate its efficacy *ad infinitum*. For our purposes, it is not so relevant because what the American experience with telecommunications has been is quite clear. All through the twentieth century, and continuing to the present, Americans considered their telecommunications to be crucial. The introduction of the digital into the world of telecommunications sped up the process of adoption and expanded enormously the variety of uses for which communications could be used. The effect on the old Telephone Industry is now clear: it is under a massive attack by new entrants, many coming from outside the world of telco—the loosely defined world of the Internet and the more closely defined community of cell phone providers.

The implied theme—that telecommunications is all about the movement and use of information—is made more explicit when looking at the media sector of the economy. The role of the digital in that sector is also extremely important because it intimately influenced the work of its industries. For that reason, the next several chapters are devoted to a discussion of the digital in these media industries, beginning with the oldest ones, those devoted to paper-based products: books, newspapers, and magazines.

8

Role of Media and Entertainment Industries in the U.S. Economy

'Tis True, There's magic in the web of it.

—William Shakespeare, *Othello*, 1621

It is tempting to describe the effect of computing and communications in many industries as revolutionary, and in fact many commentators routinely do. Yet as argued in this book and its predecessor volume, evolutionary is the more realistic descriptor of what has happened, so far. But now, we can look at a collection of industries that are rapidly revolutionizing in wide-ranging ways via the digital hand. As the rest of this book demonstrates, digital technologies have made possible three transformations across a variety of industries that are conveniently called either media and/or entertainment. First, digital technologies were used the same way as in other industries to handle accounting, data collection and distribution—in short, in support of normal business operations. Second, the digital made possible new ways of doing the core tasks of these industries: delivery of news over the Internet, digital animation in movies, games played on hand-held computers. The flexibility and diversity of applications made possible by digital technologies led to a variety of new products and services one could only dream about in 1950.

Third, these technologies directly and unequivocally led to the convergence of products, services, functions, companies, and industries to such a degree that, just as with telecommunications, it is difficult to define their boundaries. Simply put, various media industries operated relatively independently of each other in 1950, such as the newly born Television Industry and the Newspaper Industry. By the end of the century, there were firms that owned newspapers, book publishers, radio stations, movie studios, music companies, and had Internet presences. Medium-sized companies became global giant firms, redrawing the borders of industries as yet unsettled in the early 2000s. Mergers and acquisitions became a big a story in these industries much as did the role of regulators in the finance or telecommunications industries, all oiled by the same facilitative capabilities of digital technologies.[1] To describe these various industries in the late 1990s or early 2000s is to engage with a complex, fluid, and very confusing topic. To be sure, elements of that story have embedded in them not only many technologies but also economic and business considerations. Keep in mind that this convolution of companies, changing industry borders, and new offerings represent a phenomenon barely two decades old. These industries began as nearly discrete entities; by examining how they came independently to embrace the digital, we gain insight, first, into the fluid and complex state of affairs today and, second, into how technology affected their constituent enterprises.

To be sure, most commentators have been observing for two decades that convergence was taking place, within which just about every conceivable digital and telecommunication tool has been used independently, mixed, and matched, contributing to the resulting actions of organizations also mixing and matching activities and industries. One brief example of convergence illustrates the megaprocess at work at the business level that makes it possible to use the term "revolutionary" in a responsible manner. Simon & Schuster (S&S) was the largest book publisher in the United States in the 1990s; it had been in the industry under various names and forms since the 1920s. In the 1990s, Paramount Communications—a large media company—owned S&S. Paramount, originally just a producer of films and later of TV programs, was acquired by Viacom in 1994. That year Viacom participated in a variety of media industries: book publishing, films, television (cable and otherwise), theme parks, sports, and radio broadcasting. That same year it merged with Blockbuster Video, which also was in the music business but was best known as a distributor of television shows and movies. To complicate matters, even within the lines of businesses such as book publishing, there were subsectors. In the case of S&S, in the 1990s it consisted of the old S&S, Free Press, Touchstone, Scribner, Pocket Books, DC Comics, and more than a dozen other publishing companies.[2]

A similar genealogy could be laid out for any of two dozen other large firms, including those centered on the newspaper, television, and radio industries.[3] This sector was—and continues to be—difficult to define. The opportunity to leverage market convergence stimulated mergers and the crossover of markets as occurred, for example, with S&S and Paramount. To be accurate, however, the potential synergies of using the digital from combined organizations normally lagged far behind the advantages that could be gained by expanding the number and types of markets one could reach, which could be done without changing any current use of the digital. However, the technology did exist, creating an environment—indeed a mindset—that encouraged M&As. Thus on the one hand we had mergers occurring but simultaneously on the other the adoption of a broad array of digital and telecommunications applications that in combination with mergers

Table 8.1
Fifteen Largest U.S. Media Companies by Amount Spent on Advertising, 1996
(Ranked from most to least)

Time Warner	Cox Enterprises
Walt Disney Corporation	Knight-Ridder
Tele-Communications Inc.	New York Times Company
NBC TV (owned by General Electric Company)	Hearst Corporation
CBS Corporation	Viacom
Gannett Company	Times Mirror Company
News Corporation	Tribune Company
Advance Publications	

Source: AdAge.com, http://www.adage.com/page.cms?pageID=871 (accessed June 20, 2003).

made possible the substantial changes which occurred in media and entertainment industries.

As an illustration of how convoluted all these industries had become, table 8.1 lists the top 15 media companies by the amount of money they spent on advertising. What is interesting to note is the mixture of different types of companies all under the heading of "Media." Each had holdings in various industries that before World War II either operated relatively independently of each other (such as book publishing and newspapers) or did not exist (television). A list of the largest media companies, ranked by sales revenues, would be different, but the pattern would remain the same, namely, they would contain businesses originally from multiple industries that had started the process of sharing products across industries, and made possible in part by the digital.

How could digital technology contribute to such changes over time? The short answer is that this technology, along with telecommunications, allowed management to do three things differently from before. First, accounting software and the ability to collect and communicate in a timely manner the information on how businesses were doing made it possible to manage disparate lines of business in a practical and cost effective manner.[4] Second, the technologies themselves made it possible to integrate businesses, such as happened with newspapers, which made available electronic versions over the Internet or delivered exclusive sources of information and entertainment quickly and inexpensively to acquired television and radio stations.[5] Third, new offerings, even new industries, could be formed as natural extensions of prior industries and firms, such as electronic games made from software and hand-held computerlike devices. The combination of all three elements operated across most of the second half of the twentieth century, although they were most aggressive in the 1980s and 1990s as the technologies morphed into applications one could readily adopt. As in other industries, there was the cumulative effect of building up inventories of prior applications and experiences which also played a strong role.

If one had to pick which of the three factors proved most influential, the second and third appear dominant. Much as we saw with the Brokerage Industry, early on media industries were run by individuals who had little appreciation for the economic benefits of automating back-room office processes, for example, and were drawn instead to new products and ways of reaching markets. To be sure, IT use in a form that made sense to media industries did not come along until the 1960s. That is why, for example, the initial wide use of computers in the Book Publishing Industry is a story about the 1970s, instead of the 1950s or 1960s. Book publishers began using computers to manage inventory for the first time in the late 1960s and early 1970s, well over a decade or more after automotive and other manufacturing industries had already extensively utilized computers to help do that work. My story about the media industries is thus heavily skewed toward the last two decades of the century, for when the digital took off in media industries it had the same overwhelming effect as did the Internet in so many other industries—rolling over companies, industries, products, and the personal lives of customers and members of the industry.

To a large extent, this chapter introduces the industries which produce and distribute media, (content) that serve to inform and entertain. Increasingly, companies that do that are being called media firms in their industries. The digital hand helped to shape the various industries selected for examination in this and in subsequent chapters.

Emerging Definitions of Media Industries

Conventional wisdom is evolving to a point of view that these industries are largely about entertainment and are not simply purveyors of information, news, music, movies, and so forth. In recent years, economists and other commentators have focused on the expanding role of media as entertainment; as purveyors of experiences, and entertainment as part of media; and perhaps most intriguing, the growing role of entertainment in the lives of Americans.[6] That last point should be of no surprise because, as a society becomes wealthier, it has more resources to spend on entertainment. Given the enormous expansion of the U.S. economy in the second half of the twentieth century, Americans possessed the economic wherewithal to seek out information and entertainment in larger amounts than before, and their actions took a particularly high-tech tone thanks to the availability of so many digital and telecommunications tools.[7] As time passes, economists and industry analysts have been converging in their view of what comprised this sector late in the century. In a detailed study of the economics of entertainment, Harold L. Vogel listed the industries that comprised what he called the Entertainment Industry, relying extensively on media technologies—ones that did not necessarily have either a SIC or NAICS code at one time or another. Table 8.2 lists the ones he included in his definition. Vogel also analyzed live entertainment that also used digital tools, although that was not a criterion for selection

Table 8.2
Components of Media-Dependent Entertainment Industry, circa Late 1990s

Book publishing	Magazine publishing	Television broadcasting
Cable Television	Movies	Television programming
Games	Music	Toys
Internet	Newspaper publishing	

Source: Harold L. Vogel, *Entertainment Industry Economics: A Guide for Financial Analysis* (Cambridge: Cambridge University Press, 2001).

by him for analysis. These included gambling, sports, the performing arts, and amusement and theme parks.[8]

U.S. government economists also have their definitions. Although more restrictive than Vogel's, they are similar. Official definitions list network (also called broadcast) TV, cable TV, recorded music, newspapers, books, magazines, and home video; sometimes they also include the Internet.[9] Shifting definitions is not a new problem for students of the sector; those who began defining emerging sectors, such as the "Information Marketplace," or the "Information Economy," early on bumped into this problem.[10] One economist at Harvard University in the early 1980s defined media and entertainment together—as became the custom of many U.S. government economists—cataloguing these as advertising, broadcasting (radio and TV), book publishing, cable TV, news wire services, motion picture distribution and exhibition, newspaper publishing, organized sports admissions, periodical publishing, printing (book and commercial), radio and TV communications equipment, and theaters.[11] Benjamin M. Compaine, who put this list together, argued that one of the elements they shared was their reliance on electronic distribution of information. The one statement we can reliably make is that, regardless of one's definition, media and entertainment industries have collectively played a growing role in the American economy. While contributing only a small percentage to the GDP over the years, media industries were some of the largest exporters in the U.S. economy, and in no industry more so than movies, followed by television shows and books.[12]

These industries collectively caused the evolution and dispersal of a large range of technologies across the American economy. Advances in digital graphics, for example, are being extended by both the games and movie industries. Other industries have distributed digitally based media; CDs and DVDs quickly come to mind from the music and movie industries, respectively. Their use has contributed to the overall awareness and reliance on the digital across all social and economic sectors of American society. In the process, the digital has displaced old entertainment technologies such as the phonograph record, even music tapes (reel-to-reel, 8-track, and cassettes) while making it possible to see movies on demand (via videos, downloads, pay-per-view programs, cable TV, and DVDs). Newspapers in 1950 could only be read in paper editions; now one can see the same stories provided by the same newspaper publishers on the Internet or summarized by TV and radio broadcasters. True, many technologies have contributed to this broad variety of ways for delivering information and entertainment, not just the digital. But it has been the increasing reliance on various digital technologies, formats, and applications that over time has complicated (some say enriched) the diversity and flexibility of information and entertainment. In the late 1990s, two leading experts on the role of the new media in American society, John V. Pavlik and Everette E. Dennis, put the case simply:

> From virtual reality to the World Wide Web, the new media technology landscape is as diverse as it is fast changing. These new technologies are radically transforming almost every aspect of how we communicate and with whom, as well as just about any other dimension of our lives, from dating to making money to health care.[13]

Because of the effects on individuals, how these diverse industries embraced the digital is as important an experience to understand as it is itself entertaining.

So, are we to include the Consumer Electronics Industry in our mixture of media sector industries? As of this writing, they barely are by historians and economists. Yet these companies overwhelmingly make products that are for entertainment or to store and transmit information, often also serving as the physical tools by which other media industries deliver their products (for example, television sets, DVD players). The plethora of products from the Consumer Electronics Industry is stunning and clearly suggestive that it is part of the media milieu; table 8.3 lists products available for sale in the United States during the Christmas season in 2003, drawn from *Parade* magazine, a Sunday newspaper insert.[14] Companies providing media entertaining products included E-Legend, SilverDial, Palm Pilot, Compaq, Olympus, Samsung, Panasonic, Canon, Handspring, Microsoft, Sony, Nokia, Motorola, Nintendo, Apple, Dell, Fisher Price, and Gateway Digital. Note that some of these companies sit in the PC Industry (Apple, Compaq, Dell, Gateway), photography (Canon, Olympus), Software Industry (Microsoft), wireless telecommunications (Nokia, Motorola), Toy Industry (Fisher Price); the rest are in either consumer electronics (Panasonic, Sony) or in the emerging Games Industry (Nintendo).[15] It would seem that the proverbial "everyone" was participating in the media sector. One should also observe that each product listed above and in table 8.3 had, literally at its core, digital technology, making possible the functions of each device.

Note that the list above also included IT industries. Intel made chips for electronic games, and it is in the Semiconductor Industry. Microsoft and other software firms sold video games or offered access to them over the Internet. But the most interesting development involved Apple Computer. Beginning in 2001 and picking up momentum in 2002 and 2003, this IT company provided products that one could argue were squarely in the media and entertainment sectors. Its iLife suite of products, for example, included tools called iTunes, iPhoto, iMovie, and iDVD, all aimed at individuals interested in working with various types of media. But the big sensation was the introduction of iTunes Music Store by which

Table 8.3
Sample Consumer Electronics Products for Sale in the United States, 2003

Cell phones	MP3 Player
Combination cell phones and cameras	Palm Pilot
Digital cameras	Plasma TV screens
DVD home theater systems	Portable speakers
Hybrid gaming devices with text messaging and e-mail	Radio headsets
iPod	Remote controllers to integrate operations of TiVo, TV, CD players, VCR, Xbox
Laptop computers	

Source: *Parade*, November 23, 2003, entire issue.

consumers could go online and download some 400,000 different pieces of music from five music recording companies at a price of 99¢ each or $9.95 for an album. In the first week of its availability, in April 2003, customers bought over a million songs, proving that recording companies could sell products over the Internet and that not all Internet users were pirates. By Christmas of that same year, Apple had sold over 25 million songs, making it an important player in the music business. That fall, Apple partnered with AOL.com to also offer gift subscriptions and accounts that, for example, parents could pay into, in effect creating a line of credit for their children downloading music.[16]

The story itself is interesting and we will come back to it in chapter 11; however, Steve Jobs's (CEO of Apple) experience with the Music Recording Industry mimicked what Vinton Cerf had with AT&T when he went to call on its scientists and executives to explain digital communications over the Internet. In a 2003 interview with *RollingStone* magazine on how he developed iTunes, Jobs commented that "there's a lot of smart people at the music companies. The problem is, they're not technology people. . . . So when the Internet came along, and Napster came along, they didn't know what to make of it. A lot of these folks didn't use computers—weren't on e-mail; didn't really know what Napster was for a few years." Even after 18 months of trying to persuade them to participate in what became known as iTunes, Jobs still believed that as a "matter of fact, they still haven't really reacted, in many ways. And so they are vulnerable to people telling them technical solutions will work, when they won't." The final telling observation is that he saw Apple as the only company in the Consumer Electronics Industry "that go [*sic*] deep in software in consumer products."[17] In short, while still rooted in IT, he was already envisioning a move closer to the media and entertainment industries. Apple barely made a profit on iTunes, but the profit lay in iPods, onto which many iTune users downloaded their music and which it had sold over a million units just in 2003. When a reporter for *Time* magazine asked him why he bothered with online music if he was only earning 10¢ out of every 99¢, he responded "Because we're selling iPods," and grinned."[18] Thus, the story, and the two industries it involved (IT and music), clearly demonstrated the potential power of a new technology, in this case the Internet, and how powerful this continuation could be in determining the fate of an industry whose products and services could be totally digital in format.

Before moving to the media sector's industries themselves, we have to resolve the issue of what constitutes *media*. The term is widely used today to mean the information/entertainment content itself (such as news, music and video); the firms that create, distribute, and sometime use this content; and the providers of the technologies and tools used to create, distribute, or access content. By that definition, the *New York Times* is a media company, but so too is IBM. Today content is often defined as the digital data, visual, musical deliverables that are made and distributed by these industries; in other words, what resides on a CD, DVD, or film, for example, or travels over the air waves into radios and television sets. However, the meaning of the term "media" has shifted over the years. In the 1950s and 1960s, dictionaries defined the term as (a) the plural of medium, (b) as something in the middle between two things or events serving as a go-between or conveyor of something, and (c) as a publication or broadcast that transmitted advertising.[19] One would not have found a definition that included PCs, software, the Internet, telephone data transfer, or anything digital. Yet by the end of the century these latter items became major components of any discussion of media.

Whatever industries one includes under the heading of media and entertainment, they all share some commonalities with other industries. First, those that existed before the arrival of digital technologies functioned essentially independently of each other. Second, they individually and uniquely embraced digital and communications technologies at their own pace, which is why the rest of this book looks at a number of these one by one.

The fact that mergers took place at a torrid pace among these industries throughout the closing decades of the twentieth century leads to the conclusion that large conglomerates played a highly influential role in the economy, culture, and society at the time. That trend does not mean, however, that their experiences with the digital merged as well. In fact, applications of the digital remained unique to each industry and sector even at the dawn of the new century. The enormous concentration of many of these industries into the hands of a few companies transcended multiple industries. Beginning primarily in the 1970s, media executives began to think about how to exploit their content in as many ways as possible. By focusing on the content, they could begin to think about how best to deliver it to market in various forms. In the 1950s and 1960s, for example, Walt Disney essentially allowed television to broadcast cartoons originally made in the 1930s and 1940s as a way to advertise his theme park in California. Books could be turned into movies, while sounds tracks of plays and movies could be sold as music albums. Media executives thus began finding ways to present their content in multiple formats and forms, and, as digital technology increasingly became available and practical, the number of options simply expanded. Over half the options could be implemented in digital form (such as TV, movies, online, DVD, CDs, even e-books).

But to understand the paths taken to this strategy and then how it was implemented, we have to examine each industry individually, acknowledging that applications influenced their activities and their relations with each other. All the industries selected for study played major roles in the world of entertainment and in the dissemination of information. Each has also been an active user of various forms of digital technologies. All have been profoundly affected by the technology—with or without their consent.

These industries for study collectively reflect the major trends in the use of digital technologies in the modern period. However, nearly at the same time, new definitions of "media" began spreading into communities of programmers, engineers, and others closer to the technologies that underpin today's media industries. For example, one definition from the 1960s and early 1970s called media, "magnetic or punched cards, or paper tapes," in other words the physical artifacts on which data resided.[20] Today that definition includes diskettes, CDs, and DVDs. Over time the two definitions—the general and technical ones—merged into what today is commonly referred to as primarily electronic, film, or paper products, and the companies and industries that distribute them; it refers to both content and their providers. Because so much entertainment now relies on digital technologies, we are seeing entertainment activities and their companies and industries also blending into the mix.

The Media Sector in the American Economy

During the second half of the twentieth century, the media sector experienced broad expansion in all its segments. A number of factors accounted for this success: a strong economy, an expanding population, and higher levels of education in the populace. Technological innovations made it possible for the sector to operate more efficiently and in a greater scale, and to offer products and services that either did not exist before or were simply better or more entertaining. Traditional black-and-white newspapers now published color photographs; television entered our lifestyles; every profession and area of interest seemed to have its affordable magazine;[21] the Internet made vast quantities of information at the click of a button.

Book Publishing Industry

In 1950 this industry consisted of a few distinguished publishing firms and many smaller, lesser known publishers. Over the next 30 years, hundreds of new publishers sprang up in response to a growing demand for books across the American economy. From 1980 to 2000, the industry began consolidating into ever larger firms, exploiting opportunities for economies of scale and scope, making book publishing "big business" and international. Table 8.4 reflects one basic result—the number of new books published between 1950 and 2000, keeping in mind that some titles published in earlier years were also still available for sale—an American in 2000 could buy any one of over 1.8 million titles. Technological innovations

Table 8.4
Number of New Books Published in the United States, 1950–2000*

1950	11,022
1960	15,012
1970	36,071
1980	42,377
1990	46,738**
2000	96,080

*New titles and editions

**The large jump between 1990 and 2000 in the number of titles is due to a change in the way titles were counted. Prior to 1990, the U.S. Library of Congress counted titles eligible for inclusion in that library, such as publications by mainline publishers, but left out those of smaller firms and even mass paperbacks. The R. R. Bowker Company, publishers of *Books in Print*, always had a larger number and by 1990, the numbers cited by government and private sources were closer to this firm's.

Sources: U.S. Census Bureau, *Statistical Abstracts of the United States*, annual updates; Andrew Grabois, R. R. Bowker Company, April 15, 2004.

in printing in the 1960s and in writing and production in the 1970s and 1980s supported the expansion of this industry. The 1990s saw the arrival of the Internet and the potential of e-books set the industry off onto another round of transformation.

In 1950, there were just over 350 publishing houses, most of which were small. That number climbed into the thousands over the next 50 years, although almost all major and many mid-tier publishers became divisions within large media conglomerates. For example, in 1998, the largest private U.S. publisher, Random House, became part of the German publishing firm Bertelsmann, making the latter the world's largest publisher. That same year, Bertelsmann bought half ownership of barnesandnoble.com, while Barnes & Noble bought Ingram, the largest distributor of books in the United States. By the end of the century the industry was dominated by 10 companies. (To put this number in some context, the Music Recording Industry had five, and the Movie Industry six major firms.)[22]

Magazine Publishing Industry

This is another print industry that enjoyed prosperity in the same period. In almost any year between 1950 and the 1980s there were over 10,000 titles available in the United States, nearly 22,000 by the early 1990s, and even more at the start of the new century. Of the 22,000, some 8,000 were trade magazines, another 4,000 were consumer publications, and 10,000 were devoted to public relations (including college alumni magazines).[23] After the arrival of the Internet, online magazines appeared such as *HotWired*, while many print magazines began publishing online versions as well. Hypertext[24] made it possible for online magazines to acquire their own "look and feel" by the late 1990s, blending color, graphics, text, sound, video, and non-sequential reading of material to create a product quite different than a paper magazine.[25] By the mid-1990s there were at least 500 online publications originating in the United States, and in subsequent years hundreds more appeared annually, such as business-to-business publications. These statistics are most likely on the low side because there are no formal registration processes to track anyone starting an online magazine. Internet searches using Google, for example, demonstrate a rapidly growing number of this genre, including blogs. Meanwhile, pre-Internet publications continued going online from every corner of American society. Examples include *PC Magazine*, *National Law Journal*, and *Progressive Grocer*. By the of the new century, magazines that only published on the Internet numbered at least several thousand, while additional thousands of print versions were also available there.[26] At a minimum, the vast majority of U.S. magazines had websites where one could learn about the print publications, obtain contact information, and subscribe; and just like newspapers, but not books, magazine publishers began transforming into a combined set of products that were both print and electronic.

Because magazines have served niche markets for well over a century, with publications targeted at almost every conceivable interest group, it should be no surprise that, as new media technologies came along, magazines would still do well. For example, when television broadcasters served up sports programs, these events stimulated interest in sports magazines. When PCs came into existence in the 1970s and 1980s, a number of highly profitable magazines appeared to meet the needs of PC users. To be sure, magazine publishers always had to worry about how much time a consumer could give a publication, given the fact that TV, radio, and the Internet also competed for their attention. Yet, after all was said and done, the number of magazines increased from 6,600 in 1950 to more than 20,000 at the end of the century. New publishers came and went, and circulations shrank as new magazines catered to niche markets. But as the nation's population grew, the circulation numbers reflected that. For example, instead of a niche magazine only having thousands of subscribers, it was possible to have far more, for example, *PC World* with 1.1 million in the late 1990s, while all its rivals also had similar size circulations. Many other new niches were possible, although with few subscribers. In most years, roughly a third of all print media shipped into the economy came from magazine publishers.

Unlike book publishing, ownerships were less concentrated in the Magazine Industry. Nonprofits, universities, lobbyists, and media companies all published magazines. The industry remained relatively fragmented; even in the late 1990s, a period of considerable M&A activity, there were only a half dozen publishers each publishing over 50 magazines, while only a dozen or so others each published between 20 and 50 titles. Thus, many publishers were small; that fact influenced the kind of IT tools they could use and how they approached the tasks of publishing and distributing their products.[27]

Newspaper Publishing Industry

The presence of newspapers in American society is ubiquitous: dailies, weeklies, Sunday editions, school and company papers—the list is long. Throughout the period 1950 to 2000, they were major vehicles for transmitting news across America, at providing more detail than magazines or electronic media about national and local themes. Contrary to those who would argue that radio and television had caused newspapers to decline, circulations grew along with the population during the second half of the century, declining only in the 1990s as alternative sources of information became more convenient and available (for example, want ads on the Internet). Table 8.5 summarizes the number of newspapers and subscribers over the half century; this is clearly an industry that did not go away. In fact, as already mentioned, many newspaper publishers also owned radio and television stations, and had a presence on the Internet.

This industry included many tiny, often family-owned, publishers whose circulations were small. It also had its large media firms. It is an industry specifically protected by the U.S. Constitution, defended and supported by government policies and laws over the past 22 decades. It is becoming more subject to regulatory influences as it moves online (just as content delivery was over the air waves earlier) and becomes parts of large media conglomerates consisting of both print and electronic media. In short, it is an industry that allows one to probe deeply into the causes and effects of technological innovations and the interplay of public policies.

As one would expect, the largest firms often were the first to integrate new technologies to handle customer accounting, writing, and printing. Some of the better-known publishing firms included Gannett, Knight-Ridder, Newhouse Newspapers, Times Mirror, Dow Jones, New York Times, MediaNews Group, E. W.

Table 8.5
Total Number of Daily U.S. Newspapers and Number in Circulation

Year	Number of Newspapers	Number of Newspapers in Circulation/Issue (thousands)
1950	1,772	53,829
1960	1,763	58,882
1970	1,748	62,108
1980	1,745	62,200
1990	1,611	62,300
2000	1,509	56,000

Sources: U.S. Bureau of the Census, *Historical Statistics of the United States: Colonial Times to 1970*, Part II (Washington, D.C.: U.S. Government Printing Office, 1975): 809; Ibid., "Communications and Information Technology 1998," October 29, 1998, p. 580, and annual updates, http://www.census.gov.

Scripps, Hearst Newspapers, Tribune, Cox Enterprises, and Thompson Newspapers. Unlike magazines, newspapers became increasingly concentrated into the hands of fewer firms, particularly after the 1980s. While this concentration did not happen because of any miracle of modern technology, IT did make it possible to achieve economies of scale, use new channels of distribution, and, of course, integrate word processing with layout and production. By the start of the fourth quarter of the century, even small newspapers were doing the same.[28]

Radio Industry

Radio stations became a popular medium for communicating news and entertainment following World War I and remained so throughout the twentieth century. Owned by companies that also participated in television and various publishing industries, radio stations often—along with television—became the cores of large media conglomerates. Table 8.6 details how many stations there were and the number of households that owned at least one radio, evidence of a large and mature industry.

The history of radio in America has many of the same features as telecommunications: the role of technology, particularly wireless communications; the formation of companies to invent and deploy (in this case) broadcasting equipment and technologies; the extensive roles of the SEC and the FCC; the numerous mergers and acquisitions; the development of niche markets; and the expansion over the Internet. The digital served broadcasters first in office applications (such as accounting) and next as both part of the transmission equipment used to broadcast programs as well as the way to store music and other sounds on such media

Table 8.6
Licensed Commercial AM and FM Stations and Households with Radios in the United States, 1950–2000

Year	Stations	Households with Radios (millions)
1950	2,835	40.7
1960	4,224	50.2
1970	6,519	81.0
1980	7,871	128.0
1990	9,379	181.9
2000	10,500*	225.0*

* Estimated.

Sources: U.S. Bureau of the Census, *Historical Statistics of the United States: Colonial Times to 1970*, Part II (Washington, D.C.: U.S. Government Printing Office, 1975); 796; Ibid., "Communications and Information Technology 1998," October 29, 1998, p. 580, and annual updates, http://www.census.gov.

as CDs. This industry adopted every major new technology that came along: computers, terminals, PCs, diskettes, CDs, DVDs, the Internet, to mention the most obvious. Historically, it has operated in two spectrums: AM and FM; today it also delivers its offerings through the Internet. As this book was going to press (2005), satellite-delivered radio was also becoming an important channel of distribution.[29] In 1950, radio was all about AM; by the end of the 1990s, 80 percent of all listened-to stations broadcast in FM. We do not yet have hard data on broadcasting share over the Internet, but it is becoming more common.[30]

The FCC is as large an influence on this industry as it is on telecommunications. The role is the same, the intentions equally similar. No expansion by any of this industry's players occurs without FCC's approval, and all broadcasting is done over spectra licensed by the FCC. Much of what we described in earlier chapters applies here, and thus does not need to be repeated. As the half century progressed, however, the FCC supported the concentration of stations to a certain extent, doing its best to ensure there were multiple stations in each market. When it deviated from that strategy, the Congress usually intervened, as happened in dramatic fashion in 2003 when the FCC tried to loosen significantly its guidelines for media companies to own large market shares and multiple media in a market (e.g., radio and TV and newspapers in the same market).

A listing of the largest radio companies in the late 1990s demonstrates how media giants dominated the market. CBS was both a radio and TV broadcaster; Walt Disney made movies, broadcast TV programs, ran theme parks, and owned ABC Radio; Cox Enterprises was a major player in print media; the *New York Times* owned two radio stations. Not all media companies owned radio stations. In the late 1990s, for example, the following did not: Time Warner, Sony, Viacom, News Corporation, Tele-Communications, Inc., Comcast, Gannett, or McGraw-Hill. The number of radio stations continued to grow from a few thousand in 1950 to over 6,500 in 1970; and to over 10,000 in the late 1990s—at the same time that television was becoming ubiquitous and the Internet was competing for people's time and attention.[31]

Television Industry

Television is a prominent icon of the late twentieth century. There are as many as three types of television industries in the United States: broadcasting (for which viewers pay no fee and which includes the major networks such as ABC, NBC, CBS, and CNN); cable and satellite (for which viewers pay a monthly fee); and public television (funded by viewer contributions and private and public grants). However one defines the television industry, its firms all used similar technologies to create and broadcast programming, had back-office operations, and were subject to government regulations. To explore the role of the digital, I combine all of these industries into one, with a tip of the hat to the specific differences among them that affected the rate of adoption or how they used the digital.[32]

Table 8.7 provides a brief snapshot of three sets of data: number of licensed commercial TV stations (to suggest the size of the business); number of households

Table 8.7
Licensed Commercial TV Stations, Cable Subscribers, and Households with TV Receivers, 1950–2000

Year	Stations	Homes Subscribing to Cable (% of total)	Homes with TVs (% of total)
1950	98	n/a*	9.0
1960	515	1.4	87.1
1970	677	6.7	95.3
1980	734	19.9	97.9
1990	1,092	56.4	98.2
2000	1,288	68.0	98.2

*First census on this data was 1955, which determined that 0.5% were subscribers.

Source: U.S. Census Bureau, *Statistical Abstracts of the United States*, various years, http://www.census.gov/statab/hist/HS-42.pdf (last accessed 3/16/2004).

subscribing to cable (the primary source of television in homes and important because the same technology can be used to bring in Internet access); and number of households with television sets (the "infrastructure"). The table documents several trends. First, the number of television stations increased over time. Second, there is the remarkable adoption of cable (and some satellite) services, particularly in the late 1970s and 1980s, with growth in the 1990s creating a service now considered almost as essential as telephone service or electricity for over half the households in the United States. Third, almost all homes had televisions by the end of the century (over 98 percent), with deployment very extensive since the late 1970s.[33]

Television was big business and thus subject to much merger activity, particularly in the 1980s and 1990s, with the FCC playing a role similar to that in radio and telecommunications. Companies owning TV stations were well known to the public: CBS, NBC, ABC, News Corporation, Tribune, Gannett, Cox, Viacom, and Turner, to mention a few. Among the cable channels were American Movie Classics (started in 1984), Cartoon Network (1992), CNN (1980), Court TV (1991), Discovery Channel (1985), Home Box Office (1972), QVC (1986), and Turner Classic Movies (1994). Because cable TV was such a large generator of revenue, many media firms were in this market by the 1990s, including Time Warner, Tele-Communications, Viacom, U.S. West Media Group, Comcast, Walt Disney, Cablevision System, Cox Enterprises, News Corporation, and USA Network.[34]

Entertainment Sector in the American Economy

There is a logical flow in the traditional Media Industry from the three forms of paper-based, print media to the two basic electronic forms. (If the Internet ever becomes an industry unto itself, then that can be viewed as a third.) Entertainment industries, however, are not as neat to categorize as part of some broad, logical clustering of industries. In fact, newspapers, books, magazines, radio, and television are also important components of the entertainment sector. The world of entertainment is both fragmented yet increasingly interconnected. It is fragmented in the sense that, for example, circuses are a world onto themselves, with their own histories, traditions, and business models, yet they appear on television and have had magazine articles and books written about them. Entertainment is integrated, however, with movie studios filming television programs, with music studios in Nashville, Tennessee, recording symphonies, rock and roll bands, and other types of music. Distribution channels are also becoming similar to each other, moving multiple types of content.

Technology increased the sectors' dependence and influence on each other while creating new business opportunities. For example, each digital platform (or medium) affected immediately and directly the Music Recording Industry and, along the way, recording as part of movies, television, video, tapes, and diskettes. The Internet made it possible to merge the activities of many technologies, although various entertainment industries had already begun the process of integrating and mixing offerings, regardless of the arrival of the digital and sometimes because of it.[35] (Not discussed in this book are various types of live entertainment and live sports to mention two, though they are economically important forms of entertainment, particularly to television and radio.) For our purposes, a few of the large, technology-dependent industries illustrate how the digital worked in this part of the American economy.

Music Recording Industry

In the early years of the twenty-first century, the Music Recording Industry was embroiled in a highly publicized, ugly battle to protect its music copyrights. Music had become readily available for free on the Internet, with millions of people downloading files without paying for them. The fight had gone so far that the industry was reported to be suing even children 12 and 14 years of age who had downloaded music off the Internet. While I will have more to say about this controversy in chapter 11, this is an excellent example of the disruptive feature of a new technology at work on an industry and, an equally superb example of an industry not reacting effectively to a transformation. Harvard economist Joseph Schumpeter, studying the effects of disruptive technologies in the 1930s and 1940s, would have found fascinating this industry's unfolding experience with computer and telecommunications and might have concluded that he had an absolutely airtight example of his ideas at work.[36]

Like all other media and entertainment businesses, the Music Recording Industry grew, driven by the rapid expansion and popularity of rock and roll, jazz, and country music into the mainstream market. The arrival of the post–World War II baby boom, with its 76 million new customers in the United States alone (not to mention the over 20 million immigrants), generated a massive growth in demand for its products. Most of the baby boomers had enough spending money since their early teen years to stimulate this growth, and in fact, they continue to spend extensively on its products (as do their children).[37] Better recording and listening consumer products (from stereo equipment and FM radio in the 1950s to iPODs in the early 2000s) also fueled demand for recorded music. In the years just prior to the wide availability of the Internet, the industry continued expanding, from $36.64 billion in revenues in 1990 to double that volume by the end of the century.[38] Expenditures for recorded music in the 1980s had also doubled from one end of the decade to the other. The story was repeated in the 1990s, as expenditures for all types of music media doubled again.[39] For the entire modern period, "Americans have been on an entertainment spending binge, and the recording industry, more than any other consumer medium except home video" has benefited the most.[40]

Despite the problems this industry has had with pirated copies of its products on the Internet, the Net became a channel through which it began selling CDs. In 1997—the first year for which we have data on sales volumes—it appears that sales first began; in the following year, these climbed to just over one percent of all sales. In subsequent years the percentage continued to climb slowly, from 3.2 percent in 2000 to 3.4 percent in 2002. While the volumes were low, they were not so out of line with the experiences of other industries, most of which only began making their products available for sale online beginning in the late 1990s. Since we are discussing sales data, it is interesting to note that CDs dominated all channels of distribution; in fact, in 2002, they accounted for just over 90 percent of all sales, up from 25.1 percent of the total in 1989. The digital format was king.[41]

When identifying the large players in this industry, it is important to understand that there are three ways to make money here: by running live shows (concerts), by producing and selling recordings, and by writing songs. Regardless of which part of the revenue stream one examines, the industry moved toward an oligopolistic structure during the second half of the twentieth century, a process that began decades earlier. In the early 1950s, four large firms dominated some 75 to 78 percent of the charted record business: Columbia, Capitol, Mercury, and RCA Victor. Rock and roll music brought new entrants, changing the dynamics but not the pattern of the oligarchy. By the early 1970s, five firms dominated nearly 60 percent of the business (WEA, CBS, A&M, Capitol, and RCA), and by the late 1990s, the number had grown to six: Warner Music Group, Sony Music, PolyGram, BMG, MCA, and EMI. They were owned by American, Japanese, and European holding companies. Consolidations and shifting names of firms continued right into the new century.[42]

Game and Toy Industries

Electronic games and toys requiring software and computing came from a variety of sources, not just from the Toy Industry *per se*. In the early 1980s, some of these products were called entertainment software and were produced by such companies as Tandy, Texas Instruments, Atari, Apple, and Commodore—all manufacturers of computing equipment, none born or raised in the Toy Industry. Independent software firms did the same, such as Sierra On-Line, Sirius, Broderbund, Continental, and Epyx.[43] Programmers began developing computer-based games in the 1950s at universities and IT firms. These were often circulated freely in the 1950s and 1960s and included such games as *Hangman* and *Baseball*. Online games came into existence in the 1960s but did not become commercial products until the 1970s. Between 1977 and 1982, the digital game business came into its own, shipping $3 billion in products in the latter year—nearly a third of the entire value of all toys shipped to stores in the United States in 1982. It took less than a decade for this market to go from nearly zero to being a major contributor to the Toy Industry.[44] Home video games represented the heart of this business, with the entry of the now-familiar Nintendo and Sega products, later Sony with its PlayStation, and even later others with games available over the Internet. By the late 1990s in the United States, revenues from these products reached $7 billion and continued increasing into the new century.[45] A variety of industries competed in the early 2000s, such as Apple, Microsoft, and Nintendo, while even IBM worked with game developers to use grid computing.[46] What historians, journalists, and economists have yet to quantify is to what extent more traditional toys also incorporated the digital, such as talking dolls. So far, however, they have learned a great deal about the effects of the digital on games, demonstrating that once again we have another industry that found highly practical uses of the digital.[47]

Photography Industry

This industry has many angles. It consists of manufacturers of cameras, related paraphernalia, film, and supplies. We can add in photo studios, and photographers (both professional and amateur). Some of the professionals work in other industries: newspapers, magazines, and television, for example. Another group consists of a rapidly growing collection of companies in the photo market, displacing established firms with new business offerings, such as happened with Apple Computer when it moved into the music business. Examples include consumer electronics, such as Sony with digital cameras and Nokia with cell phones that took photographs. Computers came into the Photography Industry in the same way and at the same time as in many other manufacturing industries; to automate accounting, support logistics, partially automate shop floor data collection, and to perform a variety of manufacturing work. In that sense, this industry was very normal, fitting the pattern identified in the first volume of the *Digital Hand*.[48] But the real excitement in this industry was not the encroachment of Japanese competition into it that began in the 1970s or the role of the Internet.

Rather, the important event was the dramatic displacement of film with digital images in the late 1990s. This transition rapidly reduced demand for film-based photography across the economy, led to replacement of film cameras with digital versions, and increased demand for software to reside both in cameras and in PCs to handle video and still shots. This swept across households, all media industries, and many in the entertainment sector, especially newspapers, news wire services,

TV, and movies. In the process, old-line photography companies entered difficult times. Nowhere was this more so than for Kodak, an American icon as familiar as Coca Cola and IBM, which ended the century in deep trouble. Some observers questioned whether it could successfully make the transition to a digital industry in time to survive.[49]

Companies that had peripherally participated in the industry now became giants within it, such as Sony and all the major Japanese camera manufacturers, who had quickly converted their product lines to digital photography. While reliable data is difficult to obtain in the best of circumstances when changes are as rapid as we see occurring in this industry, it probably is realistic to conclude that by the end of 2004, more digital than film photographs were now being taken in economically advanced countries. Film was retreating, much as happened to telegraphy when long-distance telephone calls dropped in cost and became easier to make in the late 1960s. The digital was becoming easier to work with than film and less expensive per image snapped.

This is an excellent industry to study for digital's revolutionary effects. Its ubiquity alone—it's the rare person who doesn't have albums of pictures of relatives and friends—makes it important in our lives. This industry in its heyday employed hundreds of thousands of people in manufacturing cameras, equipment and film, while others developed and printed photographs—a task that may well disappear if film photography goes the way of telegraphy and the typewriter. Additionally, tens of thousands of individuals made their living as photographers for newspapers, government agencies, and yet other tens of millions of customers used its products. While demand for its goods and services grew throughout the twentieth century, prospects for the financial health of such traditional businesses as manufacturers of 35mm film cameras, film, and film processors remained in question as their products changed (along with who provided them), and the labor content of work shrank.

Patterns and Practices in the Adoption of Computer Applications

Can we generalize about the patterns of adoption of computing in such an array of diverse industries as are found in photographs? The evidence presented in subsequent chapters suggests that we can, but only partially. For more mature, stable industries, such as book publishing and newspapers, we can generalize with confidence and marshal substantial quantities of evidence in support of these conclusions. Like other industries that had the highest churn during the half century, or where digital technology had been changing the very nature of their work and industry structures, patterns are at best tentatively understood, as in movie making, television, music recording and now most dramatically, photography. It is important to recognize that the form of an industry often influences the type of digital adoptions that take place and their rate of integration into the fabric of its work and enterprises. But in the cases of both media and entertainment, we also see where the availability of a new technology (or application) can profoundly change how work is done very rapidly. This is especially so where economic or technological barriers to enter an industry were initially low, as in music recording, video game and film making, but higher than in such capital-intensive industries as book publishing and television broadcasting. We also have other patterns at work such as with television in all its forms: television sets that only included digital technology at the very end of the twentieth century, four decades after cathode ray tube terminals had gone into wide use in tens of thousands of companies, government agencies, schools, and PCs. It is also the case where management at the major networks resisted converting to digital broadcasting until mandated by law, and that was to take effect nearly a decade into the new century.[50]

There are several patterns, however, that emerge clearly. As happened with manufacturing, insurance, banking, and telecommunications firms, large film-related companies were the earliest to adopt computers and telecommunications for use in digital applications. Normally, these were for back-room office operations, such as accounting, finance, customer relations, logistics, and transmitting information and programs. These companies had enough scale and scope to justify adoption of tools that could drive down operating costs. Into this category one can place the print media. The smallest enterprises came later to computing after the cost of using the digital had declined sufficiently to make the technology either affordable or cost justified. Into this collection we can put radio and many movie studios, for example. However, by the end of the century, any organization that had more than a couple dozen employees used computers to handle routine office work.

When we look at the use of the digital in the development and performance of revenue-generating activities, we see a different pattern. We can observe that rates of adoption varied enormously based on the needs of different industries. Publishers required better printing machines to manufacture their products and, in time, the digital became part of the machines that produced books, magazines, and newspapers. Typewriters went the way of mechanical cash registers when word processing came into existence. At first, computers were mainframe based and entered the Newspaper Industry as early as the 1960s, though not widespread until the 1980s. At the same time news services in all media industries began using databases of information in support of their work. First, they relied on online databases of specialized information in the 1980s, such as LexisNexis. By the late 1990s they used a plethora of databases, some for a fee (such as Forrester, Gartner) and many free on the Internet (like government agency websites).[51] The gating factors for adoption were the combined availability of word processing software and hardware, and their relative costs.

In movies, television, and music recording, however, we see the adoption of the digital occurring much later, realistically not until the late 1980s and 1990s, in those portions of a company involved in the production of products, which consumers began using immediately. To a large extent the requisite technology developed later than back-room machines and software. Costs also had to come down to the point where industries that traditionally had smaller enterprises could afford them.

Another pattern that jumps out is the attitude displayed by management in these various industries. The more a digital application involved such mundane things as accounting and finance, the less concerned managers were with its effects on business practices. If an application lowered operating costs it was welcomed, especially since the technology did not fundamentally change the structure of organizations or industries. Competition was not so severe that a digital application's function was to lower operating costs. In this situation, they welcomed the digital at the same rates evident in many dozens of other American industries. Later on, however, the more a digital application affected products, sales, and profits, the greater the resistance of management to adoption as they saw the technology threatening existing profit streams and their own careers. Printers in newspaper companies resisted improvements in printing technologies in the 1960s and 1970s out of fear they would lose jobs (they did), mimicking the pattern of behavior that occurred at the same time in the Railroad Industry as computers made the requirement for so many workers on a train go way.[52] Later the same occurred in the Steel Industry.[53] When management faced changing its entire inventory of practices and equipment to take advantage of digital technologies, they balked at the cost and the potential effect on profits. This happened in television, music, and, to a lesser extent, movie making. The threat to existing business models and prior investments in these industries mirrored what occurred in the Telecommunications Industry in the way management dealt with the Internet and, earlier, with packet-switching telephony.

Management resisted fundamental changes long and hard but suffered from a unique consequence of digital technology they could not overcome: its flexibility. Most media and entertainment industries contained many cottage industry-like functions that, as occurred in the Software Industry, made it possible for individuals or very small groups to find new uses for the technology which immediately challenged the status quo. Best-selling digital games, for example, were often written by one individual. That new circumstance, much as occurred with software, could not be ignored by established firms once customers began demanding these new products. For many media and entertainment companies, a bigger threat came next from the Internet.

Once a firm could deliver sound, pictures, animation, and the like over the Internet, all established media firms were threatened because they could no longer control distribution. The key to profits in radio, television, movies, music, books, newspapers, and magazines lay in controlling when products came to market and at what price, done by controlling most aspects of the distribution process. When a consumer could go onto the Internet and download music for free or at low cost, the Music Recording Industry felt to its core the destruction of vast streams of revenues and profits. Its initial reaction was to determine how to protect itself from the Internet; only later did firms begin thinking about how to leverage it for profits. Both issues remain serious problems and opportunities for media and entertainment industries today. Prior historical experiences over many centuries suggest that if the problem is not resolved, some industries as we know them will disappear, much like the nineteenth-century ice cutters who finally vanished when refrigerators were invented that made ice cubes.[54]

The amount of time it took these industries to embrace the digital varied, but base technologies could often be used in multiple industries at the same time and thus in a *de facto* manner stimulated simultaneous changes. For example, PCs and the Internet became available to all industries at the same time, and so how each reacted to them was partially a result of the availability of these digital tools, along with circumstances unique to each industry. Something as simple as the availability of a particular medium, such as diskettes and CDs, can tell us a great deal about when new applications and products could be adopted. The same applies to how people used the Internet. That is why, for example, there were so many studies done by organizations about who used the Internet and for what, a pattern likely to continue for a long time to come.[55] Customers and users of these digital ephemera largely compel activity on the part of companies and industries, if for no other reason than that these firms are threatened by other industries or start-up companies who can provide the proverbial better mousetrap through deployment of increasingly inexpensive or novel digital products and processes. In short, there appeared to be a digital Manifest Destiny at work far broader in scope and depth than simply the Internet. As subsequent chapters demonstrate, that transformative process became evident by the end of the 1970s or early 1980s, and achieved all-encompassing proportions by the late 1990s.

A brief survey of the volume of digital media moving through the American economy begins to quantify in precise terms the evolution toward more digitally based products and services, and the revolution that it evolved into by the end of the century. Table 8.8 offers data documenting the expansion in the use of some digital recording media. To be sure, some of these digital products were shipped to consumers as blanks without anything recorded on them, such as CDs they

Table 8.8
Recording Media Shipped by Manufacturers, 1975–2000
(Unit shipments in millions)

Year	Vinyl Singles	Vinyl Albums	CDs	CD Singles	Cassettes
1975	164.0	257.0	X	X	16.2
1980	164.3	322.8	X	X	110.2
1985	120.7	167.0	389.5	X	339.1
1990	27.6	11.7	3,600	60.0	442.2
1995	10.2	2.2	9,300	110.9	272.6
2000	4.8	Y	9,400	34.2	76.0

X=Did not exist at the time.

Y=No longer tracked, volumes miniscule.

Sources: U.S. Bureau of the Census, various annual tables; data from Recording Industry Association of America, Washington, D.C., at http://www.riaa.com (last accessed December 14, 2003).

could use to record music downloaded off the Internet, much like one could copy television programs onto video in the 1980s or music onto magnetic tape in the 1950s. That is why we can view the data as a surrogate emblem of a general pattern of migration toward digital formats over time. They provided the historical background for why students of digital information in the early 2000s could argue that the vast majority of all new data in the world had already gone digital.[56]

Table 8.8 also displays data about nondigital media, such as vinyl records, to illustrate when the digital began crowding out older formats. CDs became a factor in 1984 when over 5 million were shipped; in 1987, over 100 million sold in 1987, followed by 2.6 billion in 1989. In the 1990s, billions of CDs shipped each year.[57] DVDs—a purely digital format for movies and videos—came into their own by 1998. In that year 12.2 million were injected in the U.S. economy and in 1999 another 66.3 million. Volumes simply continued climbing right into the new century, often growing year-over-year by over 28 percent. What is important to note with these two sets of statistics is how fast this particular medium gained wide acceptance in the market, displacing sales of all older formats in far less time than existing offerings had, in turn, done to earlier ephemera.[58] In short, the buying public knew good digital products when it saw them, and had enough experience in adapting to new formats to embrace them more quickly, in effect, than did the industries involved in producing DVDs. It was the same experience as occurred in the Telecommunications Industry when the public moved rapidly to the Internet and to wireless products.

The effect of the DVD on the Movie Industry and on television content producers was no less spectacular as Americans acquired the ability to record and play materials owned and sold by those two industries (the video volume as shown in table 8.9). Unlike the Music Recording Industry, however, the Movie Industry created a distribution process whereby it could generate revenue and profits from videos (later in DVD format) as well as from showing its products in theaters. Sales of DVDs rose across all product types each year right into the new century.

The distribution process of the Movie Industry is worth exploring in more detail because its model and operative experience is relevant to media industries and, more urgently, could possibly inform that of the Music Recording Industry

Table 8.9
Percent of U.S. Households with VCRs, 1980–2000

1980	1.1
1985	20.8
1990	68.6
1995	81.0
2000	85.0*

*Estimated.

Source: U.S. Census Bureau, various years and reports.

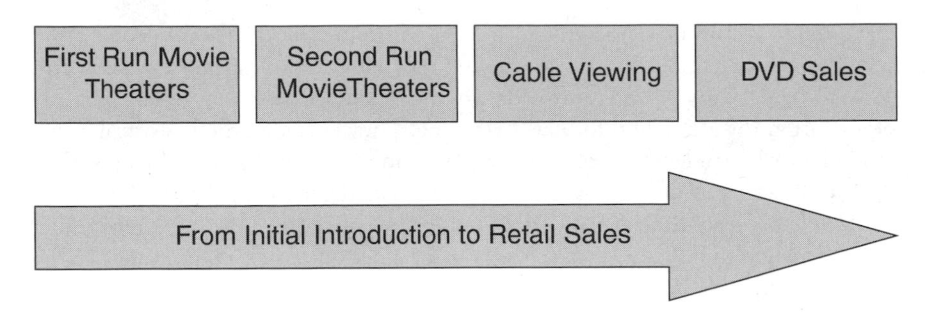

Figure 8.1
Simple distribution chain for U.S. movie industry

and such firms as Apple Computer eager to enter the entertainment or media sectors. Figure 8.1 provides a simple model of the distribution chain for the Movie Industry. In the 1950s, theaters generated all the revenue in this industry. By the 1960s, television broadcasters began contributing to that revenue flow. Then in the 1970s the system became more diverse thanks to videos (which could be rented or sold). In the 1980s and 1990s options increased again as the ability to offer pay-per-view and later DVDs added options such that, by the end of the century, revenues from DVDs far exceeded that from theaters. As this book was going to press in 2005, it was not uncommon for 70 percent or more of a major movie's revenues to come from the sale of DVDs. By then DVDs had virtually killed the video rental business, one that had been so important just a decade earlier. With the passing of each decade, digital versions of movies increasingly became important possibilities and realities, along with new ways to generate income. Finally, one should note that the process became global and that the move from one channel to another sped up or overlapped. Thus a movie in the early 1950s might show in a first-run theater for two months before moving to a second-run theater, while today a first-run theater may show a movie only for a couple of weeks, making its portion of the revenue literally over two weekends, while bootlegged DVDs can be bought over the Internet within days of a movie's release.

In short, once migration to the digital took off, it happened quickly, suggesting why media and entertainment industries faced such enormous changes by the early 1990s. The digital, however, was part of a broader set of innovations that go back to the core notion that managing the distribution of content across formats and industries was the way to increase revenues.

In hindsight, we can identify signals of the impending technological wave, cues already evident by the 1980s. For example, in training materials used within IBM in the late 1980s, employees working with customers in media industries were told:

> Perhaps no other industry marketing area has customers who are or will be more impacted by the "Age of Information" than the media industry with its

> emphasis on the value of information as a resource. Many businesses in the media industry owe their very existence to the fact that information possesses just as fundamental a value as more tangible assets such as money, capital goods, real property, labor, and raw materials.[59]

The text read like classic sales hyperbole at the time; now we know that it was an accurate statement of the emerging reality. Management would have no recourse but to use digital technology to manage their assets—although, truth be told, even these training materials still focused more on such back-office applications as publishing systems, consumer information, production and control applications, accounting, and routine office systems than on the creative and production processes.

Is the Internet an Emerging New Media or Entertainment Industry?

By the early 2000s, the Internet had become an important issue in any discussion about the role and future of media or entertainment industries. In the parlance of those two industries, it was the new channel of distribution for content. Historically distribution was the most important source of profit for many companies and industries and thus was not a trivial development. The Music Recording Industry saw it as a threat, faced file swapping and consumers downloading music off the Internet for free. Newspaper and book publishers saw an opportunity to sell products over the Internet via both their own websites and those of such distributors as BarnesandNoble.com. and Amazon.com. The Movie Industry figured out how to use the Internet to transmit, rent, and sell movies. So there are many paths being taken in this new channel of distribution, both willingly and unwillingly. What is absolutely clear, however, is that the Internet's role in the present and future of the various media and entertainment industries cannot be ignored. All of this brings us to the question: Is the Internet in the early stages of becoming an industry itself—a combined mega-media and entertainment industry that we can put alongside book publishing, music, and movies?

To be an industry, economists and business experts generally agree certain conditions should exist. There must be a group of companies that compete against each other, sell relatively similar types of products or services, and recognize each other as being in the same markets. Normally, one also sees national associations representing the interests of those collections of companies, often also industry trade magazines aimed at either audiences interested in the industry's products or at issues of interest to members of the industry. It is also helpful if government economists collect information about that economic cluster, assigning it a set of SIC or NAICS codes so those who study the industry can find a large body of useful economic data about it. As an industry matures, it acquires a style, a method, a collection of handed-down values and practices that influence profoundly the behavior of people and firms within it. Finally, if national interests so dictate, its activities might be regulated by an agency.[60]

In addition, a "business model" should exist, that is to say, a way in which

Table 8.10
Business Models Evident for Online Content Services, 1970s–2000s

Model	Features
Videotext	Used TV screens to convey text-based material; popular with some newspapers in the 1970s, recycling content after being published on paper. Not well received by consumers.
Paid Internet	Exploited pre-existing Internet; content providers offered their information for a fee over the Internet, beginning in 1980s. Access complicated; users generally did not want to pay for information.
Internet/Web Ad Push	Content providers used mailing lists to send out advertisements paid for by the advertisers themselves, targeting appropriate demographics. Spam is latest variant of this approach, and grew out of earlier mass mailing methods.
Portal and Personal Portal	Users exposed to advertising and information of interest to them or that are provided at sites of interest, building off traditional newspaper advertising model. Allows users interested in an ad to go to the advertiser's site for more information; widely deployed after 1997/98.
Digital Portal	Combines current content portals with video and audio streaming as bandwidth increased among users, allowing pay-per-view services, transmission of multimedia content. Began deployment in early 2000s.

Source: Based on material in Robert G. Picard, *The Economics and Financing of Media Companies* (New York: Fordham University Press, 2002): 27–30.

work is done, companies are organized, and profits are made. In the world of online content providers, there are currently a variety of business models, all of which are in various stages of development, flux, and experimentation. Such companies as AOL.com, Yahoo!, MSN, and Netscape Netcenter give users free access to content on the Internet but offer other services for a fee. Free services include e-mail, voice mail, access to online shopping, chat rooms, but for a fee downloading of software, and access to a variety of information using search engines. Table 8.10 catalogs various business models in evidence, circa 2003. In subsequent chapters I address specific changes to pre-Internet business models resulting from the emergence of the Net, but suffice it to say here that the Internet is an extension for other media industries, not their replacement. What table 8.10 suggests, however, is that there are nuances at work that allow us to ponder if the Internet will someday be an industry.

Researchers at the University of Texas have been some of the most aggressive and articulate in claiming that an Internet sector exists. Though they are careful, however, to call it the "Internet Economy" rather than an industry, nonetheless, they do discuss it in terms one would use to describe an industry. While their declarations have been criticized as premature, and their definition of what constitutes the Internet Industry far too broad, their typology is useful to begin the debate and to catalog results. Table 8.11 summarizes the components of their model. Essentially, it consists of four layers of companies: first, those who provide

Table 8.11
Four-Layer Model of the Internet Economy, circa 1999

Layer	Activities	Sample Firms
Infrastructure	Backbone and service providers, networking hardware and software firms, PC and computer manufacturers, security vendors, fiber optics manufacturers, line acceleration manufacturers	Qwest, MCI Worldcom, Mindspring, AOL, Earthlink, Cisco, Lucent, 3Com, Dell, Compaq, Axent, Checkpoint, Corning, Ciena, Tellabs, Pairgain
Applications	Internet consultants, commercial applications, multimedia applications, web development, search engine software, online training, web-enabled databases	USWeb/CKS, Scient, Netscape, Microsoft, Sun, IBM, RealNetworks, Macromedia, Adobe, NetObjecs, Allaire, Inktomi, Verity, Sylvania Prometric, Assymetrix, Oracle, IBM, Microsoft
Intermediary	Market makers in specific industries, online travel agents, brokerages, content aggregators, portal and content providers, ad brokers, online advertising	VerticalNet, PCOrder, TravelWeb, ITravel.com, E*Trade, Schwab, Cnet, ZDnet, Yahoo!, Excite, Doubleclick, 24/7 Media, ESPNSportszone
Commerce	E-tailers, manufacturers selling online, fee and subscription companies, airlines selling online, entertainment and professional services	Amazon.com, eToys.com, Cisco, Bell, IBM, WSJ.com

Source: Adapted from "The Internet Economy Indicators," http://www.internetindicators.com/features.html (accessed November 10, 1999).

the infrastructure of the Internet (such as AOL, PC vendors, even manufacturers of fiber optics); second, an Internet application group that uses the infrastructure to conduct business (such as consultants, application and web developers, and writers of search engine software); a third, "Internet Intermediary Layer," comprising those companies that improve the efficiency with which buyers and sellers meet on the Net (including online travel agents, brokers, portals, and advertisers); the fourth, a commercial layer, in which goods and services are sold (for example, Amazon.com, online entertainment).[61] Typically, discussions about the Internet by those in journalism, academia, government, and business focus on the first and fourth layers of this model.

While this model includes everything from IBM to Microsoft, to those firms we normally think of as 100 percent Internet online firms (pure plays), and thus may need to be modified as economic activities on the Internet become clearer, the typology allows us to view the Internet almost like an industry. For example, in 1999, it employed 1.2 million people, a third of whom worked in the first layer, over 230,000 in the second, another 250,000 in the third, and the rest—nearly 482,000—in the fourth, all generating revenue in the range of $300 billion. The research team has argued that this fourth and largest layer had nearly 11,000 companies participating. I was not able to find any economist who would concur with the revenue figures (they are too high), and the employment data is subject to much debate. The estimate of 11,000 firms illustrates the problem since they were not all pure players on the Internet; many also had brick-and-mortar operations and identified themselves with pre-existing industries, such as automotive or retail. So are they also part of this new Internet Industry? They would probably argue no; instead they might posit that the Internet was merely another channel of distribution which they chose to use. The evidence presented in the first volume of the *Digital Hand* confirms that attitude and that the practice existed in over a dozen of the nation's largest manufacturing and retail industries.[62]

Another model, more widely recognized for describing the Internet, holds that there are essentially three groups that worked to make the Internet viable as a massive new channel of distribution of information, services, and even goods. First there is a group offering information supported by subscribers and advertisers—the approach taken by many newspapers, for example, by 2000. A second model is that of subscribers and users paying for services and information obtained from the Internet. We see this today with book publishers and even with pay-per-view cable networks, and on the Internet with LexisNexis, Westlaw, and Dialog. The third group involves advertisers paying for an offering on the Internet. TV and radio broadcasting have used this business model for decades, and it began appearing on the Internet by the end of the 1990s, often with advertisers paying the website provider a set amount for every time an individual viewed their commercials. Spam was another variant of this model.[63]

The Internet also has a unique feature: nobody owns it. No one dominates it; a joke in the 1990s was that Bill Gates wanted to acquire the Internet to extend Microsoft's market share but couldn't find the person from whom to buy it. As Benjamin M. Compaine, an expert on media, so clearly observed, "the Internet is a collection of technologies, hardware, software and systems that does not lend itself to concentrated control."[64] He concluded that it could not be seen as an industry. Compaine found problems with who was "in" and "out" of the Internet, how people could make money at it, and in the ownership of its activities and assets.[65] Finally, we can point to the fact that some observers of media industries have accused various media firms of failing to leverage effectively the Internet, and as a consequence, losing traditionally secure market shares, levels of employment, and profits. As one reporter covering these industries noted: "Starting in April 2000 but becoming definitive later that year, the Internet business I was writing about simply disappeared, leaving only a few battered survivors, including AOL Time Warner, Yahoo!, e-Bay, Priceline, and a few more."[66]

The problem was not the firms cited in the quote ignoring the Internet; a great deal of attention had been devoted to it.[67] The problem stemmed from executives in various media industries dealing with a new channel for distributing their products and services that had no real precedent. "New" is the key word to use to describe the situation they faced. It is important to recognize that their products lent themselves particularly well to inclusion in the Internet because they could be digitized and distributed with telecommunications of all types, from dial-up telephones to broadband and DSL to satellites. Add in anyone's list of unanticipated consequences, and we have the makings of never-before-seen business challenges and opportunities, many so new that management had not yet had time to develop, deploy, and confirm effective strategies. Students of technological innovation often go further and argue that such changes often have a profound effect on the social structure and handed-down values and practices of organizations and social groups but, in the process, destroy some careers and social networks while enhancing those of others. The ice cutters of New England of the late 1800s suffered the negative side of this process, while those manufacturing refrigerators enjoyed the opposite. But ice cutters, refrigerator manufacturers, and late twentieth-century media executives all shared a fear of the unknown.[68] In short, our modern media executives faced some well-understood concerns.

Finally, we can ask: What do customers do and think about the Internet? There is a great deal of survey data on how Americans use the Internet, with reliable information available as far back as 1996.[69] It documents the fact that users treat the Internet as a virtual world encompassing all industries and media. They do not see it as an industry, just as they do not see a shopping mall as an industry or their community as an industry. They see the Internet as an extension of pre-existing providers of goods, services, and information, a new way to perform existing tasks. A recent study by Joel Waldfogel, an economist for the FCC, demonstrated that users in the late 1990s moved around from one media to another, simultaneously on and off the Information Highway. Using a large body of survey material, he learned that Americans easily substituted and concurrently used the Internet, TV, and newspapers for news, for example. They replaced some newspaper reading with TV, radio, and the Internet; interestingly, weekly newspapers were not displaced in part or at all by TV, radio, or the Internet. But clearly competition existed for the consumer's attention among all the media, not just the Internet. He found that use of all forms of electronic media kept increasing during the 1990s; that users relied on a variety of types of media to obtain information, news, and entertainment; and that no one media dominated. Waldfogel concluded:

> The mid-1990s has been a period of shifting media landscape. Use of radio, daily newspapers, and traditional television have declined, while the use and availability of other media have increased. Cable television use has grown, as has the availability and use of weekly newspapers and the Internet.[70]

Thus we face the same problem with consumers as that of the University of Texas study: the proverbial "everyone" seems involved, given the fact that over two-thirds of the nation had access to the Internet by the end of the century and that between 40 and 50 percent had used it. What Waldfogel's study suggests, however, is that at the level of the media sector we can identify macro patterns of economic behavior that suggest a sector exists, of which the Internet is a component, but not that the Internet itself is an industry, with all the features one would expect to see.

Tables 8.10 and 8.11 indicate that many industries were participating in important ways in the Internet. Media industries constituted only one set of participants, yet they garnished a great deal of attention and thus imposed an image of the Internet on the public that suggested it was a media industry. In fact, however, the Internet was broader. In early 2000—at the height of the world's interest in investing in dot-com companies—there were some 280 firms called "Internet companies" trading on various stock exchanges. That fall, many saw their equity values drop a minimum of 75 percent, costing those who invested in these firms a loss of some $1.7 trillion. Participating in those losses were important players in the Internet world who were not media companies, such as Cisco (who makes routers) and IBM (consulting, hardware). In short, the combined size of the economic sectors participating in the Internet was far larger than all the media industries put together.[71]

Conclusions

The next four chapters will describe how extensively computing and telecommunications technologies affected the media and entertainment industries. It will become quite obvious that several broad patterns of behavior took place. These industries used technologies both for back-room operational efficiencies and to improve the quality and form of their content. All became intimately involved with the Internet and all prior basic technologies of the age. There are also several patterns that emerged on the effects of technologies. For one thing, borders between industries blurred as companies and industries moved in and out of each other's markets. There were also institutional convergences taking place as large multimedia companies emerged over several decades, building on enterprises from multiple industries and lashing them together thanks to new capabilities of telecommunications and computing. In addition, they collectively acquired a presence in the economy and in the minds of citizens, regulators, and observers that far exceeded that of the period just following the end of World War II. The result is that today we cannot take seriously any discussion about the New Economy, let alone about the Information Age, without taking account of the role of media and entertainment industries. That last observation may be the most interesting of all because it represents a basic change in the look and feel of the economy. However, what did not happen was the displacement of long-standing sectors such as manufacturing, retailing, finance or telecommunications.

We should keep in mind that these industries are just entering a period of profound change such as they have not seen since the early decades of the twentieth century, and so how this round of transformation will end is not yet clear. We may indeed be several decades away from knowing the results. What we do know, however, is that the digital hand is very much involved. To begin understanding how, we will first turn to the most old-fashioned of media industries, those based on paper: books, newspapers, and magazines.

9

Uses of Computing in Print Media Industries: Book Publishing, Newspapers, Magazines

What can I do now to revolutionize the way my magazine is published?

—Dan Mcnamee, 1998

The book publishing, newspaper, and magazine industries have functioned as important components of American society for over two hundred years. By the American Civil War, all three types of publishing had become distinct industries in the full sense of the term. Each also had reacted to fundamental changes in technology dating back to the nineteenth century, when new printing techniques came about for the first time in nearly 400 years, and when the first electronic information-handling tools came into use, most notably the telegraph and telephone. So these industries had a long time to develop best practices and traditions, a set of values, and a strategy to deal with technological changes. They were also prosperous and precious to a nation dedicated to a democratic and republican form of government, central to which was the free flow of information. One could argue that the success of these three industries was crucial to the economic, social, political, cultural, and religious welfare of the nation. But one could also say that they are very American in their reaction to competition, economic opportunity, and the injection of new technologies that serve both as opportunities and threats.

These industries present us with a number of opportunities for learning how technologies enter industries that do have long-standing practices and traditions, technologies that are both obvious in why they should be implemented (such as typesetting and composition applications) and others that remain uncertain yet clearly important (like the Internet and electronic publications). The print industries have been around so long that there were no memories of prior transformational experiences and lessons had to be relearned, as happened with basic changes to hot-type printing in the 1800s. To what extent, therefore, did these industries respond to the digital hand when compared to other industries? To what extent did the digital come in forms evidently useful in other industries, such as in the Retail Industry for codes or had to be molded into specific forms to be relevent to this sector of the economy?

These industries teach us several things. First, the most distinctive uses of digital and other electronic applications occurred when they addressed industry-specific productivity issues, such as pre-press and printing applications. Second, adopting an application early in the life cycle of a new use provided comparative economic advantages, but doing it too soon meant one was developing better performing systems for others. Third, these industries were as cautious and prudent as any in the American economy; they wanted a clear sense of what an application of the digital could do and what the economic benefit was in exchange for investing time, money, and risk.

Of all the industries described so far in this book, these are the most interesting from yet another perspective: their products are far better understood by more people than are such things as Certificates of Deposit (CDs), IRAs, whole life insurance, or EDI. Everyone knows what a book, newspaper, or magazine is. These are also industries that few people know much about, or can even guess what takes place behind the doors of print shops or editorial offices. With the Internet now such a pervasive part of modern society, and with these industries present on the Web in highly visible and important ways, it becomes possible to look at these industries from new perspectives. The tweedy book editor, the slovenly dressed reporter of so many black-and-white movies and old TV shows, and the fashionable people whose pictures grace so many mass media magazines all take on new personas when we look at the nuts-and-bolts of how these industries do their work.

Book Publishing Industry

The American Book Publishing Industry experienced fundamental structural changes during the last half of the twentieth century, transforming from a cluster of small enterprises amid a few relatively large firms in the post–World War II period into one populated by mighty global media firms that include among their divisions newspapers, television, radio, and other information providers. Demand for books increased all through the half century, driven by three trends: (1) the influx of military veterans going to college in the late 1940s and early 1950s on the G. I. Bill; (2) the rapidly growing K–12 student population—the "baby boomers," many of whom also went to college and graduate school; and (3) an expanding population of more literate members. These events unfolded in a period when other media attracted the time of potential readers: television, radio, and the Internet. Nonetheless, the industry grew from an industry so tiny in 1950 that it was hardly measurable, its products consumed by a small, educated elite, to the behemoth we know today. To a small extent that characterization should be qualified by the fact that the industry had developed the inexpensive mass paperback for distribution to members of the military during World War II, thereby increasing a demand for books that continued right into the postwar period.[1]

Throughout most of the twentieth century and continuing to the present, the industry consisted of several communities, each of which interacted with the others and had its own requirements for computing.[2] The story of the adoption of the digital by this industry is largely about how various parts of its enterprises' departments digitized and ended up linked more closely. In the process, some remarkable technologies and applications came into use that clearly demonstrate how technologies which are intensely industry-centric, and priced to fit its economic models, can set the stage for rapid deployment of new tools. While book publishers were slow to adopt the digital, when they did they changed how they worked enormously, along with many of the fundamentals underlying their business models. One brief example illustrates the point: In 1950 all books were sold by individual orders or through bookstores. In 2004, nearly 20 percent of all new titles, and nearly as many second-hand tomes, were sold over the Internet. In 1950 most new books sold at retail; in 2004 a large percentage were available at some discount.

But first let us understand the components of the industry's ecosystem. It always consisted of seven players: authors, publishers, information providers (who also could be publishers), book manufacturers (also called printers), wholesalers, retailers, and customers. Authors provided content, and for most of the period simply sent it to publishers who had the responsibility to accept or reject the material, work to improve it, and then manage the production and sale of books. Editors lived in the publishing end of this industry. Book manufacturing was often done in the earlier period by the publishers themselves, particularly the larger houses which had integrated operations (editorial, production, manufacturing, and sometimes retailing).

Production refers to the project management duties a publisher performs to take a book from manuscript to completed bound text. As the century progressed, independent printers (manufacturers) played a growing role such that by the end of the century hardly any publisher had its own internal print shops where books were printed, bound, boxed, and shipped to wholesalers or directly to stores or customers. Printing operations often also manufactured other products, such as telephone directories for the Bells and magazines. By the end of the century there were a few giant firms in printing, such as R. R. Donnelly & Sons, who were able to become sizable organizations by implementing a continuous stream of technological innovations, resulting in lowered unit costs of books, more print runs, and higher quality, affordability, and variety of their products (think of books with color pictures, not just black and white). Within this industry, a large percentage of the digital innovations concerned printing applications.

Wholesalers took delivery of books from various publishers and sold them to retailers. That segment of the industry also concentrated more as time went on and, like wholesalers in other industries, adopted the kinds of digital technologies discussed in the prior volume of the *Digital Hand*. In time, these firms, such as the giant Ingram, would also ship new products emerging from the book publishing, IT, and media industries, such as software, videos, and CD-ROMs. Finally, there were the retailers, which at mid-century consisted of thousands of small independent book stores scattered across the nation, primarily in cities and larger towns, that over time became increasingly large book chains, such as Barnes & Noble, Dalton, and Borders. Later retailers and wholesalers adopted the same kinds of digital technologies evident in other industries—such as inventory control applications and point of sale (POS) terminals and digital systems—at roughly the same time and for the same reasons.[3] For example, as inventory control applications were embraced by large retailers, so too were they in book publishing circles. In 1950 one could buy a book only by ordering directly from a publisher or by going to a book store. By the end of the century, that was still an option, or one could simply buy it over the Internet or the phone. One of the Internet's most famous early players was an online bookstore—Amazon.com—which rapidly built on its early success with books to become a more generalized retailer selling such things as pharmaceutical products, software for PCs, video games, and music.

Information providers were those firms that produced a myriad of physical products ranging from directories, indices, annuals, magazines, newsletters, and so forth, that were often devoted to the needs of one industry, such as insurance or brokerage. Over time, much of this information became available electronically, first through dial up access (in 1970s and 1980s) and later via the Internet; LexisNexis, West, and Dialog were important early examples. The number of such services increased in the 1980s from a handful to thousands, thanks to the emergence of database management tools coupled to declining costs for CRTs and computer data storage, making such online services financially attractive to end users. It is worth focusing briefly on this community because the impression one gets from reading about the spread of the Internet is that access to online databases from commercial providers was minimal before the arrival of the World Wide Web. Nothing could be farther from the truth. Major publishers of electronic data existed in the 1970s and 1980s, some of whom went on to publish paper-based products. In the period 1975 to 1994, the number of databases grew from 301 to 8,776 worldwide, approximately half of them originating in the United States, and with some 52 million records in 1975 and 6.5 billion in 1994. One survey of 14 American providers turned up the fact that in 1974 these had 750,000 searches conducted just in the United States, with the number climbing in 1982 to 7.5 million; even before the Internet came into its own, in 1993 "hits" reached 58.3 million. We can thus note that long before the Internet was a widely available technology, Americans were accessing online commercially provided databases from firms that considered themselves part of the media or publishing industries.[4]

There are other subgroups in the industry that are not part of our discussion because they make up only small, fragmented communities. These include literary agents, who sell book ideas and manuscripts to publishers on behalf of authors; advertising agencies, which provide marketing support for books; book packagers, which are normally small firms that commission books and even do some of the writing or production work on an outsourcing basis for publishers; legal firms that specialize in copyright law; and the Paper Industry, which was so large in this period that it would require its own chapter to describe.

Over the course of the past six decades, several other trends became evident. First, "the trade," as it was called, became more capital intensive and less labor intensive. As companies grew through consolidations and hence had the size necessary to make important investments in technology, firms spent larger percentages of their cash flows on capital projects. Second, these investments, both in technology and in consolidations, were motivated in large part by a desire to drive down labor costs. This proved particularly so for craft labor, which was heavily unionized (especially in printing), and other costs since the industry was always seen as potentially more profitable than it actually was, motivating executives to actualize the potential by driving down costs. Third, computing-based skills of workers in this industry increased over time, from printers who began their adoptions in the 1960s with new printing technologies, to the wholesalers and retailers in the 1970s, the editorial departments in the 1970s and 1980s, and authors who embraced word processing in the 1980s and 1990s. Fourth, as has been evident in so many other industries, the earliest adopters of the digital were often the largest firms because they could afford the cash outlays, initial risks, and had the most to gain from the economies of scale.

Finally, the products emanating from this industry changed from being purely paper based in their development (from an author's typed manuscript) to an overwhelmingly electronic body of material, of which only one output was the paper-based book. This industry demonstrated how its products could evolve from atoms to electrons, all in essentially one working generation of its managers and employees. This is remarkable because, while there had been some important technological innovations in printing in the nineteenth and early twentieth centuries, the industry and its products had not fundamentally changed in hundreds of years. Johannes Gutenberg, who invented the world's first moveable type printing press in 1455, would have understood the production and printing processes of 1875, and would have admired books printed in 1950 and how they were sold. By 1980, he would have been confused, indeed lost, in the publishing world.

We should note here that the industry was—indeed some still think is—slow to adopt computing and new technologies. The image of the slightly rumpled editor having lunch with an *avant garde* novelist remains well entrenched in the minds of the public, movie producers, even with many within the industry itself. Yet it is not reality. To be sure, the true value-added contribution of an editor is to find authors who can write books that help the firm make money. Very few books ever fulfill that objective, but those that do make it possible for a publisher to publish the ones that don't. Today book publishing is a demanding business that requires achieving financial objectives that are more closely similar to those of newspapers, magazines, radio, and television. It is a global industry that is complicated by the growing integration of products and marketing programs across multiple media industries. Thus, for example, a book published by one division of a media company might be favorably reviewed in a mass market magazine edited by a second division of the same firm, and its author interviewed on a television program owned by the same conglomerate.

There is a softer, more idealistic thread, however, that permeated this industry throughout the twentieth century, a value system that survived the introduction of computers, the dislocation of printers from jobs, and the creation of large media conglomerates. It is the aspiration and belief that book publishing has a noble cause to it: the enrichment of civilization. This notion became so infused in the psyche of the industry's editors that at mid-century they sometimes were less concerned about making a profit on a book than they were in publishing good books. An editor from Princeton University Press, writing in the late 1960s, describes the ethos:

> A publisher is known not by the skill with which he runs his business but by the books he publishes. The history of publishing is the history of great houses that published great books; it is also the history of literary taste, which itself is made in part by publishers. . . .
>
> It has been said that printing, like architecture, is a servant art. Printing serves publishing, and publishing serves civilization.
>
> If a publisher runs his business well, he has greater opportunities to favor quality, to distribute it widely, and thus to contribute something worthwhile to mankind.[5]

This set of values often silently played a role in the adoption of some of the practices of the industry during the second half of the twentieth century, and computing was not immune from its influences.

Early IT Deployment (1950s–1970s)

Computing came to book publishing later than to so many other industries, most notably other manufacturing ones. In the 1950s, most publishers were too small to need or afford the expensive mainframes then available, and relied instead on precomputer information-handling tools, such as punch card tabulators and adding machines, primarily for accounting applications and for a few limited tape-based text-entry systems to feed some printers. Much as in the brokerage business, backroom operations were not the central focus of senior management, the industry's leaders had worked their way up the organization (when there was such a thing) largely on the editorial side. Labor was cheap, and even in printing, major innovations in its technology independent of computing did not come until the 1960s.

However, there were a few exceptions in the 1950s. For example, the Reader's Digest Association, the nation's largest book club with several million subscribers in the late 1950s, was an extensive user of information technologies to track accounts, to bill, and to ship books, fulfilling millions of orders. It used two MODAC computers (the 404 and 414) to do the necessary processing.[6] Though the industry's trade magazine, *Publishers Weekly*, did not report on any installations in the 1950s, several were deployed.[7] Reflecting the issues and concerns of the day, a reading of this publication demonstrates that the industry as a whole did not begin to pay attention to computers until the 1960s, increasing its coverage of the topic as time passed.

When computers came to book publishing, it was in the largest firms for traditional accounting applications. Daniel Melcher, president of R. R. Bowker, a publisher normally on the "bleeding edge" in the use of various technologies throughout the period, observed in 1967 that "most large publishers have computers, and many small ones use computers" for the same reasons as in other industries, "for billing, accounting, inventory control, and sales analysis. Very few report any net savings resulting from conversion to computer, and most went through agonies in the conversion process. They all hope for tangible economies in the future." He was tough on the industry's use of computing in these early years, arguing that these machines "have unmistakably lengthened the time it takes to fill an order, and have made it almost impossible to understand a royalty statement or get an intelligent answer to a complaint or query."[8] His was a common protest, similar to those in other industries during their early years of adoption, but evidence that the industry was beginning to embrace the technology. However, he noted that firms also deployed computers in typesetting, in speeding up composing activities, and in automating indexing (important to his firm). He pointed out four reasons why computers were becoming attractive to his industry: rising cost of labor, declining costs of computers and other machinery, growth in the market for books, and an impatience on the part of publishers and customers for books and data to come to market more quickly.[9]

Antidotes to Melcher's negative assessment can be found in the press releases of the early to mid-1960s. The American Bible Society, which had been distributing religious publications for over a century, ordered an IBM System/360 in 1965 to help manage its distribution of 48.6 million publications to 131 countries, using this new technology to maintain and keep current contribution lists, to perform inventory control for over 27 million volumes, and to process orders.[10] At the same time, Fawcett Publications installed an IBM 1460 computer system to process book orders, anticipating a reduction of 75 percent in the time required to serve 800 wholesalers. Fawcett's data-processing manager, Joseph Federici, had committed to using the system to "speed through large quantities of detailed data and tell us the most economic means of distributing new books" in one day as opposed to the previous four.[11] His counterpart at Hitchcock Publishing Company, Donald Hogan, three years later ordered his second generation of equipment, moving from an IBM 1440 to a System 360 Model 25 to do inventory and other applications "30 percent faster at the same monthly rental as our present 1440 computer system" and to add new applications, such as an improved reordering process and handling reader service inquiries.[12]

In the years immediately preceding these installations, members of the industry began looking at computers as a possible means of handling the growing volume of transactions and flow of products to market. Billing and accounting functions accounted for $1.17 out of every $100 spent in the industry, while some of the largest firms reported in one survey in 1962 that it was as high at $4.60. So, the hunt was on for increases in productivity. Early installations addressing these issues began drawing attention in the industry, particularly when computers were used in warehouse operations and for accounting. Beginning in the early 1960s, almost all the largest publishers built new warehouses, using the occasion to automate in part some of their prior accounting and inventory processes, each dutifully reported in the industry's trade literature.[13] In that decade, computers began appearing at all the major publishers, while smaller enterprises began relying on service bureaus for some of their data processing.[14]

Starting in the 1960s, however, an innovation emerged that ultimately would have an even greater effect on productivity in the industry during the 1970s and 1980s: computing in the composition and printing processes. Composition refers to all the activities associated with converting an author's manuscript into pages of printed text, all properly aligned to look as it will when published, while printing is the act of physically manufacturing the book (printing and binding). Telling the story can become complicated, because innovations in this part of the industry were occurring at the same time as in magazine and newspaper publishing; indeed, those two industries were pushing the technological envelope more than were book publishers. But the latter benefited from these developments because book manufacturing firms, such as R. R. Donnelly & Sons, acquired new equipment that had some computer-based controls when manufacturers that introduced new printers were selling them to all publishers. All three paper-based industries began embracing these new composition and printing technologies in the 1960s, but newspaper and magazine publishers did it more frequently and earlier than book publishers.[15] Yet at the time that new printing presses began appearing, as one member of the book publishing world put it, his remained "a craft-oriented industry of small businesses."[16]

The key technological developments beginning to affect this industry included computerized typesetting, the use of CRTs in typesetting, and the early yet limited use of optical character-reading equipment to scan manuscripts into systems that could reset the material into pages.[17] Finally, the development of various photocomposition machines continued the historic process of shifting work from craftsmen to those who could manipulate data on a screen using software. In typesetting the core task was—and still is—to set pages so that the text's margins are aligned (justified) and, more complicated from a software perspective, words are correctly hyphenated. These tasks require skill and are expensive, time-consuming tasks to perform.[18] Typesetting software first appeared in the early 1960s; by 1964, 98 such systems were installed in the United States, working in all three print industries. The number reached 663 in 1968; there was rapid adoption of the application because it worked.[19] A keypunch operator or typist, with some training in assigning codes to text, could produce a tape on an electric typewriter device, and by the early 1970s, using CRTs, that could be fed into a computer that then set type. Expensive compositing room craftsmen no longer were needed. This work could also be done more quickly since it was no longer manual and, if the software worked well, more accurately. In the Newspaper Industry, speed in particular was important since deadlines were normally within hours of a reporter drafting his text. This was the most important application in printing to appear in the 1960s and it received much attention throughout the industry.

It took a while for the Book Publishing Industry to understand its economic benefits. For example, as late as 1967, one book manufacturer noted that "we feel using the computer for justification and hyphenation is not economical," and opted for continued manual page setting. At the time some 15 percent of the Printing Industry was using the new technology, a percentage that grew over time.[20] To be sure, there were problems with the early systems: poor quality of margin justification and hyphenation, and pagination. Sometimes the rhetoric became emotional: "So much of computerized composition is an abomination that there may be cause to wonder whether high quality book work can, in fact, be achieved through the use of programs of computer typesetting."[21] However, when it worked it reduced production time, increased quality control, and was becoming economically viable. Growing consensus emerged that it was faster to compose using computers than earlier methods. As salaries rose during the late 1960s, the industry continued to experiment and adopt the application, with those who took the plunge focusing on how CRTs made the whole data-entry process so much easier and quicker.[22] As adoptions increased, led primarily by the Printing Industry, capital expenditures rose more for printers than publishers, the latter simply buying services from the former. Yet publishers remained intimately engaged in the debate and adoption of the technology because, as the president of McGraw-Hill Book Company pointed out in 1975, "manufacturing accounts for more than 20 percent of a publisher's expense, and is responsible for the physical product and to a large extent the price at which it must be sold."[23] Government economists looking at this industry in the early 1970s concluded that already it was changing as a result of computing, leading to new skills in data processing. They concluded that computerized typesetting had become the most important new technological development underway in the industry.[24]

By the mid-1960s, printers had become very concerned about the effect this new technology would have on their careers, especially in the Newspaper Industry. Large numbers of printers were unionized and generally accepted the inevitability of technological changes improving the economic and technical performance of their industry, but nonetheless felt an intense anxiety over possible loss of jobs. So long as the printing business continued to grow, the two conflicting issues could coexist, although flaring up from time to time during contract negotiations. Printing unions faced much larger problems than computers, however, with the move to photo-offset printing techniques, beginning in the 1950s and extending widely by the end of the 1970s, because the switch called for whole new sets of skills. Older hot metal type, or linotype, virtually and quickly disappeared by the early 1980s, swept away by new printing methods that increasingly integrated the use of computers to move text to plates rapidly and effectively. Unions worked to protect jobs and to keep skills current among their members.[25] While unions worried, innovations in printing technologies came to the information-processing community and to the public at large, both of whom began sensing that publishing was changing, first most evident in newspapers and magazines with new formats, but then also with books.[26]

During the mid- to late-1970s, composition (that is, typesetting) systems spread across the industry, as did CRTs for online page makeup, and word processing tools began appearing, first among the ranks of authors, and later in editorial offices, although this technology did not have a widespread effect on internal publishing operations until the 1980s.[27] Composition increasingly shifted from specialized organizations in printing companies that set pages to in-house operations, using less expensive personnel with online systems to do the work. This change constituted a fundamental and important shift in the division of labor in the industry. One *PW* survey suggested that roughly a quarter of all American publishers had established internal composition processes by late 1978, with many others reporting they soon would do the same.[28] Some were also outsourcing this function to freelancers. One report on trends in editing and composition from 1980 described the situation:

> Traditional divisions of labor among publishers and suppliers have been changing for some time, because of new technological operations and revised ideas on managing the cost of editing and design. On the one hand, publishers have increasingly employed outside freelancers or service firms to perform editing and design work; on the other, new editing and composition technology favors performance of this work in one location, coupled with the production of typeset matter.[29]

Inventory control systems and computer-based retail store systems (linking inventory reordering back to wholesalers, and the collection of POS data) spread slowly during the decade, mimicking the process evident in other retail industries where inventory control and order fulfillment became more closely linked composite processes as the years went by.[30] During the first half of the decade, links between retailing and distribution systems remained crude; in fact, one observer noted that "the book industry as a whole, which has had as ready access to computers as have the many other industries which have successfully dealt with their distribution problems, has failed so far to develop a system that would effectively link retail inventory control with publishers' order processing and shipping procedures," calling for industry standards to end the "chaos."[31] By the end of the decade the industry was on its way to addressing these issues. The flow of books and information about inventory and sales, from publisher to retailer and then from retailer back to publisher, had come to be recognized as a process that could be—indeed, had to be—integrated via computers and telecommunications.

By that time the industry was suffering from three problems: too much information on pieces of the process; too many books being returned by retailers that publishers then had to store (after going through the expense of printing them), sell at a loss (in the trade called remaindering), or destroy; and the lack of understanding of what sold in a timely fashion, hence lost sales. As one reporter pointed out, "the new vision that emerged in the 1970s, and that is now being pursued by the industry's leaders, is of truly integrated operations that form one coherent electronic publishing information system," a movement he characterized as "still in mid-stream," involving publishers, wholesalers, and retailers.[32]

It is not uncommon to think of on-demand printing as a phenomenon of the 1990s, perhaps even of the late 1980s; however, in the 1970s the industry began considering it as its ability to print profitably smaller numbers of books increased, thanks to new printing presses and the increased use of electronics, the digital, and software tools. As early as 1972, computers made the production of pamphlets, loose-leaf publications, and bound books more cost effective than in the past, suggesting that new competitors might invade the book business. The introduction of informational audio- and videotape cassettes also provided alternative ways of delivering content, creating the concept of "multimedia" in the industry. Photocopying materials also posed problems related to lost revenues and copyright issues, similar to those faced by the Music Recording Industry a quarter of a century later.[33] In the 1970s McGraw-Hill and other publishers began experimenting with on-demand processes to alleviate these problems while also engaging in the development and sale of publications in formats other than books.[34] In the 1970s, new databases became available online for bibliographies, abstracts, full-text and statistical materials. These were designed largely for use by libraries, scientists, engineers, and financial communities. Key early commercial players included Dun & Bradstreet, W. Dodge, Standard & Poor's 500, and Valueline, providing economic statistics, financial information, and facts about companies. This development caused some publishers to form alliances and launch pilot programs to enter the field, as did McGraw-Hill, which became a giant in this market by the end of the 1980s.[35]

IT Deployment Before the Internet (1980s–1994)

If we were to count the number of *Publishers Weekly* articles devoted just to IT in the 1980s and early 1990s, and compared that number to the volume from the 1960s and 1970s, we would conclude that the interest in and importance of computing in the industry had more than doubled. In fact, computing permeated each component of the industry. Early adopters of applications in accounting and typesetting moved on to second- and even third-generation versions, and to hardware and software that provided online capabilities, exploited more modern technologies, and were integrated more effectively across applications and segments of the industry. They included authors who, by the early 1990s, routinely wrote using word processing, hence providing this industry with electronic input to a growing array of editorial, typesetting, and printing systems driven by the digital. In short, the industry was well on its way to becoming a highly digitized element of the economy.

Rather than spend the next several pages documenting the increased use of computing in word processing, editorial work, composition, typesetting, printing, accounting, marketing, inventory control, and retail applications—all of which occurred in wave after wave—we should focus on those important issues which emerged in the 1980s and 1990s that proved to be either significant extensions of the earlier applications or the consequence of new technologies. The two that appeared most to fit these criteria were the role of software as products and in electronic publishing.

I want to acknowledge and dismiss a third discussion that, while noisy, did not fundamentally change, namely the future of the book as an essential medium for the transmission of information, knowledge, and pleasure. Beginning in the 1960s and continuing through the early 2000s, there was an ongoing colloquy among members of the industry and its observers about whether printed books would survive in the Information Age. The answer? More books were sold at the end of the century than in the 1960s—as were many other related products such as CD-ROMs, videos, and even software games.

In the wake of the launch of the PC in the late 1970s and its rapid expansion during the 1980s, software products to run on those systems reached the market in vast quantities. This was the period, for example, when Microsoft came into existence, but so too did many other companies, selling software products for spreadsheets, word processing, database management, accounting and finance, and later games, education, and home entertainment. In 1981, the year in which IBM introduced its PC, estimated revenues for just nine software publishers totaled $70 million; two years later, they reached $468 million. The continued growth in this new segment of the IT market can be demonstrated by one example: Microsoft, which represented just slightly less than 10 percent of the entire Software Industry, alone sold $6 billion in products in 1995, the vast majority of its revenue coming from systems software and applications. In short, software products for the PC had become a massive business in the United States and around the world, one that kept growing to the end of the century.[36] Software came in little packages, and from the beginning of the PC, it seemed the proverbial "everyone" saw these as an extension of, or addition to, the ephemera of the Information Age. Publishers quickly began asking themselves what role, if any, they had in this new world.

By the early 1980s, some had begun answering the question: they wanted to be part of it. Educational software products called "courseware" were seen as both a natural extension and threat to textbooks. In 1983, one literary agent staked out the aggressive high ground in favor of entering this market: "Publishers are sitting on treasure chests of print titles that can be adapted as software products."[37] Yet much discussion took place about how close to a book software was, and how it differed in terms of the way it was made, sold, and supported. Emerging channels of distribution were also new to the industry; for example, should software be sold through PC retailers or bookstores? Ignorance and confusion related to what was clearly a rapidly growing market led *Publishers Weekly* (*PW*) to produce a nearly 150-page primer as a special supplement in early 1984 to begin educating publishers about this new media.[38] As a reflection of concerns in the industry, this report demonstrated that publishers had already recognized the vast economic potential that lay before them, with sales figures bandied about in the tens of billions of dollars. Publishers began forming small alliances with software houses and others to begin learning about software products. For example, McGraw-Hill teamed up with Carnegie-Mellon University in early 1984 to develop courseware as companion products for college texts and began exploring with bookstores how such items might be sold.[39]

All through 1984 and 1985, the industry went to school on software and computing to understand better how to move forward. Initially, most publishers saw education as the key market, an extension of their textbook business which had so sustained the industry since the late 1940s. Closely related to this opportunity were the burgeoning database markets, in which subscribers could obtain electronic data by dialing up an information provider. Did that represent yet another opportunity to sell information in some physical format, and would the traditional Book Publishing Industry participate?[40] It was all so uncertain in the mid-1980s. Some enthusiastic publishers jumped into—and then quickly out of—the software publishing business as they realized it was a different way of making money from what they were accustomed to. Yet many experimented, soldiering on with different projects. Examples included Addison-Wesley, John Wiley, Harper & Row, Reader's Digest, Bantam, Prentice-Hall, Simon & Schuster, Random House, and Houghton Mifflin—all giants in the industry.[41] During the course of the 1980s and early 1990s, however, publishers established a partial coexistence with software publishers so that some software was sold in book stores; the Book Publishing Industry essentially got out of the business of developing software except for text enhancements, which functioned similarly to its tradition of publishing teaching aids and tests. This pattern paralleled what happened in the industry as it adopted video, audiotape, and subsequently CD-ROM products for sale through retail and online outlets.

A by-product of the "computer revolution" was the emergence of a new category: books about computing. This niche first developed in the 1960s, and the industry press commented on its characteristics from time to time.[42] However, the spread of PCs across the economy in the early 1980s caused a vast new healthy market for computer texts and "how to" volumes to come into being. In a major survey of the market conducted texts in late 1988, *PW* reported that 49 percent of the book dealers it contacted sold books about computing; 90 percent carried books about word processing, almost as many sold books on database management and operating systems, and a large percentage had books about programming languages (80 percent), spreadsheets (71 percent), and miscellaneous other topics such as IBM PCs, Macintosh machines, and telecommunications. At the same time as this survey, the stores that carried computer books were asked about their practices on selling software; most reported they had dropped these for various reasons, ranging from the high price of the products to "too much pilferage of displayed programs," and their inability to demonstrate the software. However, the big surprise was that a third of the stores did carry some software products, mainly games, children's educational packages, word processing, spreadsheets, and graphics-related items.[43] Publishers and chains continued expanding both their computer book lines in the 1990s and their CD-ROM business; the latter category kept growing right through the end of the century, with major book chains leading the way such that some large stores, such as Barnes & Noble, frequently carried several thousand book titles, or roughly 4 percent of their inventory just in this category.[44]

Meanwhile, traditional publishers, such as Random House, Macmillan, John Wiley, and Prentice-Hall, published thousands of titles, and new companies came into the business, such as IDG with its *Dummies* brand, selling often millions of copies of individual titles. We should note that the first IDG Dummies publication was *DOS for Dummies*, published in late 1991, which sold 1.5 million copies in its first 14 months. At the time (February 1993) it reached this mark, other best-selling computer books included *PCs for Dummies* (from IDG), *WordPerfect for Dummies* (IDG), *How Computers Work* (Ziff-Davis), and the *Mac Bible* (Peachpit Press).[45] So not only were computers influencing how the industry did its work, but also books about computers were beginning to have a measurable impact on revenue sources.

As early as 1992 and 1993, CD-ROMs were already a sizeable market, offering text, game and music. In 1992, 2 to 3 million CD-ROM players had been sold, a number that would climb throughout the decade, with half going into home markets and a large number into schools. Early products included reference and encyclopedic works, strategy games, and educational tools. Encyclopedias were early success stories; both Compton and Grolier sold more than 500,000 copies each of their encyclopedias in CD-ROM format—twice as many as their hard-copy equivalents. In short, CD-ROMs represented a fast growing market.[46]

Another major development of this period concerned electronic publishing, which spilled out past such traditional publishing activities as editorial, typesetting, and printing to how one should deliver information and even books and other smaller printed products in the future. These issues grew out of the fact that content was more available in digital form and thus could be produced and delivered in ways inconceivable even as late as the 1960s. The definition of what constituted electronic publishing changed over time, never becoming quite clear, and always reflecting new applications of the digital.[47] But as a macrohistorical progression it involved, first, the injection of the digital into the process by which information and publications were created and prepared for publication and printed; and, second, the end product that became available to readers. The first group of applications involved use of the digital in typesetting and composition in the 1960s and 1970s, quickly followed by creation of databases one could access for a fee in the 1970s and 1980s. Development of digital endproducts followed, such as CD-ROMs by the early 1990s. Electronic publishing came to integrate computers across the entire value chain such that by the end of the 1980s, from author to finished product, electronic publishing was in full swing.[48]

In the period from the late 1970s through the early 1990s, technological advances furthered the use of computers. First, mainframe software tools to help editors and compositors appeared and improved. Second, authors gained cost-effective access to word processing on PCs and to art and graphic tools, particularly with Apple Macintosh microprocessors (first introduced in 1984).[49] Third, and simultaneously, some publishers began producing final product in digital form that

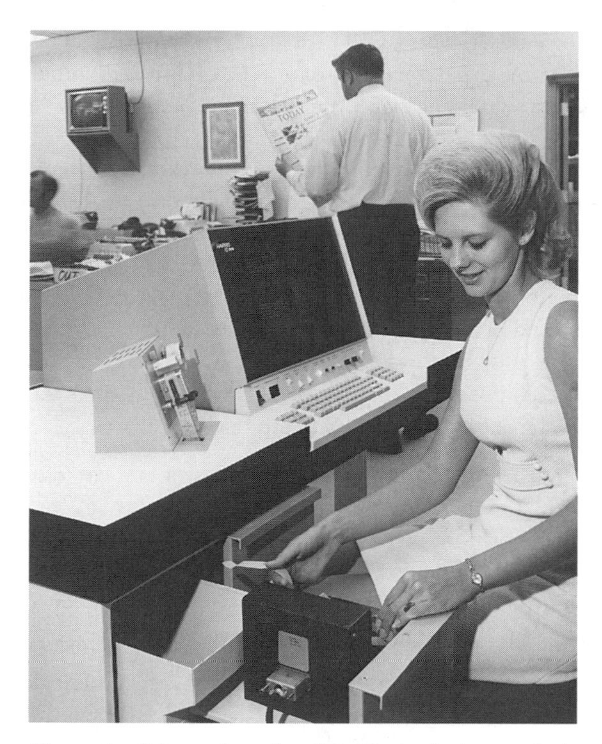

Electronic editing and proofreading using a display terminal in book publishing became widely available in the 1970s. This example is from 1974. Courtesy Archives of the Charles Babbage Institute.

one could dial up to access, such as databases from The Source, DowJones News, and CompuServe—none of whom were members of the Book Publishing Industry, thus representing a threat to the traditional paper-based publishers.[50] By the end of the 1980s, desktop publishing became a widely available application to small publishers and individuals, also posing a potential threat to book publishers. In fact, the term "desktop publishing" came into the vernacular in 1984, the creation of Paul Brainerd, president of Aldus Corporation.[51]

A casual perusal of the industry's literature clearly demonstrates that beginning in the 1970s, publishers began paying attention to electronic publishing, and that in the 1980s it was the predominant macrotechnology issue of the day. They actively debated visions of "on demand" books or the ability to publish short runs quickly and cost effectively; the subject first drew attention in the 1980s and early 1990s, but it did not become a practical reality until late in the century. One of the gating factors in its adoption was an affordable technology. Another concern involved uncertainties about what role the industry should play in disseminating information: should it be only on paper, via electronic media, through dial-up databases, or a combination? Could one make money at this or be threatened by companies traditionally not in the industry? Many writers declared the arrival of one application or another far earlier than in fact happened, much as occurred in all industries as enthusiasts declared prematurely the arrival of "electronic revolutions."[52]

By the early 1990s, most publishers were publishing content in digital form using electronic-based technologies. They had overcome problems of affordability and the lack of in-house technical knowledge of computers, and by 1990 their discussions had moved on to questions concerning electronically published books—selling whole books in electronic form that one could buy on diskettes (later CD-ROMs), downloaded from a database, or as e-books. Various experiments underway across the industry explored what these developments could mean.[53] A new mindset slowly emerged in the industry, expressed exuberantly by Richard Snyder, chairman of Simon & Schuster in 1992:

> We are not just a publisher anymore, but a creator and exploiter of copyrights. We sell information in any form, in any way you want it. We're utilizing our copyrights in the various media that other people are creating. We're not a technology company, and don't confuse us with one. But we are out of the confines of print, although that doesn't mean that we are out of the print business. We now sell the same information in various forms.[54]

The industry was driven by technological innovations, competition from within and outside the industry, authors pushing the technology at publishers, and growing evidence that using computers did drop production costs by 10 to 40 percent and cut production times by as much as half. When it made sense, composition was brought in-house, printing was outsourced, distribution was done both electronically directly from a publisher's database or, as before, using subcontracted wholesalers. In short, who did what work, and how, changed significantly.[55]

Everyone, it seemed, was aware of the changes. In a *PW* survey done in late 1993, its reporters found that publishers believed universally that the paper book would not go away any time soon, and that CD-ROM products would increase in importance, beginning with reference works moving to electronic formats, followed by the linear textbook. Technical problems remained, of course, such as the lack of industry technical standards, and the looming questions posed by the Internet. If anything, there was a resigned acceptance that books would play a smaller role in the years to come. Janet Wikler, an executive at HarperCollins responsible for creating and selling electronic products in the early 1990s, reflected the melancholy opinions of many of her peers when she told *PW* that "my generation may be the last to be predominantly verbal, to have a strong visceral affection for books."[56]

Meanwhile, inventory-control processes were enhanced much along the lines evident in other industries, with books imprinted with bar codes, with shared databases on backlists and titles on order shared between marketing and warehouses, and between publishers and printers. R. R. Donnelly and its competitors continued to invest heavily in digital technologies, which made it possible to produce shorter runs more quickly by the late 1980s, even leading some printers to wonder if they should not also be wholesalers—and wholesalers to wonder if they should print on demand if delivering electronic versions of their products. There was much interest in electronic delivery because all through the 1980s and 1990s its cost declined, while the expense of shipping books kept rising.[57]

Recent Trends (1995–2005)

1995, an artificial date for suggesting a new era, is when the Internet became an increasingly important channel for the sale and delivery of products. As tools such as Mosaic and Netscape became available to make access to the Internet easier, more information went on the Net. It became a major channel for information exchange after 1994; with the appearance of security software a couple of years later, it also evolved into an important channel for the sale of information and products of all kinds offered by publishers. Newspapers early and effectively exploited this new tool, about which more is described below.[58] Traditional distributors in the book industry, such as Ingram, quickly recognized the significance of the Internet, and moved forward to exploit it,[59] while the industry began to think, much like S&S's CEO, that it was all about content. But, as occurred with every new technological twist, some wondered and had to be reassured. Dick Harte, a software developer of book-selling tools, boldly declared in 1996, "If you're a good retailer in the real book world, you can succeed online."[60] A student of the role of computing, David Moschella, reflected back on this period and the Internet's effects on publishers:

> The Web has been a tough market to participate in. In many ways, the problems all started with the rise of Amazon.com. The ability to sell books via the Web created the opportunity for all sorts of new publishing industry value creation—recommendations, reviews, reader feedback, sales rankings, affinity marketing, expanded title availability, more efficient distribution and inventory management, chapter samples, tables of contents, and so on. Unfortunately for the traditional book publishers, just about all of this new value has been provided by Amazon.[61]

But book dealers and publishers moved to the Internet very rapidly during the second half of the 1990s, and by the end of the decade some were selling up to a third or more of their products through that channel.[62]

But we might want to ask, why books? The book business was supposed to be slow and conservative when it came to radical new business models or changes in technology, although the historical records indicate quite the opposite. So what happened? The first and most obvious answer is that when someone started selling books online and succeeded, the rest of the industry had no choice but to take notice and follow suit. Amazon.com became the poster child for that action. While more will be said about this e-tailer below, we do know how its founder came to choose books to sell online, and that provides insight into "why books." Jeff Bezos, Amazon.com's founder, while determining what kinds of products might sell well over the Internet in 1993—at a time when selling over the Net was very immature but with many new users of the network getting online quickly—he concluded that book publishers provided better terms and conditions to retailers than the Music Recording Industry (at the time dominated by six recording companies); that there were 3 million book titles available for sale versus 300,000 music CDs, for example; that the book publishing business was highly fragmented (unlike music and computers), with no dominant player to worry about; that customers knew what a book was (one did not have to describe it); that it was a product that sold well; that there were no geographic limits to the market for books; that books could be acquired quickly from two existing giant distributors in the industry (Ingram Book Group and Baker & Taylor). Finally, there already existed a limited body of insights validating the attractiveness of book selling over the Internet in the form of the few existing Internet book dealers, such as clbooks.com and books .com. So a market existed, as did products, the ability to acquire them, and the way to ship them. It was a perfect congruence.[63]

The publishing business as a whole continued to be good in the 1990s, even after the arrival of the Internet as a major channel for business transactions. The number of publishers increased from approximately 2,000 in 1997 to over 2,800 in 2001. The number of employees grew by about 15 percent from just over 100,000 to some 115,000, while the industry ended 2001 with revenues of $22.4 billion in the United States, compared to the global revenues of $80 billion.[64] The major structural change in the industry concerned another round of consolidations that resulted in global publishing behemoths. Table 9.1 lists major consolidations from just one year—1999—to illustrate the variety and size of the trend, continuing the historic pattern begun in the 1980s of intertwining this industry into the affairs of other media and entertainment industries. By the start of 2001, concentration in the industry had progressed to the point to where several firms dominated in each major segment of the industry. The giants were Bertelsmann, with worldwide revenues in 2000 of $16.5 billion; News Corporation, at $14.15 billion; and Pearson, with $5.37 billion. To put these sizes in context, the world's largest tradebook publisher—Random House—only generated $4.2 billion that year—and it was the property of Bertelsmann.[65]

Customers continued buying books, but also other media. In 1998, a milestone was reached when consumers spent more on home videos for the first time than on books.[66] But closer to home, we should ask where the industry sold books. In the years immediately preceding the predominance of the Internet, people bought the majority of their books at independent bookstores and small chains. Using 1992 as a base year, large chains sold 24 percent of all books, independents almost 25 percent, book clubs 16.8 percent, and price clubs the rest. In 2000, large chains held their own (23.7 percent), independents had shrunk a bit (14.8 percent), book clubs held on (18.6 percent), while price clubs actually increased share (6.6 percent). But the Internet was the new entrant, owning 7.1 percent of the market, outselling price clubs.[67] Over the next four years, the Internet's share of the U.S. market grew to roughly 15 to 20 percent.

In the 1990s the industry had its 800-pound gorilla to deal with: Amazon .com. One cannot discuss the Book Publishing Industry of the past decade without

Table 9.1
Major Acquisitions in the U.S. Book Publishing Industry, 1999

Acquiring Firm	Acquisitions
Amazon.com	Bookpages, Telebook, Internet Database, Junglee, PlanetAll
Bertelsmann	Random House, barnesandnoble.com, Springer Verlag
Harcourt General	Mosby
Pearson	Simon & Schuster
Reed Elsevier	Mathew Bender

Major acquisitions (those above and others) totaled $11.14 billion in 1999.

Source: Jim Milliot, "Whatever Next?" *Publishers Weekly* 246, no. 1 (January 4, 1999): 47.

understanding its role. Jeff Bezos, who worked as a financial analyst in New York, conducted research on possible Internet-based businesses in the mid-1990s and concluded for the reasons discussed above that selling books over the Internet could be a viable business and decided in 1994 to start such a company. After a year of preparations and building the software infrastructure needed, he launched the firm on the Web in June 1995. He accepted orders for books, using *Books in Print* and other digitally available bibliographies of the time as his "catalog" and off-the-shelf hardware and software to help create the offering. It became popular, business grew all through the second half of the decade, and in the process, his firm became the beacon for Internet-based businesses. Along the way he added other products to his suite of offerings: music and videos (1998), and drugs and greeting cards (1999), for example. Table 9.2 shows the revenue growth for this firm over the time period; in the latter years, approximately 70 percent of the revenue came out of the U.S. economy and the lion's share of that from books, followed by music and videos. The firm has been the subject of so much discussion that we do not need to review the issues here; what is important to note, however, is that just as in brick-and-mortar operations, online bookselling became a highly concentrated business such that by the early years of the new century the dueling titans were Amazon.com versus BarnesandNoble.com. Each expanded around the world, added new offerings, and became big businesses with multiple media products for sale. In short, much like the rest of the book business, book selling had become highly digital.[68]

The arrival of e-books represented the final innovation of the century. Though subject to multiple definitions, it essentially consisted of one of two formats. The first involved creation of a digital tablet or flat portable screen that could hold books in digital storage and that could be called up to be read, scrolling page to page. Various experiments with such devices took place all through the 1990s, but as of the early 2000s had not yet been sufficiently perfected to displace paper-based books.[69] The second version, the one that began to gain traction in the early years of the new century, involved downloading purchased digital files of a chapter or book (usually in PDF format) from either a book dealer, such as Amazon.com, or

Table 9.2
Revenues of Amazon.com, 1995–2004

1995	$511,000
1996	$15.70 million
1997	$147.80 million
1998	$610.00 million
1999	$1.60 billion
2000	$2.76 billion
2001	$3.12 billion
2002	$3.93 billion
2003	$5.26 billion
2004	$6.92 billion

Source: Annual reports, Amazon.com. Note that 1995 was not a full year as the firm went "live" at mid-year.

directly from a publisher, such as Pearson or Oxford University Press, two of many that had launched Internet book-selling initiatives. Earlier, in the late 1990s, publishers began forming alliances with smaller firms that had the capability of publishing e-books delivered over the Internet.[70] The latter tended to result in the sale of shorter publications, initially from small presses or research organizations, later from mainstream book publishers selling chapters of books (first textbooks), later novels and short stories. The industry was still learning how to make a profit from these products. Pearson, for example, lost a reported $200 million with its Internet initiative in 2003, so there was much yet to learn while also conditioning the market to accept this new type of product for a fee. E-books suffered from all the normal issues of a new format: lack of technical standards, not enough content (much material was not yet covered by electronic publishing clauses with authors), undeveloped markets, and, finally, insufficient awareness of what formats and types of material would work best in this medium.[71] But clearly, it was seen in the industry as the next great wave of change—and it was totally digital.

Twenty-First Century Challenges

There were a number of challenges facing this industry in the early 2000s, all of which involved digital issues—clear evidence that this industry had profoundly changed.

The first and foremost involved the Internet, which had rapidly become a major channel of distribution and was beginning to influence the very essence of what defined a book and the topics that would sell. But how much influence had the Internet had on this industry? As a vice president at McGraw-Hill in early 2001 said, "not enough."[72] It was the essential truth, for the industry had only worked with the Internet for one decade, and if the events of that short period of time were any indication of the volume of change and innovation yet to come, then "not enough" meant so much more would happen soon that was unpredictable. The second issue concerned books themselves: would they finally become e-books, or would they be enhanced with CD-ROMs that leveraged hypertext to create books that could be read out of sequence, to read extended sidebars, to hear voices and other sounds, or to view videos as part of the text? New technologies always seemed on the horizon; for example, in 2004 it was electronic paper or ink (digital platforms and text) while multimedia books had been an emerging collection of technologies, formats, and media for a decade.[73]

What were the economics involved in publishing with new technologies, the business models, and the skills and resources needed by a traditional book publisher to compete in the industry? The extensive reconfiguration of the industry that occurred in the final two decades of the twentieth century through consolidations and acquisitions figured into the mix of influences on what future products would look like. We are not yet at a point where sufficient time or equilibrium has been reached to answer the question. Experimentation continues. One example of that has been the creation of novel alliances and partnerships that leverage core competencies of members with the capabilities of the digital. One simple case illustrates the trend. The Wharton School at the University of Pennsylvania and Pearson Education announced in February 2004 that they would soon begin operating an alliance under the logo of Wharton School Publishing. The business school, with its ability to find meaningful material on business topics and a strong brand name, would recruit and develop book manuscripts. Pearson would publish and distribute these around the world. Deliverables would be in print, audio, and other interactive formats—books, audio books, e-books, CD-ROMs, and videos—and in various languages. What makes the project novel is that right from the launch, both announced that the end products would be in multiple formats, all made practical by use of digital technology. What is particularly interesting is that both are thinking and describing the individual pieces of content as traditional books. Thus, for example, in the press release, both spoke about 15 books being published in the near future, listing titles and authors as one might see in a publisher's catalog, yet the final products would range from printed books to tapes and electronic files.[74]

Another challenge came from the industry never being immune to the same pervasive concerns about copyright that affected both the music and movie industries. While copyright laws had evolved since the 1970s to account for software, electronic publications, and other digital formats, access to material in legal and illegal ways had improved, creating both risks and opportunities for the industry, issues not yet resolved.[75]

While many other topics could be raised, there is a fourth that extends the cultural handed-down values of the industry: the quality of the content itself. Richard Curtis, owner of an e-publishing firm and a long-time member of the industry, in late 2000 expressed a widely held view of his colleagues: "What are we in this business for? We're in this business to write content, to publish content that will last forever. Not physical books, but content."[76] In the midst of concerns about funding new distribution channels, finding new markets, doing it all globally, making books interactive multimedia products, one other editor pondered, "Who will be the Homers and the Chaucers of the digital age? Is the new technology going to free us in the way that writing helped free our thinking?"[77]

Newspaper Industry

This industry is quite different in its business operations from book publishing. For one thing, it does its own research and writing, which is the role of reporters. For another, it produces a product that looks quite different than a book. Yet a third difference is that its products are written, printed, and distributed in a very short period of time—hours in some cases—while books may take years to write, produce, and distribute. Newspapers cost much less than books. While the lion's source of revenue in book publishing comes from the sale of products to individuals, in the newspaper business the bulk comes from advertising fees. Newspapers are far more labor intensive than book publishing, particularly on the composition and printing sides, and more intensely involved with labor relations and union issues. As for similarities, both spend a great deal of money on paper, and they often use similar technology to print their products.

The major activities of a newspaper can be summarized quickly. In the physical production of a newspaper reporters gather and write up information, reporting on events of the day or more general topics; editorial staffs pick what stories to run, clean up the writing style and decide how much emphasis to give a story by where it is placed in a newspaper; composition and production set up pages and print newspapers, or today also post them to websites (today, composition activities are part of the editorial process, though not so in 1950). On the back-room side, there is an advertising department that sells two types of space for advertisements: those provided by companies (like automotive or space for furniture dealers), and those

At the New York Times, *an employee collects and delivers information to the news room using an IBM 370 Model 145, 1973. Courtesy IBM Corporate Archives.*

from individuals; ads are sold, set in pages, and then sent to production to be integrated into the newspaper. Circulation is responsible for selling copies of newspapers to subscribers or to retail outlets and for their physical distribution. Through most of the half century, about 10 percent of the workforce worked in advertising, with circulation activities accounting for roughly another 12 percent, administration and management 15 percent, editorial functions between 8 and 12 percent, production and maintenance another 48 percent. Despite extensive use of computers through most of the period, it remained a labor-intensive business.[78]

The digital hand could be seen at work in each area of the business, beginning slowly at the start of the 1960s and becoming pervasive as the decades rolled past. To a large extent, the early the story of IT in the Newspaper Industry is about the use of computing in support of business, pre-press, and production applications; and later about reporting and, finally, use of the Internet to distribute stories. Table 9.3 summarizes the key areas of use as of the mid-1980s, by which time IT existed in various forms in most departments and in both large and small newspapers.

IT Deployment (1960s–1970s)

Computing came initially to newspapers for the same reason as to so many other industries: to drive down the costs of labor and thereby improve productivity. One expert on the industry stated the problem facing the industry at the start of the 1960s: "Find a means of halting the rise of costs which are overrunning revenues in spite of the heaviest ad volume and best business climate newspapers have ever known."[79] More specifically, "the production department of the newspaper spends more money than most of the others put together—and never takes in a nickel."[80]

Table 9.3
Typical IT Applications in the U.S. Newspaper Industry, mid-1980s

Business Applications	Pre-Press Production Applications
ABC reporting (Audit Bureau of Circulation)	Classified advertising page makeup
Advertising order entry	Computer managed printing
Advertising and sales forecasting	Display advertising makeup (including graphics, page makeup)
Circulation management	Editorial systems (including news text entry/edit)
Direct mail promotions	Page layout (i.e., composition)
Distribution	Pagination
General accounting	
Payroll	
Reader demographics and marketing	

Source: IBM Corporation, *Industry Applications and Abstracts* (Rye Brook, N.Y: IBM Corporation, 1988): 16-1–16-2.

The new printing presses that were coming onto the market were thus both an attraction and a problem. They were attractive because they used less-skilled labor to produce newspapers using offset production methods that were quicker than before, but they were also capital intensive. Offset methods, which output photographed pages that are then turned into plates, cut out the old labor-intensive hot-type production methods.

Once such photographic techniques entered the process in the early 1960s, one could start thinking of using OCR technology and, as soon as online technologies became available in the mid-1960s, CRTs with combination word processing, editing, and composition functions all-in-one, also driving down costs. That change swept rapidly through the industry in the 1960s and 1970s. Press releases from various newspapers and computer vendors from the early 1960s spoke about faster, less expensive production in editorial and composition activities.[81] Major city dailies led the way, such as the *New York Times* and the *Detroit News*.[82] In fact, the latter became the first major newspaper in the United States to install word processing for all its reporters in the early 1960s. Reporters entered their stories on CRTs, which editors then edited and sent to production. At the same time, the national wire services began using computing to collect and transmit stories to newspapers and other subscribers. Associated Press (AP) and United Press International (UPI) were the earliest to do this. Since some dailies acquired nearly 65 percent of their stories from such sources, having them available in electronic format and transmitted over telephone lines made it possible to start putting this material directly into computer-based composition systems; it became widespread by the end of the decade.[83] The conversion proved relatively quick because the early years of the decade saw the majority of computing in the industry devoted to composition: hyphenation and justification of news stories and classified text. To be precise, as of 1962, there were eight major installations of computer-based composition systems in the United States.[84] Once computers were installed, however, new uses were added. For example, the Miami *Herald* installed a typesetting system in April 1963 based on an IBM 1620 system, later adding disk and memory to handle advertising, billing, circulation, payroll, and accounts payable functions.[85]

The implications became obvious early on. As one member of the industry wrote in 1963, "by providing an integrated system for scheduling of printing and production facilities it may be possible to materially improve the utilization of the normally large over-capacity of most major newspaper plants."[86] The notion of integration was important in leveraging assets (people, time, and machines) and extended across the entire industry. As happened in both book and magazine publishing, publishers had to be exposed to the technology and learn how best to deploy it. That process occurred all through the 1960s, largely driven by newspapers themselves, which could and did report on their uses of computing.[87] The effects proved positive. A survey conducted by the Research Institute of the American Newspaper Publishers Association in March 1965 reported that 38 newspapers then used computers for composition, of which 10 had also started using the technology to do layout in display advertising.[88]

Over the next few years, various manufacturers of software and hardware introduced products to do composition, editing, and word processing in the industry. IBM, DEC, Sperry, RCA, and many specialized firms became active here. By the end of the 1970s, applications operated in a variety of systems ranging from large mainframes to minicomputers, providing different price points for large and smaller newspapers. All of these systems made the technology increasingly affordable across the industry. By the late 1970s, some major suppliers had dozens of installed systems, while a few claimed over one hundred. At that time, over 60 percent of large and medium-sized newspapers used some form of computer systems; by 1975 over 80 percent used photocomposition, which made it possible to migrate to computer-based applications. Thus, "automated newsroom systems," as they were known by the late 1970s, were well understood in the industry. It led to the integration of key functions within a newspaper which had previously been carried out relatively independently of each other, in part because of the craft nature of the business, but also as a result of job content terms in union contracts.[89]

As the first wave of applications—composition and typesetting systems in the 1960s and 1970s—worked its way through the industry a second involved reporters who were forced to abandon their typewriters, some as early as the 1960s. By the 1980s, all used combined word processing/editing systems and CRTs hooked to mini- and main frame computers. In the late 1980s, some began using personal computers that could communicate stories back to centralized word processing systems. In the 1970s, and extending widely in the 1980s, scanning and transmission of photographs filled another major gap in the automation of pages and text.[90]

In the 1960s and 1970s, computing went beyond merely using technology to improve and speed up traditional operations. This was so much the case that the Newspaper Industry became the target of those critics of automation across the American economy fearing computers would destroy jobs. We now know what the results were for the number of jobs held and lost.[91] First, there were the consolidations which, regardless of the extent to which computers were used, created opportunities for improved economies of scale. In 1960, 109 firms owned 560 newspapers, or 30 percent of all newspapers and some 45 percent of all circulations. Thirty years later 136 enterprises owned 1,228 newspapers, or 75 percent of all newspapers and 81 percent of the circulations. Many of these firms also had assets in other media industries, as noted in chapter 8. That was the extent of consolidations during a period when demand for all media, including newspapers, in general went up and had a dampening effect on the job pool. One of the reasons such consolidations could be done is because digital technologies (both computing and telecommunications) facilitated realization of economies of scale and scope.[92]

Looking at union membership allows us to take the issue of jobs and technology down to another level of detail. Labor in the 1950s in this industry was highly unionized within organizations representing specific crafts and jobs in the industry. All members were affected by technologies of various types, such as telecommunications and printing and typesetting, not just computing. The main players, however, were the International Typographical Union (ITU) and the Newspaper Guild. Each merged with the Communications Workers of America

(CWA), the first in 1987, the latter in 1995. The labor side consolidated into less-specialized unions at the same time as management consolidated into less-specialized corporate conglomerates. As one union leader explained, such labor mergers into larger unions were done in part to face "the technological challenges of the remainder of the twentieth century."[93] The shift caused by technology came early, in 1961, when the *Los Angeles Times*, the nation's largest newspaper, became the first one in the country to start using computing in composition, and continued to be one of the most extensive users of computing within the industry through the 1960s and 1970s.[94] The number of craft workers declined, as did the population of print workers needed in general by this newspaper. Other newspapers followed suit, particularly those with nonunion labor since they did not have work contracts in place blocking such changes. At the start of the decade, the ITU had 94,523 members, but at the end of the decade—one that saw good growth in demand for newspapers—membership had declined to 91,848. Use of CRTs and less-skilled labor in the 1970s simply accelerated this trend of needing less-skilled craftsmen. Statistics from the Guild documented a similar process: 26,000 members in 1975, then 20,000 in 1995.[95] However, the most dramatic evidence of what happened came from the U.S. Census Bureau, which reported that between 1970 and 1980, the percentage of typesetters and compositors declined by two-thirds, yet the total number of people working in the industry actually increased, from 422,657 to 510,760. In short, as one student of the newspaper business observed, it went from being "top-heavy with machine operators to an industry largely staffed by white collar and clerical employees operating computers."[96]

Computers improved typesetting and other operations sufficiently for publishers to install them quickly in this period. They were instrumental in reducing the power of the craft unions. They did work more efficiently, lowered labor costs, and optimized tax deductions then available for companies investing in capital machinery. Table 9.4 provides data on the extent of deployment of the various new technologies in the period. We can see old technologies declining and new ones

Table 9.4
Deployment of Electronic Technologies and Decline of Hot Metal Linecasting in the United States, Selected Years (1963–1978)

	Number of Machines (Systems)				
Technology	**1963**	**1969**	**1973**	**1976**	**1978**
Hot metal linecasters	11,175	11,557	6,690	1,877	1,158
Computers	11	529	719	1,206	1,982
Photocomposition machines	265	903	2,395	3,076	3,090
Video display terminals			685	7,038	15,841
OCR readers			186	671	713

Source: Derived from data in Anthony Smith, *Goodbye Gutenberg: The Newspaper Revolution of the 1980s*, Table 9 (New York: Oxford University Press, 1980): 133.

displacing them very quickly. The data are evidence that the Newspaper Industry was also acquiring thousands of workers capable of using this new technology, representing another shift toward a different way of doing a job. The information is very revealing about the extent and speed of the switch to the digital. Photocomposition equipment quickly displaced hot metal technology. The biggest drop in use of the old took place between 1969 and 1973, when the number of installations dropped from 11,557 to just under 6,700. Meanwhile, installations of photocomposition systems rose sharply all through the period, more than tripling in number between 1963 and 1969 (265 and 903, respectively). The number of computers grew naturally on the heels of photocomposition installations, from 11 known installations in 1963 to 138 by the end of 1966, and yet another surge by the end of 1969 to 529 systems. The growth in CRTs suggests that online systems, often connecting photocomposition and editorial functions, took off in 1971, with the predominance in installations occurring between 1973 and 1975, as the number of units grew from a base of 685 to nearly 4,000. That tells us online systems and applications came into the Newspaper Industry in some significance about a half-dozen years after they began appearing in manufacturing and financial industries.[97]

Lest one conclude that what we were seeing was some inevitable march of progress, the reality proved quite different; it was a march of change afflicted by potholes. In addition to workers witnessing the end of their present careers, and responsibilities shifting across the industry, came the angst of change itself. An expert on the industry looking at the situation at the end of the 1970s described the issues facing management and their employees:

> The real problem in applying computer science to the newspaper industry was that of breaking down habitual attitudes and methods within the newspaper organization itself. To alter, for example, the method for collecting, billing, credit checking, and setting classified advertising meant that an entirely fresh view had to be taken of the industry, one that in itself necessitated outside expertise of a kind that most newspapers didn't realize they needed.[98]

Newspapers felt their way along, evolving toward a production environment in which multiple tasks were combined or more tightly coordinated than before.

Demand grew for those individuals who could operate electronic digital systems and dropped for the craftsmen of old. Composition and typesetting activities were subsumed into the writing and editorial processes. Studies done in the 1980s demonstrated that the number of IT specialists in editorial departments had steadily risen.[99] Business and technology are still human subjects, requiring a long process of learning and growth, not just collections of numbers so one final piece of data on the nature of the change underway is offered. Edward P. Hayden, a member of the ITU, wrote in the 1980s that "I used to be a printer, a member of the aristocracy of skilled labor. . . . It was a craft. . . . It took six years or more for an apprentice to become a journeyman; I learned the necessary computer codes for typesetting in about eight weeks."[100]

The reductions in and consolidations of staffs made possible by computing

Minneapolis Star and Tribune *newspaper composed and set advertising copy online, 1976. Courtesy Archives of the Charles Babbage Institute.*

continued all through the 1980s and 1990s as newer, more powerful, and less expensive systems came online, building on the inventory of earlier installed applications. But, as always, a combination of factors influenced events. As one sociologist observed, "such cut-backs are related to various factors including the desire by proprietors to increase profits and the rising costs of production associated with price increases in newsprint, they are made possible in part by technologies which allow production to occur with a smaller staff," citing the United States as the place where this was mostly the case.[101]

Recent Trends (1980s–2005)

The history of computing in the 1980s and 1990s incorporated every new digital technology that came along. When PCs finally made their major appearance in the 1980s in corporate America, they began replacing CRTs attached to mainframes. These also acquired communications capabilities which allowed reporters to access databases of information (from book publishers, information providers, and all the wire services). By the late 1980s, they also began transmitting stories from the field back to the home offices; indeed, this occurred in print, radio, and TV reporting.[102] This practice increased with the wide adoption of "lugables," then "portables," and finally laptops.

With the availability of dial-up services in the 1980s and then the Internet in the 1990s, the Newspaper Industry now faced the kinds of issues book publishers did with regard to online editions of their publications. Their questions were the same: How could they make money on the Internet by selling advertisements or subscriptions or by giving away content? How could one protect copyrights? What formats would work best? Should stories appearing in print editions be the same length and style on the Internet? What effect would one medium have on the economics of the other? This debate continues to the present. However, some patterns are observable.[103]

The industry came to the Internet with surprisingly considerable experience in delivering news in electronic ways. By the mid-1970s, newspapers in the United States and Europe had begun thinking about how to deliver news using online systems, dial-up services, subscription offerings, and so forth. Over the next two decades the industry experimented extensively with various delivery models and technologies until it and its customers coalesced around the Internet as the primary platform for delivering electronic news. As one student of the process later wrote, "the 1980s were years of exploration. Newspapers tinkered with a variety of technical and communication options, from providing directory services to personal computers to delivering news via facsimile, and learned about the commercial feasibility of these endeavors by studying how users responded to them."[104] One very early experiment that received wide exposure in the industry illustrates the process. Knight-Ridder, Bell South, and AT&T began an experiment in 1978 called the Viewtron project. Knight-Ridder established Viewdata Corporation of America for the joint venture; over the course of the next several years it ran a series of experiments with groups of customers around the United States to provide shopping and banking, data, news, and more over telephone lines. Many of the offerings shifted the reader from a passive user of information (like a newspaper reader) to an active user who could perform applications (such as online banking). The ventures tested pricing mechanisms in exchange for information and services during the early 1980s. The project failed for a number of reasons, most notably the arrival of the PC displacing earlier devices used to deliver services, the high costs, and an insufficient number of subscribers. In the 1980s, other experiments by various newspapers focused more on the delivery of news over teletext, audiotex, and fax, using stand alone or others devices connected to television sets, all during a period when PCs began entering homes and offices, eventually becoming the public's platform of choice and widely deployed by the time the Internet became available. By the time those two technologies were widely implemented, newspapers had learned a great deal about what kind of data to put online.[105]

The first major application of the Internet by the public was as a source of information, followed by e-mail and online purchasing. Newspapers were some of the first providers of information on the Internet, with all major newspapers and news services having their own websites by the late 1990s. The list is endless but includes *USA Today*, *New York Times*, *Wall Street Journal*, *Washington Post*, and *Los Angeles Times*, to mention a few. By the start of the twenty-first century it was difficult to find even a small newspaper without a website. At first, newspapers simply reprinted text from print editions, then went on to add information in support of those stories, then to modified text online to make stories shorter.

By the 1990s, the practice of producing continuous updates to stories had become fairly common, often occurring several times a day (as opposed to once or twice daily in print versions). This continuous updating of stories had been done by newspapers for years, of course, but the new factor in the 1990s was the Internet, which allowed for updates that were almost real time. Some topics, like business or sports news, which were updated only once each day in print, were updated every 15 minutes by the late 1990s.[106] But the Internet also made it possible to make archives available. For example, the *New York Times* placed over 50,000 book reviews onto its site, with material dating back to 1980. Newspapers dealt with the issues of electronic newspapers and the Internet individually, but there was also collaborative activity. The most famous involved eight newspaper companies that in 1995 came together to create the New Century Network. The newspaper firms involved in this project were important members of the industry: Advance Publications, Cox, Gannett, Hearst, Knight-Ridder, Times Mirror, Tribune, and the Washington Post. The Network established an Internet-based delivery of newspapers and information sharing among the network of newspapers. The goal was to get 75 newspapers online, making them available to their subscribers. Various other projects of a similar nature launched in the second half of the decade as well, either as a defensive tactic in case print versions of newspapers were to disappear, or as a way of enhancing their reach to the market.[107]

By the end of the century, even the content covered by newspapers had expanded, giving some newspapers a magazine-like quality. For example, instead of just publishing a news story (the traditional print approach), a website would now add side bars with additional content (what magazines had done for decades). Conversely, some magazines did the same, taking advantage of news sources owned by their conglomerate. This melding of practices actually began with newspapers in the 1980s and magazines in the 1990s with print versions, when newspapers sought to make their publications more attractive by adding non-news materials, such as advice on gardening in the spring and about plays and entertainment in the fall.[108]

In many ways, however, newspapers appeared the same. They still devoted about 60 percent of their pages to advertising and the rest to editorial content, not so different from 40 or 50 years earlier. About 65 percent of all costs went to printing and distribution, again similar to decades earlier. The industry kept slowly growing; at the turn of the century, advertising sales in the United States hovered at $50 billion, with annual growth rates in the area of 5 percent; in 1950 sales had been $2.5 billion. It was still the largest publishing industry.[109]

At the turn of the century, newspapers were available across the United States, manufactured simultaneously in various printing plants using material downloaded off satellites so that the same edition appeared in New York and Los Angeles. The *New York Times*, *USA Today* and *The Wall Street Journal* fit this category, both in print and electronic versions. *The Wall Street Journal*, with a circulation of 1.7 million in print (making it the largest in the nation) had 620,000 online subscribers as well. Today *WSJ*'s biggest competitors do battle with it on the Internet, such as Bloomberg and the *Financial Times*, both with rich Internet content.[110] The *New York Times* launched its electronic edition on October 24, 2001, delivering it over the Internet for readers to download into their own

Table 9.5
Websites of Leading Newspaper Chains, circa 2003

Gannett (owns *USA Today*)	www.gannett.com
Knight-Ridder (owns 28 U.S. newspapers)	www.kri.com
Advance Publications (owns 22 papers)	www.advance.net
Times Mirror (owns *Los Angeles Times, Newsday*)	www.tm.com
Dow Jones (owns *Wall Street Journal*)	www.dj.com
New York Times Company (owns *The New York Times*)	www.nytco.com
Cox Enterprises (owns *Atlanta Journal-Constitution*)	www.coxenterprises.com
Tribune Company (owns *Chicago Tribune*)	www.tribune.com
E. W. Scripps (owns 18 newspapers)	www.scripps.com
Hearst Corporation (owns *Houston Chronicle*)	www.hearstcorp.com

Each of these enterprises owns multiple newspapers and other media, such as magazines, radio and television stations. In 2003 the combined list owned over 220 daily newspapers in the United States, with a total of approximately 40 million subscribers.

Source: Data compiled from John V. Pavlik and Shawn McIntosh, *Converging Media: An Introduction to Mass Communication* (Boston, Mass.: Pearson, 2004): 89–90.

personal computers or laptops. This edition offered key word search capabilities of full text, a zooming function to enlarge photographs, one-click scanning from one article to another, and the ability to store content. This electronic newspaper charged 65 cents for each issue, bringing in incremental revenue beyond what the firm collected for the printed version. A quick trip through various websites proves who was on the Net with content (see table 9.5).

Demand for online news increased in the American market in the late 1990s, helping to explain why newspapers moved to that medium relatively quickly. One survey conducted by the Newspaper Association of America in 2001 reported that 37 percent of all adults in the country had gone online for news or other information during a sample one week period, up from 1997, when only 15 percent had done so.[111] Other studies confirmed similar patterns of behavior.[112] Yet the industry's senior managers were still trying to figure out what role advertising should have in their business. The *New York Times* found a system that worked; the digital edition required subscription, which made it possible to gather information about readers that in turn it used to sell advertising and reach targeted audiences. The editor of the *New York Times Digital*, Lincoln Millstein, reported in June 2002 that he was successfully selling advertising for this electronic product.[113]

But like all other industries, its members were not always clear on how the Internet would work well for their respective companies. Business models, nature and type of content, the economics involved, and issues over security and copyrights all played out in this industry. Concerns about competition also worried editors, who began realizing that other companies in different industries could compete effectively (such as electronic news services), while others in the industry wanted to learn what lessons there were to be learned from Amazon.com's experience and the failed dot-com. Various convention agendas and stories of the American Society of Newspaper Editors of the late 1990s and early 2000s makes it quite clear that much uncertainty about computing and communications still weighed heavily on the minds of publishers.[114]

But the nature of work also concerned them. In addition to all the automation of composition, printing, accounting, advertising, and the like, much still was changing and nowhere more so than in the core profession of journalism itself. To be sure, reporters had all moved to word processors by the mid-1980s, so much so that they began to experience repetitive stress injuries such as carpal tunnel syndrome and tendonitis from typing on systems that were far more effective in capturing more typed text faster through word processors than typewriters.[115] On a brighter note, reporters now could use computers to mine data of large quantities of government tax and land records, for example. Their colleagues in television and radio could conduct multiple and simultaneous interviews with people, creating a new genre called "interactive journalism." Reporters and staff clearly had to have strong digital skills,[116] but they only reinforced strategies of the industry. The use of online databases by reporters, for example, had a relatively long history. As soon as such pools of data became available, reporters began tapping into them, particularly for investigatory journalism that required analyzing large bodies of government and commercially available files. Beginning in the late 1960s and expanding over the next 20 years, reporters learned how to pore through electronic files, constructed in-house databases regarding specific research projects, and learned to analyze electronic data, in the process creating a subcategory of journalism called computer-assisted journalism (CAJ). One veteran of the industry described how CAJ changed from the 1960s to the mid-1990s:

> When journalists began conducting computer-assisted projects, it was a cumbersome and difficult task. To examine the records of a government agency with a computer, they first had to create their own database from paper documents. Moreover, they had to use a mainframe computer with neither ready-made software nor an interactive screen on which to type their instructions. More recently, as personal computers became commonplace in newsrooms, some reporters began to see the potential of employing the machines that they had been using only for word processing for counting, sorting, and comparing various kinds of information that they could obtain on magnetic tape.[117]

In one survey of 130 early CAJ projects, 40 of the reports were about politics and campaign financing; another 19 studies concerned crime and police issues, and another 18 were about minorities and their problems. Half focused on national level themes, another third on local. Of the 130 studies, 97 percent relied on various government electronic files, overwhelmingly involving analysis of magnetic tape records, though a third of all studies required converting paper records to electronic files.[118] They all had in common the practical examination of vastly larger files than might have been possible before. By the mid-1990s, such studies had become relatively common fare for most major and regional newspapers.

With so much data available, niche specialties seemed to pop up everywhere;

journalists would have to specialize—or would they? In 2000, one reporter, Frank Houston, thought about the future he and his colleagues faced:

> If the Web can be considered in cosmological terms, then right about now we're in the infinite nanosecond after the Big Bang, an inflationary moment when all that matter spreads itself out. Now things start to coalesce. New technologies and faster connectivity are at work, as is the gravity of financial pressure. In five years thousands of media jobs have come into existence. Headlines that once surprised . . . seem familiar. New journalistic content is being created everywhere, but though the audience is continually growing, it is finite. If one can read breaking news on literally dozens of different sites, for example, how many headline services are likely to prosper?[119]

It is a good question to ask, one facing the industry today as it ponders its collective next round of arm wrestling with the digital hand.

Magazine Industry

Perhaps the most distinctive feature of magazine publishing is its product; magazines tend to be published more frequently than books, less so than newspapers, have more illustrations and artwork than either of the other two, and come in different shapes and sizes. They come and go in rapid response to the changing tastes of American readers. This industry also outsources more of its internal operations than the other two publishing industries.

For decades prior to the arrival of television it was the mass medium of the United States; after the arrival of TV it gradually evolved into a vast collection of publications aimed at niche markets. While there are many large media companies in the industry, of the three publishing industries it remains the most fragmented because it has more, smaller enterprises and fewer, more specialized sets of subscribers. It is so fragmented that even the number of magazines published in the United States is subject to some dispute with estimates ranging from 12,000 to over 22,000 by the year 2000. Most enterprises publishing in this industry also had only one or just a few magazines. The point to keep in mind is that outsourcing and fragmentation are key features of this industry's structure, influencing adoption of various technologies.[120]

Yet it shares some common features with its two cousins in publishing. The editorial process of collecting, writing, and organizing material is relatively similar, as are the tools used, from composition systems to word processors to graphic arts software. Printing technologies are also similar; in fact, large printing companies manufacture magazines for multiple publishers just as they do for multiple book publishing enterprises. Like newspapers, editorial, advertising, and circulation processes are finely tuned as the core of the business, regardless of whether they're outsourced or not. Subscriptions tend to hover closer to newspaper numbers, although some have vast circulations, such as *PC Magazine*, with 1.2 million in 1998.[121] Magazine editors and their production departments are always under enormous pressure to make publishing deadlines, a requirement to publish on time made more complicated than in the newspaper business by the fact that so much more of the production work of a magazine is outsourced. That in turn requires greater levels of coordination of activities and technologies with the suppliers of various services.[122]

Magazines were not immune to the effects of digital technologies late in the century. Two distinctive ones are worth mentioning: zines and other Internet-based publications. Zines constitute a new category of publications that first appeared in the 1990s. They are low-circulation periodicals aimed at a very narrow audience, are created using desktop publishing tools running on a PC, and are "printed" either through electronic distribution or simply photocopied. Some audiences may be as few as 25 or as many as 100,000, but they are normally tiny (perhaps aimed at a neighborhood or a particular niche of teenage girls). It is not clear how many of these exist, although one estimate suggested 10,000 in the United States just in the mid-1990s.[123]

The Internet became a factor for magazines, either as websites or, increasingly more like newspapers, with electronic editions or subsets of materials originally published in hard copy. In 1998, for example, 23 of the top 25 circulating magazines in the United States had websites; all the large firms were online, such as Time Warner and Meredith. The Internet augmented exposure for print editions, but like newspapers and book publishers, it was not clear at the end of the century how one could make a profit selling periodicals over the Internet.[124] However, the industry was robust and felt few negative effects from the Internet. New magazines came and went; but business was generally good.

IT Deployment (1960s–1995)

Computers came to magazine publishing in the 1960s, The most detailed inventories of computer installations in the 1950s were maintained by Automation Consultants, Incorporated; its lists showed none for magazine publishers, and only a handful for book and newspaper firms, mirroring the types of applications and patterns of adoption seen in book and newspaper publishing.

As for the other publishing industries, printing systems were of concern to magazine publishers. Advances in technology in the 1950s and 1960s led magazine editors to pressure vendors to be more innovative because magazines needed advances in handling photography, color, graphics and other formats sooner than the other publishing industries. Also important, the quality of the end product often had to be far higher than for books or newspapers. Advances in lithography, photocomposition, and offset printing opened the door to new ways of producing magazines. As in the other two industries, the mundane issues were also important, such as the ability of photocomposition systems to hyphenate and justify well. Tape-based systems first came into this industry much as in the other two, but quickly migrated to online CRT-based approaches beginning in the 1960s. Computers transformed these early systems into highly sophisticated graphic arts operations in wide use by the 1980s. As in book and newspaper publishing, these

Early use of zip codes for marketing and distribution by a magazine publisher relied on computing, mid-1960s. Courtesy Charles Babbage Institute.

systems began reducing the requirement for expensive craft laborers in setting type, printing, and later even graphics work. The 1970s belonged to early online systems as better software and more digital storage became available and affordable. In the early 1980s, desktop publishing began seeping into the industry and by the end of the decade was in wide use. As several experts on the industry noted, the effect of these new photocomposition systems was enormous: "In particular, computerization of typography showed that specialists from other disciplines were starting to redefine many of the production tasks that had once been dictated by the skills and heritage of printing craftsmen."[125]

For the first time in some 500 years, management began changing the fundamental production processes in publishing. As the experts above pointed out in the mid-1990s, "Today desktop publishing and electronic prepress, the children of photocomposition, are the gateways through which advanced technologies continue to transform printing into a communications medium for the twenty-first century."[126]

It did not take long for management to seize upon other ways of managing work flows, for example, in managing subscriptions and other routine activities. Meredith Publishing Company, for instance, installed a mid-sized IBM system 360 Model 30 in the mid-1960s to help address the problem of meeting narrowing production and distribution deadlines. In the case of its *Better Homes & Gardens*—

with a subscription list of 6.7 million and some 70 regional editions and 140 different products (some for home delivery, others for newsstand sales)—the firm needed to make sure each one arrived where it was supposed to go on time. The IBM 360 was used to set type and to schedule production tasks in a very precise manner, while allowing for adding in last-minute advertising and editorial content. Meredith also used the system for its other magazines which, at the time, totaled over 12 million issues per month. By the late 1960s, Meredith was using it to handle subscriptions, billing, subscription analysis, and preparation of audit and other accounting reports; other firms began implementing similar changes in the early 1970s.[127] Subscription management has long proved to be an expensive and complicated process, because of turnover in lists or changes in mailing addresses. *Look*, for example, annually experienced an annual 20 percent change in addresses from existing subscribers, and turned to software and rapid data entry to change the process, speeding it up and lowering labor costs.[128]

The earliest users of computers were publishers of trade magazines, who found the digital useful in composing annual suppliers' directories. In short, the workload most mimicked the text-intensive activities of newspapers, which had led the charge in the early 1960s to adopt computer-based composition tools. The next advances came in digitally available illustrations and photography (1970s and 1980s) which could more easily be integrated with text, in effect, completing the evolution to all digitally based printing processes. Systems in the 1970s went online, and costs of computers, memory, storage, and other peripheral equipment kept dropping, making it ever more affordable for smaller firms to adopt the computer. In the 1980s, desktop systems made it easier to reduce reliance on expensive experts in composition and design; by the end of the decade, off-the-shelf software and hardware products were widely available and inexpensive enough for small publishers to use. Yet adoptions were slow, particularly for PC-based tools, since typography remained the preserve of high-end expensive systems, a circumstance that changed as time passed. The consequences were enormous; technology decimated a whole profession. Industry participants later observed that by around 1985—the year many consider the birth year of desktop publishing—professional typesetters had "failed to anticipate . . . the willingness of publishers, advertisers, and others to put quality into the scales with control—and expanded control of the publishing process that personal computers now promised to grant them."[129] Table 9.6 catalogs the building blocks of desktop publishing available in the 1980s, setting the industry off in a new direction. By the end of the decade, it was possible

Table 9.6
Technical Components of Desktop Publishing, 1980s

Personal computer (Macintosh first to offer desktop publishing)
Page layout software (e.g., PageMaker, QuarkXPress, Illustrator)
Laser printer (e.g., LaserWriter for producing hard-copy proofs)
Page description language (e.g., PostScript)

to design graphics, set colors, and lay out pages with high quality output using such software products such as Adobe Systems' Illustrator and Photoshop. There even existed a page-layout program that was being widely embraced called QuarkXPress.[130] One survey conducted in 1991 turned up the fact that 85 percent of all publishers were using desktop printing applications, an enormous increase just in a half decade.[131]

Thus, in a very short period of time, fundamental changes to work flows had occurred: photocomposition systems and automation of editorial, production, accounting, and subscription activities in the 1960s and 1970s, followed by desktop writing and publishing in the 1980s. Some who lived through the transformation recalled "that desktop computers have hammered the traditional production chain into a production mesh—a matrix of tasks linked in new patterns that have challenged nearly all of the conventional wisdom of production management."[132] As in newspaper publishing, layout and design work migrated toward editors and writers and away from production craftsmen. Increased throughput of digital systems now drove production schedules, reducing cycle times and allowing for last-minute changes. Fewer people were needed to process a given amount of pages. One could learn production skills in weeks rather than over years. Therefore, the costs per page kept dropping. A survey conducted in 1994 by *Folio*, the industry's leading trade magazine, noted that 90 percent of its respondents were extensive users of computer-based editorial and production systems.[133]

As digital systems spread across the industry, it became easier for editors either to keep the work in-house with less-skilled personnel or to outsource more functions, beginning with printing but later extending to various editorial functions. By the mid-1990s, editors were able to conclude that the digital had delivered capabilities that drove down operating costs and sped up work: "its essential promise has been fulfilled . . . nonspecialists can set type; lay out pages; add artwork and photography; specify color; make and correct color separations; and output the results for proofing, film and plate making, or even for direct imaging on certain kinds of digital printing presses."[134] Concerns about the quality of design and color production remained intense through the 1970s with its low-end systems, but was essentially resolved by using high-end software tools, many of which reside on specialized computers, general-purpose mainframes, and in high-performance desktop systems from Apple and Sun, for example. Use of PostScript, the de facto industry standard software tool, became widespread in the 1990s. PostScript continued to improve in function and performance all through the decade. The most critical quality problem of the prior two decades with desktop systems—color—began fading, a process that had started in the late 1980s. In fact, *Folio* reported in 1994 that 27 percent of editors were doing their own four-color film work in-house, while 34 percent processed their own four-color proofs in their own offices.[135]

Responsibilities and roles were thus changing, largely the product of what the digital made possible. Accounts from the period suggest that expensive work outsourced in the 1950s and 1960s began coming back in-house as digital systems made composition and layout more possible. These systems later made the work of designers easier, forcing editors, however, to increase their supervision over these employees to make sure that work was completed on time and in a cost-effective manner.[136] For example, they impressed staffs on the need to put as much material into a page's layout as possible, since paper routinely accounted for some 50 percent of all production costs of a magazine. By the mid-1990s, no step in the production process had escaped the influence of the digital hand.[137]

Recent Trends (1995–2005)

The major technological intrusion into the Magazine Industry in this period was, of course, the Internet.[138] As occurred in book and newspaper publishing, magazine publishers saw it as a double-edged sword. On the one hand, the Internet represented a new channel for distributing content, but on the other it threatened existing sources of revenues and profits, particularly from advertisers. It was already known from experience with television that new media can and do draw consumer and advertiser attention away from print publications. While magazine publishers were first wary of the Internet though later embracing it, they remained nervous all through the 1990s. As late as January 2001, long after magazine publishers had started using the Internet as a channel of distribution, *Folio* published an analysis of issues related to this media in which it aired the concern that had been lingering for over a decade:

> Free can be such an ugly word. For magazine publishers, that one disagreeable word has epitomized the threat of the Internet in the mid-nineties, making the Web the electronic equivalent of Ross Perot's "giant sucking sound." The Internet screamed "free," but in reality could only take: take away subscribers, take away advertisers, take away revenue. If publishers tried to cozy up to it, pouring in money and resources, it would only take that much more, most thought. To an extent, those fears have been warranted. . . . [T]he Web is still a profit vacuum that online advertising is unlikely to fill. And Internet users, of course, still demand free.[139]

To be sure, the author also pointed out that publishers were learning how to sell advertising for Internet-based publications and how to complement their print products with electronic versions and complimentary offerings. But issues remained: how to protect copyrights, convince advertisers to invest in this channel, and recruit subscribers; and what to publish.[140]

The story of the industry's relationship with the Internet during these years parallels that of book and newspaper publishing. First, publishers established websites to explain who they were. Second, they posted material they had published in print. Third, they posted material tailored to the Internet, including online or e-magazines, and made it possible to subscribe to periodicals. Simultaneously, they also used the channel to cross-sell print products or to encourage print readers to visit their websites. Members of the industry also shared their experiences through *Folio*, in private conversations, and at industry conferences. A major issue across the decade was how to preserve subscriptions to print publications while acquiring electronic ones or new subscriptions to print media over the Internet. Experimentation continued into the twenty-first century. Meanwhile, publishers posted themselves on the Internet, leading one observer in late 1998 to comment that "the Web-enabled magazine is fast becoming old news," although they were still struggling with what the new business models had to be in order to sustain revenues and profits.[141] By 1999, the Internet had become the most important issue facing publishers, and they were learning how to use it. For example, by late 1999, one survey demonstrated that about 88 percent of all U.S. magazine publishers were taking subscriptions over the Internet, although marketing over the Net still had not proven profitable for most firms.[142]

Had the Internet become a profitable channel? If so, when did that happen? The answer is not clear. *Folio* published various accounts of individual publishers experiencing successes here and there, but the telling evidence came from a survey it did late in 2000. A third of the respondents said their sites were profitable or on the verge of being so, while over half said it might take as long as five years for that to happen. The same survey pointed out that, in the late 1990s, many publishers felt the Internet was drawing attention away from their printed products, thus continuing to nurture their nervousness regarding the role advertising revenue would play in the years to come.[143] The dot.com shakeout in 2000 gave the industry a nervous jolt because these new firms had advertised so much through magazines. Of the $3.6 billion spent by dot.com companies between January and August 2000, 25.7 percent went to magazine publishers, representing the sum of $933 million. That figure, in turn, had grown by $719 million over revenues from dot.coms in 1999. After the dot.coms collapsed, so too did a large portion of that revenue stream.[144]

The industry also looked closely at its internal operations involving the use of the Internet as a tool for production. Management had to make decisions about what software tools to use over the Internet such as PDF; how to move information to, from, and through the Internet to its various suppliers and printers; and how best to integrate the Web into its processes. In dealing with these issues, publishers were no different from their colleagues in so many other industries. Integrating the Web into their operations became an important operational issue by the mid-1990s, one that remains ongoing.[145] In parallel to Internet-related activities, new generations of software tools enhanced use of the digital internally, particularly in production, finally making it possible for all copy (print, black and white, graphics, and color) to be digitized and managed by using software tools, from composition through proofing to printing, with nearly half of all publishers using such integrated tools by the end of the century.[146]

Another set of activities unfolded all through the 1980s and across the 1990s that were followed closely by book and magazine publishers. It involved the evolution of scholarly journals, many of which were published by book publishers (such as university presses), scholarly institutions (such as scientific and technical societies), and media and magazine publishers. These are typically published two to twelve times a year, although more normally on a quarterly basis; they rely extensively on subscriptions for their revenue (although also on limited advertising) and are intensively text oriented. They are produced like books, and experience similar cost patterns. Libraries and scholars are their primary customers. Beginning in the 1980s, and extending all through the 1990s, the cost of producing these journals rose faster than for books, magazines, or newspapers. Libraries, facing stiff budget cuts, pressured publishers either to hold down subscription fees or to migrate to electronic versions, which typically cost 30 percent less for publishers to produce and distribute. Because there were thousands of such journals, any switching of format from print to electronic posed the same questions and uncertainties that book and magazine publishers faced with their other products. The number of electronic journals increased slowly all through the 1990s, beginning with periodicals in the hard sciences and various technical subjects. Circumstances were complicated by the fact that the academic community in the United States remained conflicted over whether to recognize publications in electronic journals as suitable for meeting their requirements to publish for tenure and academic success. Many of the issues related to areas such as legitimacy, cost of production and sale, copyright, and reprints, which were faced by both book and magazine publishers, often played out first in this subset of the publishing world. For that reason this class of publications remained a subject of great interest to magazine publishers throughout the 1980s and 1990s.[147]

Conclusions

These industries teach us several important lessons about the deployment of the digital hand across the American economy while reaffirming insights teased out of the historical record of one industry after another. The discussion is complicated because deployment of digital and communications technologies into the fabric of an industry's activities have yet to reach some stable state—one where we can say "it is all done" and be able to make observations and draw conclusions. But we can learn from the experiences of these industries by looking at them from three perspectives: role of technology, organizations, and across the economy.

Briefly stated, vendors, engineers, and users demonstrated in these three industries how digital technologies could be bent to their needs. The digital could be embedded in composition and printing systems; software could be used to hyphenate, provide editorial support, and later mix and match text with pictures, then with color—all the while improving in performance and quality. We have seen flexibility and continuous transformation of the technology in all forms over time in all industries. Less dramatically, but nonetheless just as important, telecommunications became more flexible and able to carry more data in different formats, particularly when networks went digital. The convergence of computing and communications, when augmented by database tools and establishment of standards for software products and programming languages, made it possible for more work to be automated or supported by IT, and facilitated movement of work from one part of the industry to another, or from one department to another within a firm. As costs dropped for computing and its various tools, more firms used them for additional applications. In turn, that provided economic incentives to create more or better digital tools.

When desktop computing came into its own in the 1980s, the world changed for many people and industries. In publishing it caused the break-up of work into smaller units, moving it back and forth within and even outside the organization. It would be difficult to underestimate the number of options one could seize upon with the technology. Its flexibility made it possible to house applications on almost any hardware or software platform, in any department or organization. As one piece of technology improved, patterns of work altered flows first on mainframes, then later to PC-based editorial and prepress work, all of which accelerated in all three industries, beginning in the 1980s and continuing into the new century.

More important than the technology was how organizations changed. The use of craft labor declined as their skills became assimilated into various digital tools. Over time, that absorption spread across all departments in all three industries to the point where by the late 1990s there was no question but that they now functioned in a digital style. The effects were everywhere: mergers, partnerships, in-sourcing and outsourcing, shifting of work toward editorial departments and later to authors, and outputting to various formats aside from paper. The nation's most famous printer, Benjamin Franklin, would have been delighted to see that the proverbial "everyman" could be his own publisher, although just as nervous about the effects on revenues. For despite all the drama around how writing, printing, and accessing information had become profoundly easier thanks to computing, there was also high anxiety. For one thing, it raised questions about whether advertisers would spend to place their messages in new formats, such as on the Web. For another, well-understood cost structures for publishing and distribution of content were continuously challenged at each turn of the technological screw, raising uncertainties in industries that had been relatively stable for decades. Uncertainties about the effects of the digital on contemporary business models have not gone away, and new opportunities and threats appear all the time.

The rise of Kinko's Graphics Corporation illustrates the issues. Founded in September 1970 by Paul Orfalea in the shadows of a California college, his firm offered photocopying services—a crude form of printing—for those preparing term papers and later thesis and other documents. Other services were next added, such as small offset printing, film processing, and even the sale of paper, stationery, and some office supplies. Students and professors were its original customers, and by the end of the 1980s used Kinko's to put together course materials for them. By the end of the century, Kinko's also had a strong base of business customers who needed high-quality, low print run collections of white papers, proposals, training materials, booklets, and so forth. These were done quickly, often using either desktop publishing or xerographic machines that had also become highly digital. It was a remarkable story of progress. By 2003, Kinko's had over 1,200 stores in 10 countries with 20,000 employees and some 100 million customers, generating annually over $2 billion in revenue. In early 2004, it became part of FedEx. Clearly, when its history is told, a major component of the story will surely be the role of the digital and communications in making possible creation of the firm and facilitating its growth.[148] Looked at from the point of view of our three industries, and adding in the traditional Printing Industry, one can only conclude that Kinko's used technology to take away market share from other publishing industries, all made largely possible by the digital. The lesson learned from Amazon.com's role in drawing away distribution activities from at least the Book Publishing Industry, by playing a combined hand of digital and telecommunications cards, is equally telling. What made Kinko's and Amazon.com possible was more than good entrepreneurship; it was digital technology.

But Kinko's did not escape the issues that had long bothered the industry, such as what to do about copyrights. Basic Books sued Kinko's for photocopying materials copyrighted by the book publisher and repackaged into "coursepacks" for use by students. In 1991 a Federal Court ruled that Kinko's was not protected by "fair use" clauses of the copyright law.[149] In short, Kinko's wallowed in one of the central and as-yet-unfolding issues in all three industries: how best to deal with copyrights in a world of digital content.

But there is more to learn here. In 2000, Kinko's, in partnership with Forbes Special Interest Publications, launched a magazine called *Kinko's Impress*, aimed at small businesses and home office workers. It discusses technological topics and other issues of interest to this readership, and is for sale in all its stores. When we combine this event with the business of creating coursepacks, we see Kinko's either returning to its line of printing/photocopying business or mimicking the Book Publishing Industry right back at its origins; in the 1400s, the first publishers were little print shops literally located down the street from a university. It is how the commercial book-publishing industry essentially began over 500 years ago. The development of moveable type by Gutenberg provided the technology necessary to give an industry direction and wherewithal, much as we are witnessing today happening with the digital and telecommunications. The parallels are both fascinating and instructive.

We will come back to the broader issue of the role of these industries as part of a larger media sector in the United States in chapter 11. The impact is more than the fact that print industries are highly visible; they are part of a much larger flow of events in the modern U.S. economy. But first we must understand what happened in electronic media, such as radio and television, because these were just as important in the role they played in the economy and in the effects the digital had on them. For that reason, the next chapter is devoted to electronically based media industries.

Because the Newspaper Industry did more with IT than the other two print media, are there hints there of what could happen in the future for all print-based industries? Pablo J. Boczkowski has looked at the various uses of online delivery of newspaper content and services and identified three consequences of IT on this industry. First, news shifted from being centered on the ideas and creations of journalists and to being "user centered," by which readers helped determine the topics and tone of the content. Second, the one-way communication of print—writer or newspaper out to the reader—is shifting to a bi-directional movement of information, in which a "wider spectrum of voices" participates in the process of exchanging views and influences. Third, news is also adding a more fragmented layer of detail, moving from national and local news to news of interest to small communities of interest. Cumulatively, these kinds of changes are affecting the culture and content of the industry. The end result is a world in which "print people" are making way for others who produce products that, while they look similar to the old print newspapers, are also quite different.[150] Philip Meyer, long a student of the newspaper business, has reminded us that the Internet is the latest of a long trail of technologies the industry has used. However, what is different is that the Internet has fragmented audiences more than any previous technology, the fragmentation Boczkowski also noticed.[151] Lest there be too much clarity in the matter, consumers have been buying increasing amounts of content from a mixture of book, magazine, newspaper and specialized data providers; in fact, in 2003 they spent nearly $1.6 billion of content in the United States alone.[152]

Radio and television are media industries that have, of course, always lived in electronic format and thus provide additional perspectives on what happens to industries and content that are delivered electronically. For that reason, our next chapter focuses on these two important American industries.

10

Digital Applications in Electronic Media Industries: Radio and TV

We have researched ourselves almost to death, and your new electronic pets, the computers, threaten to finish us off.

—C. Wrede Petersmeyer, 1963

The radio and television industries are the oldest electronic media and entertainment components of the American economy, the telegraph and telephone industries notwithstanding. Both were the "high tech" firms of their early days, and each has variously and widely been characterized as "mature" and always squarely in the middle of America's entertainment sector. The two have a heritage of understanding electronic technologies, including telecommunications of every kind, and thus we would expect to see early examples in these industries of digital technologies entering the American economy. Furthermore, with the advent of the Internet as yet another electronic channel for distributing media, we should derive key lessons about the role of the digital in the worlds of media and entertainment. The short answers to the questions just raised are that these industries embraced the digital when it made economic sense for them to do, worried about and resisted changes that threatened the way they generated revenues and profits, and, like so many other industries, puzzled about how best to use, resist, and leverage the Internet. As happened in print industries, over time the Internet became a major force for change. Ironically, that source of change came later to these industries than to the older print-based media firms. An industry being dubbed "high tech" does not necessarily make it the first to deploy a new digital technology. Viewers, for example, still saw their football games on tube TVs in the 1980s, although all the major providers of sports programming had started to use software to provide instant replays of increasingly sophisticated sorts and small cameras that offered views of the action not available in the 1970s. In both cases, however, use of computing predated the Internet by many decades and long before consumers noticed that their end deliverables (e.g., programming or even the humble radio or "tube") had changed. In short, these two industries embraced the digital in their own ways.

Radio Industry

This industry is one of the most pervasive in the United States; well over 95 percent of the population either listens to or has access to radio at home, at work, and in their automobiles. Millions have portable radios with headsets that they wear as they work out. The ubiquitous nature of the radio dates to the dawn of the computer, but the general widespread nature of radios reaches back into the 1920s.[1]

There are essentially five sections in this industry. The first involves the production of programs, live or prerecorded. These programs can be prepared long in advance of broadcasting (such as a group of top-10 hits for music or a program broadcast to many stations from one central site) or be transmitted live (as in talk radio). The second comprises broadcasting programs over the airwaves either from a radio station or from a central site that transmits programming via satellite to one or more radio stations for local airing. Third is advertising and customers—those in a radio station selling ads and firms and organizations buying ads—which is how radio stations make money; the more one targets particular listeners, the larger the number of advertisements that can be sold at competitive or premium prices. Fourth, there are the independent organizations that determine how many listeners there are, and whose data become the basis for setting advertising fees. Finally, we have the listeners, who affect tastes in programming and, from the point of view of the digital, select how they want to receive broadcasts (through receivers, over the Internet, and the like).[2]

The Radio Industry is the oldest electronic entertainment component in the economy, dating back to the early twentieth century. It is an industry that depends directly and intensely on the use of various electronic technologies, ranging from microphones and recording equipment to broadcasting and transmission devices. Like its counterpart, television, this industry has applied every major advance in electronics that came along in the twentieth century. When satellites became available in the 1960s, for example, radio used them to transmit; when transistors became available, they were integrated into radio sets (which made them lighter, hence portable, and less expensive, so available to more people). All through the 1960s and 1970s, miniaturization of electronics made broadcasting equipment even lighter and smaller. With the introduction of computer technology into programming equipment in the 1980s and its widespread use by the 1990s, mixing, changing, and automating programming became feasible. The industry also used every sound storage technology invented, from reel-to-reel magnetic tape in the 1940s to CDs in the 1980s. And as in so many other industries, technology made it possible to operate with fewer employees.

In 1950, aside from the large networks (NBC, ABC, and CBS) and their affiliates, most stations were local affairs and independently owned. Over the next five decades, the FCC increasingly permitted companies to acquire more radio stations, often in combination with television broadcasting stations and other media, and by the end of the century large media companies owned many radio stations.[3] Management used various technologies to expand market reach and increase productivity. They achieved market reach by sending the same programming to multiple stations via satellite, and improved productivity by using computers to do normal accounting and payroll functions, and to create programming. By the end of the century, many radio stations also broadcast over the Internet.

Business and Economic Influences on the Adoption of Technology

Fundamental structural characteristics in this industry constantly influenced decisions concerning acquisition, deployment, and replacement of technologies over the course of many decades. These are economic and business environmental conditions that remain operative to the present. Robert G. Picard has probably been as effective as anyone in calling our attention to these realities, suggesting that all media industries be looked at from four perspectives: market, finance, cost, and operations, a typology useful to apply in this book. For both radio and television, his findings help us understand the environment in which the digital worked. Picard thinks in terms of firms; however, since industries are composed of individual companies, his views are just as useful in understanding the aggregation of firms.

The market for radio has been and continues to be closely tied to formats (the type of programming provided, such as country music vs. rock 'n' roll), which are linked to local (geographic) markets. This latter circumstance became increasingly important as television came to represent national markets and radio local interests. Regulatory barriers set by the FCC limit which rivals may come into a particular market, although regulators want to ensure that competition exists for listenership. As a result, radio stations experience intense competition—not only from rival stations broadcasting in their own geography, but also from television and now the Internet. Audiences have always been unstable, switching stations frequently, within even one or two minutes, so holding customers attention long enough to expose them to one's advertisements remains a constant challenge. Furthermore, in the elegant language of marketing, there has always existed a "high elasticity of demand for advertising," which means stations constantly have to sell "spots." Finally, regulators require stations to provide a variety of free "air time" to the public sector, whether rival politicians running for office, announcing local disasters, and providing local news and weather reports.

From a financial point of view, one could operate a station with minimal capital and, equally important, with low fixed costs. Since programming can always be acquired that was created by someone else, this also means a station could maintain low production costs. Regardless of who creates programs, that activity remains very expensive, as does labor-intensive production of local news. Picard characterized marketing costs as "moderate." Normally, however, sales functions are also labor intensive.

From an operational perspective, radio stations are moderately labor and equipment intensive. The introduction of various radio technologies enabled people to be more productive—or unnecessary. As computing made equipment less expensive, radio stations adopted new devices and software. The lion's share of a station's budget has always gone to core costs, thus expenses are constant, with little latitude to reduce budgets without, in effect, going off the air. Most programming is acquired from a supplier, with the exceptions of local news and some local programming, such as broadcasting town council or school board meetings. The lion's share of revenue comes from advertisers, so radio stations attempt to stabilize their revenue streams by negotiating long-term advertising agreements with local firms, giving them discounts on costs in exchange for longer term commitments. The financial performance of radio stations tends to be less cyclical than in other media industries, despite the enormous energy expended generating advertising revenue.[4]

The Special Role of the Transistor Radio

The story of the transistor radio is a useful introduction to the other parts of the Radio Industry. Scientists at AT&T initially developed the transistor in the late 1940s; their company licensed the rights to over 250 companies to manufacture and sell it in the 1950s. The transistor proved especially useful in radio, as manufacturers of radios could now replace less reliable tubes with more durable electronics. Transistors were also small, which meant radios could be placed in more automobiles, homes, or portable devices. In addition, reception and audio quality improved dramatically with this technology. So how is the transistor related to the digital? Circuits and functions of transistors eventually became components of computer chips, which could perform ever greater amounts of electronic activities. Later, listeners could preprogram stations they wanted to listen to and even when (as in a modern clock radio in a bedroom used to wake up people). Transistors used far less electricity than tubes, making it possible to use batteries to power them—hence be portable.

The first transistor radios were developed in the United States. In fact, the "first" commercially available product was unquestionably the Regency TR-1, introduced in October 1954. Within months others came onto the market; by the end of the year, Sony's first entrant, the TR-52, arrived in stores. In time Sony came to dominate this new market. Early transistor radios were expensive; for example, the TR-1 sold for $49.95 ($334 in 2003 dollars). Even so, sales were brisk in the 1950s. There were 85 million radios in the United States. In 1960 that number had reached 156 million (including some tube-based radios).[5] Yet all through the 1950s, radios became smaller and their costs declined. In the early 1960s, they finally reached a point where they had become mass-market products.

Table 10.1
Number of U.S. Households with One or More Radios, 1950–1996

	(Millions)
1950	40.0
1960	50.2
1970	62.0
1980	78.6
1990	94.4
1996	98.0

Source: Various annual statistics, Bureau of the Census, U.S. Department of Commerce.

Portability became a major feature. Instead of only being in cabinets (or nearly 9 × 9 inches square installed in automobiles), one could literally have a "pocket" radio. For example, the TR-1 measured a mere 5 inches high, 3 inches wide, and 1¼ inches deep.

During the 1960s, sales of vacuum tube radios declined rapidly, and by the end of the decade it was almost impossible to buy one. Table 10.1 documents the number of radios in the United States covering the period of the transistor and computer chip and illustrating the proliferation of this technology. (In the 1990s, the average household had over five radios, and over 96 percent had at least one, putting radios' market penetration in the same range as televisions and telephones.[6]) Texas Instruments (TI) became a leading supplier of much of the early technologies of the 1950s and 1960s. It was the same company that introduced some of the earliest computer chips and that remained a major source of digital components for all industries (including the Computer Industry) throughout the second half of the twentieth century.

Tape and radio devices continued shrinking in size all through the 1970s and 1980s, thanks to the miniaturization of the transistor, now in the form of computer chips. Perhaps no device more symbolizes the trend than Sony's Walkman, one of the first "portable stereo systems"; this tape player with headphones, was introduced in 1979 in Japan, and in the United States the following year. It proved to be a success, with some 150 million sold by 1995 around the world, a large number for a small electronic consumer product at the time. Dozens of rivals also introduced innovations all through the 1970s and 1980s. As early as 1986, Sony began selling the D-50, a portable audio product that played compact discs, which are digital. Subsequent digital sound systems and radios appeared that provided additional functions besides playing music. These individual clocks, alarms, digital tuners, rechargeable batteries, wireless headphones, and logic controls were all available on the same listening devices. By the end of the century one could buy a completely digital, portable radio with headphones for as little as $20.[7]

IT Deployment (1950s–1970s)

As in so many other industries, the larger the enterprise, the more sense it made for it to rely on some automated accounting equipment and, later, computers. Mid-sized radio stations began using service bureaus to do some of their routine accounting by the late 1950s. For example, Station WVCG in Coral Gables, Florida, used IBM keypunch equipment to collect data for program log preparation, billing and general ledger accounting, and to compile information for its publication, *WVCG Music Magazine*.[8] But through the 1960s, larger stations began finding useful ways to use computers in-house, to automate programming, that is, what stations broadcast. One company's experience illustrates the process:

> The electronics behind this is the Schafer 8000, a digital computer whose component parts can fit into a single 19-inch rack. That includes the broadcast control center, the computer, the expandable electronic memory, teletype control, system interface, and power supply.
>
> The computer turns both stations on, operates all the separate functions from music to joining the network, then turns both off at night.[9]

In the 1970s, this application spread widely across the industry. Like so many uses of computers, this one grew out of earlier, nondigital, punch card–based systems for sorting, tracking, and organizing programs.[10]

The migration from vacuum tubes to transistors, chips and their accompanying software in computers was a long process that took place from the early 1960s into the 1980s. In addition to their technical advantages over vacuum tubes—they're smaller and more reliable, and operate at lower temperatures—by the early 1960s, transistors had become far less expensive. The same process occurred with computer chips. Expensive in the 1960s when compared to transistors, by the mid-1970s they proved both technologically superior and cost effective. Printed circuits wiped away a whole generation of tubes, transistors, resistors, capacitors, and other electronic components. As electronic equipment became more reliable, newly added timing components made it possible to automate the playing of long-running tapes so that a radio station could set up considerable amounts of programming with minimal use of staff. With such devices one could broadcast programs for hours at a time with little human oversight. FM stations, in particular, were drawn to this application because they were chronically short of cash, and thus staff (the largest cost element for any station). During the 1960s and 1970s companies appeared, offering a combination of programming and automated equipment to stations, and creating a submarket of the radio business that has remained active to the present.[11]

One problem that all radio and television stations faced was how to ensure that all the advertisements they aired were broadcast according to the contract terms and instructions of advertisers. Increasingly, they turned to computers to track and schedule this. The market proved sufficiently lucrative that vendors outside the industry began offering such products. In late 1972, for example, IBM introduced a system, called the IBM System for Television and Radio. Using a System/3 (an early IBM minicomputer), software tracked unsold air time, resolved scheduling conflicts, separated competing advertisements from appearing too closely together, and scheduled other announcements.[12]

Advertisers and rating agencies took an early and even greater interest in computers for help in their relations with both radio and television stations. In the early 1960s, computers held out the promise that advertisers would be able to analyze more quickly a variety of factors to determine when, and how much, advertising to place with specific radio and TV stations and for which broadcast programs. Management in these agencies recognized quite clearly the strengths and limits of computing. In 1966, Joseph St. Georges, vice president of the advertising agency Young & Rubicam, described the role computers could play in his industry:

> The computer provides cost-per-thousand data, reach and frequency data, gross-rating-point data, homes delivered, and costs. It does not provide program evaluation, market evaluations, station evaluations. And it cannot supercede the buyer's judgment in these areas. To us it had the positive advantage of doing tremendous quantities of analytical arithmetic with great speed and complete accuracy. It enables us to make better buys faster with fresher availabilities. It does not prevent us from confirming hot opportunities the moment they are offered.[13]

The application became one of the first adopted in the world of radio and has remained an important one to the present. The quote at the beginning of this chapter, from C. Wrede Petersmeyer, president of Corinthian Broadcasting Company, provided evidence of the growing amount of data that computers began collecting on the effectiveness of broadcasting, a trend that became more rigorously developed and enforced all during the second half of the century.[14] Much as point-of-sale data in retail began influencing the kinds of decisions companies made about what merchandise to sell, determining which programming in radio and television to invest in became dependent on massive quantities of data collected by computers.

One could tell that the application was spreading quickly just by noting the issues that arose, many of which were similar to those in other industries where deployment of the digital hand had been extensive. By the mid-1960s, for example, advertisers were complaining that data about broadcasts needed to be standardized so that information could be ported from one system to another. The need to make the technology more "user friendly" for marketing led to seminars and industry meetings on the subject. Why? Because in the mid-1960s large advertising agencies were rapidly adopting computers for the first time; hardly any agency billing over $20 million operated without one.[15] In turn, radio stations subscribed to these increasingly computer-generated marketing reports, which they used to arm their sales departments in setting prices for advertising and to attract advertisers to markets that became more competitive as more radio stations started broadcasting in the same geographic areas, particularly in large urban communities.

By the late 1970s one could look back at the Radio Industry and begin identifying the effects technology had on it. Most commentators on the industry like to point out how radio's dominant mass-media position shrank due to the wide adoption of television in the 1950s and 1960s, forcing the industry to adopt strategies aimed at narrower, niche markets.[16] To be sure, that did happen, finding those markets required larger, more precise amounts of market penetration/demand information, hence the attractiveness of computers. The availability of portable, transistorized, and inexpensive radios was also another factor in the transformation of the industry as it moved toward a more specialized collection of markets.

One study conducted at Harvard University in the early 1980s placed more emphasis on the effects of computing on the industry than on the arrival of television as the transformative set of influences on radio. Historians and economists simply have missed the point that in conjunction with television, digital technologies also had important effects on the radio business. The link between advertisers and radio was the essential path by which early digital applications came to influence the industry. Richard S. Rosenbloom, the author of the Harvard study, explained what computers supported, facilitating the interaction that had been occurring:

> The radio industry's customers . . . are its advertisers—predominantly local advertisers. Its real product, offered to those customers, is its audience. The growing strength of radio has come from its ability to deliver an audience of specified characteristics at relatively low costs.[17]

In the next decade, the digital became an equally important part of the internal operations of the radio station.

IT Deployment (1980s–1990s)

The 1980s proved to be a remarkably important decade for the industry. For one thing, desktop computing came into its own as a reliable, inexpensive, software-rich technology. Larger minicomputers rapidly dropped in cost for the same reasons that PCs were inexpensive. The former, by now a mature technology, had been extensively deployed and thus was well understood and part of a company's existing infrastructure. Satellite and broadcasting hardware became increasingly digitized, driving down the cost of operation while acquiring additional functions and capacity to handle more data (such as broadcasts). Software tools designed specifically for electronic media companies in general, and to a certain extent for radio, proliferated. Meanwhile the number of radio stations continued to increase, creating incentives for suppliers of specialized technology to enter the market. By June 1987, there were 10,128 licensed radio stations in the United States. To put that number in perspective, at the same time there were 1,008 operating television stations in the country. Of these TV firms, only 100 or so operated independently; the rest were network affiliated, creating the critical mass that made possible more extensive use of IT by the television industry at an earlier time than by radio stations.[18]

Table 10.2
Typical Radio Industry Applications of Data Processing, 1980s

Billing and accounts receivable
Order processing for commercials (sales application)
Program scheduling
Traffic (commercials) control

Source: IBM Corporation, *Industry Applications and Abstracts* (White Plains, N.Y.: IBM Corporation, 1988): 16-6. Similar lists had appeared in earlier editions, dating back to 1980 and updated almost every year during the decade.

Computer Industry records from the period demonstrate that IT providers expected approximately 100 radio and television firms in the United States to be large users of data-processing technologies. Additional hundreds were prospects for smaller systems. By the early 1990s, one could not imagine any of the some 10,000 radio stations not being equipped with at least some forms of digital technology. At a minimum, all their modern broadcasting equipment now relied on computer chips and software. So what were the common early data-processing applications in the Radio Industry? Table 10.2 provides one list published in the late 1980s, but originally tabulated in the earlier part of the decade. Uses of computing for these business and operational activities increasingly were woven into the fabric of daily tasks. Thus, for example, sales departments used software tools to enter orders on CRTs for commercials; the software identified for which programs each order was intended, the frequency, and the cost. A program file tracked what programs were going to be broadcast every day, detailing this information by the minute, and identifying what commercials they were to play during the day. Creating precise schedules for each day of news programs, weather, music, and so forth, became a major application of the digital, as did identifying exactly to the minute when and where specific commercials were to be aired. Scheduling commercials was already a very precise activity, so employing computers to do this work was a natural and evolutionary application of the new technology. The reason for the precision is easy to explain: ads had to be aired during the specific time periods advertisers had paid for. The flow of data began by taking an order for a commercial, scheduling it into a program(s), recording when in fact it was aired, then transforming that piece of information into a bill, which was then tracked and collected.[19]

A development that affected radio, television, and print media involved the adoption of portable electronic tools in the 1980s and 1990s. News reporters in the 1980s embraced the PC and, in the 1990s, it seemed the proverbial "everyone" in the profession had a laptop. In fact, reporters were some of the earliest, most extensive users of laptops in both the print and electronic media industries. By the middle of the 1990s, these machines could transmit video, voice, and text back to home offices (news rooms) for immediate broadcast or publication. Portable phones in the 1980s and 1990s provided similar immediate reporting which was enhanced when cell phones could transmit video images as well, as we saw vividly during the Iraq War of 2003. Just as newspaper reporters had access to online databases for information as far back as the 1970s, so did reporters in the 1980s and 1990s, along with an extensive collection of information available over the Internet in addition to the source that existed before arrival of this new network. There is growing consensus among journalists and editors that information technologies did more to change the profession of journalism in the last two decades of the twentieth century than any other reason. As in print media, editors and reporters in electronic media had to work quickly since the technology made it possible for them to deliver text and messages rapidly and for editors to transform and change them before broadcasting. Reporters began spending less time writing in a newsroom, because they could draft or record their material closer to the physical locations of the events about which they were reporting. With the availability of massive quantities of cheaper digital storage in the 1980s, news reports could be recorded in digital formats rather than on old magnetic tape, which deteriorated in quality, thus making it possible for engineers to archive recordings which could be called up quickly by reporters and editors as either source material or for rebroadcast.[20]

The story of the late 1980s and early 1990s is also about equipment continuously becoming digitized, modular, and less expensive, which made it possible for ever smaller stations to broadcast. At the same time, large media companies continued acquiring stations as fast as the FCC allowed. In the early years of the Internet, radio employees used the Internet as a way of acquiring information, then added their own websites as places people could go to for programming information and to other community-related data and links. Like newspapers, radio and television stations later added information to augment what they presented on the air. But the big watershed event for radio, as with all electronic media and telecommunications late in the century, was less the Internet and more the Telecommunications Act of 1996.

Recent Trends

The greatest effect of the new law was not so much technological as it was organizational. It stimulated in radio, as in other media industries, another round of consolidations which, in turn, made it possible for media companies to continue leveraging the digital to improve their productivity. Deregulation dating back to the 1980s had made it possible for the number of radio stations to increase before 1996, from 10,500 in 1985 to more than 12,000 by 1997, of which 10,000 were commercial stations that sold advertising to generate revenue and profits. The law of 1996 made it possible for owners of radio stations to acquire additional properties and cluster them into chains. The other radio stations were nonprofits, such as those affiliated with the National Public Radio (NPR) network.[21] The Internet was less affected by the Telecommunications Act of 1996; however, radio stations and media turned to the Internet more for technological and marketing reasons than because of regulatory permissions or rulings.

Yet at the end of the century the industry remained relatively unconcentrated when compared to other media industries, despite a great deal of M&A activity. From 1996 to early 2001, the number of commercial radio stations increased by 7.1 percent, but the number of owners of radio stations declined by 25 percent, largely caused by mergers. How concentrated did the industry become? In 1996, the two largest owners of radio stations had fewer than 65 stations each; in March 2000, the two largest had just over 440 stations. They merged in August, creating Clear Channel Communications, added in other stations, and quickly had 1,000 stations, with another 200 or so pending acquisition. As a comparison, the second largest group owner at the time, CBS, owned only 250 stations. Concentration at the regional level took place as many smaller groups with several to a few scores of stations also formed. In the largest metropolitan areas at the turn of the century, typically one firm generated between 46 and 73 percent of all local radio advertising revenue. More dramatically, about 20 percent of all stations changed ownership in the years following passage of the 1996 law. However, large, publicly traded radio firms entered the new century with considerable debt because of their acquisitions, although profits remained strong. Profits would have been greater if firms did not have to pay out so much on interest to support their extensive debt loads[22]

All through the period many industries experienced growing concentrations through M&A activity. Most consolidations, however, were of a mega-variety, with large corporations merging together as occurred, for example, with the merger of Exxon and Mobile in the Petroleum Industry, or the string of mergers in banking. However, in the Radio and Television Industries the pattern played out in a slightly different way. To be sure, relatively larger enterprises merged for the same reasons as in other industries and because of similar environmental circumstances (such as regulatory approval), but consolidations occurred almost in two phases. There were the consolidations of relatively large companies and across wide sectors of the nation, and there were regional consolidations. The former mirrored what occurred, for instance, in the Banking Industry, creating national players, while the latter followed more along the lines of the Grocery Industry, where local (within a city, state, or region) enterprises emerged. Like the Grocery Industry, the radio and television industries comprised collections of relatively small companies, though they were often affiliated with the national networks in order to be able to acquire and broadcast national programs. Otherwise, they were small, localized enterprises, which optimized the economies of scale balanced against what the regulators would allow.

While not as high when compared to many other industries, the concentration into relatively large radio groups now existed and increased in the early years of the new century. Table 10.3, which lists the largest firms in 2001, demonstrates that concentration of the number of stations remained low, thus theoretically placed in a position to compete, and ostensibly to provide diversity of programming. Meanwhile, however, concentration of revenues had accelerated dramati-

Table 10.3
Top Eleven U.S. Radio Station Owners, 2001
(Ranked by percentage of total industry revenue)

Rank	Name	Stations	% of Total Industry Revenue
1.	Clear Channel Communications	972	26.2
2.	Infinity Broadcasting	185	18.2
3.	Cox Radio Inc.	86	3.5
4.	ABC Radio Incorporated	53	3.4
5.	Entercom	97	3.3
6.	Citadel Communications Corporation	210	2.7
7.	Radio One Incorporated	48	2.1
8.	Emmis Corporation	22	2.0
9.	Susquehanna Radio Corporation	27	2.0
10.	Hispanic Broadcasting Corporation	46	1.9
11.	Cumulus Broadcasting Inc.	257	1.9
	Totals	2,003	67.2%

Source: U.S. Mass Media Bureau, Federal Communications Commission, *Review of the Radio Industry, 2001* (Washington, D.C.: U.S. Federal Communications Commission, September 2001).

cally over just a few years. Concentration put some pressure on management to use any technology that would optimize scale economics and scope.

Meanwhile, a new form of broadcasting, called digital audio broadcasting (DAB), made its appearance in Canada and in Europe in 1992. This form of broadcasting had the ability to deliver CD-quality programming via terrestrial or satellite communications. In the early to mid-1990s little happened with DAB in the United States, but in 1999 the FCC began searching for ways to encourage its introduction in the United States. Stations were reluctant to change their AM or FM formats, waiting instead for technological advances that would allow them to transmit AM and FM programs via DAB in a cost-effective manner. Complicating the picture was the availability of the Internet, which provided listeners with the programs and music on demand, a great deal of it of CD quality. That option raised additional questions about the wisdom of moving to a DAB format. And, as often happens early in the life of most transmission technologies, there were ten different technologies and standards for implementing DAB, and none had bubbled to the surface as the most widely acceptable version.[23]

The Internet, however, provided a dramatic source of change because listeners could obtain their programming—both talk radio and music—from a new electronic channel. The exciting story here is not about radio stations using IT for accounting, or websites posting their programs and additional material. Rather, by the end of the century, the main event was the emerging potential of the Internet affecting programming content and the manner of accessing it. To be sure, the

Radio Industry uncovered the potential of this new media much like other industries.

Because music, voices, and hence programming, can be transmitted over the Internet as digital files, anyone with high-speed access to the Internet can listen to programming coming in over telephone or TV cable lines rather than through radio signals. In practice, this became possible when radio broadcasters delivered programming over the Internet. While the role of the Internet in broadcasting is more a story of the early twenty-first century than of the late 1990s, it all began when stations in the '90s first used it to communicate additional text and visual information to listeners, much like any other industry. But its broader potential as a new or additional channel for distributing programming led to its implementation in the industry by the late 1990s. By then, several thousand radio stations in the United States were simultaneously delivering their programs over the Internet and the airwaves. Some small operators (called webcasters) even broadcast their programs solely over the Internet, since it represented existing infrastructure for which they did not have to pay. This venue proved popular, as it capitalized on the habits of millions of individuals who had been downloading music off the Internet for a number of years and thus were conditioned to receive music and other audio material via their PCs, which now were equipped with speakers and sufficient memory to accept and store audio data. In essence, one could argue that radio stations were not broadcasting but instead were delivering on-demand audio programming. In other words, a listener could now begin dictating when and what programming to listen to. As Harold L. Vogel, a leading industry watcher observed, deployment became quite extensive rather quickly.

As one of the newest uses of the Net, streaming material over the Internet has, as yet, an undocumented history. Conventional accounts attribute to KLIF in Dallas, Texas, the distinction of being the first radio station to do so with live broadcasting in partnership with AudioNet (later called broadcast.com and now part of Yahoo!) in 1995. By 1999 the firm was generating capital value in the neighborhood of $5 billion. The company became a provider of Internet streaming for other firms as well, occasionally having over a million participants in some of its programs.[24] So, this use is a portent of things to happen.

Also unclear at the moment is how many pure Internet radio stations operate in the United States. Early estimates (circa 2002) placed the number between hundreds and a couple of thousand in the United States, and some 10,000 worldwide.[25] Since these stations are not as regulated as are traditional AM/FM stations, counting noses is approximate at best. Compounding the lack of information is the fact that popular global webcasters are emerging, such as Spinner.com. The only fundamental law the "globals" must abide by is the Digital Millennium Copyright Act, which does not allow advance posting of music play lists, which would allow listeners to plan ahead of time when to listen to music. However, perhaps just as popular as music broadcasts are online live broadcasts of a myriad of sporting events, for which listeners are charged a fee to access over the Internet.[26]

As stations both on and off the Internet shaped their business activities in the early 2000s, they refurbished internal infrastructures with as much digital technology as necessary, because that basic technical standard and practices allowed material to feed into a station in a relatively common format. Control rooms enhanced their automation and audio routing to match new equipment capabilities, such as downloading compressed digital[27] sound from satellites and reformatting it so that the sounds of voices and music match what humans hear as their natural sound. Peter Allen, chief engineer at a state-of-the-art radio station in 2001, stated that "ninety percent of recorded audio is now played from the AudioVault drives. We use very little tape except as backup and for the few programs that still come to us on DAT and cassette."[28] His comment was less revealing about how advanced his station was than it was about the state of the industry's use of technology. Tape had been a prime medium for decades; now digital disk drives and other digital formatted media had displaced it. The transition radios went through in the late 1990s and early 2000s was less to an Internet format than to an all-digital one.

That transition made it possible for radio stations to begin migrating to high definition radio (HD), which can be delivered over AM, FM, and Internet channels. Capital investments were quite low when compared to television, which required greater infusions of capital to swap out old equipment. It only cost about $100,000 in 2001 to 2003 for a radio station to convert to HD, compared to the over $1 million per television station. While that was good news for radio, the need to move to all digital was driven by the market where consumers already had most of their music in CD formats (all digital) and similarly received digital-quality music off the Internet. As one reporter commented, "HD radio will simply bring radio services into essential parity with the rest of consumer audio."[29] The FCC encouraged voluntary conversion by radio stations, unlike its mandate for television to complete the changeover by 2006. Yet not all the technology was adequate; one observer noted in 2003 that "American terrestrial radio broadcasters remain stuck at the starting gate, with no standard format approved and no receivers in the market."[30] As occurred so many times with so many digital technologies across so many other industries, the availability (or lack of) standards and tools often dictated the pace of adoption. For consumers, it was only in January 2003 that they could finally buy an HD radio receiver for their cars, making it possible to experience digitally based radio transmission. Unlike satellite radio—also a relative newcomer at the end of the century—listeners did not have to pay a subscription fee for HD radio. However, both formats did well as consumers acquired broadband Internet connections and HD tuners, with rapid takeoff in their use finally starting in the spring of 2004. Some 2 million listeners used satellite radio in 2004; no hard data existed on HD users since it was a new technology, but that number had to be far smaller. Meanwhile the number of listeners to conventional AM and FM radio remained high.[31]

In summary, in the early years of the new century, AM and FM radio, the stalwart formats for decades, competed or collaborated with the Internet, direct digital satellite transmission, digital terrestrial radio, and HD. The array of distribution options for listeners and media companies had expanded over time, rather than consolidating into fewer technological options. The effects were interesting and complex. Consumers were demanding delivery of media when they wanted it, not when some broadcaster chose to air it. Consumers were fragmenting into smaller niche markets, making it more difficult for broadcasters to reach them. Listeners also had become comfortable with multitasking and consuming multiple media at the same time. Radio's advertisers had to develop more complex and integrated marketing strategies that spread across various types of radio, television, Internet, and even print media. Competitors often sprang up in multimedia firms, resulting in steep learning curves, new opportunities, and angst in the Radio Industry. Observers of the media world at large noticed that starting in the 1980s, and early 1990s, in radio as in so many other media industries, management increasingly became dependent on technology for competitive advantages.

Television Industry

From the perspective of business operations, both the radio and television industries compete for the same audience's time and for the same pool of advertising dollars. Both aim to inform and entertain; both are subject to the regulatory control of the FCC. They even essentially carry out the same activities—such as programming, broadcasting, and selling advertising—and have for over a half century. But there are some differences, one very important one being that television operations require a lot more funding and staffing; TV broadcasting is not cheap and cannot be done on a shoestring, as can some radio. As a consequence, operations are more complex in terms of number of employees and activities. Being larger, television companies, therefore, adopted computers earlier than their voice rivals, following the clearly evident pattern evident in so many industries that the largest firms were often the first to embrace computing. Yet both industries were integrally affected by the availability of various digitally based equipment and tools.[32] Because of the generally larger size of the firms in television, these companies were always in a better position to acquire other media companies, for example in radio and newspapers, and so their executives and industry practices proved influential in many other media industries.

The Television Industry can be even further understood by comparing it to other industries rather than just to radio, as is normally done. Because television has run along essentially three channels of distribution—terrestrial broadcasting (free TV, like ABC, NBC, CBS, FOX) and cable and satellite (for a monthly fee)—the industry sees itself almost as two entities, operating in parallel and yet competing against each other. The Insurance Industry has its property and casualty side and also its life sector. Both television and insurance have seen mergers over the decades as a way to gain sufficient critical mass to compete effectively. The parallels are remarkably similar in terms of how management responded to their markets. But there are divergences; insurance has to contend with large volumes of data, collections, and payments, which television does not. Television has to worry about acquiring programming and broadcasting, activities obviously of no concern to insurance. Broadcast and cable television did not merge while some

Table 10.4
Major Phases in the History of the U.S. Television Industry, 1947–2005

1947–1952	Modern analog television begins spreading across American society
1952–1964	Television broadcasting spreads further, regional and local TV companies are established, cable television takes off, satellite transmissions begin
1964–1972	FCC protects broadcast television, constrains ability of cable to expand
1972–1975	FCC begins a thaw in its regulatory constraints on cable, reflecting the national movement toward deregulation of many industries
1975–1984	Broadband broadcasting begins with satellite distribution, FCC regulations decline, cable expands at the expense of broadcast television
1984–1992	Deregulation of the industry continues, as majority of homes subscribe to cable and still view broadcast television
1992–1996	Congress re-regulates cable rates while expanding competition in the industry; direct broadcast satellite services begin expanding
1996–2005	Television industry operates essentially the same as before, but Internet comes into its own (raising the possibility that it would be a rival to TV as a way of delivering programming into the home), migration to digital TV broadcasting begins, and a new generation of TV sets (various flat-screen technologies) begins

Source: Adapted and extended from Patrick R. Parsons and Robert M. Frieden, *The Cable and Satellite Teleivison Industries* (Boston, Mass.: Bacon and Allyn, 1998): 20–21.

insurance companies sell both casualty and property and life policies. But the fundamental notion that there were different, yet related, industries (or subindustries) operating side-by-side and in direct competition is insightful. Insurance companies fall into one camp or another based on the products they offer and services and skills they specialize in, while television largely divides into groups based on technological alternatives for delivering programming to the public, a practice reinforced by a myriad of FCC rulings. In other words, television illustrates how technology and organizations coevolved, offering yet additional evidence that all technologies, and their uses and users, affect each other simultaneously. Table 10.4 briefly, almost arbitrarily, provides highlights in the industry's history.

The history and description of the Television Industry has been written by many and thus need not be repeated here.[33] However, we need to account for the effects of technology, both analog and digital, on business models—that is, on how one made money in this industry.

Business and Economic Influences on the Adoption of Technology

In the 1950s and 1960s, the business model in the Television Industry was the same as in radio. Broadcasting was wireless or via satellite, and the major networks generated revenues by selling advertisements. What one could or could not do was determined largely by the FCC and by what advertisers and audiences wanted. Technology was either an enabler or a limiting influence.[34]

Cable television provides an example of how technology can enable a new business model. While advertising was still important in this corner of the industry, the ability to transmit programming directly to a home or office over a cable or telephone line created the opportunity to charge a subscription fee for a variety of combined channels. More channels, hence more programming, became possible, allowing cable companies to differentiate their offerings by varying the combination of channels subscribers could access to provide specialized programming for niche markets.[35] Cable came into its own in the 1970s, and, by the end of the century, over 60 percent of all households subscribed to cable because it offered a larger variety of programming than the "free channels." Cable increased its penetration to over 67 percent of all television households by mid-2003.[36] It delivered far better quality transmission, since programming was piped in over wire and thus was not subject to the atmospheric and electronic vagaries of both traditional radio and TV broadcasting. As technology in cable television improved in the 1970s, specalized channels emerged, such as the History Channel and Home Box Office (HBO), the latter of which offered specific programs, movies, and sporting events for a fee added on to one's monthly cable bill.[37] The quality and quantity of programming that could be transmitted down a line increased enormously with the installation of fiber optic cable. Thus, just as telephone companies installed fiber optic cable to increase the volume of data they carry, so too did cable companies, providing, as two industry analysts noted, "almost infinite, expansion of bandwidth."[38]

In the 1980s, a new competitor entered the field: satellite TV. As regulators encouraged expansion of the cable industry in the 1980s, both cable and satellite transmissions increased. Between 1976 and 1987, satellite delivery services increased from 4 to over 70, with penetration of markets rising, expanding revenues from some $900 million to nearly $12 billion in the period.[39] At first, direct satellite transmissions required large expensive dishes; by the early 1990s, these dropped in cost and size, and by 2000, could be purchased and installed for less than $400. A customer with a dish avoided monthly cable costs (although had some monthly satellite costs for programming), pulled in hundreds of channels and even programming from national and international wireless broadcasting companies, such as CBS or the BBC. By the 1990s, satellite technology was primarily digital and provided additional controls at the user's end for selecting programming.[40]

As digital technologies became more available in the 1980s and 1990s, there were various new ways of doing business. The important basic change was the gradual conversion from analog to digital frequencies, which enabled a form of broadcasting that was interactive with video and communications. This made it possible for viewers to select what programming they wanted, receive information with their programming, delete advertisements, determine when to view a program, and whether to alter the programs themselves.

IT Deployment (1950s–1970s)

In the beginning, all programs and broadcasting were analog. Digital technologies first came into the industry through traditional accounting and other business

CBS News TV correspondents analyzing voting patterns in November, 1962, collected by an IBM computer. Courtesy IBM Corporate Archives.

applications, much as occurred in so many other industries. The journey was a slow one. A careful examination of the archival records of Burroughs, CDC, and IBM reveal almost no activity in the Television Industry in the 1950s. A thorough review of the installations of IBM 650s, the most popular system of the mid-1950s, failed to turn up any listings of ABC, CBS, and NBC. Yet we know that some computing arrived early in the industry, as CBS installed a small Univac system in the mid-1950s to handle routine accounting activities.[41] On the broadcasting side of the business, however, equipment remained analog and electromechanical all through the 1950s and 1960s. One reason for the limited adoption was that, in the 1950s, television broadcasters had begun relying on service bureaus to do some of their mundane accounting work, such as payroll. The most widely used rating service, A. C. Nielsen, however, was an extensive and early user of IBM equipment, beginning in 1938 with punch-card equipment to track and analyze radio programming. In 1955, the firm installed the first of four IBM 650 computers to do the same, including analyzing its assessments of television programming.[42] Advertising agencies that worked in both radio and television were also some of the earliest users of the new technology to compile reports for the FCC and information regarding advertising activities by account. One report from 1963 listed 16 advertisers using computers to do this kind of work.[43]

It was not until the 1960s that there was a substantial deployment of computers in the Television Industry, as smaller systems became available and their economic benefits clearer. In fact, though, it was still a curiosity. One would see articles describing how these machines had been used as props in television programming, as occurred in the Movie Industry. One publication, for example, reported in 1955 that an IBM sorter "starred" in the highly popular *$64,000 Ques-*

tion TV program and that an IBM 702 computer was scheduled to appear in another.[44]

As in every other industry, management often learned initially about a new technology from the trade magazines. In this industry, the voice of *Broadcasting* became the essential medium. In February 1964, this journal published a short account of computing in television that gives us insight into expectations and actual use. It focused on the anticipated introduction of transistors into broadcasting, taping, transmitting, and camera equipment, all of which would lower operating costs and allow for smaller equipment. In the period 1963 to 1965, many television stations remodeled their studios, replacing first-generation television broadcasting equipment with newer devices that could, for example, also broadcast in color. But clearly the use of computing was something for the future: "thanks to the advance in solid state devices, the station of the future, it's believed, will have equipment that is all transistorized. And that future . . . is coming within the next two to three years."[45] Advertisers were also hearing about the future. Marion Harper, Jr., an advertising industry executive, touted the role of the "information revolution" in 1964, arguing that computers would make advertising more of a science than an art due to the vast increase in information one would have to work with.[46] But it was a hard sell because many doubted the economic value of the new technology in the early- to mid-1960s. As one advertising executive noted as late as November 1966: "Advertising agency computers haven't even begun to pay for themselves," although he acknowledged their value much along the lines articulated by Harper.[47]

The networks themselves, however, were beginning to use the technology. CBS led the way, relying on a Univac 1050-III system to provide information of use across the enterprise, such as in support of station sales and clearances. By the mid-1960s CBS also had its Morning Minute Plan for directing air-time events on the computer and was in the process of integrating that application with billing, much as was being done in the Radio Industry. CBS offset the $12,000 a month rental charge with clerical savings. NBC was also rapidly developing a program inventory and affiliate records system residing on an RCA 301. As late as the end of 1966, ABC did not use computers for programming and sales work, instead deploying precomputer punch-card systems to handle advertiser billing and station compensation, although it hoped to go "live" with its first computer-based applications in 1966. The push toward computing came from advertising agencies, to be sure, but also from the rating organizations. A. C. Nielsen Company began automating reports on audience data and rate information in the 1950s, which introduced TV stations to another advantage of computers. As these came down in cost, TV stations began embracing them.[48]

We have the example of the Meyer Broadcasting Company of Bismarck, North Dakota, which owned three TV stations, three FM radio stations, and two AM stations, and employed 140 people. In 1972, it installed an IBM System 3 Model 6, a small system by contemporary standards, to increase the accuracy and quality of data for programming and scheduling. Management also used the system for such traditional accounting activities as accounts receivable, general ledger, payroll, accounts payable, cable TV billing, product analysis, and sales analysis. But its primary purpose was to schedule commercials for both radio and TV. As a manager at the company reported, "With computer scheduling, we can offer the advertiser a better buy and we have a more flexible marketing tool." The company could rotate advertisements quickly and accurately around its stations and track what was sold, broadcast, and billed.[49]

By the end of the decade, and continuing all through the 1970s, such applications slowly made their way into various television broadcasting companies across the United States. As forecast earlier, transistorized equipment began reaching the broadcasting market, and by the end of the 1960s some vendors, such as General Electric, were introducing equipment that could automate television broadcasting much as was occurring in radio, particularly FM stations, and that could interact with a station's computer-programming applications. RCA had introduced a tape recording system that relied on integrated circuits by the end of the 1960s as well, joining GE in making the earliest computerized broadcasting equipment available to the industry. Service bureaus were reaching out; for instance, Data Communications Corporation, which in 1970 introduced a subscription service to give a station online access to the company's digital computer for help in routine station data-handling applications, such as accounting, demographic analysis, and comparison of performance of one station to another.[50] By the early 1970s, broadcasters were using computers to track their inventory of movies and other programs, much as one used inventory-control applications in other industries.[51]

Loudly approaching the industry were the implications of less expensive computing, further developments in satellites, and, of course, advances in telecommunications. These were subjects of much speculation at industry trade shows, in its publications, and at the FCC, which had to pronounce policies or points of view nearly every time some new technological development came along. One example of this examination came, in the form of a report prepared by a group of consultants for the U.S. President's Task Force on Communications Policy in 1969. Their careful analysis of demand for communications pointed out quite accurately, as it turned out, how many people would be watching television in the future and the number of channels they wanted or could possibly use. They focused on how distribution would evolve in the industry by describing what became the basis of the "free" versus cable debate and the role of transmitting wirelessly or through wired systems. They reported that industry leaders believed both approaches would continue to expand rapidly in the years to come, and that wired services would improve the quality of transmissions (that is, produce clearer pictures). The public in 1968 had on average access to six TV stations, and they predicted that number would grow without any particular stimulation on the part of regulators, thanks to market demand and technical developments. The report clearly anticipated the wide expansion of cable television, which, in fact, occurred just a few years later.[52]

These debates and innovations took place against the backdrop of massive national migration toward television. Already by 1961, 89 percent of the nation's households had at least one television set. By the end of 1976, 97 percent had TVs, but the population had grown too in that period, so the actual number of families with TVs expanded to over 22 million households. Beginning in the mid-1960s, a technical revolution occurred in the transition to color from black-and-white TV as people kept their old black-and-white TVs but bought color receivers. By around 1977, nearly half of all households in the United States had more than one TV, while viewership rose from an average of 5 hours per household in 1961 to about 6 hours and 15 minutes in 1976, with households owning color monitors hovering at 7 hours.[53] Miniaturization of electronics, introduction of computing and more advanced telecommunications, and a growing demand for television programs created an opportune circumstance in the 1970s and beyond for additional enhancements.

One other area of technological change in the 1960s and 1970s involved spectrums—VHF and UHF channels, for instance. Much as we saw with radio and in the telephone business, the FCC constantly had to determine who to encourage using specific spectrums. In the case of cable television, of course, there was the added dimension of wired transmission and later satellite dishes. It is a complicated story. But as the ability to transmit data digitally became possible, particularly in the 1970s, the dual Television Industry and the FCC faced the issues and opportunities created by technological developments. The FCC generally encouraged (even mandated) that the industry expand their offerings within a spectrum, and required that television receivers handle larger numbers of channels in various spectrums, a trend that continues to the present in response to both market and technological realities.[54]

What we can conclude from the experience of the 1950s through the 1970s is that adoption of computing by the industry proved slow and tentative. To be sure, there were three important developments. The first involved use of computers to speed up and reduce the cost of business operations, such as accounting and advertising. While the work sped up, became easier to do, and possibly cost less, essentially the fundamental work of the industry did not change as a result of the arrival of the digital. The second was the injection of the digital into transmission technologies, such as satellite and cable systems, though it was not until the 1980s that one could begin seeing its effects on this industry. Then there was the one area in which computing's technologies did lead to important changes: miniaturization of equipment. While the history of this class of devices awaits its historian, the fact remains that computer chips made video cameras smaller, hence more portable and less expensive. Meanwhile, program management and transmission equipment too also took up less space, and became cooler (hence less expensive to operate) and more reliable as the industry migrated from vacuum tubes to microprocessors. In addition, over time equipment became more sophisticated and less expensive. Just as editors at newspapers could rework their reporters' material in real time, so too would editors in television. This third set of changes led almost all commentators on the industry and its management practices to marvel at the effects of transistors and microprocessors on the daily work of the industry.

CBS news reporters collect election night data for minute-by-minute reporting on results. Note that the system is online, realtime, 1968. Courtesy IBM Corporate Archives.

Consumers, however, only experienced a few consequences of the digital. Yes, television sets integrated transistors, hence making them more reliable, less expensive, and smaller, but for many years that is all consumers were aware of. Digitally based applications, such as changing channels with a remote control or skipping scenes on DVDs, were not yet developed. The one major visible change of the period was the introduction of color television, but computing did not cause its arrival or adoption. What about cable television? Did computers affect its arrival? After all, adoption of cable has frequently been called "explosive," since the nation went from roughly 14,000 subscribers in 1952 to 4.5 million in 1970 and to even far larger numbers by 1980. That growth came because, once the public knew it could have more channels they wanted them, and those who wanted to provide them managed to get the U.S. Congress and the FCC to allow this to happen, despite enormous resistance by the wireless broadcasters. One did not need computers to make available more channels. This new offering was all about regulators and competition, not about innovations in digital technology.[55]

IT Deployment (1980s–1990s)

The major events in the Television Industry from the 1980s through the mid-1990s were less technological and more regulatory and business-oriented. The FCC continued to deregulate both the radio and television industries, as per the wishes of the U.S. Congress. In 1982, for example, the FCC deregulated many activities in the Radio Industry, then two years later Congress passed the Cable Act which deregulated cable TV rates and services. In 1992, after complaints from the public that cable companies were raising rates too high, Congress passed a law re-

regulating cable TV rates and services, which the FCC followed up with regulations that implemented the new controls. The following year, however, the FCC began planning for DAB implementation and, in 1997, allocated channels for digital television broadcasting. The major regulatory event of the 1990s, of course, was passage of the Telecommunications Act of 1996 which made it possible for media companies to increase the number of broadcast firms they could own.[56]

There were other, simultaneous events going on that one should keep in mind. During the 1980s, the number of homes subscribing to cable exceeded 50 percent and continued increasing during the decade. In addition, consumers were purchasing VCRs; by 1988 over half the homes in America had one. While not a digital technology, with it Americans could use their television sets to see shows they purchased or recorded on demand, conditioning them to the idea that they could control when and what they viewed. With the arrival of digital CD technology in the 1990s, the concept behind the VCR expanded. (Just for the record, we should note that compact discs were available in the United States as early as 1983, although they did not take off in popularity until a number of years later.) Between the regulatory changes implemented by the Congress and the FCC and the expansion of cable and satellite services directly into homes, the amount of programming available to viewers increased substantially in the 1980s.[57]

On the consumer side, one technical innovation began affecting the business model of the industry—remote control devices (RCDs). No single innovation in television in the 1980s was so directly apparent to the viewer than the rise of the mighty little "clicker." With this simple device, one could change channels and later operate VCRs relatively easily and, most important, quickly. When a commercial was being aired, one could switch to another channel without even getting up to see a different program, and avoid being exposed to the advertisement. The same held true for a program that did not hold the attention of the viewer. The implications were vast. For one thing, both programs and commercials had to become more interesting to viewers. Commercials had to alter their hard-sell sales. Now, an advertiser had no guarantee that the audience for a TV program would see the message. To keep a viewer's attention, ads had to be better and more entertaining, be linked to the themes and audiences of specific programs, and be faster paced. Power to determine what a viewer was seeing began shifting to the consumer, a trend that continued in the 1990s with the advent of the Internet and wide deployment of "programmable" VCRs. In the 1980s, the device took off, and the percentage of homes armed with an RCD grew from 16 percent at the start of the decade to 77 percent at the end. In the 1990s, one could hardly acquire a TV that did not have a remote control included. By then they were also programmable; that is, users could control both VCRs and TVs from the hand-held unit, with a large variety of functions ranging from channel selection to recording to controlling sound volume to pausing, and so forth.[58]

The device was originally conceived at Zenith as early as the 1960s but, like so many other technology-based units, evolved slowly over time; first came the primitive analog, and expensive version, then the one that was cheap, ubiquitous, rich in function, and digital. In time, the technology migrated to other devices.

In the 1990s, for example, one could acquire remote control devices to operate radios and CD players—indeed whole sound and entertainment systems—to turn lights on and off in a home, or to operate digitally based temperature control devices. By the mid-1990s, the Automotive Industry began introducing remote control clickers to lock and unlock cars, even to turn the car's lights on and off. The concept of remote controls permeated many aspects of American life.[59]

In addition to the remote control device, video became an important media platform because it gave viewers flexibility in their viewing habits; it also had a great impact on the Movie Industry. But let us start by acknowledging that in the beginning video was analog, another way of recording, in this instance a combination of sound and image, and was first used by TV production departments and companies, not consumers. The videocassette recorder (VCR) allowed the "harried" viewer, as one historian called Americans of the 1970s and 1980s, the option to pick when they could watch a program, be it a rented movie or a recorded TV program. Being able to record and not merely play back a recording was also essential. As Frederick Wasser, the VCR's historian, has pointed out, the lack of flexibility and the increasing monthly costs of cable led many Americans to spend more on video than cable by 1984, a shift in expenditures and time that cable never caught up to again.[60] Prerecorded videotapes were available to rent through independent distributors, and even others linked to the Movie Industry; when we combine the total number of retail outlets, the demand becomes apparent. In 1980, some 2,500 rental stores operated in the United States; by the end of 1985, that number had grown to 23,000, and in 1990 to 29,000. The number of households renting per store also rose, from just over 670 per store in 1980 to nearly 2,200 families per store in 1990. Helping this growth was the drop in rental prices per video, from on average $7 each to $2.31 in 1990.[61]

Because movies represented such a large percentage of content, we will have more to say about the video business in the next chapter. However, we should acknowledge that videos became an extension of the movie business, yet another channel of distribution for its products. Cable and broadcast television chose not to embrace video as much because their revenues were based on advertising and subscription.

Video technology, however, had important effects on television and consumers. For example, video cameras became small and affordable enough to replace old film systems and to increase the number of individuals using the technology so that when digital photography came into existence, for both still and moving pictures, there was a large receptive market. Second, video affected how television did some of its work, such as in electronic journalism.

Almost from the earliest days of analog video, industry experts anticipated the arrival of digitally based versions. One survey of the industry's technology, published in 1982, pointed out that disks could hold more data (hence more images) per square inch of recording space than tape, were cheaper to manufacture, and could be searched just like a disk drive on a computer. Finally, and not to be ignored, "discs wear better than tapes."[62] While the story of why video remained analog and in tape format is too long to tell here, suffice it to note that research on the digital version of the technology was well underway in the 1970s.[63] All during the 1980s, television programming departments became interested in the technology, as did some in the Movie Industry, and by the early 1990s software tools began arriving in the industry that would allow producers to create, edit, mix, store, copy to tape, and present content, relying on software and computers. Even IBM entered the market, with a series of offerings in July 1994.[64] At the same time, IBM and others (such as ICTV) were already involved in creating digital tools that would allow viewers to switch from broadcast television to various interactive services (such as games or movies on demand)—services that became increasingly available toward the end of the century.[65]

On a more mundane level, operations in the industry continued to be converted by digital tools. In addition to handling accounting and business operations, news management functions were beginning to rely extensively on PCs and software tools. Two professors writing about the industry's news operations in the early 1980s observed that as "has already happened in the newspaper, magazine, and book publishing fields . . . computers are used to create and edit material, set type, lay out pages, and control printing processes," although "it remains to be seen whether television newsroom computers will have as dramatic a cost-saving impact." What editors wanted in television was "more efficiency and tighter control in the news operation."[66]

The digital continued expanding in the Television Industry on a broader scale in the early 1990s. Two industry watchers observed in 1998 that "the digital format is displacing analog at every level of television, from the capture of images to editing to storage to transmission."[67] They recalled circumstances earlier in the decade that motivated television to move to extensive use of digital tools:

> Through the early 1990s most video had to be shot, or captured, in traditional ways, on analog film or videotape, and converted into digital form later. But major electronics manufacturers such as Panasonic and Sony are producing equipment to capture video directly to digital formats, either in tape or disc. Television production already had adopted digital, or "nonlinear" editing, using videotape shot in analog and converted to digital. Digital cameras are now even available for the home consumer market, although at premium prices.[68]

The move to digital occurred across broadcast, cable, and satellite television. Production studios built in the 1990s went all digital, such as TCI's National Digital Television Center. More important than digital production was digital storage, by which programming could be received, stored, and retrieved faster and easier than from old analog tape systems. Such digital systems also made it possible to insert messages, such as advertising.

The technical overviews of technology in the industry published late in the twentieth century only described digitally based broadcasting, programming and station operations. The battles came in the lack of technical standards. These included debates about High Definition Television (HDTV) and data compression standards, which were needed for satellite broadcasting and so various parts of the industry could transmit signals with others. Even the Movie Industry joined in at the time, itself involved in developing what became known as the MPEG-1 standard for data compression.[69]

Viewers fell in love with video because it provided the ability to collect and view material whenever they wanted. At first the industry saw it as a threat, but fears were allayed as viewing of broadcast and cable television by consumers kept growing through the entire period. Digital tools made their way into operations, program development, and broadcasting across the entire industry during the 1980s, and for all intents and purposes, the digital was ubiquitous by the early 1990s. The one unfortunate fallout? The mighty clicker led to the creation of the iconic image of the modern male media aficionado, the Coach Potato.

Recent Trends

The ten years following the mid-1990s proved to be a period of much anticipation for new digital technologies in the Television Industry. The most important source of interest was, of course, the Internet. What effect would it have on television? It was a question similar to that asked by the Radio Industry in the late 1940s and early 1950s about television. There was also the continued migration of broadcasting and programming from analog to digital formats, driven by economics and regulatory favoritism. By the early 2000s, a whole new generation of television sets came on the market, most often called "flat screens" and representing a variety of technologies contending for market preference, anticipating the day when television would be all digital, high density, programmable, and so on. It was a time when observers of the industry and computing at large spoke of technological "convergence," of independent technologies blended together. This was all thanks to the versatility of digital technology that could mix and match images, sound, text, and work on disks, video, and other media; travel via satellites, over old copper telephone lines, through cable wiring, and optical fibers; and delivered to a user by a telephone company, a satellite provider, or a cable company. Coupled to this was the parallel business convergence as the FCC and the U.S. Congress made it easier for multiple types of media firms to be a part of the new enterprises, with TV companies owning radio stations, newspapers purchasing TV and radio stations, telephone companies acquiring cable firms, and the like. More changes, as well as anticipation of changes, occurred in the industry in that 10-year period (1995–2005) than in the prior 20. Ironically during this period there was more organizational convergence underway in the form of mergers and acquisitions than there was technological.

But when conversations turned to technology, the focus of attention seemed to be the Internet. In 1998, two commentators on the industry reflected the common rhetoric and view of the day:

> Through the 1990s, technology has transformed the television business. The three-network system has toppled. Most U.S. households subscribe to cable TV and own at least one VCR. More than ten million enjoy motion pictures

> in home theaters with screens exceeding 40 inches (measured diagonally, as always, to make the set seem larger than it really is). The Internet provides an alternative form of on-screen information and entertainment; video is becoming part of the Internet. Change is happening fast, fueled in part by available investment dollars and an impetus toward an improved worldwide communications system involving telephones, television, computers, and interactive work and play devices.[70]

In one paragraph we see the mantra of convergence, the effects of prosperity, and the positive anticipation of changes just on the horizon.

Bruce M. Owen, like so many of his contemporaries, pondered whether television would be broadcast over the Internet into one's home, making it possible for either telephone or cable companies to combine TVs and PCs to create a new medium. But as he suggested in 1999, and so far has been proven right, the much anticipated convergence has not happened. Why? Economics and technology continue to influence the flow of events. For television programming to come by way of the Internet requires massive broadband capacity, far in excess of what is currently available going into homes. The cost of installing fiber optic cabling from the street into the home, which would provide the capacity, is too expensive and so far has not been done. Downloading programs and movies remains a batch application that has to be done with a PC. But other influences are at play: at the moment, digital technologies happen to be less expensive than analog; but technical standards for transmission and programming have yet to stabilize into widely accepted formats. The Internet is not yet quite right for television: "The Internet has the wrong network structure, and Internet pipelines are too small for television. The Internet, however, may eventually carry competing entertainment services that reduce the television audience."[71]

To be sure, there were technical problems to overcome in the early years of the Internet. A contemporary description (circa 1997) gives us a sense of some of these: "Full-motion video is currently a reality on the relatively young Internet medium; the picture quality is marginal and the images are jumpy, but the situation is improving rapidly, and with the fast access . . . the new paradigm may evolve quickly."[72] By the early 2000s, however, the anticipation of television and the Internet somehow coming together had not dimmed. Observers were waiting for "interactive TV," not merely broadcasting off the Net. The discussion had been enriched by the continued conversion of television from analog to digital. However, the move to the Internet had been, to put it mildly, "slow to develop in the TV medium, largely because the traditional broadcast medium does not allow for direct viewer input in the program viewers receive."[73] It was an old dream, one that dated back to 1953, when CBS had a program called *Winky-Dink and You* that asked children to help Winky-Dink escape difficult situations by drawing solutions on wax paper on their TV screens, such as a bridge to escape wild animals; it failed as too many children simply drew on the screen with crayons, not on wax paper, to the consternation of mothers who had to clean the CRTs. Other failed experiments occurred over the next several decades, including efforts by cable TV, including the widely publicized WebTV project launched by Microsoft, that combined television and e-mail using a traditional TV and a combination small keyboard and set-up box, providing e-mail and web surfing.[74]

While television and the Internet had yet to merge in some meaningful way by the early 2000s, there was universal anticipation that this would occur; debates and forecasts centered on the details of how. Would it be a real-time delivery of programming off the Net, or would people download programs to watch later?[75] Would the problem of the "last mile" be solved using existing wire coming into the home or be leapfrogged by some wireless solution? Ignored by most commentators, with the major exception of Owen, was the fact that there remained a host of business issues. The Internet is essentially free, and consumers have demonstrated enormous reluctance to pay for material coming from it, as we saw with the experience of print industries and as we will see in the next chapter with the Music Industry. There are the existing models of advertising paying for "free" broadcasting over the major networks, and of subscribers in cable television, both of which have to be reconciled with the new medium. Media companies had resolved the problems of who delivered: cable, telephone, or some other industry. In addition, as noted in prior chapters, with the increased availability of so many ways for someone to spend their time, and with technologies that have created more fragmented audiences (especially with the arrival of cable TV), questions remained about the ability of a provider to collect together enough of an audience to create an economically critical mass to run a profitable operation. An old idea born in manufacturing industries during the 1980s now came to irritate the business models of media industries: mass customization or, to use IBM's more insightful phrase from the early 2000s, "on demand." Implicit in both notions was the idea that customers wanted things, and when in the way they wanted them, which meant mass-production models of running a business had to be replaced with new processes and business templates that made it possible to provide almost made-to-order goods and services at the same price as mass-produced products. That posed a radical change to any firm in any industry.[76]

Multiple channels of distribution have become as complex with television and the Internet as with radio. Broadcast-based delivery of entertainment include 3G wireless, digital satellite, low earth orbit constellations, HDTV (high definition TV), digital sideband broadcast, digital satellite radio, digital terrestrial radio, good old AM and FM radio, terrestrial TV, and analog satellite. It is an impressive list. On the purely digital side, there are digital cable, broadband/Internet streaming, simulcast, Internet-only delivery, rebroadcast, LAN/WAN networks, VPNs, Bluetooth, broadband/Internet downloads (which can be done by subscription), and peer-to-peer (which can be legal or illegal). There was also a similarly confusing set of analog delivery methods, not the least of which was the grandfather of them all, analog cable. Looking at the issue from a device perspective, there were PC hard-drive playback systems, hard-drive HIFI, MP3 players, embedded playback in PDAs and cell phones, CDs, DVDs, DVD audio, DAT, mini-discs, CD-Rs, digital video, SACD, cassettes, and did I mention that vinyl records never fully went away? This myriad of technologies simply points out that much has yet to be resolved regarding standards and economic options for both suppliers and consumers. It becomes easier to understand why even an intellectually simple idea to understand, such as convergence, at the moment is a horribly complicated topic. What we can conclude, however, is that despite all the interest in the Internet, it had not yet fundamentally altered the Television Industry.[77]

Closer to reality and beginning to sort its way out is the more fundamental migration of analog to digital broadcasting. There are essentially three types of digital television: digital broadcasting itself, HDTV, and interactive television (ITV). It is not uncommon for commentators to mix and match all three as if they were one, but they are not. They are yet other examples of how flexible and diverse digital technology can be, adding function but also complicating business models and an industry's reaction. Simply put, digital broadcasting is terrestrial transmission of video and audio in digital formats, what the FCC wants the industry to move to as a regular format for delivering programs to the public. The Telecommunications Act of 1996 mandated that all TV stations broadcast digitally by 2006 and trade in their spectra currently used for analog broadcasts for new ones that were digital, all to live within the physical limits of available spectra. So far, the evidence suggests the deadline will not be met because of the enormous costs of conversion, with estimates suggesting that in some instances this can be as much as $20 million per firm. There are also technological difficulties and the requirement that viewers acquire either digital TVs or converters for their analog sets. As of April 2001, only 187 commercial stations were broadcasting in digital, some 15 percent of all broadcasters, though by late 2003 the number had climbed to 574. Most stations, however, agreed that at some point they would deliver digital content.[78]

Then there is HDTV, after the Internet perhaps the subject of more speculation than any other digital topic in the industry. Originally an analog technology, it originated in Japan in 1973 and provided far better picture clarity than traditional American analog television. While the Japanese remained with their analog version, digital varieties appeared in other parts of the world. But with so many standards proposed for HDTV, it has not yet received wide endorsement either by the American public or its television industry.[79] However, its current form suggests it might yet have a future: it was developed by a consortium comprised of academic and industry engineers, achieved its premier status after several other options were considered and eliminated, and leverages a great technical strength—digital technology as configured for computing. What is perhaps most interesting about the development of HDTV is that there were technical choices, ranging from simply improving analog broadcasting (what the Japanese elected to do) to building on advances in computer technology that were underway at the time (the U.S. approach), as well as a variety of other options.

The third digital version is normally referred to as interactive TV—what originated with Winky-Dink—that for many promises to be the wave of the future. The industry has sponsored various experiments with interactive TV. For example, in the early 1990s Time Warner launched a pilot in Orlando, Florida, that turned out to be an economic failure. The experiment also happened just as the World

Wide Web came into existence. Other projects, launched largely by cable companies, became victim to costs, technical difficulties, and the collapse of the dot.coms at the end of the century, including Microsoft's Web TV (although still available as of 2004).[80] But in an odd turn of events, websites that carried programming or were extensions of television survived. For example, CNN and FOX both encouraged viewers to log onto their respective websites to express their opinions and fill out surveys related to TV programs, with results reported both on their websites and on the air. Online communities on the Internet also cluster around types of TV programs, such as soap operas.[81] As with radio, magazines, newspapers, and books, the Internet is seeping into the Television Industry in an all-too-familiar manner: incrementally, cautiously, with more fanfare than fundamental effect so far, and coexisting with pre-existing technologies and applications.

While many in the Television Industry and in the larger IT community worried about convergence of TV and the Internet, the public at large was engaged in its own approach. People were treating the two sources of entertainment and information as separate options that they could use at will, one versus the other. As the Internet increased its content in the late 1990s and early 2000s, people watched less television and spent more time online similar to the practice of the people in the 1950s who increased the amount of time they spent watching TV and spent less time listening to radio. How much of a switch people have made to the Internet away from TV is not yet clear. While the evidence is anecdotal, the pattern of behavior is clear.

Yet some broadcasting over the Internet has begun. Streaming video over the Internet began sometime around 1996; data on the number of early users is sparse; but whatever time they spent on it was attention diverted from watching television. We have some evidence that by mid-1999 up to 75 percent of all American Internet users had spent up to 30 minutes a week watching video over the Internet.[82] The data is suspect, however, first because in order for that many people to have done so would have required that they have PCs connected to the Internet, and we know that access at the time was far less.[83] Second, not all individuals who logged onto the Internet were downloading video, because at that time most access to the Net was over dial-up lines, which either were too slow to see video in real time or took too long to download any substantive amount, such as a full program. However, for those who did have access to the Internet, downloading up to a few minutes of video from some news source was occurring and was not burdensome for the technology at the time.

With expanded use of broadband connections in the early 2000s, it became increasingly practical to download video. As infrastructures improved, so too did experimentation increase with broadcasting over the Net. The Alternative Entertainment Network TV (AENTV) pioneered this approach by offering original video over the Internet as well as programming first aired on television. However, the viability of its business model remained in question because, as of 2003, it had only hundreds of hours of programming in inventory and did not charge for access to them.[84] Leaving aside the fact that most people probably would like to watch video and TV programs on a bigger screen than on their laptop or PC and in a more relaxed setting, such as lying on a sofa or ensconced in a large overstuffed chair, experimentation with various configurations of technologies is occuring as people try to determine how best to use new tools.

Meanwhile, TV news organizations posted video and other information to their sites, much like newspapers and magazines. These included CNN.com, CBS News, ABC News, Fox News, NBC News, and MSNBC.com, to list the obvious. When Viacom acquired CBS in 2000, expectations rose that some new TV-Internet combined set of offerings would emerge. But all of these various activities are in early stages of evolution, along with original Internet video programming. Executives in the industry, however, have been watching the experiences of other firms in various media industries, in particular the Music Industry, which was experiencing what it argued was a significant loss of revenues from people downloading music off the Internet. The question was further complicated by the fact that the cost of developing original programming remained just as high as without the Net, while as a channel of distribution the Internet could potentially be quite inexpensive.

TV firms were left with the same problem as radio and music: How could one make money transmitting over the Internet? The answers are not clear. Early experience selling advertising has not been as positive an experience as with print or more traditional TV.[85] Carl Shapiro and Hal R. Varian, looking at the general problem of how one could build profitable businesses on the back of the Internet were quite blunt on the matter: "Technology changes. Economic laws do not."[86] These two professors also turned to the historical record for insights: "If you are struggling to comprehend what the Internet means for you and your business, you can learn a great deal from the advent of the telephone system a hundred years ago."[87] One would think that their book, *Information Rules*, was simply stating the obvious: that to thrive in business, one should generate profits, and if you cannot, then change the business plan. But since the Internet was such a compelling and simultaneously dangerous and confusing new element in the economy at large, many executives had yet to work out the details of how to make a profit. So their book, which became a best seller when it came out in 1999 and has continued to be highly influential, reached into the heart of the matter. The experiences of both the radio and TV industries reflected the problems these two professors addressed. In fairness to executives in the industry, the Internet became an important business issue during the second half of the 1990s, and thus they have had only one decade to deal with it. It took far longer for the telephone business of the nineteenth century, for example to work out the details of its business model and stabilize its operations into what eventually became known as the Bell System.

Two media experts have defined the Internet problem this way in 2004:

> Broadband digital media present both an opportunity and a threat for those in the media and communication industries. On the one hand, long-standing corporations, institutions, and entire industries are being turned upside down by the digital revolution. Businesses built on analog technologies of production and distribution are trying to figure out how to adapt to the digital age.

> New efficiencies of creating and delivering content in a digital, networked environment are emerging throughout the world. Long-held, highly profitable business models based on the analog world are less viable in a digital marketplace.[88]

Uncertainty about the wider effects of the digital, and not just the Internet, pervade the thinking of many TV executives. On the one hand, markets are expanding, indeed going truly global, but on the other, old ways of doing business cannot serve as the path forward. Like other media industries, TV is facing a world moving from spectrum scarcity to broadband abundance, from concentrated large national audiences to fragmented groups of viewers with the power to choose what to watch and when, from high to lower production costs, from limited broadband availability to increasingly rapid deployment and access, and the convergence of PC and TV technologies, the implications of which remain unclear.

Audiences increasingly are becoming aware of these fundamental sources of change in their TV world, but nothing is more evident to them now than what is happening to their television sets. What is occurring reflects a pattern repeated many times before, as in the early stages of adoption of PCs in the 1970s to early 1980s, and even earlier, with the move from vacuum tube to transistor TVs, then from black and white to color. With recent TV sets, two new technologies have emerged, not one flat-screen version. The first involves projection TVs, those very large flat screens that became popular in the early to mid-1990s as part of "home entertainment centers." But the first major leap forward in visual quality, driven in large part by a combination of new flat-panel technology and digital components, has been liquid crystal displays (LCDs) and plasma screens. LCD technology is the same as the one on laptop computers and many desktop PCs. LCDs use electricity to align crystals so that light selectively passes through them, while a plasma screen is illuminated by bits of charged gas. Both of these technologies provide a dramatically clearer picture than the old CRT-class TV and come in different shapes, usually larger and more rectangular than traditional TVs. LCDs and plasma TVs utilize the technology developed for computer screens, PDAs, and other digital devices. They appeared on the market at the dawn of the new century and by 2003 were drawing substantial attention from the public.

However, both of these technologies (LCDs and plasma) are still in early stages of deployment and remain far more expensive than the traditional cathode ray tube television sets. As late as December 2003, for example, a 27-inch flat-panel TV sold for more than $2,000, while the same-size CRT version could be purchased for less than $500. And plasma TVs cost more than LCDs. Because of the high cost of both new formats, 85 percent of all new TVs sold in the United States in 2003 were CRTs. Larger plasma TVs drew a lot of public praise, but 42-inch models cost between $4,000 and $6,000 in late 2003. Thus, the actual market for this new technology remained tiny, with the public anticipating that these would drop in price over time, much as had their PCs, PDAs, and earlier, compact stereo equipment and portable "boom boxes."[89] By this time, consumers practiced a *de facto* Moore's Law, knowing from experience that these kinds of devices dropped in price from the high to the affordable in a three-to-five year period, which almost literally tracks to the rule of thumb that capacity grows and prices drop at a predictable rate. From the Television Industry's point of view, therefore, little has changed in the family rooms of America. It too was subject to the same Moore's Law effect.[90]

More important to the industry early in the 2000s was the introduction of replay TV software like TiVo, which gained popularity with the public. TiVo is a combination digital video recorder (DVR), which is like a VCR but combines a digital hard drive (as we have on our laptops, for example) with software (the TiVo service) that allows users to download TV programs that they can store and view later. Downloads can come from cable, digital cable, satellites, or antennas, and be scheduled for viewing. Software allows a user to search the hard drive for an upcoming program by title, type, movie actor, team, and keyword, and as of 2004, its disk drive could store up to 140 hours of programs. Furthermore, one can use TiVo to pause a live broadcast.[91] The implications are extraordinary because this service continues a trend started with the Internet and earlier with radio, with consumers increasingly determining when they would listen to or view programming, made possible by software and digital technologies. As with other media, consumers were also able to determine if and when to listen or view advertising, whether or not to pay for media, and when and how they wanted to do so.[92] More than LCDs or plasma screens, software is challenging long-standing business models in the Television Industry. Not since the deployment of the remote control has the industry faced a challenge to its operating process because of a technological innovation, since all major transformations in the 1980s and 1990s came from changes in regulatory and market circumstances. Even the continued miniaturization of electronics did not hold out the same potential for changing the fundamentals of the industry's business model as did the growing ability of consumers to determine when and what to view. It remains a drama still playing out with no obvious conclusion.

Conclusions

These are two fascinating industries, important because of their roles in providing entertainment, informing the public, and stimulating political and social debates, all the while promoting the sale and use of goods and services through advertising. In a time when so many social commentators think of American society as an "entertainment economy," people today listen to more radio and watch more television than in earlier decades. So anything that goes on within these two industries is important to understand. Like so many other parts of the nation's economy, they embraced digital technologies and often were part of much discussion about the role of computing and telecommunications in defining the nature of modern society. It was a role they had played throughout the second half of the twentieth century, decades before the arrival of the Internet or the upheavals that occurred in the Telecommunications Industry in the 1980s. But because these industries increasingly have linked to other media segments of the economy, it has become difficult, if not impossible, to discuss them in isolation from such other industries as newspapers, satellite, and telephone. By the end of the century, we could see them becoming extensions or parts of the broader IT segment of the economy, whether because Microsoft had entered it through MSNBC, or AT&T due to its acquisition of TCI, or most recently, because of the changes underway in the kinds of television sets reaching the market. In short, these two industries are more connected to others than ever before, in part because of the flexibility of digital technology which has facilitated the crossover in offerings and uses (the convergence that analysts and engineers speak of).

But as occurred in so many other industries reviewed in this book, it is remarkable how important regulatory behavior has proven to be. While we may want to think that technologies are driving most changes in society, they are less influential than the FCC or the U.S. Congress. This observation might please an ex-Soviet central planner, but there is a difference. While the Soviets decided how much an industry would grow and what it had to do, American regulators favored creating environments where competition could thrive, yet in an almost contradictory manner dictating how that might occur, such as by mandating standards, setting timetables for technical conversions, or controlling access to spectrums. The role was not hidden. Reed E. Hundt, chairman of the FCC during the Clinton Administration, was quite open on the matter:

> Always government would have the challenge of writing laws and regulations that open monopoly businesses to new entrants and new innovation. Again and again, government would need to renew the community of America by creating and sharing equitably new public goods born from the confluence of technology and economics.[93]

As we saw in these two industries, and in particular with telecommunications, regulators could get in the way of the ideal and smooth adoption of technologies. It is a cautionary example illustrating that economist Joseph Schumpeter's idea of "creative destruction" was often a messy process.

Technologies often tipped the balance of economic power in an industry. In the Retail Industry, the massive collection of information at POS terminals made it possible for merchants to reverse the long-standing practice of the Second Industrial Revolution in which manufacturers told retailers what to sell; now merchants told manufacturers what to produce. In the Banking Industry we saw consumers use ATMs to have, in effect, 24/7 banking, but also the economic power to continue writing their checks. The list of examples is a long one, but the point is the same: technology has increasingly empowered consumers with the ability to pick when, where, and how they would acquire and use the offerings of scores of industries. Radio and television are simply additional examples of that process at work.

If there was a surprise with these industries, it might be how late they embraced the digital. To be sure, as noted before, they used computers to do back-office processing as early as the 1950s and more frequently in the 1960s and 1970s, and incrementally embedded it in the equipment they used to create programs and to broadcast. One might have thought that the digital would have been all over these industries in massive quantities in the 1970s. But there were economic gating factors: the costs of acquiring and converting to the technology; the compelling reliance, familiarity with, and cost effectiveness of analog alternatives; and, of course, the speed with which regulators made changes possible and practical. In hindsight, the evidence would suggest that it should hardly be a surprise that the digital did not become pervasive in either industry until late in the twentieth century. That is when a confluence of technologies, regulatory changes, and consolidations of media companies across industries created the environment leading to deployment of the digital to all corners of these two industries. But note that it was not the wonder of the technologies that made this possible. Rather, predigital, nontechnological—that is, they were economic, regulatory, and organizational—forces were at work. Firms invested when they could figure out how to make a profit, and were put at great risk with every new digital technology that came along because they challenged the viability of revenues from advertising. Cable subscription revenues, often coupled to near-monopoly status granted by local governments, mitigated the economic risks of new technologies. Organizational considerations regarding scope and scale, when combined with regulatory practices, more often than not set the pace and defined the ebb and flow of mergers, acquisitions, and entry into different (new) or additional markets. By the early years of the new century, one could see a confluence of changes perhaps equal to the churn that had earlier affected radio and television, the former in the 1920s, the latter in the 1950s.

Did the entertainment segment of the American economy begin amalgamating as a result of the arrival of so many digital options? To what extent did the collective experience of the media industries unfold in those others which began relying on the same technologies and channels of communications, such as the Internet or, earlier on, the humble computer chip? These are the issues addressed in the next chapter. The industries selected for study relied in 1950 on technologies far different from the digital, but they were so computerized and dependent on digital technologies by the early years of the twenty-first century that it is difficult to imagine a time when computers were not as vital to them. This transformation occurred relatively late in the century but perhaps more extensively than happened to radio and television, giving us a possible hint of what a more intensely digitized future might look like for so many of the industries studied in this book.

11

Digital Applications in Entertainment Industries: Movies and Recorded Music

One of the greatest misconceptions about modern movies is that visual effects are generated by computers. Nothing could be further from the truth. Human inventiveness is the most important ingredient and it always will be.

—Piers Bizony, 2001

The conventional typology of modern media links discussions regarding radio to those about music, and then that movies should closely be compared and contrasted with television. In the previous chapter I linked radio and television together, because those industries saw each other as rivals and accommodated their behavior in response to that perceived linkage. In the case of movies, music, and games, we see other perceptions and identities at work. Movie executives saw themselves locked in battle and later in alliances with television; radio was thus less of a factor to them. However, radio was not out of the picture because there were links dating back many decades. Radio personalities became movie stars as far back as in the 1930s, and in certain markets studios experimented in limited ways with owning radio stations so as to better promote films, particularly in the Los Angeles market.

Music and movies, however, dabbled in many similar technologies, and they suffered and enjoyed the same consequences of the digital by the end of the century. Radio and TV were rivals because of new technologies—mainly the emergence of "picture radio"—while movies and music shared similar copyright and technological experiences, thus making a comparison of one to the other more useful. As for links to electronic games or even digital photography, the Movie Industry has come to rely on computer-based animation and uses a variety of graphical digital tools, innovations in one industry directly influencing what the other could do. While it is still unclear about who stimulated the greatest number of innovations, movie studios or game producers, neither was shy about exploiting developments from the other source, and thus by the 1990s we could identify the relative influence of a technology or business process in one industry on the other.

To achieve those fresh perspectives, we have to look at the mix of industries in new ways. That is why, for example, our discussion travels from radio to television, then to movies, next to music, and finally to games and photography. Yet we should also recognize that none of these industries has operated in isolation. The Music Recording Industry feeds a great deal of content to radio, while the Movie Industry provides a number of senior executives to the Television Industry, along with massive amounts of programming. Many technologies, some digital, have provided a path for all of these industries to follow toward innovative practices, new markets, and new offerings. There are similar enough business practices to create incentives for companies across industries to merge, leading to the large media companies that came to dominate each of the traditionally separate media industries and firms by the end of the 1980s. In the background, but playing an extraordinarily important role, a galaxy of government regulatory agencies, government policies, and legislation has proven influential, all of them encouraging and facilitating mergers that in turn fostered expansion of oligarchies in these industries. One must understand the interaction of the three sources of influences—technology, business practices, and regulatory behavior—to appreciate how it was that digital and communications technologies came to play such important roles in American media industries and, conversely, how these industries came to be so influenced by digital technologies.

Movie Industry

Nearly 100 million times a month, people go up to a ticket counter at an American motion picture theater and buy a ticket to see a movie. Every month over three times as many people sit down in the comfort of their homes to view a movie they rented or purchased. While these are very large numbers, movie attendance today is actually smaller per capita than it was six decades ago, when movies were the primary entertainment that people went out to enjoy. Though television diverted large numbers of people away from movies, beginning in the 1950s, the Movie Industry continued to grow as the nation's population did, and once again we saw consumers enjoy more than one form of media at the same time, for example, movies and TV.[1] It would be difficult to exaggerate the popularity of movies in American society. Indeed, much of modern American culture can be attributed to or has been shaped by the values and tastes reflected in the products of this industry. To be sure, it is another chicken-and-egg conundrum of which influenced the other, movies or cultural norms. And information technology facilitated making the media both entertaining and available in various forms and venues. Outside the United States, the world's perception of the United States has also been profoundly influenced by Hollywood while other local movie industries were equally aggressive in deploying various information technologies.[2]

This industry has long been the subject of intense examination by economists and business professors eager to understand how it functioned. The patterns of oligopolistic behavior, the ebb and flow of control over distribution, the interactions of antitrust lawsuits, and the practice of creating companies just to produce one movie (in which all work is essentially outsourced) drew this group's attention.[3] A century-old industry, moviemaking inspired along the way a vast body of publications about it, with the result that it is often easier to study just one aspect of it, such as movies as an artistic form,[4] movies as a business model,[5] movies as a collection of technologies,[6] or movies as a social component of American life.[7]

For the historian examining the role of computing and technology in the American economy, there is much to learn from this industry. In particular, there are three aspects of its experience relevant to our understanding about the role of the digital hand and communications in modern America. First, there is the nearly unknown story of how computers were used in normal business operations, such as in accounting and marketing. Second, there are the effects computing had on how the industry operated, such as in distribution of movies before and after the arrival of VHS videos and DVDs. Third, and the one aspect that receives the most attention, is the role of computing in the creation of movies. In 1950 no director used computers to produce a movie; by the early years of the twenty-first century no entertainment movie was without at least a modicum of computer-generated graphics. In the 1950s, though a great deal of moviemaking technology was deployed almost as fast as it was invented, it still took years to see the digital in movie production, because technology evolved slowly into effective forms useful to the industry. By the end of the twentieth century, however, the Movie Industry was intensely digital and its uses of computing were pushing the technological boundaries of many aspects of IT, from how to use massive computing for modeling to creative deployment of realistic graphics. It proved difficult to tell who was "pushing" the technology further into new applications, the video game business or movies.

The Movie Industry also has much to teach us about the timing of when a business adopts a technology, because this industry does not have its core processes profoundly influenced by regulators. To have experienced what radio, TV, and telecommunications underwent, the FCC (or some other agency) would have had to determine the size and format of film, for example, the functions one could put into a camera or lens, and even what percent of the artistic output had to be devoted to specific themes, (politics, nonprofit, and community events), and so forth. Leaving aside the accounting, tax, and environmental regulations that affected all industries in the United States, the two forms of regulations this industry faced were requirements from time to time to warn audiences if movies had sexual content, bad language, or violence; and important court rulings about whether or not studios could own or control channels of distribution for their products, rulings which were variously applied. In short, American regulators played a relatively benign role in this industry when compared to the financial, electronic media, and telecommunications industries.

The Movie Industry comprises a complicated galaxy of studios, distribution companies, theaters, and many specialized firms (such as those which produce animation, others manufacturing or leasing equipment used in the industry), and nearly 40 craft unions. However, for our purposes we will look at three of this industry's components: production, distribution, and (to use an industry term) "presentation," because all three came to use computers over time. Production consists of companies that make movies, such as the Big Six studios or the independent film companies that often rent space from the larger studios in which to do their work. Distributors promote films—to make arrangements to place them in theaters, to sell advertisements and movie-related products (such as action figures), and to cascade movies through various stages of exposure from first-run theaters all the way to DVD sales in stores. Presentation is the exhibition of the movie to the public. It is important to recognize that there is a set series of ways in which the public is exposed to movies. The premiere weekend release of a new movie (known as first-run), what we normally hear about, is only the beginning of the cycle of events. A good launch allows "word of mouth" and media "buzz" to promote the movie as it proceeds through other forms of presentation. After being in a first-run theater, a movie normally moves to pay TV and home video at the same time it shifts from first-run to second-run theaters, where tickets cost less. After pay TV, movies become available on DVD, VHS, both for rent and for sale. One can easily see where IT comes into play. Satellite communications and TV cable make it possible for people to elect to pay to see a movie and for a cable company to transmit it into the home. Videos, while analog, taught the consumer the habit of renting or, more important, to view movies when they wanted. When DVDs became available in the 1990s the habits learned from videos in the 1980s simply continued and were amplified by the greater opportunity to purchase a high-quality copy often with additional content, such as unused scenes and interviews.[8]

The industry is routinely characterized as an oligopoly because many of its activities are concentrated in the hands of very few companies, a circumstance that has essentially remained the case since its beginnings. Five, sometimes as many as seven, but normally about six firms have dominated the creation and pattern of distribution (from theaters to home video, pay-per-view, pay cable, broadcast TV, basic cable; and to end product, such as DVDs and action figures). The names of the Big Six (also called the Majors) have changed somewhat over the years, but they grew out of prior groupings, some of whose names have been familiar to the American public for three-quarters of a century. In the 1990s, for instance, they were Disney, Paramount Pictures, Sony Pictures, Twentieth Century-Fox, Universal Pictures, and Warner Brothers. In various ways they also dominated key distribution channels and the timing of when movies moved through the various presentation "windows" to the public. In 1950 the studios saw television as a threat;

Table 11.1
Big Six U.S. Movie Studios and Their Owners, circa 1998

Studio	Owner	% Box Office Market
Disney	Walt Disney Corporation	21.9
Paramount Pictures	Viacom	15.8
Sony Pictures	Sony	10.9
Twentieth Century-Fox	News Corporation	10.6
Universal Pictures	Seagram's	5.5
Warner Brothers	Time Warner	18.7

Source: Various data from Douglas Gomery, "The Hollywood Film Industry: Theatrical Exhibition, Pay TV, and Home Video," in Benjamin M. Compaine and Douglas Gomery (eds.), *Who Owns the Media? Competition and Concentration in the Mass Media Industry*, 3rd ed. (Mahwah, N.J.: Lawrence Erlbaum Associates, 2000): 360, 382.

by the 1980s they had figured out that TV was yet another channel for distributing their work and a new source of revenues. By about 1990, movie revenues from cable and video equaled those from theaters. At the end of the century, revenues from DVD sales and rentals outpaced those from theaters.[9]

The oligopolistic nature of the industry solidified over time—as in so many other industries—through consolidations and mergers. The history of mergers is complicated, but as table 11.1 shows, the Big Six were owned by a variety of large media companies, with the interesting exception of the liquor company Seagram's, which acquired MCA in the mid-1990s and changed the studio's name to Universal. Using 1998 data, we can see that these few studios owned 83.4 percent of the theater market as measured by ticket sales. Similar data could be provided for any decade to demonstrate how centralized the industry remained throughout the second half of the twentieth century. All six firms were extraordinarily profitable enterprises in the 1990s, hence their attraction as M&A candidates; indeed; they were very profitable over the course dating from the 1950s.[10] To be sure there were also other smaller independent firms that strived to join the Big Six club. The most recent publicly recognized aspirant is DreamWorks SKG, put together in 1993 by Steven Spielberg, Jeffrey Katzenberg, and David Geffen and partially funded by Microsoft's cofounder, Paul Allen.[11]

This industry's operations are only marginally understood by other industries or even by IT vendors selling to its members. As documented by traditional economic metrics, the industry is quite small given the enormous visibility it has on the American scene. In 1954, the U.S. Bureau of the Census reported that it employed 20,800 people and generated worldwide revenues of nearly $2.4 billion, all in an era when industry executives and observers feared television could destroy sources of vast sums of revenues. Over the next 10 to 15 years, employment declined (in 1967, for example, to 16,700) yet revenues continued growing (in 1967, the industry generated nearly $3.5 billion).[12] But the data are not as clearcut as one would like. For example, the 20,800 in 1954 were for employees on the payroll of the industry, but the figure does not include freelance workers who were and are such an important source of labor. The 1954 revenue data also demonstrate what an important export business movies was for the United States, with half the revenue coming from the non-U.S. market. It is in the domestic revenue data, however, that we can see the inroads made by television in the 1950s and 1960s; in those two decades, ticket sales actually declined by roughly 20 percent and only came back to 1954 levels late in the 1960s before growing steadily from the 1970s to the end of the century.[13] Keep in mind, too, that the U.S. population grew during the entire period from nearly 151.7 million in 1950 to over 294 million in 2004, and the vast majority had access to television.[14]

The industry continued expanding, however. In 1990 the industry reported employing 152,500 in production and another 111,100 in theater services. The first number doubled by the end of the century, while the second only increased by about 7,000. As noted earlier, admissions to theaters remained constant between 1.2 and 1.3 billion in any given year in the 1990s, with an uptick to 1.4 billion at the end of the century.[15] Revenues from theaters in the 1990s grew also, from $5 billion in 1990 to over $7 billion late in the decade.[16]

So what does this brief overview teach us about this industry? First, over the half century its value chain expanded from merely generating revenues in theaters to a plethora of income sourcing made possible largely by the continuous arrival of new technologies, beginning with television, followed by satellites, cable, video, and most recently DVDs. As this book went to press in 2005, the industry was in the midst of yet another round of technological opportunities and problems, this time posed by the growing capacity of individuals to download digital copies of movies into their homes, much the way they had music a few years earlier. Second, because of the good profit margins in the industry (often in the 15 percent range), companies were attractive takeover candidates. Third, the industry has remained highly concentrated in just a few players. While we have not discussed the degree of concentration of distribution companies and theaters, suffice it to acknowledge that they too followed the example of the large studios, although not achieving as intense a level. As Kristin Thompson and David Bordwell, two distinguished scholars who have extensively studied the industry pointed out, its generally oligopolistic stance contributed to its conservative penchant in the kinds of projects it took on, while retaining an iron hold on its practices, most notably over production and distribution.[17]

We can put too fine a point on the oligopolistic description, however, because the industry also comprises a loose confederation of specialized firms (such as those that only do animation or digital effects) and the dozens of craft unions that provide actors, make-up people, photographers, and so forth, all whom remain tightly linked to each other within their own organizations, more loosely within the industry, and as part of a highly integrated team while working on a specific project. The industry operates in a highly unpredictable environment, as no one has ever been able to predict reliably how to make a financially successful movie. Increasingly, evidence is mounting that relying on the economic performance of a movie is less of a success factor for the industry and its major players than the industry's resilience, which allows it to adapt quickly and embrace new ideas. In short, the Movie Industry is an ideal case for teaching other industries how to be adaptive and resilient in the face of extreme uncertainties and great financial risks.[18]

Computers as Movie Stars

Interest in computers on the part of actors, directors, scriptwriters, and studio heads predated the invention of these machines. Computers were the subject of so many movies that one can generalize about the collective attitude of Hollywood toward the social and technical implications of this technology as it evolved over time. Many have argued that the attitudes reflected in movies (and TV, for that matter) on a myriad of topics, not just computers, conditioned to one degree or another how the public viewed those same subjects.[19] Because attitudes regarding the digital influenced decisions people made about using this technology at work, what they learned from the movies influenced how the public responded to computers in general.

The concept of computers and their role in society received attention beginning in the 1950s. One theme portrayed computers as giant brains which were large machines that could displace workers. The theme ran through movies for decades, beginning with *Desk Set* (1957) in which Katharine Hepburn, a corporate librarian, is threatened with the possibility of being displaced by a computer (though in the end she learns to use rather than fight the efficiency move). As a larger theme, for decades computers were portrayed as part of the workplace. A second theme concerned the role of computers in crime and other evil acts. Who can forget the megalomaniacal HAL in *2001: A Space Odyssey* (1968), or *Tron* (1982), where the hero becomes a prisoner in a computerized world and can only survive by being an outstanding electronic game player? The evil theme came back again most recently in the highly popular movie *Matrix Reloaded* (2002). Other themes have appeared, such as what happens when computers take over (*Colossus: The Forbin Project* [1969] or *Red Alert* [1977]), the role of technology in the hands of computer geniuses or in crime (*Deadly Impact* [1985] or the earlier *Hot Millions* [1968]), and computers doing good deeds (*The Computer Wore Tennis Shoes* [1969]). Computers were active in love and sex, of course, as in *Hunk* (1987), playing on an old American theme dating back to the nineteenth century in which, in its modernized form, a computer operator transforms into a muscular blond hunk by striking a deal with the devil. There also were dozens of movies in which computers facilitated love by helping socially awkward people meet attractive members of the opposite sex.[20]

A common theme in literature and movies concerning computers was the existence of some omnipotent artificial intelligence (AI), some of it directed toward evil intents, while in other cases toward the potential betterment of humankind. The 1950s have often been called the Golden Age of Science Fiction, and movies were there. *Gog* (1954), *Tobor the Great* (1954), *Forbidden Planet* (1956), and *The Invisible Boy* (1957) were emblematic of the period. The 1960s saw movies that were far more optimistic both about technological progress and the motives of artificial intelligence. This decade saw the introduction of notably obscure movies, such as *Cyborg* (1966) and *The Billion Dollar Brain* (1967), but it also introduced *2001: A Space Odyssey* (1968), perhaps the most famous computer/robotic movie of the century. Many movies involving AI appeared in this decade—the same one in which the United States launched its project to land people on the moon. In the 1970s, movies swung back to a growing technophobia, moving from government and Cold War scenarios into daily life. This was the decade of *Star Trek: The Motion Picture* (1970) and *Star Wars* (1977). In any given year one could count on the industry to introduce up to six AI-related movies, though most were minor productions. The 1980s provided both pessimistic movies, such as *Blade Runner* (1982) and *The Terminator* (1984), as well as romantic comedies and nonthreatening computers, including *Heartbeeps* (1981), *Electric Dreams* (1984), and *Making Mr. Right* (1987), the latter ostensibly one of the first overtly high-tech "chick flicks." In the 1990s, more movies of the darker genre appeared, such as *Terminator 2*. The industry devoted more attention to the broader role of information in modern society, often with threatening implications of spies and serial killers on the Internet and so forth. There were no stellar productions, but one could see the change just by the titles of some of these films: *Circuitry Man* (1990), *The Net* (1995), and one movie based on a wonderful book concerning how computers were designed, *Ghost in the Machine* (1993).[21]

An interesting question is: How real were the computers seen in the movies? In the 1950s and 1960s, the computers in the movies were rented from IBM, Burroughs, and Sperry Univac, beginning with *Desk Set*. By the end of the 1980s, routine office scenes showed computer terminals everywhere; the typewriter no longer played a supporting role. The unreal also coexisted, such as the computers in *Star Trek*. As every new communications technology appeared, if too made it into the movies. These were not the central characters, merely a reflection of what modern offices and work habits were like late in the century (such as omnipresent cell phones and laptops). As graphics and animation tools became more powerful, beginning in the 1970s, movie producers gained a wider flexibility in presenting computer themes. In *Terminator 2: Judgment Day* (1991), an actor plays the role of a robotic creature that takes on liquefied features. Perhaps one of the most widely seen uses of special effects, graphics, and the notion of the computerized being was in the character Jabba and various storm troopers who appeared in *Star Wars* movies, particularly in *The Empire Strikes Back*.[22] But by then, robotic/computerized creatures had been a long-standing fixture in movies.

Thanks to the Movie Industry, millions of Americans were initially introduced to the concept of a computer, and then to a variety of perspectives about what these systems looked like and what they did. For those who were not exposed to computers at work, as was probably more so the case in the 1950s and 1960s than in the 1970s forward, movies were profoundly important in creating images for people. In time, as people became familiar with the technology, they could put movie themes regarding computers into a context that had not existed before, thereby making computers in the movies less awesome than they had been in earlier decades. Nonetheless, both Hollywood and the Video Game Industry continued to introduce new visions of computers.

How Computers Were Used in Making Movies

The most important theme we can explore regarding the role of digital technology in this industry clearly has ties to the movies themselves. To be sure, IT affected how movies were distributed, as the media used by consumers forced the industry to recognize and later profit from the public's desire to see movies when they wanted and where. These themes, however, have not drawn as much attention as how computers came to dominate the nature of the output of the industry through the effects it had on the artistic work of its members. To understand how technology affected the work of the industry one needs to appreciate all of these issues, but the use of computers in movies made it possible to tell stories and to entertain in so many new ways that we should begin our analysis there. When the technology reached the point where it could enhance the artistic mission of the industry, everyone from producers and directors to technical staffs embraced IT to such an extent that today it is nearly impossible to tell where old-fashioned photography of real people and places ends and where computer-based graphics begin. Even how film is shot and edited has become profoundly high tech, with extensive use of digital computing. All of this began slowly in the second half of the 1960s and proved essentially ubiquitous by the end of the 1990s. How did such a transformation occur in such a short period of time?

We know several things. First, digital tools were being used to create television programs and advertisements for TV, often earlier than in movies. So the people who did that work were a natural conduit of information to colleagues working in the Movie Industry. Indeed, as studios did more work for television and advertisers, knowledge about the digital spread through the business. Second, as technology became more feasible, for example had enough computer memory to hold images, hardware and software vendors busily pushed their wares. Third, while members of the industry described their embrace of the digital as a major event that occurred in the 1990s, we know business uses of IT had begun appearing in the industry years earlier, for instance, as a way for directors to watch their budgets and model economic options for staffing. We can assume that bureaucratic practices led to interactions among IT people and the creative side of the business, while we are certain developments in other industries seeped into this industry. Finally, we should acknowledge that as the decades passed, creators of movies, from actors to directors and producers, were more formally trained in their subject areas at various colleges and universities. Institutions of higher learning exposed them to new technologies and created the necessary links by which one community could reach out to the other and form yet another path for the digital to enter the Movie Industry.

In fact, that last observation is a logical starting point for describing the arrival of the digital to the world of moviemaking. In 1962, Ivan Sutherland, then a graduate student at the Massachusetts Institute of Technology (MIT), developed what soon came to be known as computer graphics, writing a software package called Sketchpad. This tool allowed him to draw cubes, lines, and other forms on a CRT screen. Two years later, he established with David Evans what may have been one of the first, if not the original, computer graphics departments at the University of Utah. Over the next few years they developed a variety of tools and techniques that blended art, graphics, and computing. They began on the path toward the Movie Industry when they formed a company in 1968 called Evans & Sutherland (E&S) to exploit their software tool, introducing their first software product in 1969. It could create wire frame images in real time, much along the lines of the CAD/CAM systems used earlier by the military and such large industries as aerospace and automotive. These images could then be painted in with flesh tones and backgrounds and made to look real.[23] The U.S. Department of Defense funded various projects at the University of Utah in the 1970s that made it possible to enhance graphical software tools. Work also occurred at Ohio State University, Cornell University,[24] the University of Pennsylvania,[25] and at the University of Illinois at Chicago in the 1960s and 1970s. When Alexander Schure at the New York Institute of Technology (NYIT) wanted to create computer-animated films, in the 1970s, University of Utah veterans worked with NYIT but were stymied by the high cost of the technology and the lack of sufficient machine-processing power to handle the large data content of images. Furthermore, software had not progressed enough. But it was all a start, because computer graphics firms began forming in the late 1970s and early 1980s, portending new developments to come. IBM also became involved in developing software tools and even won an Academy Award in 1973, the first ever given for a software tool in the Movie Industry; its program monitored light in printing motion pictures, in other words, controlling the precision and uniformity of color film.[26]

One other important development that required software was motion control. This function allowed a camera to repeat the exact same movements and functions multiple times, using automation which made it possible to have multiple filmings, as for example, some of the space scenes in *Star Wars* (1977) with ships passing through space with some backgrounds moving or remaining constant. Crude mechanical means had been used since 1950, but by the 1970s software could control the physical movement of equipment and later contribute to animation that could be mixed with actual filming. This capability allowed an actor to play multiple roles in the same scene, to film several actions that all had to be superimposed on each other, and to create some spectacular footage.[27]

The next big jump to movies came in 1978 when George Lucas (best known as the creator of the *Star Wars* and *Indiana Jones* trilogies) hired Edwin Carmull (a Utah Ph.D.) from NYIT to run Lucasfilm Computer Development Division. Soon after, some half-dozen other graphics studios opened in the United States, all providing services to various media industries. Initial projects included developing "flying logos" (logos that moved) for television (including *NBC Monday Night at the Movies* and *ABC's Wide World of Sports*). Almost no work was done in these early years for movies; rather it focused on television where one could charge far more to develop a commercial than for a longer animation sequence in a movie. A frame of film still required more data (hence more computing calculations) than a frame of television image, making it more practical to do work at that time for TV. However, early attempts began to apply this technology to movies. The "first" use of the digital in movie graphics is normally attributed to a film called *Future World* (1976), which had a segment lasting only a few seconds of a computer head of Peter Fonda and of a hand. However, the one event that caused the industry to take computer-based graphics very seriously came in 1982 with the release of *Star Trek II: The Wrath of Khan*, which included a stunning 60 seconds of full-color graphics.[28] An expert on movie graphics described the event as a collection of "firsts" in the use of computing:

> It required the development of several new computer graphics algorithms, including one for creating convincing computer fire and another to produce realistic mountains and shorelines from fractal equations. Fractal equations are often simple-looking mathematical formulas that can create infinitely complex images in computer graphics. In addition, this sequence was the first time computer graphics was used as the center of attention, instead of being used merely as a prop to support other action. No one in the entertainment industry had seen anything like it, and it unleashed a flood of queries from Hollywood directors seeking to find out both how it was done and whether an entire film could be created in this fashion.[29]

The industry now began using computer graphics in more films, causing the technology to evolve all through the 1980s.

The amount of available computing also increased. For example, Digital Productions, formed in 1982, acquired a Cray X-MP supercomputer for $10.5 million for the purpose of producing computer-based graphics. Digital imaging no longer was a minor, even primitive activity; it now was becoming part and parcel of movies. All through the 1980s, graphics firms acquired more computing power and advanced software tools, adding color and more intense and detailed images, and used all the various mini- and mainframe systems that the Computer Industry introduced in the 1980s. High-powered workstations and PCs also made it easier in that decade for artists to work in real time to create images and to share files and knowledge. Computer technology added more processing speed, capacity, and storage, all of which were needed to advance the use of the digital in movies. Costs of computing dropped, causing those graphics studios saddled with large mainframes and equally large debts to collapse as others replaced them armed with the new, less expensive computing technologies of the late 1980s. Over the next several years, the studios and graphics firms deployed software tools in making movies, television programs, and, of course, ever more sophisticated TV commercials. The first full-length television cartoon appeared in 1993 (*The Incredible Crash Dummies*), signaling that now one could put together an entire program using computers. The mid-1990s saw computer animation continuing to benefit from developments in all three markets—movies, television programs and commercials—while also achieving technical and artistic milestones.

One problem directors and producers faced concerned whether to record on film or video. The issue revolved around the fact that a more detailed image (pixels) could be put on a film frame than in a video frame, making film still a very attractive medium in the 1980s and 1990s. Then there was the question of how to create systems that had the capability of handling larger amounts of color data, since more data meant a better image. The capacity of computers to do this kind of work kept growing. In the highly watched case of Disney's restoration of *Snow White and the Seven Dwarfs* (1937), this first feature-length animated movie required massive data handling. A single film frame has approximately 40 megabytes of data. An average length movie might have 130,000 frames, generating a collection of five terabytes of data, or the amount of information needed to fill some 25,000 PCs, each having a disk in the 200 megabyte range. That is all technical language to suggest a lot of information needed to be processed by a computer. Kodak had developed various software tools in the late 1980s to help the process along, such as programs to eliminate scratches and particle images from the surface of film frames so they would not appear in a final, digitally remastered version. Kodak also developed software to facilitate the restoration of the color (called digital painting) as it originally existed in *Snow White and the Seven Dwarfs*.[30] It was a fabulous success. One observer noted at the time: "Never before had every frame of an entire film been converted to digital imagery, manipulated in some way, and then printed back out to film."[31]

As the ability to place movies into digital formats increased in the 1990s, it became easier for directors to use digital technology to speed up the editing process that occurs after a film has been shot. That capability had arrived in the industry in 1985, and it encouraged editors to work with video copies of movies and to cut or move frames around quickly without having to literally cut and connect pieces of film and then later, as with *Snow White*, put the end result back on film.[32] With this technique, editing could be done faster and even earlier in the production process. Images could be altered (enhanced), and a film editor could now jump back and forth from one frame to another, mixing and matching to tell a story.[33] A veteran editor of the industry reflected on the changes of the past two decades:

> It has been a relatively short period since audio tape was edited with a razor blade and videotape was edited by using tracing powder to locate track pulses. It has been a remarkably short cycle that has brought three distinct approaches to electronic nonlinear editing.
>
> The basic quest of nonlinear editing systems in the past has been singular: How fast can the footage be displayed to the editor? All three approaches—videotape, laserdisc, and digital—have sought to reduce the time required to get to the different pieces of footage.[34]

The financial well-being of the industry in the 1960s facilitated experimentation with new forms of technology in the 1970s and 1980s. The successes of television and movies that appealed to the public, at a time when a growing number of viewers provided the economic wherewithal around the world to support more extensive and expensive uses of computing, created what clearly became a Golden Age in the 1990s.[35]

Conversion from analog to digital had slowly been underway since the 1980s. The speed of conversion to the digital was driven by a variety of factors: the quality of the software tools needed to do the work, the capacity of computers to store enough data and to manipulate data quickly, the declining costs of computing, the advancing skills of the technicians and editors, in the affordability of the tools, the industry's willingness to use them, and the public's reaction to such animation. In short, it was not just about new digital tools appearing in the market or about directors and others being for or against the use of digital technologies. They embraced these as fast as they became available, in fact much faster than engineers in other industries. They borrowed lessons learned from television programming and advertisements, and whole companies appeared that specialized in particular types of animation. By the end of the century, members of the industry were asking: why continue using film? Why not distribute movies to theaters on DVDs or download them from satellites, just like radio and TV programming? The technology had advanced sufficiently by the late 1990s so that these questions could be asked because the image quality of a totally digital format was now sufficiently acceptable for viewing in theaters.[36]

By the early years of the new century, cameras and high quality video had so lowered the costs of film production that individuals could make full-length movies, produced for thousands of dollars instead of millions, and get them into the market. But as the quote at the start of this chapter reminds us, there is more to filmmaking than good technology. What technology has done, however, is to provide the opportunity to improve the artistic side of the process by more than just good graphics. For example, by using the Internet for communications, scriptwriters began communicating and sharing ideas, with the result that websites have appeared in support of this activity. These include Internet Filmaker's FAQ (www.filmmaking.net), Writer's Exchange (www.writers-exchange.com), and Cinematography Mailing List (www.cinematography.net). Other sites made it possible for writers to submit scripts and stories for review, such as Script.com, Zoetrope Virtual Studios, and Screenwriter's Connection. Use of these Internet-based facilities remained embryonic in the early 2000s, but parallels the overall increased use of this communications medium evident in so many industries.[37]

The most important development of the early 2000s came in the increased use of digital technologies in the postproduction phase of filmmaking—in all the work done after film or video has been shot, in editing, changing, and altering images. While use of digital tools to do these tasks dates back to 1992 with the arrival of nonlinear editing tools from Avid Technology, it took filmmakers a few years to become familiar with these kinds of software aids. By the end of 2002, three companies offered software products in support of postproduction work (Avid, Media 100, and Pinnacle Systems). One report on the use of such tools noted:

> Today, non-linear editing is the norm for commercials, television shows, miniseries, features films, independent films, corporate videos, and multimedia software. The changeover is all the more remarkable given the fact that any eco-

nomic advantages of the benefits (speed and ease) were not immediately apparent.[38]

Why did this happen? Over time, costs had come down for the necessary software, from hundreds of thousands of dollars in the early 1990s to less than $10,000 by the early years of the new century. That trend helped and mirrored what happened in so many other industries, when adoption rates increased as costs decreased for digital tools. But as the same report noted, "In part it was the pressure to be at the cutting edge of post-production technology that drove most post-houses to make the financial plunge."[39] Add in the industry's penchant to adopt tools that enhance artistic expression of filmmakers, and we have a prudent understanding of why the technology was adopted and what dictated the speed of adoption.

How Digital Technologies Affected the Business Practices of the Movie Industry

Not surprisingly, the digital made its way into the business and managerial practices of the industry, resulting in altered work activities. The largest institutions in the industries—the studios—were the earliest users of computers for the same reasons as in any other industry, to control costs of accounting and other back-room data management. Outsourcing payroll was a common application in the 1950s and 1960s, while accounting operations remained fairly primitive until the late 1960s. During the 1970s, however, the studios were using computers to facilitate distribution and marketing, such as modeling where to exhibit movies. At 20th Century-Fox Film Corporation, for example, management began using a mid-sized IBM computer (IBM System 370 Model 135) in 1977 to help select which theaters to use. The studio's vice president, Peter Myers, described the rationale: "Profit from a film can be significantly increased if we place it in theaters with proper revenue histories, and arrange contracts that favor the studio's interests." This studio routinely released 15 to 20 movies each year to thousands of theaters. Myers reported, "before we used the IBM system, it was just an impossible clerical task to manually examine theater history every time we reviewed a contract." His expectation was straightforward: "We expect the IBM system to be a major factor in helping us take better control of film placements and, as a result, ensure us more profits."[40]

It was in the 1970s that studio management finally became fully aware of the potential business uses of computing and began implementing applications across all its major functions, beginning with accounting and finance, then moving to sales and marketing. By the late 1970s, all the major studios were using in-house computers and had installed online systems to manage personnel, production, and payroll internally, all for the purpose of driving down costs of operations. Managing budgets for movie projects became a major application. Dick Gallagher, a DP manager at Paramount, told one reporter, "our production people nowadays are so budget-minded they take the IBM 3741 workstation with them on location."[41] He described what was done:

> Our production cost system begins with a budget: for the script and what the lead talent contracted for in compensation. But there's a multitude of variables. How many extras will be hired? How much film? How many trucks and honey [makeup] wagons? Motor crews? How many one-camera scenes versus three-camera? How many light bulbs? At the end of each day, the actual costs are entered into the 3741, which may be plugged into a hotel room wall socket, on location. The diskette is sent to the studio. The producer can learn the actual costs the next day. If it's a TV production he may cut scenes 5 and 22. He can use fewer extras. He can save air fares by choosing the next location closer to home. He can cut time out of his prep or his shoot."[42]

This description points out that creative activities in this fundamentally creative industry were just as subject to normal business managerial factors as in any other. It is an important (if obvious) comment to make because so much of the literature on the role of the digital in movies is about the creative side—special effects and animation—without a tip of the hat to the owners of the studios who demand fiscal responsibility, especially as the studios became the properties of nonmovie companies. In the case of Paramount, for example, its holding company, Gulf+Western, required the studio to report its expenditures and other financial data on a daily basis in conformity to G+W's corporate reporting standards; hence, accountants dragged their heavy desk-sized 3741s around to filming locations.

Lest one think that Paramount was the only one doing this, IBM's archival records document that at the same time Universal Studios (which made *Jaws*) also was required by its holding company to account for expenditures and for that purpose had installed one of IBM's largest computers, a System 370 Model 158, and a network of 3270 CRTs. As one manager at the firm reported in 1977, "bookkeeping accounts for 85 percent of our computer activities." In fact, "payroll checks alone would drown us without the computer."[43] Hollywood may have had the most complicated payroll system one could find in the American economy:

> [Checks] can range from over a million dollars for a theatrical star, to under five dollars for a TV rerun residual payment. If an extra uses her own ice skates, she gets paid more. If an extra gets wet as part of the job, he also gets more. We have to deduct and pay extra according to the rules of the 39 unions and craft guilds. We have between 4,500 and 7,000 on the payroll at Universal on any one day. Last year we issued 42,000 Federal W-2 forms, so at least that many people worked for us during the year. We've issued as many as 30 checks in one day to the same person, in different amounts, for different shows. Our system tracks the tax structures of all the states. If we're shooting in another state, that state has first tax claim on a cast member's earnings there; his home state, second. Last year we issued more than 300,000 checks. If the vouchers are in by 1 p.m., the checks will be ready by 5 p.m.[44]

It is almost exhausting reading that description, but it reflected what was happening at all other studios—and to a similar extent in the Music Recording Industry.

Other accounting uses were more typical of what many industries had, such as inventory control—that is, movies. At MGM in the 1970s, software kept records on 1,500 films, providing information on their availability and features (such as whether in color, or if silent). At Warner Brothers, another large IBM 370 Model 158 was simply the latest in the use of information-processing tools. As early as 1941, the studio had installed its first card-tabulating equipment for accounting applications, and a decade after commercial computers had become available it started migrating these applications to mainframes. By the early 1970s, online access to accounting and financial data on a daily basis began affecting decisions regarding production of movies and distribution channels in which to place new movies. In the 1980s, deployment continued spreading across the industry,[45] and by the 1990s, laptops had become as familiar on the set as a good set of lights and a coffee cart.

The arrival of the Internet in the mid-1990s did not fundamentally alter the accounting and production uses of computers. Rather, the Internet, when combined with the increasing availability of movies in digital format, created a whole new set of problems and opportunities that are now beginning to have as profound an effect on this industry as elsewhere, most notably in the Music Recording Industry. DVDs as a technology also influenced the Movie Industry, particularly distribution—the cash-generating part of it. We need to understand the effects of both DVDs and the Internet if we are to appreciate how the work of the Movie Industry began another round of changes in the late 1990s which, as of this writing (2005), have yet to play out fully.

The hype about the importance of DVDs is being borne out by experience. DVDs, which are essentially digital CDs for movies, have all but supplanted analog videotape. Widely available since 1998, they provide a very clear image and often include trailers, alternative endings, scenes that did not make it into the final movie, interviews with directors and stars, and stories concerning production, often amounting to two to six additional hours of content. They have become extremely popular, embraced as an extension and enhancement of the consumers' practice of the 1980s and 1990s of renting or buying videos and seeing them in the comfort of their own homes. Americans embraced DVD technology faster than they had previous digital consumer goods, outpacing even cell phones, music CDs, and the Internet. A few statistics suggest the magnitude and rate of adoption. Table 11.2 documents the adoption of DVDs and, for comparative purposes, other digital media. Notice the explosion in consumer consumption of music videos and DVDs, in the case of the latter, increasing by over 100 percent in 2001 over 2000. In a survey conducted in 2004, the Pew Research Center for the People & the Press reported that 76 percent of the public owned a DVD player, more than quadruple the percentage reported in 2000 (16 percent). Interestingly, over 60 percent of the public in 2004 also owned VCRs, although it was not clear how often they still used them.[46] To be sure, all consumer digital media enjoyed enormous growth rates in their early years of availability, with people clearly very comfortable with each new technology as they built on their prior experiences with earlier digital mediums. Rapid adoption and comfort with the digital proved to be a worldwide phenomenon, not limited to the United States.

Table 11.2
Expenditures for Digital Media Products, Selected Years (1990–2001)
(Millions of dollars)

Medium	1990	1995	1997	1998	1999	2000	2001
CDs	3,451.6	9,377.4	9,915.1	11,416.5	12,816.3	13,214.5	12,909.4
Music videos	72.3	220.3	323.3	508.0	376.7	281.9	329.2
DVD videos				12.2	86.3	80.3	190.7
DVD audio							6.0

Source: U.S. Census Bureau, *Statistical Abstract of the United States*, various years, all available online at its website.

Initial reaction of many in the industry to the arrival of the DVD was that people would start copying movies and not pay for them. But that notion was quickly put aside when management recognized that DVDs, like video before, introduced millions of people to older movies and to repeat business, all the while providing an additional source of revenue. Then in 1997 in the United States, and in the next year in Europe, DVDs became widely available, quickly becoming the most rapidly accepted home entertainment format in history. VCRs took ten years to reach the same rates of penetration that the DVD achieved in its first six years. All the while, the cost of DVD players kept dropping (by over half between 1999 and 2001), making it possible for more people to acquire the necessary hardware. The number of movies available on DVD also kept growing.

DVD sales around the world in 2003 exceeded $11 billion and in the following year climbed to over $15 billion. For those selling DVDs, the business model proved quite attractive because the profit margin on a DVD was about 66 percent, versus 45 percent for a videocassette. Wholesale prices for DVDs in the early 2000s hovered around $16. After deducting $2.75 for marketing expenses, $1 for manufacturing, another 90 cents for packaging, and 80 cents for distribution, the average gross profit on a DVD was $10.55. Put in more meaningful terms, if a movie did well and sold, for example, 11 million DVDs—not an unrealistic number—that alone would spin off $121 million in profit, far more than most movies make in the theater.[47] By the early 2000s, DVDs often kept studios afloat financially since only 1 in 10 feature films recoups its costs at the box office, though roughly 4 out of 10 ultimately turn a profit from all sources of income (overseas box office sales, DVDs, rentals, etc.). To put DVDs into further perspective, compare the enormous growth evident in table 11.2 with the data in table 11.3, which show revenues from the sale of tickets at theaters in the United States. While the numbers remained large, they also were flat or only grew slightly in the early 2000s. Clearly the future seems to point toward DVDs. One simple comparison of revenues of box office sales versus DVD sales for a handful of movies from 2002 illustrates the relative importance of this new digital technology (see table 11.4).

While it is still early in the life of the DVD, it is not yet clear if VHS sales

Table 11.3
Annual U.S. Theater Ticket Sales, Selected Years (1990–2003)
(Billions of dollars)

1990	1995	1997	1998	1999	2000	2001	2002	2003
5.0	5.5	6.4	6.9	7.4	7.7	8.4	9.2	9.2

Source: "Yearly Box Office," http://www.boxofficemojo.com/yearly/ (last accessed 6/13/2004).

Table 11.4
Revenues from Box Office and DVD Sales for 5 Movies, 2002
(Millions of dollars)

Movie	Box Office	DVD
Spider-Man	215.3	403.7
The Fellowship of the Ring	313.4	257.3
Monsters, Inc.	255.9	202.0
Harry Potter and the Sorcerer's Stone	317.6	166.7
Attack of the Clones	303.2	144.8

Source: "Top DVD Sales 2002," http://www.boxofficemojo.com/dvd/2002/sales/ (last accessed 3/12/2005).

have suffered significantly because of the adoption of DVDs. That would be consistent with the broader pattern evident in the United States across many industries and consumer products of more than one medium in use. However, the video rental market has begun suffering as people simply buy a DVD of a movie. A second, very recent phenomenon is the introduction of DVD-ROM, making it possible for a personal computer either to be a screen or a platform from which to acquire movies off the Internet. Given the large number of PCs installed in the United States, this medium holds a high potential for expansion. Regardless of that fact, the Movie Industry has now embraced DVDs as a major deliverable.[48]

The role of the Internet in the Movie Industry can quickly be summarized. In the mid-1990s, studios began creating websites to post information about their movies. By the end of the century, they were establishing websites for each major movie so that fans could get information about them, provide commentary, and even download screen-savers and place orders for products.[49] Smaller distribution and exhibition firms used the Internet to compete more forcefully with the majors. By 2002, marketing over the Web had become a crucial component of what a studio did. But the big story regarding the Internet may only just now be emerging.

The major issue is the growing availability of movies on various Internet sites that people can download—often for free—as pirated copies. It is essentially the same problem the Music Recording Industry faced beginning in the late 1990s. The later timing of it coming to the Movie Industry was due to, first, the time it takes to make and post digital copies of movies; and, second, the need for massive broadband capability to download a 90-minute movie, which represents massive quantities of data far in excess of a 3-minute song. Most consumers do not yet have sufficient bandwidth to download movies, but the studios see it coming. As one industry report in 2002 assessed the situation:

> When it comes to the Internet, the major Hollywood studios are determined not to make the same mistakes they have watched the music industry make. The challenges are really two-fold. Firstly, there is the emergence of new Internet companies employing radically new business models that might overtake the existing industry giants and their traditional business practices. Secondly, there is the challenge of piracy.[50]

Executives in the Movie Industry knew that whatever happened to music companies would eventually become part of their experiences and problems as well. Some industry observers predicted the end of Hollywood and the majors in the face of the challenge of streaming and downloading. However, the Movie Industry thinks it has learned several lessons from the Music Industry's response to the Internet. First, music companies simply ignored for too long the threat posed by the Internet, so when the industry reacted it did so by striking back at downloaders through the courts, with mixed results. Second, in the process it attracted a great deal of negative press coverage as newspapers began reporting 12-year-old children and college students being sued for all they were worth. Third, the Recorded Music Industry was unable to launch the Secure Digital Music Initiative (SDMI), its effort to provide end-to-end secure distribution of music over the Internet. Fourth, the solution may lie in providing legitimate ways for viewers to download movies for a fee.[51]

In 2002 and 2003, downloading pirated copies of movies began, although it is not clear how widespread this practice had become because of the limitation of sufficient bandwidth. A few movies even began appearing over the Internet before the movies opened at the theaters. Unauthorized high-quality DVD copies had floated around for several years, often made in the studios or by advertising agencies associated with the films and often appeared on the Internet. While the industry probably exaggerated the number of pirated downloads, nobody could be sure. Industry-wide meetings in the summer of 2004 devoted time to analyzing this growing global problem. Seizure of illegal DVDs in 2003 and 2004 were approaching or exceeding the number of legally available DVDs, and stories of police raids on shipments of 20,000 to 30,000 illegal DVDs at a time alarmed the industry, feeding its concern about the growth of the problem.[52] The industry responded, much like the Music Recording Industry, by suing people they caught downloading, mainly from websites that rebroadcast television shows and movies.[53] The industry was also exploring ways to distribute movies online for a fee in 2004 and 2005, but that is still an embryonic activity, with firms experimenting with various business models such as licensing, subscriptions, sponsorships, and advertising.

In 2002, the industry faced yet another problem, the "hacking" of films, by which someone used software to alter a pirated copy and devise different endings, change some lines, add pornography, and so forth. But as of this writing (2005), the major counterinitiative taken by the industry is the same as music's: legal action. As one industry report observed, "consumers have quickly learnt to expect the media content they want when they want it. This demand for content over the Internet is a powerful force that legal, technological and other tactical measures may not fully be able to curb."[54] A lawyer in the industry, Peter Dekom, was blunt, "It's clear that if you don't give the consumers what they want, they'll find a way to get it their own way."[55] This issue remains the central market reality that both industries have yet to address satisfactorily.

In defense of the industry, the majors began reaching out to consumers directly. They are attempting to establish an Internet VoD platform that all studios can use to sell or rent movies over the Internet, although the business model at the moment is leaning toward a monthly subscription or pay-per-view approach. This would, of course, bypass distribution companies and theater owners, representing a monumental shift in the fundamental structure of the industry if it is successful. What the industry's executives learned from music is that they have to provide a secure, commercially viable alternative to being Napsterized, and they have to do it quickly; they cannot afford to wait to act until households have sufficient bandwidth.

By 2003, Hollywood had arrived at a point where the business model surrounding a movie had changed. Kristin Thompson observed this change with *The Lord of the Rings: The Fellowship of the Ring*. *Fellowship* is one of three movies in the series, and it is what the industry calls a "franchise," consisting of various products such as action figures and clothing with logos, and videos. The following practices were routinely being applied to all other major movies. First, the movie went into DVD, the preferred medium over VHS since its production costs were so low in comparison and the market larger. Second, the movie's producers used elaborate special effects to create spectacular scenes in this fantasy movie. That in turn provided additional explanatory material that could be included in the DVD product. Third, the movie spun off the now-common assortment of action figures, T-shirts, posters and other items. Thompson noted the effective use of websites, both those established by the producers and others created by fans, in which movie clips, commentary, and news stories generated additional word-of-mouth promotion for the movie. One could also download off the Internet screen savers, wallpaper, interviews, and trailers. How popular was this channel of promotion? In the first week in which trailers of the movie became available on the Internet (April 2001), over 8 million downloads took place. Continuously placing more material on the Web allowed fans to follow the progress on filming and rollout of the movie, giving fans a "virtual keyhole" into the production process, a fundamentally new dimension in the movie world. Fans discussed scenes as they were being developed and offered suggestions for changes. Polls on various issues were taken before and after the the movie's opening, making for an effective way of sustaining interest in the movie after launch.[56] Nothing like that had ever occurred before in the industry.

By mid-2004, computer experts were developing ways to download movies quickly. One estimate placed the number of downloads occurring at the time of between 400,000 and 800,000 *each day*. The same week that the press reported the high-speed downloading of film projects (over 30 gigabytes per minute), the Motion Picture Association announced it was launching another antipiracy campaign. A reporter from *The Hartford Courant*, John M. Moran, expressed the frustration of many when he ended his report with the statement, "Hollywood should learn a lesson from the music industry's troubles. The best defense against pirating is to give consumers a legal way to purchase movies and TV shows online." He concluded that "most will pay a reasonable price for legal downloads, if only they are available."[57] Movie studios were reluctant to move too fast in endorsing video-on-demand (VoD) because of the effect that it would have on its highly profitable sale of DVDs, a major cash cow for the industry.[58] It presented a classic business problem faced by all industries moving from one generation or type of technology to another, where costs, sales, and profit margins are different. How does one replace stable, predictable revenue and profit streams from one type of product with another without seeing a decline in either cash flows or profits? Every industry that has been affected by digital technologies continues to face that issue, as have all industries over the centuries as they have moved from one base technology to another. It just has become more intense since the arrival of digital technologies.[59]

New possibilities also emerged in the world of digital (video) games. Because the movie *Lord of the Rings* was a fantasy, it lent itself to a connection to electronic games. The link between movies and digitally based games had become evident in the 1990s but not until the end of the century had that symbiosis begun widely manifesting itself as part of the business opportunities facing both the movie and video game industries. To be sure, games linked to themes explored by movies had been around for over two decades, including one related to *Lord of the Rings* (LOTR) (*The Hobbit*, released in 1982).[60] In 2001 Vivendi Universal Games acquired licensing rights to produce various video games based on the *Lord of the Ring* books, and *The Fellowship of the Ring* appeared in October 2002. The company also enhanced its repertoire of game-development technology both in support of this and other fantasy games. At least one other firm (Electronic Arts [EA] Games) introduced a game directly related to the three LOTR movies. The studio chief at EA, Don Mattrick, reported that "we got all of Peter Jackson's computer drawings and images to base our work on. So you're not looking at our interpretation of the movie, it *is* the movie."[61] As Kristin Thompson observed, this was the "recent trends [*sic*] in movie-based games. In the past two years or so [2001–2003], designers have tied their film-based computer games more closely to the characters, situations, and settings of the original sources. Film companies, seeing the advantages, have cooperated in many ways."[62] She reported that film-related computer games now generated more revenues than theatrical ticket sales, and the digital media were converging in interesting ways for the consumer. At the end of 2002, Americans had over 37 million DVD players, but there were 55 million DVD-playing video game consoles as well, demonstrating that digital games had become "a major factor in the home-video market."[63] As an example, in 2002 total DVD sales income as a proportion of total domestic ticket sales was 82.1 percent just for EA's

Fellowship of the Ring game. Theatrical performances for major movies were increasingly evolving into large marketing events in order to prop up the major sources of revenues: games, DVD copies of the movies, and products. What a contrast that was to 1950.

There is one final segment to comment on: the exhibitors (theaters) themselves, which work with distributors to schedule films into their locations. As movie theater chains came into existence throughout the second half of the twentieth century, they too standardly used computers to handle accounting, personnel, and movie-scheduling activities. The important story, however, is the emerging issue of what will happen to these companies now that movie studios are beginning to develop movies in digital formats. Theaters use projectors that project analog film. Many of these projectors are fully paid for and last for decades. The advent of digital media, say downloaded to a theater off a satellite or off a DVD, would require new projectors. Given that technology changes so fast, theater owners worry that they would have to replace these new projectors with even newer ones in a few years. The industry's practice so far is for theaters to bear the burden of expense in acquiring new equipment.

In the early 2000s, and even beginning in the late 1990s, there has been a lot of ballyhoo about moving to digital exhibition, much as there was about viewers doing the same at home. The data on deployment demonstrates that once again, we are dealing with an industry that moves to a new technology when it can afford to and when it makes economic sense. In 1999 (the first year for which there are data), there were possibly 10 screens in the United States that could show digital movies; that number grew to 19 in 2001, then jumped to 76 in 2002, where it has remained. Globally, there were some 250 screens that showed digital movies in late 2004. If we look at how many movies there were to exhibit in such theaters, we see that the numbers are small as well. In 1999, Hollywood produced 4, independents and the rest of the world none. Hollywood made 9 the following year, independents 1, and the rest of the world 7. By mid-2004, Hollywood was up to 13, independent studios another 2, and the rest of the world 13 for a total of 26. If we add up the entire inventory from 1999 through 2004 from around the world, the grand total is 142 films.[64] Thus, both the total number of available films and screens was miniscule, especially when we consider that in any of those years the worldwide industry produced about 4,000 films for 141,000 screens, with 500 films coming out of the United States to play on 31,000 screens. Hype aside, this next reaching out to the digital hand appears far away.

Music Recording Industry

Companies in this industry not only record, manufacture, and sell recorded music; they also are involved in handling the representation of music writers and performers, managing copyrights, and organizing live performances. Some companies only specialize in one aspect of the industry, while the majors, just as in movies, dominate the industry and focus on both distribution of its products and protection of its copyrights.[65] In the early 2000s, this industry was very publicly involved with trying to preserve its historically high levels of revenues and profits by protecting its copyrights in the face of what it called piracy of its music off the Internet. While this issue is discussed below, the industry's history demonstrates that issues related to emerging digital technologies have been reaching out to this industry since the 1950s in many ways—some helpful, and some not so helpful. It is an industry that has experienced so many changes, driven by changing technologies, markets, and organizations, that it provides yet another excellent window through which to see how the digital has been affecting American society and work.

The industry plays two major roles in the economy: it produces and sells recorded music, and its products are the mainstay of radio broadcasts. In the case of the former, it was always about sound recordings, from records to tapes to CDs and now DVD videos. As for radio, recall that prior to the arrival of television, radio stations broadcast talk shows, comedies and dramas, live performances, and recorded music. Television ate into radio's dominance in all but the broadcasting of recorded music, which by the end of the 1950s was the major programming activity of radio stations, leading to a closer link to the Recorded Music Industry than ever before, one that has remained to the present. The volume of business in this industry always placed it in the top five sources of entertainment and media expenditures, competing with television, magazines, books, home videos, and movies in theaters for all manner of revenue: advertising, tickets, CDs, T-shirts, and so on for the public. Americans spent more time listening to recorded music than with any other media, with the exception of television. By the 1990s at least half the American population purchased at least one album each year.[66]

Yet the industry has essentially three other sources of income as well. The first derives from a string of activities related to live performances: career managers, talent agents, promoters, venues for performances, and ticket sales. The second income stream comes from the recordings themselves—the production, manufacture, distribution, and sale of records, tapes, CDs, and DVDs, for instance. The third involves songwriting, with publishers acquiring copyright permissions to market a song, thus allowing them to negotiate the use of music with producers (who make the recordings) and with those companies that manage live appearances of the artists who sing the songs.[67] All three have involved computers in varying degrees over the decades. Like so many other media industries, the Recorded Music Industry exhibits a high level of oligopolistic behavior. It too has a few dominating firms that control distribution of music and who make the majority of their profits off a few, highly popular artists. As early as 1952, for example, four companies controlled 78 percent of all charted records: Columbia, Capitol, Mercury, and RCA Victor.[68] The enormous growth in rock 'n' roll music in the 1950s altered the configuration of the major firms as new entrants (mainly labels) introduced a heightened round of competition into the industry. But oligopoly returned by the early 1970s; in 1972, for example, five corporations controlled 58.2 percent of the album charts. Control further concentrated all through the 1970s and 1980s, involving various rounds of consolidations, with the same level of heightened activity in the late 1980s and early 1990s evident in all other American media industries. In the early 1990s, four distributors controlled nearly 62 percent of the market in the United States: WEA, Sony, PolyGram, and BMG. On the eve of the Internet's entry into the life of the industry, the major global firms were Warner Music Group (the largest), Sony Music, PolyGram, BMG (part of Bertelsmann), and MCA.[69] As for other media industries, this one aspired to vertical integration, giving its major firms the ability to control all aspects of production, from finding the music to recording, distributing, and selling it.[70]

This industry has been highly successful over the past six decades because of a combination of factors: the profitability of the products (largely thanks to the technology), the industry's ability to control so many aspects of its work, and the enormous popularity of recorded music across all generations in a period when the nation could afford to spend billions of dollars on entertainment. Other forms of media have also proved enormously influential for the well-being of the industry. Over the entire period nearly half its income had come from copyright fees paid by radio stations to broadcast music, another 30 to 40 percent from actual recordings sold to consumers, and the rest from sales of sheet music and other ancillaries.[71] Combined revenues from all sources grew from the 1950s almost continuously into the new century. By the time of the Internet (1994), sales of recordings alone had reached $12 billion annually. But more important was the shift that took place over the years as the major firms increasingly managed copyrights and the granting of permissions to others in the industry to use and sell their products. A combination of technological innovations and changes in copyright law led the industry to focus its attention on the management of its ever-expanding portfolio of assets by licensing its properties. That is why in all probability the industry came to be so intensely concerned about copyright infringements on the Internet, which it labeled "piracy," and less effective (or concerned) at finding successful ways to sell products to consumers that leveraged the technological realities of the day. Indeed, in studying over two dozen industries, I did not find as much discussion about copyright issues as in this industry in the 1990s and 2000s.[72]

Nowhere is this attention so evident over so many decades as in radio. It is important to understand the basic business of radio, which is to sell advertising, not music; stations use music to attract listeners who then are exposed to advertising. Music companies want exposure for their offerings on the radio in order to generate royalty income and create market demand for specific songs and artists.

Most new music is introduced to the market through radio, giving stations profound influence over the buying tastes of the American public.[73] When music videos came onto the market in the 1980s, TV began to play a growing role in influencing music tastes as well, but not to the same degree as radio. Furthermore, the massive surge in the sale of music videos is a 1990s story. In 1991, the industry sold 6.1 million videos and in 1993, nearly 11 million, which grew to over 12.6 million in 1995, the first year the Internet became widely accessible to the public.[74] Sales kept growing right into the new century, along with those for CDs and DVD versions of musical videos. But sales for music had built over time, giving CDs momentum. Two industry analysts observed that "for 25 years, beginning in the mid-1950s, sales of sound recordings grew an average of 20 percent a year. The most dramatic growth came in the 1970s when sales rose from less than $2 billion at the beginning of the decade to over $4 billion in 1978."[75] The arrival in the American market of CDs in 1984 had pushed up overall sales, leading to combined sales of $6.25 billion in 1988 for records, tapes, and CDs. A decade later (1998) revenues reached $13.7 billion, with the fastest growth as a percent of sales coming from CDs.[76] By 2001, 2.9 percent of all sales of music had taken place over the Internet, almost all of it for CDs.

Did the growth in sales revenues in the final two decades of the twentieth century occur for the same reasons as in the 1950s to 1970s? In the earlier period, new recording media (stereophonic LP records and tapes) made possible better quality and more accessible music. In addition, the arrival of rock 'n' roll, R&B, and even TV helped sales—TV because it forced radio stations to rely more on music (rather than dramas) for programming. Industry observers noted that in the second half of the century, technology played an increasing role in the financial success of the industry:

> New technology has been responsible for much of the industry's growth. In 1948 the LP was introduced; stereo dominated toward the close of the 1960s. Prerecorded cartridges and cassettes, which were introduced in the mid-1960s, opened an entirely new market for record manufacturers. In 1983, cassettes became the configuration of choice, but by 2000 CDs dominated unit sales, with cassettes constituting only about 35 percent of unit sales and vinyl less than 1 percent.[77]

But it did not end there, because digital audio tape (DAT) came along, serving as a counterpart for tape as CDs had been for earlier formats. VCRs also played music and were widespread (over 95 million U.S. households had players in 2001). But our observers also noted that marketing played an important role, right alongside the expanded use of technologies. Rack jobbing (a third party supplying shelves in a store)—so evident across all industries supplying retailers and record clubs—and ordering music over the Internet from retailers proved essential to the overall success of the industry.[78] So in answer to our question, circumstances had changed in how music was sold.

Finally, we need to acknowledge the special role copyright laws played in making this industry so reliant on the management of its assets. That in turn led to an obsession with any activity that threatened its control over musical properties. Table 11.5 lists the key U.S. legislation that supported digital and other media copyrights. What jumps out from the table is how active the U.S. Congress was from 1988 on in creating a network of legal protection for digital recordings, in which it was spurred on by many media industries. These laws were intended to protect the copyrights of firms owning music and other electronic media while aligning legislation with various international agreements. For example, the No Electronic Theft Act (1997) articulated what constituted criminal theft of digital recordings, while others defined the length of copyright periods and who was ex-

Table 11.5
Key U.S. Copyright Legislation, 1988–1999

1988	Berne Convention Implementation Act
1992	Audio Home Recording Act
1995	Digital Performance Right in Sound Recordings
1997	No Electronic Theft Act
1998	Technical Correction to 1976 Copyright Act
1998	Digital Millennium Copyright Act
1998	Copyright Term Extension Act
1999	Fairness in Music Licensing Act

empt from acquiring permission to use recordings.[79] All of this legislation came into being at the same time as digital and communications technologies made it increasingly easier for individuals to inexpensively and illegally obtain copies of music.

The History of Technology in the Industry

This industry leveraged a variety of technological innovations over the course of a full century. Its managers were no strangers to the opportunities and dangers of disruptive technologies. Notwithstanding the fact that in the early years of the twenty-first century most observers of the industry criticized the way its management dealt with the issue of the Internet and what their trade associations called "piracy," it did not take away from the historical record that this industry has long concerned itself with digital technologies. It is easy to think linearly about technological changes in the industry, moving from LP records to tapes to CDs and then to the Internet and on to DVDs. However, if we look at any list of all the other digital activities underway in the course of the last half of the twentieth century, we can begin understanding how complicated, even confusing, all these innovations were to all media industries, creating in the process some ill-served thinking and many real concerns. While not a rigorous explanation of recent industry responses to Internet downloads, it nonetheless reinforces an age-old lesson well understood by historians, that things are never quite as simple as they seem.

Thanks to a substantial research project underway on media history at the University of Minnesota, there is a growing list of innovations in chronological order that reminds us that, in any given year, there were normally several digital or other technological events occurring with either a direct (or subsequent indirect) bearing on the Recorded Music Industry, not to mention other media industries, ranging from the technologies mentioned above to the whole series related to the evolution of PCs, computers, and telecommunications. The list averages over a dozen events per year. Table 11.6 extracts key technological events

Table 11.6
Timeline of Major Technological Events Affecting Media Industries, 1950–1975

1950	Vidicon camera tube introduced, improving television pictures
1951	Color television sets first sold in the United States Commercial computers enter the market
1952	Early transistor radios sold by Sony Univac used by CBS to project winner in U.S. presidential election
1953	Pre-recorded reel-to-reel tape at 7½ lps introduced
1954	Regular color TV broadcasts begin in the United States Commercially available transistors appear on the market First major year of sales of transistor radios
1955	First tests of fiber optics in telecommunications
1956	Transistors first used in car radios
1957	USSR launches first space vehicle, *Sputnik*, alarming United States about the currency of its technology
1958	Color video tape first appears Stereo LP records first go on sale First use of cable to carry FM radio programs
1959	All-transistor radios that fit into shirt pockets introduced
1960	Electronic music first created
1961	Wireless microphone first used in a movie Early computer chips first introduced
1962	High-speed digital telecommunications lines first used in telephone services FCC requires UHF tuners on TV sets
1963	Emergency Broadcast System launched in United States
1964	Picturephone tested, unpopular with public
1965	Computer-based telephone digital switching first used Intelsat I communications satellite launched Cartridge audiotapes first go on sale
1966	No major event
1967	Newspapers and magazines digitize production on a wide scale Prerecorded movies on videotapes first sold A hypertext system developed at Brown University
1968	TV photographers start using portable videotape recorders
1969	Audio music tapes first sold using Dolby Noise Reduction
1970	FM radio stations began "narrowcasting" AP begins transmitting news by computer
1971	ARPANET launches what eventually becomes Internet Wang 1200—world's first computerized word processor
1972	Digital television first emerges from laboratory
1973	AP, UPI start using online systems Noncompatible video players drive manufacturers out of the market
1974	Arcade video game *Tank* uses ROM chips Early use of international digital voice transmission
1975	Optical videodisk system demonstrated by Philips Microcomputer construction kits become available in U.S. markets

Source: Extracted from "The Media History Project Timeline," http://www.mediahistory.umn.edu (last accessed 7/2/2004).

Table 11.7
Timeline of Major Technological Events Affecting Media Industries, 1976–1990

Year	Event
1976	Dolby stereo first used in movie theaters A still camera (Canon AE-1) first to use a microprocessor
1977	Atari introduces programmable home video game in a cartridge Nintendo begins selling computer-based games
1978	RCA introduces Selectravision videodisk Cellular radio assigned spectrum
1979	First news groups appear on the Internet Game players discover multiuser game possibilities on Internet
1980	*Pac-Man* introduced, exposing large segments of population to computer-based games Lasers first used to set print type
1981	Nintendo introduces computer-based game *Donkey Kong*
1982	*Tron*, movie from Disney, also an arcade game
1983	First CDs sold Laser disk used in *Dragon's Lair*, an arcade video game
1984	First portable compact disk player sold Newspapers start offering online text versions CDs sold widely in United States
1985	Nintendo enters home video game market TV networks begin satellite distribution to affiliate stations
1986	Digital audio tape (DAT) invented First international standards set for audio, video, and digital recording Japanese firm introduces *Game Boy*, a hand-held game using microprocessor First encyclopedia published on CD-ROM (*American*)
1987	IBM introduces PC with VGA Cortada household acquires its first *Game Boy*, daughters declare *Pac-Man* outdated
1988	Digital Disk Playback (DDP) used for uncompressed digital sound Prodigy introduces dial-up network service CDs outsell vinyl records
1989	First manipulation of photographs using personal home computers
1990	Second edition of *Oxford English Dictionary* made available on CD-ROM Kodak introduces Photo CD player *Dick Tracy* first 35 mm film with digital soundtrack New generation of videodisks introduced

Source: Extracted from, The Media History Project Timeline," http://www.mediahistory.umn.edu (last accessed 7/2/2004).

from the first quarter-century that had a direct link to the industry. Table 11.7, covering the subsequent 15 years, demonstrates how the pace picked up, particularly with regard to digital developments. Table 11.8, on the most recent period, gives a picture of the current complex and tumultuous period.

Technological developments came from many sources, such as from computer

Table 11.8
Timeline of Major Technological Events Affecting Media Industries, 1991–2004

1991	Semi-user-friendly Gopher provides navigation tool for accessing Internet files Recordable compact disk drives for CD-R introduced to the market
1992	Sony introduces Mini-Disk, a recordable magneto-optical disk Compact disk music sales surpass cassette tape sales
1993	MPEG-2 standard for television pictures adopted at world conference IMAX 3D digital sound system first installed in a theater in New York City Computer-generated dinosaurs used in movie *Jurassic Park* Movie studio uses nonlinear editing computer system for first time Graphical user interface, Mosaic, developed for World Wide Web
1994	Digital satellite TV service, DirectTV, first offered in United States Netscape Navigator replaces Mosaic as Web browser, opens Internet to millions of users
1995	*Toy Story* first totally digital feature-length movie Experimental CD-ROM disk carries full-length movie Sony PlayStation and Sega Saturn use 32-bit system for home video games Amazon.com begins selling books online Direct Broadcast Satellite (DBS) transmits first digital programs to home dishes Audio of live events now heard over the Internet
1996	Thirty million Americans use the Internet, 45 million globally Nearly 30 percent of all U.S. libraries have access to Internet
1997	DVD players and movies sell well in the United States Over 50 million North Americans (U.S. and Canada) use the Internet Palm introduces first hand-held computer device Ultima Online, a multiplayer online role-playing game, begins operation First point-and-shoot camera introduced by Kodak Streaming audio and video available now over the Internet Almost all U.S. newspapers have Internet sites and stories Over 40 percent of all U.S. homes have a PC
1998	Megapixel cameras made available to general public HP introduces first personal digital assistant (PDA) product Music Industry upset over downloading of sound files for free with MP3 HDTV broadcasts begin in United States Cable TV channels begin using compression technology 150 million users of the Internet, half in the United States First digital TV programs broadcast in the United States V-chips first installed in American television sets
1999	Millions download music free off the Internet; Music Industry faces a crisis
2000	U.S. court limits Napster's Internet music file-sharing activities U.S. Congress passes Children's Internet Protection Act VCRs enter market rapidly, reaching some 85 percent of U.S. households More than 3 million blank, recordable CDs sold monthly Film *Quantum Project* produced for distribution over the Internet, not theaters Nine-minute video, *George Lucas in Love*, premieres on the Internet Stephen King's novel *Riding the Bullet* becomes a best-seller over the Internet, not for sale in bookstores

(*continued*)

Table 11.8
Timeline of Major Technological Events Affecting Media Industries, 1991–2004 (continued)

2001	Personal headsets display movies, video games, and spreadsheets for first time Americans spend more on electronic games than on movie tickets iPod music player introduced More than half the nation uses the Internet Special computer-based animation used in all major movies, notably *Shrek*, *Harry Potter*, *Crouching Tiger*
2002	Government reports 9 out of 10 children have access to the Internet First feature film fully developed in digital format, *Attack of the Clones* DVD sales surpass those for VCRs; estimated 40 million U.S. homes have DVD player Apple computer system introduced that can create movies in DVD format
2003	More DVDs than videotapes are rented in the United States for the first time iTunes online music store offers music at 99 cents each
2004	Apple introduces iPod which can hold 10,000 music files; fits in shirt pocket HD car radios first go on sale, delivering digital AM and FM broadcasting at near CD quality

Source: Extracted from "The Media History Project Time Line," http://www.mediahistory.umn.edu (last accessed 7/2/2004).

and software manufacturers,[80] television and telecommunication manufacturers, games and electronics firms, but none directly from the Recorded Music Industry. In any year, therefore, it would have been difficult to spot innovations and rates of adoption of what might normally seem to be disparate developments that had little or no implications for one's own industry. Since managers in most industries did not pay close attention to technological developments in other industries, music executives would not likely have noticed not only that their technological base was changing profoundly, but also that the underpinning business model and practices of their industry were being altered. By the time the industry reacted (1999), the detailed lists make it obvious that the proverbial "train had left the station," with other industries and sources of innovation affecting the Recorded Music Industry. Furthermore, the industry had not introduced major technological innovations; it had simply adopted them as they were created, primarily for recording music, running normal business operations, and delivering their products. It is no wonder, therefore, as Steve Jobs noted at the time, that this industry did not understand what was happening in the market, particularly in regard to the role of the Internet.[81] Yet the signs had been there for years, forecasted and discussed by professors of media and business management in academia and industry experts residing in various information technology industries. One has to wonder if their articles and books were read by management teams in American music companies.[82]

In fairness to the industry, however, it did adopt digital technologies as they came along. One area in which computers in time came to be used increasingly was by songwriters and musicians in creating music itself. Much as in the Movie Industry, where computer-based graphics and animation enhance directors' work, so too in the world of music did digital technology made it possible to create and produce music in new and more efficient ways.

In the case of electronics in music, the earliest known use dates back to 1897 to Thaddeus Cahill's Teleharmonium, which weighed seven tons and was about the size of a boxcar. As new electronic components became available in the twentieth century, people adopted these to make sounds, for example, by building devices that mimicked percussion instruments, often called rhythm machines. Early devices were, in effect, preprogrammed to create certain beats, such as for the tango or mambo. Beginning about 1980, a musician could program to any beat. By the end of the century, most of these machines had been supplanted by software embedded in other devices, including PCs.

The major tool in electronic music was the synthesizer, a device (typically a keyboard) designed to reproduce sounds, usually of recognizable instruments (such as, piano, guitar, or drum) by manipulating electrical currents. As early as 1958, engineers at RCA had tinkered with devices to reproduce music and voice, building an instrument called the Mark II Music Synthesizer. In the 1960s, large systems appeared in studios that could perform some of this work, but credit for creating the first modern configurable music synthesizer goes to Robert Moog, who built his first device in 1964. In 1968, Walter Carlos (now Wendy Carlos) published an album, *Switched-On Bach*, based on music played on Moog's device, which became a million-plus best-seller record. Moog's synthesizer was used that decade by hundreds of musicians. Other devices were developed by various individuals in the 1960s, but all were analog and predigital. In the 1970s, new machines relied on solid-state components, which made them smaller (hence portable) and less expensive, but even these were monophonic, broadcasting only one tone at a time. Polyphonic sound finally came in 1976 with the introduction of the Yamaha CS-80, which used computer technology and was large and expensive. Then in 1983 a standardized digital control interface was created, setting the stage for all-digital synthesizers and another class of devices called samplers.[83] By the 1990s, Musical Instrument Digital Interface (MIDI) software existed, making it possible to mimic electronically a growing variety of musical instruments and sounds that a composer could use to create music from a combination of keyboards and CRTs.[84]

In the Recorded Music Industry, the story is told that the well-known blind musician, Stevie Wonder, approached Raymond Kurzweil (the inventor of a reading machine for the blind) and challenged him to make an instrument that could create a far richer sound than before. Kurzweil's company developed the K250, the first ROM-based sampling keyboard that reproduced the sounds of a variety of instruments. Introduced in 1983, it impressed many in the industry with the quality of the sounds of piano, strings, choirs, drums and other instruments.[85] This early start was built upon by other vendors all through the 1980s and 1990s to the point where costs for equipment and their sizes declined, and functions increased. In the 1990s, one could acquire commercially available software to write and play music on a humble PC. To be sure, musicians had experimented with punch-card software-based systems as far back as the 1950s and 1960s, but not until nearly the end of the century had the technology evolved far enough to make this an economically affordable and practical application. Systems that cost nearly $100,000 in the 1970s could be had for less than $1,000 by the end of the century—some, in fact, for less than $500.[86]

Because of this technology, single-person bands became a recognized subset of the music business, called "bedroom bands." The wide use of PCs, even as early as the 1980s, had led to the emergence of a new genre of music, called "chip music" or "bitpop." The one technology that proved essential to the process comprised hard-disk recording systems on PCs that made it possible for both an amateur musician and the professional to create, store, improve, and play music on hardware. Thousands of bands and recordings employed digital technology to create and present music.[87]

Business Applications in the Recorded Music Industry

This industry shares many of the same patterns of operations and managerial concerns as the Movie Industry. The most obvious examples include the widespread use of subcontractors and outsourcing of work that led to the kinds of complex accounting practices noted, for example, in the number of checks an actor or performer might receive. In music, the challenge was the accounting for royalties to collect and pay for use of the industry's copyrighted products. In fact, there were more individual products to track in the Recorded Music Industry than in movies because there were more copyrighted songs than movies and many more radio stations to bill and collect from over time, as well as more performers to track and pay. But the nature of the work was similar and lent itself to early uses of computers to partially automate accounting and clerical work. In addition, both industries were anxious to protect copyrights in the face of the increasing capabilities of consumers to acquire productions without paying for them. This led to a near-chronic frustration over the lack of sufficient capacity of computers to handle incremental volumes of work or new uses, particularly on the creative side where the amount of processing power and memory, along with the availability of software tools, were often the technical issues determining their rate of adoption and the cost of such technology. In both industries, it did not seem that technical expertise was as much of an issue because there were competent individuals available, often running their own specialized firms, to whom various sectors of an industry could outsource work, be it tracking and paying royalties or performing creative work.

The earliest uses of information technologies in the Recorded Music Industry involved accounting applications; these dated back to the period of World War II, and often depended on adding and other calculating machinery whose base technologies were first introduced into the American market in the early twentieth century. As new tools came along, these were slowly embraced. For example, the industry had an organization responsible for keeping track of music performances owned by its members. Broadcast Music, Inc. (BMI), was the clearinghouse for collecting royalties on performances and then paying performers and others who held copyrights. In 1958, BMI installed a Friden Flexowriter with an IBM 47 Tape-to-Card Converter to create a quasi-automated tool to track the growing number of musical selections being aired, for example, by the ever-increasing number of radio stations. It relied on an early sampling algorithm to calculate payments:

> Before it can pay songwriters, composers and publishers for the performances of their music, BMI must tabulate these performances before calculating and making payment. These basic functions are handled by the Logging Department, which uses a scientific sampling procedure and an integrated data processing system built around Taller & Cooper paper tape and edge-punched card reading and punching unit.[88]

BMI first started automating its logging operations in 1941, using IBM punch-card equipment and "tub files," each holding 60,000 cards listing names and composers of BMI-licensed tunes. In the 1950s BMI accumulated 30 tub files. In 1953, the company began falling behind in its work because of the volume, hence its move to more modern equipment, a process BMI continues to the present. A press release dated February 1, 1978, in which IBM and BMI announced the installation of a computer system (IBM System 370 Model 148), for example, described the same application as implemented in the 1940s and 1950s, just with larger volumes of transactions: 47,000 songwriters and publishers, nearly a million song titles, 8,000 radio stations, over 750 television broadcasters, and tens of thousands of night clubs, discotheques, jukeboxes, symphony orchestras—and, yes, even elevators.[89] Radio stations also implemented logging applications, first with electromechanical punch-card equipment and later, as they could afford it, various generations of computers.[90]

As computers came into the Recorded Music Industry, additional services were added. Take the case of BMI in 1978, for example:

> Today BMI affiliates can receive bonus payments for all song titles that remain popular over a sustained period. The new schedule is based upon the cumulative number of performances per song title, excluding network television. Formerly, the bonus payment rate was based upon the total number of performances for an entire catalog of song titles by each writer or publisher. Now any BMI licensed song title can earn a bonus payment after 25,000 performances.[91]

The new system provided for "faster distribution and more detailed payment records of all monies due BMI writers and publishers" than before; it would "more efficiently collect performance royalty payments from broadcasters and other organizations that use BMI-licensed song titles," and tracked "for the first time the performance of all motion picture and syndicated television show themes used by broadcasters." Long before Internet piracy issues, BMI worried about it and hoped that its system would "improve control over possible copyright infringements."[92] It was a logical, early, and effective use of computing, addressing a core function in the industry that built on prior use of information-handling technologies. But computing also spread to other functional areas of the industry.

A core element of this industry is the recording studio. Studios are fixed expenses and constantly install the most advanced technical recording equipment. Musicians, singers, and bands rent time by the hour at these facilities to record their creations. For decades, recording equipment was electromechanical and analog, much like TV equipment. As digital recording equipment became available, beginning in the late 1970s, studios adopted the new tools. By the mid-1990s, over half of all recording equipment was digital, driven by PCs, specialized computers, and other devices. Digital studios had a more nuanced capability of recording many sounds more precisely than analog (for example, they could record more tracks) and charged anywhere from 65 to 100 percent more per hour.[93] As recording equipment became less expensive, artists could do their own studio work. Geoffrey P. Hull, who studied the industry's practices in the mid-1990s, reported:

> Technical advances are making it more possible for musicians and bands to record and distribute their music to wider audiences. On one front, the advances in low cost, high quality "home recording" equipment mean that many bands and musicians can afford their own recording gear or go to a low-cost studio. The equipment available enables them to make high quality recordings without having to go through any record company or pay studio rentals of hundreds of dollars per hour. Lower costs in the manufacturing of compact discs now mean that these same bands can make CDs in small numbers.[94]

He went on to say that musicians and lesser-known bands could then sell their products over the Internet worldwide, an option that did not exist earlier when the lack of a record contract with an established firm essentially meant bands could not make or sell their own music.[95]

This industry's use of modern marketing techniques became evident by the early 1970s. As they created more organized marketing plans for specific songs, singers, and bands, marketing managers needed more information about the make-up of their markets. The lack of data about sales patterns and buying preferences presented a particularly frustrating problem, one that the industry began resolving by placing Universal Product Codes (UPCs) on all their products, beginning in 1979. The industry joined so many other product industries, such as producers of packaged foods and book publishers, in collecting data from POS systems from retail establishments. That made it possible to begin accumulating substantial amounts of useful data in the 1980s. As rack jobbers and retailers expanded their use of UPC and POS systems in the 1980s, the individual stores or labels were now able to collect data on actual sales across the nation and to build marketing plans based on that information. Then in 1989, a company called SoundScan started collecting POS data nationwide and selling them to the Recorded Music Industry's firms, retailers, and distributors. These were reports on sales by artist, title, market, and retail establishment. Within a few years, SoundScan was an integral part of the industry as BMI had become decades earlier, and both for the same reason; they collected and reported data essential to the operations of the industry; each was essentially an information-processing firm—high-tech and effective users of IT.[96] By 1997, SoundScan was routinely collecting sales data from over 14,000 stores and providing *Billboard* with its rankings of best-selling music. In short, by the early 1990s—less than a decade after its founding—SoundScan had changed how the Music Recording Industry made decisions regarding sales and marketing, making it one of the most data-intensive media or entertainment industries.

Another major area of management's interest concerned knowing what music radio stations actually broadcast, or actual radio airplay. That concern linked back to what fees to collect, of course, but also connected to marketing considerations. Radio airplay charts had long been developed by gathering information from program directors at radio stations; specialized firms compiled such charts, for example R&R, *Billboard*, Gavin Report, BMI, and CashBox. BMI asked stations what they performed while other firms used various alternative sampling techniques. Then in 1990, *Billboard*'s BPI Communications established Broadcast Data Systems (BDS) to use a computer-based system to monitor actual broadcasts. Music would be loaded into a computer and then the system would monitor radio broadcast waves, essentially executing a pattern recognition process of comparing what was broadcast to what music was in its system, establishing what had been played by whom and when.[97] The data were compiled then resold to the industry, advertisers, and others. The system was improved all through the 1990s as the industry came to rely on its data. By around 1998, BDS could identify several million patterns (songs). Retailers installed kiosks loaded with BDS data on store floors for consumers to identify Top 40 or other best-selling music. One survey provided evidence that by 1996 over 40 percent of music buyers in the United States were using kiosks either as sources of information about music or to actually listen to potential purchases.[98]

These various business applications continued to be refined and expanded across the American economy and into other countries, particularly in East Asia and Western Europe and represented many of the core applications of computing and telecommunications in the industry, all integral to its operations today.

The Internet and the Piracy Issue

But what role has the Internet played in the operations of this industry? If we had not reviewed the uses of computers discussed earlier in this chapter, one might have thought that the industry's only concern and use of computing involved the Internet. However, as should be clear by now, the Internet was the latest in a long string of digital and telecommunications technologies to come along, and the problems it posed were not completely new to the industry.[99] In earlier times, when new technologies had come along to threaten the industry's large profit streams, the industry persuaded the U.S. Congress to pass laws to protect copyrights. Record rental shops and home copying represented threats to the industry dating back many decades that were worked out. The Internet, while resurfacing many of the same concerns, proved to be a larger threat that also had a very public face. The speed with which this new technology presented dangers to the industry outpaced earlier experiences; it took less than a half decade for problems to surface.

Forecasts of future business conducted over the Internet, circulating in the industry by the early 2000s, pointed out how large the issues had become. Jupiter Media Metrix predicted that the market for direct electronic delivery of music would reach $5.5 billion by 2006, with a potential customer base of nearly 211 million people. As early as 1999, Jupiter had started recommending to companies in the industry that they move rapidly to electronic distribution of their products. By the end of 2002, annual sales of music online had already reached the $1 billion mark, with legal downloads approaching $30 million. But the industry reacted cautiously to the suggestion that it move rapidly to online sales and delivery for fear of disrupting existing channels of distribution. To be sure, online sources emerged, such as N2K's Music Boulevard, CDnow.com, and, of course, Amazon.com, the latter enjoying 65 percent of the online music sales market by the end of 2001.[100]

It did not take long for a number of technological formats to emerge that made it possible for people either to legally or illegally download music off the Internet directly to their PCs or to blank CDs. The formats involved tools for compressing digital copies of music so that the amount of data that needed to go over a telephone or dedicated line was minimized, thereby making it possible for music to arrive quickly. The most widely recognized standard was Moving Picture Experts Group's collection of standards, better known as MPEG, used for movies, video, and music in digital compressed formats, which worked with PCs using software from Microsoft and Apple and with Unix. MP3 was the most widely used version of MPEG's formats and became the tool of choice for those who wanted to download music files.[101] In 2000, an organization called MP3.com placed 45,000 copyrighted CDs onto its website and sold a service called MyMP3.com to users in order to access this music. The Recording Industry Association of America (RIAA) the industry's trade association, quickly sued for copyright infringement, however, and the service was stopped.

There were other providers, however, with Napster the best known of them. Beginning in 1999, Shawn Fanning, a student at Northeastern University taking a programming class, organized a search engine and wrote software that made it possible for people to search for music over the Internet by title, communicate with the files that had copies of the music, and download that material onto their computers. It instantly became popular with college students, who used the large mainframe systems at their universities (processors and telecommunications) to copy this material. By February 2001, at its peak, Napster had over 26 million users who in total were downloading hundreds of thousands of music titles. The industry's five largest record companies, and some music artists, sued Napster for making it possible for people to violate copyright laws. To make a long story short, Napster was forced to shut down its service in 2001.[102] Users moved immediately to other services, the most popular of which was KaZaA, and between March and August of that year, the number of users of such systems skyrocketed from 1.2 million to 6.9 million. By the end of 2002, users all over the world had downloaded music some 37 million times using KaZaA's software. The Recorded Music Industry was alarmed at the rapid expanded use of such downloading tools and launched a series of lawsuits against various service providers.[103]

As early as 1998 and 1999, the controversy over downloading was drawing wide attention, both in the press and in other industries. For example, book publishers were concerned about the implications for them. It did not help that members of the music industry were themselves unable to figure out what to do. They were told by one music industry lawyer, David Lesser, that getting control over the issue was "like trying to kill cockroaches in a tenement apartment."[104] Newspapers reported that the industry was trying to develop software to sabotage people's computer systems if they pirated music.[105] And the president of the RIAA, Cary Sherman, opined that "we will see an entire generation who grow up thinking they don't have to pay the people who create music, referring primarily to children and young adults."[106] It did not help that the highly regarded Pew Foundation, in one of its many reports on how Americans were using the Internet, reported in July 2003 that "the number of downloaders who say they don't care about copyright has increased since July-August 2000," from 61 to 67 percent of all Internet users who downloaded music. Just as disconcerting was its conclusion that, for a total of 35 million people in the United States, "It is clear that millions of Americans have changed the way they find and listen to music." Pew reported that every time the industry shut down a service provider, within days people moved to others.[107]

In an effort to discover the most egregious downloaders, the industry sent subpoenas requesting information from Internet Service Providers (ISP) in July 2003.[108] That same month, Ted Bridis, AP's technology writer, began to report on the growing crisis of the industry, alleging that "parents, roommates—even grandparents—are being targeted in the music industry's new campaign to track computer users who share songs over the Internet, bringing the threat of expensive lawsuits to more than college students."[109] Meanwhile, some judges ruled that subpoenas of ISP records were illegal.[110] The problems continued unabated right into 2004. Headlines in newspapers all over the nation did not help the industry's image: *Mercury News* reported that "music suits aim at schools," while *CNNMoney* announced: "Recording industry group has sued about 3,000 people since September for copyright infringement."[111]

The RIAA, which was doing most of the suing, flooded the media and its website with dramatic rhetoric against piracy:

> Old as the Barbary Coast, New as the Internet—No black flags with skull and crossbones, no cutlasses, or daggers identify today's pirates. You can't see them coming; there's no warning shot across the bow. Yet rest assured the pirates are out there because today there is plenty of gold (and platinum and diamonds) to be had. Today's pirates operate not on the high seas but on the Internet, in illegal CD factories, distribution centers, and on the street."[112]

RIAA had expanded its accusations from pirated records and "online piracy"—the Internet issue—to include counterfeit and bootleg recordings (a major problem in Asia and particularly in China, especially in the 1990s). It argued that on a worldwide basis the industry was losing $4.2 billion to piracy, crimping its economic health.

But as far back as the summer of 2002, observers had started noting that perhaps the industry's 15 percent slump in business was not caused by illegal downloads. Forrester Research reported that "we see no evidence of decreased CD buying among frequent digital music consumers," rather, that the industry had to find a way to make "it easier for people to find, copy, and pay for music on their own terms."[113] In early 2004, economists began weighing in on the issue.[114] Felix Oberholzer at the Harvard Business School and Koleman Strumpf at the University of North Carolina at Chapel Hill conducted the most comprehensive economic analysis. In the arcane language of their discipline, they reported that "downloads have an effect on sales which is statistically indistinguishable from zero, despite rather precise estimates."[115] They observed that downloads of music actually stimulated sales if the listener liked what he or she heard. Moreover, since file sharing did cause the price of music to drop, it stimulated increased sales of music. They calculated that it would take 5,000 downloads to displace the sale of one CD. "High selling albums actually benefit from file sharing" and "while downloads occur on a vast scale, most users are likely individuals who would not have bought the album even in the absence of file sharing."[116] They blamed other factors for declining sale of music products in 2002 to 2003: "poor macroeconomic conditions, a reduction in the number of album releases, growing competition from other forms of entertainment such as video games and DVDs," and "a reduction in music variety stemming from the large consolidation in radio along with the rise of independent promoter fees to gain airplay, and possibly a consumer backlash against record industry tactics."[117] Various surveys in 2003 and 2004 presented evidence that the anti-piracy campaign was not effective and that if anything, downloads were increasing.[118]

As noted previously, others were trying to deal with the problem, such as Steve Jobs with his iPod, which allowed for fee-based downloads at under a dollar per song through Apple's iTunes Music Store. Between 2001 and 2005, Apple Computer sold millions of the devices and introduced new models that could store thousands of songs, with software to search and play music. Soon after, even the legal community was endorsing the concept that finding legitimate ways to provide music made more sense than court cases. Lawrence Lessig, a lawyer and professor at Stanford University, reacted positively to Jobs's approach and negatively to the industry's notion of digital lockboxes and other security measures: "I think the Apple system will begin to teach this lesson: Perfect control produces less profits than imperfect control. Always."[119]

The industry experimented with online subscription services, but these were not terribly successful since consumers still wanted to own the music on which they spent money.[120] Jobs commented that since executives in the industry did not use computers or e-mail themselves, "they were pretty doggone slow to react. Matter of fact, they still haven't really reacted."[121] After 18 months of trying to get the industry to consider his option of purchased legal downloads, he launched his business: "People don't want to buy their music as a subscription. They bought 45's; then they bought LP's; then they bought cassettes; then they bought 8-tracks; then they bought CD's. They are going to want to buy downloads. People want to own the music. You don't want to rent your music—and then, one day, if you stop paying, all your music goes away."[122] This is a perfect quote for a book on the history of any technology and business, because no new technological innovation can fully trump existing patterns of behavior, at least not overnight. The failure of the industry's treatment of the online sales issue between 1998 and 2005 was neatly summarized by Jobs: "the subscription model of buying music is bankrupt."[123] As this book went to press in 2005, the industry was still suing people and still trying to sell subscriptions—and people were still downloading music.[124]

Conclusions

Both the Television and Music Recording Industries faced a remarkably wide range of issues churned up by various digital technologies, particularly since the late 1970s. These varied from fundamental concerns about the protection of copyrights to shifting sources of competition to new channels for distributing their products. The basic tools of the industries were also part of their transformation, whether a TV camera or television set, or the humble 45 rpm record that ended the century as a museum piece, replaced by CDs and DVDs. How both industries did their daily tasks changed, thanks to the increased use of digital technologies either to enhance creativity and variety, on the one hand, or productivity on the other. Yet their roles and what they had to do remained essentially the same; Hollywood still made movies; and musicians still composed, sang and recorded music. They just did these things quite differently from how they did them in the 1950s. What is remarkable is how fast changes came.

While manufacturing companies and banks began using computers in the 1950s, we did not see that happening at equivalent rates in the television and recorded music industries until two decades later. To a large extent that time lag was caused by the need for various technologies to evolve into different forms so that they could be used in ways specific to each industry. Automotive companies did not need the kind of animation tools that the Movie Industry required, although automotive firms did have their advertising agencies use them in the 1990s to produce advertisements that were striking and entertaining, just as in the movie, music, and game businesses. But as in other industries, once the digital hand could contribute to the core work of either industry, they did not hesitate to reach out and grab it.

Adoption of the digital was not inevitable or smooth. The installation of new applications often is slow and complicated, because early adopters either run into technical problems or because management does not always know what to adopt or when. We saw the first pattern unfold with animation tools; it took over a decade for these to finally be useful and effective. In the second case, we have the Internet, which also illustrates more emphatically in this industry another truth about the digital experience: it is possible for an industry to have a deep understanding of one form of digital technology and little appreciation of yet another variety and have both situations be true at the same time. Record companies had no problem moving from LPs to cassettes to CDs and then to DVDs, but they also have not figured out how to deal with downloads and "piracy" over the Net. This dual circumstance has appeared in one fashion or another and at one time or another in all industries; in these two industries, it is particularly intense.

While much has changed in these industries, much has not. The fundamental missions have remained the same. Essentially, the same firms have stayed in charge from one decade to another, along with the types of players, whether production companies, distributors or theaters and stores. As in so many other industries, digital technologies have not slowed the pace of consolidation and oligopoly in either one. As in other media industries, however, computing has facilitated consolidation of multiple media companies together—and every once in a while an odd combination, such as a liquor manufacturer owning a media company. But why did the fundamental structures of industries change the least? The answer probably lies in the fact that digital technologies have always remained subservient to the business interests of the firms. No technology was willingly adopted if it threatened a firm's business structure or its revenue and profits streams. Technologies were embraced if they supported the existing base. Thus a songwriter would use a synthesizer if it made him or her a better musician; a studio would use a computer to track operating expenses and help directors control their budgets; DVDs would be used to create net new sources of revenues, not to replace movies exhibited in theaters. The list of examples is endless but the point is simple: the digital hand was welcomed if it supported the core purposes of an industry and if it was effective at doing that.

But that does not mean that new business models were not possible, indeed we have examples all over these two industries. We only need look to Napster. File-sharing applications teach us that new business models could appear rapidly with the potential to significantly undermine entrenched ways of doing business. If there is one conclusion we can reach, given the state of technology and what is going on in all media industries, is that movies and recorded music remain prime candidates for new business models that could well threaten the fundamental organization of their respective industries.

In addition, we have the influence of consumers on the whole process, for they have been known to shift their expenditures—voting with their wallets, so to speak. We saw it in banking where they loved their ATMs, but wanted their branches; in telecommunications, with their adoption of cell phones; in movies, radio, and music, in their desire to enjoy these when and where they wanted. Movies and music purchases are affordable to nearly everyone and thus represent literally billions of microdecisions that affect these industries. The "voice of the customer" that so many business commentators speak about is quite loud in the movie and recorded music industries. So what have these industries done? They have embraced the technologies that attracted customers, whether animation and special effects in movies, or synthesized music and CDs and DVDs. Beginning in the late 1970s both industries also accumulated a great deal of marketing and sales data about the buying habits of their customers, making these two industries at least as data intensive as any in the media sector of the economy. One IBM report from 2002 concluded that "information about the content that is distributed and who consumes it will be the new strategic high ground."[125]

In the last decade of the twentieth century and in the early years of the new one, both industries struggled with copyright issues. Leaving aside noisy lawsuits, the strategies deployed to protect the rights were quite traditional. Primarily, they turned to the U.S. Congress to update copyright laws to reflect the digital realities of the new times; then they followed up with the time-honored strategy of suing violators. The Recorded Music Industry, however, has demonstrated that these two traditional approaches may no longer be sufficient. Mergers and acquisitions continued to be a favorite tool of both industries as a way to protect and enhance business. However, the Internet may have created a new circumstance beyond the control of law and prior practice. It is difficult to assess to what extent the Internet is stimulating the use of M&As, for example, although we can conclude that as the public embraced new digitally based products from either industry, new business opportunities presented themselves—the classic reason for M&As in the first place. But the Internet still remains a highly visible central concern to both industries. MIT Press thinks it sells more books because it places so many of them out on the Internet for people to sample and even download. Steve Jobs of Apple Computer makes the point that perhaps music producers should do the same. As in earlier times, emerging digital capabilities pose challenges to existing business models and to management teams, who are just as conservative as so many managers in so many other industries.

We have two more industries to examine in order to appreciate the broad effects of digital technologies on modern America; the Games Industry, and photography. The first may be the newest high-tech industry in the United States—it was born in the 1980s—and one of the least understood. Yet it generates billions of dollars worth of business. The second one, the photography business, consists essentially of the behemoth Kodak and its rivals. In less than a decade, digital photography has become an almost ubiquitous phenomenon. We need to understand why that is and how it happened. Do these two industries represent old patterns of behavior in new places or new circumstances and business realities? They are the subjects of the next chapter.

12

Digital Applications in Entertainment Industries: Video Games and Photography

The game industry is not a technology business, but an entertainment business with a technology component.

—Ernest Adams, 2003

Michael J. Wolf, a thoughtful expert on the media and entertainment industries, has convincingly argued that "we are living in an entertainment economy"[1]—an economy that had become quite digital by the 2000. Nowhere has that become more visible than in the video game and photography industries. The technologies underpinning them have profoundly transformed both their products, who came to dominate their respective industries, and the practices evident in each. Looking at these two industries after all the others demonstrates the melding of various technologies and products in their two areas. Consumers could play video games over the Internet, or take and send photographs (even video) using a wireless telephone. In short, the functions of these industries' gadgets all acquired a highly futuristic feel to them, but they were real; millions of Americans have made them the sources of many hours' worth of entertainment. So, what do these industries have to teach us about how the digital hand was changing the nature of industries, work, and play in America?

Video Games Industry

Ernest Adams, whose quote opened this chapter, is a game producer and designer and likes clearly set priorities. He takes the position so evident in all entertainment industries, namely, that the digital hand is subservient to the core role of its members. However, the Video Games Industry could not exist if it were not for a myriad of digital and telecommunications technologies that came into existence since the 1960s. At the heart of its technology is a combined system made up of computer chips and software. This industry is an example of essentially a new class or form of entertaining and economically important activities made possible precisely because of the digital hand. It serves as a case study of what emerging industries look like in our age of computers.

It is also an industry still in the process of assuming its form. One of the characteristics of a proto, or new, industry is that people have not yet agreed on what to call it. Ernest Adams calls it the Game Industry. Melissa A. Schilling, who studies the economics of the industry, refers to it as the Video Game Console Industry. Harold L. Vogel avoids the problem by merging his discussion of the topic into the larger theme of toys and games. The Toy Industry co-opts all this industry's products in a similar fashion by treating them as yet another category of products (video games) right alongside dolls, bicycles, and sports toys. IBM and Microsoft think in terms of the Online Games Industry.[2] Invariably, when a new technology appears, early names of an industry are linked to it, such as the Computer Industry or the Software Industry. As the functions of the industry become more set, we see the industry goes through stages of naming. In the case of the Computer Industry, it first fragmented into various sub-industries (e.g., software, disk drives, semiconductors) then into the macro Information Processing Industry (1980s) and recently the Information Technology (IT) Industry (1990s). The same has been happening to this gaming industry, which at the moment appears to be settling on Video Games Industry as a name on its way to some other as yet to be established more permanent label.[3]

But why do we care about names? For the historian, the evolution of names suggests how groups of people who would become part of a common industry came together, identified with shared values and practices, and clustered into companies that competed and collaborated. For consumers of an industry's products and services, names indicate their offerings. For economists and members of the industry, names serve up both clarity and confusion about what's being discussed. Adams demonstrated this last point in a book on how to become a participant in the industry:

> When the term was first used, "video game" always meant a coin-operated arcade game—they were much more common than the early console machines. Then people began porting text-only games from mainframe computers to personal computers and calling them "computer games"; then personal computers got graphics and computer games began to be called "video games" as well.

> Now there are games available on mobile telephones and built into airplane seats; there are web-based games, handheld games, and electronic gambling machines. The whole situation is a real muddle.[4]

The mega trend was the migration, merger indeed, of computers and various forms of screens to create the various platforms on which video games are played today.

Regardless of what we call it, this industry has some remarkable features. First, it is a large, indeed, global industry; in 2003 it generated some $20 billion in revenue; selling 200 million copies of games annually just in the United States in the late 1990s, and employing over 200,000 people. In 2004 the industry grew an additional 8 percent, selling $6.2 billion in products just in the United States.[5] It reached this size over the course of barely three decades, growing out of an earlier pre-digital games market (such as for electromechanical pinball and arcade games, and of course, paper-based games) but most recently and directly based on software written to run on mainframes, then smaller computers, and consoles (which are essentially single purpose microcomputers). It is also an industry that has only recently caught the attention of historians, economists, and others looking at the expanse of the entire IT world.[6]

A Short History of the Industry, 1960–2005

Like so many industries, this one emerged out of several prior lines of business. People have made mechanical games for centuries and adapted many technologies to them, ranging from mechanical devices in the eighteenth century (automatons) to electromechanical equipment for coin-operated arcade games before World War II. Analog and digital computers in the 1930s and 1940s implemented simple games, such as *Baseball* and *Hang Man*. Scientists and mathematicians also have a long heritage of using game theory to study mathematics and other bodies of scientific knowledge, and when computers became available to them in the 1950s and 1960s, they did not hesitate to incorporate computing technology.[7] The emergence of cybernetics and artificial intelligence led to numerous attempts to use computers to play chess,[8] which led to the recent highly publicized contests between an IBM supercomputer and a world chess champion.[9] These events, however, were more research exercises intended to enhance scientific knowledge of how the brain works, and less about a desire to create games for fun or marketing hype.[10] But for both the industry and the public at large, these more intellectual exercises introduced the idea of the machine as a player.

The video game business came out of many developments, not the least of which was the computer, and most specifically, computer chips and software programming languages. While other sources existed (such as television technology), I want to focus on the digital path in this brief history of the industry. The wide deployment of computer chips in the 1970s was the fundamental enhancement to digital technology that made possible modern games, because regardless of whether we are talking about arcade video games, those that reside on specially made consoles (such as GameBoy), or those that are played on PCs, they all depend on these microprocessors. In the 1950s and 1960s, the only computer-based games were those written to work on large mainframes, usually at universities. But in the 1970s, two lines of modern games began, the first involving arcade video games, and later in the decade, others residing on personal computers. Arcade games evolved into home game consoles and video arcade games all through the 1970s to the present, with such famous milestones as the Atari 2600, Genesis, the rise of Nintendo in the 1980s, the highly popular PlayStation in the 1990s, and PlayStation 2 and Microsoft's Xbox in the early 2000s. This crowded line of development spilled over to related equipment, such as games on cell phones and PDAs (beginning in the late 1990s) and video poker games at casinos (as early as the 1970s).[11]

PC games emerged in the 1970s for the Apple II and Commodore 64. The widespread availability of Apple and IBM machines in the 1980s, with their continuously increasing processing and data-storage capabilities, attracted many game designers, often young men, all with extensive computing skills. Many started their own companies or became game writers for others, while various enterprises sold subscriptions to games over dial-up telephone lines (for instance, CompuServe and America Online). As tools came along that enhanced the gaming experience, developers quickly adopted them. These included music and voices and VGA (video graphics array) tools for improved graphics, in the 1980s. In the 1990s, 3-D graphics hardware and CD-ROMs made games more realistic, and sound blasters for noise came into their own; at the same time, game playing over the Internet began. By 2000, games also resided on DVDs, and the primary constraint to Internet-based games was the lack of broadband connections into homes.[12]

To a large extent, improvements to games were made possible by innovations in technology, such as CD-ROMs. A game programmer working in the 1990s described the effect on his world:

> CD-ROMs changed the PC game landscape enormously and, a little later, the console game landscape as well. It was now possible to create really large games. About the largest number of floppy disks ever shipped with a game was 12, for *Ultima Underworld II*, which together amounted to 25 megabytes of data once they were all decompressed. Today, a single compact disc can hold 26 times that much information. . . . CD-ROMs allowed games to include photorealistic graphics, high-quality sound, and even small movies. One of the first games to take advantage of this new technology was called *The 7th Guest*, and it was so spectacular in its day that people bought CD-ROM drives just to be able to play it.[13]

To a very large extent, my account of the video games industry relies on the primary sources of its origins and composition—first, information technology and, later, firms in the consumer electronics industries that moved in and were capable of expanding this industry to levels not achieved by the first generation of games providers. However, we need to recognize that existing board game producers too made video games and, like Kodak in the Photography Industry, had to adapt to the arrival of computing. The experience of Parker Brothers illustrates how traditional participants in the games business responded to the new technology. Parker Brothers is perhaps best known for its highly popular board games, such as *Monopoly*, *Clue*, *Trivial Pursuit*, and, for even longer, *Tiddley Winks*. Founded in 1883, in the same era as Kodak, it quickly became a major player in the board game business before World War I. In 1975, management was wrestling with the issue of what role should computing technology play when they were approached about the possibility of using a microprocessor in a game that would allow it to react to what a player did. Management did not like the idea of a game that could be played by one individual, since their products were designed as family games requiring three or more people to play together. In addition, the company had no internal skills to develop electronic games. Furthermore, high-tech games cost $30 each, while Parker Brothers sold *Monopoly* for $10 and could not imagine anyone wanting to pay much more for a game; at the time, the firm was generating $20 million a year just from this one product. In fairness to management, it should be noted that they saw the arrival of electronic games as inevitable, as did their competitors. At the New York Toy Fair of 1976, three electronic games were displayed by competitors, which created a certain amount of consternation among some senior executives.

So the firm dipped its toe into the digital waters and began developing an electronic game in the late 1970s. Initial forays into the electronic games market proved successful, especially with its early game *Merlin*, a spin-off from the *Star Wars* movie. Like other game manufacturers, however, sales declined in 1982 and 1983 for their electronic games. Some executives saw the decline in sales as another fad ending, perhaps not realizing that a shift in the types of games people would play was just beginning. The first step was the move away from electronic arcade-like games to video games, the great story of the 1980s and 1990s. The memoirs of Philip E. Orbanes, an executive at the firm at the time, reprinted a conversation with the company's president, Randolph Barton, illustrating how so many companies reacted when a fundamentally new technology threatened their industry, in this case the video game:

> While believing that video games was indeed a promising field, he felt that Parker Brothers should enter it by creating a system that would play board games, perhaps an upward-pointing TV console built into a coffee table's surface. The cost of developing such an idea, at a time of belt-tightening, overwhelmed Barton. He came back to me and asked how much it would cost to decode the Atari system.[14]

Despite the expense, Barton approved the project. It was an excellent example of transition concerns, adoption of a reverse engineering strategy, and the anxiety over changing markets and patterns of behavior. The company courted arcade game makers in the early 1980s and worried about Japanese video game manufacturers in the late 1980s. In time, and after spending vast sums of money—far beyond what the company had historically had to in order to create paper-based board games—the firm acquired expertise in electronics and video game design. After the video game crash, Parker Brothers reverted to making board games.

Orbanes recorded that "henceforth, all new product effort would be concentrated on traditional games, some of which might have electronic chips inside, but which were mainly to be made of good old-fashioned cardboard, printing, and plastics."[15] In the 1990s Parker Brothers was taken over by another toy and game company, Hasbro Games, bringing to an end the firm that had for over 100 years sold board games; this story illustrates how a firm can react to new technologies, even reject the modern, and prove that the "old" could coexist with the "new" technologies and succeed.

The history of this industry consists of various individuals introducing one game after another.[16] It is a history of hits and failures. The first widely popular video game was *Pong*, developed by Nolan Bushnell in the early 1970s and marketed by his company, Atari. Bushnell sold his company to Warner Communications in 1976 for approximately $28 million, which was roughly a typical year's worth of revenue for his firm.[17] *Pong* launched the modern era of gaming, operating first as an arcade game in bars, and later in conjunction with home TVs. In 1977, Atari came out with the Atari Video Computer System (VCS, later renamed the 2600), which signaled a second generation of games.[18] In the mid-1980s, two Japanese firms Nintendo and Sega, entered the American market, and in time dominated the global video game business.[19] Consumers, meanwhile, became attracted to home versions of arcade games that resided initially on consoles and later on PCs.

Rather than review the arrival of so many thousands of games, which would be like writing a history of major songs or movies, Table 12.1 lists a few of the great games of the late twentieth and early twenty-first centuries. As innovations in technologies occurred, new games did too. Table 12.2 lists major genres of games, much as the Book Publishing Industry has categories of books (novels,

Table 12.1
Popular Video Games in the U.S., 1961–2004

Year	Game	Year	Game
1961	*Spacewar*	1987	*Legend of Zelda*
1966	*Periscope*	1990	*Super Mario Bros. 3*
1972	*Pong*	1993	*Doom, The 7th Guest*
1976	*Telstar, Death Race*	1994	*Donkey Kong Country*
1978	*Football, Space Invaders*	1996	*Super Mario 64, Quake*
1979	*Lunar Landing*	1998	*Pokemon*
1980	*Pac-Man*	2001	*Matrix, Grand Theft Auto, Mario Kart Advance, Super Mario Advance, Madden 2002*
1981	*Donkey Kong*		
1982	*Ms. Pac-Man*	2003	*The Sims Makin' Magic* (expansion pack)
1983	*Dragon's Lair*	2004	*Doom 3*

Sources: Rusel Demaria and Johnny I. Wilson, *High Score! The Illustrated History of Electronic Games*, 2nd ed. (New York: McGraw-Hill/Osborne, 2004); Steven L. Kent, *The Ultimate History of Video Games* (Roseville, Calif.: Prima, 2001): xi–xvi; various Amazon.com listings of best-sellers, http://www.amazon.com. Amazon had 14,302 games for sale during the summer of 2004.

Table 12.2
Major Video Game Genres

Genre	Examples
Action	*Pong, Mario Bros.*
Fighting	*Street Fighters, Legend of Zelda*
Strategy and war	*Command & Conquer*
Sports	*Tony Hawk, PGA Tour Golf*
Vehicle simulators	*F-16 Falcon, Flight Simulator*
Construction and management simulations	*Roller Coaster Tycoon*
Online role-playing games	*Drakkar, Empire Builder*
Puzzles and software toys	*Mind Rover, Breakout*

Sources: Ernest Adams, *Break into the Game Industry: How to Get a Job Making Video Games* (New York: McGraw-Hill/Osborne, 2003): 42–47; Martin Campbell-Kelly, *From Airline Reservations to Sonic the Hedgehog: A History of the Software Industry* (Cambridge, Mass.: MIT Press, 2003): 282; J. C. Herz, *Joystick Nation: How Videogames Ate Our Quarters, Won Our Hearts, and Rewired Our Minds* (Boston, Mass.: Little, Brown and Company, 1997): 24–31.

biographies) and the Movie Industry's has categories of movies (comedies, westerns). As one would expect, by the late 1980s, the industry was competitive, with companies timing introductions of new games either to trump a rival's anticipated product or to optimize for the Christmas season, which was when most products were sold.[20] To a very large extent, pressure to introduce new games on a continuous basis reflected the fact that the shelf life of a video game was often less than a year, much as with songs from the Recorded Music Industry. As with the latter, the former needed to introduce new titles and hope that a few of them would become best sellers.[21] Furthermore, by the 1990s, development of games had moved from the lone programmer or two writing games in the 1970s to large staffs—replete with producers, directors, writers, graphics experts, sound producers, publicists, marketing, distributors—working on myriad aspects of games, leading to the adoption of the same approach used by the Movie Industry to create its products, known as the Hollywood Model.[22]

As games started being played mostly on PCs, or on dedicated consoles, two patterns became evident. First, selling consoles mimicked the marketing of razors: one did it inexpensively in order to sell the blades, or in this case, additional games that could be played on either device. Various technologies were introduced, such as cartridges to hold games (1970s–1980s) and one that plugged into game consoles, making it possible to play various games on one machine (but one at a time). Consoles were designed to attach to TVs (1970s–1990s), with the latter providing the screen needed for games. Over time, digital storage was added to hold games, creating in the process a whole new class of hardware that emerged at the same time as PCs but were not PCs *per se*. The significance of this development is hard to overstate, because consoles were some of the earliest digital, or computer-like, devices designed for specific purposes that relied extensively on digital technologies much as occurred with early word processors. Until then all digitally based devices were computers, such as mainframes, minis, and PCs. Now we had the emergence of new uses of the digital, applications that in time, for example, became the technological bedrock of today's consumer electronics, devices that accomplished specific things by using computing and telecommunications technologies. As with razor blades, profits were in the games, not the consoles or PCs. So selling one's platform was essential to a firm's success by the late 1980s. For companies that just wrote software, picking the right platform (technology base, such as an operating system) for its products was an equally important strategic decision as which game to produce.

A second pattern concerned the entrance of traditional computer companies into this industry, such as IBM in the early 2000s. The founders of the future Apple Computer became interested in video games in the 1970s as it became evident to them that this represented an expanding new market. Both Steven Jobs and Stephen Wozniak created video games; Jobs even worked for Atari. But the two men established Apple as a computer company, not as a video game provider. Atari collapsed as a result of the video games crash of 1982–1983 discussed later, but Apple's path did not depend on games for revenues, instead offering a variety of business, home, educational, and graphics applications. When the firm developed the Apple II, however, the machine had the ability to accept various input devices to make it possible to play games on it. In fact, the machine came with a game, *Little Breakout*. Future machines also had games, and over the years the company worked on and off on consoles and, of course, games themselves, introducing a variety of technical innovations.[23]

Perhaps more important over time was Microsoft's role because it had more customers than Apple ever did. While at the end of the century the giants of the gaming industry were consumer products companies, such as Sony with its highly popular PlayStation 2, Microsoft's prior experience at the time was limited to *Flight Simulator* (actually not developed at Microsoft) and the series of games called *Age of Empires* (also developed outside the firm), both played using PCs. Executives at Microsoft increasingly became concerned when Sega and Sony began exploring ways to deliver their products over the Internet and for ways that players could access games through their products because Microsoft saw that development as a direct threat to its market base. As one industry observer explained, "If video consoles became a primary portal for accessing the Internet, they could undermine the foundation of Microsoft's empire. This possibility was strengthened by the fact that nearly 25 percent of PC households surveyed indicated that the primary use of the PC was for playing games."[24] In November 2001, Microsoft introduced its own console, the Xbox, with a raft of games that could only be played on it, using its existing software distribution channels to sell the console and games.[25] By the new century, rivalries had extended beyond "best games" to include platforms, with Sony PlayStation 2 (introduced in March 2000), Microsoft's Xbox (November 2001) and Nintendo's GameCube (November 2001) providing the key players. PlayStation had over a thousand games that could be played on its platform by the end of 2002, Microsoft 215 games, and Nintendo 306 games.[26]

This evolution toward standard—indeed dominant—platforms and products mimicked exactly what happened in so many other high-tech industries. In PCs, there are IBM and Apple standards; in PC software, Microsoft is the most widely used standard/platform for operating systems, with over 90 percent of all personal computers using its products; IBM won the mainframe standards battle in the 1960s, never relinquishing that position. Battles over Beta and VHS standards for video are the stuff of legends; the same occurred with bar codes and DVDs, among others. In fact, we can take it as a sign of technological maturity when an industry devolves to product and marketing strategies dependent on widely accepted platforms and technical standards. By the early 2000s the Video Games Industry had yet to fit this pattern of behavior, although increasingly the number of different platforms had shrunk, but not yet reaching the level of conformity to platforms as we saw with Windows, IBM and the compatible PCs, and Apples providing de facto technical standards, often as a result of effective marketing that crowded out alternatives.

The Video Game Industry also acquired more definition. In the 1970s, there were thousands of start-up firms around the world and, as the industry formed and matured, the number of surviving firms that developed and published products shrank to a few, not unlike the pattern evident in the Software Industry.[27] The weeding out process occurred just as rapidly. By 1983, substantive firms existed, each generating millions of dollars in revenues every year. The industry had acquired some of the "look-and-feel" elements of the Recorded Music Industry just before its conversion into a global media industry. Key players included Adventure International, Infocom, Broderbund, Datamost, Sierra Online, and Synapse. These firms faded into history, replaced by a wave of new entrants from entirely different backgrounds in the second half of the 1980s, mostly from Japan (Nintendo, Sega, Sony). The conventionally held view about key historical periods in the industry labels the 1960s and 1970s as the birth of the industry. This was followed by a major decline in business in 1983, when companies and magazines in the industry folded, largely the result of too many low-quality games having saturated the market in 1982, turning off customers. One study of the period estimated that of the 135 significant firms in the industry, a mere 5 to 6 survived the crash.[28] This second period was then followed by a third, post-crash era, characterized by global growth and prosperity.

The Japanese brought a new generation of technology, games, and ways of doing business. These included better use of licensing and outsourcing techniques, already being honed by the Software Industry, and adoption of marketing and distribution methods widely deployed by the Consumer Electronics Industry. The largest American firms in the late 1980s were Atari, Coleco, and Intellivision; all three were severely pummeled by Japanese firms that produced outstanding games, priced well, and marketed effectively in the 1990s. Martin Campbell-Kelly, the leading authority on the history of the Software Industry, explained the success of the Japanese in the late 1980s this way:

> It is fair to say that Nintendo was solely responsible for the renaissance of the worldwide videogame industry. However, this had less to do with Mario Bros. as a "killer app" than with Nintendo's control of the software market for the NES. When the console was being designed (around 1983, at the height of the videogame crash), it was recognized that it was necessary to control the supply of third-party software. This was achieved by incorporating inside each game cartridge a chip that acted like a key, unlocking the NES's circuits. Only Nintendo-manufactured cartridges had the chip.[29]

The key idea to draw from this description was the role of control over the platform and other intellectual properties that went into the making of games. It proved crucial. Nintendo won market dominance, providing over 80 percent of the consoles in use by the mid-1990s, analogous to what happened with Microsoft's operating systems in the PC market. Plus, it took Nintendo fewer years to achieve its dominance than it did for the PC industry to standardize on IBM compatibility with Microsoft in control of key software products.

By the late 1980s, one could see the establishment of a set pattern of behavior in the industry that has extended to the present. (While one could present several chapters describing these, others already have.[30]) First, over time this industry increasingly viewed itself as an entertainment industry, not as part of the world of IT; it just happened to be an extensive user of digital technology which it deployed to carry out its mission. Second, because nearly half its sales came during the Christmas shopping season in most industrialized countries, it was (and is) very much attuned to what retail and other consumer product companies worry about and how they market their goods. Third, because its products are based on engineering and other technical issues, complex problems have to be solved that might not interrupt a book writer, for example—such as how to address issues in creating innovative graphics, or limitations of hardware. Fourth, pricing a product to be competitive is also a sensitive issue:

> It's about the same price as a trip to the movies for the whole family, if you buy them all popcorn and soft drinks as well. It's cheaper than taking them to a baseball game, and much cheaper than taking them to a football game. On the other hand, it's much more expensive than watching broadcast TV, which is free at the point of delivery (the TV set).
>
> Video games normally give about 20–40 hours' worth of entertainment—some many more than that—which means that the cost of the entertainment is $1–2 an hour. This is a pretty good rate of return, given that the movie or the baseball game will be over in two or three hours but you can go on playing the video game for weeks if you want to, and your roommates and friends can play it, too.[31]

The industry sold its games, although stores emerged around the country renting them as well. Increasingly in the late 1990s and early 2000s, players bought games directly from publishers, increasing the profit margins for the publisher, although distribution remained important. The example of Electronic Arts (EA), a major producer since the 1980s, illustrates the process. EA produced games (such as, *Archon*, *Pinball Construction Set*), then launched various distribution initiatives in the late 1980s by forming distribution alliances, though it stayed out of the business of manufacturing consoles or running its own retail outlets. In the next decade, the firm acquired other publishers of games and licensed games others had produced, such as those with leading movie brands (*Harry Potter*, *Lord of the Rings*, *James Bond*). As of 2004, the firm remained the largest independent game publisher in the United States. Another trend of the time was subscription-based games available over the Internet.

By the early 2000s, the industry competed more over platforms than over who had the best-selling games on the market. The rivalries over consoles mimicked what occurred routinely in the software and computer hardware markets. For example, during the Christmas season of 2002, the great battles were between Sony with its PlayStation 2 and Microsoft and the Xbox. He who had the most widely embraced platform had the largest opportunity to sell games that only ran on that console. But there were other side battles and opportunities that spilled out. For example, magazines devoted to games appeared, such as *Electronic Gaming Monthly*, and also for particular platforms too, while book publishers introduced the "official" guides to various platforms and games. Ziff Davis Media and Future Network were key players in the publishing arena.[32] The Book Publishing Industry had long debated whether to sell games in book stores, a dialogue that dated to the late 1980s.[33] As games became increasingly more attractive to the American public at large, bookstores had to worry about whether to stock games that played on one platform or another. Amazon.com became one of the first major online booksellers to do so.

As the amount of memory and speed increased in consoles and PCs, so did graphics, such as 3-D graphics. Playing games interactively on the Internet also became a rapidly expanding pastime, with some games achieving the status of international best sellers, so too game products such as Rockstar Games' newest versions of *Grand Theft Auto* (introduced in 2001) which generated hundreds of millions of dollars in revenues.

The crucial new development was the emergence of online gaming. Sony, for example, provided tools to game writers over the Internet; then these people made games available over the same medium. In 2003, Sony and others began experimenting with grid computing, which allowed unused computer power in multiple mainframes and PCs all over the world to be used to power games.[34] Online gaming remains a story just unfolding as this book went to press.

Business events in this industry reflected those in other high-tech industries: the rise and fall of firms, the growing and shrinking market shares, and the effects of continuous innovations. Some of the movement was quite dramatic. For instance, Nintendo had placed its consoles in one-third of all American homes by

1992. A decade later, its market share of the games business had shrunk from 90 to 15 percent. In the early 2000s, half its profits came from its handheld GameBoy, a decade earlier from console-based video games. Yet at the same time, Americans kept buying games; in fact, half the homes in the country had at least one video game console by the end of 2002. The global market of $27 billion had doubled in just seven years from less than $14 billion, with Sony the big winner. Many reasons for the dramatic changes in fortunes have been offered up, ranging from Nintendo not watching the details of its business, to shifts in customer demographics (in 1990 two-thirds of all players were younger than 18, in 2003 the average player was 29 years old), to better consoles and games. We have insufficient evidence to understand definitively what has happened; we can reasonably surmise, however, that poor execution of strategy and insufficient speed may have accounted for the problems faced by Nintendo. By 2003, both Sony and Microsoft were transforming their products to fit into all-in-one home entertainment systems and to incorporate music, TV, and other functions, thereby expanding their economic opportunities. The industry still relied on sales of GameCube (15 percent market share in 2002), Microsoft's Xbox (20 percent) and Sony's PlayStation 2 (65 percent).[35]

By 2003, it was becoming clear that traditional computing firms were becoming part of the games mix, from Microsoft to IBM. As games moved to the Internet it was almost inevitable that these firms would too since online gaming required substantial amounts of high-performance, high-capacity, computationally intensive systems. When a new game was published, it was not uncommon for tens of thousands of people to access it within minutes of its availability, even hundreds of thousands within a week. That required far more sophisticated computing and telecommunications than a humble little console cartridge could provide. Gaming was becoming another form of e-business. Companies in the computer games industry still worried about pirating and how best to protect their patents and copyrights, which seemed to take on a greater urgency as the industry's total supply of games increased over time, providing a larger target of things to steal. Gaming companies joined others such as software and music in attempting to enforce the law, but also looked to the digital hand for help.[36] In July 2004, for example, IBM and seven other firms announced they would establish a consortium to create standards and technologies to serve the dual purpose of enhancing digital entertainment and protecting copyrighted material, buttressed by the passage of the Digital Millenneum Copyright Act of 1998 (DMCA). The consortium, called the Advanced Access Content System Licensing Administrator (AACSLA), brought together major players from three industries: entertainment, consumer electronics, and computing and software; its members included IBM, Intel, Microsoft, Panasonic, Sony, Toshiba, Walt Disney Company, and Warner Brothers. The consortium outlined its intent and the trend:

> This cross-industry consortium represents the first time an effort of this magnitude has been made across the entire "value chain" of digital content. Each group is playing an integral role in the creation, management and distribution

> of next-generation content and has a mutual interest in the vitality of the content and its delivery platform.[37]

The first step involved developing technical standards for technology and licensing of content, a process underway as this book was published in 2006.

By 2004, the landscape had become quite cluttered with offerings. There were thousands of games that ran on consoles, on PCs, and over the Internet. Consoles dominated, however, and were even becoming networked. One marketing description of Microsoft's Xbox illustrates what was happening with products in the industry:

> The Xbox offers a truly jaw-dropping experience. Its Darth Vader-like housing is supercharged with features such as an Ethernet port (for LAN parties and broadband Internet games), an 8 GB hard drive (for saved games, downloaded enhancements, and characters), and four controller ports (for accessories such as a microphone headset or additional game controllers). The optional DVD kit even lets you use your Xbox as a DVD player, making it a complete entertainment system.[38]

The industry had experienced a turbulent evolution, but not one atypical of all emerging industries based on a collection of technologies that were still evolving and had not yet settled down to common standards, platforms, and configurations. Behind the science and the entrepreneurial spirit that drove so much that happened in this industry was the revenue side of the equation, what economists so clinically describe as the "supply side." By moving from the "demand side" of the story to deployment we get a better understanding of the dynamic and fast growing importance of this industry to American life.

Extent of Deployment of the Video Game Industry

The extent of this industry's entrance and role in the U.S. economy and in the lives of Americans can be gauged by a review of its revenues and an estimation of how many people play with its products. Before tracing the evolution of growth and deployment, a few recent numbers suggest the scope. In 2002, this industry hovered at $10.3 billion in the United States; that year, Americans spent $9.5 billion at movie theaters. How many new games were released each year? In 1995, publishers introduced 550 games; in 2002 the number had climbed to 850. On a global basis, Sony had been the star in the industry since 1995. Its video games business generated $36 billion in revenues between 1995 and the end of 2002, while in the same period Nintendo accumulated $32 billion in revenues. It was not uncommon for successful games to generate anywhere from $7 million in revenues to nearly $100 million.[39] In short, this industry had become big business.

It was not obvious to the early developers of games in the 1970s that their playful activities would end up forming an industry in its own right. Early accounts had "gamers" being universally surprised at the popularity of their creations installed on university computers.[40] As we know, it all seemed to begin with a little game called *Pong*:

> If one were to list American cultural phenomena of the 1970s, alongside CB radio, hot pants, and Charlie's Angels one would have to include the video-game Pong, a game of electronic table tennis that briefly mesmerized America and much of the developed world. It is not an overstatement to say that Pong, produced by Atari, was the springboard for today's vast computer entertainment industry.[41]

In its first full year of sales (1972), it brought in over $1 million, a massive amount of money given the fact that there was no established Video Games Industry. Students of the industry agree that the world of gaming changed with the arrival of *Pong*.

With the arrival of Atari's Video Game Computer System (VCS) in 1977, one could begin to think of a second generation of video games now reaching the market. By 1983, Atari had sold over $5 billion of systems worldwide, $3 billion worth in the U.S. market that year. When one adds in various competitors, we can say there was a real, indeed new, industry operating in the American economy. Campbell-Kelly has estimated that in 1981 some 4.6 million video game consoles were introduced into the American economy; in the following year, the number grew to 7.95 million units, and the following year saw 5.7 million—a total of over 18 million units in three years. These devices generated $577 million in revenue in 1981, $1.3 billion in 1982, and $540 million in 1983. Shipments of game cartridges also were substantial: 34.5 million in 1981, another 60 million the following year, and an additional 75 million in 1983. People were acquiring multiple games that they could play on the same consoles. By combining sales of consoles and cartridges, we see that in 1981 the industry generated $1.38 billion in the United States, an additional $2.8 billion the following year, and $1.9 billion in 1983, turning in a massive growth rate from 1981 through 1982, and healthy revenue streams the following year.[42] The last year—1983—is the all important year the industry refers to as the crash, when sales declined precipitously. Sales data for the period remain imprecise; however, we know the number of consoles and cartridges (games) shipped dropped and just as alarming, the cost per console did too, from $166 on average to roughly $95.

Business turned around after 1983, however, and the volumes reflect what happened. Over the course of the next 15 years, sales of video games doubled and, in fact, came to represent a third of all sales by the Toy Industry in the United States.[43] Economist Melissa A. Schilling has clearly demonstrated that growth in sales represented a variety of successfully implemented strategies, combining the introduction of a large stream of new technologies and products with skillful deployment of standards and lockout strategies all timed to optimize the ever-important shopping seasons, product introductions relative to those of competitors, and so forth. She has also collected data on volumes, pointing out, for example, that in the late 1980s and early 1990s, new consoles were acquired by the hundreds of thousands in the United States; 600,000 were sold by Sega in 1989, while NEC added 200,000 of its own into the market. Some of the newer consoles remained expensive, as high as $700 each, although more typically they cost anywhere from under $130 to $200, with the latter closer to the norm. Cartridges were usually under $50 each. Each new generation of technology led to another round of increased sales in the U.S. economy—a process she has tracked carefully. For example, after Sega introduced a new generation of consoles in 1991, its shipments into the United States went from about 1 million per year to 5.5 million in 1991 and 1992, only declining in 1995 as it achieved its maximum market penetration for this generation of products.[44] And in that year, a new generation of consoles came into the market (32-bit systems), such as Sony's PlayStation. By the end of the following year, Americans had acquired 2.9 million of them.

Because the Internet did not become a major factor in this industry until the late 1990s or early 2000s, the prior patterns of product development, introduction, and sales (evident in the early 1990s) continued deep into the decade. U.S. government economists began tracking industry activities and reporting various statistics beginning with 1996. Table 12.3 offers details on consumer spending on video games from 1996 through 2001. To illustrate the relative extent of these expenditures, the table includes data on several other types of entertainment that had become digital during the same period, such as CD recordings of music. Several patterns become obvious. First, the nation's appetite for such products proved extensive and grew faster than inflation, (less than 5 percent/year). In part that can be accounted for by the appeal of the products themselves, including the quality of the technologies on which they were based, but also by a thriving economy making these discretionary expenses affordable. The second pattern is the growth rate in expenditures for video games, which is one of the fastest listed in the table; only the Internet comes even close to that kind of expansion. In effect, Americans doubled their per capita annual expenditures on games in the period. Even allowing for price fluctuations, it is an impressive performance.

Another way to look at the data is to ask: How many hours did the so-called

Table 12.3
Consumer Spending Per Person for Video Games and Other Forms of Digitally Based Entertainment, 1996–2001
(U.S. dollars)

Type	1996	1997	1998	1999	2000	2001
Video games	11.47	16.45	18.49	24.45	24.85	27.96
Home video	85.98	92.38	92.58	97.33	102.46	109.60
Consumer Internet services	13.24	20.87	27.63	41.77	50.63	62.08
Recorded music	57.47	55.51	61.67	65.13	62.80	60.57

Source: U.S. Census Bureau, *Statistical Abstracts of the United States*, various tables, usually titled "Media Usage and Consumer Spending."

Table 12.4
Hours Per Person Per Year Spent Playing Video Games and Using Other Forms of Digitally Based Entertainment, 1996–2001

Type	1996	1997	1998	1999	2000	2001
Video games	25	36	43	61	75	78
Home video	54	53	36	39	46	56
Consumer Internet services	10	34	54	82	106	134
Recorded music	292	270	283	290	264	238

Source: U.S. Census Bureau, *Statistical Abstracts of the United States*, various tables, usually entitled "Media Usage and Consumer Spending."

per capita person spend with various media in the same time period? In table 12.4, we see that listening to music remained the most popular activity, but we also see a direct correlation to table 12.3 in the growth in the amount of time spent playing video games and in using the Internet. One has to be cautious in accepting the data as absolute since an "average" or "per capita" American can include anyone from a child to a senior citizen; but the data is suggestive of a fundamental shift. Furthermore, since people only have a fixed amount of time they can spend using all media (theoretically up to 24 hours in any given day, but in practice less than five), as new media usage rose we should expect to see a decline in time spent with other media. People did reduce the number of hours they spent reading books, though they spent about the same amount of time reading magazines and newspapers (as in prior years), but increased the number of hours they spent watching television and going to the movies. These generalizations have to be qualified by acknowledging that there were different patterns evident by age. For example, in the early 2000s, TV viewing by children was dropping while that of men in their late 50s was increasing. But all that said, it appears that in general a shift away from paper-based media and toward digitally based forms of entertainment and content may have started. We have much anecdotal evidence to suggest that this definitely was the case with teenagers and young adults.

Our picture of deployment would not be complete without understanding the number of people who played games. Data on this community are also some of the most tenuous we have, but they clearly suggest trends. The number of players has increased steadily over the years and broadened out by age groups and gender, so the stereotypical image of a young white male high school or college gamer of the 1970s does not fit the profile of the player in the twenty-first century. In the United States in the 1960s and early 1970s there were probably several thousand hardcore "gamers"—people who wrote and played games on computers, mostly at universities and government data centers—but since there were no contempory studies on these players, this is simply an educated guess. By the end of the 1970s, that number had grown to many thousands, possibly tens of thousands, and they began meeting for gaming contests and conventions.[45] Very quickly in the 1980s, however, the numbers grew to several million players, and into several thousand game writers. Hard-core gamers never went away, but they were joined by younger and older, mostly male, game players in the early 1980s, and by women and girls by the late 1980s. Sega's products tended to appeal to girls more than Nintendo's, and GameBoys were used by children and young adults of both genders.[46]

By the early 1990s, it appears that half the players in the world were located in the United States; 20 years earlier, the proportion was probably over 90 percent. As the population of users increased, demographics changed, particularly between the early 1990s and the early 2000s. A driving factor in game development has become the hard-core PC and console players living in the United States—by 2002 their numbers were estimated at between 7 and 10 million. A hard-core player has come to mean an individual who buys at least 10 games a year; games' firms learned that these users became bored quickly and thus sought out new games to add to their existing favorites. (There are some hard-core gamers who stay with the same games for years, but this practice is the exception.). Hard-core gamers were and still are, in the language of marketing people, "the most influential demographic," so if they like games with violence and warfare, the industry produces that, rather than gentle games for children who do not buy 10 new games each year. (Nintendo has managed to be an exception to this generalization.)

When one adds in young players (under age 25), the number of players approaches 30 million. Survey data suggest that casual and adult players amount to another 25 to 30 million. In the late 1990s–early 2000s, the age of a gamer increased from 18 to 19 years of age to 28, probably representing the hard-core players of the early years. We know that 70 percent of all console players in the early years of the new century were at least 18 years old. Since the late 1990s, the market for teenage players has continued to represent a growth demographic with no signs of abating. Given the fact that a surge occurred in birth rates in the nation from 1989 through 1993, one can understand how teenagers began influencing fashion and preferences. We know also that women increasingly came to gaming and that by 2002 they comprised about 35 percent of the total number playing on consoles and 43 percent on PCs.[47] The lateness of girls and women reflects a change in whether games being sold appealed to them or not. Violent "shoot 'em up" games had always appealed less to girls and women than to boys and men, and that circumstance hardly changed after the arrival of video games. This generalization, however, may also be shifting; extant data is just not precise enough to say with certainty.[48]

As gaming added online versions in the early to mid-1990s (such as *Doom* in 1993), industry analysts and others began collecting data on its demographics. We know that the first to move to online gaming were hard-core gamers, who spent more time online than others. One survey suggested that this group spent 30 to 40 hours a month online, consisting of a community of between 2 and 3.5 million players just in the United States. But gamers played against each other around the world, so it is difficult to tell how many were in the United States. Asian players were coming into the market rapidly in the early 2000s, a trend that had begun in the mid-1990s, particularly in Korea where broadband connections were extensive. In the United States, as of 2002 less than 8 percent of all homes had broadband connections, so the market for online gaming there remained constrained. At the time, broadband was not yet a factor because there were alternative channels of delivery of high-speed playing, such as stand-alone consoles, for example, or online at Yahoo! and such. Online gaming took place simultaneously with console and PC-based gaming; and what limited evidence we have suggests that players tended to use one platform rather than many (for example, only Nintendo or only Microsoft). People could access Internet games with later version consoles although most people used PCs to join in these games via subscriptions to game sites or via free game sites. Even games no longer available for sale could be played over the Internet.[49] About half the players, and sales in the late 1990s and early 2000s, originated in the United States. Thus for 2001, for example, global sales reached $26 billion, of which $11 billion originated in the United States, and includes both hardware and software. Console sales that year hovered at $9.4 billion globally of which $4.4 billion came from the United States. There were, in addition, over $500 million in sales of handheld consoles (e.g., GameBoys), again about half the world volumes.[50] In the following year, some 60 percent of U.S. residents age 6 and older (roughly 145 million people) played with computer games; that same year over 221 million computer and video games were sold in this country. Satisfaction levels with this form of entertainment proved quite high as well, and online gaming revenues climbed to over $1 billion.

The Pew Internet and American Life Project reported that 66 percent of all teenagers in the United States downloaded games off the Internet that they played on their own devices. This is not the same as saying they were playing a game on the Internet with rivals around the world; these are two different kinds of activities. The former is more like what we see with downloading of music and movies. Pew's research also noted that 57 percent of girls played digital games, 75 percent of boys played games on the Internet. The same report called out the fact that teenagers on average spent 4 hours a week playing video games, while other researchers presented evidence that college-age men were spending over 15 hours per week doing the same.[51]

To put an industry-centric point on all this data, sales in 2002 for gaming software (the latest year for which we have comprehensive, firm U.S. numbers) rose by 21 percent over the prior year, not including the sale of hardware—a huge surge by any standard. While the sale of consoles and other hardware grew more slowly, leading to a total aggregate increase in industry sales of 7 percent year over year (largely caused by a weak U.S. economy at the time), nonetheless the industry was still expanding across many demographic groups in American society and the number of hours spent gaming by hardcore and moderate users was still rising.[52] Furthermore, for nearly three decades, the United States remained the largest single market for such products, even though by the end of the 1980s the industry was global, with major suppliers located in the United States, Europe, and Asia.

We might ask, what did American gamers play in the early years of the new century? Action games accounted for 25 percent of all sales of software products; sports another 19.5 percent, racing games 16.6 percent, and general entertainment software 7.6 percent.[53] The ambiguous "other" amounted to 31.2 percent, comprising many of the types of games listed in table 12.2. "Other" continued growing in volume and use as the demographics of players expanded to older players and more numbers of girls and women.

What do all these data on trends and demographics suggest? First, as all the historians and observers of digital gaming have stated, is that these games are fun and exciting. On a more scientific note, they stimulate and engage various cognitive functions, require intense focus, and appeal to the competitive nature of many players. Gamers are tuned in to "market forces" and to the possibilities presented by technological innovations. For example, we are seeing the ability of gamers on online games to change plot lines and even to impose their own likenesses on the characters. Second, more than half the population under the age of 30, and over the age of 6, has played video games at one time or another. The amount of time people are reporting playing video games is sufficiently high to conclude that this is a new and important addition to the nation's entertainment repertoire. To be sure, parent groups have bemoaned the violence in action games, but so too did people in earlier decades protest sex in novels, calling for the banning of books.[54]

Video games have integrated prior bodies of experiences and pleasures in American society into games. This is actually quite a broad activity, ranging from games tied to movies,[55] popular books, all major sports, even professions like space travel and military careers. In fact, one of the most popular games of the early 2000s was the U.S. Army's official recruiting video game; indeed, all the armed services now use video games for training.[56] Most commentary about video games places them in the context of the "New Media" category; the historical record demonstrates, however, that the industry was also the child of the Information Age, spawned by its technology and nurtured by a population increasingly comfortable with all manner of digital tools and toys.

Photographic Industry

Photography has existed for over 160 years; for most of its existence, the industry's products consisted of cameras and their related paraphernalia, film, and film development and printing supplies. These have been sold through retail outlets, such as specialized photography stores and department and specialty stores. Photography stores, drugstores, and other retail outlets have also provided film-processing services. Professional photographers and newspaper, magazine, and electronic media employees make their livings taking photographs. The lion's share of the business, however, has rested with the amateur photographer, with a second, smaller market comprising specialized services, such as medical imaging. In the United States for most of the twentieth century, the Eastman Kodak Company has controlled 80 percent or more of the film market; it has also been a player in both the camera and film-development market. Its key competitors for film have consisted of a few yet varied rivals, but for the most part it has been challenged by the Japanese firm Fujitsu from the mid-1960s on (which took more than 10 percent of market share away from Kodak) and the European manufacturer Agfa, which holds a smaller share. Competition over various types of cameras has existed since the early 1900s, with European rivals dominating prior to the 1960s and Japanese firms, such as Canon and Nikon, afterwards.[57]

The industry had remained both profitable and essentially stable for decades until the 1960s, when Japanese competition began making serious inroads into Kodak's business. By looking at film sales, we can establish that interest in photography kept growing. In 1983, for example, 594 million rolls were sold in the United States, climbing to over a billion roles in 1999, but subsequently sliding over the next half dozen years as digital photography claimed market share from the traditional business. In that same period, sales of cameras by all vendors also did well, with over 18 million sold in 1983 and between 14 and 18 million sold annually over the next 15 years.[58] Photographic equipment of all types also did well. Taking a broader historical view, from the 1950s to the end of the century, one can conclude that business had been good. By 1970, total annual sales were above $4.4 billion; sales doubled by the end of 1976, and doubled yet again by 1982.[59] While many new products were introduced over the half century, and continuous improvements made to the capabilities of lenses and film, the base technologies remained essentially the same until the early 1990s. Film was based on a large body of chemical processes and knowledge, cameras on mechanical and electrical mechanisms that used film.

Then the early effects of the digital began to be felt in the industry, resulting in a massive transformation in the base technologies and subsequently in who became the key suppliers of new products. Nowhere is this more evident than in the sale of digital cameras in the United States. The first digital products for the amateur market appeared in 1996; prior to that, high-end, expensive, specialized digital cameras represented a tiny, indeed negligible market. The general consumer embraced these new cameras rather quickly, particularly as they obtained better quality, and were easier to use and less expensive. Within 18 months, sales had exceeded 1 million units, reaching 2 million in 1999, and became 4 million in 2000. That last year represented the historic high-water mark for traditional film-based camera sales in the United States; thereafter, sales began a continuous decline, while those for digital cameras increased.[60] U.S. government economists began tracking the volume of sales in 1996, reporting $177 million, climbing to $483 million the following year, then to $524 million in 1998, reaching $1.2 billion in 1999.[61] In short, in the early years of the new century technology, sales revenues grew fivefold in four years.

Kodak's experience during this period tells a revealing story. This firm responded slowly to the industry's transformation and to other competitive pressures (for example, on both traditional cameras and film). Its story is an account of the decline in size and success of a senior American corporation. Senior management at Kodak continued striving to enter the digital market with bridge products that could take it from older base technologies to the new, though it stumbled in its introduction of the Photo CD product. Copy print machines, for example, represented an initiative, as did scanning equipment later developed for use by the medical profession and, as late as 1995, a relaunch of Photo CD. So it was not an issue of the company not wanting to try nor a lack of understanding of what was happening across the industry. What is crucial to understand is that Kodak and all its rivals understood perfectly well that a profound and rapid transformation was taking place in what their industry sold and by whom. We can debate the rate of execution of new strategies, and lament the internal bickering and faulty execution; but others have started to dissect this company's activities and those of its rivals.[62] What the tale below reveals, however, is that changes are as dramatic and complex as the move to the digital and are anything but easy or quick to accomplish.

Early Uses of Digital Technologies in the Photographic Industry, 1950s–1980s

The origins of the modern digital camera can be traced back to 1951, when television video images were recorded by capturing live images from TV cameras and converting them into digital files. These videotape recorders (VTRs) became widely available in the Television Industry after 1956. In the 1960s, the National Aeronautics and Space Administration (NASA) embraced digital photography over analog methods for use with its space probes mapping the surface of the moon. Scientists working for NASA took these digital files and enhanced the images with new software tools. Meanwhile, the U.S. government added capabilities to digital photography in use by spy satellites, and spread its overall understanding of digital imaging into other industries, such as computing and defense-related companies. Texas Instruments, a leading manufacturer of computer chips, introduced the first digital camera in 1972; soon after, Sony brought out a digital video camera that took video freeze-frame pictures. Kodak next entered the market with digital products. The breakthrough to a mass commercial market, however, did not begin until the mid-1980s and into the amateur photographic market not until the 1990s.[63]

Before moving to that story, we should recognize several patterns at work. First, digital cameras emerged over a long period of time, as occurred with so many other digitally based products. In this case, its evolution took over a quarter of a century just to reach highly specialized, barely post-experimental products, and then another decade before one could consider them consumer electronics. Second, space programs and Cold War applications pushed the engineering forward, with government agencies funding much of the basic development, which later spilled over into the consumer market in the form of high-end products.

For many decades, the Photographic Industry also lived in a world of transition, and nowhere was this more evident than with the camera itself. Beginning in the late 1970s, and extending rapidly in the 1980s, all major camera manufacturers put electronics into their cameras. They still used film, but they also started to use microprocessors to make their products easier to use (giving them, for example, point-and-shoot capabilities, light controls). In the process, companies like Kodak, Canon, and others, acquired considerable expertise in the use of electronics. By the end of the 1980s, it was almost impossible to find a camera that did not have some complex electronic components. The one big exception was

Polaroid's product line and later disposable cameras.[64] It was not until the mid-1980s, however, that the industry made the leap to digital cameras aimed at the professional.[65] Often considered the first camera to address this market, Canon's RC-701 became available in 1986 at a cost of about $3,000, making it clearly a product aimed at professional users, such as newspaper photographers who needed to transmit quickly images back to a newspaper.[66] Video cameras continued to be a source of innovation; for example, in 1987 Kodak introduced a still video system that included seven products for storing, manipulating, transmitting, and printing electronic still images.[67]

An easy way to follow the trail of the digital into the American industry is through the experience of its largest member, Kodak. In the 1950s, this firm in many ways reflected the same good and bad practices as other large, high-tech, science-based manufacturing companies of the day. It embraced the use of computing for similar reasons: to control inventory, to reduce operating budgets for data collection and accounting, to facilitate sales support, and for product R&D. Its first computer was an IBM 705, which came into the firm in early 1956 as a pilot project, "to work experimentally in order to determine its potentialities and whether it can be of economic usage in the Company." Like most early users in large corporations, "this long-range research project" was to test the system to see if it could do "sales analysis, scientific calculations, and" deliver "data previously not readily available."[68] In May 1956 Kodak installed a second IBM 705 computer system, initially to do centralized billing for its sales organization, but with the intent of adding many other accounting applications. Both systems were large by the standards of the day.[69] Billing alone justified the systems because there were seven sales districts in the United States, each intended to transmit daily sales information to the computers via IBM transceivers. Computer systems could speed up the timeliness of the work, reduce the number of clerks required to accomplish it, and provide management with more current information.[70] Over the next half-dozen years, additional applications were added to the system, requiring larger computers. In 1962, Kodak replaced the 705s with larger twin IBM 7080s. These new systems ran payroll and inventory-control applications. Production control and scientific work spread knowledge about computing into new corners of the firm.[71] Data-transmission equipment, using Bell System Dataphones, wide-area telephone service (WATS) lines, and IBM 1013 data transmission card terminals were added in the early 1960s, making larger numbers of employees and departments dependent on computing to do their work.[72]

Scientists and product developers began using computing in the 1950s and by the mid-1960s had a solid appreciation of the technology. Experiments in building small computers took place in addition to scientific calculations using IBM's systems.[73] Calculations regarding the behavior of light, modeling construction of buildings, study of chemical properties (such as of magnetic oxide for recording tapes), and other experiments with color film all demonstrated growing sophistication in the use of this equipment by those in the firm in the best position to understand its potentials.[74] All during the 1970s and 1980s, these technical communities continued using every new form of computing that came along.

Senior managers, and their scientists and product engineers, understood the coming of the digital. In the 1980s, in particular, they scaled back R&D on film—although they continued to introduce innovations in film products right to the end of the century—and began investing more in digital photography. Digital photography was about capturing an image that did not use film, processing chemicals, and photographic paper; rather, storing pictures in discs which could then be displayed on either a TV or computer screen and thus also printed on paper. Kodak focused its research on digital cameras, an initiative already launched by Japanese camera manufacturers.[75]

Digital Photography Comes of Age, 1990s–2005

In 1991, Kodak introduced the Photo CD, a blank compact disc to which one posted images taken from a traditional role of film, which could then be printed on traditional paper or be viewed on a TV screen or PC. It was a classic example of a product transitioning from an old to a new technology. Management initially aimed it at the commercial market where high volumes of picture-taking took place—advertising agencies, publishers, and others who needed to store and use many images. Commercial interest exceeded that of the consumer, but the product failed to meet the company's sales expectations.[76] Some in the firm resisted doing much more in this area, fearing it would hurt their end of the business.

Kodak's entry into the digital image market was ailing. Managers saw digital photography as a threat to their traditional film business. Nonetheless, Kodak continued experimenting and developing products. By the mid-1990s, digital cameras were available, and commercial users were beginning to embrace the new technology, such as insurance companies documenting accidents and other claims, and real-estate agents photographing homes and commercial properties. All camera manufacturers expected digital cameras to continue dropping in price, and in 1996 they reached a point sufficient to create a consumer market. All through the early to mid-1990s, Kodak shifted resources to the emerging new world of digital photography. In 1994, Kodak introduced an important consumer product, called Digital Print Station, which made it possible for consumers to reprint and enlarge prints using a scanner; the images could be transferred to many surfaces, from photographic paper to coffee mugs or clothing. Its first digital camera came out in 1995, but faced intense competition from 25 rivals; this segment of the market has yet to turn a profit at Kodak.[77]

Since the camera was the digital battleground of the industry, it is necessary to understand its emergence in some detail. But first, what is a digital camera? It has an LCD screen, which one looks at to frame what is to be photographed. Light going through a lens with a shutter reaches light-sensitive material. Then a digital camera senses the colors of the lights coming into it and their intensity, converts them to digital format (the binary 1s and 0s of digital data), then passes the information to a processor that adjusts the image, contrast, and so forth, to form a picture. The picture is a record, which is then compressed and finally stored in the camera's memory as a file. The more computer power and memory a camera has, the greater the clarity of the picture one can take, and the more pictures that can be stored in the camera.

The industry's press recognized that a new chapter in the history of photography was being written. In 1990, in describing Kodak's introduction of its Photo CD system, Dan Richards at *Popular Photography* framed his story as the "electronic bridge to the future."[78] By then digital cameras and products were quickly beginning to reach the market. In 1992, Kodak introduced its Digital Camera System (DCS) for use by professional photographers, whereby one could take pictures with a Nikon camera modified by Kodak, and the following year a writeable CD which MCI used to create corporate telephone bills and could be used to capture images. A year later it brought to market over 30 products to enable digital photography.[79]

The staff at *Popular Photography* provided its first assessment of digital cameras in 1991, an analysis and ranking that they subsequently did annually. In their first assessment, the staff looked at 17 electronic cameras (not yet called digital) and acknowledged that the right name had yet to be established: "some call it still video, others electronic photography, but whatever the term, there's something magical and exciting about recording pictures without film."[80] They noted, however, that film-based images were still of better quality. Those companies who provided the 17 cameras included Canon, Fuji, Konica, Nikon, Olympus, Panasonic, and Yashica, clear evidence that Kodak was in a major battle for market share in cameras. Many in the industry considered 1991 the first major year in the life of the digital camera because of the introduction of so many new cameras and picture printing equipment.

Over the next several years, the industry exposed the public to the nuts-and-bolts of digital photography.[81] As the technology improved, professional photographers adopted the new type of camera, but it took this community nearly a decade before it had broadly accepted it; its rate of adoption was paced as much by issues of quality of digital versus film as by simply becoming familiar with a new tool.[82] In the early- to mid-1990s, waves of new products from all the key players in the industry reached the market, including Kodak. By the mid-1990s, one could access sites on the Internet that included information on how to use digital photographic equipment, where one could download images, and where to find data on products; of course, Kodak had its own site (launched in 1995 as Kodak.com).[83]

Photographers debated the quality of digital versus film photography during these years, but theirs was mainly a discussion among enthusiasts and professionals, not the public at large. In one discussion, published in April 1996, *Popular Photography* complained about the exaggerations made by computer vendors and others that digital photography was as good as film images. For those who did not need high-resolution pictures, however, "digital cameras can actually provide acceptable images and be cheaper and more convenient to use."[84] Software used to move images from camera to PC, however, varied in ease-of-use and function, leading to mixed reviews of this class of products.[85] Three years latter, however, a writer in *Popular Photography* expressed amazement at how quickly digital photography had taken off; the answer was better digital images that were good enough for the public and at lower prices. The number of new film-based cameras introduced in 1999 declined, signaling that a shift was now underway from an old to a new form of photography.[86] However, the film versus digital debate did not end. While the number of U.S. households owning digital cameras doubled from 4.4 percent to 9.3 percent between 1999 and 2000, commentators still debated the issue.[87] In March 2001, *Popular Photography* carried an extensive article, complete with tables summarizing the pros and cons of digital and film-based systems. Meanwhile, vendors kept introducing and selling a range of products from the very expensive (over $3,000) to less-expensive portable cameras in the $200 to $500 range. The debate also included a discussion that had gone on for over three decades about the quality and value of digital still versus digital video photography. There were also related issues concerning printers, paper, inks, storage, and the like because these were collections of products that provided large sources of revenues and profits.[88]

What are we to make of these continued debates? While the number of households with digital cameras kept increasing, so did use of film. We are observing a technological transformation that is still very much in its early stages. Even the ways pictures are handled remain in flux. Online photo processing began at the end of the 1990s; image-editing software tools, which were quite numerous in the early years of the twenty-first century, were rapidly changing in function and costs. By 2003, cell phones with digital cameras from Motorola, Sanyo, Nokia, Samsung, Sony Ericsson had become some of the most popular consumer electronics around the world.[89]

Not only was the technology changing within the industry but so too were the demographics of its users. Professional photographers no longer were the major users of digital photography by the late 1990s; the proverbial "average" consumer now dominated the market. Early mass-market users of digital photography essentially comprised men attracted to new technologies. At the end of the 1990s, users' ages began dropping from an average of 48 in 1999 to 45 the following year, a trend that continued into the new millennium. Women increasingly began using digital cameras in the late 1990s, as well, with one survey indicating 41 percent of all users. (Women were also users of 70 percent of all film cameras.) Both genders bought digital cameras to be used in conjunction with and in addition to traditional cameras (more than two-thirds in 2000). The big spurt into digital photography by the general public may have started between 1999 and the end of 2000, when acquisitions doubled year over year. Thus, while 1.8 percent of households reported having at least one digital camera in 1999, by the end of the following year that number had more than doubled to 4.6 percent, and by the end of 2003 to 31 percent, as noted earlier. Mothers in particular were the newest customers for cameras that were relatively inexpensive and convenient to use in snapping pictures of their children, representing a major shift in the customer base for the industry.[90]

Of growing concern was what was happening with the photograph printing business, whose volumes were declining. For decades this was a cash cow. In the United States during the early 1990s—before the effects of digital photography were evident—it was not uncommon for consumers to purchase 640 million rolls of film, representing between 16 billion exposures, generating roughly $5 billion in revenues. Volumes in all three elements increased throughout the decade, reflecting the growth of the industry at large, regardless of technology. The high water mark was 2000, when the industry sold to amateur photographers 781 million rolls of film, which generated 18.5 billion images and $6.2 billion in revenues. Beginning in 2001, however, volumes of all three began declining slowly.[91] It wasn't the number of photographs being taken by people that was diminishing; rather, it was that digital was increasing. In the beginning of digital photography most people printed their own pictures or simply stored them on their PCs. To counter this, the industry began offering digital-printing services, and Kodak introduced various products to move film images to disk and equipment to print digital images inexpensively at drug stores and other specialty retailers, so that the decline in film printing revenues could be replaced with digital print sales.

An industry organization, the Photo Marketing Association International, estimated that in 2004 the number of digital images printed by the industry would hover at 5.4 billion. It projected that by 2006, half of all prints would come from digital cameras, mostly from parents photographing children and not having time to print these at home.[92] Survey data from the turn of the century suggests, however, that there were additional uses made possible by digital technology. Table 12.5 summarizes the uses of digital photography in 2001, based on a reliable industry survey. Note that while there were some uses that were long evident with film photography, some of the most extensive uses were unique to the digital (such as to e-mail pictures or to use with computers).

Another major shift only becoming evident just as this book was going to press concerned cell phone photography. While sales of digital cameras in general

Table 12.5
What U.S. Consumers Did with Digital Photography, 2001

Activity	Percent Doing This
Send photos by e-mail	77.3
Preserve memoirs	67.8
Share later with others	66.9
Pure enjoyment	63.2
Enjoy taking photographs	41.2
Use photos in a computer (hobby reasons only)	31.5
Use photos in a computer (business reasons only)	18.6
Use photos for business purposes	16.7
As gifts	16.7
Master skills of digital photography	15.0
Artistic expression	12.0

Source: Survey conducted by PMA Marketing Research Department, "Digital Makes its Debut," 2001, p. 7, http://www.pmai.org (last accessed 8/4/2004).

grew year over year in 2003 by over 60 percent around the world, the sales of cell phones expanded dramatically as well. In 2003, some 80 million cell phone cameras were sold worldwide, 6 million alone in the United States. Industry observers argued that the reasons for the relatively slow sales in the United States concerned the poor quality of the images and the lack of features available on regular cameras (such as zoom and flash). Yet the market was on track to sell around 11 million units in the United States, in part thanks to a new generation of cell phones, but also to an awareness of their existence and growing popularity. So far, most pictures are shared online, not printed. Sprint reported that in 2003 its customers only took 66 million digital pictures with their phones. Since people tend not to forget to take their cell phones wherever they go, but often do not think to bring a camera, the industry expected the number of spontaneous photographs taken to increase sharply as consumers replaced older generation cell phones with these new ones.[93]

What was going on with the Internet? To be sure, camera and film manufacturers had their websites; online retailers sold cameras, equipment, and film; and there were the usual collections of chat rooms, photography sites, and so forth. By the early years of the new century, digital photography was making possible yet a new use of the Internet: online family albums, used in tandem with blogs (personal diaries and websites). People set up family websites beginning in the 1990s, but over the next few years began populating these with essays; mothers were posting pictures of their children and maintaining online diaries, for example. Estimates of how many of these sites existed are hardly reliable, ranging from 300,000 to 3 million; the Pew Internet & American Life Projects has collected some data suggesting that parents are online more than nonparents, and that the former use it the most to communicate with relatives and friends.[94] But that is essentially all we know at the moment.

The changes in the industry described over the last few pages have been intrinsic. Yet the industry has taken a bifurcated view of the matter. For example, *Popular Photography* began covering digital matters on a regular basis in the 1990s and after 2000 divided its monthly publication into film and digital sections, though coverage of film-based photography remains extensive. Kodak has been accused of moving to the digital world too slowly. To be sure, the company experienced terrible business conditions during the 1980s and 1990s, as Japanese camera and film competitors took away market share with attractive, less expensive products. Kodak introduced many new products for the consumer market while beginning a slow process of developing offerings for commercial and health industry customers. In its annual report for 2003, published in the spring of 2004, it stated that "our strategy is firmly rooted in Kodak's core businesses: digital and traditional imaging. Continued success in both components of the business is vital to the Company's future as the merger of information and imaging technologies (infoimaging) accelerates the demand for digital imaging solutions."[95] In 2003, the firm generated over $13 billion in net sales, of which some $4 billion came from digitally based products and services, with growth driven "primarily by consumer digital cameras, printer dock products, inkjet media and motion picture print films," along with commercial uses of digital offerings.[96] Yet later on in the letter, we read also that the firm was still introducing film-based products, in this instance "breakthrough technology in silver halide film systems for entertainment and consumer imaging."[97] The headlines on Kodak in 2004 read like those one would have expected to have read five years earlier, for example, "Film company to take $1.7 billion charges as it moves from traditional film to digital imaging," and would "slash its work force by 20 percent" as "the company accelerates a painful shift away from the waning film photography market."[98] Its great rival in film sales, Fujifilm, while recognizing that the film market was now mature, also remained squarely committed to serving this market while simultaneously selling audio and video media, cameras, computer products and photofinishing services.[99]

If we look at where sales revenues came from, we can conclude that Kodak and others were responding to actual market conditions when in the late 1990s and early 2000s they did not abandon completely the promotion of film-based photography for what they had acknowledged was the future, namely digital technology. In its annual reports to the industry, the PMA (Photo Marketing Association) documented the transition over the years. Regarding sales in 2000, after cataloguing where sales came from for all cameras, it reported to its members that "despite the growth of digital camera sales, conventional camera unit sales continued the strong growth they exhibited in 1999. Conventional camera unit sales increased by 11.1 percent in 2000, to 19.7 million units."[100] For prior years, PMA's reports did not mention digital photography as a major component of the market. In short, when executives looked at their pie charts and the hard data, they concluded that while digital photography was gaining momentum, they would have to continue selling film-based products to sustain their current levels of revenues as required by shareholders and prior levels of performance.

It was a sign of the changing times, however, when in April 2004 the Dow Jones industrial average dropped Kodak from its list of components along with AT&T—both of which had been part of the core index since the 1930s, icons of Corporate America. In part, as the press reported, Kodak's performance contributed to this change, citing the growth of digital photography as having eaten away at Kodak's business.[101]

Conclusions

Both the video game and the photography industries exhibit patterns of behavior evident in many other corners of the national economy. Perhaps the most obvious is the length of time the gestation of a new technological base took. It is quite common to see 10 to 40 years pass before a new technology matures to the point where it is useful for an industry to adopt, exploit, and deploy. In the case of the video game industry, though people flirted with games on mainframe computers as far back as the 1950s, it was not until microprocessors and personal computers first became available in the 1970s that digitally based gaming took off. Furthermore, it took another decade for the new industry to acquire definition and sufficiently deploy its products to draw wide attention. This industry's experience followed similar paths to those of newly emerging industries: *ad hoc* entrepreneurial activities that evolve into companies, which in turn grow or are acquired by older firms or by other new ones; the evolution toward common technical standards (e.g., making games IBM PC compatible); battles for control of platforms; clearly defined channels of distribution; and so forth.

The Photographic Industry remained quite fragmented all through the second half of the twentieth century. In fact, this may be the most fragmented industry reviewed in this book. Economists working for the U.S. government list 35 different sets of economic activities that one could say make up this industry. These include photojournalists, photographers, photography and photofinishing services, wholesale and retail outlets, as well as component and equipment manufacturers, including specialized digital camera manufacturers. What bonds them together, however, is the use of photography as the medium for their work and, increasingly over time, digital as their base technology. Because the technology had the flexibility to be used in many ways, it made it possible for disparate parts of the fragmented industry to identify and even be linked to other clusters within it, as we saw in other media industries. Simultaneously, it also diffused into other industries, such as PC manufacturing, influencing the way television advertising and movies were made and the nature of the content created in those industries. As a consequence, although the Photography Industry has been around a very long time, this is an industry that is very loosely defined and undergoing considerable change. Indeed, when compared to the Video Games Industry, one could conclude that the latter is already in a far more stable circumstance than the former. While this may seem counterintuitive, because gamers have to produce new products all the time and rely on ever-changing technologies, their methods of writing games, what kinds of products they need to produce, and how they distribute these are relatively stable when compared to what is going on in the world of photography. Moreover, game customers are also relatively comfortable with what to do with their products and how frequently they acquire and dispose of them. In photography, we are witnessing a fairly rapid migration toward digital photography and the use of other new tools, such as blogs on the Internet. What we did not discuss in this chapter are the developments in industrial and medical-imaging systems, which are massively transforming into digitally based tools as well, but which also took time to evolve to a useful stage.

These two industries share one feature: the need to present images to customers in highly attractive forms. So far this has meant essentially a static picture. Rapidly emerging is a new, additional requirement: to provide a high-quality image or images in real time, which represents a dramatic shift from the traditional batch frame-based pictures of the past. Meeting these two requirements requires pressing digital technology to evolve toward greater capabilities to collect, process, and store data for both still and moving pictures—to do this with animation in games but to create real-time visual effects in photography. In both instances, we see consumers pressing both industries to keep innovating, and supporting that pattern by buying their products. I have been most impressed by both the number of products acquired from each by the American consumer and by how quickly that happened once the public became aware of the products' existence. These two industries have joined so many others in the world of consumer goods in using digital technologies, most particularly consumer electronics. Just as we saw the convergence of digital technologies across many media industries, so too are we seeing with games and photography that the producers have begun melding and converging with other media companies. This is so much the case that we can reasonably speculate that movies someday may also be games; that photography, games, and movies will merge into one vast sector that provides hardware, software, content, and entertainment.

If Michael J. Wolf is correct in arguing that large portions of the American economy are increasingly operating as entertainment industries, then the two industries reviewed in this chapter are becoming the face of that new world. To what extent, then, do these two industries link into the other dozen or so discussed in this book? To address that question, and to explain the overall effects of digital and telecommunications technologies on those industries, we turn to one final chapter.

13

Conclusions: Lessons Learned, Implications for Management

There is inherent in the capitalist system a tendency toward self-destruction.

—Joseph A. Schumpeter, 1942

Computers and telecommunications played important roles in the affairs of each industry discussed in this book. While each became an extensive user of an array of computing technologies, each was affected in different ways. Some embraced computing early, such as insurance and banking, others later, as did magazine publishers and the Recorded Music Industry. But once it started down the path of using IT, an industry saw dramatic change in how it conducted its work. The effects led to the transformation of industry structures—which companies dominated or were displaced—and what new threats and opportunities surfaced. In general, industries that provided services were more sensitive to the effects of technology than those in manufacturing, retailing, or public sector. Computing also affected services industries quicker, although they used this technology later than companies that manufactured or sold goods.

Most of what we know about the role of computing and communications in the American economy is based on studies done of manufacturing and retailing industries. Therefore, some of the patterns of behavior documented in this book are not yet reflected in mainstream economic or business literature concerning services industries. To a large extent we attribute that pattern to the lack of economic and industry data.[1] The two very large exceptions are banking and telecommunications, which have long been the subject of much attention and continue to be vital components of today's economy. But the relative paucity of material—

particularly related to the uses of IT—on the industries studied in media and entertainment, for instance, was striking when compared to what is available for manufacturing, transportation, and retail industries. That should be of no surprise since manufacturing and retail industries had long been major components of the economy. Only in the last three decades of the twentieth century did media and entertainment industries begin to expand their share of activity within the American economy. Despite much hyper-rhetoric regarding the "Information Age" and the "Services Economy," we are only just beginning to understand all industries in their contemporary forms.

This may seem a surprising statement to make, since the proverbial "everyone" thinks they understand what the Newspaper Industry does, or the Television Industry, and certainly the Movie Industry. But appearances and perceptions can be deceiving. For example, given the enormous fragmentation and practice of outsourcing in the Movie Industry, trying to understand who is part of that industry as opposed to being part of the Video Games Industry or the Television Industry, and subsequently how they all used digital tools, proved challenging to describe. Yet there is a growing body of literature on each of these that all but ignores the activities of others but their own, much as we encountered in the Insurance Industry where for decades life and property/casualty operated apart from each other as if they were two entirely different industries. Television studies are about TV, with hardly a nod to the role Hollywood plays today in providing content. Discussions about copyright problems in the Recorded Music Industry usually do not include analysis of similar issues experienced by producers of movies, books, or videos, although that is beginning to change. In some instances, there is the difficult problem of identifying what constitutes an industry, compounded by the fact that the rapidity of changes in digital technologies may cause further redefinition before one can arrive at a fixed answer; this was most evident in the Photography Industry.

The range of industries in the services sector of the American economy far exceeds that of manufacturing, transportation, and retail. In the first place, some industries could be viewed as manufacturing, such as telecommunications or book publishers. The Recorded Music Industry has operated stores, and one could argue that bank branches are retail operations because they share similar managerial and operational practices. However, as the half century passed, it became increasingly clear that, while distinctions between industrial and services sectors were not always crisp and clear, the industries described in this book were in the services sector. The senior members of the New Economy had also been distinguished participants in the Old Economy: banking, insurance, brokerage, book publishing, newspapers, magazines, and telecommunications. They were large players on the stage of the American economy all through the twentieth century, with no end in sight for their pivotal roles in the twenty-first. But others gained in stature both in terms of size and the interest with which consumers approached them, particularly when they were clustered together in a late twentieth-century descriptor, the media industries. That group of industries made very good sense because as time passed, they participated in each other's industries, often became the building blocks of large media corporations, and frequently integrated their products and services, thanks to the capabilities of computers and communications.

That consolidation has not occurred to the same extent with entertainment industries. Here we still have problems of definition. For example, is TV an entertainment industry or a media industry? Both descriptions make sense. What about newspapers and magazines that also appear on the Internet? Are they media or entertainment? What about the video games business, which seems to be moving toward the Movie Industry as fast as movie producers are incorporating games into their suite of entertainment products? Finally, what are we to say about the growing observation of many that most American industries are evolving into organizations that mimic the entertainment industries (such as the Hollywood Model) and that are adopting those practices and processes?

What lessons can we learn from the adoption of computing by services industries? There are several observations that can quickly be made. In all the industries looked at, the more their output was information, the more they were profoundly affected by the evolution of digital and communications technologies. Those that produced the fewest physical products had the greatest opportunity to leverage technology to lower operating costs (as we saw in insurance) and to offer new services (banking) or new nonphysical products (Internet-based newspapers) and services (Internet-based stock purchases, ATMs in banking). Even when there were physical products, these often took forms that were new to the last half century, based on digital technologies; we saw this happen with video games and digital photography. In addition, over the half century, all these industries were able to provide a greater variety of services and products as technologies expanded their capabilities. The Internet, of course, is the stunning example of a collection of technologies that allowed firms in all industries to transmit data, text, images, sound, and services.

These industries often could not exploit technologies in their earliest incarnations as effectively as did manufacturers and retailers. It is no surprise, for instance, that insurance companies became massive users of computing at the end of the 1950s and 1960s, because the essential application was data collection, sorting, and analysis using computers, capabilities that the technology made possible in those years. But until the emergence of good graphics and animation software tools in the 1980s, many media and entertainment industries could not use computers to improve the quality of their work. They did not have the armies of data-collection clerks and salespeople that existed in banking, insurance, brokerage, and even telephony. Put simply, as computing in particular, but telecommunications as well, expanded its capacity to store data, improved functions and reduced costs the more they were embraced by the media and entertainment industries. Thus computing became a major influence on most of the industries described in this book beginning in the 1980s. As these technologies improved rapidly, all these industries embraced them quickly. The emergence of a reliable Internet, rich in capacity and function by the end of the century, simply propelled these industries into yet a new style of operation that has nearly obliterated our memories of how they operated a mere two decades before. However, for most of these industries the Internet was not the first digital paradigm transformer because mainframes, online computing, and PCs had started the process of conversion to a new style of operation years earlier.

This leads us to perhaps the most important observation about the role of IT in these industries: technology affected them faster and more dramatically than in manufacturing and retailing. The reasons are not hard to discern. The first and perhaps most obvious is that the products and services generated by these industries could be made and delivered electronically. In contrast, automobiles are still made out of metal and someone still has to pump petroleum out of the ground and convert it into usable forms. A second reason lies in the fact that, unlike many "Old Economy" industries, these service industries did not have armies of workers in thousands of large buildings and factories whose productivity had to be improved in order for their employers to remain competitive. The issue in financial, telecommunications, media, and entertainment industries revolved more often around how to improve, enhance, or change services provided to the public, not just how to lower the costs of labor and raw materials. To be sure, there were exceptions. Western Electric, and later Cisco, were just as obsessed with improving manufacturing as was General Motors; but in general the motivations for using technology were more varied in services industries. Finally, we can observe that as new technical capabilities became available, companies could either expand to fairly large units to exploit it, as we saw in banking, or could move quickly to use it because they were small and specialized, as we saw with games. But the final analysis has to rest on one fundamental fact, namely, that digital technology and telecommunications were at the core of what these industries did to generate revenues and profits. At their centers were information, entertainment, and data sharing with customers, all of which could increasingly be provided in digital forms. To do that required companies to change their organizations and transform their inventories of human capital, and to do so at speeds that mimicked the transformations occurring in the technologies themselves.

Was it all worth it? It is a question that is addressed throughout this book. One can argue that on the whole the answer was yes. Each of these industries deployed digital technologies to help generate profits, and most did. The one great exception which has yet to play out fully is telecommunications. Here, exuberance over investing in new technologies, in combination with important regulatory intrusions into the affairs of this industry, led to such a large amount of overcapacity, soon-to-become rapidly ageing communications technologies in the 1990s (such as switches) and too much fiber optics that whole companies were crushed by declining profit margins and mounting debts. Financial industries, however, lowered their operating costs thanks to computing, formed larger enterprises, and sold new products. And media industries grew into a media sector with large and profitable publishers of books, magazines, and newspapers who increasingly delivered their products electronically and also operated in radio and television. The technologies of the age also made it possible for firms to flirt with other industries, as we saw telecommunications companies do with cable television, radio companies with television and movies, and now movies with video games. The one thing we can be sure of is that, as a general rule, management does not adopt a new technology unless it thinks it makes good economic sense for the company. Not a single industry reviewed resisted technologies for long, often adopting them faster than manufacturing, process, transportation, or retail industries. Frequently, the benefits were clear to a decision maker, and because many organizations were smaller they could thus make decisions faster than very large enterprises. To be sure there were exceptions, but they were essentially minor. For example, movie studios were slow to adopt accounting software, yet their animators quickly utilized leading-edge software to enhance movies.

Finally, we can ask, as the work of these industries changed as a result of the use of new technologies, did work forces also transform to support new ways of doing tasks and managing circumstances and assets? While the obvious answer is yes, it must be recognized as a nuanced response. In the financial sector, workers bolted computing onto existing processes in the 1950s and 1960s, much as did manufacturing companies. When computing became pervasive with a particular offering or service, often the work force involved shrank or was displaced, such as occurred with the arrivals of ATMs and integrated customer files making multiple offerings possible to a single customer. Computers and machines, driven by software tools, led to fewer workers and the de-skilling so decried by many observers, most of whom generalized about what was happening within services industries based on their knowledge of manufacturing industries.[2] The same observation about the role of computing and the work forces in financial industries can be made of telecommunications, although the latter acquired vast quantities of digital expertise in the 1980s and 1990s comparable to what occurred in other high-tech industries, such as in the Software Industry or the Video Games Industry. All the paper-based media industries acquired digital expertise, thereby transforming the nature of some of their work forces, first in printing their products (because computing came first to this part of their value chains), then in the editorial and writing sides of their businesses, causing workforces to learn more about technologies and requiring greater amounts of formal education.[3] In industries that already used electronics, such as radio and TV, workforces changed as their work became based on the use of software; this was especially the case beginning in the 1980s. Workers responsible for creating content were perhaps the most affected by the arrival of the digital. These included animators in movies—even movie directors by the late 1990s—writers of video games, and producers in the music business. They all had to learn how to use digital tools, generally beginning in the 1980s and absolutely so by the early 1990s.

To go back to a theme introduced in the first volume of the *Digital Hand*, the *style* in which work was done had changed substantially in less than a quarter of a century in these industries. This held true also for how companies were organizing, although that transformation occurred at a slower pace than did the changes in their work practices and offerings.[4] Much has been written about how organizations flattened, workers given more responsibility for decision making, while new firms, partnerships, and alliance were created, all of which can be tied back to consequences growing out of the use of computing and telecommunications in massive quantities in a very short period of time.[5]

Role of IT in Services Industries

So much of the discussion about the role of IT in modern economies is rooted in the experiences of manufacturing, transportation and retail industries that we should ask whether services sector industries had a different experience. Issues raised by industries that were large and well-established before the arrival of computers have long interested economists attempting to understand the transformation of the American economy in the late twentieth century, so the experiences of financial, telecommunications, media, and entertainment industries are important to appreciate as part of the larger discussion. Economist Richard N. Langlois has looked at IT applications, technological innovation, and the role of capitalism, but almost as if through the eyes of a business historian. He has argued that the Chandlerian construct of modern corporations—the visible hand of management controlling the throttle of modern economic activity—is not the end game of economic development. Rather, he sees it as a way station on a journey back to a fundamental idea originated by Adam Smith, namely, that over time enterprises specialize. He argues that economies were de-emphasizing the vertical integration of companies that had taken place since the late nineteenth century, supported by the introduction of various computing and telecommunication technologies that facilitated managerial control of events, companies, and industries. He explains that as technologies became increasingly available and diverse, organizations and technologies co-evolved, one affecting the other. It is an elegant and widely held notion that technology is affecting how organizations function—clearly a theme supported in both volumes of the *Digital Hand*. The majority of his work is based on examples from financial and manufacturing industries.

However, Langlois has studied one example that illustrates what could be done in nonmanufacturing industries. In telecommunications, he has observed that as PBX switches became available, enterprises could move from centralized information flows over telephone lines to decentralized distribution of data, such as conversations. In the elegant language of economics, this change created the opportunity in the Telecommunications Industry to seek new "rents" by unbundling prior services. In other words, in selling phone services delivered through the local Bell company's network one could buy PBXs from upstart firms to do the same, such as Cisco. The earlier development of microwave transmission was yet another example of breaking the monolithic control on the markets held by AT&T, making it possible for MCI to come into existence. In turn, that development caused yet another crack in the wall of the classic Chandlerian corporation, in this case, within the highly regulated Telecommunications Industry.[6]

What occurred in all the Industries described in this book is exactly what Langlois documented, the unbundling of services and markets within industries. Niche players appeared in all industries, and the Hollywood Model could be found scattered across many of them to one degree or another. Subcontracting, outsourcing, unbundling, partnering, and formulation of alliances were all manifestations of basic changes occurring in the structure and activities of markets and industries. It was, in short, a return to the specialization of labor and firms that Adam Smith described as the work of the invisible hand in the 1770s. This broad pattern of diversification began in the 1960s and 1970s in financial and telecommunications industries, spread to the print and electronic media industries beginning in the late 1970s, and across all the other industries that we looked at in media and entertainment in the 1980s and 1990s. All were abetted by the extensive deployment of computing and communications that made it possible to distribute and disseminate information and managerial control, the process I describe as the digital hand at work. In fact, it is a process still underway with no discernible direction because of the large amount of ongoing evolution in all the base technologies. The process suggests that we may still be early in some historic transition comparable in scope to what Alfred D. Chandler, Jr., described with the creation of corporations in the first place, a process that took over five decades to unfold.

There seems to be one experience from the services sector that we could point out as a contradiction, namely the enormous amount of M&A activity, particularly late in the century. As we have seen, many causes contributed to M&As and, more specifically, to the consolidation of many firms into larger enterprises. The majority of studies of the phenomenon are based on manufacturing companies.[7] Langlois calls the shift in organization and work "vertical disintegration," whereby work shifts to outsourced organizations such as temporary manpower, and with an emphasis on high throughput in work flows. This could happen either within the context of a large corporation or smaller specialized firms, mixing and matching a combination of narrowly focused functions with general capabilities. As technical and operational standards increased, the need for centralized dissemination of information declined; one now had to make sure that smaller organizations and groups of people conformed to standards. These standards regarded how a process worked, often referred to what IT goods and services were used, and even to regulatory practices that transcended firms within industries.[8] Langlois argues that technology made specialization so possible that one could move companies in and out of conglomerates or hollow out these large enterprises, and do it frequently. The churn experienced in the Brokerage Industry and the Movie Industry are examples of that process at work.

The experiences of industries in finance, telecommunications, media, and entertainment have other lessons to teach. They aspired by accident or intent to mimic some managerial practices of the industrial sector. These included automation of routine repeatable tasks, which some observers went so far as to criticize as the Taylorization of work.[9] This, in turn, led to changed demographics in work forces. For instance, data collection by insurance companies either shifted to computerized tools or were outsourced to countries with lower labor costs. However, it proved difficult to apply routine to many of the activities in media and entertainment industries. It became possible to automate many services in telecommunications and banking, for example, but not always as possible to do so in many other industries. This can be largely explained by the iterative, constantly unique nature of the work done. Writers never wrote the same things twice; all songs were unique; no two movies were exactly alike, not even remakes.

Furthermore, there were so many technological innovations that directly affected the tasks and content of work in these industries that it became difficult to stabilize practices. To exaggerate a bit, every time a game designer sat down at a terminal to create a new game, something was different: the software tool used, the CRT they sat at, the plot of the game to be created. In the Television Industry we see this pattern on a large scale with different forms of transmissions constantly affecting what kinds of programs to offer and how many as, for instance, when the industry responded to the displacement of some broadcast TV by cable, satellite, and interactive TV based on use of the Internet. From three national networks to hundreds of niche channels in one generation, we see that the ability to stabilize work and processes was nearly impossible to accomplish, as too much was changing. In short, the more one moved from the industries studied earlier in this book to those examined later, the greater the departure from the routine of more stable work processes and organizations, and the more we see direct and immediate effects of technological innovation on the work of individuals, the organization of enterprises, and the configurations of industries. The churn that inevitably grew out of the macro process of injecting IT into these industries proved far greater than what economists and business historians had noted in either the industrial or retail sectors.

On top of all that was the ability of companies in one industry to move into those of another, or to co-opt markets and products. We are beginning to see that happen with movies and games; today no major movie appears without at least one companion video game. Experts in movies work in the video games business, and game writers now work for firms that serve the Movie Industry. Telecommunications flirted with cable television for a while, and there has been a historic migration of offerings from all three financial industries toward each other. The one exception, and it is only a partial exception, is in the print media industries. While technology facilitated integration and standardized production processes in book and magazine publishing, much as we saw in many manufacturing industries in the same period of time, the print media proved less fluid in transforming their way of working than those in the electronic and entertainment industries. However, this observation is all about relative changes, because they all transformed at one speed or another.

This leads to a basic theme of the work of Richard R. Nelson, who has studied how economies grow. He argues that differences in how companies are structured and operate affect how they thrive. In effect, Nelson was in a debate with Joseph A. Schumpeter, who gave us the notion that technology stimulates the constant recreation of businesses and industries through a gale of "creative destruction." Both scholars acknowledged that technologies evolved and that their adoption was evolutionary in practice. The experience of the industries studied in this book confirms this pattern; however, it also seems that it operated more quickly than evident in manufacturing, for example, or in our industries in the years prior to the arrival of the computer. Nelson concluded that differences in firms do matter within an industry.[10] What Schwab did in its industry clearly demonstrated that fact; it was also an excellent example of how IT made that all possible. The real insight, however, is that Schwab was not an exception; every industry we studied contained multiple examples of that at work.

But technologies alone do not confer uniqueness on a firm. Nelson stated the proposition clearly:

> It is also the case that to be effective a firm needs a reasonably coherent strategy, that defines and legitimatizes, at least loosely, the way the firm is organized and governed, enables it to see organizational gaps or anomalies given the strategy, and sets the ground for bargaining about the resource needs for the core capabilities a firm must have to take its next step forward.[11]

In short, sound managerial practices were just as necessary in services industries as in manufacturing or retailing.

We have seen examples in every industry of new technologies either reinforcing existing competencies of firms or disrupting them. Those who have looked at this process argue that younger firms tend to leverage new technologies more effectively than older ones; and that for older enterprises to respond effectively they often have to change both strategies and senior management.[12] That observation rests largely on the experience of industrial firms, but the examples of the industries studied in this volume confirm the relevance of that finding to services industries, and nowhere more so than with movies, music, and video games. The problem for existing enterprises is also compounded by firm differences that operate in a global context, for example, the role of Japanese consumer electronics manufacturers entering both the video game and digital photography markets, bypassing entrenched American providers of these goods because of other, broader, strategic considerations. Sony, for instance, viewed digital photography and cell phones not as parts of the photographic or telecommunications industries, but rather within the context of its consumer electronics business. Steve Jobs, CEO at Apple Computer, approached recorded music not as entertainment, but as content that could stimulate market demand for his iPod digital hardware.

As one moved toward the latter decades of the twentieth century, another of Nelson's conclusions increasingly became evident in finance, telecommunications, media, and entertainment industries:

> I want to put forth the argument that it is organizational differences, especially differences in abilities to generate and gain from innovation, rather than differences in command over particular technologies, that are the source of durable, not easily imitatable, differences among firms. Particular technologies are much easier to understand, and imitate, than broader firm dynamic capabilities.[13]

Put more precisely, "from one point of view it is technological advance that has been the key force that has driven economic growth over the past two centuries, with organizational change a handmaiden."[14] This is at the core of the role played by the digital hand in the American economy, with its presence felt in all industries. The experience of media and entertainment industries, in particular, is that the digital hand was heavy and constantly in motion, and nearly equally so in telecommunications and financial industries as well.

My findings also align with Nelson's on the effect of organization on technology—namely, that some of the innovations would not have been possible without innovations in how organizations were structured. The resistance of AT&T's senior management to the initial creation of the Internet provides a textbook example of the important effect firm structure has on the acceptance and evolution of a technology. Financial and telecommunications firms acted more like industrial enterprises in that they were large and often encumbered with highly defined, even stiff, corporate cultures that made only certain types of technological innovations more attractive. However, industries populated with smaller enterprises that were responsible for creative products such as movies, music, and video games appeared more receptive to the kinds of innovation that came along in the last three decades of the century. They were capable of forcing the technology to innovate in specific ways, as we saw in the push for software tools to do animation and special effects.

When looking at the industrial and retail sectors, I observed that most innovations came from large firms that could afford the expense and risks associated with the adoption of new uses of computing and telecommunications. That stood in sharp contrast to what many economists and business observers were writing in the 1980s and 1990s. However, when we look at the industries presented in this book, we see that it often was the small firms in a media or entertainment industry, or even in brokerage and telecommunications, that introduced important changes in what their industries sold and did; how they performed and organized—and that changed fundamentally the actions within a industry. This observation must then lead us to the conclusion that generalizing about patterns of change (or adoption of technologies) across the entire economy is not a reliable way to arrive at an understanding of how economies change or provide "rules of the road," which managers so frequently want. In short, the experiences of all these industries are sufficiently different when looked at from individual perspectives that we have to accept the notion that the transformation of work and of the economy in the United States was and is profoundly complicated and nuanced. Economists have long studied this process at the firm level and concluded that individual enterprises have unique experiences; the same holds true for industries. To understand how an economy experiences a transformation, we need to understand very precisely what was happening under the surface waves of the national economy, at the industry level.

Patterns of IT Adoption

All the industries studied varied in their uses of computing and communications; yet, they shared some common features that offer insights into how the American economy worked and evolved over time. Recognizing these patterns helps inform our later discussion about the broader issues concerning old versus new economies. Some of the patterns of adoption are not new; nonetheless, they are pervasive enough to be called out before discussing some interesting differences in behavior. There are six elements in the pattern of adoption in financial, telecommunications, media, and entertainment industries worth noting as general trends, with the understanding that there were many exceptions and variations in their manifestation.

First, these industries initially used new hardware, software, and telecommunications to continue previously set ways of doing work or to support current organizational structures. In all industries except video games, processes and organizations existed prior to the arrival of digital technologies and the rich variety of telecommunications tools of the late twentieth century. Companies looked continuously at how these tools could be used in support of existing operations, most notably to improve operational (labor) and economic (profits) productivity. The video games business was the exception because it came into existence after digital tools were widespread. Grafting a new technology onto that which already existed was a common practice prevailing in the half century, beginning with computer mainframes and continuing into the early twenty-first century with portable IT devices. The approach was a logical starting point because it was practical, reactive, and cautionary. Once firms learned what they could do with a technology, it was difficult to imagine an alternative approach to adoption.

Second, firms in all industries experimented with various forms of digital technologies to learn how to reinforce existing work patterns to drive down operating costs or to support novel offerings and services. Banks looked to eliminate branch offices; newspapers sought numerous electronic distribution channels and formats; even conservative book publishers tried new printing methods, editorial tools, and even e-books. There was a broad willingness to experiment, to have an open mind. There were exceptions of course, none more notorious perhaps than at the New York Stock Exchange, where employees on the floor frequently displayed Luddite behavior, possibly even destroying equipment that had the potential to displace them. The Brokerage Industry was also quite slow to innovate and apply IT, but even here there were notable exceptions, such as Schwab with its aggressive deployment of online systems. But in general, we can conclude that services industries were just as curious about the digital hand as management in manufacturing and retail industries. Competitive pressures to improve productivity often stimulated interest, but so did consumer demand for additional products and services. In the cases of both the media and entertainment industries, digital tools made it possible to expand creative expression.

Third, industries tended to gravitate toward common uses of technology, such as for applications (e.g., accounting), technical standards and platforms (e.g., Apple's graphics tools or Microsoft Word), and shared practices and policies (e.g., for check and credit card clearing). Newspapers, for example, moved to delivering material over the Internet as opposed to using such alternative electronic channels as videotext or private dial-up services. A number of reasons accounted for this. In the case of those industries in which companies had to interact, the move to standard interfaces and practices proved essential, such as in banking with checks or later in brokerage with transactions. In some instances the influence of regulators could be seen, most notably in telecommunications, radio, and television, where optimizing existing bandwidth for wireless communications was vital. In other cases, consumers expressed their preferences by what they purchased. In time, historians will probably come to the conclusion that consumers were so attracted to the Internet that this preference drove many industries to embrace the Net when they might otherwise have continued experimenting with other electronic alternatives. Once a standard channel for delivering services or products became widespread, companies mixed and matched services over the Internet. The same held true for digital media, such as CDs and CD-ROMs, where consumer behavior had much to do with their standardization and even what kind and volume of content was added to them. Standards made it possible to add services that were unique, such as digital photography to cell phones, which in turn could communicate over telephone networks, creating new markets and sources of revenues. In short, standards stimulated business.[15]

Fourth, once industries began embracing various applications and technologies, they all entered a period in which technological innovation proved relentless, with the result that the accumulation of incremental changes in work and organizations stimulated considerable changes at all levels in an industry. It is often the continuing result—the accumulation of incremental transformations—that is so often the subject of commentators about the computer revolution or the new economy. In every industry discussed in this book, changes were frequent but also evolutionary—applications and new adoptions always came one upon the other in short steps, not by leaps and bounds, and each was different from the one before. However, the intervals in time between one incremental move and the next one became shorter, as if a head of steam had built up, caused by the ability of various industries to introduce new technologies and tools, of user industries to embrace and deploy them, and of markets more willing to buy them. Costs of technology kept dropping throughout the half century, while the expense of labor kept rising. Entertainment and financial products, though often expensive, in fact, came down in cost. The fees one paid to buy 100 shares of stock at the end of the century were far less than 50 years earlier; some services were even free, such as a few online newspapers.[16]

Fifth, the adoption of technologies, the adaptation of work practices and organizations, and their consequences occurred in a progressive manner as management at the firm level embraced new uses of IT in reaction to potential threats, as a way of reducing operating expenses, or in response to new opportunities. The one great exception that almost can stand as a sixth trend is that industries in media and entertainment were generally far more willing to experiment and innovate using digital technologies, while financial and telecommunications industries were less so. In particular, knowledge of digital technology seeped into many media and entertainment industries. We have the wonderful examples of how the Movie Industry acquired knowledge of the digital through individuals experimenting with software to do animation, and of people who were interested in writing computer-based games ultimately creating their own industry. Whatever reason one wants to ascribe as to why people tinkered with technology, we can say that at both individual and firm levels, people learned about various new tools, used them, acquired skills and insights, then attempted to find more new ways to use them and added to those, in a continuous process of evolution. Brokers were slow to tinker, while video game writers were quick to change. Institutional barriers to innovation were the greatest in telecommunications, although corporate cultures in insurance and book publishing also slowed innovation.

Sixth, consumer tastes gave industries direction in incorporating the digital. When a new service was offered electronically, consumers were quick to accept or reject it (they liked downloading music in the 1990s, but dropped pagers quickly after cell phones arrived giving them a more appealing alternative). They had an appetite for digitally based products that grew voracious over time. We can see that with each new round of less expensive or functionally richer digitally based goods and services, consumers were open to adopting them and did so in a shorter period of time than 30 years early. While it took over a decade and a half for television to become relatively ubiquitous, the Internet achieved that status in a third less time. We are observing a similar pattern today with the adoption of digital photography. The migration from long-playing records to tape then to CDs also illustrates this pattern at work. Various reasons account for the behavior: declining costs of digital media and telecommunications, equipment that works better, is easier to operate or is more fun (such as PCs, PDAs, cell phones), or that offers better quality entertainment (such as CDs, CD-ROMs); more disposable income; prosperous economy; growing familiarity—the reasons vary, but the pattern is clear.

Furthermore, consumers almost subconsciously embraced and applied Moore's Law, demonstrating two interesting patterns of behavior. As consumers accumulated experience with one technological product or service, they seemed to learn about the next one quicker and embraced it with growing confidence, such that by the end of the century they were quite comfortable with new devices and willing to use them. The old jokes about not being able to program a VCR (often because the devices really were complicated to use) gave way to an era of digital disk drives, CD-ROMs, flat panel TVs, DVDs, and cell phones that took pictures, all easier to use and rich in functions. Second, while most people probably never heard of Moore's Law, they came to understand almost instinctively that digital devices would increase in function and decline in cost, often over a period of 18 months to 3 years or so. As price points in consumer electronics and declining costs of long-distance phone calls demonstrated, one could almost see the law being applied. It may not be too early to argue that by the 2000s, consumers of all ages had learned how to acquire and use digital tools, creating in the process a new style of consumer behavior. Looking at the reaction of consumers to digital tools is a subject awaiting serious study, but since we know industries respond to consumer tastes and behavior, the analysis of what the industries presented in this book did serves up a large quantity of information about the influential role of an ever more digitally savvy public. Just as industries were increasing the proportion of their investments in IT to run their operations and to sell digital goods and services, so too did consumers increase the proportion of their expenditures devoted to high-tech goods and services. The dozens of statistics we looked at from American government agencies document this significant shift in consumer buying patterns over the half century, particularly after 1985 and expanding right into the new century.

In summary, industries embraced IT to improve operations, to add new functions, and (for those selling to the public) to offer a variety of digital goods and services. For their part, consumers became increasingly comfortable with the innovative goods and services and accepted them as they became available. The failed experiments and offerings were significantly outnumbered by the successes, demonstrating that both industries and consumers had moved to a new style of doing, selling, buying, and using.

There is one fundamental consequence that must be kept in mind. Across more than the two dozen industries discussed in both volumes of the *Digital Hand* IT was used so intensively in both operations and in products and services that the declining costs of computing, coupled to increases in functional capacity and performance, fundamentally enhanced the attractiveness of new products and services or lowered their costs, creating a nearly deflationary economy by the late 1990s. Things were becoming faster, cheaper, and better over time. We have seen whole sets of products and services transforming: In the brokerage business, use of online systems became ubiquitous; digital cameras are rapidly displacing film; and consumers are voting with their dollars that they want their music and video digital, on demand, and less expensive. In looking at the effects of IT on the economy, Graham Y. Tanaka reached the same conclusion: "All *kinds* of things were getting faster, better, and cheaper."[17] My evidence indicates that those three characteristics of the New Economy affected internal processes, organizations (which became more responsive to the market and quicker to transform), and the products and services.

Financial, Telecommunications, Media, and Entertainment Industries in the New Economy

Since the 1970s, there has been a growing discussion among economists about whether the economies of highly industrialized nations have been evolving out of the First or Second Industrial Revolution into a New Economy. The debate has been spurred on by discussions of the Productivity Paradox in the 1980s and 1990s and the massive investments made in IT in the 1990s. There is growing consensus that some sort of structural change has occurred but is not complete. There is also wide acceptance among economists that IT has played a significant role in that process, although the particulars remain ill-defined and still subject to debate. Economists agree generally that the traditional economic measurements are not adequate for describing the changes in the services sector; in fact, one economist has devoted a whole book to the problem of measuring the American economy as a result of its shift away from primarily manufacturing to services.[18] Thus, we are left with many imprecise definitions of the New Economy, begging the question as to whether there even is such a thing.

The evidence presented in both volumes of *The Digital Hand* strongly demonstrate that important, indeed fundamental, transformations are underway that have already affected such traditional measures of economic performance as productivity, employment patterns, availability of new products and services, and sources of wealth creation and destruction. Nowhere does this seem more so than in the bulk of the industries reviewed in this book. By examining how IT was used in so many industries at the industry level, we now know that important changes took place at the "street level" in firms and industries, and that these changes are not over yet. As a result, our findings contribute important specifics to the debate about the New Economy and do so in ways that transcend the dependence economists have on U.S. government statistics on trends. In sum, the evidence favors the argument that we are moving to a New Economy.

But first, we need to understand what comprises the New Economy. Most definitions include at their core the large injection of digital and communications technologies across most industries in all manner of products and operations. It is crucial to keep in mind that both are evident in the economy of 2005, for example, and were not there in 1950 or even as late as 1970. Since the use of information to do the work of selling and buying has increased over the decades and across most industries, much of that information has been collected, analyzed, and used by people relying on computers and communications in integral ways. But as economist Roger Alcaly reminds us, computing and communications alone did not solely stimulate the transition:

> New business methods have evolved over the last thirty years in response to the pressures and opportunities presented by not only the new technologies but also increased global competition, deregulation of many industries, and important financial innovations, including the development of the junk-bond market and the spread of hostile takeovers. Compared to the rigid Galbraithian economy, the business arrangements of the new economy are leaner, more flexible, and more entrepreneurial.[19]

More than earlier economists, Roger Alcaly emphasizes the changes in how companies operate, which is confirmed by our evidence when looked at in tandem with the new products and services. Furthermore, the digital hand has altered the composition and macro practices of whole industries. Some have been for the better, as we saw in banking, while others have threatened fundamental desegregation and restructuring as the Recorded Music Industry now faces, and as movie studios and the Photographic Industry are on the verge of confronting.

Tanaka has suggested that the New Economy has four important features:

1. It leverages new technologies to "deliver more value per dollar" through a continuous round of new products and services.
2. It produces and uses increasing amounts of IT (both computers, and other

products that have IT embedded in them, e.g., automobiles or ATM services).
3. Its companies deliver continuously increasing productivity and products and services that are predictable, based on the technologies upon which they depend.
4. Its companies charge less for the same function with every new offering, while Old Economy enterprises tend to charge more; the former contributes deflationary influences to the economy, the latter inflationary effects.[20]

It is not a bad list. Alcaly adds greater emphasis on the effects of all these new goods, services, and practices on productivity: "Whatever else we might wish it were, a new economy is one that changed significantly through the adoption of innovative new technologies and business practices, leading to a meaningful and sustainable increase in the rate of productivity growth."[21] One tends to turn for insights to David Landes, the premier economic historian of the First Industrial Revolution, who argues that one needs to see rapid deployment of a new technology and its effects on increased productivity to declare that a new economy was emerging.[22]

Because productivity can be measured, it is a popular metric used by economists and economic historians. By that measure, many economists have documented fundamental shifts in American productivity.[23] Beginning with Joseph A. Schumpeter in the 1930s and 1940s, and extending through the work of Nobel laureate Robert Solow, Moses Abramavitz, Paul David, Nathan Rosenberg, David Landes, and Richard R. Nelson, a large body of evidence has accumulated showing that technological advances and a wide variety of innovations were major determinants in the increase of both productivity and standards of living in all advanced economies.[24] Economists have recorded the sharp increase in productivity that the United States enjoyed through most of the twentieth century, and the particularly sharp increase that came in the second half of the 1990s (averaging 2.5 percent per year). There is mounting evidence that the way technology contributed to that growing productivity made it possible for producers of goods and suppliers of services to generate increasing amounts of value for every dollar or hour expended on them by producers and providers.[25] In economic terms, the real output per person-hour increased continuously with every round of introductions of new goods and services. The evidence in this book suggests that this process was made largely possible by innovations in how companies worked, which made them more productive, and by the products and services they offered, which made their offerings attractive in the marketplace. As Nelson has reminded us, the role of individual companies in pushing forward these changes affects both their industries and the economy.[26] He makes very good sense because as churn mounts, new competitive opportunities are created, particularly for innovative early entrants, though not necessarily for all. Economic rewards follow, as we saw with the early online brokerage firms and now within the digital photography market.

Most commentators on changes in American economic productivity focus on what happened in the late 1990s, although we are learning more about what influenced productivity all through the second half of the twentieth century as government agencies collect better data and more economic studies are done. We are developing a better understanding of the specific role of technology, most precisely about computing and telecommunications. There are essentially three insights. First there is the IT sector of the economy itself, which produces the hardware and software of the New Economy such as computers and cell phones, boosting productivity as Moore's Law plays out in the products and services they offer. Second, the technologies themselves become less expensive over time (Moore's Law again), thereby conferring additional productivity to companies that now have to spend less for such tools. Third, and as a consequence of investments in IT, companies can afford to invest in improving internal operations which lower operating costs, hence either preserving existing profit margins or potentially adding some.

In hindsight, everything seems clear. What our case studies indicate is that when the economy did well, industries thrived and some firms exploded with growth and profits. However, there were unintended consequences in every industry that contributed to the nascent New Economy, three of which are worth calling out. First, most of these industries were able to grow, increasing their percent of the total economy, because they used IT to reach markets with better or more cost-effective products. Put another way, it would be difficult to explain why the services sector of the economy grew so much as a proportion of the total economy after 1970 without acknowledging the supporting role played by IT. Other than for some futurists, such as Alvin Toffler or Daniel Bell, management in industries generally did not consider, let alone forecast, such a consequence.

Second, more than in manufacturing, an industry could be severely threatened or hurt by IT and its resultant consequences. Telco's exuberance in overinvesting in fiber optics and in aging technologies nearly destroyed major players in the industry, saddling them with a colossal debt overhang that is taking many years to flush out of their balance sheets. We are currently witnessing the ravaging effects of IT on the Recorded Music Industry, and the beginning of a similar, if milder, effect on the Movie Industry. The lesson here is that once technology begins to affect a firm or an industry, management does not have the option of ignoring it, but they must respond. That response includes investments in technology, its effective use, and changes in the structure of both firms and industries. Those changes are major contributors to the creation of the New Economy. No industry is exempt from the process.

Third, technology broke down barriers between firms and industries that were often unpredictable, making movement from one economic sector to another swift and usually surprising. It often took less than a decade to occur. What our industry case studies teach us is that changes came from all manner of technologies—and not just from the Internet, which seems to be the favorite example cited. The effects were broader and deeper, emanating from less expensive telecommunications; from the humble diskette and CD that in ever newer versions could hold more data; and from all manner of PCs, mainframes, disk drives, and other media that could collect, store, analyze, and produce information.

The implications and consequences are not always positive. As change occurs in an economy, the ability to predict what one should do is compromised. As one economist noted when lamenting how difficult it was to predict the future, "economic predictions are tolerably accurate when progress is steady and they are least needed, but badly miss turning points when, for example, a business expansion is soon to become a contraction, or a bear market in stocks is about to begin, and prescient forecasts are most needed."[27] But all the news is not bad. There are patterns of success evident in the New Economy. In the case of the Movie Industry, we saw that the ability of firms to form and dissolve, using specialized work forces, made it adaptable to changes in their business and in the economy at large in a timely fashion. Technologies came and went, but in the greater context of what the industry was attempting to do—make movies and profits—firms thrived. The lesson is clear: when firms combine use of IT and constantly changing business processes and practices, they are successful in the New Economy. When both are missing, firms are damaged. When the NYSE and brokerage houses slowed down their use of computing and new methods, they were hurt. Music recording companies are struggling with the issue today of how to leverage the new technologies so as to preserve current levels of profits and market shares. That intention will probably collapse under the weight of new circumstances, because invariably industries seem eventually to make changes that prove advantageous in the face of new circumstances. Services sector industries are more susceptible to the consequences of new technologies arriving on the scene, but also vary in speed in which they have to respond to thrive.

While Schumpeter wrote extensively about this issue, I would add that we now know that the "gale of creative destruction" was felt most intensely within specific industry sectors and even within specific communities within firms. The experience of the Recorded Music Industry hints that the real storm is most evident within the communities of various players in an industry who stand to win or lose in a transformation; for example, musicians may capture a higher percentage of profits in the future at the expense of producers and distributors.

Wrestling with Old Problems and New Issues in the New Economy

One of the most important problems faced by industries discussed in this book concerns copyright protection across all media and entertainment for the simple reason that their copyrighted material can all be translated into digital form and thus is susceptible to copying, legally or illegally. Illegal copying is today a global issue because the consequence is denying an industry's firms the revenues and profits that they are legally entitled to collect. In turn, the loss of those revenues has the potential of destroying firms or at least causing them to be smaller or less profitable; on a grander scale, it can force changes in the structure of an industry itself. We have already discussed the problems faced by the Recorded Music Industry, which are beginning to appear in the Movie Industry and in the Video Games Industry. From an artistic view, all three also are beginning to experience molestation of their artistic works as people change lyrics in songs and scenes and endings in movies, though the Video Games Industry's products are less susceptible because the games themselves often contain the ability for game players to create alternative outcomes. The old paper-based industries—books, newspapers, and magazines—have yet to experience these problems to the same degree, but in all probability they will as their products become available in digital formats and are sold directly to consumers or through subscription services. Television and radio content is copied all the time, but since advertisers provide the revenues, illegal copying has less of an economic impact on those two industries. Cable TV has always worried about illegal taping of its transmissions, lost revenues, and copyright violation. Banks, insurance, and brokerage firms worry about data security and access to their systems and the increasing ease with which customers can change their suppliers of financial services, so their issues are more distinct than those shared by media and entertainment industries. Because copyright problems are so pervasive, they constitute the first and most immediate structural issue in the New Economy.

So what is being done to work through these issues? In every prior case of a technology threatening copyrights in the United States, solutions were developed that combined changes in law, the legal actions to enforce them, and industry practices that leveraged contemporary protections and optimized revenues and profits. Increasingly, the legal community has become involved in the discussion, primarily in support of actions to enforce existing copyright laws.[28] Most proposals to fix the problem, however, are coming from outside. For example, we saw Apple Computer offering a legal downloading service to help sell its iPods; by mid-2004, other rivals had entered the market to sell similar services.

First we should ask, why do we care about the issue, other than that it violates current copyright laws? Many commentators argue that the Recorded Music Industry is charging too much for its products anyway, and that perhaps movie CD-ROMs are also far more expensive than they need be. Yet, as the portion of the economy that comprises media and entertainment grows, it becomes important for those industries to operate in an economically healthy way for the good of the nation. There are also more specific reasons to fix the problems. To begin with, if we can resolve open issues, industries will be able to deliver their products at lower cost so that consumers can acquire them for less. Second, consumers want more of their entertainment products delivered digitally and quickly. Third, as more forms of entertainment become less expensive, the variety of materials will increase, offering a culturally diverse set of products and experiences difficult to access today. Fourth, over time one could expect consumers and entertainers to meld, for example, leading the former to be able to modify legally plot lines of movies (this already occurs legally with some games) or change or add lyrics to music. Today all that is illegal.

These reasons for improving delivery of entertainment over the Web also define an impressive list of limitations. The most obvious problem is that it threatens to reduce or eliminate revenues that should be going to the creators of media and entertainment content. Second, if one made it possible for consumers to alter content, the originators of the material would find that troublesome; it is already a problem that has emerged in the Movie Industry, discussed earlier in Chapter 11. Imagine if, in *Gone With the Wind*, the Confederacy prevailed in the Civil War, or in the *Terminator* the bad guy won.[29]

The most comprehensive set of proposals to address the various issues of the Internet, the digital, and these industries has come from William W. Fisher III, a professor at the Harvard Law School. The first option could involve what the Recorded Music Industry has been doing for some time, namely asking the courts to enforce existing copyright laws. As of September 2004, the industry had filed 110 lawsuits against 4,297 people in the United States; the problem was now even more severe than even a year or two earlier, made so by the escalation of legal action.[30] The Movie Industry has also embraced this strategy. In addition to lawsuits, one could write more effective copyright laws suitable for the age of the New Economy, and possibly support various subscription offerings to allow consumers access to products. Various industries and the U.S. Congress have been implementing this approach since the early 1990s. However, one could reasonably expect that the concentration of firms into just a few, more powerful ones, already underway across media and entertainment industries, would continue, narrowing the range of products available to less risky offerings by known popular musicians and bankable movie stars.

A second set of proposals would draw from the experience of regulated industries, such as the Telephone Industry, relying on the fact that there would only be a few industries and that regulators could force them to provide a broad array of intellectual content, as is done in the Television Industry and in the Radio Industry. Regulators could mandate that content producers license their material to distributors and then regulate what producers could charge consumers for movies and music. The industries would make money, the Internet would become the delivery channel of preference (because it would be the cheapest) using broadband communications, but creativity might be constrained; firms and consumers would have high transaction costs, along with the inevitable problems of economic distortion that comes with government regulations in a capitalist economy. Furthermore, since these industries are global, how would one implement such an approach across more than 200 countries?[31]

Fisher has proposed what he calls "an administrative compensation system." Because his proposal is the most thoroughly thought out at the moment, it deserves to be understood:

> The owner of the copyright in an audio or video recording who wished to be compensated when it was used by others would register it with the Copyright Office and would receive, in return, a unique file name, which then would be used to track its distribution, consumption, and modification. The government would raise the money necessary to compensate copyright owners through a tax—most likely, a tax on the devices and services that consumers use to gain access to digital entertainment. Using techniques pioneered by television rating services and performing rights organizations, a government agency would

> estimate the frequency with which each song and film was listened to or watched. The tax revenues would then be distributed to copyright owners in proportion to the rates with which their registered works were being consumed.[32]

He further suggests that once in operation, Congress could modify existing copyright law to allow reproduction and use of published film and music. For that matter, one could do the same for printed materials.

While Fisher's book is both thoughtful and imaginative, the experiences described in the *Digital Hand* suggest that problems would still exist. The first and most difficult one is the involvement of the U.S. government as a clearinghouse. When the Clinton administration attempted to inject the government into the operations of the payer systems in the Insurance Industry, for example, it was rejected by all the industries affected (health, insurance, even industries that carried insurance for its employees), as well as by many politicians and government officials. It was a case of the nation being in no mood for the involvement of the government for a variety of reasons, ranging from concern about profits and bureaucratic complexities to ideological reluctance to expand the scope of government's influence in the private sector of the economy. Second, there are organizations in operation in the private sector that could perform cross-firm case management, a lesson we learned from the Radio Industry. Third, most regulated industry executives—and even regulators at the FCC, Food and Drug Administration, and Federal Reserve, to mention a few—would argue that government is best suited in establishing rules of the game, not in participating in the game itself.[33] Bankers and telephone company executives would argue that more freedom of action, rather than less, stimulates innovation and technological progress, not to mention increases the number of offerings to the market.[34] Yet some sort of copyright clearinghouse seems very compatible with historical experience; the final version would probably be a private one, as in the Recorded Music Industry. In addition, practices similar to those allowed to consumers by producers of printed content might be effective.

Meanwhile, we are left with an important historical legacy. Where novel digital storage and playback devices came into being and could also access the Internet, those industries most likely to lose income fought back, as Fisher has readily admitted, often winning most of the battles in courtrooms or in legislative hallways. The battles intensified as did innovation in recording technologies and communications. The list reads like a prizewinning catalog of major technological innovations crucial to the operation of the New Economy: digital audiotape recorders, software to circumvent encryption, music lockers, Webcasting, centralized file sharing (often on university and college computers or through Internet service providers), decentralized file sharing, and recordable CDs. Fisher is correct in arguing that this sad tale is not about villains, yet we still are left with "tens of millions of Americans, frustrated by the foregoing costs and restrictions, who regularly violate the law—an unstable and culturally corrosive state of affairs."[35] Newspapers, magazines, and book publishers have been relatively immune to the growing problem, but history predicts that as their content goes digital, they too can expect to face the same issues and anguish over what solutions to support. It is very much a game in play, with even thoughtful students of the problem not yet getting the right strategies defined, and all at a time when industry associations and corporate executives are in no mood to consider fundamental reforms. So far, only Steve Jobs at Apple Computer seems to have devised a workable approach that holds out economic incentives for all concerned, though it also requires refinement, as rivals to his service are demonstrating.

A second issue related to the New Economy and the industries reviewed in this book concerns the broad questions of innovation and transformations, which were served up as early as the 1930s by Joseph A. Schumpeter and which have continued as a dialogue among economists and business managers ever since. What do the experiences of these industries with IT have to teach us? Schumpeter's observation that advances in technology—and, I would add, the refinement in the use of technologies within an industry—is an evolutionary process borne out by these industries. Incremental implementation of a machine or software, followed by modifications to both, and then the adoption of new uses has been more the norm than not. As companies learned about the capabilities of technologies, experienced their consequences, and had access to additional changes in the technology, they made further changes. What confuses us, however, is the fact that in some industries change was such a ground swell that using his idea of evolutionary change sounds discordant. To be sure, the cumulative effect of changes in all these industries in just one to two lifetimes has been dramatic, but nonetheless it has been evolutionary. Furthermore, the rate of adoption has varied from one industry to another, with applications often making a difference. For example in banking, old legacy systems from the 1950s and 1960s proved difficult to change in the 1980s and 1990s, reinforcing habits of business silos (lines of business), which often resisted the new capabilities of technology. And in music, the lack of a large inventory of old IT applications blew Schumpeter's gales of "creative destruction" across the industry. So, his characterization of the arrival of technologies in an economy remains a relevant perspective.

Richard R. Nelson lamented that the evolutionary approach to the transformation of technology and use is an "inherently wasteful" process.[36] One can build an excellent argument in support of that proposition, particularly if one takes into account that productivity increases and costs of operation decline if a productive technology is implemented earlier rather than later. However, I would offer a different perspective, namely that because management is intrinsically cautious, radical changes to a base technology or application put a firm or industry at sufficient risk that we should expect important economic disasters to result in negative productivity within a company, industry, or even economy. Those readers who have had to implement quickly a major system without careful preparation and experience understand the risk, as so many ERP installations in the 1980s and 1990s proved.[37] However, what Nelson calls "generic aspects of new technology" are quickly learned across multiple industries, and thus the risk of asymmetrical ignorance of a technology or application is minimized.[38] Put in less prosaic language, as firms gained experience with an application, that knowledge spreads quickly first through the rest of their industry and then across industries. Furthermore, that dispersal of insight and experience knew no national boundaries, particularly as firms and industries globalized their uses of technologies, beginning in the 1970s and early 1980s.

In addition, the concept of productivity varies, depending on who is talking. In the case of management—the people who actually make the decision to install and use specific pieces of technology, software, and new processes—their benchmark of productivity relies on their firm's competitive weaknesses and strengths, corporate- and division-level strategies, and business model and budgets. Those realities mean that every enterprise in an industry or economy has a unique definition of what productivity means to it. Only after hundreds of thousands or millions of adoptions of various technologies and applications do economists come in and sweep up data to create composite views of an industry's or a nation's productivity. My point is that managers do not knowingly use a technology that they think is unproductive unless they are forced to do so or out of some ignorance about the tool under consideration. One of the features of a capitalist economy is that management is given considerable latitude to decide how best to proceed, and their personal and organizational self-interests dominate their judgment.

A great deal of knowledge about the use of IT was also industry specific, not generic. Many observers have observed this phenomenon under many circumstances in various centuries.[39] The patterns observed in the use of computing and telecommunications in our services industries confirms that this pattern of behavior continues, with applications tailored to the particular needs and experiences of individual firms and industries. The database management software used by the Banking Industry, for instance, was not the same as the database software packages used by animators in movies. The list is endless, the point is obvious. Let Nelson have the final word: "While I have written as if there were a sharp distinction between generic knowledge and particular technique, of course the line is blurry."[40]

Innovation as a fundamental characteristic is, of course, a major feature that economists point to when describing the New Economy. Schumpeter already pointed out that innovation led to investments in new uses and activities, displacing capital expenditures in older technologies and practices.[41] As this book was being written, innovation as a theme once again was spreading out from the circle of economists to public policy makers, consultants, and executives in all industrialized countries as if enjoying a renewed sense of fasion.[42] Our industries teach us that innovation occurs on a continuous basis in each industry with important consequences usually underappreciated until they take effect. Responding to these consequences proves to be a sloppy, inefficient process that management muddled through. We have seen in all industries compelling cases for innovation, and surprises in what happened. The ability of an industry's companies to respond (usually react) to innovations and their consequences often was the crucial ingredient for the success of a firm or industry. Yet that capability was not evenly distributed; for example, in the Movie Industry, animators were excellent in embracing new technologies, "pushing the envelope" of applications, but the accounting and marketing sides of their firms were slow to adopt new applications and did so unimaginatively, playing instead a follower's role. Print media were quick to embrace innovative printing technologies, but the marketing, accounting, and editorial applications of IT progressed more slowly. In each instance, we saw that no firm willingly adopted a technology unless it made good economic sense to do so and unless management had a reasonable level of confidence that they could succeed in using a new tool and receiving some benefits in exchange.

The third issue related to the New Economy that attracts much attention concerns the changing nature of the skills workforces have or need. One of the features of the New Economy that so many have pointed out is the requirement for workers to know about computing and other high-tech tools tailored to the specific tasks of their industries. Students of the issue cite the greater amounts of formal education that workers today have versus those in earlier times, and the movement to have employees participate in continuous training and "upgrading" of skills.[43] To be sure, that clearly has been the case in all our industries if for no other reason than that people needed to learn how to use computers in their daily lives (such as for writing e-mail or using a cell phone). But as economists Frank Levy and Richard J. Murnane have pointed out, record keeping in the Brokerage Industry shifted so much to computers that the remaining work was moved from clerks to those who managed mutual funds. In addition, workers migrated from one industry to another in response to new opportunities and to the decline of others. While they cite the example of manufacturing jobs going away as others increased in services industries,[44] the same occurred within and across services industries not only as opportunities presented themselves (such as, merging games and movies) but also as skills became portable (for example, the use of animation in movies, TV shows, and video games). In short, the mix of jobs changed over the years. We also know that the number of jobs did not shrink over the half century; rather, there are more jobs today in aggregate in the services sector of the economy than there were in 1950, whether measured as absolute number of jobs or more relevantly, as a percent of total number of positions. In the services sector the number of jobs grew from 11.6 percent of all employees in the economy in 1969 to 13.9 percent in 1999. Blue-collar and office clerical employees suffered the most as IT made the greatest number of inroads into their work, reducing the total number in these areas from 56 percent of the nation's workforce in 1969 to 39 percent in 1999; at the same time, however, the economy continued to grow. Technicians increased as a percent of the workforce from 4.2 to 5.4 percent, while professionals (including teachers, lawyers, consultants) moved from 10 to 13 percent of the total workforce. Management across the entire economy jumped from 8 percent to 14 percent of the total. The remaining roughly 15 percent were scattered in other industries, such as agriculture.[45]

Each industry studied in this book experienced variations of these patterns, some more dramatically than others. The vast armies of clerical personnel in insurance and banking shrank, for instance, but more complex jobs also came into these industries. Media industries required individuals who could work with both print and electronic tools. Telecommunications shed the most number of workers in this half century, even though the amount of telephony use increased overall. However both the Video Games Industry and Software Industry were new to our half century, with all jobs being net additions to the economy. In many industries not studied in this book, services sector jobs also increased, such as for janitors and security guards, paid for by firms in other industries that could afford them. Levy and Murnane concluded that the number of menial jobs increased, "but the general shift of occupations is toward higher-end jobs." Furthermore, they argue that "while computers are not responsible for all these changes, they do play a major role in bringing them about."[46] The evidence from our industries reinforces their conclusion and also demonstrates that complaints and lamentations about the de-skilling of the American workforce just do not hold up in the face of what we know today. In 1969 such professions as management, administrators, professionals and technicians accounted for 23 percent of the workforce; in 1999 they comprised 33 percent of the workforce. Put another way, the rate of growth in percents was greater than the rate of shift to the services sector (from 11.6 percent to 13.9 percent between 1969 and 1999).[47] Economists have moved on to the more substantive issue of what skills and cognitive behaviors are required of the modern workforce, a discussion outside the scope of my book, but nonetheless indicative that old debates are just that, old, and that we need instead to move on to more current concerns.

Is the United States Becoming an Entertainment Economy?

The question is both fascinating and relevant. It is fascinating because of the implications for how firms would be organized and managed, on the one hand, and how technologies, could, indeed will, differ from prior experiences, on the other. It is relevant because so many industries reviewed are in the entertainment business: movies, radio, TV, video games, books, magazines, even newspapers. We could also stretch the logic to include cell phones and playing the stock market. But on a more serious note, how products and services are developed and sold is profoundly affected by the evolution of digital and communications technologies. One of the leading proponents of the argument that the nation's economy is focusing more on entertainment in its style of operation is Michael J. Wolf.[48] This expert on American media has pointed out that the percent of consumer spending on entertainment has been rising steadily for years, approaching 5.5 percent by 2000. He has also observed how many media activities have acquired an entertainment style to them, such as news stories, and TV magazines, whose features are more entertaining than necessarily newsworthy. He points out that the business celebrities of our age are well publicized people like Ted Turner (founder of CNN), Michael Eisner (CEO of Disney), and Rupert Murdoch (TV and newspaper companies). Wolf also notes that even banking and brokerage advertising contains entertainment qualities, including emphasizing fun and recreation. Finally, he has commented on the growing appetite for content being provided increasingly on an on-demand basis, that is to say, when a consumer wants to "enjoy" a product or service, be it news, a movie, music, magazine, or financial service.[49]

What are we to make of his observations? What do our industries teach us? At a fundamental level, the expansion in both the entertaining technology (TV, cable television, Internet delivered radio, CDs, DVDs, and so on.) and the infrastructure for delivering these goods and services to us cheaply and quickly (cable, Internet, iPods) has grown steadily and quickly since the 1980s. The types of materials have also grown, beginning in the 1960s, ranging from more titles per year published by the Book Industry to more television channels and radio stations one can go to for entertainment. Even the financial sector has expanded its ability to offer a wider variety of products today, thanks to the capabilities of computers and telecommunications, than was the case in the 1950s and 1960s. While we may see technological conversions take place, the content has become more varied and thus made it possible for consumers to be highly particular in what they want and to gain access through an integrated technological infrastructure. Thus, PCs work with the Internet, cell phones with video games, video games with the Internet, and so on.

But this cornucopia of content and services has also created its own problems. For one thing, along the way the American consumer has acquired a healthy appetite for free content. Though a consumer is prepared to buy a book or a paper-based magazine, that same individual is reluctant to pay for similar content that is in digital form over the Internet; hence the problem of file sharing faced by print, music, and movie companies. Consumers still want to own what they pay for, but the digital hand makes it easy for content providers to move away from offering physical products. The consumer continues to want the tangible object—physical copies of music, movies, and books, for example. Perhaps this is a pre-New Economy behavior, but one nonetheless. The dilemma for so many industries is that they do not want to manufacture or deliver physical products when it is so much cheaper to use electronics, yet companies want to maintain the same prices for electronic offerings as for physical ones. So far, the consuming public is resisting this to the point of violating copyright laws because they want both digital copies and physical copies.

The bigger problem for all the industries discussed is how to make a profit in a world where the costs of copies of the content itself are dropping rapidly while the expenses of producing the original forms are increasing sharply. For instance, a video game created for $100,000 in the late 1970s now costs millions of dollars to produce. A CD full of music costs the music distributor less than a half-dollar to manufacture and distribute, but the same content can be transmitted over the Internet for a fraction of that expense. With different industries moving in and out of others with different business models, the problem becomes more accentuated. For example, Apple Computer would probably be willing to sell music to iPod users for 5 cents instead of 99 cents (in 2004) if it could acquire the music for something approaching a nickel in order to sell iPods, which is how the firm makes a profit. That would mean the Recorded Music Industry would have to suffer severe losses in profits if Apple proved successful. Bankers have noticed the same phenomenon with brokerage firms offering free checking and credit card companies also becoming pseudolenders, extending lines of credit so far that one can buy cars and houses and put them on a Visa card. So Visa's profit becomes a traditional bank's lost opportunity. Even stalwart telephone companies face this problem today with Internet-based long-distance telephone calls that essentially knock the legs out from under a traditional major source of both revenue and profit to the industry. Viacom and Time Warner began eying the video games business as a source of new income in late 2004; in short, every industry is subject to this sort of rivalry.[50]

After a decade of experience with the Internet as a vehicle for delivering content and services, and as a source of revenue, profits, and losses, we are learning a few things. First, advertisers cannot and will not foot the total bill to deliver free content to consumers, the Radio Industry business model notwithstanding. Second, consumers are not terribly enamored of rental and subscription services; they prefer to have access to content and experiences the same way they have always had with records and books: they want to buy, enjoy, and sell them, and photocopy or magnetically copy them.[51] What we are learning on the positive side, however, is that consumers are willing to pay for content, financial services, communications, and experiences when they are delivered to them at a time of their choosing (on demand). Banks have figured this out with 24-hour services, as have cable television companies. Print media is just beginning to experience the online world on a substantive basis, while music, movies, and video games are in the eye of the New Economy's storm.

Since no business can long survive without being profitable, one can expect in time that all the industries described in this book will transform their practices and offerings to conform to three market realities. First, they will have to deliver goods and services in forms customers are willing to pay for; often that will result in fundamental changes in prices and the costs of producing and delivering content. Just as I have argued in this book that industries embraced slowly new styles of doing business as a result of the deployment of IT, so too are customers doing the same; it is just that we understand their transformation less than we do what has happened to companies and their industries. Second, firms will use content from one industry or another to bolster their own existing business models, hoping to extend those out longer in time before having to face the inevitable fundamental restructuring. To accomplish that task, companies will have to continue learning new ways to leverage the digital hand effectively. Simultaneously, they will have to borrow from other industries' methods and experiences. Third, just as happened in earlier decades to other industries, the economics of profits, losses, and sources of revenue will change for most. This is a process well underway across all the industries presented in both volumes of the *Digital Hand*. Will it be toward the Hollywood Model?[52] Or will firms increasingly look like consumer electronics or cable television? Brick-and-mortar business models are finding it difficult to coexist with digital or virtual markets, so while one could expect the former to continue, there is evidence that the economy is in some transition toward an as yet ill-defined future state that already economists have labeled the New Economy and that reporters and social scientists have named the Information Age.

Those who would argue that services industries are adopting the personnel and work practices of the old Industrial Age may only be momentarily right, because even those older industries are transforming. Simon Head, who directed the Project on Technology and the Workplace at the Century Foundation, has argued that Taylorist methods of organizing work are appearing all over the services sector of the economy, citing many examples.[53] But at the same time, services sector practices are also appearing in manufacturing industries. In short, while the end result remains far from clear, what is obvious is that new styles of organizing work, companies, and industries are emerging and are moving back and forth across many industries, none of which would have been possible without the involvement of the digital hand.

Global Patterns and Implications

Every industry discussed in this book is global in scope and organization. American culture, and particularly music, movies, and television programming, provide some of the largest sources of exports from the United States to the world. This is not a new story; such exports have been taking place since the early decades of the twentieth century. Information technology, however, has sustained the process by increasing the ease with which it can be done. Because distances have virtually eroded due to telecommunications (telephony and the Internet in particular), what is new is how companies can operate globally, as if vast portions of the earth were simply extensions of their national markets. To be sure, one can overstate the case. Regulators in various nations have much to say about telecommunications, for example, while eBay can sell certain goods in the United States that would land its corporate officers in legal trouble if sold in France; yet consumers in each nation can access eBay. Perhaps one of the most famous examples of globalization was the use of CNN by both Iraq and the United States during the Gulf War in 1991 to see what the other side was doing, or the worldwide popularity of American TV programs or Japanese video games. While markets are segmented abroad to account for local cultural differences, it mirrors what is done for customer segmentation within a national economy. Local offices are often set up to reach local customers and conform to local legal requirements for a presence in the markets served. One can still have a staff meeting with colleagues from around the world using a telephone conference line.[54]

Globalization of industries has led to consolidations and M&A activity with different patterns than were evident in the first two to three decades following World War II. Rather than just American firms acquiring overseas components, the reverse is also happening. For example, European book publishers have been very aggressive in acquiring American publishers. The British Pearson Longman, for instance, owns Prentice-Hall, long one of the most important American publishers of books on IT. European and Australian firms have acquired American newspapers. Japanese consumer electronics companies have bought American movie studios, music recording companies, and video games firms. Their business models and practices, along with their uses of IT, are as sophisticated as and often more advanced than those of American enterprises. One criticism that can reasonably be made of my book is that I have focused only on the American experience; and since every industry discussed in this book had the same kinds of experiences in other countries, limiting the story to the United States provided an unduly narrow focus. But that would be a reaction to today's circumstances, not to those of the 1950s, 1960s, and 1970s, when our digital odyssey began most forcefully in the United States. In addition, space limitations and just the quantity of material to cover required that I limit the study to the United States. But we need to recognize that by the end of the 1980s, all these industries operated in a global fashion at operational, marketing, and strategic levels. Thus today one can read online the *London Times* or *The Wall Street Journal*, as well as download illegally European, Asian, and American music, use a PC made in Asia or the United States, while easily (legally) calling friends using a cell phone manufactured in Europe. American television and radio programs have competed with offerings from around the world for some time, thanks to satellites, cable, and short-wave radio. The BBC's programming is quite commonly available in the United States, both for radio-based news and in the form of Masterpiece Theater productions on television.

Financial systems were some of the earliest to go global, particularly ATMs. First introduced in the United States in the 1970s, nearly ubiquitous across the country by the end of the 1980s, users had access to almost any bank from nearly all ATMs regardless of vendor. By the mid-1990s, an American in Europe could go to any ATM machine in Western Europe and in many East Asian cities and access their checking accounts. American Express, Visa, and MasterCard had earlier made their cards usable around the world. To be sure, wire transfer services dated back to World War I, but it didn't provide the scale of online banking worldwide. These days, not being able to conduct a routine financial transaction virtually in any advanced economy in the world would be seen as odd, out of the ordinary.

The implications for management are numerous and serious. History shows that entire firms and industries have in some cases been turned upside down directly as a result of emerging digital technologies and management's reaction to them. Simultaneously, new industries and opportunities for economic advances have also appeared. In the former category we can put AT&T and its industry, in the latter, the Video Games Industry. The swings in fortunes are massive, at the industry level amounting to billions of dollars, whole percents of GDP, for instance, and with important implications for how other industries, firms, and individuals function. When I looked at manufacturing and retail industries, I was impressed with the effects of the digital on them. However, to a large extent those effects paled when compared to what happened to banking, telecommunications, most media, and to those entertainment industries examined in this book. New forms of digital technology and applications often took only between less of a year and three years to work their effects on an industry in the services sector, while in manufacturing and retail the process usually took closer to a decade or more. If ever there was a time and a sector where management has to be diligent in understanding emerging technologies and new sources of competition (firms, industries, and products), these industries provide the examples. When commentators speak about the effects of the Internet, no industry is immune, although media and entertainment were affected more rapidly than so many others. Even book publishers were affected, although so far more in the use of traditional IT than of the Internet. But that is all changing as this paper-based book goes to press in 2005. Even my publisher, Oxford University Press, the oldest continuously operating academic publisher in the world (for more than 500 years), now routinely places content on the Web.

Even the Insurance Industry is now undergoing basic changes in how it operates. Work is shipped to other countries with lower wages, while scanning technologies are in wide use—both made possible by relatively cheap and effective IT and telecommunications. Insurance firms are now active on the Internet, using that vehicle as a channel to communicate with business partners and clients. Payer systems in use by the U.S. government, the Health Industry, and insurance firms are today likely to be digital, although the volume of the paper trail left behind is almost embarrassing in this age of information; but even that is changing. In short, no industry seems immune from the effects of the digital on a global basis.

Work continues to be automated, while delivery of goods and services in some electronic form is increasing—and again, on a global basis. For management, thinking strictly in terms of national markets is becoming problematic as customers and business partners come from around the globe. At the moment, the most serious impediments are usually local laws, some customs, and the activities of regulators. However, regulators from across the world are also standardizing their practices, or at least making them regional. The U.S. Food and Drug Administration's practices have been widely adopted around the world; the European Union's regulations apply to 25 European countries; some telecommunications standards are converging on a global basis. Already one can use cell phones around the world (if signed up for a plan that offers the service); the technology exists. Standardization at the regulatory level is also being matched by the kinds of employees that are needed. They include individuals who can work with various technologies across multiple cultures, are multilingual, and can solve problems and communicate well. These may sound like soft skills, but they are in short supply across all industries around the world.

Digital technology has knocked the legs out from under many industries' business plans and sources of profits—often quickly, as is happening in music recording. Richard N. Langlois and others have suggested that the result will be a deconstruction of some of the Chandlerian corporate models; we will see if that happens. He is correct in pointing out, however, that new ways of making money are rapidly emerging, including new business models.[55] Simultaneously, we have seen in our case studies the creation of ever larger firms, right into the new century. Today we have large multimedia firms encompassing all the media and entertainment industries discussed; financial empires are still being built with ever-larger banks in the United States and now M&As are crossing borders in all three financial industries and also in telecommunications. So, while Langlois is correct about the direction of new firms, it may take longer than he implies—no matter, since neither of us claims to be a forecaster. There will remain the requirement to coordinate activities of expensive projects, which will require large enterprises or conglomerates of well-coordinated partnerships, alliances, and subcontracting. They may mimic the Hollywood System in the beginning and later move to a more sophisticated, complex, more global-centric set of business organizations.

Managers who deal with technological issues ultimately worry about when to introduce a technology into their organization, why, and when to replace it with something different. Much of this book has been about the displacement of older ways of doing things with newer processes, using new technologies. As historian Thomas J. Misa has noted, displacement is all about "how societies, through their decisions about technologies, orient themselves toward the future and, in a general way, direct themselves down certain social and cultural paths rather than other paths."[56] When we look at what happened at the firm and industry levels, decisions to acquire and dispose of various processes and technologies was all about how enterprises wanted to do their work in the future.

At a more tactical level, what advice might one give a manager operating in these services industries about orienting their firms to the future? Recognizing that industries do not necessarily make such decisions, rather it is firms on a case-by-case basis, managers are nonetheless influenced by the collective views held within an industry, and increasingly by the work of industry committees that establish technological standards for software, applications, hardware, telecommunications, business policies, and operating practices. Thus, we see at the "street level" Misa's idea of societies doing things with technology.

Management's Role in the Age of the New Economy

So, the first advice to a manager is the most obvious: make decisions about how to use specific technologies and how best to direct the digital hand. A corollary is that managers must make such decisions; they cannot be put off because rivals in and outside the firm will always be looking at ways to leverage technology for personal and institutional gain.[57] These decisions do not need to be first of a kind; they can trail a bit behind what others have done. Since the future arrives unevenly from one industry to another, usually a novel application of a technology has already been tried somewhere before, a manager just may not know where. That is where vendors, industry publications, various professional associations, and academics can help, continuing a process in evidence across the New Economy: the elimination of asymmetry in ever-larger bodies of knowledge. The most widespread formula for successful business books in the last 30 years of the twentieth century is to collect case studies of how firms are doing something new or different, and to name specific companies, in the belief that if managers know how someone else did something, they will have the confidence, knowledge, and shared experience to do almost the same with minimal personal or economic risk to the firm. The concept of benchmarking was built on that idea, as were the publishing programs of so many distinguished business-book publishing enterprises, most notably the Harvard Business School Press.

Second, in every industry looked at, we saw that there was a continuous influx of new forms of digital and communications technologies, working their way in. So, managers must continuously be students of how that is occurring within their firms, in their industry, and across the fence in other industries. The first two are obvious, but why the third source? We have seen that industries are porous and that digital technologies have to be facile in support of business strategies that allow management in one industry to poach on markets in others. The cross-industry movement of services seems to have a much larger potential for radically changing the business prospects of a firm in services industries than have occurred in manufacturing, transportation, or retail sectors of the economy, although to be sure they were not immune from the phenomenon either. In addition, technologies and the management of the digital hand are shifting quickly and can profoundly affect practices across industries.

Joseph A. Schumpeter had as his central point that new ways displace old ways in our economy. I would add, however, that management needs to understand that this is not some fine theoretical economic pronouncement or law, but an intense, urgent, visceral, positive, and negative process. The key corollary I would add to Schumpeter's idea is the emphasis that the process is urgent and intense. This is broader than the managerial practices of cycle-time improvement—an issue of great interest to management, particularly in manufacturing and in all industries managing large inventories in the 1980s and 1990s.[58] It sits squarely at the center of an age-old issue of management: should one react to a situation or be proactive, seizing the moment? The experiences of our industries with digital technologies suggest that both are appropriate responses, but decisions and actions must be executed quickly.

Third, the opportunity to be the poacher in someone else's industry seems greater today than even 30 years ago. Churn, even chaos and growing economic risks create opportunities. In each industry we saw clever managers move in novel ways: Charles Schwab in the way he traded securities, banks moving toward securities, newspapers moving to the Internet with more content, movies edging toward the games business, radio to TV, TV to movies, movies to TV—the list is nearly endless. It has occurred in every decade where there was a substantial presence of the digital at work in the core processes of a firm and its industry. One can reasonably expect that to continue. So, management needs to look for the opportunity to enhance its firm's capabilities by co-opting practices in other industries or leveraging one's own goods and services linked to those in another, much as Apple Computer is doing with music and iPods. Thinking through the implications is a nice intellectual exercise that management often does not have the time or the inclination to conduct. So, managers should focus on the very tactical issue of how best to deploy their existing or newly arrived digital tools to enhance current goals and capabilities. They will, however, have to be on the watch for the inevitable evolution of the organization, culture, and activities of the firm. That will happen; we just do not always know the form these will take. The formal study of the coevolution of technology and organizations as a body of knowledge and practice in industry is only just beginning. Current work on this theme is also linked closely to attempts to anticipate how industries are evolving so that management can have a strong voice in articulating the future it wants for firms. Interestingly, despite the turmoil the digital hand has caused over the past half century, it is also a body of technology that has given management enormous influence over the course of their futures. The evidence for this phenomenon is only just beginning to reach credible levels, but it should not be ignored.[59]

Fourth, don't always resist trends. Nowhere is this more evident at the moment than with copyrights. Every industry running into piracy and other copyright issues is suing customers and requesting the U.S. Congress to provide additional legislative protection. In the end, those strategies will only have minimal positive effects; management has to find alternative solutions. Some attempts to use the digital hand by adding encryptions to content, requiring pass codes to access material, and so forth, may only be stop-gap measures, although they probably will provide the greatest number of positive and negative unintended consequences rather than litigation and legislation, because the legal strategies do not directly affect most customers. The solutions will lie more likely in changed business practices, terms and conditions for offering goods and services, and appropriate economic incentives for customers to alter their patterns of acquisition and behavior.

Finally, given the enormous threat to traditional sources of profits in these industries posed by a constant stream of new digital applications, what is management to do? The record makes clear that holding on to old business models and value propositions is not a sustainable strategy. As new technologies become available, managers have to face the fact that they may have to alter how they are organized, how costs are created and covered, and the way profits are distributed. Using myself as an example, I could publish my own book by using a variant of the Hollywood Model by outsourcing production, printing, and distribution, thus cutting out Oxford University Press; so Oxford may have to find ways to motivate me to continue using its services. In years to come, that will probably require different contractual terms from what we have today. In turn, that will alter how OUP generates revenue. The Recorded Music Industry is beginning to experience a similar phenomenon whereby artists are starting to self-publish but need to find new channels of distribution which, if successful (and probably will be to a certain extent), will deny the traditional industry enormous amounts of revenues. At a minimum, we should see the album of old (now in CD form) disappear as consumers buy individual pieces of music rather than collections of 12 to 15 songs on a CD. One can easily imagine the profound affect that transformation alone will have on the industry at large. We are seeing these kinds of issues across all the industries studied.

Management will not be able to avoid consequences and, therefore, they must be willing to destruct creatively their existing cash cows, business models, and old ways of doing things, replacing them with new ones that align more with the capabilities of the digital hand and to the tastes of their market segments. To be sure, incremental approaches will work. However, most managers are reluctant to do even that; those who are more imaginative or braver stand a greater chance of being successful, and the subject later of a good business book. The bottom line is that these industries are currently in the process of moving to new styles of doing business, much the way manufacturing firms in the 1970s to 1990s went through a transformation away from the old Fordist, mass-production approach to a new model characterized more by mass-customization products made by a carefully organized galaxy of business partners and through alliances than in highly integrated manufacturing firms. The point to remember is that the process of change in services industries appears to be just starting and the end of the cycle is still not in sight. If history is any guide, the cycle may take several more decades.[60] In short, senior management has to embrace a way of managing today that differs from that used by those individuals who hired them into their industries.

Arrival of a New American Economy?

Americans love technology. Given half a chance, they will invariably turn to it to solve problems and exploit opportunities. Nowhere is this truer than in business, but it also extends to all walks of life.[61] Americans often accept the notion that technology almost has a life of its own; it keeps coming, changing and improving. Historians have noted, however, that the digital hand is not a living creature, merely a collection of tools that has been forged out of the social values, business practices, consequences of prior use, and scientific knowledge. There is no inevitable "march of progress" underway when it comes to technology. However, because technological innovations in computing and communications have generally rewarded their users in positive ways and the nation's economy in a productive manner, it is easy to jump to the conclusion that the United States is a nation filled with pro-technology Americans.

But the truth lies elsewhere. The economy and society of the United States are capitalist and open, facilitating the relatively free flow of ideas, innovations, and business development.[62] Without those attributes, it would be difficult to imagine how the digital hand could come to play a prominent role in any industry in the short period of a half century. Our industries also demonstrate that the adoption of digital technologies was not merely an American phenomenon; it happened all over the world. Elsewhere I have argued that one of the ways technology seeped into many economies was by way of American firms exporting their ways of doing business to branches in other countries and, secondarily, by foreign rivals adopting those that made the best sense for them.[63] We saw examples of that practice in banking, brokerage, telecommunications, movies, and video games. However, what I did not dwell on was the traffic coming the other way, with the exceptions of the cell phone and digital photography. For the process of deploying digital technology is an international practice with direct effects on the United

States. As digital technology spread around the world beginning in the mid-1960s, it built up a head of steam by the late 1980s that suggests that with specific applications, Europeans or Asians were ahead of the Americans (cell phone development and use in Europe and Asia, digital photography in Japan, and so forth). While American technology got a fast start, thanks to Cold War military funding of R&D and sales, other nations were able to take many initiatives, most notably Japan. As time passes, the distinctiveness of one nation's practice of embracing specific uses of IT seems to be diminishing in the sense that the proverbial "everyone" simultaneously learns about a new tool and adopts it more or less at the same time, although often in ways unique to a nation's culture and business practices.

What a contrast 2005 is when compared to 1950. A half century ago there were no cell phones, PCs, CDs, DVDs, Internet, cable television, satellites, computer chips online banking, websites, laptops, pagers, PDAs, stereo equipment, color TV, ATMs, video games, camcorders, digital cameras, tape cassettes, transistor radios, boom boxes, GameBoys, or iPods. Ma Bell dominated American telecommunications; long-distance telephone calls were major events. Checks were ubiquitous as financial instruments, and local branch offices existed in almost every town in America for banking, insurance, telephone service, and utilities; it was how one did business. Cash was king; credit cards were a thing of the future. So much has happened in such a short period of time that it is difficult to imagine an era when all our digital devices were not present. For Americans under the age of 30, that is their reality; for those of us over that age, we have to work hard to remember a pre-digital time in our lives. It is difficult to exaggerate the scope of the changes wrought by the digital hand.

Yet finally we still must ask, is the United States evolving to some new economy as a result of its extensive use of the digital hand? The answer is yes. However, we can go one step further and add that large swaths of the world's economy are on the same path, most notably the majority of the countries in the OECD. As this chapter was being completed (2005) one of the consequences of 9/11 was the resurgence of nationalism in many countries and the recognition of the continued importance of national governments. So, the question becomes, is globalization as we had come to know it in the 1990s dead? If so, what role will IT play in the future? The historical truth is that globalized economic activity has been with us for several thousand years. Its form ebbs and flows, changing along the way; in the past two centuries the charges have altered society, thanks to technological innovations in transportation, weapons, and information handling.[64] The digital genie has been let out of the bottle. When a terrorist in Afghanistan can use a cell phone and a laptop to coordinate activities around the world, one realizes that the digital hand has truly worked its way into the very fabric of the world. The digital hand will do what its firms, institutions, societies and governments want from it. IT will be remade in accordance with the wishes of that complex set of players, just as it always has.

We have devoted a great deal of space in this book to technologies, applica-

Table 13.1
U.S. Employment by Select Sectors and Select Industries, 1992, 2002, 2012

	Employment (millions)		
Industry/Sector	**1992**	**2002**	**2012**
Total	**123**	**144**	**319**
All Manufacturing	16.8	15.3	15.1
All Services	87.5	108.5	129.3
Financial	6.5	7.8	8.8
Professional & business services	11.0	16.0	20.9
Education & Health	12.0	16.1	21.3

Source: U.S. Department of Labor, "Table 1. Employment by Major Industry Division, 1992, 2002, and 2012," http://www.bls.gov/news.release/ecopro.t01.htm (last accessed 9/11/2004). The table includes in the total services such other industries as utilities, wholesale and retail trade, information, transportation, and all forms of government (federal, state, local).

tions, firms, industries, and the national economy. How are the workers in this economy now deployed, and where are they headed? It is a complex topic, more suitable for others to write about, concerning how their roles will change, but the issue is germane to our time. The U.S. Department of Labor has put together some statistics that describe the distribution of the American workforce in 1992 (just before the advent of the Internet), 2002 (the last year for which complete data was available before this book went to the publishers), and its prognostications for 2012 (see table 13.1). Three trends are evident. First, the total number of employees increased between 1992 and 2002 and, despite lamentations to the contrary, will continue to grow, solid indication that the New Economy will be a healthy one, creating new jobs. Second, the manufacturing sector remained essentially flat all through the period, a clear reflection of either massive increases in productivity or the loss of work to other nations. Third, the services sectors grew dramatically between 1992 and 2002, by 24 percent, and are expected to continue to expand between 1992 to 2012, by 19.2 percent. All of the manufacturing sectors constituted 13.6 percent of the work force in 1992, which dropped to 10.6 percent in 2002, and is expected to shrink to 9.2 percent by 2012. In the services sector, the influx of workers will continue, from 71.3 percent in 1992 to 78.2 percent in 2012. Furthermore, the proportions of jobs within specific industries in the services sector going out to 2012 remain essentially unchanged. The forecasts by the labor economists tend to be fairly reliable, but the actual numbers are not as important as the trend: the services sector will continue to grow as a percent of the total New Economy. The news is essentially positive, though there is some shrinkage in manufacturing, mining, utilities, and agriculture.[65]

Adam Smith in 1776 argued that businesses tend to specialize; today we call those businesses industries and see his pattern continuing. Alfred D. Chandler, Jr., in the 1970s and 1980s argued the importance of large enterprises that can support the scale and scope of many economic activities; that, too, remains true in most industries, although in ever-changing forms. I believe both would agree that digital technologies offered a helping hand.

APPENDIX A

Role and Use of Industries, Sectors, and Economic Models as Concepts for Understanding Business and Economic Activities

Beginning in the 1970s and extending to the early 1990s, the highly regarded expert on business strategy, Michael E. Porter, argued the case for understanding the structure and nature of industries and the role they played in creating firm and national-level competitive advantages.[1] His ideas had a strong influence on thinking about the design of the research for this book and its predecessor. Indeed, in the earlier volume, I summarized his thinking about how best to study industries and how to leverage his ideas to understand the role of technology within an industry. In this volume, I offer an expanded view, in that we have industries that are part of the larger economic cluster involving the services sector, which makes up the largest component of the U.S. economy. They may actually represent multiple economic sectors, global industries as tightly intertwined as a national industry, or some new emerging industry that may be a replacement for a sector either in fact or name. Such possibilities can influence the work of the business historian and, indeed, even the thinking of strategists and management at the firm level. Moreover, they cause historians and executives to confront the whole issue of economic models. Economists love them, particularly American economists, but historians find them too restrictive and artificial.[2] Business managers study them less frequently and are instead normally exposed to pale replicas of the work of economists.

By deconstructing some of the issues involved in clustering and defining industries and sectors, business historians and managers concerned with exploiting an industry-changing technology can frame their thinking, research, and decisions in ways that utilize historical perspective in practical ways and that derive results that are productive. The act of leveraging historical insight to understand a firm, an economy, or modern business history is a basic purpose behind this book and its predecessor. In this book we have seen how banks, insurance companies, and brokerage houses evolved into a highly interconnected network of financial offerings and firms such that referring to all of them as the Financial Industry became fashionable, even convenient. A similar process is unfolding in the Telecommunications Industry (also called a sector). If a banker had a sense, say in 1975, that such an integrated possibility existed, that realization might have influenced his or her business strategy; indeed, regulatory bodies at the state and federal level did realize it. For the economist, if a model had existed that demonstrated convincingly that technologies tended to cause industries to aggregate or merge, then he or she would have advised firms, industry opinion makers, and public officials about the blessings and curses of the process. Others, particularly those selling technologies, would have known how to alter their marketing strategies as IBM did occasionally, for example, when it sought to provide the plumbing for intra-industry communications (for example, for regulatory data in the Pharmaceutical Industry or when it created IVANS in the Insurance Industry). Finally, there is the business historian who could have used such a model as an initial test bed from which to raise questions about the past, either to disprove the validity of the model or to enrich our understanding of its features and foibles.

But let us begin with sectors. Simply put, sectors are groups of industries that have some relation to each other, sharing common features. In the financial sector, we think of banking, insurance, and brokerage; for a long time, the U.S. government also included real estate, although by the end of the century it was removed from this specific collection. Other sectors include manufacturing, made up of industries that make things; and, for most of the twentieth century in the United States, there was the services sector, which included all industries that performed services, ranging from running hotels to the legal profession. Public sector refers to all government agencies and educational institutions, and the agricultural sector involves all industries that grow crops, raise animals, or sell these items wholesale. And in recent years, U.S. government economists have begun to talk about an information sector, much along the lines that economist Fritz Machlup suggested as far back as the 1950s.[3] So there are numerous sectors, but not as many as there are industries. Industries within sectors share some common practices, activities that may be independent of each other (wine growers do not work with cattle ranchers, for example) or that are highly integrated (with financial industries). Sectors also allow one to borrow ideas and practices from another industry, or to conveniently catalog a large body of economic data deemed useful in rationalizing a description of a national or the international economy. In short, it is a convenient tool for collecting perspectives.

Let us return to Porter. In 1990 he published a major study, *The Competitive Advantage of Nations*, in which he described the emerging view of sectors.[4] Specifically, since all the industries described in my book are considered to be in the megaservice sector, let us focus there, beginning with his useful definition of services sector activities: "The term services encompasses a wide range of industries that perform various functions for buyers but do not involve, or only involve incidentally, the sale of a tangible product."[5] He is quite right in adding that because of the breadth of services available in an economy, there really is no agreement on a "taxonomy of services." Porter postulated that by cataloguing the activities of a services industry along his now famous "value chain," one could begin understanding the role of a services industry in the economy. As services slide in and out of industries and even sectors (such as the maintenance of equipment provided by a manufacturing company, which places this maintenance in the manufacturing sector), his concept of the value chain allows us to understand where activities and firms sit within a sector and thus within the economy. Disintegration of services can be better understood as services become more specialized over time. We need not pursue his discussion about how services industries and sectors operate in a competitive climate; it is enough to acknowledge the value of thinking in terms of sectors and not be confused by the movement of activities from one sector to another, or more frequently, from one industry to another.

The Digital Hand has as an implied subtext the notion that there may be emerging in the American economy multiple services sectors, not just the one normally thought of by economists, managers, and historians. Specifically referring to the services sector, as opposed to the manufacturing or agricultural sectors, has long proven too general a descriptor of the economy. That proved so especially after World War II, when economists noticed that the services sector of the U.S. economy was rapidly becoming the largest component of one of the world's biggest national economies.[6] Yet so long as industries maintained their traditional borders, one could always turn to a description of an industry to understand a portion of the economy. But as this book demonstrate, industry borders began changing across many industries, beginning as early as the 1970s and picking up momentum in the 1980s; they show no signs of reaching new stages of stability of definition as this book is published. Leaving aside for the moment my argument that technology served as an important, but not sole, stimulus for such transformations, the problem of perspective remains that even industry-based definitions of clusters of economic activity are not as stable and convenient to rely upon as in prior decades. Hence, we may have to begin thinking about placing new industries, new clusters of industries into sectors and even possibly retiring the now supralabel *services sector*, since we now frequently call whole economies by such labels as services economies, or even information economies.[7]

There is also the issue of economic models, but first, I must betray my biases. Historians in general do not like abstract, absolute descriptions of patterns of behavior, because every situation is seen as different from any other. I am a historian and so view models as convenient intellectual shorthand, not thorough reflections of actual patterns of behavior. As one historian years ago noted, "the economic model redefines history with dangerously convenient assumptions, with the hypothesis of an unchanging external environment, while attributing to its agents stereotyped and excessively general purposes."[8] In short, models can be false, even misleading, views of reality because every situation is unique. Relying on long strings of mathematically based descriptions of conditions, often using massive databases of information that we do not know to be accurate or thorough, simply compounds the suspicions of any historian. However, if one could speak about a long period of time as relatively stable, either because things did not change that much or we don't think they did (as in Paleolithic societies which existed for hundreds of thousands of years), then the temptation exists to use some sort of model with which to describe the behavior of many people.

What are we to do in the situation where firms, industries, even whole sectors, were transforming even as these words were being written? If nothing else, this book demonstrates that a great number of changes occurred not only in the last three decades of the century, but continue today. The evidence presented in support of that observation suggests economic models that either generalize about the past or worse, attempt to describe the present (a fleeting moment at best, based on what we are seeing in the industries described in this book), are inaccurate, and probably would have to be labeled misleading.

Yet a large number of sources I have used to underpin this book came from economists, many of whom either wholeheartedly created and used economic models or, at a minimum, flirted with them in some quasi-delightful abandon. Economic models of firm and even industry-level behaviors are useful in suggesting questions to ask of extant evidence of a period (what historians call primary sources), and they present ideas of how one firm or industry might be acting that might not otherwise have occurred to the historian. They offer up a great deal of raw data that historians can study as part of their work. This same comment applies to managers who use models to ask if there are alternative paths to profits they had not thought of before, or who enrich their understanding of the their own firms and industries, in turn inspiring future behavior.[9] For example, one can demonstrate that the Banking Industry did improve operational efficiencies when it worked collectively with its competitors to standardize certain practices (such as in the case of MICR coding of checks). Defining levels of efficiencies and the necessary patterns of cooperation would delight an economist in pursuit of an economic model, while a historian would see this as one piece of evidence of what happens in an industry where high levels of cooperation among rivals became necessary in order to perform mundane tasks (cashing and crediting checks). For the business manager it would be a useful signal that perhaps some of the most complex operational inefficiencies might be partially mitigated by looking outside the firm through some cooperative initiative with rivals, as bankers did, and as occurred in the Grocery Industry with the deployment of the bar code. So models have their uses, but within limits.

Finally, there is the question of technology and its role. The history of computing and telecommunications in the industries in this and the previous book seems to suggest that no definition of an industry or sector of the late twentieth century in the American economy can ignore the role and influence of technologies. I think it is an understatement to say that technology has profoundly influenced daily work and the configuration of products, services, firms, and industries. The evidence presented also demonstrates that strategies, government regulations and laws, wars, weather, and other exogenous factors in most instances continue to be still more influential than the mighty computer. Thus, to include them in the activities which make up the definition of an industry, sector, or economic model continues to make good sense.

In summary, concepts of industry, sectors, and economic models are generally useful mental tools with which to catalog events and patterns of behavior. They help stimulate new avenues of research, thought, and action, but they also have their limits, not the least of which is that they are not thoroughly accurate, never absolute truths. At best, they are moderately reflective of reality. Describing the specifics of a situation or event as it occurred or existed at any one time is the unique work of the historian.

However, effects of IT and telecommunications across the American economy did raise some fundamental questions about the definition of industries and sectors. Traditionally, both were defined by the principle of "birds of a feather stick together" in that they occupied certain parts of the economic landscape: they shared markets, competed against each other, and did many things in common. But economists and historians have discovered that as boundaries between industries and sectors shift (as a consequence of using these technologies), their notions of industry and sector become less precise. If General Motors is selling loans, can we still only think of that company as being a manufacturer of cars and trucks? What do we do with its financial arm (GMAC), which is now larger than some of its rivals in the Banking Industry? As services are extracted from one industry and moved to another, thanks to the capabilities of computers, the Internet, and other technologies, we can expect such questions of definition to arise. At the sector level, if whole new industries are formed, when is it time to call a new configuration a sector? Let's be ridiculous and extreme for a moment to illustrate a problem. What if an entire national economy were to become a services economy? One might say, we would invent new sectors, as seems to be slowly developing with the U.S. government's use of the NAICS taxonomy as it replaces the old SIC codes.[10]

The point is that this problem is thorny but crucial to resolve. As demonstrated in both books, identification of firms with industries (and, one can assume, also with sectors) is important because a whole collection of infrastructures and relationships develop around those clusters of economic units, which in turn alter behaviors of member firms, even causing new sets of technologies to emerge in support of those. Issues related to measuring economic productivity, national policies toward economic development and taxation, and regulatory scope of responsibilities for agencies all come into play. While this is not the place to propose fundamental answers to address these problems, one purpose of these books is to help managers, economists, historians, and others sort out how work is changing in America and to define some of its implications. Figure A-1 is a simple attempt to illustrate the changing configuration of a sector that may be emerging. It is not

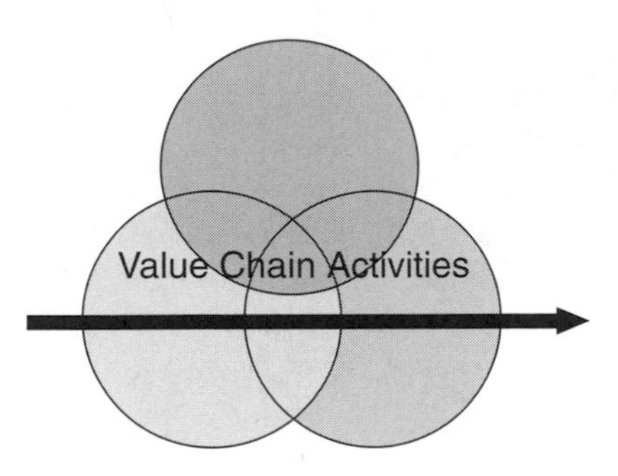

Figure A-1
Interactions of multiple industries within a sector. Circles represent industries.

the clean, engineering, even Newtonian picture one might have produced to describe a sector in the past (such as a large box with little boxes representing all the industries neatly ensconced in it), or even Porter's comforting and possibly aging value chains, but it begins to suggest a more dynamic, fluid, ill-defined reality characterized less by individual firms within industries or industries within sectors, than by the functions or lines of business that share rivalries, common borders and dependencies.

I am disturbed at having to present such a nebulous figure because like others interested in the economy and as a business manager, I prefer absolute definitions that categorically tell me, for example, that I am in the Computer Industry or in the Banking Industry, not that I lease digital hardware, some of which may have been manufactured on behalf of my company by a third party in another industry, and which also is sold through the Retail Industry in some computer store in a shopping mall. But that is the reality we face, compliments of IT and telecommunications. Others will have to take up the question of how to reconcile traditional definitions of sectors and industries with this emerging reality. It also pains me to suggest that Porter's depiction of the value chain is perhaps becoming out-of-date, because it has proven so useful to so many business people for over a quarter of a century, including myself. However, since the notion of a value chain remains very relevant, perhaps the next step in its conceptualization is to impose it on multiple industries, sectors, or streams of activities that transcend various industries.

APPENDIX B

How a Telephone Works and the Basics of Telecommunications and Networking

There are two fundamental reasons for having a basic understanding of the technology underpinning telephony. First, the infrastructure comprising a network is complex and massive and thus is a major driver, along with the regulatory agencies, in the construct of most portions of the telecommunications world and its practices. Second, there has been a fundamental shift in the base technologies of telephony from analog to digital over the last two decades of the twentieth century. These have important implications for the its future. The changes are as profound as, say, moving from making ships out of wood to metal. Of course people cannot often actually see the difference in telephonic technology, but they know the difference in the clarity of calls and the functions they can perform with the digital versus analog. Most people think of the telephone as the device on their desk or mounted on a kitchen wall. In the Telecommunications Industry that device is called a handset or terminal. It is merely the tip of the iceberg, tip of the pencil, the starting or ending point of a long chain.

Think of a telephone as part of a system, a network of many technologies. This network includes telephones, transmission media, call-and message-switching equipment, and signaling technologies that control the traffic of voice and data from one point to another. All telephone communication systems have essentially four basic elements: the telephone itself; the transmission equipment that carries a signal (voice or data) from the handset into the network; switching hardware and software; and human operators, who used to receive calls and switch them to lines that direct the message to their intended destinations but now are more often troubleshooters. These all involve various techniques for connecting one telephone circuit to another. Coordinating all these is signaling, which is all about how telephone networks are controlled and instructed to connect specific calls.

The basic concept of the telephone is that a handset converts the sound waves produced by someone's voice into analogous electrical waves, which are then transmitted over wire to the intended recipient, via the four parts of the network listed above, and then converted back into sound that can be heard by the listener. Traditional telephones have carbon granules through which an electrical current flowed, with sound creating pressures on the granules using a thin diaphragm, thereby affecting the intensity of the electrical current flowing through the handset out over the telephone line; newer phones used digital means to do the work of carbon granules. At the opposite end the reverse occurs. Switching involves connecting the signal transmitted by a speaker into the line toward the intended recipient. Originally, a telephone operator did this by physically plugging one line into a switchboard to connect to that person. That process was continuously automated all through the twentieth century; first, by using precomputer technologies, later computers, and by the late 1980s fully digital technologies. AT&T devoted decades and hundreds of millions of dollars to improving switches. In the process, it produced a vast body of telephonic technology.[1] Local control offices were located all over the nation, connecting local calls to transmission lines that allowed conversations to go long distance to other local control offices, which then passed calls to intended recipients. Over the decades the capacity of the nation's telephone network to handle more calls rose nearly exponentially.

In the "old" world of telecommunications (pre-1984), the switch was the key value component of the network and was the property of telecommunications companies; businesses and homes had telephone lines, but none had a switch. The switch has a long history dating back to the mid-nineteenth century and was the innovative technology that allowed telecommunications as a business to evolve from telegraphy to telecommunications. A century later, when switching became digital, switching could be handled by various proprietary systems, and the key value component of the telecom business began evaporating. Today, physical transport and switching are rapidly being commoditized, raising serious questions about how telecommunications companies can add value to the network now that their more than century-long key technology—the switch—is no longer under their sole control.

So far, I have described what is normally called analog signals. In other words, the voice is converted to electrical impulses—known as the voice signal—and has the same oscilloscope readings as the original signal (the voice). By the late twentieth century, analog also came to mean transmissions that were digital. When implementation of digital telephony became possible, the voice went into the phone as before—they still are analog signals—but then was translated into digital impulses (the binary code of digital signals) wherein the voice is sampled many times a second and each sample is assigned a binary series of numbers, through a process called modulation. The digital signals now look like standard computer-based digital data, and are carried through the line, switches, and so forth, to the intended recipient.

Digital switching (also called packet switching) was first tried in 1962, and by 1969 over a half-million voice channels were in use across the nation. Digital telephony enjoys two fundamental advantages over prior approaches. First, signals can be reproduced exactly as originated, as opposed to analog transmissions, which lose signal strength and pick up distortions over long distances which is why long-distance telephone calls were often hard to hear and scratchy until late in the twentieth century. Second, digital circuitry is much less expensive to install and maintain than the pre-digital circuitry, which consisted of a great mass of copper wires. In the digital era, optical cables can carry many times more calls in a fraction of the number of lines. Furthermore, digital circuits can be fully automated and controlled by computers. The one great negative in using digital is the cost of converting from analog systems, an expense borne by American telephone companies.

There are other considerations with the digital to keep in mind. Once a telephone company converted sound to digital signals, then it could also transmit data (such as pictures, text) over telephone lines, because the technology required to convert text to digital forms (the on/off electrical impulses in a computer) had been used for decades as the central technology of computing. Since data that could be read by a computer were electrical impulses and electricity flows through a telephone line, one could now send all types of information over a telephone line. So data, voice, pictures, and so forth, were converted to digital signals, transmitted to the recipient, and then converted back into images or sound so that the person could understand the messages.[2] Once that was possible, then the name of the game was to speed up the transmissions and increase the capacity of the telephone systems to handle ever larger amounts of data over a wire. Improving modulation techniques was important and more feasible in a digital world. Prior to the arrival of digital telephony, existing technologies were more or less adequate to handle the volume of telephone calls, but with crude data transmission starting in the 1960s, the pressure to handle greater volumes more quickly drove technologists toward the digital.

We should recognize that there is a vast difference between optimizing a network for voice calls versus data calls. In a voice network—such as what existed for a century—the intelligence to manage the flow of conversations lay in the network. These are called "smart networks" because switching technologies control the flow of events. In a data network, intelligence for managing the flow of information resides in the data, which contains information, and in the devices and applications that package the data and tells it where to go. A data network does not need a traffic policeman or stoplight as might have existed in a traditional analog switch. It also does not matter for data networks whether the data travels via main roads, highways, or back roads as long as the data reaches its destination. In voice networks, however, all traffic is carried on main roads and highways because these kinds of networks segment their bandwidth into dedicated lines or time slots, which is akin to having a limit on the number of lanes and cars per lane. In the digital scenario, data traffic can always get to where it is going on time.

We should understand the effects of the Internet on telephony. In the early days, one connected a PC to the Internet by dialing a telephone number using an existing telephone network, going through an existing network of switches. Data transmissions tend to carry more electrical signals on a line (data never pauses the way we do in conversation; it only slows down if the line is crowded with other data), resulting in enormous traffic jams on the network. Eventually, the length of time users were connected live on the network exceeded the amount of time of a typical conversation. As time went by, users simply acquired a second telephone line dedicated to Internet access, which is why by the mid-1990s we saw a dramatic increase in the volume of new telephone numbers and area codes. So the hunt for more capacity and speed was on. Part of the capacity could be handled by the use of optical fiber, of which massive amounts were installed across the nation, but also in speeding up the transmissions themselves, hence the 1990s phrase, "broadband." Specific speeds of transmission determined if a line was a narrowband (able to handle one voice channel, as in old POTS), wideband, which was faster and used to move data in the 1980s and 1990s, or broadband, which handled even more traffic (to be technical, capable of handling more than T-1 rates, or roughly 24 channels at wideband speeds).

It is important to understand the role of increasing demand for telephone numbers and area codes. This happened first because of growing demand (described in Chapter 7), but also because government regulations established zones to make sure the country would not run out of telephone numbers. The government also sells blocks of telephone numbers to carriers that they can issue within their zones. These two regulatory practices have a tremendous impact on local and regional switching because each switch needs to be programmed to recognize the zone from which the call originates and the zone to which it is being directed. Therefore, the system for handing off calls from zones to regions to national networks has become dramatically more difficult to do over time. All switches had to be reprogrammed, particularly in the 1990s.

One way to speed up transmission speed was by taking signals (a conversation) and separating them into bunches (packets), then shipping these down a line with other packets of partial conversations—much like highway traffic, which is all intent on moving things and people in the same direction. The ability to ship pieces of transmissions, called packet switching, grew out of digital technologies. In effect, packet switching made it possible to keep lines fully loaded with transmissions but organized so that all the pieces of, for example, your conversation, could be put back together in a timely and correct manner at the other end of the line. That meant one could simultaneously use the same line at home both to take a telephone call and to log onto the Internet. This ability was the true genius of the innovations that Vinton Cerf and Paul Baran had developed in the early 1960s. Add in the hunt for faster transmission speeds and one begins to see the continued complexity of the telephone network in the new century.

Finally, let's discuss the lines themselves. Twisted pair, which is the copper telephone wires that have been used for a century and continue to be deployed in every home and office, can handle about 7.7 million bits of information per second, or roughly 120 telephone conversations. Fiber optics can handle 45 million bits of information per second, or just over 700 telephone conversations. By using packet-switching approaches to transmit waves of light through such networks (called Wavelength Division Multiplexing, or simply WDM), one could, as of 1996, push through a line some 40 billion bits of information per second, which translated into 625,000 telephone conversations. By 2000, that combination of WDM and fiber optics could handle 1.6 trillion bits of data per second, or 28 million conversations. Now one can understand why the telephone companies in the United States made vast investments in optical networks. Between 1996 and 2000, installation of fiber optic capacity grew at a 158 percent compound rate.

In Chapter 6, I noted that as the century progressed, data transmissions became an increasingly important part of the traffic over a telephone line. Then came widespread availability of the Internet, which is based on digital technology. So we need to understand further the differences between the pre- and post-Internet telephonic technologies. Prior to the arrival of the Internet, switching consisted of a body of technologies called circuit switching, which dated back to the first decade of the twentieth century (albeit with improvements over time) and was the basis of POTS. Two devices connected through a dedicated line on the network (called the circuit), and voice or data moved back and forth in the same order as originally transmitted. Consumption of bandwidth was fixed, in other words, the two parties owned the entire capacity of the line between them regardless of whether they used all its capacity or not; nobody else could use the line. The origins of this approach date back to the days of telegraphy. When telephony first came into use, dedicated wires were strung from town to town, and each line was a party line, so one had to schedule specific times to talk. Since the original concept of voice switching was based on this concept, it survived through the years even though technologies now afford us more effective ways to create a channel of communications between two distant parties. Digital approaches represented a radical departure from the old model.

In the world of packet switching, digital is now the basis of the Internet and most data communications services. Software manages the networks and determines routes. One's use of a telephone line varies, since there can be other packets of data also using the same line, so dedicated access is never guaranteed. Of course, if a network were down, packets can bypass a bad line and arrive at their destinations via alternative paths.[3]

In the 1990s, a technical debate, tantamount to a holy war, was in full swing in the Telecommunications Industry. To put it bluntly, on one side were the traditional telephone industry representatives, pejoratively called "Bell Heads," and on the other side the "Net Heads." The Bell Heads argued that switching networks needed to be highly intelligent, that is, smart enough to know that one wanted a telephone line and who to get the call to. They argued that the telephone itself should be "dumb" with no processing capability, leaving computing intelligence in the network itself. Thus, the network is the service of value to the user. The Net Heads argued the opposite, specifically that networks should be dumb and that the telephone (or terminal) should be smart and have computing capability. In their model, networks are simply utilities that, like the street in front of one's home, merely provide facilities for "smart" drivers to direct their vehicles wherever they want to go. As the Internet became a preferred channel for handling the movement of data, activities, voice, and sound in the 1990s, one could quickly realize how important such debates had become.

Telephone and other telecommunication provider companies are built around one view or the other. Many of the tens of thousands of Internet Service Providers (ISPs), so prevalent in the 1990s, sided with the Net Heads, while the Bells were trapped into maintaining preexisting, older technologies and infrastructures. In the early 2000s, as more traditional telephone and cable companies began rapidly to displace ISPs in the fast-growing market for high-speed Internet connections and telephone services, one could again quickly see the debate taking on an urgency. The business dilemma all this presents is that no service provider has been able to figure out a viable business model to transition from one operating model to another. Nor has any determined when it makes most sense operationally to transition—that is, when one can make the most money or create the most value for end users. Meanwhile, traditional phone systems continue their historic transformation to fully digital, high-speed, packet-switching approaches, leaving behind long-standing technologies of the old Bell System. As with so many other technologies, therefore, we see that how a telephone works is as much a discussion about business issues as it is about technology, electronics, and physics.[4] Telephony and its industry were originally about physics, then electronics and technology, but now are all about fully developed businesses, and that governs the other three issues.

NOTES

Chapter 1

1. The literature is now vast on the topic. However, two studies recently have summarized the economic perspectives: Erik Brynjolfsson and Brian Kahin (eds.), *Understanding the Digital Economy: Data, Tools, and Research* (Cambridge, Mass: MIT Press, 2000); and Robert E. Litan and Alice M. Rivlin (eds.), *The Economic Payoff from the Internet Revolution* (Washington, D.C.: Brookings Institution Press, 2001). For a sociologist's perspective that also reviews other points of view, see Manuel Castells, *The Internet Galaxy: Reflections on the Internet, Business, and Society* (Oxford: Oxford University Press, 2001).

2. Harry E. Mertz, "Electronics Down to Earth," *Banking* 48 (December 1955): 49, 138, 140.

3. The American Bankers Association (ABA) routinely conducts surveys on various uses of IT, publishing the results in the *ABA Banking Journal*. At the end of the century one important survey showed that 22 percent of all bank customers had used the Internet to interact with their bank, although only 5 percent did that as their primary channel of communication with their bank. Traditional branches remained the most widely used channel (51 percent), followed by ATMs (29 percent); see "What We Have Here Is Channel Evolution," *ABA Banking Journal* (September 2001): S14.

4. Jack Weatherford, *The History of Money: From Sandstone to Cyberspace* (New York: Crown Publishers, 1997): 191–263; Elinor Harris Solomon, *Virtual Money: Understanding the Power and Risks of Money's High-Speed Journey into Electronic Space* (New York: Oxford University Press, 1997): 3–34.

5. Lauren Bielski, "Year of the Wallet?" *ABA Banking Journal* 99 (December 1999): 56, 58, 62–63; Bill Orr, "E-banking What Next," Ibid., 93 (December 2001): 40, 42, 44, 46.

6. Alan Gart, *An Analysis of the New Financial Institutions: Changing Technologies, Financial Structures, Distribution Systems, and Deregulation* (Westport, Conn.: Qurum Books, 1989): 245–272.

7. Coopers & Lybrand, "Expansion of Banks into Insurance," in Everett D. Randall (ed.), *Issues in Insurance*, vol. 1 (Malvern, Penna.: American Institute for Property and Liability Underwriters, 1987), 225.

8. Ibid., 248.

9. Beginning in the early to mid-1990s, credit card providers (such as Visa) began making available to nonbanking organizations credit cards that displayed on the plastic the name and images of the sponsoring organization. A percentage of all profits earned from cards sold through an organization would be returned to the branded enterprise. Thus, for example, a university or a charity, could raise additional funds by encouraging their members to adopt a credit card with the logo of the sponsoring institution. A second trend involved

organizations issuing their own credit cards and either doing the processing themselves or outsourcing the back-room processing to a credit card company. The most widely known example of a nonfinancial institution issuing a card is AT&T's Universal Card, launched in 1990. For details about this, see David Evans and Richard Schmalensee, *Paying with Plastic: The Digital Revolution in Buying and Borrowing* (Cambridge, Mass: MIT Press, 1999): 219–221, 232, 257–258.

10. Note that Real Estate Industry has always employed a large number of people. Much of its information processing revolves around mortgages, however, which the Banking Industry handles and which is discussed in this book.

11. Charles J. Woelfel, *Encyclopedia of Banking and Finance* (Chicago, Ill.: Probus Publishing Company, 1994): 69.

12. Ibid.; for a useful discussion of the role of exchanges, see pp. 1087–1089.

13. U.S. Bureau of Economic Analysis, U.S. Department of Commerce.

14. Bureau of Economic Analysis, "National Income and Product Account Tables," available at http://www.bea.doc.gov/bea/dn/nipaw/Ta.

15. Sherlene K. S. Lum and Brian C. Moyer, "Gross Domestic Product by Industry for 1998–2000," *Survey of Current Business* (November 2001): 20.

16. J. Brooke Willis, "United States," in Benjamin Haggott Beckhart (ed.), *Banking Systems* (New York: Columbia University Press, 1964): 841; U.S. Bureau of Labor Statistics, *Technological Change and Manpower Trends in Six Industries*, Bulletin No. 1817 (Washington, D.C.: US GPO, 1974): 41, 50; U.S. Bureau of Labor Statistics, *Technology and Its Impact on Labor in Four Industries*, Bulletin No. 2242 (Washington, D.C.: US GPO, May 1986): 35; Sandra D. Cooke, "Structural Changes in the U.S. Banking Industry: The Role of Information Technology," June 1997, unpublished paper, Office of Business and Industrial Analysis, U.S. Department of Commerce, p. 34.

17. Other contenders for that claim, however, include electricity and the telephone; both, however, entered the economy in the late nineteenth century and spread across the nation during the first half of the twentieth century.

18. George K. Darling and James F. Chaston, Jr., *The Business of Banking for Bank Directors* (Philadelphia, Penna.: Robert Morris Associates, 1995): 11.

19. Evans and Schmalensee, *Paying with Plastic*, is as useful as any source for an introduction to the issue. The literature on this topic, however, is huge.

20. On life insurance, see Gart, *An Analysis of the New Financial Institutions*, 245–272, and on property and casualty insurance, see 273–293.

21. H. Thomas Krause et al., *Insurance Information Systems*, vol. 1 (Malvern, Penna.: Insurance Institute of America, 1990): 1–31.

22. Introductory literature on this industry is numerous. A good start for the pre-deregulated era is L. E. Davids, *Dictionary of Insurance* (Totowa, N.J.: Rowman & Allanheld, 1983); and for the past two decades of the century, Woelfel, *Banking and Finance*, 598–606.

23. A useful way to be introduced to this industry—the most obtuse of the three financial ones we are discussing—is to read an introductory textbook on its services and, ideally for historical perspective, one that has been through multiple editions over time. For a good example, see Charles P. Jones, *Investments: Analysis and Management* (New York: John Wiley & Sons, 1985, also published in 1988, 1991, the 8th edition published in 2001). These kinds of texts exist for many industries. For insurance, for example, there is Emmett Vaughan, *Fundamentals of Risk and Insurance* (New York: John Wiley & Sons, 1972), which was published in multiple editions into the 1990s; and perhaps the longest-running text on insurance, Kenneth Black, Jr., and Harold D. Skipper, Jr., *Life Insurance* (New York: Meredith Corporation, 1915) and which continued publication through various publishers throughout the century.

24. For introductions to the role of exchanges, see F. E. Block, *Security Analysis* (New York: McGraw-Hill, 1987); K. Park and A. W. van Agtmael (eds.), *The World's Emerging*

Stock Markets (Chicago, Ill: Probus, 1993); and for a series of articles on modern trading habits and technologies, see Jessica Keyes (ed.), *Financial Services Information Systems* (Boca Raton, Fla: Auerbach, 2000): 667–753.

25. For government "grand strategy" regarding policy, see Lewis M. Branscomb (ed.), *Empowering Technology: Implementing a U.S. Strategy* (Cambridge, Mass.: MIT Press, 1993); regarding telecommunications (including radio, TV, and Internet), see Alan Stone, *How America Got Online* (Armonk, N.Y.: M. E. Sharpe, 1997); but also see the memoirs of the Chairman of the Federal Communications Commission from 1993 through 1997, Reed E. Hundt, *You Say You Want a Revolution: A Story of Information Age Politics* (New Haven, Conn.: Yale University Press, 2000); on the special issues of television, Don R. LeDuc, *Cable Television and the FCC: A Crisis in Media Control* (Philadelphia, Penna.: Temple University Press, 1973); George H. Quester, *The International Politics of Television* (New York: Lexington, 1990): 5–28, 109–124; William F. Baker and George Dessart, *Down the Tube: An Inside Account of the Failure of American Television* (New York: Basic Books, 1998): 23–24, 27–31, 103–104.

26. Allen N. Berger, Anil K. Kashyap, and Joseph M. Scalise, "The Transformation of the U.S. Banking Industry: What a Long, Strange Trip It's Been," *Brookings Papers on Economic Activity* 2 (1995): 59–65.

27. For a series of studies on banking regulations through the early 1970s, see William H. Baughn and Charles E. Walker (eds.), *The Bankers' Handbook* (Homewood, Ill.: Dow Jones–Irwin, 1978): 1049–1102.

28. Berger, Kashyap, and Scalise, "The Transformation of the U.S. Banking Industry;" 55–201; Frances X. Frei, Patrick T. Harker, and Larry W. Hunter, "Retail Banking," in David C. Mowery (ed.), *U.S. Industry in 2000: Studies in Competitive Performance* (Washington, D.C.: National Academy Press, 1999): 179–212.

29. Alan Gart, *Regulation, Deregulation, Reregulation: The Future of the Banking, Insurance, and Securities Industries* (New York: John Wiley & Sons, 1994): 103–114.

30. Coopers & Lybrand, "Expansion of Banks into Insurance," 246–247, 252–263; Gart, *Regulation, Deregulation, Reregulation*; 59–79.

31. Geisst, *Wall Street*, 231–236, 251, 259, 263–264; 332–336.

32. On the earlier period, see Milton Friedman and Anna Jacobson Schwartz, *A Monetary History of the United States, 1867–1960* (New York: National Bureau of Economic Research, and Princeton, N.J.: Princeton University Press, 1963): 240–296, 362–363; Clay Anderson, *A Half Century of Federal Reserve Policymaking* (Philadelphia, Penna.: Federal Reserve Bank of Philadelphia, 1965); Thomas K. McCraw, *Prophets of Regulation* (Cambridge, Mass.: Harvard University Press, 1984); Joel Seligman, *The Transformation of Wall Street: A History of the Securities and Exchange Commission and Modern Corporate Finance* (Boston, Mass.: Houghton Mifflin, 1982).

33. "S.900: The Gramm-Leach-Bliley Act," *ABA Banking Journal* 91 (December 1999): 6–8, 8A–8B, 12.

34. A useful introduction to regulatory issues is Gart, *Regulation, Deregulation, Reregulation*.

35. Raymond C. Kolb, "The Paper Work in Banks," in William H. Baughn and Charles E. Walker (eds.), *The Bankers' Handbook* (Homewood, Ill.: Dow Jones-Irwin, 1966): 158.

36. The history of how IT vendors sold their products has yet to be written; indeed, most records needed with which to prepare such studies are still locked up in the archives of those companies that maintained such records, most of which are not accessible by historians. Major exceptions are the Burroughs and CDC collections at the Charles Babbage Institute at the University of Minnesota. Both companies were major players in the financial industry. IBM was too; however, these sorts of records are few. I have commented on how various companies sold in general, in *Before the Computer: IBM, NCR, Burroughs, and Remington Rand and the Industry They Created, 1865–1956* (Princeton, N.J.: Princeton University

Press, 1993); and also in *The Computer in the United States: From Laboratory to Market, 1930 to 1960* (Armonk, N.Y.: M. E. Sharpe, 1993). For passing comments on how other vendors sold, see Franklin M. Fisher, James W. McKie, and Richard B. Mancke, *IBM and the U.S. Data Processing Industry: An Economic History* (New York: Praeger, 1983).

37. On the history of check automation, see the next chapter.

38. James W. Cortada, *The Digital Hand: How Computers Changed the Work of American Manufacturing, Transportation, and Retail Industries* (New York: Oxford University Press, 2004): 89–160. Japanese firms, however, demonstrated levels of coordination with their suppliers that far surpassed those of American firms in the 1960s and 1970s, comparable to what U.S. companies did in later decades. This was especially so with regard to the two national automotive industries. For more on the Japanese model, see Richard J. Schonberger, *Japanese Manufacturing Techniques: Nine Hidden Lessons in Simplicity* (New York: Free Press, 1982); Hiroyuki Hirano, *JIT Factory Revolution: A Pictorial Guide to Factory Design of the Future*, J. T. Black, trans. (Cambridge, Mass.: Productivity Press, 1989); Robert E. Cole, *Managing Quality Fads: How American Business Learned to Play the Quality Game* (New York: Oxford University Press, 1999).

39. Steven M. H. Wallman, "The Information Technology Revolution and Its Impact on Regulation and Regulatory Structure," in Robert E. Litan and Anthony M. Santomero (eds.), *Brookings-Wharton Papers on Financial Services 1999* (Washington, D.C.: Brookings Institution Press, 1999): 207–248; Frederic S. Mishkin and Philip E. Strahan, "What Will Technology Do to Financial Structure?" Ibid., 249–287; Zvi Bodie, "Investment Management and Technology: Past, Present, and Future," Ibid., 343–389; Robert E. Litan, *The Revolution in U.S. Finance* (Washington, D.C.: Brookings Institution, 1991): 1–55.

40. For example, Anthony Gandy and Chris S. Chapman, *Information Technology and Financial Services: The New Partnership* (Chicago, Ill.: Glenlake Publishing and Fitzroy Dearborn, 1997): 3–11; Sandra D. Cooke, "Structural Change in the U.S. Banking Industry: The Role of Information Technology," June 1997, paper number ESA/OPA 97-6, Washington, D.C.: Economics and Statistics Administration Office of Policy Development; William M. Randle, "Banking: The Biggest Risk Is to Do Nothing," in Christina Ford Haylock, Len Muscarella, and Ron Schultz (eds.), *Net Success: 24 Leaders in Web Commerce Show You How to Put the Internet to Work for Your Business* (Holbrook, Mass.: Adams Media Corporation, 1999): 209–237.

41. Ruth Schwartz Cowan, *A Social History of American Technology* (New York: Oxford University Press, 1997): 78–82, 230; Carroll Pursell, *The Machine in America: A Social History of Technology* (Baltimore, Md.: Johns Hopkins University Press, 1995): 91, 92, 239; and most useful, David A. Hounshell, *From the American System to Mass Production, 1800–1932* (Baltimore, Md.: Johns Hopkins University Press, 1984): 218, 249–253.

42. Cortada, *The Digital Hand: How Computers Changed the Work of American Manufacturing, Transportation and Retail Industries*, 78–80. The standard work on the Toyota approach is Taiichi Ohno, *Toyota Production System: Beyond Large-Scale Production* (Cambridge, Mass.: Productivity Press, 1988); see also B. Joseph Pine II, *Mass Customization: The New Frontier in Business Competition* (Boston, Mass.: Harvard Business School Press, 1993); Richard J. Schonberger, *Building a Chain of Customers: Linking Business Functions to Create the World Class Company* (New York: Free Press, 1990): 4–6, 45–47, 218–219.

43. Cortada, *The Digital Hand.*

44. Andrew Tylecote, *The Long Wave in the World Economy: The Current Crisis in Historical Perspective* (London: Routledge, 1991): 36–70. Another earlier, similar analysis is by C. Perez, "Structural Change and Assimilation of New Technologies in Economic Social Systems," *Futures* (October 1983): 357–375.

45. Tylecote, 36. He describes a new style (as opposed to a pre-existing style) "as the most efficient and profitable, change in response to the appearance of new key *factors of*

production which are: (a) clearly very cheap by past standards, and tending to get cheaper, and (b) potentially all-pervasive." Italics are Tylecote's.

46. Jo Anne Yates, "Coevolution of Information Processing Technology and Use: Interaction Between Life Insurance and Tabulating Industries," *Business History Review*, no. 1 (1993): 1–51; Aetna Life & Casualty, *Information Systems at Aetna Life & Casualty: From Punched Cards to Satellites, and Beyond* (Hartford, Conn.: Aetna Life & Casualty, 1984); George E. Delehanty, "Office Automation and the Occupation Structure: A Case Study of Five Insurance Companies," *Industrial Management Review* 7, no. 2 (Spring 1966): 99–109. For dozens of bibliographic citations on early uses of computing in the Insurance Industry, see James W. Cortada, *A Bibliographic Guide to the History of Computer Applications, 1950–1990* (Westport, Conn: Greenwood Press, 1996): 46–48, 174–177.

47. This has been a central issue of my prior research, *Digital Hand: How Computers Changed the Work of American Manufacturing, Transportation, and Retail Industries*; see Cortada, "A Framework for Understanding Technological Change: Lessons from Information Technology, 1868–1997," in Gary D. Libecap (ed.), *Entrepreneurship and Economic Growth in the American Economy* (Amsterdam: JAI, 2000): 47–92; and for an application-centered discussion, "Commercial Applications of the Digital Computer in American Corporations, 1945–1995," *IEEE Annals of the History of Computing* 18, no. 2 (1996): 16–27.

48. Mary J. Cronin (ed.), *Banking and Finance on the Internet*; (New York: John Wiley & Sons, 1997); Jessica Keyes (ed.), *Financial Services Information Systems*. (New York: Auerbach, 2000). Both summarize much deployment data. Industry magazines routinely conduct surveys and publish the results.

49. George Basalla, *The Evolution of Technology* (Cambridge: Cambridge University Press, 1988): 1–25. Basalla does for the history of technology and how it changes what Thomas S. Kuhn did for scientific change, see *The Structure of Scientific Revolutions* (Chicago, Ill.: University of Chicago Press, 1962; 2d edition, enlarged, 1970).

50. There is no formal history of ATMs; however, one historian, Paul E. Ceruzzi, has commented on their origins in *A History of Modern Computing* (Cambridge, Mass.: MIT Press, 1998): 80–81. The best history of the bar code is by Stephen A. Brown, *Revolution at the Checkout Counter: The Explosion of the Bar Code* (Cambridge, Mass.: Harvard University Press, 1997).

51. Historians have not studied the security issue and the Internet; however, the industry trade press has covered the story. The Banking Industry was the first to interact with regulators and technology firms to address the issue, and thus its industry publications are a useful source of material on this topic: Bill Orr, "Banking on the Internet," *ABA Banking Journal* 86 (November 1994): 67, 71, 73, 75; Penny Lunt, "Payments on the 'Net. How Many? How Safe?" Ibid., 87 (November 1995): 46–48, 50, 52, 54; Bill Orr, "Security: What Everyone's Wondering About," Ibid., 87 (December 1995): 45, 47; "Issues and 2001," Ibid., 92 (December 2000): 33–34, 36; Mary Rose, "Internet Security Analysis Report: An Executive Overview," in Keyes, *Financial Services Information Systems*, 395–401.

52. Good research has been done on how military requirements influenced the development of computers in the 1940s through the 1960s. See Kenneth Flamm, *Targeting the Computer: Government Support and International Competition* (Washington, D.C.: Brookings Institution, 1987); and his *Creating the Computer: Government, Industry, and High Technology* (Washington, D.C.: Brookings Institution, 1988); see also Paul N. Edwards, *The Closed World: Computers and the Politics of Discourse in Cold War America* (Cambridge, Mass.: MIT Press, 1996); about the Internet, there is Arthur L. Norberg and Judy E. O'Neill, *Transforming Computer Technology: Information Processing for the Pentagon, 1962–1986* (Baltimore, Md.: Johns Hopkins University Press, 1996); and Janet Abbate, *Inventing the Internet* (Cambridge, Mass.: MIT Press, 1998).

53. On productivity paradox, Brynjolfsson and Kahin, *Understanding the Digital Economy*,

16–18, 76–77; Cortada, *The Digital Hand: How Computer Changed the Work of American Manufacturing, Transportation, and Retail Industries*: 35–40.

54. Carl Shapiro and Hal R. Varian, *Information Rules: A Strategic Guide to the Network Economy* (Boston, Mass.: Harvard Business School Press, 1999).

55. Philip Evans and Thomas S. Wurster, *Blown to Bits: How the New Economics of Information Transforms Strategy* (Boston, Mass.: Harvard Business School Press, 2000): 1–21.

56. Castells is the latest of a long list of voices, *The Internet Galaxy*: 64–115.

57. Recent examples from the Federal Reserve's own economists include Allen N. Berger, Seith D. Bonime, Daniel M. Covitz, and Diana Hancock, "Why Are Bank Profits So Persistent? The Roles of Product Market Competition, Information Opacity, and Regional/Macroeconomic Shocks," March 24, 1999; Allen N. Berger, "The Effects of Geographic Expansion on Bank Efficiency," undated (circa 1999); Allen N. Berger, "Explaining the Dramatic Changes in Performance of U.S. Banks: Technological Change, Deregulation, and Dynamic Changes in Competition," Working Paper No. 01–6, April 1991; Allen N. Berger, "Technological Progress and the Geographic Expansion of the Banking Industry," June 2002; Allen N. Berger, "The Economic Effects of Technological Progress: Evidence from the Banking Industry," September 2002.

58. IBM, *Industry Applications and Abstracts* (White Plains, N.Y.: IBM Corporation, 1988): 7-1–7-29.

59. Ibid., p. 7-1.

60. Ibid., pp. 14-1–14-2.

61. Ibid., pp. 19-1–19-3.

62. David S. Pottruck and Terry Pearce describe this process at Charles Schwab, *Clicks and Mortar: Passion Driven Growth in an Internet-Driven World* (San Francisco, Calif.: Jossey-Bass, 2000); and for a series of case studies, see Mary J. Cronin (ed.), *Banking and Finance on the Internet* (New York: John Wiley & Sons, 1998); and Jessica Keyes (ed.), *Financial Services Information Systems* (Boca Raton, Fla.: Auerbach, 2000): 667–754.

63. "Community Bank Web Census 2001," *ABA Banking Journal* 93 (April 2001): 32–38; "What We Have Here Is Channel Evolution," *ABA Banking Journal* 93 (September 2001): S14–S17; Evans and Schmalensee, *Paying with Plastic*, 86–91, 94.

64. For the United States, the U.S. Bureau of Economic Analysis tracks the nation's inventory of housing by number and cost; the U.S. Department of Commerce tracks major products, including cars, while the Automotive Industry publishes monthly production and sales figures.

65. In the population, 41.2 million did not have health insurance (14.6 percent). The data are for 2001, U.S. Census Bureau, Press Release, September 30, 2002, http://www.census.gov/Press-Release/www/2002/cb02-127.html. Pension data are from Leonard Wiener, "Pining Away for Pensions," December 24, 2001, *USNews.Com*, http://www.com/usnews/biztec/articles/011224/24pension.htm.

66. Wiener, "Pining Away for Pensions."

67. My unpublished estimate places the number of American investors who invested in dot.com stocks at over 50 million, largely through 401(k) and NASDAQ transactions.

68. Lee S. Sproull, "Computers in U.S. Households Since 1977," in Alfred D. Chandler, Jr., and James W. Cortada (eds.), *A Nation Transformed by Information: How Information Has Shaped the United States from Colonial Times to the Present* (New York: Oxford University Press, 2000): 257–280.

69. On the slow adoption of debit cards, Evans and Schmalensee, *Paying with Plastic*, 297–319; "What We Have Here Is Channel Evolution," *ABA Banking Journal* 93 (September 2001): S14, which did not even list use of TV-based banking anymore, but did include telephone and PC banking.

70. Glenn R. Carrroll and Michael T. Hannan, *The Demography of Corporations and Industries* (Princeton, N.J.: Princeton University Press, 2000): 413.

71. But to keep the record straight, buying "on credit" was not invented in the 1950s, rather, as historian Lendol Calder has demonstrated, that practice dated back to the nineteenth century, as did the concepts about linking one's reputation to their ability to repay a loan, in *Financing the American Dream: A Cultural History of Consumer Credit* (Princeton, N.J.: Princeton University Press, 1999): 3–33. Calder focuses this book on the period from 1890 to 1940.

72. Daniel J. Boorstin, *The Americans: The Democratic Experience* (New York: Random House, 1973, Vintage Books ed., 1974): 427.

73. Ibid., 428.

74. Solomon, *Virtual Money*, 201–218; Evans and Schmalensee, *Paying with Plastic*, 85–107; Weatherford, *The History of Money*, 219–232.

75. Cronin, *Banking and Finance on the Internet*, vii.

76. Ibid., viii.

77. Weatherford, *The History of Money*, 219–263; Evans and Schmalensee, *Paying With Plastic*, 207–246.

78. The issue of using history to provide management with context is intriguing, one hardly explored. I have made an initial attempt to define practices for business management that wants to use history, in James W. Cortada, "Learning from History: Leveraging Experience and Context to Improve Organizational Excellence," *Journal of Organizational Excellence* 21, no. 2 (Spring 2002): 23–29.

79. I leave it to business managers to draw conclusions from the work of historians and economists; I leave it to economists to exploit the historic record, while I hope other historians will search the past of American industries to understand the course of events of American society in the last half century.

80. David S. Landes, *The Wealth and Poverty of Nations: Why Some Are So Rich and Some So Poor* (New York: W. W. Norton, 1998): 516, and for a brief discussion of the views of historians, economists, and businessmen, see 514–516.

81. Their seminal works include Thomas H. Davenport and Laurence Prusak, *Information Ecology: Mastering the Information and Knowledge Environment* (New York: Oxford University Press, 1997); Davenport and Prusak, *Working Knowledge: How Organizations Manage What They Know* (Boston, Mass.: Harvard Business School Press, 1998); Don Cohen and Laurence Prusak, *In Good Company: How Social Capital Makes Organizations Work* (Boston, Mass.: Harvard Business School Press, 2001).

82. I reported on earlier research I had done on the changing nature of work, play, and other leisure activities in the United States in previous books, but had not focused on the specific issue of *style*. My earlier findings indicated that Americans were extensive users of information and its technologies in all facets of their lives. Their collective propensity to be early adopters of new information tools was reinforced by the research emerging out of the *Digital Hand* project, which focused on industries but, because those industries were points of interaction with aspects of people's lives (e.g., media for entertainment), reflected the tastes and practices of the public at large. That propensity to use new technologies facilitated the emergence of a new style of interacting with the financial industries, a trend still unfolding but sufficiently far enough along to be something different from what existed in 1950, as Daniel Boorstin suggested. For my earlier research, see Cortada, *21st Century Business: Managing and Working in the New Digital Economy* (Upper Saddle River, N.J.: Financial Times/Prentice-Hall, 2001); and Cortada, *Making the Information Society: Experience, Consequences, and Possibilities* (Upper Saddle River, N.J.: Financial Times/Prentice-Hall, 2002). Also useful for gaining a rich context by taking the story back to the eighteenth century is Alfred D. Chandlers, Jr., and James W. Cortada (eds.), *A Nation Transformed by*

Information: How Information Has Shaped the United States from Colonial Times to the Present (New York: Oxford University Press, 2000).

Chapter 2

1. John A. Kley, "Are You Getting Ready for Electronics?" *Banking* 46 (May 1954): 38. The author had already given the topic some thought because he was, at the time, a member of the American Bankers Association's Bank Management Commission Committee on Mechanization of Check Handling, and the chairman of that committee's Technical Subcommittee. For a description of this committee's work in the mid-1950s, see Ibid., 46 (June 1954): 34.

2. C. M. Weaver, "The Age of Faster Service," Ibid., 46 (June 1964): 133.

3. Allen N. Berger, "The Economic Effects of Technological Progress: Evidence from the Banking Industry," p. 1, unpublished report, September 2002, available from the Federal Reserve Board.

4. Ibid., pp. 3–4; Emma S. Woytinsky, *Profile of the U.S. Economy: A Survey of Growth and Change* (New York: Frederick A. Praeger, 1967): 416.

5. J. Brooke Willis, "United States," in Benjamin Haggott Beckhart (ed.), *Banking Systems* (New York: Columbia University Press, 1964): 839–916; Arnold A. Heggestad and William G. Shepherd, "The Banking Industry," in Walter Adams (ed.), *The Structure of American Industry*, 6th ed. (New York: Macmillan, 1982): 319–347; Allen N. Berger, Anil K. Kashyap, and Joseph M. Scalise, "The Transformation of the U.S. Banking Industry: What a Long, Strange Trip It's Been," *Brookings Papers on Economic Activity* 2 (1995): 55–201; Sandra D. Cooke, "Structural Change in the U.S. Banking Industry: The Role of Information Technology," June 1997, ESA/OPA 97–6 white paper; X. Frei, Patrick T. Harker, and Larry W. Hunter, "Retail Banking," in David C. Mowery (ed.), *U.S. Industry in 2000: Studies in Competitive Performance* (Washington, D.C.: Academy Press, 1999): 179–214; Berger, "The Economic Effects of Technological Progress."

6. Willis, "United States," 839–916; Woytinsky, *Profile of the U.S. Economy*, 414–421.

7. U.S. Bureau of Labor Statistics, *Technological Change and Manpower Trends in Six Industries*, Bulletin 1817 (Washington, D.C.: US GPO, 1974): 41–53.

8. Eric N. Compton (who was the vice president of the Chase Manhattan Bank), *Inside Commercial Banking* (New York: John Wiley & Sons, 1980): 173.

9. Heggestad and Shepherd, "The Banking Industry," 322.

10. The monthly issues of ABA's publication, *Banking*, for the entire period provides hundreds of comments concerning the potential of IT.

11. Heggestad and Shepherd, "The Banking Industry," 329.

12. Ibid., 329–333.

13. Ibid., 342.

14. Berger, "The Economic Effects of Technological Progress," Table 1, p. 37.

15. These are essentially the points made by U.S. government economists looking at the industry, U.S. Bureau of Labor Statistics, *Technology and Its Impact on Labor in Four Industries*, Bulletin No. 2242 (Washington, D.C.: US GPO, May 1986): 35–45.

16. Frei, Harker, and Hunter, "Retail Banking," p. 185.

17. Ibid., pp. 183–188; Cooke, "Structural Change in the U.S. Banking Industry," 15–16.

18. Data from Cooke, "Structural Change in the U.S. Banking Industry," 15–16.

19. Dwight B. Crane and Zvi Bodie, "The Transformation of Banking," *Harvard Business Review* (March–April 1996): 109.

20. Ibid., 111.

21. The data for this and the next paragraph were drawn from Crane and Bodie, "The Transformation of Banking," 109–117.

22. See, for example, Cooke, "Structural Change in the U.S. Banking Industry," 1–41.

23. For an excellent synopsis, see Berger, "The Economic Effects of Technological Progress, Evidence from the Banking Industry."

24. This unintended positive consequence for the economy has not been fully quantified and properly studied, although economists recognized the phenomenon as occurring. For discussion, see Ibid., 2–3.

25. IBM for example, received nearly 100 percent of all its invoices from vendors electronically and paid them the same way. The majority of its employees have been paid electronically since the 1970s.

26. Crane and Bodie, "The Transformation of Banking," 110.

27. Raymond C. Kolb, "The Paper Work in Banks," in William H. Baughn and Charles E. Walker, *The Bankers' Handbook* (Homewood, Ill.: Dow Jones–Irwin, 1966): 159.

28. Ibid., 159.

29. Ibid., 159.

30. Ibid., 160.

31. James L. McKenney, with Duncan C. Copeland and Richard O. Mason, *Waves of Change: Business Evolution through Information Technology* (Boston, Mass.: Harvard Business School Press, 1995): 41.

32. Paul Armer, *Computer Aspects of Technological Changes, Automation, and Economic Progress* (Santa Monica, Calif.: The RAND Corporation, 1966), unpaginated, reprinted by RAND Corporation; National Commission on Technology, Automation, and Economic Progress, *Technology and the American Economy*, Appendix Volume I, *The Outlook for Technological Change and Employment* (Washington, D.C.: U.S. Government Printing Office, February 1966), copy in CBI R, Quarto HD6331.2.US.A766, Charles Babbage Institute Archives, University of Minnesota, Minneapolis.

33. Robert S. Alsom et al., *Automation in Banking* (New Brunswick, N.J.: Rutgers University Press, 1962); "Around the World 70 Times," *Check Clearings*, no. 236 (February 1963): 2; "Automatic Banking," *Data Processing*, no. 3 (July–December 1960): 156–169; John J. Cooley, "Bank Checks and Bank Automation," *Banking* 55, no. 7 (January 1963): 51; Walter Dietrich, "Optical Handling of Checks," *Datamation* 10, no. 9 (September 1964): 39–46; Edward T. Shipley, "The Case for Uniform Punched Cards," *Audigram* (February 1954): 4–5; C. M. Weaver, *Check Operations in the Banks of Tomorrow* (Chicago, Ill.: First National Bank of Chicago, 1954) is an early memoir of automation.

34. For a clear description of the process, including illustrations of the machines, see American Institute of Banking, *Principles of Bank Operations* (New York: American Institute of Banking, 1956): 29–30, 66, 70–71, 86–90, 259, 360; see the same title for check-routing symbols used in the 1950s, which are essentially the same today, p. 359. For general descriptions of precomputer–based uses of information handling technologies, see "Automatic Banking," *Data Processing*, no. 3 (July–December 1960): 156–160. An important series of reports emerged from the industry in the early years regarding MICR and the modern check that captured much of the debate over productivity using technology, justification, and the role of computing. The most important of these are American Bankers Association, *Automation of Bank Operating Procedures* (New York: American Bankers Association, 1955); ABA, *Magnetic Ink Character Recognition: The Common Machine Language for Check Handling* (New York: American Bankers Association, 1956); ABA, *Placement for the Common Machine Language on Checks* (New York: American Bankers Association, 1957); ABA, *National Automation Conference Proceedings* (New York: American Bankers Association, 1963); and ABA, *Automation and the Small Bank* (New York: American Bankers Association, 1964).

35. McKenney, *Waves of Change*, 47; David Rubinfien, "Automation of Bank-Check Accounting," *The Journal of Accountancy* 103 (March 1957): 41–65; "Automatic Banking," *Data Processing* no. 3 (July–September 1960): 156–169; John A. Kley, "Mechanization of Check Handling," *Banking* 48 (February 1956): 60–62. For a description of how the tech-

nology worked, see Robert H. Church, Ralph P. Day, William R. Schnitzler, and Elmer S. Seeley, *Optical Scanning for the Business Man* (New York: Hobbs, Dorman & Company, 1966): 1–66. For a more detailed description of the technology applied to checking, complete with process flowcharts and description of hardware, see IBM, *IBM 1401 Data Processing System with Tapes and the IBM 1419 Magnetic Character Reader for Demand Deposit Accounting* (White Plains, NY: IBM Corporation, 1961), Box B-116-3, DP Application Briefs, IBM Corporate Archives; Digital Data Corporation, *New OCR Methods Speed Data to Bank Computers* (Minneapolis, Minn.: CDC, 1969), CBI 80, Box 1, Series 1, Serial Publications, Charles Babbage Institute, University of Minnesota; John V. Vardealas, *The Computer Revolution in Canada: Building National Technological Competence, 1945–1980* (Cambridge, Mass.: MIT Press, 2001): 145–149.

36. "Automatic Handling of Checks," *Banking* 50 (February 1958): 42–43; John L. Cooley, "Magnetic Ink's Imprint on Banking," Ibid., 52 (November 1959): 42–44; Richard L. Kraybill, "Magnetic Ink and Brass Tacks," Ibid., 52 (January 1960): 48, 121; "What the Annual Reports Say About Electronic Banking," Ibid., 52 (March 1960): 43–44, 111, 115; Richard L. Kraybill, "MICR: What's in It for Smaller Banks?," Ibid., 53 (June 1961): 52–53, 107, 110. For a history of checks with a strong testimonial for adopting the MICR, see William R. Kuhns, "Checks," Ibid., 53 (January 1961): 45–51.

37. Herbert Bratter, "Progress of Bank Automation," *Banking* 55 (September 1962): 47.

38. On the Federal Reserve survey, Ibid., 47–48.

39. Aaron Lebedow, "MICR: Current and Future Applications," *Banking* 55 (December 1962): 60.

40. The history of ERMA has been well-documented, See: McKenney, *Waves of Change*, 41–95; A. Weaver Fisher and James L. McKenney, "The Development of the ERMA Banking System: Lessons from History," *IEEE Annals of the History of Computing* 15, no. 1 (1993): 44–57; James L. McKenney and A. Weaver Fisher, "The Growth of Electronic Banking at Bank of America," Ibid., 15, no. 3 (1993): 87–93; David O. Woodbury, *Let ERMA Do It: The Full Story of Automation* (New York: Harcourt, Brace & World, 1956): 158–180; Homer R. (Barney) Oldfield, "General Electric Enters the Computer Business—Revisited," *IEEE Annals of the History of Computing* 17, no. 4 (1995): 47–55 and his longer study, *King of the Seven Dwarfs: General Electric's Ambiguous Challenge to the Computer Industry* (Los Alamitos, Calif.: IEEE Computer Society Press, 1996): 1–124. For a detailed, contemporary description of the applications done with ERMA, see "D21 The Bank of America's 'Friend ERMA,' " undated report, CBI 55 Market Reports, Box 70, Folder 1, Archives of the Charles Babbage Institute, University of Minnesota.

41. GE's role has recently been described in detail, along with the long-term effects on that firm's role in the Computer Industry, Oldfield, *King of the Seven Dwarfs*, 239–244.

42. McKenney, *Waves of Change*, 79–82, and for subsequent history of the bank's use of computing to the early 1990s, pp. 82–93.

43. Quoted from A. R. Zipf in Hershner Cross, *Computers and Management: The 1967 Leatherbee Lectures* (Boston, Mass.: Graduate School of Business Administration, Harvard University, 1967): 56.

44. IBM, *Banking Operations at Data Center, Inc.* (White Plains, N.Y.: IBM Corporation, undated [1966]), Box B-116-3, DP Application Briefs, IBM Corporate Archives; IBM, *Data Processing Services at the American National Bank of Beaumont* (White Plains, N.Y.: IBM Corporation, undated [1962]), Box B-116-3, DP Application Briefs, IBM Corporate Archives).

45. The first detailed report published by IBM on these applications appeared in 1962, IBM, *General Information Manual, Automated Deposit Proof and Transit* (White Plains, N.Y.: IBM Corporation, 1962), Box B-116-3, DP Application Briefs, IBM Corporate Archives; but see also IBM, *Central Information System at the Bank of Delaware* (White Plains, N.Y.:

IBM Corporation, undated [1966]), Box B-116-3, DP Application Briefs, IBM Corporate Archives); E. A. Casnellie, "Automated Proof and Transit System," in DPMA, *Data Processing*, 9, *Proceedings 1965* (Dallas, Tex.: DPMA, 1965): 91–106.

46. Dale L. Reistad, "The Impact of Automation on the Nation's Banks," *Banking* 57 (October 1964): 51–52.

47. Dale L. Reistad, "The Impact of Automation on the Nation's Banks," Part Two, *Banking* 57 (November 1964): 106–107.

48. John W. Riday, "The 'Checkless Society'," *Banking* 61 (September 1968): 49–50, 145–146.

49. Charles C. Smith, "Misconceptions Impede a Start toward Electronic Funds Transfer," *Banking* 66 (October 1973): 95–96, 98.

50. Peter J. Brennan, "Better Resting Place for Bank Checks?" *ABA Banking Journal* 72 (May 1980): 47.

51. "Turnpiking effect" is IT slang for how computing causes more work rather than less, using the analogy of what happens when a major highway replaces a country road. For example, if someone drove to and from work on a country road and it took 45 minutes each way, then was given the option of taking a superhighway that only required 10 minutes per trip, that individual might drive on the road at lunch time to do some errands or even to go home for a while. Another example is when a city or state adds another lane to a highway and then a couple of years later the road is just as crowded as before. Why? Because now more people feel they can commute from farther distances on this road, as opposed to a smaller road. In effect, turnpiking increases traffic, then slows it down. The same often occurs with computerized applications that are more convenient and thus are used more frequently.

52. Brennan, "Better Resting Place for Bank Checks?" 47.

53. Brennan, "Better Resting Place for Bank Checks?" 47.

54. Bill Orr, "Check Processing the 'Star Wars' Way," *ABA Banking Journal* 72 (May 1980): 56, 58; Joseph Asher, "Keep the Checks, Send the Data," Ibid., 73 (April 1981): 57, 60; "Roundtable of Experts Ponders Payment Systems of the Future," Ibid., 73 (September 1981): 48–49, 53, 55–56, 58, 61–62; Richard Gilgan, "Deregulation's Impact on Check Processing," Ibid., 74 (June 1982): 35, 37; "Checks in Transition: Why, and How Fast?," Ibid., 74 (June 1982): 38, 40, 43, 45, 47; "Bar Code Technology Comes to Check Processing Arena," Ibid., 78 (March 1986): 38, 40.

55. Gilgan, "Deregulation's Impact on Check Processing," Ibid., 74 (June 1982): 35, 37.

56. Gilban, "Bar Code Technology Comes to Check Processing Arena," Ibid., 78 (March 1986): 38.

57. "How Do You Compare on Prime Pass Rejects," Ibid., 79 (May 1987): 62, 64, 66.

58. Orla O'Sullivan, "The Check Is in the e-Mail," Ibid., 90 (January 1998): 52–55.

59. Jeffrey Selingo, "Tear Out a Check, Then Watch It Vanish," *New York Times*, October 9, 2003, p. E6.

60. Berger, "The Economic Effects of Technological Progress," 7–8.

61. Kolb, "The Paper Work in Banks," 168–169.

62. On this earlier technology, see James W. Cortada, *Before the Computer: IBM, NCR, Burroughs, and Remington Rand and the Industry They Created, 1865–1956* (Princeton, N.J.: Princeton University Press, 1993, 2000).

63. Raymond Rogers, "The Great Growth of Time and Savings Deposits," *Bankers Monthly* 79, no. 9 (September 15, 1962): 11–16; "Time and Savings Deposits at Member Banks," *Federal Reserve Bank of New York Monthly Review* 42, no. 7 (July 1960): 118–123.

64. "D22 Bank Posts Benefit from Ramac Investment," CBI 55 Market Reports, undated [circa 1958] Box 70, Folder 1, p. III D20–1, Archives of the Charles Babbage Institute, University of Minnesota.

65. "D24 Ramac at PSFS," undated [circa 1959/60], CBI 55 Market Reports, Box 70, Folder 1, Archives of the Charles Babbage Institute, University of Minnesota.

66. "D20 Applying a Computer to Various Banking Operations," CBI 55 Market Reports, undated [circa 1958] Box 70, Folder 1, p. III D20-1, Archives of the Charles Babbage Institute, University of Minnesota; "D24 Ramac at PSFS," undated [circa 1959/60], CBI 55 Market Reports, Box 70, Folder 1, Archives of the Charles Babbage Institute, University of Minnesota. The latter is a detailed description of the application, complete with a photograph of the system in the bank.

67. For an application brief on the use of two Burroughs 220s at the First Pennsylvania Banking & Trust Company, see "D25 Old Bank Learns New Tricks," undated [circa 1960], CBI 55 Market Reports, Box 70, Folder 1, Archives of the Charles Babbage Institute, University of Minnesota. Like other large banks installing computing, this bank chose to implement multiple applications on these systems: revolving credit plan, retail banking, mutual funds, and demand-deposit accounting.

68. "Automation—With the Old, Familiar Forms," 52 *Banking* (February 1962): 49–50.

69. Herbert Bratter, "Progress of Bank Automation," *Banking* 55 (September 1962): 48.

70. Dale L. Reistad, "The Impact of Automation on the Nation's Banks," *Banking* 57 (October 1964): 52.

71. U.S. BLS, *Technological Change and Manpower Trends in Six Industries*, 43.

72. For an early statistical study of the types and number of tasks performed by customers using ATMs, see "Latest Figures for ATMs and Other Retail Services, *ABA Banking Journal* 74 (March 1982): 110–111.

73. BLS, *Technology and Its Impact on Labor in Four Industries*, 36–37, quote, p. 37.

74. William Friis, "Electronic Journaling Helps Truncate Paper in Branches," *ABA Banking Journal* 73 (January 1983): 82.

75. Penny Lunt, "What Will Dominate the Home?" *ABA Banking Journal* 87 (June 1985): 37.

76. Kolb, "The Paper Work in Banks," 161.

77. For descriptions of loans in the late 1950s and early 1960s, see Ibid., 160–164; Neal J. McDonald, "An On–Line Savings and Loan System," *Datamation* 11, no. 10 (October 1965): 81–84; Horace H. Harrison, "Punch Card Accounting for Installment Loans," in William R. Kuhns and William P. Bogie (eds.), *Present Day Banking 1957* (New York: American Bankers Association, 1957): 153–163.

78. "D22 Bank Posts Benefit from Ramac Investment," p. III D22-3.

79. For a description of computer-based mortgage processing using an IBM 1401 system, see IBM, *General Information Manual, IBM 1401 Tape System for Savings Accounting and Mortgage Accounting* (White Plains, N.Y.: IBM Corporation, 1960): 17–39, Box B-116-3, DP Application Briefs, IBM Corporate Archives.

80. John N. Raleigh, "How a Computer Earns Its Keep," *Banking* 53 (August 1960): 51.

81. Bratter, "Progress of Bank Automation," 48.

82. Dick Brandon, "What Should Bank Management Expect from Automation?" *Banking* 56 (August 1963): 134.

83. Reistad, "The Impact of Automation on the Nation's Banks," *Banking*, 53; Reistad, "National Automation Survey of 1966," 59 (September 1966): 38.

84. For an early explanation of credit-scoring as an operations research application, see H. Martin Weingartner, "Concepts and Utilization of Credit-Scoring Techniques," *Banking* 58 (February 1966): 51–53.

85. Dale L. Reistad, "New Technological Developments," *Banking* 58 (April 1966): 34–35. For case studies of loan management on a computer, see John J. Cross, Jr., "How We Put Commercial Loans On-Line," Ibid., 58 (May 1966): 39–40. Henry J. Coffey and Francis R. Murphy, "Imagination and Automation Speed Installment Loan Handling," *Banking* 61 (February 1969): 83–84.

86. When federal insurance became a reality, thereby eliminating the possibility of bank runs, it had the unintended consequence of making it easier for banks and savings and loans to make riskier loans than they would have otherwise, causing the system to be abused. That resulted in the failure of many banks in the 1980s. For details, see Robert E. Litan, *The Revolution in U.S. Finance* (Washington, D.C.: Brookings Institution, 1991): 2–19.

87. Cooke, "Structural Change in the U.S. Banking Industry," 11–15. On the balance of power tipping toward the consumer, see David Siegel, *Futurize Your Enterprise: Business Strategy in the Age of the E-Customer* (New York: John Wiley & Sons, 1999); and Philip Evans and Thomas S. Wurster, *Blown to Bits: How the New Economics of Information Transforms Strategy* (Boston: Harvard Business School Press, 2000).

88. As early as the late 1980s, expert systems were coming into the banking industry. See, for example, Carol McGinn, "Computer Mentors Give Loan Officers a Hand," *ABA Banking Journal* 82 (November 1990): 45, 49, 51, 53, quote p. 45; "New Automated 'Experts' Ready for Lenders," Ibid., 84 (January 1992): 61–62; Mark Arend, "Outsource Consumer Lending?," Ibid., 85 (August 1993): 43–44; "What Will Dominate the Home?," Ibid., 87 (June 1995): 36–38, 40, 42, 45; Steve Cocheo, "Mortgage Machine," Ibid., 87 (October 1995): 53–54, 56; Steve Cocheo, "Scoring Gains Ground While the Web Waits," Ibid., 89 (January 1997): 52, 54.

89. For descriptions of the credit card process, see D. Dale Browning, "Bank Cards," in Baughn and Walker, *Bankers' Handbook*, 879–894; Charles T. Russell, "Charge Cards: Where Do We Go From Here?" in *Data Processing: Proceedings 1967* (Boston: Data Processing Management Association, 1967): 277–282.

90. David Evans and Richard Schmalensee, *Paying with Plastic: The Digital Revolution in Buying and Borrowing* (Cambridge, Mass.: MIT Press, 1999): quote and data, p. 25.

91. Data from U.S. Census Bureau, http://www.census.gov/prod/2004pubs/03statab/domtrade.pdf (last accessed 3/14/2004).

92. Universal refers to the fact that a card could be used in any state or, for that matter, anywhere in the world, as opposed to credit cards of earlier years that could only be used in a defined area, such as within a state or only with merchants who subscribed to a limited size network of a credit card service provider.

93. Evans and Schmalensee, *Paying with Plastic*, 61–65.

94. Ibid., 66.

95. Ibid., 85–110.

96. Ibid.

97. Robert D. Breth, "How to Make Credit Cards Profitable," *Banking* 60 (February 1968): 51–53, quote, p. 51; Charles T. Russell, "Charge Cards—Where Do We Go from Here?" *Data Processing*, 12, *Proceedings 1967* (Boston, Mass.: DPMA, 1967): 277–282; Robert A. Johnson, "Credit—and Credit Cards," in Paul F. Jessup (ed.), *Innovations in Bank Management: Selected Readings* (New York: Holt, Rinehart and Winston, 1969): 101–109.

98. Thomas N. Overton, "How a Computer Solved Our Credit Card Problems," *Banking* 61 (October 1968): 100.

99. Ibid., 100–101.

100. William B. Stevens, "Expanding Opportunities in the Bank Card Market," *Banking* 63 (July 1970): 47–48, 83; Milton O'Neil, "Charge-card Networks Are Working," Ibid., 65 (September 1973): 116, 118, 120; Kenneth M. Adams, "The Bank Card: Yesterday, Today, Tomorrow," Ibid., 66 (September 1974): 118–120; "Special Report: Bank Cards," Ibid., 68 (August 1976): 50, 54, 56, 60, 97; Joe Asher, "Five Cards for One Bank?," Ibid., 69 (May 1977): 39–41, 130; Henry J. Schaberg, "Story of a First Generation Bank Card," Ibid., 69 (September 1977): 116; Charles W. McCoy, "Bank's Card Has Big Side Effects," Ibid., 70 (September 1978): 66, 68.

101. Adams, "The Bank Card," 119.

102. "Changing Face of Automation," *Banking* 67 (September 1975): 32.

103. Donald P. Jacobs, "Taking a Look at Banking in the '80s," Ibid., 72 (August 1980): 66.

104. "Visa and MasterCard: Self-Evaluations," Ibid., 74 (September 1982): 48, 50, 52, 55–56, 58, 60, is an interview with the presidents of both organizations who made it perfectly clear they intended to do more than just process credit card transactions. The story is updated for the 1980s in Daniel Hall, "Going for the (Plastic) Gold," *ABA Banking Journal* 81 (September 1989): 102, 104, 106.

105. See, for example, Steve Cocheo, "Bank Cards at the Crossroads," *ABA Banking Journal* 79 (September 1987): 66, 68–69, 71–72, 75, 77, 79.

106. Thomas D. Steiner and Diogo B. Teixeira, *Technology in Banking: Creating Value and Destroying Profits* (Homewood, Ill.: Business One Irwin, 1990): 129.

107. Ibid., 112–143.

108. Mark Arend, "Uneasy at the Top," *ABA Banking* Journal 83 (December 1991): 87.

109. Ibid., 88.

110. Mark Arend, "Card Profits: How Far Will They Slide?" Ibid., 84 (September 1992): 79, 81–82; "Visa's Payment Network Girds for Growth," Ibid., 102, 104.

111. Mark Arend, "Smart Cards: Are Banks Missing the Boat?" Ibid., 85 (July 1993): 68, 70–71; Penny Lunt, "The Smart Cards Are Coming! But Will They Stay?" Ibid., 87 (September 1995): 46–48, 50, 52; Bill Orr, "The Great Card Question: Will It Be Smart or Debit?," Ibid., 90 (September 1998): 54–56, 58; Lauren Bielski, "Smart Cards, Coming Up to Bat," Ibid., 90 (November 1998): 57, 58, 60, 62, in which 11 pilot projects are reviewed; Evans and Schmalensee, *Paying with Plastic*, 321–322.

112. On First Data Corporation, the largest credit card transaction processor of the 1990s, see Alan Levinsohn, "The Rise and Sprawl of a Card Services Empire," *ABA Banking Journal*, 90 (July 1998): 45.

113. Ibid., 45–47.

114. Evans and Schmalensee, *Paying with Plastic*, 297.

115. Ibid., 298–300.

116. Ibid., 314.

117. Ibid., 315–316; Jeannine Aversa, "Electronic Payments Lead Checks," *Wisconsin State Journal*, December 7, 2004, p. C9.

118. Evans and Schmalensee, *Paying with Plastic*, Ibid., 308.

119. Patricia A. Murphy, "Regional Networks Drive Debit Growth," *ABA Banking Journal* 84 (September 1992): 6, 73, 75–76.

120. An early tagline on an article by Robert E. O'Neill captured the essence of the business problem: "Who pays for what?" in his article, "What's New in EFT?" *Progressive Grocer* (August 1986): 59.

121. Not to be confused with e-business or B2B (business to business), which involves transactions among organizations and usually were not done using credit cards for payments except for small transactions and even then on an exceptional basis.

122. Solomon, *Virtual Money*, 219–245; Weatherford, *The History of Money*, 245–247; Steven M. H. Wallman, "The Information Technology Revolution and Its Impact on Regulation and Regulatory Structure," in Robert E. Litan and Anthony M. Santomero (eds.), *Brookings-Wharton Papers on Financial Services 1999* (Washington, D.C.: Brookings Institution Press, 1999): 207–248.

123. Solomon, *Virtual Money*, ix.

124. Charles J. Woelfel, *Encyclopedia of Banking and Finance*, 10th ed. (Chicago, Ill.: Probus Publishing, 1994): 62.

125. IBM, *Industry Applications and Abstracts* (New York: IBM Corporation, 1988): 7-21–7-22.

126. For a clear, early description of the ATM application, see "24-Hour Automatic Tellers: How Big a Boom?" *Banking* 65 (February 1973): 17–19, 78.

127. Steiner and Teixeira, *Technology in Banking,* 91–92.

128. Ray M. Hayes, "The ATM at Age Twenty: A Productivity Paradox," *National Productivity Review* 9, no. 3 (Summer 1990): 273–280; Linda Fenner Zimmer, "ATM Installations Surge," *The Magazine of Bank Administration* (May 1980): 29–33.

129. "24-Hour Automatic Tellers," 78.

130. "Changing Face of Automation," 33.

131. "Our ATMs Now Are Busy Around the Clock," *Banking* 69 (October 1977): 90.

132. Ibid., 90.

133. One could speculate that the larger the bank, the more administrative overhead existed, adding costs that had to be recovered by requiring higher revenues from an ATM. Technology long affected corporate sales, general, and administrative (SG&A) expenses and is always subject to measurement by management in large enterprises.

134. "ATMs and Travelers Checks: Worthwhile for Bankers?" *ABA Banking Journal* 73 (January 1981): 82, 84. For an anecdotal example, see Bill Treeter, "Midwest Bank Proves an all-ATM Office Can Work," Ibid., 73 (February 1981): 37–38. For the survey data, see "Latest Figures for ATMs and Other Retail Services," Ibid., 74 (March 1982): 110–112.

135. "What's the Real Significance of National ATM Networks?" Ibid., 74 (September 1982): 38, 40, 43–44, 47.

136. Paul Strassmann, *The Business Value of Computers* (New Canaan, Conn.: The Information Economics Press, 1990): 324–325, 384–385.

137. "Technology Catches On, But People Still Like People," *ABA Banking Journal,* 78 (May 1986): 111.

138. For additional deployment data from the same period, see "Crashing the ATM Wall," Ibid., 78 (September 1986): 82–83.

139. Meena Modi, "ATMs: At What Cost?," Ibid., 79 (November 1987): 44. See also pages 46, 48.

140. Ibid.

141. Steiner and Teixeira, *Technology in Banking,* 93.

142. Ibid., 94. For discussions on how one might improve the economic efficiency of ATMs, see "It's Time to Start Thinking About Redeploying ATMs," *ABA Banking Journal,* 77 (September 1985): 53, 56; "Bankers Forum: Attempting to Scale ATMs' 33% Wall," Ibid., 77 (December 1985): 34–35. The percentage refers to the fact that for many years only a third of a bank's customers used ATMs.

143. "Reshaping the Networks," *ABA Banking Journal* 79 (November 1987): 50.

144. Ibid., 95.

145. "High-tech Branches Streamline Customer Service," Ibid., 83 (July 1991): 38, 40, 44; Mark Arend, "Are Your Bank's ATMs Earning Their Keep?" Ibid., 83 (November 1991): 37–38, 42; Mark Arend, "Technology Shatters 'Bankers' Hours," Ibid., 85 (May 1993): 57–58, 60–61.

146. Judge W. Fowler, "The Branch Is Dead! Long Live the Branch!," Ibid., 87 (April 1995): 41.

147. Ibid.

148. Beverly K. Wayne and Curtis B. Wayne, "Branches for a Small Planet," Ibid., 87 (August 1995): 43; but see the entire article for examples, pp. 43–45. See also Lauren Bielski, "Rethinking the Branch Experience," Ibid., 91 (August 1999): 28–30, 32, 34, 46.

149. For his views and the actions taken by him within his own bank, see Steve Cocheo, "Main Street Meets the 21st Century," Ibid., 87 (October 1985): 44–45, 47, 49, 51; quote, p. 44.

150. Joe Asher, "The Second ATM Revolution," Ibid., 90 (May 1998): 51–52, 54, 56.

151. "What We Have Here Is Channel Evolution," Ibid., 93 (September 2001): S14.

152. For a short, excellent description of automated clearing house applications, see "Automated Clearing House," in Woelfel, *Encyclopedia of Banking and Finance,* 61–62.

153. Mark J. Flannery and Dwight M. Jaffee, *The Economic Implications of an Electronic Monetary Transfer System* (Lexington, Mass.: Lexington Books, 1973): 1.

154. Ibid., for quoted statements by Federal Reserve officials, pp. 1–2.

155. Ibid., 2; Richard M. M. McConnell, "The Payments System: How It Works Now and Why It Is Changing," *Banking* 66 (May 1974): 35; Federal Reserve Board of Governors, *The Federal Reserve System: Purposes and Functions* (Washington, D.C.: Board of Governors, 1974).

156. Richard M. M. McConnell, "Fed's Mitchell Says Expanding RCPCs Are Doing Well—And Getting Better," *Banking* 65 (May 1973): 30–32; "FRCS '80: What Is It, and Is It Important?" Ibid., 74 (February 1982): 82, 84; Brown R. Rawlings, "Historical Perspective on Fed's Check Clearing Role," Ibid., 74 (October 1982): 43, 45–46, 48.

157. Ibid., 42–44; "Fed Takes Aim at Delayed Disbursement," *ABA Banking Journal* 80 (September 1988): 22, 24; "Consolidation, Truncation Mark Fed Operations Strategy," Ibid., 84 (August 1992): 58, 60; Alan Levinsohn, "Fed Taps Small Firm for EFT Translation Software," Ibid., 90 (May 1998): 76.

158. Mark G. Bender, *EFTS: Electronic Funds Transfer Systems Elements and Impact* (Port Washington, N.Y.: Kennikat Press, 1975): 39; and for a detailed description of the application, pp. 37–83.

159. For case studies on use, see Eugene M. Tangney, "Electronic Funds Transfer (EFTS) in Depository Institutions," in Baugh and Walker, *Bankers' Handbook* (Homewood, Ill.: Dow Jones-Irwin, 1978): 241–254; Robert M. Klinger, "Electronic Delivery Systems for Retail Services," Ibid., 909–920.

160. Charles Gardner, "Progress in Atlanta: Automated Payments Plan May Set U.S. Pattern," *Banking* 64 (April 1972): 22.

161. Ibid., 22–26.

162. Robert F. Clayton, "Electronic Funds Transfer Is Coming," Ibid., 65 (September 1972): 42, 44, 46.

163. "Electronic Funds Transfer Systems: One, Two or More? Bank-Run or Fed-Run?" Ibid., 66 (July 1974): 26–27, 52–54, 57, 62–63; Peter Brennan, "For Automated Clearing Houses, the Greatest Need Is Volume," Ibid., 68 (June 1976): 37, 96, 98; "Stating the Case for Community Banks' Role in EFTS," Ibid., 67 (May 1975): 31–33; "Correspondents Examine Economics of EFTS," Ibid., 67 (November 1975): 31–33, 80, 82–83, 86–100; "Automated Clearing House Network: Its Potential for Growth," Ibid., 71 (April 1979): 94, 96, 99, 101, 103; "After a Decade, Where Is EFT Headed?" Ibid., 72 (May 1980): 82, 84, 86, 88, 91.

164. For a history of TYME, see Paul F. P. Coenen, "Four Banks and TYME," Ibid., 71 (May 1979): 40–42. On an important network of the time, CHIPS, including a flowchart of its transaction process, see "CHIPS Clears Same-Day, at $200 Billion a Day," Ibid., 73 (November 1981): 84–86.

165. Bankers were bombarded with studies and monographs describing EFTS and their business justification at the dawn of the new decade. See, for example, Allen H. Lipis, Thomas R. Marschall, and Jan H. Linker, *Electronic Banking* (New York, N.Y.: John Wiley & Sons, 1980); Kent W. Colton and Kenneth L. Kraemer (eds.), *Computing and Banking: Electronic Funds Transfer Systems and Public Policy* (New York: Plenum Press, 1980); and an earlier study, Thomas Kleinschmit, *Electronic Funds Transfer: The Future Is Now* (Minneapolis, Minn.: Federal Reserve Bank of Minneapolis, September 1976). See also "Private-Sector ACH Movement Is Rolling Along," Ibid., 77 (April 1985): 101, 105, 107, 109 "ACH: Has Its Time Arrived?" *ABA Banking Journal*, 79 (May 1987): 70–73; Alec Smith, "ECH Ahead of Its Time," Ibid., 80 (May 1988): 69–70, 72; "Use of ACH on Rise in the Midwest," Ibid., 80 (December 1988):66–68.

166. "Selling the ACH," Ibid., 81 (March 1989): 46.

167. Edward E. Furash, "Payments System Under Siege," Ibid., 86 (June 1994): 55.
168. Ibid., 55, 56.
169. BLS, *Technological Change and Manpower Trends in Six Industries*, 47–48; BLS, *Technology and Its Impact on Labor in Four Industries*, 36–37.
170. Penny Lunt, "Payments on the 'Net: How Many? How Safe?" *ABA Banking Journal* 87 (November 1995): 46–48, 50, 52, 54; Bill Orr, "Security: What Everyone's Wondering About," Ibid., 87 (December 1995): 45, 47; "Could e-Cash Threaten Payment Integrity?" Ibid., 89 (November 1997): 57, 62, 64–66, 68.
171. Clay Hamlet, "Community Banks Go Online," Ibid., 92 (March 2000): 61–64.
172. Many commentators took this point of view. One of the most distinguished and thorough researchers who embraced this notion is Manuel Castells, *The Rise of the Network Society* (Oxford, Eng.: Blackwell, 1995): 151–326.

Chapter 3

1. James W. Cortada, *The Digital Hand: How Computers Changed the Work of American Manufacturing, Transportation, and Retail Industries*, vol. 1 (New York: Oxford University Press, 2004).
2. For an early account of the bank, see Penny Lunt, "Welcome to *sfnb.com*: The Paradigm Just Shifted," *ABA Banking Journal* 87 (December 1995): 40–41, 43, 45.
3. John A. Kley was vice-president of The County Trust Company in White Plains, New York, at the time he made the comment, published in his article, "Are You Getting Ready for Electronics?" *Banking* 46 (May 1954): 38.
4. C. M. Weaver, "The Age of *Faster Service*," *Banking* 46 (June 1954): 34.
5. "Univac Cuts Accounting Expense," *Banking*, 47 (May 1955): 104.
6. Periodically, these would be described in *Banking*, and beginning in the 1970s in IT publications, such as *Computerworld*.
7. Paul E. Ceruzzi, *A History of Modern Computing* (Cambridge, Mass.: MIT Press, 1998): 143–306; Emerson W. Pugh, Lyle R. Johnson, and John H. Palmer, *IBM's 360 and Early 370 Systems* (Cambridge, Mass.: MIT Press, 1991) provide a massive history of the technology of the period.
8. Robert V. Head, "Banking Automation: A Critical Appraisal," *Datamation* (July 1965): 24–28; Frederick S. Hammer, "Management Science in Banking," in Paul F. Jessup (ed.), *Innovations in Bank Management: Selected Readings* (New York: Holt, Rinehart, and Winston, 1969): 285–292; Dale Reistad, "Bank Automation," in Alan D. Meacham (ed.), *Data Processing Yearbook* (Detroit, Mich.: American Data Processing, 1962): 162–165, 203.
9. C. Russell Cooper, "Management Information Systems for Banking," *Data Processing*, 9, *Proceedings 1965* (Dallas: DPMA, 1966): 68.
10. Forde Steele, "Automation's Acid Test: Does It Pay?" *Banking* 50 (May 1958): 42.
11. Dale L. Reistad, "The Impact of Automation on the Nation's Banks," *Banking*, 57 (October 1964): 51–56, second part of the same report, Ibid. (November 1964): 106–109; Dale L. Reistad, "National Automation Survey of 1966," Ibid., 59 (September 1966): 35–38, and part two of same survey, Ibid. (October 1966): 50–55.
12. Reistad, "National Automation Survey of 1966," Ibid., 59 (September 1966): 52.
13. James A. O'Brien, "How Computers Have Changed the Organizational Structure," *Banking*, 61 (July 1968): 93–95; and for a case study of internal lobbying, see Atushi Akera, "Engineers or Managers? The Systems Analysis of Electronic Data Processing in the Federal Bureaucracy," in Agatha C. Hughes and Thomas P. Hughes (eds.), *Systems, Experts, and Computers* (Cambridge, Mass.: MIT Press, 2000): 191–220.
14. Willis J. Wheat, "Bank Management and the Computer," *Banking* 61 (August 1968): 60.

15. Ibid., 61.

16. Horst Brand and John Duke, "Productivity in Commercial Banking: Computers Spur the Advance," *Monthly Labor Review* (December 1982): 19.

17. Ibid., 22. The issue of productivity is also well discussed by Allen N. Berger, "The Economic Effects of Technological Progress: Evidence from the Banking Industry," undated [2002], Federal Reserve System; he includes all the key bibliography.

18. There exists a massive body of material from the period on this subject, prepared by the publishers of *Office Automation* and *Office Automation Applications* handbooks, now housed at the Charles Babbage Institute. These include case studies of implementation and descriptions of specific products, such as check-processing equipment from Burroughs and other vendors; CBI 55, Computer Product and Market Reports Collection, Archives of the Charles Babbage Institute, University of Minnesota.

19. "How Banking Tames Its Paper Tiger," *Business Review* (Federal Reserve Bank of Philadelphia) (June 1960).

20. Brand and Duke, "Productivity in Commercial Banking," 23.

21. One of the most comprehensive surveys on deployment of computers by industry for the year 1977 reported that 17.7 percent of all systems went into the financial sector, second after all of manufacturing, which claimed 25 percent, Philip S. Nyborg et al., *Information Processing in the United States: A Quantitative Survey* (Montvale, N.J.: AFIPS Press, 1977): 4.

22. U.S. Bureau of Labor Statistics, *Technological Change and Manpower Trends in Six Industries*, Bulletin 1817 (Washington, D.C.: U.S. GPO, 1974): 42–44.

23. Arnold Kaplan and Per Lange, "Highlights from the '79 Bank Telecommunications Survey," *Banking* 72 (February 1980): 92.

24. "Latest Figures for ATMs and Other Retail Services," Ibid., 74 (March 1982): 111.

25. Ibid., 112.

26. "How Do You Compare on Prime Pass Rejects?" *ABA Banking Journal* 79 (May 1987): 62, 64, 66.

27. U.S. Bureau of Labor Statistics, *Technology and Its Impact on Labor in Four Industries*, Bulletin 2242 (Washington, D.C.: U.S. GPO, May 1986): 35–41.

28. For additional data on bank expenditures for IT, see William C. Hunter and Stephen G. Timme, "Technological Changes in Large U.S. Commercial Banks," *Journal of Business* 64, no. 3 (1991): 339–362.

29. Those expenditures were actually down by between 5 percent and 7 percent from the previous year's, but rose again later in the decade. "Technology Report: Smarter Bankers, Fewer Vendors," *ABA Banking Journal* 85 (June 1993): 71–72. On the effects of M&As, see Osamu Hirokado, "Competition in the Financial Industry: Who Can Survive?" white paper released by Program on Information Resources Policy, Center for Information Policy Research, Harvard University, 1995, pp. 52–57.

30. "Who's Right," *Banking* 65 (October 1972): 19–24, 80–81; Paul W. Fields, "Can a $23-Million Bank Be Happy With Its Own Computer? This One Is," Ibid., 65 (May 1973): 33–34.

31. "24-Hour Automatic Tellers: How Big a Boom?" Ibid., 65 (February 1973): 17–19, 78; "ATMs and Travelers Checks: Worthwhile for Bankers?" Ibid., 73 (January 1981): 82, 84; Bill Streeter, "Midwest Bank Proves an All-ATM Office Can Work," Ibid., 73 (February 1981): 37–38; "What Role for ATMs?" Ibid., 80 (November 1988): 45, 48.

32. Milton O'Neal, "Point-of-Sale Systems: 'Still Testing'," Ibid., 66 (January 1974): 21–23, 88–89; Joe Asher, "What the Point-of-Sale Revolution Means to Banks," Ibid., 66 (August 1974): 32–34, 77–79; "POS: Is the Future Now?" Ibid., 78 (September 1986): 66, 68, 70; Alec Smith, "POS: Where Have We Been? Where Are We Going?" Ibid., 80 (September 1988): 104, 106, 108; Mark Arend, "Are Your Bank's ATMs Earning Their Keep?" Ibid., 83 (November 1991): 37–38, 42.

33. "Contrasting Viewpoints on How to Attain Electronic Funds Transfer," Ibid., 66 (May 1974): 29–32, 82, 84, 86, 88, 90, 93–94; "Electronic Funds Transfer Systems: One, Two or More? Bank-run or Fed-run?" Ibid., 66 (July 1974): 26–27, 52–54, 57, 62–63; "Stating the Case for Community Banks' Role in EFTS," Ibid., 67 (May 1975): 31–33; "Correspondents Examine Economics of EFTS," Ibid., 67 (November 1975): 31–33, 80, 82–83, 86, 88, 90, 92, 94, 96, 98, 100; "The Electronic Era: Greater Insurance Risks," Ibid., 77 (January 1985): 36–38.

34. "Automated Clearing House Network: Its Potential for Growth," Ibid., 71 (April 1979): 94, 96, 99, 101, 103; James F. Lordan, "More Changes Needed to Reach ACH Potential," Ibid., 75 (April 1983): 38–40, 42; Robert L. Caruso, "New Look at ACH Cost/ Benefit Details," Ibid., 75 (April 1983): 44, 46; "ACH: Has Its Time Arrived?" Ibid., 79 (May 1987): 70–72.

35. Peter J. Brennan, "The Pros and Cons of Stand-alone Minicomputers for Trust Work," Ibid., 72 (January 1980): 92; David M. Ahiers, "Personal Computers: You Can't Afford to Ignore Them, But Be Cautious," Ibid., 73 (October 1981): 101–102, 104; "Micro–Decision Time," Ibid., 75 (August 1983): 61, 64, 66–67, 69–70, 72, 77, 79–80, 82, 85, 89; Lawrence H. Cooke, Jr., "PCs Are Better Than 'Dumbies,' " Ibid., 78 (October 1986): 20, 22; Jeffrey M. Lynn and Linda J. Peters, "Why Don't All Banks Go for Databases?" Ibid., 76 (June 1984): 106, 111; Alec Smith, "EDI: Will Banks Be Odd Man Out?" Ibid., 80 (November 1988): 77, 79, 81.

36. Joe Asher, "With Pressure on Profits, Banks Must Cut Their Operations Costs," Ibid., 68 (June 1976): 33–35, 148; Peter J. Brennan, "How Cheaper and Abler Computers Are Affecting Correspondent Banking," Ibid., 68 (November 1976): 41, 56, 58; Bill Orr, "More Banks Shifting to Outside Data Processing," Ibid., 72 (October 1980): 60, 63–64, 67; John E. P. Borden, Jr., "Private Networks Help Turn Costs into Assets," Ibid., 76 (January 1984): 44–46; "Will the Pendulum Swing Back to Service Bureaus?" Ibid., 76 (December 1984): 76; Donald S. Atkinson, "Is an In-House Computer Right for Your Bank?" Ibid., 77 (October 1985): 44–45; "Cutting Costs While Calling Bonds," Ibid., 82 (September 1990): 120, 123.

37. See, for example, Charles W. McCoy, "Bank's Card Has Big Side Effects," Ibid., 70 (September 1978): 66, 68; "Roundtable of Experts Ponder Payment Systems of the Future," Ibid., 73 (September 1981): 48–49, 53, 55–56, 58, 61–62; Bill Streeter, "Bankers Rethink Key Payment Systems Issue," Ibid., 74 (April 1981): 37–38, 40; Donald J. Dunaway, "Automating Check Returns: What's the Payback?" Ibid., 76 (May 1984): 46–48; Richard Gilgan, "Deregulation's Impact on Check Processing," Ibid., 74 (June 1982): 35, 37; "Checks in Transition: Why, and How Fast?" Ibid., 74 (June 1982): 38, 40, 43, 45, 47; Steve Cocheo, "Bank Cards at the Crossroads," Ibid., 79 (September 1987): 66, 68–69, 71–72, 75, 77–78; M. Williams Friis, "Is POS Approaching Critical Mass?" Ibid., 77 (September 1985): 49–51; Mark Arend, "Card Profits: How Far Will They Slide?" Ibid., 84 (September 1992): 79, 81–82.

38. "Technology: Ally or Enemy of Customer Service?" Ibid., 84 (September 1992): 88, 90, 92; Mark Arend and Penny Lunt, "Rethinking the Branch," Ibid., 85 (October 1993): 38–40, 42–43.

39. Blaire C. Shick, "Privacy—The Next Big Issue in EFT," Ibid., 68 (March 1976): 70, 72, 74, 76; Bill Streeter, "People, More Than Technology, Are Still Key to EFT Security," Ibid., 74 (July 1982): 29, 33–34, 37; Donald G. Miller, "Beware of the 'Hacker' Attack," Ibid., 76 (November 1984): 50, 53, 56; "How Safe Are ATMs?" Ibid., 79 (November 1987): 58, 60; "War Prompts Banks to Review Security," Ibid., 83 (March 1991): 76, 80. The issues were generally resolved by economists in favor of significant productivity gains for the industry using computers, see Berger, "The Economic Effects of Technological Progress."

40. Peter J. Brennan, "Changing Technology Dictates Changing Management Style," Ibid., 69 (March 1977): 44. The literature on the changing nature of bank management and organizations, while not rich, is sufficient to reflect the issues of the day. See, for ex-

ample, John W. Spiegel and Robert M. Horton, "Systems for Producing Bank Services—Concepts and Needs," in William H. Baughn and Charles E. Walker (eds.), *The Bankers' Handbook* (Homewood, Ill.: Dow Jones–Irwin, 1978): 221–229; Craig G. Nelson, "Managing the Data Processing System for Banks," Ibid., 230–240; Bertrand Olivier and Thierry Noyelle, *Human Resources and Corporate Strategy: Technological Change in Banks and Insurance Companies in France, Germany, Japan, Sweden, and the United States* (Paris: O.E.C.D., 1988); Arnold A. Heggestad and William G. Shepherd, "The Banking Industry," in Walter Adams (ed.), *The Structure of American Industry*, 6th ed. (New York: Macmillan Publishing, 1982): 319–347, but especially pp. 338–340; Thomas D. Steiner and Diogo B. Teixeira, *Technology in Banking: Creating Value and Destroying Profits* (Homewood, Ill.: *Business One* Irwin, 1990): 241–262.

41. *Banking* 69 (March 1977), 46.

42. Donald P. Jacobs, "Taking a Look at Banking in the '80s—I," Ibid., 72 (August 1980): 65–66, 70.

43. Bill Streeter, "New Players to Push for More Competition," Ibid., 73 (January 1983): 75.

44. Ibid., 70, 72, 75.

45. Frances X. Frei, Patrick T. Harker, and Larry W. Hunter, "Retail Banking," in David C. Mowery (ed.), *U.S. Industry in 2000: Studies in Competitive Performance* (Washington, D.C.: National Academy Press, 1999): 195.

46. Ibid., 210.

47. Ibid., 190–195, 207–210.

48. JP Morgan signed a seven-year outsourcing contract with IBM, valued in excess of $5 billion. Part of the arrangement calls for the bank to pay fees to IBM for the amount of computing it uses, including services on a pay-as-you-go basis, a relatively new set of terms within both the computing and banking industries.

49. Discussed in detail by Berger, "The Economic Effects of Technological Progress."

50. *Proceedings, National Automation Conference, American Bankers Association, New York, July 13–16, 1964*, p. 38.

51. Berger, "The Economic Effects of Technological Progress," 28.

52. Steiner and Teixeira, *Technology in Banking*, 3–9.

53. Ibid., 14–16. Citicorp's portfolio continued to shift toward credit cards and consumer loans into the 1990s.

54. Ibid., 59–62.

55. Ibid., 64–66.

56. Sandra D. Cooke, "Structural Change in the U.S. Banking Industry: The Role of Information Technology," unpublished paper, June 1997, U.S. Department of Commerce, BEA, ESA/OPA 97–6, pp. 31–34. She concludes IT did generally make banks more profitable and allowed them to increase revenues.

57. And they were profitable. See Berger, "The Economic Effects of Technological Progress."

58. Ibid.,

59. Allen N. Berger, "The Consolidation of the Financial Services Industry: Causes, Consequences, and Implications for the Future," unpublished paper, November 1998, Federal Reserve Board.

60. For example, Mark J. Flannery and Dwight M. Jaffee, *The Economic Implications of an Electronic Monetary Transfer System* (Lexington, Mass.: Lexington Books, 1973); Thomas Kleinschmit, *Electronic Funds Transfer: The Future Is Now* (Minneapolis, Minn.: Federal Reserve Bank, 1976); Kent W. Colton and Kenneth L. Kraemer (eds.), *Computers and Banking: Electronic Funds Transfer Systems and Public Policy* (New York: Plenum Press, 1980); Allen H. Lipis, Thomas R. Marschall, and Jan H. Linker, *Electronic Banking* (New York: John Wiley & Sons, 1985).

61. Allen N. Berger and Loretta J. Mester, "Explaining the Dramatic Changes in Performance of U.S. Banks: Technological Change, Deregulation, and Dynamic Changes in Competition," Working Paper No. 01-6, April 2001 (Philadelphia, Penn: Federal Reserve Bank of Philadelphia, 2001), first page (unpaginated).

62. Ibid., 1.

63. Allen N. Berger and Robert DeYoung, "Technological Progress and the Geographic Expansion of the Banking Industry," unpublished paper, June 2002, available from Federal Reserve Board of Chicago.

64. Robert E. Litan, *The Revolution in U.S. Finance* (Washington, D.C.: Brookings Institution, 1991): 1. For a tutorial on how banking worked in the 1980s, and at the same time a less critical view of the industry, see Peter S. Rose, "The Economics of the Banking Firm," in William H. Baughn, Thomas I. Storrs, and Charles E. Walker (eds.), *The Bankers' Handbook*, 3rd ed. (Homewood, Ill.: Dow Jones-Irwin, 1988): 187–206.

65. Bertrand Olivier and Thierry Noyelle, *Human Resources and Corporate Strategy* (Paris: OECD, 1988): 36.

66. The most useful sources of information on usage of the Internet are the ongoing studies conducted by the Pew Foundation and those of the U.S. Department of Commerce; however, additional studies by academics and industry associations (such as the ABA) appeared at the rate of dozens per month in the early years of the new century, most of which were posted to the Web and could be found by looking at the Internet sites of their various business schools, academic departments, and trade associations.

67. For a sense of the interest by a highly respected student of the Internet, see Manuel Castells, *The Internet Galaxy: Reflections on the Internet, Business, and Society* (Oxford: Oxford University Press, 2001). He opens his book with a simple sentence that lays out the case for this interest: "The Internet is the fabric of our lives," p. 1.

68. See, for example, A. Gandy and C. S. Chapman, *Information Technology and Financial Services* (Chicago, Ill.: Glenlake Publishing, 1997); Cronin, *Banking and Finance on the Internet*; Keyes, *Financial Services Information Systems and* also her *Handbook of Technology in Financial Services* (Boca Raton, Fla.: Auerbach, 1999) and *Internet Management* (Boca Raton, Fla.: Auerbach, 2000).

69. Government agencies and universities had used the Internet since the early 1970s. For details, see Janet Abbate, *Inventing the Internet* (Cambridge, Mass.: MIT Press, 1999): 83–145.

70. The role of feedback in knowledge creation and technology is an important one to understand, as it is recognized as an important phenomenon at work in all manner of innovations. For a recent, very lucid discussion of its role, see Joel Mokyr, *The Gifts of Athena: Historical Origins of the Knowledge Economy* (Princeton, N.J.: Princeton University Press, 2002): 20–22, 96–98; Linsu Kim and Richard R. Nelson (eds.), *Technology, Learning, and Innovation: Experiences of Newly Industrializing Economies* (Cambridge: Cambridge University Press, 2000). Most of the research has focused on manufacturing industries; we have so little on services industries. For an excellent set of examples from the manufacturing sector, see David C. Mowery and Richard R. Nelson (eds.), *Sources of Industrial Leadership: Studies of Seven Industries* (Cambridge: Cambridge University Press, 1999).

71. Lee S. Sproull, "Computers in U.S. Households Since 1977," in Alfred D. Chandler, Jr., and James W. Cortada (eds.), *A Nation Transformed by Information: How Information Has Shaped the United States from Colonial Times to the Present* (New York: Oxford University Press, 2000): 264–265.

72. Arthur B. Kennickell and Myron L. Kwast, "Who Uses Electronic Banking? Results from the 1995 Survey of Consumer Finances," unpublished paper, July 1997, available from the Division of Research and Statistics, Board of Governors of the Federal Reserve System.

73. 2002 was chosen because it was the latest period for which we had data at the time this book went to press.

74. John B. Horrigan and Lee Rainie, "Getting Serious Online" (Washington, D.C.: Pew Internet and American Life Project, March 3, 2002): 2, 21.

75. Deborah Steinborn, "System Builds on Branch Strategy," *ABA Banking Journal* 86 (December 1994): 72.

76. Jude W. Fowler, "The Branch Is Dead!" Ibid., 87 (April 1995): 40.

77. Suzanne Donner and Cathy Dudley, "Balancing Customer Contact and High-tech Delivery," Ibid., 89 (January 1997): 18.

78. Ibid., 18, 20.

79. Ibid., Tom Flynn, "The Supermarket Branch Revisited," Ibid., 89 (October 1997): 24, 26, 28; Steve Cocheo, " 'Tech' Weds 'Touch' in a Marriage for Convenience," Ibid., 89 (November 1997): 26, 28; Lauren Bielski, "Rethinking the Branch Experience," 91 (August 1999): 28–30, 32, 34, 46; Joe Shalleck and Mark Underwood, "What's the Right Back Office Solution?" Ibid., 93 (December 2001): 49–50, 52.

80. See, for example, Everett M. Rogers, *Diffusion of Innovations*, 3rd ed (New York: Free Press, 1983): 1–86; George Basalla, *The Evolution of Technology* (Cambridge: Cambridge University Press, 1988). Both are useful for understanding when and how people change technologies.

81. Coverage in the industry press, replete with interviews with various types of bankers, demonstrated the ambiguity of the issue. See, for example, Penny Lunt, "The Smart Cards Are Coming! But Will They Stay?" Ibid., 87 (September 1995): 46–48, 50, 52; and a cover story by Penny Lunt, "Is It First and Goal for Debit Cards?" Ibid., 88 (September 1998): 44–47; another cover story, Tom Anderson, Greg Case, Tim Gokey, and Zubin Taraporevala, "Who Will Survive the Bank Card Shakeout?" Ibid., 89 (September 1997): 55–56, 61–62, 64; on a competitor not a bank, Alan Levinsohn, "The Rise and Sprawl of a Card Services Empire," Ibid., 90 (July 1998): 45–47; Bill Orr, "The Great Card Question: Will It Be Smart or Debit?" Ibid., 90 (September 1998): 54–56, 58; Lauren Bielski, "Smart Cards, Coming Up to Bat," Ibid., 90 (November 1998): 57–58, 60, 62.

82. Orla O'Sullivan, "The 'Super ATM' Comes of Age," Ibid., 89 (May 1997): 72, 74; Joe Asher, "The Second ATM Revolution," Ibid., 90 (May 1998): 51–52, 54, 56.

83. Lauren Bielski, "Security Gets a Higher Profile," Ibid., 93 (December 2001): 54, 56, 58, 60, 61; Edmond Blount, "Reengineering Customer Expectations," Ibid., 87 (March 1995): 36–38, 42; "Adding Life to Bank Brokers' Sales," Ibid., 93 (March 2001): 44, 46, 48.

84. Lauren Bielski, "Did Y2K Deep Freeze Data Processing?" Ibid., 91 (May 1999): 38–41. Y2K was not completely to blame in the downturn of IT industries after 2000; the global economic recession had more effect because it caused so many firms to cut back on their investments in new applications.

85. Alan Levinsohn, "The Search Continues for the Elusive Integrated 'Solution,' " Ibid., 90 (June 1998): p. 73.

86. Orla O'Sullivan, "The Profitability Riddle," Ibid., 90 (February 1998); 78, 80, 82; Alan Levinsohn, "Citibank Recharts Its Technology Course," Ibid., 90 (May 1998): 40–41, 44, 46, 48; "It's Not About Banking It's About Money," Ibid., 91 (December 1999): 43, 46, 48, 50, 52, 54.

87. Bill Orr, "Banking on the Internet," Ibid., 87 (November 1994): 67. See also pp. 71, 73, 75.

88. Penny Lunt, "What Will Dominate the Home?" Ibid., 87 (June 1995): 36.

89. Daniel Schutzer, "Get Ready for Electronic Commerce," Ibid., 87 (June 1995): 47–48; Beverly K. Wayne and Curtis B. Wayne, "Branches for a Small Planet," Ibid., 87 (August 1995): 43–45; Penny Lunt, "The Virtual Bank Needs a Back-Office Alignment," Ibid., 87 (September 1995): 82, 84, 88, 90; Penny Lunt, "Payments on the 'Net: How Many? How Safe?" Ibid., 87 (November 1995): 46–48, 50, 52, 54; Bill Orr, "Security: What Everyone's Wondering About," Ibid., 87 (December 1995): 45, 47; Ed Blount, "Rise of the Virtual

Trust Company," Ibid., 88 (March 1996): 46–47; Penny Lunt, "Groupware Gets Webfied," Ibid., 88 (June 1996): 70–72; Bill Orr, "Smaller Banks Move into Internet Banking," Ibid., 89 (August 1997): 66–67.

90. Orla O'Sullivan, "The Check Is In the E-mail," Ibid., 90 (January 1998): 52–55; Bill Orr, "Web Banking—Now in Stage Two," Ibid., 90 (July 1998): 64; Lisa Valentine, "VANs vs. the Internet: May the Best Man Win," Ibid., 90 (November 1998): 52, 54; David W. Medeiros, "The Revolution Has Just Begun," Ibid., 90 (December 1998): S3–S6, S9–S11, S16–S17, S21.

91. Penny Lunt, "Welcome to sfnb.com The Paradigm Just Shifted," Ibid., 87 (December 1995): 40–41, 43, 45; J. Christopher Westland and Theodore H.K. Clark, *Global Electronic Commerce: Theory and Case Studies* (Cambridge, Mass.: MIT Press, 1999): 91–120; Kim Humphreys, "Banking on the Web: Security First Network Bank and the Development of Virtual Financial Institutions," in Mary J. Cronin (ed.), *Banking and Finance on the Internet* (New York: John Wiley & Sons, 1998): 75–106.

92. For a massive anthology of descriptions of Internet-based applications, see Keyes, *Financial Services Information Systems*.

93. Donal O'Mahony, Michael Peirce, and Hitesh Tewari, *Electronic Payment Systems for E-Commerce*, 2nd ed. (Boston, Mass.: Artech House, 2001): 128–143, and for a list of such services by supplier (circa 2000), pp. 128–129.

94. Bill Orr, "Community Bank Guide to Internet Banking," *ABA Banking Journal* 90 (June 1998): 47–49, 52–53, 56–58, 60.

95. "You and Your Bank on the 'Net: Evolution of a 'Wall'?" Ibid., 91 (February 1999): S8–S9, S13–S14.

96. See, for example, the discussions held in the *ABA Banking Journal* in this period, Lauren Bielski, "Home Equity Loans on the Web," Ibid.,91 (March 1999): 42–44; "E-commerce Is Here. Do You Have a Plan?" Ibid., 91 (May 1999): 49–50, 52; Lauren Bielski, "Modern Call Center Handles Web Mail, Too," Ibid., 91 (May 1999): 60, 62; Bill Orr, "One Bank's Response to the Online Trading Boom," Ibid., 91 (May 1999): 72; Bill Orr, "E-banks or E-branches? Both Are in Play as Early Adopters Make Them Work," Ibid., 91 (July 1999): 32–34.

97. "At Last Internet Banking Takes Off," Ibid., 91 (July 1999): 36; Bill Orr, "Who's Afraid of the Web?" Ibid., 91 (August 1999): 44–45.

98. "Who's on First?" Ibid., 91 (July 1999): 40.

99. Alan Levinsohn, "Online Brokerage, The New Core Account?" Ibid., 91 (September 1999): 35.

100. Ibid., 34–35, 38, 40–42.

101. The ABA published many articles in every issue of its journal in this period, discussing issues and also patterns of adoption. It also published a white paper on the Internet, Clay Hamlet, "Community Banks Go Online," Ibid., 92 (March 2000): 61–64.

102. William Streeter, "Issues for 2001," Ibid., 92 (December 2000): 34. See an additional study on deployment by Karen Furst, William W. Lang, and Daniel E. Nolle, "Who Offers Internet Banking?" *Quarterly Journal* 19, no. 2 (June 2000): 29–48, which is rich with data and bibliographic citations.

103. Data taken from one of the most important surveys done about the Internet and banking and the industry's customers, "Community Bank Web Census 2001," Ibid., 93 (April 2001): 32–38; for its predictions, in part a result of this and other surveys, see Lauren Bielski et al., "Tech Trends: A 5-Year Scan," Ibid., 93 (December 2001): S3–S23. See also, Karen Furst, William W. Lang, and Daniel E. Nolle, "Who Offers Internet Banking?" *Quarterly Journal* 19, no. 2 (June 2000): 29–48; Karen Furst, William W. Lang, and Daniel E. Nolle, "Internet Banking: Developments and Prospects," Economic and Policy Analysis Working Paper 2000-9 (September 2000).

104. Star Systems, Inc., *Web Aggregation: A Snapshot* (August 2000), http://www.star-system.com (last accessed 10/20/03).

105. This new battlefield is well described by Mary J. Cronin in her edited volume, *Banking and Finance on the Internet*, 1–18; but see also in the same volume, Ravi Kalakota and Frances Frei, "Frontiers of On-Line Financial Services," Ibid., 19–74; Karen Furst, William W. Lang, and Daniel E. Nolle, "Internet Banking: Developments and Prospects," (Cambridge, Mass.: Center for Information Policy Research, Harvard University, 2002).

106. William M. Randle, with Jeffrey Kutler, "Banking: The Biggest Risk Is to Do Nothing," in Christina Ford Haylock, Len Muscarella, and Ron Schultz (eds.), *Net Success: 24 Leaders in Web Commerce Show You How to Put the Internet to Work for Your Business* (Holbrook, Mass.: Adams Media Corporation, 1999): 223.

107. Eric K. Clemons and Lorin M. Hitt, "Financial Services: Transparency, Differential Pricing, and Disintermediation," in Robert E. Litan and Alice M. Rivlin (eds.), *The Economic Payoff from the Internet Revolution* (Washington, D.C.: Internet Policy Institute and Brookings Institution Press, 2001): 115.

108. Ibid., 87–128.

109. Randle, "Banking," 229.

110. Ibid., 230.

111. "Is Internet Banking Profitable? A Study of Digital Insight's Offering" (Cambridge, Mass.: Celent Communications, October 2000), available from the ABA's website.

112. Federal Reserve Staff, *The Future of Retail Electronic Payments Systems: Industry Interviews and Analysis*, Staff Study 175 (Washington, D.C.: Federal Reserve System, December 2002).

113. Doug Johnson, "Internet Banking: Is It Offense or Defense?" (New York: ABA, April 2002), available on ABA's website. With regard to bill payment profitability, it appeared not to have been achieved yet as of late 2002, Michael S. Derby, "Online Bill Payment Faces Obstacles," August 28, 2002, Yahoo! News.

114. Barbara Kiviat, "Bye-Bye, Paycheck," *Inside Business* (November 2003), unpaginated.

115. "Ask the Expert: Check Imaging: Big Changes in Banking," *InformationWeek*, May 19, 2003, p. 8; "Bush Signs Electronic Banking Bill," *The Mercury News*, October 28, 2003, http://www.bayarea.com/mld/mercurynews/news/world/7120555.htm (last accessed 10/29/2003); Associated Press, "Bush Signs Electronic Banking Bill," *Los Angeles Times Latimes.com*, October 28, 2003, http://www.latimes.com/news/nationworld/politics/wire/sns-ap-bush-electronic-checks,1,2 . . . (last accessed 10/29/2003); Jeannine Aversa, "Greenspan Urges Financial Flexibility," *Wisconsin State Journal*, October 30, 2003, p. E3.

116. National Telecommunications and Information Administration, *Report to Congress: Electronic Signatures in Global and National Commerce Act*, Section 105 (a) (Washington, D.C.: U.S. Department of Commerce, June 2001): 5.

117. "ComScore Analysis Reveals Usage of Online Banking and Bill Paying Have Grown Dramatically in the Past Year," press release, June 17, 2004, ComScore Corporation, http://www.comscore.com/press/release.asp?press=467 (last acccessed 6/17/2004).

118. The subject of banking productivity based on the digital has drawn the attention of the economic community's most illustrious members, such as Nobel laureate Lawrence Klein, who along with Cynthia Saltzman and Vijaya G. Duggal joined the ranks of so many others to report "we find evidence of increasing returns to scale and also of the large contribution to overall productivity in finance that comes from the economic process of delivering output from computer and data processing services to the financial sector," "Information Technology and Productivity: The Case of the Financial Sector," *Survey of Current Business* (August 2003): 37.

Chapter 4

1. For his quote and the paper from which it came, see Norton E. Masterson, "Statistics for Management," in G. F. Michelbacher and Nestor R. Roos (eds.), *Multiple-Line Insurers: Their Nature and Operations* (New York: McGraw-Hill, 1970): 224.

2. James W. Cortada, *Before the Computer: IBM, NCR, Burroughs, and Remington Rand and the Industry They Created* (Princeton, N.J.: Princeton University Press, 1993): 59–60.

3. JoAnne Yates, "Co-Evolution of Information Processing Technology and Use: Interaction Between the Life Insurance and Tabulating Industries," *Business History Review* no. 1 (1993): 1–51; JoAnne Yates, "Early Interactions Between the Life Insurance and Computer Industries: The Prudential's Edmund C. Berkeley," *Annals of the History of Computing* 17, no. 1 (1997): 60–73; Henry N. Kaufman, "Some Uses for the Hollerith Machines," *Transactions of the Actuarial Society of America* 11 (1909–1910): 276–295; Percy C.H. Papps, "The Installation of a Perforated Card Systems with a Description of the Peirce Machines," *Transactions of the Actuarial Society of America* 15 (1914): 49–61.

4. I make the same points in Chapters 2 and 3 regarding how banks shared common needs with insurance yet also had differences. Right to the end of the century, observers of both discussed them together. See, as one recent yet not untypical example, Jessica Keyes (ed.), *Financial Services Information Systems* (Boca Raton, Fla.: Auerbach, 2000).

5. Kenneth Black, Jr., "A 90-Year Retrospective," *Best's Review*, Life Edition 91, no. 6 (June 1990): 47–48, 50, 52, 54, 156, 158; Kailin Tuan, "Life Insurance in the Stages of Insurance Development," Ibid., 73, no. 9 (January 1973): 26, 28, 30, 63–66; Melvin L. Gold, "The Future of the Life Insurance Industry," Ibid., 83, no. 5 (September 1982): 20, 22, 24, 26.; George T. Stewart and Robert Chaut, "Insurance," in Lester V. Plum (ed.), *Investing in American Industries* (New York: Harper & Brothers, 1960): 343–381.

6. The best way to document the ebbs and flows of technologies is by reading the monthly issues of *Best's Review*, which came in two editions throughout most of the period—one aimed at life insurance companies, the second for property/casualty firms.

7. Emma S. Woytinsky, *Profile of the U.S. Economy: A Survey of Growth and Change* (New York: Praeger, 1967): 421–425.

8. Ibid., 425–426.

9. U.S. Census Bureau, 1997 Economic Census, "Bridge Between SIC and NAICS SIC: Menu of SIC Divisions," http://www.census.gov/epcd/ee97brdg/INDXSIC2.htm; Sherlene K. S. Lum, Brian C. Moyer, and Robert E. Yiskavage, "Improved Estimates of Gross Product by Industry for 1947–98," *Survey of Current Business* (June 2000): 42.

10. I am deeply grateful to Professor J. David Cummins, of the Wharton School, for sharing this information with me. Dr. Cummins has long been an expert on the industry, and the author of numerous studies that comment on the role of technology. His most recent book is J. D. Cummins and A. M. Santomero (eds.), *Changes in the Life Insurance Industry* (Norwell, Mass.: Kluwer Academic Publishers, 1999).

11. Edward I. Altman and Irwin T. Vanderhoof (eds.), *The Strategic Dynamics of the Insurance Industry: Asset/Liability Management Issues* (Burr Ridge, Ill.: Irwin, 1996): x.

12. Robert E. Litan, *The Revolution in U.S. Finance* (Washington, D.C.: Brookings Institution, 1991): 29–33. The author provides an excellent analysis of the entire U.S. financial sector during the 1980s.

13. Alan Gart, *Regulation, Deregulation, Reregulation: The Future of the Banking, Insurance, and Securities Industries* (New York: John Wiley & Sons, 1994): 187.

14. Charles H. Cissley, *Systems and Data Processing in Insurance Companies* (Atlanta, Ga.: FLMI Insurance Education Program, Life Management Institute LOMA, 1982).

15. For an introduction to the Internet and the financial sector, see Mary J. Cronin (ed.), *Banking and Finance on the Internet* (New York: John Wiley & Sons, 1998); passing comments in Keyes, *Financial Services Information Systems*; Eric K. Clemons and Lorin M. Hitt, "Finan-

cial Services: Transparency, Differential Pricing, and Disintermediation," in Robert E. Litan and Alice M. Rivlin (eds.), *The Economic Payoff from the Internet Revolution* (Washington, D.C.: Brookings Institution Press, 2001): 87–128, especially pp. 112–115; see also the ongoing publication of white papers by economists at the U.S. Federal Reserve and routinely posted to their various websites.

16. However, others were commenting on the potential of computing in insurance. The author of what most historians consider to be the first book about computers, Edmund C. Berkeley, commented on it early, "Electronic Machinery for Handling Information, and Its Uses in Insurance," *Transactions of the Actuarial Society of America* 48, Part 1 (1947): 36–52.

17. On this first company's experience with computers, see U.S. Bureau of Labor Statistics, *The Introduction of an Electronic Computer in a Large Insurance Company*, Report Number 2, Studies of Automatic Technology (Washington, D.C.: U.S. GPO, 1955); Adelbert G. Straub, Jr. (ed.), *Examination of Insurance Companies*, 6 vols. (New York: New York State Insurance Department, 1954), see vol. 4.

18. John D. De Nault, "The Impact of Computers on Multiple-Line Operations," *Journal of Machine Accounting, Systems and Management* 9, no. 11 (November 1958): 8–10; John D. Finelli, "Use of Electronics in the Insurance Business," in Adelbert G. Straub, Jr. (ed.), *Examination of Insurance Companies* (New York: New York State Insurance Department, 1954): IV, 649ff; James W. Everett, "The Effects of Electronics," *Best's Insurance News*, Life Edition, 63, no. 3 (July 1962): 67–70; Paul Kircher, "Study of a Successful Computer System," *Journal of Accountancy* 104 (October 1957): 59–64; Stephen F. Loughlin, "Direct Access Processing," *Data Processing: Proceedings 1967* (Boston, Mass.: DPMA, 1967): 145–148.

19. Cissley, *Systems and Data Processing*, 5–7; John D. De Nault, "The Impact of Computers on Multiple-Line Operations," *Journal of Machine Accounting* 9, no. 11 (November 1958): 8–10; David M. Irwin, "New Directions in Insurance Data Processing," *Data Processing Proceedings 1963* (Detroit, Mich.: DPMA, 1963): 189–196; P. Adger Williams, "On-line Real-time Processing in Insurance," *Data Processing Proceedings 1964* (New Orleans, La.: DPMA, 1964): 88–92; Philip J.F. Smith, "Life Insurance Program Augments CFO Package," *Journal of Data Management* (March 1965), reprinted article, copy in RC 2500, Box A-1245-1, IBM Corporate Archives, Somers, N.Y.; J. W. Cannon, "Building and Managing An Insurance Processing System," *Data Processing Proceedings 1965* (Dallas, Tex.: DPMA, 1966): 112–130.

20. There is one study of a company's long-term experience with data processing over many decade, *Aetna Life & Casualty, Information Systems at Aetna Life & Casualty: From Punched Cards to Satellites, and Beyond* (Hartford, CT: Aetna Life & Casualty, 1984). On how the industry moved from service bureaus to in-house computing, see Elmira L. Del Col, "A Quarter Century of Automation History: From Batch Systems to In-House Equipment," *Rough Notes* 132, no. 3 (March 1989): 80–81.

21. Survey data is rather limited for this industry, however see Raymond Dash, *Data Processing Users Survey*, LOMA Special Release no. 8 (Atlanta, Ga.: LOMA, November 1978); U.S. Bureau of Labor Statistics, *Technology and Labor in Five Industries*, Bulletin 2033 (Washington, D.C.: U.S. Department of Labor, 1979): 41–49; L. N. Fuchcar, "Interloc: A Real-Time Management Control System," *Data Processing Proceedings* 1965 (Dallas, Tex.: DPMA, 1966): 131–138.

22. Cissley, *Systems and Data Processing in Insurance Companies*, 14–16.

23. "Punched Cards: A Step Along the Road to Tape," undated [1957–58?], *Office Automation Applications*, III, D1, pp. 1–9, Market Reports 55, Box 70, Folder 1, Archive of the Charles Babbage Institute, University of Minnesota.

24. "Transceiver Network Facilities Premium Billing," undated [1958?], *Office Automation*

Applications, III, D2, pp. 1–11, Market Reports 55, Box 70, Folder 1, Archives of the Charles Babbage Institute, University of Minnesota.

25. There are numerous case studies of specific insurance firms using computers, complete with detailed descriptions and flowcharts of their applications and with photographs, in the Market Reports 55, Box 70 in particular, at the Archives of the Charles Babbage Institute, University of Minnesota.

26. All through the half century, various insurance associations hosted computing or automation conferences in which dozens of presentations were made about the experiences of individual firms with digital and telecommunications applications. LOMA, in particular published its proceedings, beginning in 1959, extending through 1989. Normally entitled LOMA, *Proceedings of the Life Office Automation Forum, 1959*, and for all of the 1960s–1980s these are available at LOMA's headquarters and library in Atlanta, Georgia. The conferences were held in 1959, 1962, 1965, 1968, 1971, 1974, 1977, 1980, 1983, 1986, 1989.

27. JoAnne Yates, "Perils of Reengineering," *Knowledge Management* (December 2000): 80.

28. John Diebold, an early proponent for the use of computers by American companies in 1952, described what the digital could do for insurance companies:

> A large life insurance company will use ten or fifteen stories of its skyscraper headquarters merely to keep records of insurance policy information. These are usually in filing cabinets distributed evenly throughout the floors. File clerks are stationed every fifty feet and are equipped with headphones connected to the clerk in charge of the appropriate section, [who] tells her the policy number, and waits until she walks to the filing cabinet containing the policy and reads him the information. By use of magnetic tapes, the storage space for the policy information of a large insurance company can be reduced from ten or fifteen floors of files to 350 or 400 spools of magnetic tape which, with control gear, would occupy one medium-sized room. This magnetic tape could be kept in a low rental area far from central headquarters.

John Diebold, *Automation: The Advent of the Automatic Factory* (New York: Van Nostrand, 1952): 94, but see also p. 95 for more details.

29. Robert C. Jacobs, "Insurance Accounting," *Data Processing Annual* (Detroit, Mich.: Gille Associates, 1961): 142.

30. Robert A. Greenig, "Electronics for Faster Policyholder Service," Ibid., 148.

31. David H. Harris, of the Equitable Life Assurance Society of the United States, quoted in a panel discussion at the 1959 LOMA Automation Forum, LOMA, *Proceedings of the Life Office Automation Forum, 1959* (N.C.: LOMA, 1959): 95.

32. U.S. Bureau of Labor Statistics, *Impact of Office Automation in the Insurance Industry*, Bulletin No. 1468 (Washington, D.C.: U.S. Government Printing Office, 1966): 17. For additional evidence, see OECD, *Integrated Data Processing and Computers* (Paris: OECD, 1961): 179–183, which describes computing at the Minnesota Mutual Life Insurance; Leonard Rico, *The Advance Against Paperwork: Computers, Systems, and Personnel* (Ann Arbor, Mich.: Bureau of Industrial Relations, Graduate School of Business Administration, University of Michigan, 1967): 62–63; Peter B. Laubach, *Company Investigations of Automatic Data Processing* (Boston, Mass.: Division of Research, Harvard Business School, 1957): 93–95; Paul Armer, "Computer Applications in Industry and Services," in Irene Taviss (ed.), *The Computer Impact* (Englewood Cliffs, N.J.: Prentice-Hall, 1970): 52; George E. Delehanty, "Computers and the Organization Structure in Life-Insurance Firms: The External and Internal Economic Environment," in Charles A. Meyers (ed.), *The Impact of Computers on Management* (Cambridge, Mass.: MIT Press, 1967): 61–106..

33. For a balanced account of the S/360 story, see Martin Campbell-Kelly and William

Aspray, *Computer: A History of the Information Machine* (New York: Basic Books, 1996): 137–150; for a technically centered account, Paul E. Ceruzzi, *A History of Modern Computing* (Cambridge, Mass.: MIT Press, 1998): 143–173; and for a business-centric review, Alfred D. Chandler, *Inventing the Electronic Century: The Epic Story of the Consumer Electronics and Computer Industries* (New York: Free Press, 2001): 91–94.

34. JoAnne Yates has done the most extensive research on how the Insurance Industry used computing, particularly in the early years, "Co-Evolution of Information Processing Technology and Use: Interaction Between the Life Insurance and Tabulating Industries," pp. 1–51, and "The Structuring of Early Computer Use in Life Insurance," *Journal of Design History* 12, no. 1 (1999): 5–24. Her major study appeared too late for me to consult for this chapter; however, it is *Structuring the Information Age: Life Insurance and Information Technology in the 20th Century* (Baltimore, Md.: Johns Hopkins University Press, 2005).

35. For examples of these applications of cutting-edge technologies of the day, see Karl F. Eaton, "The Impact of Integrated Systems on Life Company Organization," *Toward Total Systems for Total Service: Proceedings of Automation Forum 1965* (New York: LOMA, 1965): 166–168; W. H. Olischar, "The Use of a Subset of COBOL as a Programming Language," Ibid., 239–241; John C. Henchel, "Service to Administration Through Integrated Subsystems," Ibid., 317–320.

36. For early reports on the ongoing innovation, see, for example, Frank V. Wagner, "Operating Under Third-Generation Concepts," *Information Please: Proceedings of the Automation Forum* (New York: LOMA, 1968): 5–8; "Time Sharing Concepts," Ibid., 48–59; Frank A. Brooks, "A Direct-Access Master File for Ordinary Insurance," Ibid., 60–66; J. R. Campbell, "Experiences With Pl/1," Ibid., 127–129; George Perry, "A Real-Time Communications Network," Ibid., 159–161; and a case involving the IBM S/360, Ernest F. Woodward, "Insurance Management and Information Systems," *Data Processing: Proceedings 1965* (Dallas, Tex.: DPMA, 1966): 131–138. Each author worked in the Insurance Industry.

37. Erie S. Anderson, "Path-Dependency and Competitive Advantage: The Adoption of EDP Systems at Northwestern Mutual Life, 1954–1964," in *Essays in Economic and Business History: The Journal of the Economic and Business History Society* 14 (1996): 291–304.

38. Jay Kobler, "Computers—A Gap Between Can't and Do?" *Best's Review*, Life/Health Edition 73, no. 2 (June 1972): 26, 28, 30, 65–69; George Schussel, "Data Base: A New Standard for Insurance EDP," Ibid., 73, no. 7 (November 1972): 42, 44, 90, 92, 94–95.

39. For inventories and descriptions of this new generation of applications, see International Business Machines Corporation, *Group Insurance Application Description and System Planning Guide* (White Plains, N.Y.: IBM Corporation, May 1975); and also IBM, *Individual Life Insurance: DB/DC Information Systems Design Concepts* (White Plains, N.Y.: IBM Corporation, January 1975).

40. Cissley, *Systems and Data Processing*, 29. The term e-mail did not come into use until the 1990s, but I have used it for convenience. The correct term from the period would have been something like "electronic data communication."

41. Ibid., 125.

42. Ibid., 91–109.

43. For a brief history of IVANS, published on the occasion of its tenth anniversary, see Charles C. Ashley, "IVANS: A Vigorous Decade," *Best's Review*, Life/Health Edition 94, no. 1 (May 1993): 67–68, 70, 72.

44. Everett Karels, "What a Minicomputer Can Do for Life Insurance Agents," Ibid., 80, no. 1 (May 1979): 36, 38, 40, 42.

45. Mary L. Wright, "Where We Are Today: The State of Office Automation in Selected Insurance Companies," Ibid., 81, no. 7 (November 1980): 62.

46. Ibid., 62, 66, 68.

47. A series of articles in the industry's leading publication documented these activities: James F. Foley, "Issues in the 1980s," *Best's Review*, Life/Health Edition 83, no. 7 (November

1982): 74, 76, 78, 80, 82, 84; John M. Driscoll, Jr., "A Computer in the Agency: Key to Productivity," Ibid., 83, no. 10 (February 1983): 66, 68, 70, 72; Patricia K. Mortenson, "Financial Planning by Computer," Ibid., 85, no. 2 (June 1984): 38, 40; Edwin E. Hightower, "Systems Implications for Asset/Liability Matching," Ibid., 85, no. 2 (June 1984): 86, 88, 90; William D. McClain, "Computer-Generated Insurance Policies," Ibid., 85, no. 5 (September 1984): 68, 70, 72, 74; Raul A. Cruz, "A Painless Approach to Agency Automation," Ibid., 87, no. 5 (September 1986): 130, 132, 134; C. Thomas Whitehouse, "Electronic Field Support for Financial Planners," Ibid., 88, no. 5 (September 1987): 116, 118, 120, 122; Donna A. Goldin, "Electronic Efficiencies of Scale," Ibid., 90, no. 12 (April 1990): 103–105.

48. Paul M. Howman, "Freeing Underwriters for the Tough Jobs," Ibid., 89, no. 7 (November 1988): 54–56; Arthur DeTore, "Catching Up with Underwriters' Needs," Ibid., 90, no. 2 (June 1989): 96, 98–99.

49. LOMA, *Field Computing: A Study of Industry Practices and Plans, Operations and Systems*, Report No. 83 (Atlanta, Ga.: LOMA, March 1986); see also the various presentations made at the 1986 LOMA automation conference, LOMA, *Leaders . . . Goals . . . Technology, Operations and Systems*, Report No. 86 (Atlanta, Ga.: LOMA, September 1987).

50. Jennifer Cranford Rankin, "Insurers Sound Off on Field Computing," *Best's Review*, Life/Heath Edition 92, no. 7 (November 1991): 102, 104, 106, 108.

51. Patrick McCarthy, "Infoglut and the Agent," Ibid., 94, no. 1 (May 1993): 84.

52. On the debate, see Richard K. Berry and Roger R. Heath, "Reinventing Agency Distribution," Ibid., 94, no. 4 (August 1993): 33–36, 110; Thomas J. Skelly, "Technology Transforming Distribution Channels," Ibid., 95, no. 1 (May 1994): 90; Paula V. Ingrassia, "Harnessing Technology to Remain Competitive," Ibid., 95, no. 1 (May 1994): 96–97; Paula V. Ingrassia, "Selling Insurance on the Superhighway," Ibid., 95, no. 5 (September 1994): 84–85.

53. Information for both this and the previous paragraph drawn from Kenneth Huggins, Dani L. Long, and Vicki Mason Pfrimmer, *Information Management in Insurance Companies* (Atlanta, Ga.: FLMI Insurance Education Program and LOMA, 1995): 172–190. This source includes illustrations of sample screens from online systems.

54. One colleague of the author, a principal in IBM's insurance industry consulting practice, was more blunt. He told me in March 2003 that the industry was "scrambling" to provide more customer-friendly applications.

55. Michael D. Gantt, Alan J. Turner, and James Gatza, *Computers in Insurance* (Malvern, Penn.: American Institute for Property and Liability Underwriters, 1985): 102–110.

56. Huggins, Long, and Pfrimmer, *Information Management in Insurance Companies*, 2–7, 12–14, 93–94.

57. Kenneth M. Hills, "Insurance Data Processing," *Best's Fire and Casualty News* 68, no. 9 (January 1968): 75.

58. "Ramac Aids Company Expansion," August 1961, *Office Automation Applications*, III, D27, pp. 1–6, Market Reports 55, Box 70, Folder 7, Archives of the Charles Babbage Institute, University of Minnesota.

59. R. Hunt Brown, *Office Automation Insurance* (New York: Automation Consultants, Inc., February 1960), in Market Reports 55, Box 70, Folder 19, Archives of the Charles Babbage Institute, University of Minnesota.

60. Gantt, Turner, and Gatza, *Computers in Insurance*, 112–115; Hal Trimble, "The Computer—An Underwriting Tool," *Best's Review*, Property/Liability Edition 72, no. 12 (April 1972): 88–89.

61. Joe S. Tressler, "Using Real-Time Data Processing," *Best's Fire and Casualty News* 68, no. 3 (July 1967): 87–89.

62. Hills, "Insurance Data Processing," 75–77; Carl Marquardt, "Automation and the Insurance Industry," Ibid., 68, no. 12 (April 1968): 40, 42, 45.

63. William J. Killen, "I.A.S.A. and the Computer—A Survey of the State of the Art," *Best's Review*, Property/Liability Edition 71, no. 2 (June 1970): 41.
64. Thomas B. Kennedy, "The Effect of Paperwork on Company-Agent Relations," Ibid., 71, no. 2 (June 1970): 54, 56.
65. For early discussions of how computing could help reinforce relations and productivity, see "On-Line Computer Service to Branches Saves Money, Improves Customer Service," Ibid., 71, no. 3 (July 1970): 60–61; Robert J. Davis, "Assisting the Agent Through Automation," Ibid., 71, no. 5 (September 1970): 88–90.
66. Raymond P. Keating, "Automation and the Agent," Ibid., 71, no. 11 (March 1971): 68.
67. Jay Kobler, "Computers—A Gap Between Can and Do?" Ibid., 73, no. 2 (June 1972): 30. Similar complaints were voiced by an independent agent two years later, William G. Ervin, "An Independent Agent's View of Automated Accounting Systems," Ibid., 75, no. 2 (June 1974): 22, 24.
68. For a thoughtful and thorough discussion of the issues discussed in this paragraph, see Ken Williams, "Remote Office Automation in the Property/Casualty Insurance Industry," Ibid., 81, no. 10 (February 1981): 32, 36, 38.
69. Norman L. Vincent, "Trends in the Insurance Industry and Their Effect on Data Processing," Ibid., 73, no. 3 (July 1972): 26, 28, 30; Gantt, Turner, and Gatza, *Computers in Insurance*, 115;
70. J. S. Smith, "Computer Time-Sharing's Emerging Role in the Property/Liability Insurance Environment," *Best's Review*, Property/Liability Edition 76, no. 7 (November 1975): 98, 100–101; "Telecommunication Speeds Settlement of Auto Accident Claims," Ibid., 76, no. 7 (November 1975): 102; Gary L. Cantrell, "Remote Job Processing as an Alternative," Ibid., 77, no. 2 (June 1978): 88–90; "The Big Payoff in Productivity," Ibid., 77, no. 5 (September 1976): 111–113; "Database Management System Improves Operating Efficiency," Ibid., 77, no. 11 (March 1977): 79–80, 82–83.
71. James McCauley, "Market Orientation and Information Systems in Property/Casualty Insurance," Ibid., 78, no. 7 (November 1977): 26, 102; Virgil L. Pittman, Jr., "Equitable General's Serv-U Center: Insurance Processing the 'Paperless' Way," Ibid., 78, no. 10 (February 1978): 70–73; Max Stillinger, "A Discussion of Policy Rating, Issue and Data Entry via Mini-Computer," Ibid., 79, no. 1 (May 1978): 101–111; Patricia Ancipink, "Space Age Communications: An Inter-Office Tool," Ibid., 79, no. 1 (May 1978): 98–100.
72. Pamela Broome, "Word Processing at Fireman's Fund," Ibid., 79, no. 6 (October 1978): 124–125; Patricia Viale, "INA's Office of the Future—In Operation Today," Ibid., 80, no. 1 (May 1979): 26, 28, 30.
73. Joseph T. Dyer, "Data: An Allocable Resource," Ibid., 81, no. 1 (May 1980): 90.
74. G. Larry Wilson, "Agency-Company EDP: Interfacing in the '80s," Ibid., 81, no. 12 (April 1981): 40, 44, 46. For another report with a similar perspective and which is an important analysis of the entire agency structure and role, see Doris Fenske, "The Agency System at the Crossroads," Ibid., 83, no. 2 (November 1982): 12, 14, 16, 130, 132, 134–137.
75. Written by Matthew H. Marder, Ibid., 82, no. 12 (April 1982): 53–54.
76. For an introduction to the issue, see Doris Fenske, "The Banks Are Coming!" Ibid., 84, no. 3 (July 1983): 14, 16, 86–88.
77. For an inventory of these, see IBM, *Applications and Abstracts 1985* (White Plains, N.Y.: International Business Machines Corporation, 1985): 14-1–14-16; H. Thomas Krause et al., *Insurance Information Systems*, vol. 1 (Malvern, Penn.: Insurance Institute of America, 1990), pp. 1–32.
78. Edward R. Olsen, "Calling on the Experts," *Best's Review*, Casualty/Liability Edition 90, no. 1 (May 1989): 44, 46, 50; Henry Rodriguez, "Boosting Profits with Expert Systems," Ibid., 91, no. 10 (February 1991): 66, 68, 70, 102.

79. Ralph Dyer, James E. Hoelter, and James A. Newton, *Optical Scanning for the Business Man* (New York: Hobbs, Dorman & Company, 1966): 172–173.

80. Charles A. Plesums, "An Image Worth a Thousand Files," *Best's Review*, Property/ Liability Edition 90, no. 1 (May 1990): 55–56, 58.

81. Lynda E. Kissack, "A 'Paper Last' Environment," Ibid., 90, no. 9 (January 1990): 68, 70, 84–85; Thomas V. McKernan, Jr., on the Automobile Club of Southern California, see "An End to Paper Clutter," Ibid., 91, no. 1 (May 1990): 66, 68, 70, 74; and another case study, Paul McKeon, "The Imaging Advantage," Ibid., 94, no. 8 (August 1993): 65–66, 68.

82. *Best's Review*, Casualty and Liability Edition 91, no. 12 (April 1991).

83. For a sampling of the issues and debates, see Mark J. Payne, " 'Exclusive' Attention to Automation," Ibid., 91, no. 2 (June 1990): 87, 89, 92, 94; Matthew H. Marder, "Up to Snuff in the Agency," Ibid., 91, no. 7 (November 1990): 75, 76, 133; R. Michael Shassere, "Rethinking Agency Automation," Ibid., 92, no. 11 (March 1992): 82, 85; Gregory A. Maciag, "Agents Pull The Strings With EDI," Ibid., 94, no. 1 (May 1993): 50, 52, 91; Paula V. Ingrassia, "Placing Underwriting Back into Agents' Hands," Ibid., 95, no. 9 (January 1995): 82–85; John Macauley and Jim Rogers, "Tools Bring Together Carriers and Agents," Ibid., 95, no. 11 (March 1995): 72–73, 75, 77.

84. Charles C. Ashley, "Rearranging the Distribution System," Ibid., 92, no. 6 (October 1991): 74.

85 Ellen J. Jefferies, "A Strategy for Agency Automation," Ibid., 93, no. 7 (November 1992): 87–88, 108.

86. Sean Armstrong, "New Agent Associations Convene in Cyberspace," Ibid., 95, No. 7 (November 1994): 104–105. I believe this is the first article published by this magazine in the property/casualty side of the industry to describe an Internet-based application.

87. Gerald D. Stephens, "The Customer-Driven Market," Ibid., 92, no. 7 (November 1991): 35.

88. David Poppert, "The Alternative Market: How Big a Threat?" Ibid., 92, no. 3 (July 1991): 18–21, 102. On laptops, John L. Beltran, "Shifting into High Gear," Ibid., 93, no. 1 (May 1992): 49–50, 52; and Paula V. Ingrassia, "Policyholders Get Roadside Assistance," Ibid., 95, no. 4 (August 1994): 86–90. On the channels, William A. Sherden, "A New Way to Deliver Products," Ibid., 93, no. 5 (1992): 35–36, 126, 128; Thomas J. Skelly, "Technology Transforming Distribution Channels," Ibid., 95, no. 2 (June 1994): 96; Donald J. Hurzeler, "New Technology Offers Competitive Advantages," Ibid., 95, no. 10 (February 1995): 50; John Gaulding, "Technological Advances Bolster Client Service," Ibid., 96, no. 4 (August 1995): 92–93.

89. James F. Karkosak, of Ernst & Young, made the charge in "The New Technology Wave," Ibid., 93, no. 6 (October 1992): 70, 73–74.

90. Helmut J. Tissler, "The Best Defense Is a Good Database," Ibid., 90, no. 1 (May 1989): 61–62, 64; George T. Van Gilder, "Technology Is Vital To Good Underwriting," Ibid., 96, no. 1 (May 1995): 42; Kristin L. Nelson, "Underwriters Go High-Tech," Ibid., 95, no. 12 (April 1995): 30–34.

91. The economics of IT has been studied by many economists and historians. For the early period, see William F. Sharpe, *The Economics of Computers* (New York: Columbia University Press, 1969); for a manager's perspective covering the period of the 1970s and 1980s, see Paul A. Strassmann, *Information Payoff: The Transformation of Work in the Electronic Age* (New York: Free Press, 1985); for a formal economic analysis, and an excellent bibliography on the topic, see Daniel E. Sichel, *The Computer Revolution: An Economic Perspective* (Washington, D.C.: Brookings Institution Press, 1997).

92. Yates, "The Structuring of Early Computer Use in Life Insurance," 5–24.

93. U.S. Bureau of Labor Statistics, *The Introduction of an Electronic Computer in a Large Insurance Company*, Studies of Automatic Technology 2 (Washington, D.C.: U.S. GPO,

October 1955). For additional information on deployment in this period, see Yates, "The Structuring of Early Computer Use in Life Insurance."

94. I documented instances of this pattern in over a dozen industries in *The Digital Hand*. Every BLS study on the deployment of technology discussed the issue.

95. Coopers & Lybrand, "Expansion of Banks into Insurance," in Everett D. Randall (ed.), *Issues in Insurance*, vol. 1 (Malvern, Penna.: American Institute for Property and Liability Underwriters, 1987): 233.

96. Jay Kobler, "Computers—A Gap Between Can and Do?" *Best's Review*, Life Edition 73, no. 2 (June 1972): 37. This same article summarized the Diebold study.

97. David J. Blackwell, "Emerging Insurance Technology—What Are the Implications?" Ibid., 80, no. 1 (May 1979): 22.

98. Raymond Dash, *Data Processing Users Survey*, Special Release No. 8 (Atlanta, Ga.: LOMA, November, 1978).

99. Ibid., BLS, *Technology and Labor in Five Industries*, 41.

100. Clericals no longer needed were often deployed into other jobs within their companies.

101. Ibid., 41–45. Also has extensive data on employment levels and skills in the industry, demonstrating a sharp rise in the population of IT experts and the requirements for more of them, pp. 46–48.

102. Mary L. Wright, "Where We Are Today: The State of Office Automation in Selected Insurance Companies," *Best's Review*, Life Edition 81, no. 7 (November 1980): 62, 66, 68; James F. Foley, "Issues in the 1980s," Ibid., 83, no. 7 (November 1982): 74, 76, 78, 80, 82, 84; Jennifer Cranford Rankin, "Insurers Sound Off on Field Computing," Ibid., 92, no. 7 (November 1991): 102, 104, 106, 108; "The Impact of Automation on Independent Insurance Agencies," Ibid., Property/Liability Edition 83, no. 9 (January 1983): 58–59, 62, 64.

103. Sidney E. Harris and Joseph L. Katz, *Capital Investment in Information Technology: Does It Make A Difference?* (Atlanta, Ga.: LOMA, October 1987): 13, 20, 24.

104. U.S. Bureau of Labor Statistics, *Technology and Its Impact on Employment in the Life and Health Insurance Industries*, Bulletin 2368 (Washington, D.C.: U.S. GPO, September 1990): 4.

105. Ibid.

106. Ibid., 5–11.

107. Barbara Tzivanis Benham, "Multiple-Company Interface: The Continuing Saga," *Best's Review*, Property/Casualty Edition 88, no. 4 (August 1987): 96.

108. Paula V. Ingrassia, "Placing Underwriting Back Into Agents' Hands," Ibid., 95, no. 9 (January 1995): 82.

109. Ellen J. Jefferies, "A Strategy For Agency Automation," Ibid., 93, no. 7 (November 1992): 87–88, 108. An earlier survey provides useful insights on speed of adoption, demonstrating that the rate of adoption of computing actually picked up speed in the early 1990s, "Agency Automation: The ACORD Survey," Ibid., 91, no. 3 (July 1990): 70, 72, 74, 76.

110. Paula V. Ingrassia, "Harnessing Technology to Remain Competitive," *Best's Review*, Life Edition 95, no. 1 (May 1994): 96–97.

111. For example, Huggins, Long, and Pfrimmer, *Information Management in Insurance Companies*, 2–7, 12–14; LOMA, *Data Processing Practices: 1993* (Atlanta, Ga.: LOMA, 1993): 44–45.

112. One survey offered considerable evidence that this pattern was global in the industry and not limited to the U.S. experience by the early 1990s; The Economist Intelligence Unit, *Global Insurance to the 21st Century* (New York: The Economist Intelligence Unit, 1996): 3–17.

113. Paula V. Ingrassia, "Selling Insurance on the Superhighway," *Best's Review*, Life/

Health Edition 95, no. 5 (September 1994): 84–85; Evelyn Hall, "Insurers Venture onto the Internet Web," Ibid., 96, no. 2 (June 1995): 40–42, 45, 96; Gregory Hoeg, "New Distribution Channels Are Beginning to Emerge," Ibid., 96, no. 6 (October 1995): 74, 76; Robert W. Stein, "Industry Needs to Improve Electronic Sales Methods," Ibid., 96, no. 11 (March 1996): 66; Suzanne E. Stipe, "Underwriters Should Go 'Site-Seeing' on Internet," Ibid., 96, no. 11 (March 1996): 94, 96; Jim Bolton, "Will Technology Change Agents' and Brokers' Roles?" Ibid., 96, no. 11 (March 1996): 98.

114. Suzanne E. Stripe, "CEOs Express Skepticism About Selling on The Internet," Ibid., 96, no. 12 (April 1996): 24–26, 28–29.

115. Economist Intelligence Unit, *Global Insurance to the 21st Century* (New York: Economist Intelligence Unit, 1996): 4–6, 26–29, 54–58.

116. Gerald H. Peters, "Insurers Still Climbing Internet Learning Curve," Ibid., 97, no. 6 (October 1996): 112, 114, 116; Bill Coffin, "Adapting to the Web," Ibid., 97, no. 11 (March 1997): 104, 106; and for the most detailed analysis of Internet activities published by this magazine to date, see Lee McDonald and Marilyn Ostermiller, "A Report From the E-Commerce Frontier," Ibid., 99, no. 1 (May 1998): 5–6, 9–12, 14–16, 18–20.

117. Dory Devlin, "Click Till You Drop," *Best's Review* 99, no. 1 (May 1998): 23–25; for a large list of insurance websites with what they have in content and company case studies, see "The Web They Weave," Ibid., 101, no. 1 (May 2000).

118. Lynna Goch, "What Works Online," Ibid., 103, no. 1 (May 2002): 24–26, 28, 30–34; Kiran Rasaretnam, "All Things Considered," Ibid., 103, no. 4 (August 2002): 104–106; Lynna Goch, "Power of the People," Ibid., 101, no. 7 (April 2001): 63–64; Keyes, *Financial Services Information Systems*, 769–771.

119. Actually founded in 1937, but launched its website in 1995 which, by the end of the century, was a major source of revenue, with sales (premiums) of $1.8 billion through electronic means.

120. Gordon Stewart, "New Age in Communication Is Challenging the Industry," *Best's Review*, Property/Casualty Edition 96, no. 11 (March 1996): 72; Suzanne E. Stipe, "Internet Offers Executives New Tool for Managing Risk," Ibid., 96, no. 11 (March 1996): 80, 82; Suzanne E. Stipe, "High-Tech Cure Is at Hand For Internet Insecurities," Ibid., 97, no. 5 (September 1996): 98, 100; Bill Coffin, "Adapting to the Web," Ibid., 97, no. 11 (March 1997): 114–115; Lee McDonald, "Back to the Future," Ibid., 97, no. 12 (April 1997): 80, 83; Lee McDonald and Marilyn Ostermiller, "A Report from the E-Commerce Frontier," Ibid., E-Commerce Supplement 99, no. 1 (May 1998): 1–30; "Have E-Commerce, Will Profit," Ibid., 99, no. 5 (September 1998): 85–95.

121. Bill Coffin, "Agents' Favorite Software," *Best's Review* 98, no. 1 (May 1997): 64.

122. Lee McDonald, Taking Insurance to the Bank," Ibid., 98, no. 10 (February 1998): 83.

123. Ibid., 83–86; "Covering the World With One Umbrella," Ibid., 99, no. 1 (May 1998): 32–34; Sally Whitney, "Banks Loom Large," Ibid., 103, no. 8 (December 2002): 100.

124. Eric K. Clemons and Lorin M. Hitt, "Financial Services: Transparency, Differential Pricing, and Disintermediation," in Litan and Rivlin, *The Economic Payoff from the Internet Revolution*, 112. On the continuing issue of agents as a channel of distribution and the Internet, see Ron Panko, "Revenge of the Bricks," *Best's Review* 102, no. 1 (May 2001): 27–32, 34–36.

125. Ibid., 113.

126. Ibid., 115.

127. "Hang Up and Make Money," *Best's Review* 103, no. 8 (December 2002): 104.

128. J. Anne Yates, "The Structuring of Early Computer Use in Life Insurance," 5–7. On structuralism, see A. Giddens, *The Constitution of Society: Outline of the Theory of Structuration* (Berkeley, Calif.: University of California Press, 1984); JoAnne Yates, "Using Giddens' Structuration Theory to Inform Business History," *Business and Economic History* 26, no. 1

(1997): 159–183. On the effects of IT on organizations, the literature is vast, but I, like Yates, have been influenced by the thinking of W. J. Orlikowski, "The Duality of Technology: Rethinking the Concept of Technology in Organizations," *Organization Science* 3, no. 3 (1992): 398–427.

129. Don Chase, "New Business Model for the Insurance Industry Demands a New Automation Model," in Keyes, *Financial Services Information Systems*, 764.

Chapter 5

1. See, for recent examples, Philip Evans and Thomas S. Wurster, *Blown to Bits: How the New Economics of Information Transforms Strategy* (Boston: Harvard Business School Press, 2000); but see also Alex Lightman with William Rojas, *Brave New Unwired World: The Digital Big Bang and the Infinite Internet* (New York: John Wiley & Sons, 2002), and Frances Cairncross, *The Death of Distance: How the Communications Revolution Will Change Our Lives* (Boston: Harvard Business School Press, 1997). For a sober opposing view that argued the case that laws of economics, not technology, still ruled business behavior, see Carl Shapiro and Hal R. Varian, *Information Rules: A Strategic Guide to the Network Economy* (Boston: Harvard Business School Press, 1999).

2. Charles W. Calomiris, "Banking and Financial Intermediation," in Benn Steil, David G. Victor, and Richard R. Nelson (eds.), *Technological Innovation and Economic Performance* (Princeton, N.J.: Princeton University Press, 2002): 285–313.

3. John Kador, *Charles Schwab: How One Company Beat Wall Street and Reinvented the Brokerage Industry* (New York: John Wiley & Sons, 2002): 54.

4. There a number of good histories of the NYSE and Wall Street in general. The latest is Charles R. Geisst, *Wall Street: A History* (New York: Oxford University Press, 1997). See also Francis L. Eames, *The New York Stock Exchange* (New York: Thomas G. Hall, 1984); Robert Sobel, *The Big Board: A History of the New York Stock Exchange, 1935–1975* (New York: Weybright and Talley, 1975); Walter Werner and Steven T. Smith, *Wall Street* (New York: Columbia University Press, 1991).

5. For a brief introduction to the American Stock Exchange, including a list of its key IT milestones, see Charles J. Woelfel, *Encyclopedia of Banking and Finance*, 10th ed. (Chicago, Ill.: Probus Publishing, 1994): 38–39. For histories of the exchange, see Robert Sobel, *Amex: A History of the American Stock Exchange, 1921–1971* (New York: Weybright and Talley, 1972); and Stuart Weems Bruchey, *Modernization of the American Stock Exchange, 1971–1989* (New York: Garland, 1991).

6. U.S. Census Bureau, "1997 Economic Census: Bridge Between SIC and NAICS," http://www.census.gov/epcd/ec97brdg/INDXSIC2.HTM; Sherlene K. S. Lum and Brian C. Moyer, "Gross Domestic Product by Industry for 1997–99," *Survey of Current Business* (December 2000): 29.

7. See, for example, U.S. government statistics, e.g., Bureau of Economic Analysis, *The National Income and Product Accounts of the United States, 1929–82: Statistical Tables* (Washington, D.C.: U.S. Government Printing Office, September 1986): 254–255. For a useful collection of papers on this industry, see Robert R. Litan and Richard Herring (eds.), *Brookings-Wharton Papers on Financial Services 2002* (Washington, D.C.: Brookings Institution Press, 2002).

8. Litan and Herring, *Papers on Financial Services*, vii.

9. If one adds in 401(k) and pension fund transactions, the percentage of the population dependent on stocks rises to over two-thirds of the workforce and their dependents by the last decade of the twentieth century.

10. Federal Reserve Flow of Funds Data.

11. Charles W. Calomiris, "Banking and Financial Intermediation," in Steil, Victor, and Nelson (eds.), *Technological Innovation and Economic Performance*, 285–313.

12. Ibid., 308.

13. Ibid., 312–313.

14. Ian Domowitz and Benn Steil, "Automation, Trading Costs, and the Structure of the Securities Trading Industry," *Brookings-Wharton Papers on Financial Services* (Washington, D.C.: Brookings Institution Press, 1999): 33–92.

15. Ian Domowitz and Benn Steil, "Securities Trading," in Steil, Victor, and Nelson, *Technological Innovation and Economic Performance*, 317.

16. Ibid., 324.

17. Alan Gart, *Regulation, Deregulation, Reregulation: The Future of the Banking, Insurance, and Securities Industries* (New York: John Wiley & Sons, 1994): 284.

18. Marshall E. Blume, "The Structure of the U.S. Equity Markets," Brookings-Wharton Papers on Financial Services 2002, 35–43. For more detailed accounts of the SEC, see Joel Seligman, *The Transformation of Wall Street: A History of the Securities and Exchange Commission and Modern Corporate Finance* (Boston: Houghton Mifflin, 1982); see also Geisst, *Wall Street: A History*.

19. Ibid., 43–45.

20. ECNs are electronic venues through which buyers and sellers trade in stocks, widely used by electronic day trades.

21. Geisst, *Wall Street: A History*, 47–49.

22. Charles Amos Dice and Wilford John Eiteman, *The Stock Market* (New York: McGraw-Hill, 1952): 25.

23. George L. Leffler, *The Stock Market*, 2d ed. (New York: Ronald Press Company, 1957): 201. In this widely available account of the stock market, the author made no mention of computers. However, he acknowledged the wide adoption in recent years of office-accounting machines; "however, the essential principles upon which the entire process of clearing is based remain unchanged" (421).

24. Richard J. Teweles, Edward S. Bradley and Ted M. Teweles, *The Stock Market*, 6th ed. (New York: John Wiley & Sons, 1992): 306.

25. Alec Benn, *The Unseen Wall Street of 1969–1975: And Its Significance for Today* (Westport, Conn.: Quorum Books, 2000): xi.

26. Domowitz and Steil, "Automation, Trading Costs, and the Structure of the Securities Trading Industry," 39.

27. Developed in 1867 at the NYSE and continuously upgraded over the next century to handle larger numbers of stocks and transactions, its purpose is to report trades on the exchange. Originally it printed data on tape; today that information is displayed electronically. For a brief description of its function and evolution, see Richard J. Teweles and Edward S. Bradley *The Stock Market* (New York: John Wiley & Sons, 1992): 140–142.

28. "Calculators to Aid Brokerage Firms," *Journal of Commerce*, January 15, 1955, clipping in Record Group 8, Public Relations, Subject Clipping 1955–1957, IBM Corporate Archives, Somers, N.Y.; " 'Brain' Apportioning Ford Stock," *Journal of Commerce*, January 17, 1956, p. 3.

29. "Wall Street Enlists Electronic Tools," *New York Times*, November 22, 1955, clipping in Record Group 8, Public Relations, Subject Clipping 1955–1957, IBM Corporate Archives, Somers, N.Y.; "Brokers Turn to Automation," *Business Week* (April 16, 1955): 118; Press release from Burnham and Company, June 6, 1966, Public Relations/Subject Clippings/ DP Applications—Stock Exchange, IBM Corporate Archives, Somers, N.Y.; "System/360 to Quote Stock," IBM, *DPD News*, June 10, 1965, p. 2. See also dozens of press articles collected by the IBM Press Review Service for the mid-1960s, all in the same IBM files.

30. For example, Benn, *The Unseen Wall Street of 1969–1975*, 10–14.

31. "Merrill Lynch Installs 'Electronic Brain' Here," *Los Angeles, California Herald and Express*, February 10, 1956, clipping in Record Group 8, Public Relations, Subject Clipping 1955–1957, IBM Corporate Archives, Somers, N.Y.; "Machines at Work in Wall Street," *Investor's Reader*, February 22, 1956.

32. "Machines at Work in Wall Street," *Investor's Reader*, February 22, 1956.

33. IBM Press Release, August 3, 1959, *Data Processing News* file, IBM Corporate Archives, Somers, NY.

34. "Big Stock Volume Keeps Lights on Late in Wall Street, But Machines Cut Overtime," *Wall Street Journal*, January 10, 1955, clipping in Record Group 8, Public Relations, Subject Clipping 1955–1957, IBM Corporate Archives, Somers, N.Y.

35. Ibid.

36. IBM Press Release, undated [1959]; IBM Press Release, February 18, 1960; IBM Press Release, August 1960; IBM Press Release, November 16, 1960; *Data Processing News* files, IBM Corporate Archives, Somers, N.Y.; "Brokerage Firm Installs 705 Model III," *Data Processing News*, September 10, 1959, p. 11; "Brokerage Firm Grows With IBM 7070 'Partners,' " Ibid., April 17, 1961, p. 9.

37. "Applications Roundup," *Data Processing News*, September 11, 1961, p. 3.

38. L. E. Clark to Frank T. Cary, October 20, 1964, C. G. Ruykhaver to M. E. Clark, October 20, 1964, Frank Cary to G. F. Kennard, December 14, 1964, NYSE file, IBM Corporate Archives, Somers N.Y.; Press Release by New York Stock Exchange, October 5, 1962 and March 8, 1965, March 1, 1966, copies in Record Group 8, Public Relations, Subject Clipping 1955–1957, IBM Corporate Archives, Somers, N.Y.

39. NYSE, *Fact Book 1964* (New York: NYSE, 1964): 39.

40. Marshall E. Blume, Jeremy J. Siegel, and Dan Rottenberg, *Revolution on Wall Street: The Rise and Decline of the New York Stock Exchange* (New York: W. W. Norton, 1993): 193–214.

41. Benn, *The Unseen Wall Street of 1969–1975*, 10–22.

42. Lawrence Shepard, *The Securities Brokerage Industry: Nonprice Competition and Noncompetitive Pricing* (Lexington, Mass.: Lexington Books, 1975): 23–29.

43. The role of IT in the regional exchanges has not been examined by historians in even a cursory manner, but we know that their activities spurred the NYSE and the SEC to action.

44. Benn, *The Unseen Wall Street of 1969–1975*, has an excellent example that he cites throughout his book.

45. Descriptions published in the 1960s and early 1970s of how the various brokerage processes worked document how labor and paper intensive these activities had become. See, for example, George L. Leffler, *The Stock Market*, 3rd ed. (New York: Ronald Press, 1963): 148–165, 180–187; Richard F. Kearns, "Processing and Executing Orders," in Frank G. Zarb and Gabriel T. Kerekes (eds.), *The Stock Market Handbook: Reference Manual for the Securities Industry* (Homewood, Ill.: Dow Jones–Irwin, 1970): 629–639.

46. For an account of those days, see Benn, *The Unseen Wall Street of 1969–1975*, 10–15.

47. "Wall Street vs. Paper Work," *The Exchange* (September 1969): 15–16, Archive of the New York Stock Exchange, New York.

48. Ibid., 11–35.

49. My account of the crisis in the previous, current, and next paragraph is drawn from Blume, Siegel, and Rottenberg, *Revolution on Wall Street*, 116–127.

50. "The SIAC Story: A Technological Evolution," *SIAC News* 5, no. 6 (July 1987): 3–7, 10; "Milestones," undated [circa 1982] inventory of SIAC's applications, Archives of the NYSE, New York. Both describe the applications in place and provide a chronology of their implementation.

51. "Milestones," 126.

52. Hans von Gelder, "On-Line Stock Quotation," *Datamation* 10, no. 3 (March 1964): 37–41.

53. IBM's press files are filled with accounts of various exchanges installing their first-or second-generation of computers. The exchanges also publicized their use of computers. For example, a press release from the Boston Stock Exchange, dated June 26, 1969, bragged about how an IBM 360 would be used to process "all buying and selling transactions for the day in as little as 3½ hours. This task formerly took six hours." The Midwest Stock Exchange Clearing Corporation had earlier done the same in a press release dated November 10, 1959, announcing it was installing an IBM RAMAC 305 system, but in the release described the key application: "As trades are processed daily at the exchange, RAMAC will absorb each transaction and automatically update each member's account. As securities are exchanged, current balances will be immediately affected by RAMAC, establishing each broker's position for daily settlement with the Exchange."

54. Zarb and Kerekes, *The Stock Market Handbook*, 26.

55. See, for example, the comments by a member of the Computer Systems Division at RCA, Henry Staehling, "Electronic Data Processing," in Ibid., 603–616; "Automation Comes to Wall Street," *Financial World* (March 6, 1968): 5, 26; "Joint Firm Tooling Up for 'Street,'" *Electronic News*, December 14, 1970, p. 36.

56. See prior note but also the example of "The Street's New Paper Cutter," *Business Week*, November 1, 1969, unpaginated as a reprint of the original article distributed across the economy, copy in CBI 90, Control Data Corporation Newspapers and Magazine Articles, Archives of Charles Babbage Institute, University of Minnesota, Minneapolis.

57. Anthony A. Barnett, "Automated Stock Market Services," CDC Corporation Newspaper and Magazine Articles, 215–228.

58. George H. Sutton, "Communications Systems," Ibid., 617–628; "Battling the Big Board to Serve Big Traders, A. F. Kay of AutEx Computer System," *Business Week*, June 14, 1969, pp. 104ff; "Find Out How Your Investments Are Really Doing: New Changing Times Computer Service," *Changing Times* 24 (March 1970): 47–49; Joseph C. Marshall, "Distributed Processing on Wall Street," *Datamation* 19, no. 7 (July 1973): 44–46; C. Saltzman, "Apple for Ben Rosen; Use of Personal Computers by Securities Analysts," *Forbes* 124 (August 20, 1979): 54–55; George Schussel and Jack May, "Wall Street Automation: A Primer," *Datamation* 16, no. 4 (April 1970): 109ff.

59. NYSE, *The Stock Market Under Stress: The Events of May 28, 29 and 31, 1962* (New York: NYSE, 1963): 49.

60. SEC, "SEC Concept Release: Regulation of Market Information Fees and Revenues," Release No. 34–42208, [undated 1999], http://www.sec.gov/rules/concept/34–42208.htm.

61. Lawrence Shepard, *The Securities Brokerage Industry: Nonprice Competition and Noncompetitive Pricing* (Lexington, Mass.: Lexington Books, 1975): 23–31.

62. Geisst, *Wall Street: A History*, 306–310.

63. The Brokerage Industry established NASD as a self-regulatory body, acting as both a complement and buffer to the SEC. Almost all brokers and their firms in the United States are members. For details on NASD see its website, www.nasd.com/copr_info. On NASDAQ's system, Teweles, Bradley and Teweles, *The Stock Market*, 6th ed., 204–206; John H. Hodges, Jr., "How the Over-the-Counter Market Place Functions," in Richard D. Kirshberg (ed.), *The Over-the-Counter Market Place: Myths, Mysteries and Mechanics* (New York: Practising Law Institute, 1974): 11–60.

64. Most of the material for the past two paragraphs were drawn from Blume, Siegel, and Rottenberg, *Revolution on Wall Street*, 128–129, 161–171, 175–180, 187–192; Geisst, *Wall Street*, 299–320; Gart, *Regulation, Deregulation, Reregulation*, 74–79.

65. R. Stovall, "NASDAQ We Love You," *Forbes* 107 (April 15, 1971): 84–85; W. Frederick Goodyear, "The Birth of NASDAQ," *Datamation* 18, no. 3 (March 1972): 42–46; "Higher Meaning of NASDAQ: Market for Over-the-Counter Stocks," *Fortune* 83

(April 1971): 141–142ff. Meanwhile press coverage of what the brokerage firms were doing also appeared. For example, see George Schussel, "Wall Street Automation: A Primer," *Datamation* 16, no. 4 (April 1970): 109ff, which concerned developments at Dean Witter & Company, or "Computer Help for Commodity Funds," *Business Week*, July 7, 1975, pp. 57–58.

66. See a raft of articles published in the mid-1960s from the *Wall Street Journal, New York Times, Journal-Courier of New Haven* (Conn.), and other papers in Public Relations/Subject Clippings, 1964/New York Stock Exchange, and for 1966 Record Group 8, Public Relations/Subject Clippings, 1966/DP Applications—Stock Exchange, IBM Corporate Archives, Somers, N.Y. See also press releases from the NYSE, December 20, 1967, July 26, 1967, and February 25, 1969, copies in Record Group 8, IBM Corporate Archives, Somers, N.Y. The NYSE published a series of documents—primarily for circulation within the industry, among regulators, and the press—taking great pains to describe the digital applications being implemented. See, for example, NYSE, "Automation in the Nation's Marketplace: The Story of Technological Advance at the New York Stock Exchange, 01/1966–10/1966," Archives of the NYSE, New York; NYSE, *New York Stock Exchange Market Data System Quotation Service: A Fully Automated Information System You Can Dial* (New York: NYSE, 1965); NYSE, *The NYSE Block Automation System* (New York: NYSE, 1968), both in the Archives of the NYSE, New York.

67. Gerard P. Lynch, "Electronic Transfer of Insurance Securities," *Best's Review*, Liability/Casualty Edition 73, no. 8 (December 1972): 26, 28; Benn, *The Unseen Wall Street*, 10–22. Benn made the comment that "without the establishment of the Central Certificate System and the efficient, remarkably swift processing of orders by computers, it would be impossible for the New York Stock Exchange to process the hundreds of millions of shares traded everyday in the 1990s," 21–22.

68. Data drawn from an internal IBM review of its selling efforts at the NYSE, March 26, 1973, at which time it was reported that the firm had just signed a contract with the NYSE to provide a floor terminal system. The account team reported that the new system would automate the specialist function "by providing him with an electronic book in which his open orders are maintained. As buy and sell orders are entered by the specialist, they will be logged by the system, which in later implementation phases will result in a 'locked in trade.' With both buying and selling brokers identified by computer, the result will be fewer errors and failures." The little scribbled slips of papers announcing orders were to be eliminated. Account Review in Record Group 8, "Trip Files," IBM Corporate Archives, Somers, N.Y.

69. Benn, *The Unseen Wall Street*, describes the environment that was very hostile to change, 1–22.

70. Blume, Siegel, and Rottenberg, *Revolution on Wall Street*, 193–199.

71. For a description of the new application, see NYSE, *Super Dot 250: The Electronic Pathway to the New York Stock Exchange* (New York: NYSE, October 1984), Archives of the NYSE, New York.

72. For a detailed discussion of the history of DOT, see Ibid., 201–209; for DOT and SuperDOT, Teweles, Bradley, and Teweles, *The Stock Market*, 6th ed., 198–199. On the growing demand and use of automated brokerage information and trading systems of the early 1970s, see William P. Rogers, "Instant Financial Information," *Datamation* 19, no. 2 (February 1973): 65–68.

73. Teweles, Bradley, and Teweles, *The Stock Market*, 6th ed., 114–119.

74. Each is described in David M. Weiss, *After the Trade Is Made: Processing Securities Transactions* (New York: New York Institute of Finance, 1986).

75. Stephen P. Rappaport, *Management on Wall Street: Making Securities Firms Work* (Homewood, Ill.: Dow Jones–Irwin, 1988): 107–108. Much of the material in this paragraph is drawn from this same source, pp. 104–108.

76. Ibid., 109.

77. For a listing and assessment of these software packages, see *PC Magazine* 5, no. 7 (April 15, 1986).

78. Charles P. Jones, *Investments: Analysis and Management*, 2d ed. (New York: John Wiley & Sons, 1988): 104–108.

79. IBM, *Industry Applications and Abstracts* (White Plains, N.Y.: IBM Corporation, 1988): 19-1–19-3.

80. Blume, Siegel, and Rottenberg, *Revolution on Wall Street*, 199.

81. Ibid., 199–201.

82. Teweles, Bradley and Teweles, *The Stock Market*, 6th ed., 198–200, 202–208, 218–229.

83. Ezra Solomon, *Wall Street in Transition: The Emerging System and Its Impact on the Economy* (New York: New York University Press, 1974): 122–123.

84. Teweles, Bradley, and Teweles, *The Stock Market*, 6th ed., 202–211. For a brief overview of what was happening with the regional exchanges in the 1970s and 1980s, see Ibid., pp. 218–230.

85. One economic study has argued that the breakers were used too frequently in the 1990s when volatility of stocks was not out of range with historical patterns, G. William Schwert, "Stock Market Volatility: Ten Years After the Crash," NBER Working Paper Series, no. 6381 (Cambridge, Mass.: National Bureau of Economic Research, January 1998), available at http://www.nber.org/papers/w6381.

86. Ibid., 6.

87. See, for example, Charles J. Jacklin, Allan W. Kleidon, and Paul Pfleiderer, "Underestimation of Portfolio Insurance and the Crash of 1987," *Review of Financial Studies* 5 (1992): 35–64; Allan W. Kleidon and Robert E. Whaley, "One Market? Stocks, Futures, and Options During October 1987," *Journal of Finance* 47 (1992): 851–878; Sanford J. Grossman, "Program Trading and Market Volatility: A Report on Interday Relationships," *Financial Analysts Journal* 44 (July–August 1988): 18–28; Teweles, Bradley, and Teweles, *The Stock Market*, 6th ed., 229–230. For the recommendations on breakers, see Presidential Task Force on Market Mechanisms, *Report of the Presidential Task Force on Market Mechanisms* (Washington, D.C.: U.S. Government Printing Office, 1988). For background on the NASDAQ, see Woelfel, *Encyclopedia of Banking and Finance*, 780–782.

88. U.S. Congress, Office of Technology Assessment, *Electronic Bulls & Bears: U.S. Securities Markets and Information Technology* (Washington, D.C.: U.S. Government Printing Office, September 1990); U.S. General Accounting Office, *Financial Markets: Tighter Computer Security Needed* (Washington, D.C.: U.S. Government Printing Office, 1990).

89. SEC, *Third Report on the Readiness of the United States Securities Industry and Public Companies to Meet the Information Processing Challenges of the Year 2000* (Washington, D.C.: U.S. Securities and Exchange Commission, July 1999), available at http://www.sec.gov/news/studies/yr2000−3.htm.

90. Tower Group, *The Information Technology Capacity Study Securities Industry 1998–2000* (New York: Securities Industry Association, April 1998).

91. This and the previous paragraph are based on *The Information Technology Capacity Study Securities Industry*.

92. Thomas A. Meyers, *The Dow Jones–Irwin Guide to-on-Line Investing: Sources, Services and Strategies* (Homewood, Ill.: Dow Jones–Irwin, 1986): vix, xv–xvi, 2, 127, and passim.

93. The executive who established this service wrote his recollections: K. Blake Darcy, "Brokerage Service: Perspectives from a Decade in Online's Most Mature Segment," in Christina Ford Haylock, Len Muscarella, and Ron Schultz (eds.), *Net Success: 24 Leaders in Web Commerce Show You How to Put the Internet to Work for Your Business* (Holbrook, Mass.: Adams Media Corporation, 1999): 238–256.

94. John Kador, Charles Schwab: *How One Company Beat Wall Street and Reinvented the*

Brokerage Industry (New York: John Wiley & Sons, 2002): 54–55, 87–90, 106–108; David S. Pottruck, *Clicks and Mortar: Passion Driven Growth in an Internet Driven World* (San Francisco, Calf.: Jossey-Bass, 2000): 198–212; Mary J. Cronin, "Charles Schwab & Company," in Mary J. Cronin (ed.), *Banking and Finance on the Internet* (New York: John Wiley & Sons, 1998): 231–250.

95. For historical perspective on day trading by a day trader, see Gregory J. Millman, *The Day Traders: The Untold Story of the Extreme Investors and How They Changed Wall Street Forever* (New York: Times Business, 1999).

96. Alan Levinsohn, "Online Brokerage, the New Core Account?" *ABA Banking Journal* 91, no. 5 (September 1999): 34–35, 38, 40–42; Office of Compliance Inspections and Examinations, SEC, *Examinations of Broker-Dealers Offering Online Trading: Summary of Findings and Recommendations*, January 25, 2001, http://www.sec.gov/news/studies/online.htm.

97. Levinsohn, "Online Brokerage, the New Core Account?" 40.

98. Office of New York State Attorney General Eliot Spitzer, "Online Brokerage Industry Report," undated [1999], http://www.oag.state.ny.us/investors/1999_online_brokers/execsum.html.

99. U.S. Securities and Exchange Commission, *Report to the Congress: The Impact of Recent Technological Advances on the Securities Markets* (Washington, D.C.: U.S. Securities and Exchange Commission, 1997): 27, 60. This report is a lengthy catalog of various applications and services provided by the Brokerage Industry.

100. Ming Fan, Jan Srallaert, and Andrew B. Whinston, "A Web-Based Financial Trading System," *Computer* (April 1999): 64–70, http://crec.bus.utexas.edu/work.

101. Office of New York State Attorney General Eliot Spitzer, "Online Brokerage Industry Report," see chapter entitled "The Market Storm of 1999—The Outages and Customer Complaints of Online Trading," pp. 1–3.

102. Ibid.

103. Eric K. Clemons and Lorin M. Hitt, "Financial Services: Transparency, Differential Pricing, and Disintermediation," in Robert E. Litan and Alice M. Rivlin (eds.), *The Economic Payoff from the Internet Revolution* (Washington, D.C.: Brookings Institution Press, 2001): 107.

104. Ibid., 107–108.

105. U.S. Securities and Exchange Commission, Division of Market Regulation, *Technology Roundtable* (Washington, D.C.: U.S. Securities and Exchange Commision, April 1998): 5.

106. Ibid., 3.

107. James Marks, "The Impact of the Internet on Users and Suppliers of Financial Services," in Litan and Santomero, *Brookings-Wharton Papers on Financial Services 1999*, 147–205.

108. Securities Industry Association, "Technology Trends in the Securities Industry: Spending, Strategies, Challenges, and Change–2001," undated [2001], http://www.sia.com/surveys/html/tech_trends_sample.html.

109. The chairman reported in detail the events that took place to bring the industry back online in his comments; Press release SEC, "Testimony Concerning The State of the Nation's Financial Markets in the Wake of Recent Terrorist Attacks," September 26, 2001, http://www.sec.gov/news/testimony/092601tshlp.htm.

110. John Kelly and David Stark, "Crisis, Recovery, Innovation: Responsive Organization after September 11," Working Paper WP 2003-02, April 12, 2002, Wharton School, University of Pennsylvania; Daniel Beunza and David Stark, "The Organization of Responsiveness: Innovation and Recovery in the Trading Rooms of Lower Manhattan," *Socio-Economic Review* 1 (2003): 135–164; Jeffrey W. Greenberg, "September 11, 2001: A CEO's Story," *Harvard Business Review* (October 2002): 3–7.

111. U.S. Securities and Exchange Commission, "Testimony Concerning Recovery and

Renewal: Protecting the Capital Markets Against Terrorism Post 9/11," press release, February 12, 2003; Ibid., "Interagency Paper on Sound Practices to Strengthen the Resilience of the U.S. Financial System," April 8, 2003; Ibid., Office of Compliance Inspections and Examinations, SEC, "Examinations of Broker-Dealers Offering Online Trading: Summary of Findings and Recommendations," January 25, 2001, Ibid.

112. New York Stock Exchange, *Annual Report 2002* (New York: New York Stock Exchange, 2003): 4.

113. For descriptions of these services, see John Buckner, "The Future of the Securities Industry: Convergence of Trust and Brokerage," in Keyes, *Financial Services Information Systems*, 669–678; and John McLeod, "The Broker Desktop: The Future of Trading Has Arrived," Ibid., 745–753.

114. McLeod, "The Broker Desktop," Ibid., 746.

115. For a brief summary of trends, see Daniel Latimore, Vikram Lund, Ian Watson, John Raposo, and Greg Robinson, *An Old Games with New Rules: Creating Value in Today's Securities Industry* (Somers, N.Y.: IBM Corporation, 2002).

116. Michael J. Martinez, "NYSE to Hear Automation Plan," *Wisconsin State Journal*, February 5, 2004, p. F7.

117. Blume, "The Structure of the U.S. Equity Markets," 54.

118. Domowitz and Steil, "Securities Trading," in Steil, Victor, and Nelson, *Technological Innovation and Economic Performance*, 314–326. The issue of cost has come up again in the wake of 9/11 because the NYSE and others began working through the issues involved in setting up back-up sites for their IT, and one of the issues involves people, not just hardware. For details, see Litan and Herring, *Brookings-Wharton Papers on Financial Services 2002*, 57–58.

Chapter 6

1. Manuel Castells, *The Internet Galaxy: Reflections on the Internet, Business, and Society* (Oxford: Oxford University Press, 2001): 1.

2. U.S. Department of Commerce, *Historical Abstracts of the United States: Colonial Times to 1970* (Washington, D.C.: U.S. Government Printing Office, 1975), Part 2: p. 783.

3. My point is informed more by the findings of those studying the role of knowledge management practices in the economy than from classical economic analysis. See, for example, the collection of essays by various researchers in Eric L. Lesser (ed.), *Knowledge and Social Capital: Foundations and Applications* (Boston, Mass.: Butterworth-Heinemann, 2000); and a similarly important collection more focused on economic issues from Dale Neef, G. Anthony Siesfeld, and Jacquelyn Cefola (eds.), *The Economic Impact of Knowledge* (Boston, Mass.: Butterworth-Heinemann, 1998).

4. Cerf told the story of how he tried to present the case to AT&T's scientists for why digital technology made for better telecommunications than existing analog approaches in David Pitchford, "Is There a Future for the Net? David Pitchford Finds Out from the Man Who Invented It, Vinton Cerf," *Internet*, June 19, 1996, p. 75. Paul Baran, who also participated in those meetings, commented that the problem was not the scientists but rather the executives. He and Cerf were next sent to in New York, where executives saw no reason to change the technology in place in a stable monopoly. The two men went to AT&T in the first place because the U.S. Department of Defense sent them to the national telephone monopoly to implement the network; only after AT&T turned its back did DOD create its own network, ultimately called the Internet; Conversation with Paul Baran, February 18, 2004.

5. The Music Industry might also qualify as the industry most in flux, although it is a smaller industry.

6. William J. Baumol points out that the transformations that technology can have on

industries are very important, "Innovation and Creative Destruction," in Lee W. McKnight, Paul M. Vaaler, and Raul L. Katz (eds.), *Creative Destruction: Business Survival Strategies in the Global Internet Economy* (Cambridge, Mass.: MIT Press, 2001): 21–38. The same volume has a series of other papers dealing with the role of the Internet in this process. For a recent, firm-level view of the role of the notion that technology causes creative destruction, see Richard Foster and Sarah Kaplan, *Creative Destruction: Why Companies That Are Built to Last Underperform the Market—and How to Successfully Transform Them* (New York: Currency, 2001): 25–89.

7. For a series of papers on this theme, Edward L. Hudgins (ed.), *Mail © the Millennium: Will the Postal Service Go Private?* (Washington, D.C.: Cato Institute, 2000).

8. U.S. Department of Commerce, *Historical Abstracts*, Part 2, pp. 775–791.

9. Quote drawn from Peter Temin with Louis Galambos, *The Fall of the Bell System: A Study in Prices and Politics* (Cambridge: Cambridge University Press, 1987): 11.

10. U.S. Census Bureau, "1997 Economic Census: NAICS 5133 Telecommunications," p. 1, http://www.census.gov/epcd/ec97/industry/E5133.HTM. For a description of NAICS, see U.S. Census Bureau, "North American Industry Classification System (NAICS)," http://www.census.gov/epcd/www/naics.html. For a description of the comparisons of SIC and NAICS, Ibid. "1997 Economic Census: Bridge Between SIC and NAICS. SIC: Menu of SIC Divisions," http://www.census.gov/epcd/ec97brdg/INDXSIC2.HTM.

11. Ibid., "1997 Economic Census: NAICS 513322 Cellular and Other Wireless Telecommunications," http://www.census.gov/epcd/ec97/industry/E513322.HTM.

12. For example, John R. Meyer et al., *The Economics of Competition in the Telecommunications Industry* (Cambridge, Mass.: Oelgeschlager, Gunn & Hain, 1980): 23–53; Mitchell L. Moss (ed.), *Telecommunications and Productivity* (Reading, Mass.: Addison-Wesley, 1981); Benjamin M. Compaine, "Shifting Boundaries in the Information Marketplace," in Benjamin M. Compaine, *Understanding New Media: Trends and Issues in Electronic Distribution of Information* (Cambridge, Mass.: Ballinger Publishing Company, 1984): 97–120; John F. McLaughlin, "Mapping the Information Basiness," Ibid., 19–67; Heather E. Hudson, *Global Connections: International Telecommunications Infrastructure and Policy* (New York: Van Nostrand Reinhold, 1997): 35–63, 89–100; Jean-Luc Gaffard and Jackie Krafft, "Telecommunications: Understanding the Dynamics of the Organization of the Industry," unpublished paper, October 2000, University of Nice–Sophia Antipolis and Institut Universitaire de France, http://www/TelecomVisions.com.

13. See the Bureau's reports called "Communications and Information Technology," http://www.census.gov/prod/2/ge.PDF, for the period of the 1980s and 1990s.

14. Frances Cairncross, *The Death of Distance: How the Communications Revolution Will Change Our Lives* (Boston, Mass.: Harvard Business School Press, 1997): 87–118; Alex Lightman with William Rojas, *Brave New Unwired World: The Digital Big Bang and the Infinite Internet* (New York: John Wiley & Sons, 2002): 75–81; Martin Fransman, *Telecoms in the Internet Age: From Boom to Bust to . . . ?* (Oxford: Oxford University Press, 2002): 34–80.

15. U.S. Department of Commerce, *Historical Statistics of the United States: Colonial Times to 1970* (Washington, D.C.: U.S. Government Printing Office, 1975), Part 2: 785.

16. Telephone and Data Systems, better known as TDS, is an excellent example of this process at work. A local telephone company established in the mid-west, it grew over the decades primarily by acquiring smaller local telephone companies, usually at the rate of 3 to 5 per year, for example, each year throughout the second half of the century, K. C. August, *TDS: The First Twenty Years* (Chicago, Ill.: Telephone and Data Systems, Inc., 1993): 148–149, 152.

17. U.S. Department of Commerce, *Historical Statistics of the United States*, 786.

18. Ibid., 788–789.

19. Data drawn from tables in Peter Temin with Louis Galambos, *The Fall of the Bell System* (Cambridge: Cambridge University Press, 1987): 4.

20. U.S. Department of Commerce, *Historical Statistics*, 783.

21. U.S. Census Bureau, *Statistical Abstract of the United States 1998* (Washington, D.C.: U.S. Census Bureau, September 25, 1998), table 918, p. 574.

22. Ibid., Table 920, p. 575.

23. Ibid., Table 917, p. 574. A major source of information on the use of the Internet in the United States is the Pew Foundation, which has had an initiative underway to study the issue since the late 1990s. It publishes all its frequent studies at its website, http://www.pewinternet.org.

24. U.S. Census Bureau, *Statistical Abstract of the United States 2000* (Washington, D.C.: U.S. Census Bureau, 2000), table 913, p. 568.

25. Ibid., table 915, p. 569; for cell phone data see table 919, p. 571.

26. U.S. Census Bureau, *Statistical Abstract* 1998, table 921, p. 576.

27. "CTIA's Semi-Annual Wireless Industry Survey," Cellular Telecommunications and Internet Association, http://www.wow-com.com/industry/stats/surveys/ (last accessed January 4, 2004).

28. Reed E. Hundt, *You Say You Want A Revolution: A Story of Information Age Politics* (New Haven, Conn.: Yale University Press, 2000): 214–216.

29. On motives for deregulatory movement, see Temin and Galambos, *The Fall of the Bell System*, 344–346.

30. Max D. Paglin (ed.), *The Communications Act: A Legislative History of the Major Amendments, 1934–1996* (Silver Springs, Md.: Pike & Fischer, 1999); Richard H. K. Victor, *Contrived Competition: Regulation and Deregulation in America* (Cambridge, Mass.: Harvard University Press, 1994); Inge Vogelsang and Glenn Woroch, "Local Telephone Service," in Larry L. Deutsch (ed.), *Industry Studies* (Armonk, N.Y.: M. E. Sharpe, 1998): 254–291. On the effects of the 1996 law, Robert W. Crandall and Jerry A. Hausman, "Competition in U.S. Telecommunications Services: Effects of the 1996 Legislation," in Sam Peltzman and Clifford Winston (eds.), *Deregulation of Network Industries: What's Next?* (Washington, D.C.: AEI-Brookings Joint Center for Regulatory Studies, 2000): 73–112.

31. Temin and Galambos, *The Fall of the Bell System*, 346.

32. For the quotes and their arguments, Ibid., 346–353.

33. McMaster, *The Telecommunications Industry*, 121–128; Temin and Galambos, *The Fall of the Bell System*, 217–306; Alan Stone, *How America Got On-Line: Politics, Markets, and the Revolution in Telecommunications* (Armonk, N.Y.: M. E. Sharpe, 1997): 81–105.

34. Crandall and Hausman, "Competition in U.S. Telecommunications Services," 85–110.

35. The FCC chose to allocate part of the space on the air waves to other firms so as to keep the market competitive and to reserve other space for potentially other wireless uses.

36. Philip L. Cantelon, The *History of MCI: The Early Years, 1969–1988* (Dallas, Tex.: Heritage Press, 1993): 93–170; Larry Kahaner, *On the Line: The Men of MCI—Who Took on AT&T, Risked Everything, and Won!* (New York: Warner Communications, 1986): 65–137.

37. McMaster, *The Telecommunications Industry*, 98–110.

38. Ibid., 135; Paul W. MacAvoy, *The Failure of Antitrust and Regulation to Establish Competition in Long Distance Telephone Services* (Cambridge, Mass.: MIT Press, 1996): 83.

39. McMaster, *The Telecommunications Industry*, 139–140.

40. Hundt, *You Say You Want a Revolution*, 159.

41. McMaster, *The Telecommunications Industry*, 153–175. A vast amount of literature now exists on this topic; I relied extensively on Crandall and Hausman, "Competition in U.S. Telecommunications Services," 73–110; and on Dale E. Lehman and Dennis L. Weisman, *The Telecommunications Act of 1996: The "Costs" of Managed Competition* (Boston, Mass.: Kluwer Academic Publishers, 2000): 23–48.

42. McMaster, *Telecommunications Industry*, 164.

43. Fransman, *Telecoms in the Internet Age*: 260–266; David J. Collis, P. William Bane, and Stephen P. Bradley, "Winners and Losers: Industry Structure in the Converging World of Telecommunications, Computing, and Entertainment," in David B. Yoffie (ed.), *Competing in the Age of Digital Convergence* (Boston, Mass.: Harvard Business School Press, 1997): 159–200; Kevin Maney, *Megamedia Shakeout: The Inside Story of the Leaders and Losers in the Exploding Communications Industry* (New York: John Wiley & Sons, 1995): 59–104; 185–218; 283–307; Om Malik, *Broadbandits: Inside the $750 Billion Telecom Heist* (New York: John Wiley & Sons, 2003).

44. Hundt, *You Say You Want a Revolution*, 224.

45. Joseph E. Stiglitz, *The Roaring Nineties: A New History of the World's Most Prosperous Decade* (New York: W. W. Norton, 2003): 93–94.

46. Key investors in R&D in North America included AT&T, Cisco, Lucent, and Nortel. For details, circa 1999, see Fransman, *Telecoms in the Internet Age*, 218.

47. Ithiel de Sola Pool, edited by Eli M. Noam, *Technologies Without Boundaries: On Telecommunications in a Global Age* (Cambridge, Mass.: Harvard University Press, 1990): 7.

48. Ibid., 21–22.

49. In particular, on the occasion of the Lab's 50th anniversary in the mid-1970s, AT&T began publishing a series of extensively researched histories that have become the standard works on the role of Bell Labs; Prescott C. Mabon, *Mission Communications: The Story of Bell Laboratories* (Murray Hills, N.J.: Bell Telephone Laboratories, 1975); M. D. Fagen (ed.), *A History of Engineering and Science in the Bell System: National Service in War and Peace (1925–1975)* (Murray Hills, N.J.: Bell Telephone Laboratories, 1978); A. E. Joel Jr. with G. E. Schindler Jr. as editor, *A History of Engineering and Science in the Bell System: Switching Technology (1925–1975)* (Murray Hills, N.J.: Bell Telephone Laboratories, 1982); S. Millman (ed.), *A History of Engineering and Science in the Bell System: Communications Sciences (1925–1980)* (Murray Hills, N.J.: AT&T Bell Laboratories, 1984); F. M. Smits (ed.), *A History of Engineering and Science in the Bell System: Electronics Technology (1925–1975)* (Murray Hills, N.J.: AT&T Bell Laboratories, 1985).

50. George P. Oslin, *The Story of Telecommunications* (Macon, Ga.: Mercer University Press, 1992): 437–455; Larry Macdonald, *Nortel Networks: How Innovation and Vision Created a Network Giant* (Toronto: John Wiley & Sons, 2000): 51–70; and for a perspective from a Bell Labs employee of the 1960s and 1970s, see A. Michael Noll, *Introduction to Telephones and Telephone Systems*, 3rd ed. (Boston, Mass.: Artech House, 1998): 331–335; but see also two articles by the same author, "Bell System R&D Activities: The Impact of Divestiture," *Telecommunications Policy* 11, no. 2 (June 1987): 161–178, and "The Effects of Divestiture on Telecommunications Research," *Journal of Communications* 37, no. 1 (Winter 1987): 73–80.

51. Macdonald, *Nortel Networks*, 51–69.

52. Arthur L. Norberg and Judy E. O'Neill, *Transforming Computer Technology: Information Processing for the Pentagon, 1962–1986* (Baltimore, Md.: Johns Hopkins University Press, 1996): 153–196.

53. Paul E. Ceruzzi, *A History of Modern Computing* (Cambridge, Mass.: MIT Press, 1998): 291–292; Edgar H. Schein, *DEC Is Dead Long Live DEC: The Lasting Legacy of Digital Equipment Corporation* (San Francisco, Calif.: 2003): 262–263.

54. Stephen B. Adams and Orville R. Butler, *Manufacturing the Future: A History of Western Electric* (Cambridge: Cambridge University Press, 1999): 189.

55. Ibid., 187–193.

56. Ibid., 199.

57. On Western Electric's manufacturing agenda, Adams and Butler, *Manufacturing the Future*, 205–213. Their coverage of the issue is thin, so see also Noll, *Introduction to Telephones and Telephone Systems*, 175–182, 190–196, 333–335.

58. David Bunnell, *Making the Cisco Connection:The Story Behind the Real Internet Superpower* (New York: John Wiley & Sons, 2000): 15–42, 113–135.

59. Noll, *Introduction to Telephones and Telephone Systems*, 200–201.

60. John P. Burnham, *The Essential Guide to the Business of U.S. Mobile Wireless Communications* (Upper Saddle River, N.J.: Prentice Hall PTR, 2002): 41–51.

61. Ibid., 58–65.

62. Ibid., 65–74.

63. Ibid., 75–76.

64. David J. Whalen, *The Origins of Satellite Communications, 1945–1965* (Washington, D.C.: Smithsonian Institution Press, 2002): 1–18, 149–165; Oslin, *The Story of Telecommunications*, 387–411; Noll, *Introduction to Telephones and Telephone Systems*, 73–83. By the late twentieth century, satellite communications had become an integral part of the story of television and thus will be discussed further in Chapter 10.

65. In the case of the telegraph, for example, as early as 1992 that sector of the Telecommunications Industry employed less than 10,000 people, U.S. Bureau of the Census, "Communications and Information Technology," *Census of Transportation, Communications and Utilities, UC92-A-1* (Washington, D.C.: U.S. Government Printing Office, 2000): 560.

66. The process has begun to address these issues; see Alfred D. Chandler Jr. and James W. Cortada (eds.), *A Nation Transformed by Information: How Information Has Shaped the United States from Colonial Times to the Present* (New York: Oxford University Press, 2000); James W. Cortada, *Making the Information Society: Experience, Consequences, and Possibilities* (Upper Saddle River, N.J.: Prentice Hall PTR, 2002).

67. Ithiel de Sola Pool provided an early call to the future; see his *Technologies without Boundaries*, but Manuel Castells proved equally effective with the same message in a series of books that began with *The Informational City: Information Technology, Economic Restructuring and the Urban-Regional Process* (Oxford: Blackwell, 1989). For a more current statement, see Lightman and Rojas, *Brave New Unwired World*. Even members of the Information Technology Industry participated in the dialogue, most famously Bill Gates, Chairman of Microsoft Corporation, *The Road Ahead* (New York: Viking, 1995), and *Business @ The Speed of Thought: Using a Digital Nervous System* (New York: Warner Books, 1999). The body of literature on this point is massive, with tens of thousands of articles and thousands of books; there is no comprehensive bibliography of this material.

68. Montgomery Phister, Jr., developed the most detailed inventory of computing hardware in use in the United States from the 1950s through the 1970s. Volumes of data-processing storage devices (such as tape and cards) were so small that he barely accounts for them in his inventory, indirectly suggesting how small the volume of data in digital form was in 1950. For his inventory, see *Data Processing Technology and Economics* (Santa Monica, Calif.: Santa Monica Publishing Company, 1976). He opined that any digital data in 1950 would have been in punch-card form, not on tape (and, of course, not on disk drives since they were not invented until later in the 1950s). But since there is currently no information available on how frequently even that data was refreshed, simply looking at the volume of punch cards sold in the United States would be misleading; Montgomery Phister to James W. Cortada, e-mail, October 13, 2003.

69. Remember, however, that capacity installed did not equal volume of digital data since some of the disk and tape capacity was always empty, and some information could have been housed in punch cards. What can be said, in general, is that if storage capacity grew by 40 percent a year, as it did in many years, then the volume of digital data probably did too. One should keep also in mind that much data were created and simply destroyed within any given year, as happens, for instance, when you delete an e-mail message from your PC. For an overview of the storage story, see Paul E. Ceruzzi, *A History of Modern Computing* (Cambridge, Mass.: MIT Press, 1998); James W. Cortada, *Historical Dictionary of*

Data Processing: Technology (Westport, Conn.: Greenwood Press, 1987): 142–148; 261–269; Phister, *Data Processing Technology and Economics*, 66–71.

70. Michael R. Williams, *A History of Computing Technology* (Englewood Cliffs, N.J.: Prentice-Hall, 1985): 228–229, 411.

71. F. J. Corbató et al., "An Experimental Time-Sharing System," *Proceedings, Spring Joint Computer Conference* 21 (1962): 335–344; but see also their *The Compatible Time-Sharing System: A Programmer's Guide* (Cambridge, Mass.: MIT Press, 1965).

72. For an introduction to these early online systems, see James L. McKenney, Duncan C. Copeland, and Richard O. Mason, *Waves of Change: Business Evolution Through Information Technology* (Boston, Mass.: Harvard Business School Press, 1995).

73. For an introduction to the notion, see Harry Newton, *Newton's Telecom Dictionary* (New York: Flatiron Publishing, 1998): 255.

74. Oslin, *The Story of Telecommunications*, 321–322, 370, 383–384, 438, 439.

75. All the key cable and telephone vendors operating in the United States provided this capability, and by the early 2000s it had become a large growth market for all. To cite one example, Madison, Wisconsin, where I live, as of October 2003, had 25 major providers in a community of less than 400,000 residents. Some of these included AOL, AT&T, Charter, Microsoft, SBC Verizon, and a large local company, TDS. For details, see Judy Newman, "The Trend for Internet Users Is Speed," *Wisconsin State Journal*, October 19, 2003, pp. A1, A5. Her data, drawn from the FCC, also point out how broadband demand had increased from 1.8 million users at the end of 1999 to 11 million at the end of 2001, 14 million by the following June, and an estimate of 17.4 million by the end of 2004, p. A5.

76. Linda Blake and Jim Lande, *Trends in the International Telecommunications Industry* (Washington, D.C.: U.S. Federal Communications Commission, 2001): 2–3; conversation between Jim Lande and author, October 21, 2003.

77. Peter Lyman and Hal R. Varian, "How Much Information?" www.sims.berkeley.edu/research/projects/how-much-info/.

78. John B. Harrington and Lee Raine, *Counting on the Internet* (Washington, D.C.: Pew Internet & American Life Project, December 29, 2002): 2.

79. Lee S. Sproull, "Computers in U.S. Households Since 1977," in Chandler and Cortada, *A Nation Transformed by Information*, 257–280.

80. U.S. Department of Commerce, *A Nation Online: How Americans Are Expanding Their Use of the Internet* (Washington, D.C.: U.S. Department of Commerce, 2001): 5. One of the pioneering studies of the subject was also published by the U.S. Department of Commerce, *Digital Economy 2000* (Washington, D.C.: U.S. Department of Commerce, 2000), which includes extensive data on business uses of this technology.

81. In addition to the materials cited in the previous note, see an ongoing study being conducted to measure use of the Internet by the University of Texas Center for Research in Electronic Commerce. For a description of this work see http://www.internetindicators.com.

82. Both quotes are from Randolph J. May, "They Just Want to Be Free," undated [late 2002], http://www.pff.org/RandysPOVsinLegalTimes/MaysPOV111402.htm.

83. Various data cited in Stiglitz, *The Roaring Nineties*, 92.

84. Paul R. La Monica, "Cleaning Up the Wireless Mess," March 3, 2003, *CNNMoney*, http://money.cnn.com/2003/03/03/technology/wireless/index.htm.

85. Darryl C. Sterling and Charles Gerlach, *Successfully Weathering the Telecommunications Storm: Is the Forecast Bright or Gloomy?* (Cambridge, Mass.: IBM Corporation, 2002): 1.

86. Ibid., 7.

87. "Millions Getting Rid of Landline Phones," Associated Press, August 4, 2003, *ABCNews*, http://abcnews.go.com/wire/Business/ap20030804_905.html.

88. Paul R. La Monica, "Cleaning Up the Wireless Mess," March 3, 2003, *CNNMoney*, http://money.com/2003/03/03/technology/wireless/index.htm.

89. *Fortune* 500 list.

90. Analysis by IBM Institute for Business Value.

91. Om Malik, *Broadbandits: Inside the $750 Billion Telecom Heist* (New York: John Wiley & Sons, 2003): ix.

92. I have analyzed the effects of the digital (especially the Internet) on the work and play of the American public over time in *Making the Information Society: Experience, Consequences, and Possibilities* (Upper Saddle River, N.J.: Prentice Hall PTR, 2000). One recent exception to that general statement are the views of Nicholas G. Carr, who dismissed IT's importance in general. See his article, "IT Doesn't Matter," *Harvard Business Review* 81, no. 5 (May 2003): 41–49.

93. Paul Baran and Andrew J. Lipinski, "The Future of The Telephone Industry, 1970–1985," R-20, Institute for the Future (1970), CBI 32, National Bureau of Standards Papers, Box 551, file 15, p. 42, Archives of the Charles Babbage Institute, University of Minnesota.

94. Ibid., 50.

95. Ibid., 53.

96. Temin and Galambos, *The Fall of the Bell System*, 336–353.

97. Quotes in Fransman, *Telecoms in the Internet Net*, 268.

98. Ibid.

99. Robert J. Chapuis and Amos E. Joel, Jr., *Electronics, Computers and Telephone Switching* (New York: North-Holland Publishing Company, 1990): 340.

100. William Aspray and Peter Freeman, "The Supply of IT Workers in the United States," in B. L. Hawkins, J. A. Rudy, and W. H. Wallace, Jr. (eds.), *Technology Everywhere: A Campus Agenda for Educating and Managing Workers in the Digital Age* (San Francisco: Jossey-Bass, Inc., 2002); with Peter Freeman, *Supply of Information Technology Workers in the United States* (Washington, D.C.: Computing Research Association, 1999).

Chapter 7

1. Claude S. Fischer, *America Calling: A Social History of the Telephone to 1940* (Berkeley, Calif.: University of California Press, 1992): 258.

2. Ibid., 255–272.

3. K. C. August, *TDS: The First Twenty Years* (Chicago, Ill.: Telephone and Data Systems, 1989): 29. Over the years, TDS often replaced entire local telephone networks with more modern technologies or leveraged other parts of its own infrastructure. In short, some acquisitions were less about acquiring networks than they were de facto franchises of additional customers.

4. Walter B. Wriston, *The Twilight of Sovereignty: How the Information Revolution Is Transforming Our World* (New York: Scribner's, 1992): 26. See also Frances Cairncross, *The Death of Distance: How the Communications Revolution Will Change Our Lives* (Boston, Mass.: Harvard Business School Press, 1997): 5.

5. Ronald Abler, "The Telephone and the Evolution of the American Metropolitan System," in Ithiel de Sola Pool (ed.), *The Social Impact of the Telephone* (Cambridge, Mass.: MIT Press, 1977): 319–321.

6. Henry M. Boettinger, "Our Sixth-and-a Half Sense," Ibid., 200.

7. M. M. Irvine, "Early Digital Computers at Bell Telephone Laboratories," *IEEE Annals of the History of Computing* 23, no. 3 (July–September 2001): 22–42.

8. Ibid., 28.

9. S. Millman (ed.), *A History of Engineering and Science in the Bell System: Communications Systems, 1925–1980* (Murray Hills, N.J.: AT&T Bell Laboratories, 1984): 364–365.

10. Irvine, "Early Digital Computers at Bell Telephone Laboratories," 28; J. Meszar, "Basic Features of the AMA Center," *Bell Laboratories Record* (February 1952): 70–74.

11. Millman, *A History of Engineering and Science in the Bell System*, 369–371.

12. "Ultra-Fast Communications Systems Links Computer Centers," IBM press release, August 8, 1961, IBM Corporate Archives, Somers, New York; K. L. Hammer, press release from Southwestern Bell Telephone Company, December 16, 1964, IBM Corporate Archives, Somers, New York.

13. For nontechnical descriptions of these kinds of transmission systems, see R. F. Rey (ed.), *Engineering and Operations in the Bell System* (Murray Hill, N.J.: AT&T Bell Laboratories, 1984): 132–133, 373–385.

14. Susan E. McMaster, *The Telecommunications Industry* (Westport, Conn.: Greenwood Press, 2002): 111–114; Richard S. Millman, *Communications Sciences, 1925–1980* (Murray Hill, N.J.: AT&T Bell Laboratories, 1984): 379–382; John Bray, *The Communications Miracle: The Telecommunication Pioneers from Morse to the Information Superhighway* (New York: Plenum Press, 1995): 208–209, 230–235.

15. For both quote and observation, William O. Baker, Ian M. Ross, John S. Mayo, and Daniel C. Stanzione, "Bell Labs Innovations in Recent Decades," *Bell Labs Technical Journal* 5, no. 1 (January–March 2000): 7.

16. American Telephone and Telegraph Company, *The Bell System's Approach to Business Information Systems* (Murray Hill, N.J.: American Telephone and Telegraph Company, 1965): 8, copy in Archives of the Charles Babbage Institute, University of Minnesota.

17. Yet they were deploying computing to handle routine operational activities; see for example, J. A. Armstrong, "Time and Work Reporting by Data Communications," in *Data Processing: Proceedings 1964* (New Orleans, La.: Data Processing Management Association, 1964): 168–175, which reports on activities at Illinois Bell Telephone Company.

18. American Telephone and Telegraph Company, *The Bell System's Approach to Business Information Systems.*

19. Iowa-Illinois Telephone Company press release, September 1, 1965, IBM Corporate Archives, Somers, New York. This firm first used digital computers as far back as 1962 with an IBM 1401 system.

20. IBM press release, March 18, 1965, IBM Corporate Archives, Somers, New York.

21. IBM press release, November 10, 1967, IBM Corporate Archives, Somers, New York. Use of voice response systems began in the 1960s; for an early case study, see James W. Proctor Jr. "The Voice Response System," *Datamation* 12, no. 8 (August 1966): 43–44, which describes its use by the American Stock Exchange.

22. American Telephone and Telegraph Company press release, June 1969; Ohio Bell Telephone Company press release, June 4, 1970; Bell Labs press release, November 6, 1971; IBM press release, November 22, 1971; ITT Telecommunications press release, May 3, 1972; IBM press release, July 18, 1972; Commonwealth Telephone Company of Virginia press release, February 1, 1972; all at IBM Corporate Archives, Somers, New York. Published articles covered similar issues; see "Big Bell System Order," *Think* (August 1973): 4–7; Geoffrey D. Austrian, "Telecommunications," Ibid. (January–February 1984): 20–31; "Seven Million Service Orders, That's a Lot," *Data Processor* (October–November 1970): 4–6; "Joining the Electronics Production Team," Ibid. (May 1972): 6–8.

23. "The Right Number Every Time," *Data Processor* (March 1976): 8–11; "Network Hastens Service Orders," Ibid. (February–March 1981): 16–17; "Center Serves Walk-In Clients," Ibid. (May–June 1981): 9–10.

24. This process has been well documented in Bell Labs' own four-volume history of its work, *A History of Engineering and Science in the Bell System* (Murray Hill, N.J.: AT&T Bell Laboratories, 1978–1985).

25. "Final Report: Survey of Current Communications Technology (Present–1976), Sub-

task II-A," Auerbach Associates, 1972, CBI 30 Auerbach Papers, Box 130, folder 10, Archives of the Charles Babbage Institute, University of Minnesota.

26. Ibid., 9.

27. Ibid., 22.

28. But because 70 percent of these calls were paid for through a monthly flat-fee billing, only 30 percent had to be tracked; Rey, *Engineering and Operations in the Bell System*, 445.

29. Paul Baran and Andrew J. Lipinski, *The Future of the Telephone Industry, 1970–1985, A Special Industry Report R-20* (Palo Alto, Calif.: Institute for the Future, 1970), National Bureau of Standards (NBS) Collection, CBI 32, Box 551, folder 15, p. 53.

30. August, *TDS*, 102–104. The reader might wonder why this small company had so many subsidiaries with autonomous names. That was (and continues to be) done on purpose. Small companies made less money and therefore were less subject to regulations of the FCC; plus, they qualified for government support since they operated in rural areas and had prices for their services set by state regulators, who were more sympathetic to local companies than to the large Bells.

31. IBM, *Applications and Abstracts* (White Plains, N.Y.: IBM Corporation, 1985): 1-1–1-11.

32. James Martin, *Telecommunications and the Computer*, 2nd ed. (Englewood Cliffs, N.J.: Prentice-Hall, 1976) but see also his earlier book, *Introduction to Teleprocessing* (Englewood Cliffs, N.J.: Prentice-Hall, 1972); Larry A. Arredondo, *Telecommunications Management for Business and Government* (New York: The Telecom Library, 1980); Don L. Cannon and Gerald Luecke, *Understanding Communications Systems* (Dallas, Tex.: Texas Instruments, 1980); J. H. Alleman, *The Pricing of Local Telephone Service* (Washington, D.C.: U.S. Department of Commerce, Office of Telecommunications, April 1977); Frank K. Griesinger, *How to Cut Costs and Improve Service of Your Telephone, Telex, TWX, and Other Telecommunications* (New York: McGraw-Hill, 1974); Richard A. Kuehn, *Cost-Effective Telecommunications* (New York: AMACOM, 1975); James Martin, *Future Developments in Telecommunications* (Englewood Cliffs, N.J.: Prentice-Hall, 1977).

33. Rey, *Engineering and Operations in the Bell System*, 446.

34. Ibid., 445–451.

35. Stephen B. Adams and Orville R. Butler, *Manufacturing the Future: A History of Western Electric* (Cambridge: Cambridge University Press, 1999): 112–183.

36. Ibid., 188–195, 198–201.

37. Rey, *Engineering and Operations in the Bell System*, 133.

38. Ibid., 463.

39. Robert J. Chapuis and Amos E. Joel, Jr., *Electronics, Computers and Telephone Switching* (Amsterdam, N.Y.: North-Holland Publishing Company, 1990): 340–342, 534–536, 538.

40. Ibid., for over 600 pages of process and application descriptions, circa 1977–1984.

41. Norman R. Holzer, "Office Automation in Insurance," *Best's Review* 82, no. 6 (October 1981): 72.

42. Ibid.

43. Ibid.

44. Bernhard E. Keiser and Eugene Strange, *Digital Telephony and Network Integration* (New York: Van Nostrand Reinhold, 1985): 4.

45. Jane Laino, *The Telecom Handbook: Understanding Business Telecommunications Systems and Services* (New York: CPM Books, 2002): 75–80.

46. The number of 800 accounts grew in the 1990s at a very fast rate, going from over 2 million in 1993 to roughly 8.5 million in 1996; Frances Cairncross, *The Death of Distance: How the Communications Revolution Will Change Our Lives* (Boston, Mass.: Harvard Business School Press, 1997): 127–129.

47. For a description of the application, see Laino, *The Telecom Handbook*, 95–113.

48. The literature on these applications is immense. One can begin with Ibid., for a nice introduction, 173–199. See also, J. M. Nilles, F. R. Carlson Jr., P. Gray, and G. J. Hanneman, *The Telecommunications-Transportation Trade-Off: Options for Tomorrow* (New York: Wiley Interscience, 1976); Ronald R. Thomas, *Telecommunications for the Executive* (New York: Petrocelli Books, 1984); Frederick Williams, *The Communications Revolution* (Beverly Hills, Calif.: Sage, 1982); John E. McNamara, *Technical Aspects of Data Communications*, 3rd ed. (Bedford, Mass.: Digital Press, 1988); for an early discussion of e-mail, circa 1970s, Jacques Vallee, *Computer Message Systems* (New York: McGraw-Hill, 1984).

49. John R. McNamara, *The Economics of Innovation in the Telecommunications Industry* (New York: Quorum, 1991): 116–117.

50. Keiser and Strange, *Digital Telephony and Network Integration*, 421.

51. Ibid., 423.

52. Chapuis and Joel, *Electronics, Computers and Telephone Switching*, 563.

53. August, *TDS*, 191.

54. The story of this period was told by Ian M. Ross, the president of Bell Labs from 1979 to 1991, in the anniversary issue of *Bell Labs Technical Journal* 5, no. 1 (January–March 2000): 7.

55. Ibid., 9.

56. For an outstanding study of the economics of this industry for the period prior to 1980, see James W. Sichter, "Profits, Politics, and Capital Formation: The Economics of the Traditional Telephone Industry," a white paper published by the Program on Information Resources Policy, Center for Information Policy Research, Harvard University, 1987. This is a large study—over 200 pages—and is available through the Center's website, http://www.pirp.hardvard.edu. See also Mitchell L. Moss (ed.), *Telecommunications and Productivity* (Reading, Mass.: Addison-Wesley Publishing Co., 1981): 93–130.

57. Robert J. Saunders, Jeremy J. Warford, and Bjorn Wellenius, *Telecommunications and Economic Development*, 2nd ed. (Baltimore, Md.: Johns Hopkins University Press, 1994): 44. See also the entire chapter on this topic, 37–63.

58. Ibid., 10.

59. Ibid.

60. Ibid., 10–11.

61. McMaster, *The Telecommunications Industry*, 143–151.

62. Carol A. Cassel, "Demand and Use of Additional Lines by Residential Customers," in David G. Loomis and Lester D. Taylor (eds.), *The Future of the Telecommunications Industry: Forecasting and Demand Analysis* (Boston, Mass.: Kluwer Academic Publishers, 1999): 43–59; M. Landler, "Multiple Family Phone Lines, A Post-Postwar U.S. Trend," *The New York Times*, December 1995, A-1, D2; N. Allen, K. Delhagen, and S. Eichler, *Young Adults Get the Net* (Cambridge, Mass.: Forrester Research, Inc., 1996); and an important study of the period, P. Rappoport, "The Demand for Second Lines: An Econometric and Statistical Study of Residential Customers," paper presented at NTDS conference, Monterey, Calif., September 1994.

63. The key studies published during the 1990s are Arthur L. Norberg and Judy E. O'Neill, *Transforming Computer Technology: Information Processing for the Pentagon, 1962–1986* (Baltimore, Md.: Johns Hopkins University Press, 1996); and Janet Abbate, *Inventing the Internet* (Cambridge, Mass.: MIT Press, 1999). In the same period, other sources on the subject appeared, such as U.S. Library of Congress, *Spinning the Web: The History and Infrastructure of the Internet* (Washington, D.C.: Congressional Research Service, Library of Congress, 1999); S. R. Hiltz and M. Turoff, *Network Nation* (Cambridge, Mass.: MIT Press, 1995); Nancy Baym, "The Emergence of On-Line Community," in Steve Jones (ed.), *Cybersociety 2.0: Revisiting Computer Mediated Communications and Community* (Thousand Oakes, Calif.: Sage, 1998): 35–68; Robert H. Reid, *Architects of the Web: 1,000 Days That Built the Future of Business* (New York: John Wiley & Sons, 1997); a very early memoir by

Howard Rheingold, *The Virtual Community: Homesteading on the Electronic Frontier* (New York: Harper Perennial, 1994); and another memoir by Michael Wolff, *Burn Rate: How I Survived the Gold Rush Years on the Internet* (New York: Simon & Schuster, 1998); Susan B. Barnes, "Douglas Carl Englebart: Developing the Underlying Concepts for Contemporary Computing," *IEEE Annals of the History of Computing* 19, no. 3 (July–September 1997): 16–26; on Usenet there is Michael Hauben and Ronda Hauben, *Netizens* (Los Alamitos, Calif.: IEEE Computer Society, Press, 1997); Katie Hafner and Matthew Lyon, *Where Wizards Stay Up Late: The Origins of the Internet* (Carmichael, Calif.: Touchstone Books, 1998); Stephen Segaller, *Nerds: A Brief History of the Internet* (New York: TV Books, 1998); David A. Kaplan, *The Silicon Boys and Their Valley of Dreams* (New York: William Morrow and Co., 1999); Tim Berners-Lee and Mark Fisschetti, *Weaving the Web: The Original Design and Ultimate Destiny of the World Wide Web by Its Inventor* (New York: HarperCollins, 1999).

64. The literature includes tens of thousands of articles and hundreds of books. However, for an introduction to how the Internet works from a technical perspective, heavily weighted with historical insight, see Anthony Ralston, Edwin D. Reilly, and David Hemmendinger, *Encyclopedia of Computer Science*, 4th ed. (London: Nature Publishing Group, 2000): 915–927, 1867–1874; and for some of the early technical guides see, Paul Gilster, *Finding It on The Internet: The Essential Guide to Archie, Veronica, Gopher, WAIS, WWW & Other Search Tools* (New York: John Wiley & Sons, 1994); D. E. Comer and R. E. Droms, *Computer Networks and Internets*, 2d ed. (Upper Saddle River, N.J.: Prentice-Hall, 1999); D. C. Lynch and M. T. Rose, *Internet System Handbook* (Reading, Mass.: Addison-Wesley, 1993).

65. This too has been well covered with thousands of articles and hundreds of books; however, some of the more important titles of the period include Mary J. Cronin, *Doing Business on the Internet* (New York: Van Nostrand Reinhold, 1994); John Verity and Robert D. Hof, "How the Internet Will Change the Way You Do Business," *Business Week* (November 14, 1994): 80–88; Geoffrey A. Moore, *Inside the Tornado: Marketing Strategies from Silicon Valley's Cutting Edge* (New York: HarperBusiness, 1995); Bill Gates and Collins Hemingway, *Business @ The Speed of Thought: Using a Digital Nervous System* (New York: Warner Books, 1999); Walid Mougayar, *Opening Digital Markets: Battle Plans and Business Strategies for Internet Commerce* (New York: McGraw-Hill, 1998); Jeff Papows, *Enterprise.com: Market Leadership in the Information Age* (Reading, Mass.: Perseus Books, 1998); a best-seller, Carl Shapiro and Hal R. Varian, *Information Rules: A Strategic Guide to the Network Economy* (Boston, Mass.: Harvard Business School Press); see also David Siegel, *Futurize Your Enterprise: Business Strategy in the Age of the E-Customer* (New York: John Wiley & Sons, 1999); Don Tapscott (ed.), *Creating Value in the Network Economy* (Boston, Mass.: Harvard Business School Press, 1999); Peter Fingar, Harsha Kumar, and Tarun Sharma, *Enterprise E-Commerce: The Software Component Breakthrough for Business-to-Business Commerce* (Tampa, Fla.: Meghon-Kiffer Press, 2000); Mary Modahl, *Now or Never: How Companies Must Change Today to Win the Battle for Internet Consumers* (New York: HarperBusiness, 2000); also a best-seller, Philip Evans and Thomas S. Wurster, *Blown to Bits: How the New Economics of Information Transforms Strategy* (Boston, Mass.: Harvard Business School Press, 2000); Michael D. Smith, Joseph Bailey, and Erik Brynjolfsson, "Understanding Digital Markets: Review and Assessment," in Erik Brynjolfsson and Brian Kahin (eds.), *Understanding the Digital Economy* (Cambridge, Mass.: MIT Press, 2000): 99–136; Alan E. Wiseman, *The Internet Economy: Access, Taxes, and Market Structure* (Washington, D.C.: Brookings Institution Press, 2000); Bob Davis, *Speed Is Life: Street Smart Lessons from the Front Lines of Business* (New York: Currency and Doubleday, 2001); Herbert Meyers and Richard Gerstman (eds.), *Branding @ The Digital Age* (New York: Palgrave, 2001); John R. Patrick, *Net Attitude: What It Is, How to Get It, and Why Your Company Can't Survive Without It* (Cambridge, Mass.: Perseus, 2001). Publications continued to appear at a torrential rate right into the new century.

66. Also the subject of much study by economists and others interested in the economic

effects of the Internet. See, for example, such crucial works as Heather Manzies, *Whose Brave New World? The Information Highway and the New Economy* (Toronto: Between the Lines, 1996); J. Isz Quarterman, " 'Com' Primarily U.S. or International?" *Matrix News* 7 (1997): 8–10, with one of the first studies on the physical location of dot.com domains, reporting 83 percent were in the United States; Kevin Kelly, *Rules for the New Economy* (New York: Viking Press, 1998); Dan Schiller, *Digital Capitalism: Networking the Global Market System* (Cambridge, Mass.: MIT Press, 1999); Martin Dodge and Narushige Shiode, "Where on Earth Is the Internet? An Empirical Investigation of the Geography of the Internet Real Estate," in James Wheeler, Yuko Aoyama, and Barney Warf (eds.), *Cities in the Telecommunications Age: The Fracturing of Geographies* (London: Routledge, 2000): 42–53; Catherine L. Mann, Sue E. Eckert, and Sarah Clealand Knight, *Global Electronic Commerce: A Policy Primer* (Washington, D.C.: Institute for International Economics, July 2000); Brynjofsson and Kahin, *Understanding the Digital Economy*; Matthew A. Zook, "The Geography of the Internet Industry: Venture Capital, Internet Start-Ups, and Regional Development" (unpublished Ph.D. dissertation, University of California, Berkeley, 2001), which became the basis of series of articles published in 2000–2001; Robert E. Litan and Alice M. Rivlin (eds.), *The Economic Payoff from the Internet Revolution* (Washington, D.C.: Brookings Institution Press, 2001); Michael Mandel, *The Coming Internet Depression* (New York: Basic Books, 2000); Chris Benner, *Flexible Work in the Information Economy: Labor Markets in Silicon Valley* (Oxford: Blackwell, 2001); a noneconomic study essential to our understanding of the dot.com story of the late 1990s is by John Cassidy, *Dot.com: The Greatest Story Ever Told* (New York: HarperCollins, 2002).

67. These subjects too have a growing literature. Some of the early contemporary works include Lawrence K. Grossman, *The Electronic Republic: Reshaping Democracy in the Information Age* (New York: Penguin, 1995); William H. Dutton, *Society on the Line: Information Politics in the Digital Age* (New York: Oxford University Press, 1999); on the California Democracy Network (DNET) there is Sharon Docter, William H. Dutton, and Anita Elberse, "An American Democracy Network: Factors Shaping the Future Online Political Campaigns," in Stephen Coleman et al. (eds.), *Parliament in the Age of the Internet* (Oxford: Oxford University Press, 1999): 173–190; Edward L. Hudgins (ed.), *Mail @ the Millennium: Will the Postal Service Go Private?* (Washington, D.C.: Cato Institute, 2000); Janet Abbate, "Government, Business and the Making of the Internet," *Business History Review* 75, no. 1 (Spring 2001): 147–176; Cortada, *Making the Information Society*, 294–368; W. Russell Neuman, Lee McKnight, and Richard Jay Solomon, *Political Gridlock on the Information Highway* (Cambridge, Mass.: MIT Press, 1997); on copyright issues, begin with Jonathan Rosenoer, *CyberLaw: The Law of the Internet* (New York: Springer, 1997); and Jessica Litman, *Digital Copyright* (Amherst, N.Y.: Prometheus Books, 2001).

68. One almost gets the impression that every social commentator of the late 1990s and early 2000s wrote a book on the subject and held strong opinions about the Telecommunications Industry. Crucial works include Scott Ralls, *Integrating Technology with Workers in the New American Workplace* (Washington, D.C.: U.S. Government Printing Office, 1994); Douglas Schuler, *New Community Networks: Wired for Change* (New York: Addison-Wesley, 1996); Steve Jones (ed.), *Virtual Culture* (London: Sage, 1997); Robert Kraut et al., "Internet Paradox: A Social Technology That Reduces Social Involvement and Psychological Well-Being?" *American Psychologist* 53 (1998): 1011–1031; Don Tapscott, *Growing Up Digital: The Rise of the Net Generation* (New York: McGraw-Hill, 1998); Barry Wellman (ed.), *Networks in the Global Village* (Boulder, Colo.: Westview Press, 1999); D. Bolt and R. Crawford, *Digital Divide: Computers and Our Children's Future* (New York: TV Books, 2000); Martin Carnoy, *Sustaining the New Economy: Work, Family and Community in the Information Age* (Cambridge, Mass.: Harvard University Press, 2000); Cheskin Research, *The Digital World of the U.S. Hispanic* (Redwood Shores, Calif.: Cheskin Research Report, 2000); Thomas A. Horan, *Digital Places: Building Our City of Bits* (Washington, D.C.: The Urban Land

Institute, 2000); Joel Kotkin, *The New Geography: How the Digital Revolution Is Reshaping the American Landscape* (New York: Random House, 2000); Robert Putnam, *Bowling Alone: The Collapse and Revival of American Community* (New York: Simon and Schuster, 2000); Manuel Castells, *The Internet Galaxy: Reflections on the Internet, Business and Society* (Oxford: Oxford University Press, 2001) which includes a rich bibliography; Benjamin M. Compaine (ed.), *The Digital Divide: Facing a Crisis or Creating a Myth?* (Cambridge, Mass.: MIT Press, 2001); Michael L. Dertouzos, *The Unfinished Revolution: Human-Centered Computers and What They Can Do For Us* (New York: HarperBusiness, 2001); Casey Hait and Stephen Weiss, *Digital Hustlers: Living Large and Falling Hard in Silicon* (New York: Regan Books, 2001). On hacking see Pekka Himanen, *The Hacker Ethic and the Spirit of the Information Age* (New York: Random House, 2001); now a minor classic on the subject is Steve Levy, *Hackers: Heroes of the Computer Revolution*, rev. ed. (New York: Penguin-USA, 2001, first ed. 1984); Dan Verton, *The Hacker Diaries: Confessions of Teenage Hackers* (New York: McGraw-Hill, 2002).

69. All the surveys of use of the Internet begin tracking usage in 1997–1998. Some early data can be gained, however, from Woo Young Chung, "Why Do People Use the Internet?" (unpublished Ph.D. dissertation, Boston University School of Management, 1998); but see also D. Hoffman, W. Kalesbeck, and T. Novak, "Internet and Web Use in the U.S.," *Communications of the* ACM 30, no. 12 (1996): 36–46. For who used it, see Ed Krol and Mike Loukides (eds.), *The Whole Internet: Users' Guide & Catalog* (Sebastopol, Calif.: O'Reilly & Associates, 1994).

70. Transmission Control Protocol/Internet Protocol, although the "P" early-on was often called "program." The definitive early work is by D. Comer, *Internetworking with TCP/IP*, 3 vols. (Upper Saddle River, N.J.: Prentice Hall, 1994–1997). For an introduction to the subject, see Adrian Stokes, "TCP/IP," in Ralston, Reilly, and Hemmendinger, *Encyclopedia of Computer Science*, 1745–1747. For an excellent analysis of the effects of TCP/IP, see Martin Fransman, "Evolution of the Telecommunications Industry into the Internet Age," (2000/01), 38–39.

71. David H. Brandin and Daniel C. Lynch, "Applications of the Internet," in Ralston, Reilly, and Hemmendinger, *Encyclopedia of Computer Science*, 917–919.

72. Ibid., 919–921.

73. Hauben and Hauben, *Netizens*; Segaller, *Nerds*; Abbate, *Inventing the Internet*; Castells, *The Internet Galaxy*; Rita Tehen, *Spinning the Web: The Internet's History and Structure* (Washington, D.C.: Congressional Research Service, 2001, available at http://www.pennyhill.com); Nirmal Pal and Judith M. Ray, *Pushing the Digital Frontier: Insights into the Changing Landscape of E-Business* (New York: AMACOM, 2001); but also consult the dozens of studies conducted by the Pew Internet & American Life Project.

74. Soumitra Dutta and Arie Segev, "Business Transformation on the Internet," Working Paper 98-WP-1035, January 1999.

75. R. Yehling, "We've All Got Mail," *The Year in Computing* (Tampa, Fla: Faircount International, 1999): 81–84.

76. Prior notes are rife with citations, circa 1998–99; but see also, Mark Stefik, *The Internet Edge: Social, Technological, and Legal Challenges for a Networked World* (Cambridge, Mass.: MIT Press, 1999).

77. Michael D. Smith, Joseph Bailey, and Erik Brynjolfsson, "Understanding Digital Markets: Review and Assessment," unpublished paper, September 29, 1999, p. 1, http://ecommerce.mit.edu/papers/ude. This paper was published, along with other material, in Erik Brynjolfsson and Brian Kahin (eds.), *Understanding the Digital Economy* (Cambridge, Mass.: MIT Press, 2000): 99–136.

78. Manley R. Irwin and James McConnaughey, "Telecommunications," in Walter Adams and James Brock (eds.), *The Structure of American Industry*, 10th ed. (Upper Saddle River, N.J.: Prentice-Hall, 2001): 305.

79. Susannah Fox and Lee Rainie, "Time Online: Why Some People Use the Internet More Than Before and Why Some Use It Less," released July 16, 2001 (Washington, D.C.: Pew Internet & American Life Project, 2001), available at http://www.pewinternet.org.

80. These activities were documented in a series of studies conducted by the Pew Internet & American Life Project, "The Internet and Education," September 1, 2001; "Wired Seniors: A Fervent Few, Inspired by Family Ties," September 9, 2001; "Cyber-Faith: How Americans Pursue Religion Online," December 23, 2001; "The Rise of the e-Citizen: How People Use Government Agencies' Web Sites," April 3, 2002; "Holidays Online-2002," January 7, 2003; "Congress Online," 2002; "Email at Work," December 8, 2002; "Untuned Keyboards," March 21, 2003; all available at http://www.pewinternet.org. See also Harrell Associates, "The Internet Travel Industry: What Consumers Should Expect and Need to Know, and Options for a Better Marketplace," June 6, 2002; Patricia Buckley and Sabrina Montes, "Main Street in the Digital Age: How Small and Medium-Sized Businesses Are Using the Tools of the New Economy," February, 2002, released by the U.S. Department of Commerce, available from Patricia.Buckley@esa.doc.gov or Sabrina .Montes@esa.doc.gov.

81. National Telecommunications and Information Administration and the Economics and Statistics Administration, *A Nation Online: How Americans Are Expanding Their Use of the Internet* (Washington, D.C.: U.S. Department of Commerce, 2001): 3–4.

82. Lee Rainie and Berte Kalsnes, "The Commons of the Tragedy," Pew Internet for American Life Project, October 10, 2001, http://www.pewinternet.org/pdfs/PIP_Tragedy _Report.pdf.

83. The Internet proved to be the most resilient telecommunication network in the nation on the occasion of this national crisis, James W. Cortada and Edward Wakin, *Betting on America: Why the U.S. Can Be Stronger After September 11* (Upper Saddle River, N.J.: Financial Times/Prentice Hall, 2002): 227.

84. Lee Rainie, "How Americans Used the Internet After the Terror Attack," September 15, 2001, Pew Internet & American Life Project, http://www.pewinternet.org. See also a subsequent study by the same author, "The Commons of the Tragedy: How the Internet Was Used by Millions After the Terror Attacks to Grieve, Console, Share News, and Debate the Country's Response," October 10, 2001, http://www/pewinternet.org.

85. Nathan Kommers and Lee Rainie, "Use of the Internet at Major Life Moments," undated, early 2002, Pew Internet & American Life Project, press release from the same agency, August 2002; "The Growth in Online House Hunting: 40 Million Wired Americans Have Used the Internet to Search for Houses or Apartments;" Katherine Allen and Lee Rainie, "Parents Online," November 17, 2002, Ibid.; Lee Rainie, Susannah Fox, and Deborah Fallows, "The Internet and the Iraq War"; undated (mid-2003), Ibid., available at http://www.pewinterent.org.

86. I have discussed the role of the Internet in American life in more detail in *Making the Information Society: Experience, Consequences, and Possibilities* (Upper Saddle River, N.J.: Financial Times/Prentice-Hall, 2002): 97–135.

87. Princeton Survey Research Associates, *A Matter of Trust: What Users Want From Web Sites: Results of a National Survey of Internet Users for Consumer WebWatch* (Princeton, N.J.: Princeton Survey Research Associates, January 2002).

88. "The Ever-Shifting Internet Populations," April 16, 20003, Pew Internet & American Life, http://www.pewinternet.org.

89. Because terms are used so loosely when it comes to telecommunications, one should understand the differences among the types of services. In the United States in the early 2000s, high-speed data services are defined as those that provide more than 200kbps connectivity in one direction, while advanced data services are those that provide 200kbps in both directions. Broadband is defined as operating at 45Mbps, but more commonly is called T-1 speed of at least 1.5Mbps. U.S. cable companies do not offer classic DSL; rather, they

provide cable modem services which use an entirely different communication medium. Although both technologies are capable of providing true bidirectional data, neither are designed to do this in the United States. For the most part, providers in the United States as of 2005 were still operating as high-speed or advanced data-service providers. These are subtle technical points lost on the American public, but not to regulators or the industries involved. Hence why this endnote was needed.

90. Press release from Pew Internet & American Life Project, May 2003; press release from Federal Communications Commission, June 10, 2003; Judy Newman, "The Trend for Internet Users Is Speed," *Wisconsin State Journal*, October 19, 2003, pp. A1, A5.

91. "How Much Information? 2003," http://www.sims.berekely,edu/research/projects/how-much-info-2003.

92. Ibid.

93. Ibid.

94. Ibid.; see also Michael K. Bergman, "The Deep Web: Surfacing Hidden Value" (BrightPlanet White Paper), http://www.brightplanet.com/technology/deepweb.asp; IDC Report on e-mail, http://emailuniverse.com/list-news/2002/10/01.html; Pew Internet & American Life Project, "Email at work," http://www.pewinternet.org/reports.asp?Report=79&Section=ReportLevel1&Field=Level1ID&ID=346.

95. Daniel Berninger, "The End of Telecom," undated [September 2004], http://www.danielberninger.com/endoftelecom.html (last accessed 9/12/2004).

96. The conclusion of all surveys of the Internet in the United States between 1997 and 2005.

97. There is a concept at work here that cannot be ignored by telco experts called the "Negroponte Switch," which loosely holds that all voice communications go to wireless and all data communications go to wires—this is what makes the most sense. However, as wireless gains capacity—and it surely will—this rule of thumb may become irrelevant, but at least in the 1990s it was right on the mark.

98. George Gilder, *Telecosm: How Infinite Bandwidth Will Revolutionize Our World* (New York: Free Press, 2000): 101. He argued in this book that fiber optics would provide the broadband. As of this writing (2005) this view was being seriously challenged by the move to wireless Internet and the massive use of cell phones across the entire nation.

99. For background on this commentator, see Om Malik, *Broadbandits, Inside the $750 Billion Telecom Heist* (New York: John Wiley & Sons, 2003): 275–279, 281–284, 283, 288–289.

100. James B. Murray, Jr., *Wireless Nation: The Frenzied Launch of the Cellular Revolution in America* (Cambridge, Mass.: Perseus Publishing, 2001); Alex Lightman with William Rojas, *Brave New Unwired World: The Digital Big Bang and the Infinite Internet* (New York: John Wiley & Sons, 2002): 27–40; Om Malik, *Broadbandits*.

101. Reed E. Hundt, *You Say You Want A Revolution: A Story of Information Age Politics* (New Haven, Conn.: Yale University Press, 2000): 92.

102. For one company's (Nortel's) experience with this transformation, see Larry Macdonald, *Nortel Networks: How Innovation and Vision Created a Network Giant* (Toronto: John Wiley & Sons, 2000): 167–174.

103. Michelle Kressler, "Tide of Wireless Customers Recedes," *USA Today*, April 25, 2001, p. B1.

104. "Bell Labs Innovations in Recent Decades," 15.

105. For details, see George I. Zysman et al., "Mobility/Wireless," 112–116. There is a fundamental difference between the architecture of a wireless network and a wired one. In wireless, the device is smart, because in order to maintain a connection it has to report information back to the network to maintain the link. In wired telephony, devices are still dumb (with the exception of VoIP handsets). This means a wireless network is designed much more like a data network, which has a computer (server) with attached PCs, than a wired network. As computers improve in capability, one could expect the role of computers

to affect whether customers move quickly or slowly toward use of wireless devices. Note that this technological concept flies in the face of the Negroponte Switch.

106. John P. Burnham, *The Essential Guide to the Business of U.S. Mobile Wireless Communications* (Upper Saddle River, N.J.: Prentice-Hall PTR, 2000): 168.

107. Data collected by CTIA, 2000, and reported in Ibid.

108. Ibid., 168–177.

109. Judy Newman, "Landlines Not Needed, Some Phone Users Decide," *Wisconsin State Journal*, August 10, 2003, pp. A1, A5; "Cry Baby Bells?" *Time*, December 9, 2002, pp. A1–A6; Paul R. La Monica, "Cleaning Up the Wireless Mess," *CNNMoney*, March 3, 2003, http://money.cnn.com/2003/03/03/technology/wireless/index.htm; "Millions Getting Rid of Landline Phones," Associated Press release, August 4, 2003, http://abcnews.go.com/wire/Business/ap20030804_905.html. The latter reported there was one cell phone for every two people living in the United States and noted that it took land lines nearly a century to reach the same level of penetration; both price and convenience have driven the shift.

110. Walter Mossberg, "European Phones Fall Down Handling e-Mail," *Wisconsin State Journal*, March 5, 2004, p. C9.

111. Bard Smith, "Making Money in M-commerce: Partnering, Platforms and Portals," *Wireless Week*, February 28, 2000, http://www,wirelessweek.com.

112. Cathy Booth Thomas, "Wi-Fi Gets Rolling," *Time*, November 3, 2003, p. A16; but see the entire issue for other articles on the same subject.

113. For examples of early adopters and for an overall account of this new technology, Jouni Paavilainen, *Mobile Business Strategies: Understanding the Technologies and Opportunities* (London: Wireless Press/Addison-Wesley in partnership with IT Press, 2001): 73–107.

114. Consumer Expenditure Survey, Bureau of Labor Statistics, updated each year.

115. Manley R. Irwin and James McConnaughey, "Telecommunications," in Adams and Brock, *The Structure of American Industry*, 304–305.

116. Heather E. Hudson, *Global Connections: International Telecommunications Infrastructure and Policy* (New York: Van Nostrand Reinhold, 1997): 19.

117. *2001 Annual Report, Verizon*, p. 8.

118. Cortada, *Making the Information Society*.

119. An important exception was the handling of complaints for such things as repairs, slow initiation of service, and billing disputes. These are carefully documented by all state regulatory agencies and are often posted to the agency's website and annual results reported by the press.

120. The literature is growing rapidly on the 1996 law; however, for an introduction to the issues, see Dale E. Lehman and Dennis L. Weisman, *The Telecommunications Act of 1996: The "Costs" of Managed Competition* (Boston: Kluwer Academic Press, 2000); McMaster, *The Telecommunications Industry*, 153–175; Robert Crandall and Jerry A. Hausman, "Competition in U.S. Telecommunications Services: Effects of the 1996 Legislation," in Sam Peltzman and Clifford Winston (eds.), *Deregulation of Network Industries: What's Next?* (Washington, D.C.: AEI-Brookings Joint Center for Regulatory Studies, 2000): 73–112.

121. Paul Davidson, "Is Telecom Act 'a complete failure'?" *USA Today*, February 8, 2001, p. 3B.

122. Demonstrated by Charles Johnscher as far back as 1981, "The Economic Role of Telecommunications," in Mitchell L. Moss (ed.), *Telecommunications and Productivity* (Reading, Mass.: Addison-Wesley, 1981): 68–92.

123. For analysis of the situation faced by the Telecommunications Industry on a global basis, including in the United States, see Darryl C. Sterling and Charles Gerlach, *Successfully Weathering the Telecommunications Storm: Is The Forecast Bright or Gloomy?* (Portsmouth, U.K.: IBM Corporation, 2002); and their *The Contact Centre of the Future: Spanning the Chasms* (Bunnian Place, U.K.: IBM Corporation, 2002); both are available at http://www.ibm.com/bcs.

Chapter 8

1. No survey of any media industry of the late twentieth century ignores the topic, as consolidations have been massive and the number of megafirms few by the end of the century. For a useful well-done introduction, albeit hostile to the trend, see Erik Barnouw, *Conglomerates and the Media* (New York: New Press, 1998); but also, Edward S. Herman and Robert W. McChesney, *The Global Media: The New Missionaries of Corporate Capitalism* (London: Cassell, 1997); and Ben H. Bagdikian, *The Media Monopoly* (Boston, Mass.: Beacon Press, 1992).

2. Albert N. Greco, *The Book Publishing Industry* (Boston, Mass.: Allyn and Bacon, 1997): 57–59.

3. Almost every contemporary survey of the various industries devotes considerable attention to mergers and acquisitions. See, for example, Albert N. Greco (ed.), *The Media and Entertainment Industries: Readings in Mass Communications* (Boston, Mass. Allyn and Bacon, 2000); Eric Barnouw et al., *Conglomerates and the Media*, 7–30. See also citations on each industry scattered across the next four chapters (published by Allyn and Bacon in the late 1990s).

4. H. Thomas Johnson and Robert S. Kaplan, *Relevance Lost: The Rise and Fall of Management Accounting* (Boston, Mass.: Harvard Business School Press, 1987): 125–151.

5. The topic is attracting much attention and is useful for understanding the technological imperatives at work. Useful studies include John V. Pavlik, *New Media Technology: Cultural and Commercial Perspectives*, 2d ed (Boston, Mass.: Allyn and Bacon, 1998, 1st ed. published 1996); and for a short introduction there is Walter Oleksy, *The Information Revolution: Entertainment* (New York: Facts on File, 1996); for historical perspective consult Randall Packer and Ken Jordan (eds.), *Multimedia from Wagner to Virtual Reality* (New York: W. W. Norton, 2001).

6. One of the most typical and widely influential examples of this view is by Michael J. Wolf, *The Entertainment Economy: How Mega-Media Forces Are Transforming Our Lives* (New York: Times Book, 1998). The book was published by a company established by the *New York Times* newspaper but owned by Random House, a book publisher, at the time of publication.

7. A key finding of an earlier study I did on the role of information in American society, *Making the Information Society: Experience, Consequences, and Possibilities* (Upper Saddle River, N.J.: Financial Times/Prentice-Hall, 2002): 11–19, 107–108, 139–145; but see also the various studies by the Pew Foundation, located at http://www.pewinternet.org.

8. Harold L. Vogel, *Entertainment Industry Economics: A Guide for Financial Analysis* (Cambridge: Cambridge University Press, 2001).

9. See, for example, U.S. Bureau of the Census, "Communications and Information Technology," available at http://www.census.gov/prod/2/ge, published periodically, but at a minimum annually.

10. For an anthology of this literature see, James W. Cortada (ed.), *Rise of the Knowledge Worker* (Boston, Mass: Butterworth-Heinemann, 1998).

11. Benjamin M. Compaine, "Shifting Boundaries in the Information Marketplace," in Benjamin M. Compaine (ed.), *Understanding New Media: Trends and Issues in Electronic Distribution of Information* (Cambridge, Mass.: Ballinger Publishing Co., 1984): 98.

12. U.S. Department of Commerce, *Historical Abstracts of the United States*, 2 vols. (Washington, D.C.: U.S. Government Printing Office, 1975).

13. John V. Pavlik and Everette E. Dennis, *New Media Technology: Cultural and Commercial Perspectives* (Boston, Mass.: Allyn and Bacon, 1998): 1.

14. Like most magazines, this one is also on the Internet, http://www.parade.com.

15. Scattered across the entire issue of *Parade*, November 23, 2003. Each major American airline makes available to its passengers a catalog of items one can order in flight; these

always offer a similar list of the latest digital devices, and serve as useful sources on these. Included in each of these catalogs in 2003 were other products not included in *Parade*: portable digital foreign language translators, radios, and alarm clocks, to mention a few.

16. Peter Lewis, "Apple iTunes Music Store, December 10, 2003, *Fortune*, http://www.fortune.com/fortune/print/0,159335,558792,00.html (accessed December 25, 2003). This reporter's assessment of the accomplishment: "Apple is almost single-handedly dragging the music industry, kicking and screaming, toward a better future."

17. Jeff Goodell, "Steve Jobs: The *Rolling Stone* Interview," December 3, 2003, http://www.rollingstone.com/features/featuregen.asp?pid=2529.

18. Chris Taylor, "The 99 Cent Solution: Steve Jobs' New Music Store Showed Foot-Dragging Record Labels and Freeloading Music Pirates That There Is a Third Way," *Time*, undated [late 2003], http://www.time.com/time/2003/inventions/invmusic.html (accessed December 25, 2003).

19. For example, see *Webster's Seventh New Collegiate Dictionary* (Springfield, Mass.: G. & C. Merriam Company, 1963): 526. This was one of the most widely available dictionaries in the 1950s through the 1970s in the United States, with editions published approximately every two years.

20. Charles J. Sippl and Charles P. Sippl, *Computer Dictionary and Handbook* (Indianapolis, Ind.: Howard W. Sams & Co., 1972): 268.

21. John Tebbed and Mary Ellen Zuckerman, *The Magazine in America, 1941–1990* (New York: Oxford University Press, 1991): 64.

22. Douglas Gomery, "The Book Publishing Industry," in Benjamin M. Comaine and Douglas Gomery, *Who Owns the Media?* (Mahwah, N.J.: Lawrence Erlbaum Associates, 2000): 62–63.

23. Charles P. Daly, Patrick Henry, and Ellen Ryder, *The Magazine Publishing Industry* (Boston, Mass.: Allyn and Bacon, 1997): 10–12.

24. Hypertext is the concept of interlocking pieces of information and names used to describe these, or collections of them, with digital tools to allow a reader to access information and text in various orders, rather than read material sequentially as occurs in printed text. For a description and history of hypertext, see Michael Bieber, "Hypertext," in Anthony Ralston, Edwin D. Reilly, and David Hemmendinger (eds.), *Encyclopedia of Computer Science*, 4th ed. (London: Nature Publishing Group, 2000): 799–805; E. J. Conklin, "Hypertext: A Survey and Introduction," *IEEE Computer* 20, no. 9 (1987): 17–41; I. Snyder, *Hypertext: The Electronic Labyrinth* (New York: New York University Press, 1997).

25. By the early 2000s, it seemed that every organization and many departments and divisions within companies also were publishing electronic newsletters. These were directed to constituents, customers, and colleagues. There have been no formal demographic studies of these to establish the number in existence, their lengths of life, and so forth. I have personally seen many hundreds of these, beginning in 1995.

26. Ibid., 235–268.

27. Data for last two paragraphs drawn from Benjamin M. Compaine, "The Magazine Industry," in Compaine and Gomery, *Who Owns the Media?* 147–191.

28. Ibid., 1–60.

29. For a useful introduction to the industry, see Alan G. Albarran and Gregory G. Pitts, *The Radio Broadcasting Industry* (Boston, Mass.: Allyn and Bacon, 2001).

30. The major source for those kind of data are the various reports produced by the FCC, all available at its website: fcc.gov.

31. Compaine and Gomery, *Who Owns the Media?* 285–357.

32. Normally one would think that commercial enterprises would lead in technological innovations; in the case of this industry, that was not so. One recent example is the Public Broadcasting System (PBS), a nonprofit TV network in the United States, which led the industry in investing in cameras and in transmission equipment to broadcast in digital

format. It actually became the first to do so in 2002, while major commercial broadcast networks put off investing in that transformation until required to do so by law (to take place by 2006).

33. Two studies provide industry/technology coverage: James Walker and Douglas Ferguson, *The Broadcast Television Industry* (Boston, Mass.: Allyn and Bacon, 1998); and Patrick R. Parsons and Robert M. Frieden, *The Cable and Satellite Television Industries* (Boston, Mass.: Allyn and Bacon, 1998).

34. Compaine and Gomery, *Who Owns the Media?*, 206–213.

35. Vogel, *Entertainment Industry Economics*, 113–114, 164, 351–354; Wolf, *The Entertainment Economy*, 117–154, 221–252.

36. Key modern works exploring Schumpeter's ideas were published just as the Music Recording Industry faced its crisis of downloading of their products; thus its experiences are not yet reflected in the main publications on creative destruction. However, his basic principles are well described and thus the literature is informative about this industry's problems. See in particular Lee W. McKnight, Paul M. Vaaler, and Raul L. Katz (eds.), *Creative Destruction: Business Survival Strategies in the Global Internet Economy* (Cambridge, Mass.: MIT Press, 2001); Richard Foster and Sarah Kaplan, *Creative Destruction: Why Companies That Are Built to Last Underperform the Market—and How to Successfully Transform Them* (New York: Currency, 2001).

37. John B. Horrigan, "Consumption of Information Goods and Services in the United States," Pew Internet & American Life Project, November 23, 2003, http://www.pewinternet.org; Mary Madden, "America's Online Pursuits: The Changing Picture of Who's Online and What They Do," Ibid., December 22, 2003.

38. Table No. 914, "Media Usage and Consumer Spending: 1990 to 2001," U.S. Census Bureau, *Statistical Abstract of the United States: 1998*, October 29, 1998, p. 572, available at http://www.census.gov/prod/3/98. In the early 2000s, when the U.S. economy was experiencing a recession, revenues for recorded music declined, down 4.7 percent in 2003 over 2002, while 2002 was down 10.8 percent over the prior year. CD sales, which accounted for 96 percent of all album sales, were down 3 percent in 2003, and in 2002 were also down by 8.8 percent over the prior year. However, music video (on DVD) sales were up in 2003 over 2002 by 75.2 percent, while DVD sales in 2003 in general were up over 100 percent from the prior year; Nielsen SoundScan data, *Milwaukee Journal Sentinel*, December 20, 2003, p. 6B. One can conjecture that the decline in traditional sources of revenue put enormous pressure on the industry to block downloading off the Internet which, as it argued, deprived it of revenues.

39. "2000 Yearend Statistics," available at Recording Industry of Association of America's website, http://www.riaa.com (accessed December 14, 2003).

40. Geoffrey P. Hull, *The Recording Industry* (Boston, Mass.: Allyn and Bacon, 1998): 4.

41. "2000 Yearend Statistics," available at Recording Industry of Association of America's website, http://www.riaa.com (accessed December 14, 2003).

42. Geoffrey P. Hull, *The Recording Industry* (Boston, Mass.: Allyn and Bacon, 1998): 27–35.

43. Martin Campbell-Kelly, *From Airline Reservations to Sonic the Hedgehog: A History of the Software Industry* (Cambridge, Mass.: MIT Press, 2003): 225–226.

44. Vogel, *Entertainment Industry Economics*, 259–260.

45. Ibid., 260.

46. IBM's definition of this technology: "Grid computing links disparate hardware and software resources, often in geographically distant offices, into a cohesive whole," making it possible to process an application using unused capacity available on multiple computers linked together. For details, see Michael Kanellos, "Big Blue Tackles New Grid Computing Services," December 11, 2002, *CNET News.com*, http://news.com.com/2100-1010_3-5120601.html?tag=nefd_top (accessed December 16, 2003). For a study of the technical

implications of this new form of computing on organizations, see "Next Generation On Demand," at www3.ca.com/Files/WhitePapers/next_generation-ondemand-wp.pdf, published December 2003.

47. Rusel DeMaria and Johnny L. Wilson, *High Score: The Illustrated History of Electronic Games* (New York: McGraw/Osborne, 2002); J. C. Herz, *Joystick Nation: How Videogames Ate Our Quarters, Won Our Hearts, and Rewired Our Minds* (Boston: Little, Brown and Company, 1997); Steven L. Kent, *The Ultimate History of Video Games: From Pong to Pokemon, The Story Behind the Craze That Touched Our Lives and Changed the World* (Roseville, Calif.: Prima Publishing, 2001); Brad King and John Borland, *Dungeons and Dreamers: The Rise of Computer Game Culture from Geek to Chic* (New York: McGraw-Hill/Osborne, 2003).

48. James W. Cortada, *The Digital Hand: How Computers Changed the Work of American Manufacturing, Transportation, and Retail Industries* (New York: Oxford University Press, 2004): 89–127.

49. For example, see Alecia Swasy, *Changing Focus: Kodak and the Battle to Save a Great American Company* (New York: Times Business, 1997): 9–60, 225–251.

50. The Internet has been seen as posing a threat to television broadcasters since at least the mid-1990s; Bruce M. Owen, *The Internet Challenge to Television* (Cambridge, Mass.: Harvard University Press, 1999): 327–333.

51. An IT industry expert and freelance writer described the use of databases in newsgathering operations, circa 1980s–early 1990s, Timothy Miller, "Data Bases: The Retrieval Revolution," in Pavlik and Dennis, *Demystifying Media Technology*, 136–141.

52. On the experience of the Newspaper Industry, see "Survey Shows Computer Use More Than Doubles in Year," *Publishers Weekly* 188 (November 8, 1965): 125; Edward K. Yasaki, "The Computer & Newsprint," *Datamation* 9, no. 3 (March 1963): 27–31; Bruce Gilchrist and A. Shenkin, "Disappearing Jobs: The Impact of Computers on Employment," *Futurist* 15 (February 1981): 44–49; Theresa F. Rogers and Nathalie S. Friedman, *Printers Face Automation: The Impact of Technology on Work and Retirement Among Skilled Craftsmen* (Lexington, Mass.: Lexington Books, 1980). On the railroads, I explain the situation in *Digital Hand*, 254–257.

53. Cortada, *Digital Hand*, 143–160.

54. For an account of the ice cutters' experience, see James M. Utterback, *Mastering the Dynamics of Innovation: How Companies Can Seize Opportunities in the Face of Technological Change* (Boston, Mass.: Harvard Business School Press, 1994): 146–151, 155–157.

55. The most useful sources of ongoing studies are those conducted by the Pew Foundation, the U.S. Department of Commerce, and, increasingly, the University of Texas. Periodically, other organizations will publish surveys, such as the FCC.

56. For a detailed collection of data on how much information had been digitized at the end of the century, see Peter Lyman and Hal R. Varian," "How Much Information," unpublished paper dated 2003, http://www.sims.berkeley.edu/how-much-info (accessed November 21, 2003).

57. Recording Industry Association of America, http://www.riaa.com.

58. The U.S. Census Bureau began collecting data on shipments of DVDs for the first time in 1999, see U.S. Census Bureau, *Statistical Abstracts of the United States, 2000* (Washington, D.C.: U.S. Government Printing Office, 2000): 573.

59. IBM Corporation, *Industry Applications and Abstracts* (White Plains, N.Y.: IBM Corporation, 1988): 16–1.

60. On what makes for an industry, see Michael E. Porter, *Competitive Strategy: Techniques for Analyzing Industries and Competitors* (New York: Free Press, 1980): 3–33, 191–274; on how to study an industry, Cortada, *Digital Hand*, 389–394.

61. "The Internet Economy Indicators," http://www.internetindicators.com (accessed November 10, 1999). For a discussion of Internet-only companies listed on stock exchanges

in 2000 (280 of them), see David Kleinbard, "Dot.Com Shakedown," November 9, 2000, http://cnnfm.cnn.com/2000/11/09technology/overview (accessed November 10, 2000).

62. Cortada, *Digital Hand*, 126–127.

63. Compaine, "The Online Information Industry," in Compaine and Gomery, *Who Owns the Media?* 446, but see the entire article too, 437–480.

64. Ibid., 439.

65. Ibid., 437–480.

66. John Motavalli, *Bamboozled at the Revolution: How Big Media Lost Billions in the Battle for the Internet* (New York: Viking, 2002): xiii. The author was both a media consultant and columnist writing about the Internet and media topics at the turn of the century.

67. Pavlik and Dennis, *New Media Technology*, 11–20, 29–50, 133–148.

68. For a useful discussion of the problems of innovation, see Everett M. Rogers, *Diffusion of Innovations*, 3rd ed. (New York: Free Press, 1983): 371–410.

69. The best source remains the Pew Foundation's Internet & American Life Project, which publishes surveys on a regular basis on how Americans use the Internet, and which are available at its website, http://www.pewinternet.org.

70. Joel Waldfogel, "Consumer Substitution Among Media," September 2002, white paper sponsored by the U.S. Federal Communications Commission, p. 17.

71. Kleinbard, "Dot Com Shakeout"; Anthony B. Perkins and Michael C. Perkins, *The Internet Bubble* (New York: HarperBusiness, 2001); but see also economist Joseph E. Stiglitz, who has provided the most balanced account so far of the conditions that created the "bubble," *The Roaring Nineties: A New History of the World's Most Prosperous Decade* (New York: W. W. Norton, 2003).

Chapter 9

1. Chandler B. Grannis (ed.), *What Happens in Book Publishing* (New York: Columbia University Press, 1957); John Tebbel, *Between Covers: The Rise and Transformation of American Book Publishing* (New York: Oxford University Press, 1987): 352–438; Lewis A. Coser, Charles Kadushin, and Walter W. Powell, *Books: The Culture and Commerce of Publishing* (New York: Basic Books, 1982): 8, 25–28; Clive Bingley, *Book Publishing Practice* (Hamden, Conn.: Archon Books, 1966): 1–18.

2. Albert N. Greco, *The Book Publishing Industry* (Boston, Mass.: Allyn and Bacon, 1997): 2–19.

3. James W. Cortada, *The Digital Hand: How Computers Changed the Work of American Manufacturing, Transportation, and Retail Industries* (New York: Oxford University Press, 2004): 283–316.

4. Martha E. Williams, "Database Publishing Statistics," *Publishing Research Quarterly* 11, no. 3 (Fall 1995): 3–9. For a detailed analysis of McGraw-Hill's role, with information concerning many of the key players, see "McGraw-Hill, Inc.: New Information Products and the Changing Role of IS&T (A)," Case study 9-187-170 (Boston, Mass.: Harvard Business School Press, January 1988).

5. Herbert S. Bailey, Jr., *The Art and Science of Book Publishing* (New York: Harper & Row, 1970): 195.

6. "Mechanization of a Subscription Fulfillment Operations," Richard Hunt Brown report, [1957?], CBI 55 "Market Reports," Folder 2, Box 70, Archives of the Charles Babbage Institute, University of Minnesota, Minneapolis.

7. The most detailed inventory of the period listed an IBM 650 at Curtis Publishing Company, Univac 60s or 120s at Concordia Publishing House and Doubleday and Company, and an unspecified computer (probably an IBM 7070) at F. H. McGraw & Company, Automation Consultants, Inc., *Office Automation Applications* (New York: Automation Con-

sultants, Inc., undated, circa 1957–1960, CBI 55, Market Reports, Box 70, Folder 2, Archives of the Charles Babbage Institute, University of Minnesota, Minneapolis.

8. All quotes, Daniel Melcher, "The Role of Computers," in Kathryn Luther Henderson (ed.), *Trends in American Publishing* (Champaign, Ill.: University of Illinois Graduate School of Library Science, 1968): 50.

9. Ibid., 51–53.

10. "American Bible Society Plans Computer to Modernize Operations," press release, June 29, 1965, American Bible Society News Release; "To Distribute Millions of Bibles With the Aid of a Computer," press release, December 22, 1966, Box "Press Reviews 1965," IBM Corporate Archives, Somers, N.Y.

11. "Fawcett to Speed Book Orders with New Computer," press release, [1965], Fawcett Publications Incorporated, Box "Press Reviews 1965," IBM Corporate Archives, Somers, N.Y.

12. Press release [1968], Hitchcock Publishing Company, "Press Reviews 1969," IBM Corporate Archives, Somers, N.Y.

13. "Computer Experts Briefed on Publishing Procedure," *Publishers Weekly* 181, no. 8 (February 19, 1962): 52; "Simplified Procedures at Little, Brown's New Warehouse," Ibid., 186, no. 10 (September 7, 1964): 31–34; "Scott, Foresmann Consolidates Warehousing at Pinola, Indiana," Ibid., 188, no. 24 (December 13, 1965): 25–28; "Wholesaler Switches to Computerized Inventory," Ibid., 194, no. 19 (November 4, 1968): 41–42; "A New Distribution Center for Lippincott," Ibid., 196, no. 14 (October 6, 1969): 29–30; "Stock Control Is More Than Counting," Ibid., 196, no. 2 (July 14, 1969): 155–157; and the second part of the same article, Ibid., 196, no. 3 (July 21, 1969): 49–50.

14. Seymour Turk, "Order Fulfillment and Other Services," in Chandler B. Grannis, *What Happens in Book Publishing* (New York: Columbia University Press, 1967): 216.

15. There are numerous accounts comparing one industry to another; however, for convenience, see U.S. Bureau of Labor Statistics, *Outlook for Technology and Manpower in Printing and Publishing*, Bulletin 1774 (Washington, D.C.: U.S. Government Printing Ofice, 1973); for early bibliography see Lowell H. Hattery and George P. Bush (eds.), *Automation and Electronics in Publishing* (Washington, D.C.: Spartan Books, 1965): 179–203.

16. Robert E. Rossell was the Managing Director of the Research and Engineering Council of the Graphic Arts Industry; for his comment, see his "Trends in Printing Technology," in Hattery and Bush, *Automation and Electronics in Publishing*, 73.

17. Reader's Digest and Time, with their large subscription lists, were some of the earliest publishers to demonstrate that scanning data with OCR equipment was less expensive that do data entry using punched cards, Ralph Dyer, *Optical Scanning for the Business Man* (New York: Hobbs, Dorman & Co., 1966): 131–132.

18. Alan S. Holliday, "Computers in Book Composition," in Hattery and Bush, *Automation and Electronics in Publishing*, 39–46; Victor M. Corrado, "Experience in Development of an Electronic Photocomposer," Ibid., 81–90; Arthur E. Gardner, "Economics of Automated Printing," Ibid., 123–127; John Markus, "Computer Experience: Directory Production," Ibid., 47–60; Bailey, *The Art and Science of Book Publishing*, 187–194.

19. *CIS Survey of Computerized Typesetters* (Los Angeles: Composition Information Services, October 16, 1968).

20. Quote and data, "Using the Computer for the Appropriate Job," *Publishers Weekly* 192, no. 23 (December 4, 1967): 64.

21. John W. Seybold, "Aesthetic Considerations in Computerized Photocomposition," *Publishers Weekly* 195, no. 14 (April 7, 1969): 54.

22. One can trace the growing industry through the pages of *Publishers Weekly*, see "Design and Technology in Computer Composition," Ibid., 190, no. 14 (October 3, 1966): 96–97, 100; "The New Composition Technology: Promises and Realities," Ibid., 195, no. 18 (May 5, 1969): 62–65; "Computer-Based Composition and Editing: An Exchange," Ibid.,

196, no. 1 (July 7, 1969): 94–95; "AAUP: Computer-Aided Composition," Ibid., 192, no. 1 (July 3, 1967): 78, 81; "Everybody's Hot for It," 197, no. 8 (February 23, 1970): 146–148; "Computer Potentials for the Graphic Designer," Ibid., 198, no. 6 (August 10, 1970): 40, 42–44; "Lannon's Fluxions for a Technology in Flux," Ibid., 198, no. 15 (October 12, 1970): 46–48, in which it states, "The consistent rise in wages is one of the main pressures behind the relentless drive for an even higher, more advanced technology in the printing industry," p. 46; "Technology's Challenge to Publishers," Ibid., 201, no. 5 (April 10, 1972): 115; Edward McSweeney, "The New Realities in Book Manufacturing," Ibid., 208, no. 22 (December 1, 1975): 27–29, 32, 34.

23. Alexander J. Burke, Jr., "The Publisher's Responsibility to Communicate," *Publishers Weekly* 208, no. 22 (December 1, 1975): 34.

24. Bureau of Labor Statistics, *Outlook for Technology and Manpower in Printing and Publishing*, 34.

25. Frank Cremonesi, "Employee Unions and Automation," in Hattery and Bush, *Automation and Electronics in Publishing*, 105–121; "Forecast 1981: Making Way for the Coming Electronic Revolution," *American Printer and Lithographer* (December 1980): 43–58; B. Gilchrist and A. Shenkin, "Disappearing Jobs: The Impact of Computers on Employment," *Futurist* 15 (February 1981): 44–49; Theresa F. Rogers and Nathalie S. Friedman, *Printers Face Automation: The Impact of Technology on Work and Retirement Among Skilled Craftsmen* (Lexington, Mass.: Lexington, 1980); Frank Romano, "You Should Understand the Method of Typesetting by Digitation," *Inland Printer/American Lithographer* (May 1977): 70.

26. W. B. Kerr, "Revolution in Printing," *Saturday Review* 51 (August 10, 1968): 54–55; "Phototypesetting: A Quiet Revolution," *Datamation* 16, no. 15 (December 1, 1970): 22–27; R. L. Tobin, "Publishing by Cathode Ray Tube," *Saturday Review* 53 (October 10, 1970): 61–62; G. O. Walter, "Typesetting," *Scientific American* 220 (May 1969): 60–69.

27. Paul D. Doebler, "Making Up Pages on Video Screens: The Last Frontier," in Arnold W. Ehrlich (ed.), *The Business of Publishing: A PW Anthology* (New York: R. R. Bowker Co., 1976): 147–150 and another article by the same author, "Video Editing Comes to Publishing," Ibid., 155–156; Seldon W. Terrant, "The Computer in Publishing," in Carlos A. Cuadra, Ann W. Luke, and Jessica L. Harris (eds.), *ASIA Annual Review of Information Science and Technology*, vol. 10 (Silver Springs, Md.: American Society for Information Science, 1975): 273–301, and by the same author an update to this article, "Computers in Publishing," Ibid., vol. 15 (1980): 192–219.

28. For the survey, "PW's Annual Technology Review," *Publishers Weekly* 214, no. 23 (December 4, 1978): 25; however, for more details from the period see two other annual surveys done by *PW*, Ibid., 212, no. 23 (December 5, 1977): 45–47, 50–52, 54, 57–60, 62, and 216, no. 23 (December 3, 1979): 17–18, 21–25.

29. Paul Doebler, "Editing, Comp Technology Integrated by Supplier," *Publishers Weekly* 217, no. 14 (April 11, 1980): 57.

30. Pat M. Welsh, "Accentuate the Positive with EDP," *Publishers Weekly* 196, no. 23 (December 8, 1969): 37–39; Pat M. Welsh, "RAP-TAG and EDP: Happy Marriage?" Ibid., 198, no. 12 (September 21, 1970): 46; Leonard Schwartz, "The Computer and the Book," Ibid., 204, no. 12 (September 17, 1973): 39–40; Sandra K. Paul, "Book Distribution and Technology: A 1980 Forecast," Ibid., 216, no. 15 (October 8, 1979): 35, 37.

31. John P. Dessauer, *Book Publishing: What It Is, What It Does* (New York: R. R. Bowker Company, 1974): 109.

32. Paul D. Doebler, "The Computer in Book Distribution," *Publishers Weekly* 218, no. 1 (September 12, 1980): 28. See also the entire article for many details, 25–28, 30, 32, 34–36, 40–41.

33. Stephen Breyer, "The Uneasy Case for Copyright: A Study of Copyright in Books, Photocopies, and Computer Programs," *Harvard Law Review* 84, no. 2 (December 1970):

281–351; for the bibliographic record of the period, George P. Bush (ed.), *Technology and Copyright: Annotated Bibliography and Source Materials* (Mt. Airy, Md.: Lomond Systems, Inc., 1972).

34. "The Information Industry: More Significant Than Ever," *Publishers Weekly* 201, no. 18 (May 1, 1972): 28–31; Paul Doebler, "The Role of Non-print in Publishing," Ibid., 202, no. 12 (September 18, 1972): 60–63; Paul Doebler, "New Computerized Services Provide Speedy (But Not Free) Copies 'On Demand,' " Ibid., 206, no. 21 (November 18, 1974): 34–36.

35. Paul Doebler, "Profile of an 'Information Buyer'," *Publishers Weekly* 204, no. 8 (August 20, 1973): 71–74; "Data Base Publishers Vying for Key Roles as Rapid Growth Looms in Business Uses," Ibid., 206, no. 12 (September 16, 1974): 38–41.

36. Statistics and for an excellent analysis of this new sub-industry within IT, see Martin Campbell-Kelly, *From Airline Reservations to Sonic the Hedgehog: A History of the Software Industry* (Cambridge, Mass.: MIT Press, 2003): 201–301.

37. Quoted in Richard Loftin, "Publishers Set Out on the Software Adventure," *Publishers Weekly* 223, no. 18 (May 6, 1983): 34.

38. "Software Publishing and Selling," *Publishers Weekly* 225, no. 6 (February 10, 1984): 28–178. It is an important source of information about the Software Industry in the United States during the early 1980s.

39. "McGraw, University to Create Software," *Publishers Weekly* 225, no. 13 (March 30, 1984): 17; Joanne Davis, "On the Software Front," Ibid., 225, no. 22 (June 1, 1984): 41–42.

40. "Software Publishing and Selling," *Publishers Weekly* 226, no. 12 (September 21, 1984): 35–62; "Surveying Software's Educational Landscape," Ibid., 226, no. 17 (October 26, 1984): 45–74; "Electronic Publishing," Ibid., 226, no. 21 (November 23, 1984): 32–55.

41. John F. Baker, "1984 The Year in Review: The Year in Publishing," *Publishers Weekly* 227, no. 11 (March 15, 1985): 28–30.

42. Books about computers began appearing in considerable numbers in the 1950s, but were aimed at specialized technical audiences not normally reached through retail outlets. In the 1960s, this same audience was frequently reached by mail-order marketing. For bibliographies citing early publications on computing, see James W. Cortada, *A Bibliographic Guide to the History of Computing, Computers, and the Information Processing Industry* (Westport, Conn.: Greenwood Press, 1990): 7–15.

43. "Selling Computer Books," *Publishers Weekly* 237, no. 4 (January 26, 1989): 396–397.

44. Conversation with Rachel McCloud of Barnes & Noble, March 31, 1998.

45. Paul Hilts, "1.5 Million Dummies Can't Be Wrong," *Publishers Weekly* 240, no. 8 (February 22, 1993): 49–50.

46. Pamela Byers, "Consumer-Base Expansion in the Digital Market," *Publishers Weekly* 240, no. 40 (October 4, 1993): 86.

47. Oldrich Standera, *The Electronic Era of Publishing: An Overview of Concepts, Technologies and Methods* (New York: Elsevier, 1987): 6–9; Geffrey Nunberg, "The Place of Books in the Age of Electronic Reproduction," *Representations* 42 (1993): 13–37; Abigail J. Sellen and Richard F. R. Harper, *The Myth of the Paperless Office* (Cambridge, Mass.: MIT Press, 2002): 14–15.

48. Michael M. A. Mirabito and Barbara L. Morgenstern, *The New Communications Technologies: Applications, Policy, and Impact* (Boston, Mass.: Focal Press, 2001): 121–130; John V. Pavlik and Shawn McIntosh, *Converging Media: An Introduction to Mass Communication* (Boston, Mass.: Pearson, 2004): 74–75.

49. Gay Courter, "Word Machines for Word People," *Publishers Weekly* 219, no. 7 (February 13, 1981): 40–43; Robert Dahlin, "Consumer as Creator," Ibid., 219, no. 13 (March 27, 1981): 21–27; Robyn Shotwell, "How Publishers Can Use Personal Computers," Ibid., 221, no. 6 (February 5, 1982): 284, 286; Beth Bird Pocker, " 'User Publishing': A New

Concept in Electronic Access," Ibid., 221, no. 16 (April 16, 1982): 38–39; Robert Dahlin, "Electronic Publishing: Steps Forward—And Back," Ibid., 221, no. 23 (June 4, 1982): 26–31; Wesley T. Carter, "Merging Text and Graphics With a New Electronic System," Ibid., 227, no. 1 (January 4, 1985): 40–41; "Electronic Publishing Systems Come of Age at Massive Print 85 Exhibition," Ibid., 227, no. 23 (June 7, 1985): 52, 55–57, 60, 62; Daniel N. Fischel, "Two Cheers for Computers," Ibid., 229, no. 14 (April 4, 1986): 20–23; Jesse Berst, "The Latest In Desktop Publishing," Ibid., 232, no. 20 (November 13, 1987): 26, 28–29; "Desktop Publishing for Independent Presses," Ibid., 232, no. 20 (November 13, 1987): 30, 32, 34–35; Robert Weber, "The Clouded Future of Electronic Publishing," Ibid., 237, no. 26 (June 29, 1990): 70, 78–80; "Upscaling DTP—New Electronic Tools Give Publishers In-House Capabilities," Ibid., 237, no. 22 (June 1, 1990): 39–40, 42; Sally Taylor, "Desktop Today," Ibid., 238, no. 25 (June 7, 1991): 19–21; Don Avedon, *Introduction to Electronic Publishing* (Silver Springs, Md.: Association of Information & Image Management, 1992).

50. Haines B. Gaffner, "PC Software and Online Databases," *Publishers Weekly* 226, no. 21 (November 23, 1984): 44–48.

51. Taylor, "Desktop Today."

52. Robyn Shotwell, "Books on Demand," *Publishers Weekly* 219, no. 7 (February 13, 1981): 47–48, and Ibid., 219, no. 23 (June 5, 1981): 50–51; Sally Taylor, "The Potential of On-Demand Printing," Ibid., 238, no. 25 (June 7, 1991): 38–40; Judith Booth, "The Digital Original," Ibid., 238, no. 49 (November 8, 1991): 43–44; Paul Hilts, "Donnelley's Digital Production Vision," Ibid., 241, no. 34 (August 22, 1994): 24–25; Paul Hilts, John Mutter, and Sally Taylor, "Books While U Wait," Ibid., 241, no. 1 (January 3, 1994): 48–50. For a balanced contemporary analysis, see Czeslaw Jan Grycz, "Everything You Need to Know About Technology," *Publishing Research Quarterly* 7, no. 4 (Winter 1991/92): 3–12.

53. Weber, "The Clouded Future of Electronic Publishing," 76, 78–80; but see also, Nathaniel Lande, "Toward the Electronic Book," *Publishers Weekly* 238, no. 42 (September 20, 1991): 2, 30; "Donnelley, Compton Join In Electronic Publishing Push," Ibid., 239, no. 18 (April 13, 1992): 9; "Thompson Pays $210M for Electronic Database, Journals," Ibid., 239, no. 19 (April 20, 1992): 6; Clyde Steiner, "The Future of Electronic Publishing: Two Views," Ibid., 240, no. 1 (January 4, 1993): 38; John F. Baker, "Reinventing the Book Business," Ibid., 241, no. 11 (March 14, 1994): 36–40.

54. Thomas Weyr, "The Wiring of Simon & Schuster," *Publishers Weekly* 239, no. 25 (June 1, 1992): 33.

55. Robyn Shotwell, "Computer Typesetting—Reflections on a Revolution," *Publishers Weekly* 222, no. 14 (October 1, 1982): 91–92; "Computers: Manufacturing's Newest Route to Productivity," Ibid., 224, no. 23 (December 2, 1983): 65, 68, 70, 72, 74; "Toward the Total Publishing System," Ibid., 226, no. 23 (December 7, 1984): 32, 36–37, 40, 42; Kimberly Olson Fakih, "Computers in Bookstores: The Menu of Possibilities," Ibid., 236, no. 14 (October 13, 1989): 15–16, 18; Paul Hilts, "The American Revolution in Book Production," Ibid., 239, no. 41 (September 14, 1992): S3–S4; "Reshaping the Flow of Production," Ibid., S8–S10, S12; "The Changing Role of Suppliers," Ibid., S15–S17, S24–S25; Margaret Langsstaff, "The Digital Traveler," Ibid., 240, no. 37 (September 13, 1993): 46–47, 52, 54; Bella Standler, "Computerizing Inventory: A Menu of Choices," Ibid., 241, no. 12 (March 21, 1994): 18–22; Paul Hilts, "The Changing Face of Print," Ibid., 242, no. 1 (January 2, 1995): 44–45; Karen Angel, "Bookselling's Brave New Web World," Ibid., 243, no. 51 (December 16, 1996): 26–29.

56. Carol Robinson, "Publishing's Electronic Future," *Publishers Weekly* 240, no. 36 (September 6, 1993): 47. See also Pamela Byers, "Consumer-Base Expansion In the Digital Market," Ibid., 240, no. 40 (October 4, 1993): 36.

57. Studies on the economic dynamics of paper versus electronics are rare; most commentators acknowledge the differences, but provide little data. One very recent collection of studies is useful; Brian Kahin and Hal R. Varian (eds.), *Internet Publishing and Beyond:*

The Economics of Digital Information and Intellectual Property (Cambridge, Mass.: MIT Press, 2000).

58. Paul Hilts, "Web Publishing Isn't Just a Hobby Any More," *Publishers Weekly* 243, no. 12 (March 18, 1996): 2, 30.

59. "Ingram Retrenching Its Multimedia Efforts," *Publishers Weekly* 243, no. 24 (June 10, 1996): 30.

60. *Publishers Weekly* 243, no. 43 (October 21, 1996), p. 14.

61. David Moschella, *Customer-Driven IT: How Users Are Shaping Technology Industry Growth* (Boston, Mass.: Harvard Business School Press, 2003): 158.

62. "Online Bookstores: Progress in Web Selling," *Publishers Weekly* 243, no. 24 (June 10, 1996): 30, 40; Bridget Kinsella, "Book Groups Get Wired," Ibid., 243, no. 47 (November 18, 1996): 46–47; Karen Angel, "Independents Make Online Alliances," Ibid., 244, no.14 (April 7, 1997): 28; James Lichtenberg, "Online Commerce: The Virtual Bookstore," Ibid., 244, no. 50 (December 8, 1997): 23 and see a second article by him, "Inching Toward E-Commerce," Ibid., 35–37; "Borders Online at Last with Books, CDs, Videos," Ibid., 245, no. 20 (May 18, 1998): 12; Paul Hilts and James Lichtenberg, "Redefining Distribution," Ibid., 245, no. 51 (December 21, 1998): 23–24, 26–27; John Mutter, "You've Got Sales," Ibid., 246, no. 1 (January 1, 1999): 50–52; "More Maneuvering Among Online Booksellers," Ibid., 246, no. 3 (January 18, 1999): 184; John High, "Web Sales Rising for Independents," Ibid., 246, no. 5 (February 1, 1999): 25–26; "New Age Clicks-and-Mortar," Ibid., 247, no. 7 (February 14, 2000): 88–89; Anita Hennessey, "Online Bookselling," *Publishing Research Quarterly* 16, no. 2 (Summer 2000): 34–51.

63. The best explanation of Bezos's rationale was written by Robert Spector, *amazon.com: Get Big Fast* (New York: HarperBusiness, 2000): 24–30.

64. U.S. Bureau of the Census.

65. Jim Milliot, "The Land of the Giants," *Publishers Weekly* 248, no. 1 (January 1, 2001): 61–63; Pavlik and McIntosh, *Converging Media,* 75; Mark Crispin Miller, "The Publishing Industry," in Erik Barnouw (ed.), *Conglomerates and the Media* (New York: New Press, 1998): 107–133.

66. Pavlik and McIntosh, *Converging Media,* 75.

67. Ibid., 75–76.

68. The best introduction to the firm is Spector, *amazon.com*; but see also Rebecca Saunders, *Business the amazon.com Way* (Dover, N.H.: Capstone 1999).

69. "Publishers Back Microsoft Call for E-Book Standard," *Publishers Weekly* 245, no. 42 (October 19, 1998): 10; John Mutter, "The Inevitable Future," Ibid., 246, no. 46 (November 15, 1999): 26–27; "NIST eBook 2000 Conference Expands Scope," Ibid., 247, no. 41 (October 9, 2000): 910; Roxane Farmanfarmaian, "Beyond E-Books: Glimpses of the Future," Ibid., 248, no. 1 (January 1, 2001): 56–57; James Lichtenberg, "Rising From the Dead," Ibid., 250, no. 40 (October 6, 2003): 50–52; Phred Dvorak, of the *Wall Street Journal,* surveyed a new generation of products in 2004 and found them functionally less superior to paper-based books, "Digital Books Not Quite Page-Turners," *Wisconsin State Journal,* July 16, 2004, p. C-9.

70. The industry had long expressed intense interest in e-publishing: "Publishers Still Searching for Profits in New Media," *Publishers Weekly* 243, no. 1 (January 1, 1996): 22, 31; Paul Hilts, "The Road Ahead," Ibid., 244, Special Anniversary Issue (July 1997): 125–128; Jim Milliot, Calvin Reid, and Steven M. Zeitchik, "Tomorrow's Publishers Today," Ibid., 247, no. 10 (March 6, 2000): 42, 44; "Random Buys Share of E-Publisher Xlibris," Ibid., 247, no. 15 (April 10, 2000): 9–10; "Barnesandnoble.com Grabs Stake in MightyWords, Plus Content Distribution Pact," Ibid., 247, no. 24 (June 12, 2000): 18, 20; "Barnes&Noble.com to Buy Fatbrain.com for $64m.," Ibid., 247, no. 38 (September 18, 2000): 9; Paul Hilts, "Looking at the E-Book Market," Ibid., 247, no. 47 (November 20, 2000): 35–36; Paul Hilts, "BookTech Looks At E-Publishing," Ibid., 248, no. 10 (March

5, 2001): 46; "Retrenchment Hits Electronic Publishing Industry," Ibid., 248, no. 19 (May 7, 2001): 9; Calvin Reid, "All e-Publishing News Isn't Bad," Ibid., 248, no. 47 (November 19, 2001): 32, and his "(Mostly) Good E-Publishing News," Ibid., 249, no. 4 (January 28, 2002): 147; "E-book Users Surveyed Online, in College," Ibid., 249, no. 36 (September 9, 2002): 12.

71. *Publishers Weekly* routinely covers this story, providing considerable detail during the course of a year's issues of the journal, reviewing the subject in one form or another in the early 2000s at least once a month.

72. Paul Hilts, "21st-Century Publishing," *Publishers Weekly* 248, no. 21 (May 21, 2001): 44.

73. Farmanfarmaian, "Beyond E-Books," 56.

74. "Pearson Education and Wharton School Launch Wharton School Publishing, New Force in Business Publishing," joint press release University of Pennsylvania and Pearson Education, February 18, 2004, http://home.businesswire.com/portal/site/google/index.jsp?ndmViewId=news_view&newsId=20040218005167&newsLang=en (last accessed 2/26/2004).

75. Greco, *The Book Publishing Industry*, 285–286; Jonathan Rosenoer, *CyberLaw: The Law of the Internet* (New York: Springer-Verlag, 1997):1–20, 243–254, 269–340; Steven M. Zeitchik, "The Great Ether Grab," *Publishers Weekly* 246, no. 24 (June 14, 1999): 28–30; Jessica Litman, *Digital Copyright* (Amherst, N.Y.: Prometheus Books, 2001): 151–191.

76. John F. Baker and Michael Coffey, "To E or Not to E . . . and other Questions," *Publishers Weekly* 248, no. 1 (January 1, 2001): 51.

77. Ibid.

78. Robert G. Picard and Jeffrey H. Brody, *The Newspaper Publishing Industry* (Boston, Mass.: Allyn and Bacon, 1997): 74–98.

79. Allan Woods, *Modern Newspaper Production* (New York: Harper & Row, 1963), p. 172.

80. Ibid., 173.

81. "Associated Press Will Install Special IBM System to Compile Stock Tables," IBM press release, September 1, 1961; IBM press release, February 11, 1963, covering AP's use of computing; "Computer Typesetting at the *Miami Herald*," IBM press release, undated [early 1960s]; press release, *The Daily Oklahoma-City Times*, March 5, 1963; *Kansas City Star* press release, February 21, 1964; "Computer to Control Typesetting at Vincent B. Fuller," Fuller Organization press release, April 23, 1965; copies of all these press releases at IBM Corporate Archives, Somers, N.Y.

82. Frank W. Rucker and Herbert Lee Williams, *Newspaper Organization and Management*, 4th ed. (Ames, Ia.: Iowa State University Press, 1974): 85–89.

83. Ibid., p. 88. The IBM Corporate Archives has an extensive collection of such material.

84. Donald F. Blumberg, "The Impact of Information Technology on Newspaper Publishing Operations," in Edith Harwith Goodman (ed.), *Data Processing Yearbook* (Detroit, Mich.: American Data Processing, Inc., 1963): 212–213.

85. Ibid., 92–94; Rucker and Williams, *Newspaper Organization and Management*, 85–103.

86. Blumberg, "The Impact of Information Technology on Newspaper Publishing Operations," 217.

87. "Computers on Deadline," *Data Processor* (September 1967); 3–7, and "Electronic Newspaper Production," Ibid. (1974): 8–9, both in Box "Press Reviews," IBM Corporate Archives, Somers, N.Y.; Ronald A. White, "Computer Systems and Newspaper Production," in Lowell H. Hattery and George P. Bush (eds.), *Automation and Electronics in Publishing* (Washington, D.C.: Spartan Books, 1965): 17–25; Wilmott Lewis, Jr., "Electronic Computer Experience: A Daily Newspaper," Ibid., 9–15.

88. Victor Strauss, *The Printing Industry* (New York: Printing Industries of America, 1967): 143.

89. Arthur H. Phillips, *Handbook of Computer-Aided Composition* (New York: Marcel Dekker, 1980): 244–298.

90. "IBM System a Fast Election Reporter," *DP News* (November 1962): 11; "Tomorrow's Newspapers," *Business Machines* (December 1963): 9–11; "K. C. Star Calculates Savings in Computer," *Editor and Publisher* (July 10, 1965): 11, 65, copies in Box A-1245-1, IBM Corporate Archives, Somers, N.Y.

91. Anthony Smith, *Goodbye Gutenberg: The Newspaper Revolution of the 1980s* (New York: Oxford University Press, 1980): 131–132.

92. Timothy Marjoribanks, *News Corporation, Technology and the Workplace: Global Strategies, Local Change* (Cambridge: Cambridge University Press, 2000): 77–90.

93. Robert S. McMitchen, "Letter to All ITU Officers and Members, re Merger and ITA and CWA," *Typographical Journal* 189, no. 2 (August 1986): 17.

94. Smith, *Goodbye Gutenberg*, 97–101, 109–112.

95. Marjoribanks, *News Corporation, Technology and the Workplace*, 77–79.

96. Ibid., 79.

97. Smith, *Goodbye Gutenberg*, 132–133.

98. Ibid., 95.

99. Gene Goltz, "Special Report: The Workforce Reorganisation," *Presstime* 11, no. 9 (September 1989): 18–23; Marjoribanks, *News Corporation, Technology and the Workplace*, 80.

100. Edward P. Hayden, "The Luddites Were Right," *Guild Reporter* 47, no. 22 (December 12, 1980): 8.

101. Marjoribanks, *News Corporation, Technology and the Workplace*, 84.

102. Carl Hausman, *Crafting the News for Electronic Media: Writing, Reporting and Producing* (Belmont, Calif.: Wadsworth, 1992): 75–77; for an update on the pattern, circa 1990s, see Joel Simon and Carol Napolitano, "We're All Nerds Now," *Columbia Journalism Review* (March/April 1999), http://archives.cjr.org/year/99/2/nerds.asp (last accessed 12/10/2003).

103. If an industry lasts long enough, it probably experienced early in its history issues which came back again but, because of the passage of time, were forgotten. Between 1876 and 1912, the Newspaper Industry had its first encounter with electronic news. "Telephone Newspapers" were conceived of at the dawn of the telephone both in Europe and in the United States, with various experiments undertaken to make that happen. A newspaper person would read the news into an open phone line, which was a customer's dedicated line to the paper. A subscriber could pick up their line at times when the news was being read over the phone and get updates on stories. Stock quotes, sports scores, and so forth were popular topics, Pavlik and McIntosh, *Converging Media*, 84. On early concerns about electronics in the post–World War II period, see the various bibliographies in Bush, *Technology and Copyright*.

104. Pablo J. Boczkowski, *Digitizing the News: Innovation in Online Newspapers* (Cambridge, Mass.: MIT Press, 2004): 19.

105. Ibid., 25–27, for the Viewtron study, but see the entire book for the early experiments.

106. James W. Cortada, *Making the Information Society: Experience, Consequences, and Possibilities* (Upper Saddle River, N.J.: Financial Times/Prentice Hall, 2002): 174–175.

107. Picard and Brody, *The Newspaper Publishing Industry*, 152–153.

108. All major news publications on the Web had these features by the early 2000s. For examples, see ABC News, CNN/Money, *New York Times Digital*.

109. Pavlik and McIntosh, *Converging Media*, 77–78.

110. Ibid., 79.

111. Rob Runett, "Changing Role for News Sites," Newspaper Association of America, *Electronic Publishing*, July 2001, http://www.naa.org/artpage.cfm?AID=2196&SID=109 (last accessed 1/8/2004).

112. Scott B. Anderson, "Pew Study Says Internet Audience Is 'Going Ordinary,' " press

release, American Society of Newspaper Editors, January 10, 2000, http://www.asne.org/kiosk/reports/99reports/1999nextmediareader/p3-5_Pew.html (last accessed 2/14/2004). Pew Internet & American Life is a project that continuously reports out how Americans are using the Internet and its findings are widely reported by the press. For its reports, see http://www.pewinternet.org/reports/index.asp.

113. Garbo Cheung-Jasik, "In the Know: Q&A with Lincoln Millstein," June 2002, http://www.namme.org/intheknow_millstein.asp (last accessed 2/11/2004).

114. Samples from 1996: Scott Whiteside, "Web Redefines Who an Editor Is," September 29, 1996. From 1997: Chet Raymo, "Learning to Navigate a Sea of Information," October 1, 1997; David Demers, "Reader-driven Marketing in the Electronic Era," July 1, 1997; Joann Byrd, "Questions Greatly Outnumber the Answers in Exploring the Ethical Issues and Values of the New Media," March 23, 1997; Howard Finberg, "You May Have to Rethink Your Whole Organization," March 1, 1997. From 1998: Kevin M. Goldberg, "Watch Out or Real-time Coverage Will Be Shot Down," May 20, 1999; Scott B. Anderson, "Technology: The Story That May Affect Us Most," June 9, 1999; "The Future of Technology and Its Impact on Newspapers," October 27, 1999. From 2000: Kevin Goldberg, "Fear of the Internet Leads to Clampdown," March 27, 2000; Bonnie Bressers, "Connecting with Readers—E-mail Shakes Up Editorial 'Feedback Loop,' " July 1, 2000. From 2001: Kurt Greenbaum, "Sifting the dot.com Wreckage for Lessons," April 1, 2001; Ellen Kampinsky, Shayne Bowman, and Chris Willis, "What Newspapers Can Learn from Amazon.com," July 1, 2001. All of these, and other stories, are available at its website, http://www.asne.org.

115. Picard and Brody, *The Newspaper Publishing Industry*, 116–119.

116. Jan Schaffer, "Interactive Journalism," speech, March 31, 2001, http://www.pewcenter.org/doingcj/speches/s_pittsburghsapj.html, and her "Interactive Journalism: Clicking on the Future," speech, undated [fall 2001], http://www.pewcenter.org/doingcj/speeches/a_apmefall2001.htm (both accessed 12/10/2003).

117. Margaret H. DeFleur, *Computer-Assisted Investigative Reporting: Development and Methodology* (Mahwah, N.J.: Lawrence Erlbaum Associates, 1997): 71.

118. Ibid., 105–109.

119. Frank Houston, "Enjoy the Ride While It Lasts," *Columbia Journalism Review* (July/August 2000): http://archives.cjr.org/year/00/2/houston.asp (last accessed 2/26/2004).

120. Benjamin M. Compaine, "The Magazine Industry," in Benjamin M. Compaine and Douglas Gomery, *Who Owns the Media? Competition and Concentration in the Mass Media Industry* (Mahwah, N.J.: Lawrence Erlbaum Associates, 2000): 147–191.

121. Ibid., 149.

122. Charles P. Daly, Patrick Henry, and Ellen Ryder, *The Magazine Publishing Industry* (Boston, Mass.: Allyn and Bacon, 1997): 43–49, 73–75.

123. Compaine, "The Magazine Industry," 153.

124. Ibid., 187.

125. Daly, Henry, and Ryder, *The Magazine Publishing Industry*, 159.

126. Ibid.

127. "Meredith Installs Printing Industry's First IBM 360 Computer," press release, Meredith Publishing Company, September 8 [1965?], copy in IBM Corporate Archives, Somers, N.Y.

128. IBM's archives contain a variety of press releases and articles from the period concerning management of subscription lists via computers. See, for example, "Media in a Time of Change," *Data Processor* (March 1971): 4–6; "Computer Helps Coin Collectors Get Hobby Magazine," press release, *Numismatic News Weekly*, July 22 [1970?]; "Publishers Clearning House Uses Electronic Device to Speed Magazine Subscriptions," press release, Publishers Clearing House, June 16, 1972, IBM Corporate Archives, Somers, N.Y. For *Newsweek*'s experience, see Auerbach Corporation, "Technical Review of Data Processing Installation." Final Report 1315-TR-1, December 1965, Auerbach Corporation Papers, Market

and Product Reports, CBI 30, Box 53, folder 19, Archives of the Charles Babbage Institute, University of Minnesota, Minneapolis.

129. Daly, Henry, and Ryder, *The Magazine Publishing Industry*, 168.

130. Ibid., 168–169.

131. Liz Horton, "Charting the Pre-press Revolution," *Folio* (August 1991): 57.

132. Daly, Henry, and Ryder, *The Magazine Publishing Industry*, 171.

133. Paul McDougall, "Desktop Survey: Publishers Are Doing It for Themselves," *Folio* (September 1, 1994): 66ff. For an example of these applications, there is the experience of *TV Guide*, Auerbach Corporation, "TV Guide System Review and Proposal Evaluation Study," Final Report 1548-TR-1, March 15, 1968, Market and Product Reports, CBI 30, Box 76, folder 13, Archives of the Charles Babbage Institute, University of Minnesota, Minneapolis.

134. Daly, Henry, and Ryder, *The Magazine Publishing Industry*, 173.

135. McDougall, "Desktop Survey: Publishers Are Doing It for Themselves," 63.

136. Daly, Henry, and Ryder, *The Magazine Publishing Industry*, 182–194.

137. For a description of the modern process, see Ibid., 206–234, hardly any paragraph avoids discussing the role of the digital.

138. This medium played less of a role in magazine publishing than in book publishing and thus is not discussed in this chapter. For details about how this medium works and its role in publishing, see Devra Hall, *The CD-Rom Revolution* (Rocklin, Calif.: Prima Publishing 1995), and for a brief discussion of early uses of the technology in magazine publishing, 158–163.

139. Andrew Mariatt, "Your Magazine and the Web," *Folio* (January 1, 2001), http://foliomag.com/magazinearticle.asp?magazinearticleleid=41328&magazineid=125&site (last accessed 2/21/2004). The phrase "giant sucking sound" was first used by Ross Perot during his U.S. presidential campaign in 1992, and referred to the loss of American jobs to Mexico that he predicted would occur if the United States implemented the North American Fair Trade Agreement (NAFTA). The agreement was implemented by the United States, Canada, and Mexico, resulting in migration of jobs across all three borders, with almost no impact on the publishing industries.

140. Mary Harvey and Paul McDougall, "Net Sub Sales: More Questions Than Answers," Ibid. (November 1, 1998); Tony Silber, "The Internet Leaders," Ibid. (June 1, 1999); Jo Bennett, "From Magazine to Media Company," Ibid. (August 1, 1999); Mark Miller, "Will Customers Pay for Online Content?" Ibid. (April 1, 2001); Caroline Jenkins, "Online Ad Exchanges Struggle for Traction," Ibid. (August 1, 2001).

141. Ellen Cavalli, "A Deeper Net," *Folio* (November 30, 1998). All citations of *Folio* below are all from the same website, http://foliomag.com/magazine (all last accessed 2/21/2004).

142. Mary Harvey, "Internet Marketing: Moving Past the Test Phase," Ibid. (September 30, 1999).

143. Survey conducted by Folio Staff, Ibid. (January 1, 2001).

144. Dale Buss, "Magazines Snare More Dot-Com Ad Dollars," Ibid. (January 1, 2001).

145. Useful for understanding this complex story, see Dan McNamee, "It's Not the Technology, It's How You Publish," Ibid. (September 30, 1998); Dzintars Dzilna, "Digital Content Control," Ibid. (September 1, 1999); Michael Weinglass, "Digital Workflow, Defined," Ibid. (June 1, 2000); Bert Langford, "E-Production Is E-Volving," Ibid. (July 1, 2000); Folio Staff, "Design Takes to the Web," Ibid. (October 1, 2000); Julian S. Ambroz, "Portal Solutions," Ibid. (December 15, 2000); Hal Hindereliter, "Edit Once, Publish Everywhere," Ibid. (September 1, 2003).

146. Caroline Jenkins, "CTP Approaches the 50% Mark," Ibid. (October 1, 2000); Michael Weinglass, "Digital Workflow, Defined," Ibid. (January 1, 2001).

147. Andrew M. Odlyzko, "Tragic Loss or Good Riddance? The Impending Demise of Traditional Scholarly Journals," November 6, 1994, available from amo@research.att.com; Dietrich Gotze, "Electronic Journals—Market and Technology," *Publishing Research Quarterly* 11, no. 1 (Spring 1995): 3–19; Andrew M. Odlyzko, "The Economics of Electronic Journals," *First Monday* (1997), http://firstmonday.org/issues/issue2_8/odlyzko/index.html (last accessed 12/2/2003), but see also his white paper, "Competition and Cooperation: Libraries and Publishers in the Transition to Electronic Scholarly Journals," amo@research.att.com; Richard Ekman and Richard E. Quanolt (eds.), *Technology and Scholarly Communications* (Berkeley, Calif.: University of California Press, 1999). All the citations have extensive bibliographies on this subject.

148. Kinko's, *Locations* (Dallas, Tex.: Kinko's Inc., undated [circa 2003]); "Kinko's History," http://www.fedex.com/us/about/news/update/kinkos/companyhistory.html (last accessed 2/25/2004).

149. *Basic Books Inc. v. Kinko's Graphics Corp.*, *Federal Supplement* 758 (1991): 1522–47; for a summary of other cases related to photocopying, see "What Are Some Landmark Fair Use Court Cases?" http://www.ninch.org/ISSUES/COPYRIGHT/FAIR-USE-EDUCATIONAL?FAIRUSE-Intro (last accessed 2/25/2004).

150. Boczkowski, *Digitizing the News*, 185–187.

151. Philip Meyer, *The Vanishing Newspaper: Saving Journalism in the Information Age* (Columbia, Mo.: University of Missouri Press, 2005). On the fragmentation of audiences, see also Deborah Fallows, "The Internet as a Unique News Source: Millions Go Online for News and Images Not Covered in the Mainstream Press," July 8, 2004, http://www.pewinternet.org (last accessed 7/9/2004).

152. "U.S. Consumer Spending for Online Content Totals Nearly $1.6 Billion," press release, Online Publishers Association, May 11, 2004, ComScore Networks, http://www.comscore.com/press/release.asp?press=455 (last accessed 6/17/2004).

Chapter 10

1. U.S. Department of Commerce, *Historical Statistics of the United States: Colonial Times to 1970* (Washington, D.C.: U.S. Government Printing Office, 1975): Part 2, 796; U.S. Census Bureau, *Statistical Abstract of the United States* (Washington, D.C.: U.S. Government Printing Office, October 1998): 573, http://www.census.gov/prod/3/98 (last accessed 8/1/2003).

2. For an excellent introduction to how the industry works, see Alan B. Albarran and Gregory G. Pitts, *The Radio Broadcasting Industry* (Boston, Mass.: Allyn and Bacon, 2001). Includes excellent bibliographies.

3. Jeffrey S. Close, "Spectrum Utilization in Broadcasting," Leonard Lewin (ed.), *Telecommunications in the U.S.: Trends and Policies* (Dedham, Mass.: Artech House, 1981): 101–162; Susan E. McMaster, *The Telecommunications Industry* (Westport, Conn.: Greenwood Press, 2002).

4. Material for the previous three paragraphs drawn from Robert G. Picard, *The Economics and Financing of Media Companies* (New York: Fordham University Press, 2002): 14.

5. Eric Rothenbuhler and Tom McCourt, "Radio Redefines Itself, 1947–1962," in Michele Hilmes and Jason Loviglio (eds.), *Radio Reader: Essays in the Cultural History of Radio* (New York: Routledge, 2003): 378.

6. U.S. Census Bureau, various reports (1999–2003).

7. The massive trend toward digital devices small enough to carry on one's person coming from various industries, not just radio and music, is explored in Nicholas D. Evans, *Consumer Gadgets: 50 Ways to Have Fun and Simplify Your Life with Today's Technology . . . and Tomorrow's* (Upper Saddle River, N.J.: Financial Times/Prentice-Hall, 2003).

8. "Punched Card System Makes Good Music," *Office Automation Applications* (undated late 1950s), Part III, Section E, pp. 1–2, Market Research CBI 55, Box 71, Folder 11, Archives of the Charles Babbage Institute, University of Minnesota, Minneapolis.

9. "New Computer System Automates AM-FM," *Broadcasting* 79 (November 16, 1970): 61.

10. Press release, Plough Broadcasting Company, "New Development in Radio Programming," August 3, 1962, IBM Corporate Archives, Somers, N.Y.

11. Christopher H. Sterling and John M. Kittross, *Stay Tuned: A Concise History of American Broadcasting* (Belmont, Calif.: Wadsworth Publishing Co., 1978): 376.

12. IBM press release, "Makes Every Minute Count," October 2, 1972, IBM Archives, Somers, N.Y.

13. "Computers Work Praised," *Broadcasting* 64 (April 8, 1963): 66.

14. "What Food for Computers," *Broadcasting* 64 (April 29, 1963): 28.

15. "A World of Computers Plugged in by Humans," Ibid., 66 (June 1, 1964): 30, 32, 34; "A Basic Redesign of Local Nielsens," Ibid., 67 (September 7, 1964): 60–61; "Computers in the Paper Jungle," Ibid., 70 (November 13, 1966): 42; "Marketing I.Q. at Westinghouse," Ibid., 71 (September 19, 1966): 38–40; "Nielsen Takes Wraps off TNT," Ibid., 74 (December 30, 1968): 49–50; IBM press release, November 19, 1968, IBM Archives, Somers, N.Y.

16. For example, see James Walker and Douglas Ferguson, *The Broadcast Television Industry* (Boston, Mass.: Allyn and Bacon, 1998); Albarran and Pitts, *The Radio Broadcasting Industry*.

17. Richard S. Rosenbloom, "The Continuing Revolution in Communications Technology: Implications for the Broadcasting Business," Incidental Paper, Program on Information Resources Policy (Cambridge, Mass.: Center for Information Policy Research, Harvard University, 1981): 11.

18. U.S. Federal Communications Commission.

19. IBM Corporation, *Industry Applications and Abstracts* (White Plains, N.Y.: IBM Corporation, 1988): 16–6.

20. Carl Hausman, *Crafting the News for Electronic Media: Writing, Reporting and Producing* (Belmont, Calif.: Wadsworth Publishing Co,, 1992): 52–53; Stephen Quinn, *Knowledge Management in the Digital Newsroom* (Oxford: Focal Press, 2002).

21. Douglas Gomery, "Radio Broadcasting and the Music Industry," in Benjamin M. Compaine and Douglas Gomery (eds.), *Who Owns the Media? Competition and Concentration in the Mass Media Industry* (Mahwah, N.J.: Lawrence Erlbaum Associates, 2000): 287.

22. Mass Media Bureau, Federal Communications Commission, *Review of the Radio Industry, 2001* (Washington, D.C.: U.S. Federal Communications Commission, September 2001): 2–3.

23. Michael M.A. Mirabito and Barbara L. Morgenstern, *The New Communications Technologies: Applications, Policy, and Impact* (Boston, Mass.: Focal Press, 2001): 162–164; Sydney W. Head, Thomas Spann and Michael A. McGregor, *Broadcasting in America: A Survey of Electronic Media*, 8th ed. (Boston, Mass.: Houghton Mifflin, 1998): 144–145.

24. Albarran and Pitts, *Radio Broadcasting Industry*, 166–167.

25. For the global inventory, see www.radio-locator.com/cgi-bin/nation.

26. John V. Pavlik and Shawn McIntosh, *Converging Media: An Introduction to Mass Communication* (Boston, Mass.: Pearson Education, 2004): 200.

27. The term "compressed" refers to transmitting only pieces of a file that can later be extended to its full original form, thereby reducing the amount of data that either has to be transmitted or saved. It would be as if one wanted to reduce the amount of typing and thus chose to write the word *online* as *o/l*, knowing later to expand that out to its full original spelling, *online*.

28. Ty Ford, "Digital Success Story at WRBS(FM)," *RW Online* (*Radio World Newspaper*),

March 14, 2001, http://www.rwonline.com/reference-room/trans-2-digital/rwf-wrbs.shtml (last accessed 3/28/2004).

29. Skip Pizzi, "Radio Broadcasters Can Learn a Lot From the DTV Transition," Ibid., April 7, 2003, http://www.rwonline.com/reference-room//skippizzi-bigpict/01_rwf_pizzi_april_7.shtml (last accessed 3/28/2004).

30. Skip Pizzi, "Digital, Everywhere But Here," Ibid., September 1, 2003, http://www.rwonline.com/reference-room;/skippizzi-bigpict/04_rwf_pizzi_sept_1.shtml (last accessed 3/28/2004).

31. Daren Fonda, "The Revolution in Radio," *Time*, April 19, 2004, pp. 55–56.

32. Douglas Gomery, "The Television Industries: Broadcast, Cable, and Satellite," in Compaine and Gomery, *Who Owns the Media?* 193–283; Bruce M. Owen, *The Internet Challenge to Television* (Cambridge, Mass.: Harvard University Press, 1999): 15–42; Pavlik and McIntosh, *Converging Media*, 159–175.

33. Owen, *The Internet Challenge to Television*; Erik Barnouw, *Tube of Plenty: The Evolution of American Television*, 2nd ed. (New York: Oxford University Press, 1990). For an extensive bibliography of other works, see Walker and Ferguson, *The Broadcast Television Industry*, 215–223.

34. A point continuously explored and documented by Owen, *The Internet Challenge to Television*, 83–88, 114–118; Thomas W. Hazlett, "The Rationality of U.S. Regulation of the Broadcast Spectrum," *Journal of Law and Economics* 33 (1990): 133–175; Roger G. Noll, M. J. Peck, and J. J. McGowan, *Economic Aspects of Television Regulation* (Washington, D.C.: Brookings Institution, 1973); Kevin Werbach, *Digital Tornado: The Internet and Telecommunications Policy*, Working Paper 29 (Washington, D.C.: FCC Office of Plans and Policy, 1997).

35. Parsons and Frieden, *The Cable and Satellite Television Industry*, 204–254.

36. In mid-2003, there were 71.9 million cable customers in the United States out of a total of 106.4 million households with television. Cable wiring passed by 103.7 million homes, making it possible to increase the number of subscribers by over a third with only minimal expense on the part of cable providers. For data on cable usage, "Industry Statistics," National Cable & Telecommunications Association, December 18, 2003, http://www.ncta.com/industry_overview/indStats.cfm?indOverviewID=2 (last Accessed 12/18/2003).

37. Bruce M. Owen and Steven S. Wildman, *Video Economics* (Cambridge, Mass.: Harvard University Press, 1992); see also Michael J. Wolf, *The Entertainment Economy: How Mega-Media Forces Are Transforming Our Lives* (New York: Times Books, 1999).

38. Parsons and Frieden, *The Cable and Satellite Television Industry*, 74.

39. Ibid., 59.

40. Mirabito and Morgenstern, *The New Communications Technologies*, 69–85.

41. Almost all data for this paragraph come from *Office Automation Applications* (New York: Automation Consultants, undated [circa 1957–1959]) CBI 55, Market Reports, Box 70, Folder 2, Archives of the Charles Babbage Institute, University of Minnesota, Minneapolis.

42. "They're Watching You Watch TV," *Business Machines* (July 18, 1955): pp. 6–7.

43. "Status Report: Agency Automation," *Broadcasting* 65 (July 1, 1963): 32.

44. "The Electronic Stars of Television," *Business Machines* (October 7, 1955): 3–5.

45. "Transistors Bring a Revolution," Ibid., 66 (February 17, 1964): 60.

46. Harper, "Need Answers? Ask the Computer," Ibid., 66 (May 18, 1964): 40. Harper, CEO of an advertising agency (Interpublic Group of Companies) already used a computer to understand listeners, circulations, and market definitions.

47. "Are Computers Worth What They Cost?" Ibid., 70 (November 13, 1966): 42.

48. "Computers: A Special Report," Ibid., 70 (November 13, 1966): 42–44, 46, 48–49.

49. "Computer Schedules Television Programming," IBM press release, September 11, 1972, IBM Archives, Somers, N.Y.

50. "Automation Keynote of NAB Exhibits," *Broadcasting* 78 (April 13, 1970): 82–83; "Computer System Shown for Broadcast Stations," Ibid., 83.

51. "What's the Plot? Who's In It?" *Data Processor* (October 1973): 6–7, IBM Archives, Somers, N.Y.

52. Eugene V. Rostow, "A Survey of Telecommunications Technology, Part 2, Appendix I," U.S. Department of Commerce/National Bureau of Standards, June 1969, CBI 32, National Bureau of Standards Papers, Box 462, Folder 13, NBS No. 69232003. The same report has an extensive discussion of the Picturephone about which it was quite optimistic but that we now know failed to attract consumer demand. For a similar view of an advanced technological future for television, see Ward L. Quaal and James A. Brown, *Broadcast Management: Radio-Television*, 2nd ed. (New York: Hastings House, 1976, first ed. 1968): 298, 417–423.

53. Christopher H. Sterling and John M. Kittross, *Stay Tuned: A Concise History of American Broadcasting* (Belmont, Calif.: Wadsworth Publishing Co., 1978): 418.

54. For the complex story of spectrum issues in the 1950s to early 1980s see, Jeffrey S. Close, "Spectrum Utilization in Broadcasting," in Leonard Lewin (ed.), *Telecommunications in the U.S.: Trends and Policies* (Dedham, Mass.: Artech, 1981): 130–157. On cable television specifically, see George P. Bush (ed.), *Technology and Copyright: Annotated Bibliography and Source Materials* (Mt. Airy, Md.: Lomond Systems, 1972): 29–33; see also Robert Pepper, "Competition in Local Distribution: The Cable Television Industry," in Benjamin M. Compaine (ed.), *Understanding New Media: Trends and Issues in Electronic Distribution of Information* (Cambridge, Mass.: Ballinger Publishing, 1984): 147–194.

55. For the data and a brief summary of the issues during the early days of cable, see E. Stratford Smith, "The Emergence of CATV: A Look at the Evolution of a Revolution," Ibid., 344–351.

56. Susan E. McMaster, *The Telecommunications Industry* (Westport, Conn.: Greenwood Press, 2002): 153–175.

57. The FCC periodically publishes data on numbers of channels, while *TV Guide* provides listings of programs by channels. American newspapers have had the practice for decades of listing local broadcasts and those provided by cable and broadcast stations. The cable TV companies also publish catalogs of their programs as well.

58. Bruce C. Klopfenstein, "From Gadget to Necessity: The Diffusion of Remote Control Technology," in James R. Walker and Robert V. Bellamy, Jr. (eds.), *The Remote Control in the New Age of Television* (Westport, Conn.: Praeger, 1993): 33.

59. Walker and Bellamy, *The Remote Control in the New Age of Television.*

60. Frederick Wasser, *Veni, Vidi, Video: The Hollywood Empire and the VCR* (Austin, Tex.: University of Texas Press, 2001): 80.

61. Ibid., 101.

62. Sydney W. Head and Christopher H. Sterling, *Broadcasting in America: A Survey of Television, Radio, and New Technologies* 4th ed. (Boston, Mass.: Houghton Mifflin Co., 1982): 82.

63. Ibid., 82–87, for a description of how digital versions of the technology worked.

64. "IBM Entertainment Systems Introduces New Software Products for Film and Video Production Professionals," IBM press release, July 26, 1994, IBM Archives, Somers, N.Y.

65. "IBM, ICTV to Join in Interactive Home Video System," IBM press release, November 15, 1994; "Interactive CD-ROM from IBM and Saturday Morning Cartoon from CBS Produce Cool Science Adventures in 1995," IBM Press Release, February 14, 1995, IBM Corporate Archives, Somers, N.Y.

66. Richard D. Yoakam and Charles F. Cremer, *ENG: Television News and the Old Technology* (New York: Random House, 1985): 316.

67. Parsons and Frieden, *The Cable and Satellite Television Industries*, 83.

68. Ibid., 83.

69. On the standards debate of the 1990s, see Parsons and Frieden, *The Cable and Satellite Television Industries*, 85–86; for a contemporary account of storage in digital formats, Yoakam and Cremer, *ENG*, 318–319; and for a typical discussion of TV technology from the period, Sydney W. Head, *Broadcasting in America*, 119–150.

70. Howard J. Blumenthal and Oliver R. Goodenough, *This Business of Television* (New York: Billboard Books, 1998): 109.

71. Owen, *The Internet Challenge to Television*, 328.

72. Blumenthal and Goodenough, *This Business of Television*, 115.

73. Pavlik and McIntosh, *Converging Media*, 172–173.

74. For a history see John Carey, *Winky Dink to Stargazers: Five Decades of Interactive Television* (Dobbs Ferry, N.Y.: Greystone Communications, 1998); see also http://www.tvparty.com/requested2.html. At its height, WebTV had some 800,000 household subscribers.

75. A. M. Odlyzko, "Internet TV: Implications for the Long Distance Network," July 27, 2001, http://www.research.att.com/amo or at amo@research.att.com (last accessed 10/22/2002).

76. The best description of mass customization is still B. Joseph Pine II, *Mass Customization: The New Frontier in Business Competition* (Boston, Mass.: Harvard Business School Press, 1993). For a description of the On Demand concept, see Craig Fellenstein, *On Demand Computing: Technologies and Strategies* (Upper Saddle River, N.J.: Prentice-Hall PTR, 2005): 3–36.

77. One problem just beginning to cause concern in television was the pirated downloading of programs, much like the Music Industry had been facing since the late 1990s, and that the Movie Industry began to experience by the early 2000s. However, as of late 2004, there was no hard data on how extensive a problem this was for the Television Industry.

78. Pavlik and McIntosh, *Converging Media*, 171–172; "DTV Stations on the Air," November 19, 2003, list issued by FCC, http://www.fcc.gov/mb/video/files/dtvonair.html (last accessed 12/18/2003).

79. Heather E. Hudson, *Global Connections: International Telecommunications Infrastructure and Policy* (New York: Van Nostrand Reinhold, 1997): 31; on details of FCC's role and decisions, Advisory Committee on Public Interest Obligations of Digital Television Broadcasters, *Charting the Digital Broadcasting Future* (Washington, D.C.: U.S. Department of Commerce, December 18, 1998): 5–10.

80. Head et al., *Broadcasting in America*, 74, 191; see also for description of service, http://www.webtv.com/pc/whatis_default.aspx (last accessed 6/2/2004).

81. Described by Nancy K. Baym, *Tune In, Log On: Soaps, Fandom, and Online Community* (Thousand Oaks, Calif.: Sage Publications, 2000); see also her bibliography for other examples.

82. The study was conducted by Arbitron News Media Internet, reported in Pavlik and McIntosh, *Converging Media*, 203.

83. The U.S. Bureau of the Census has regularly monitored the population of users as has the Pew Foundation, both since the late 1990s. Each posts this kind of data to their home pages.

84. For details, see www.aentv.com On this website we see the following self description: "Formed in 1995, AENTV is the leading producer, aggregator, and syndicator of streaming media content on the Internet" (last accessed June 1, 2004).

85. Pavlik and McIntosh, *Converging Media*, 203–207.

86. Carl Shapiro and Hal R. Varian, *Information Rules: A Strategic Guide to the Network Economy* (Boston, Mass.: Harvard Business School Press, 1999), 1–2.

87. Ibid., 2.

88. Ibid., 210.

89. Chris Isidore, "Sony Finds TV World Is Flat," *CNNMoney*, October 23, 2003, http://

money.com/2003/10/21/news/international/sony_outlook/index.htm (last accessed October 23, 2003); "Want a Cheaper Flat-Panel TV?" Ibid., December 4, 2003, http://money.com/2003/12/04/technology/techguide_flatpanel.reut/indexhtm (last accessed 12/8/2003).

90. I found no direct evidence that the American consumer knew Moore's Law, although obviously a tiny percent would have if they were historians or economists with expertise on the industry. But one could track prices of high-tech devices and compare that against sales data to document the process. Moore's Law holds that the number of transistors per square inch on a computer chip (integrated circuit) double every eighteen months.

91. Home page for TiVo, http://www.tivo.com (last accessed 6/2/2004).

92. TiVo illustrates what the confluence of competition, consumer interest, and technology can do to the business model of a media firm. Competitors came into the same market, while demand for DVRs increased; there were 3.5 million users by the end of 2003, with industry experts predicting that number would double by the end of 2004. Cable providers were offering capabilities similar to TiVo, representing the fastest growing segment of the DVR market after TiVo had dominated it in 2003, leading one industry observer to question its future; "TiVo is in a fight for relevance and finds itself at a crossroads in its brief but high-profile history." Eric Hellweg, "TiVo at the Crossroads," *Business 2.0*, August 9, 2004, http://www.business2.com/b2/web/aticles (last accessed 8/12/2004).

93. Reed E. Hundt, *You Say You Want a Revolution: A Story of Information Age Politics* (New Haven, Conn.: Yale University Press, 2000): 226.

Chapter 11

1. I have discussed this idea of concurrent use of multiple technologies in *Making the Information Society: Experience, Consequences, and Possibilities* (Upper Saddle River, N.J.: Prentice-Hall PTR, 2002).

2. Douglas Gomery, *Shared Pleasures: A History of Movie Presentation in the United States* (Madison, Wis.: University of Wisconsin Press, 1992): ix–xiv, 294–299; Kristin Thompson and David Bordwell, *Film History: An Introduction* 2nd ed. (Boston, Mass.: McGraw-Hill, 2003): 680–681; Alvin Toffler cited movies to help him describe the evolving American culture, particularly in *Future Shock* (New York: Random House, 1970).

3. Recent examples include Barry R. Litman, *The Motion Picture Mega-Industry* (Boston, Mass.: Allyn and Bacon, 1998): 7–95; Tino Balio (ed.), *The American Film Industry* (Madison, Wis.: University of Wisconsin Press, 1985); Douglas Gomery, "The Hollywood Film Industry: Theatrical Exhibition, Pay TV, and Home Video," in Benjamin M. Compaine and Douglas Gomery (eds.), *Who Owns the Media? Competition and Concentration in the Mass Media Industry* 3rd ed. (Mahwah, N.J.: Lawrence Erlbaum Associates, 2000): 359–435, and which includes an extensive bibliography.

4. Standard overviews are David Bordwell, *On the History of Film Style* (Cambridge, Mass.: Harvard University Press, 1997); Kristin Thompson, *Storytelling in the New Hollywood: Understanding Classical Narrative Technique* (Cambridge, Mass.: Harvard University Press, 1999); David Bordwell and Kristin Thompson, *Film Art: An Introduction*, 6th ed. (New York: McGraw-Hill, 2001).

5. Harold L. Vogel, *Entertainment Industry Economics: A Guide for Financial Analysis*, 5th ed. (Cambridge: Cambridge University Press, 2001): 35–102; D. Lees and S. Berkowitz, *The Movie Business* (New York: Random House, 1981).

6. John V. Pavlik and Everette E. Dennis (eds.), *Demystifying Media Technology* (Mountain View, Calif.: Mayfield Publishing Co., 1993); Thomas A. Ohanian, *Digital Nonlinear Editing: New Approaches to Editing Film and Video* (Boston, Mass.: Focal Press, 1993): 1–6, 307–316; Mark Cotta Vaz and Patricia Rose Duignan, *Industrial Light & Magic: Into the Digital Realm* (New York: Ballantine Books, 1996); Matt Hanson, *The End of Celluloid: Film Futures in the Digital Age* (Mies, Switzerland: RotoVision SA, 2003).

7. Thompson and Bordwell, *Film History*, contains rich lists of bibliographies throughout that touch on this theme.

8. Frederick Wasser, *Veni, Vidi, Video: The Hollywood Empire and the VCR* (Austin, Tex.: University of Texas Press, 2001): 77–103, 200–202; Walter Oleksy, *Entertainment: The Information Revolution* (New York: Facts on File, 1996): 73–78; Kristin Thompson, "Fantasy, Franchises, and Frodo Baggins: *The Lord of the Rings* and Modern Hollywood," *The Velvet Light Trap* 52 (Fall 2003): 45–63.

9. Gomery, "The Hollywood Film Industry," 360.

10. The exception, and it is only in relation to the better performance of the industry in the 1950s and from the 1970s forward, was the 1960s, but even then, business was good as noted later on in this chapter.

11. Thompson and Bordwell, *Film History*, 525, 687, 702.

12. U.S. Bureau of the Census, *Historical Statistics of the United States: Colonial Times to 1970* (Washington, D.C.: United States Government Printing Office, 1975), vol. 2, p. 855.

13. Ibid., vol. 1, p. 401.

14. U.S. Census Bureau.

15. Litman, *The Motion Picture Mega-Industry*, 58.

16. Gomery, "The Hollywood Film Industry," 364.

17. Thompson and Bordwell, *Film History*, 680–716; and more fully explained in Thompson, *Storytelling in the New Hollywood*, 1–49, 335–352.

18. P. H. Longstaff, Raja Velu, and Jonathan Obar, "Resilience for Industries in Unpredictable Environments: The Movie Industry," unpublished paper, March 2004 (Cambridge, Mass.: Center for Information Policy Research, Harvard University); Arthur De Vany and W. David Walls, "Bose-Einstein Dynamics and Adaptive Contracting in the Motion Picture Industry," *The Economic Journal* 106, no. 439 (November 1996): 1493–1514; see also Mark Litwak, *Reel Power: The Struggle for Influence and Success in the New Hollywood* (New York: William Morrow & Co., 1986). I have been most influenced about how networks function by Duncan Watts, *Six Degrees of Separation: The Science of the Connected Age* (New York: W. W. Norton, 2003); and Steven Strogatz, *Sync: The Emerging Science of Spontaneous Order* (New York: Hyperion, 2003).

19. For a chronological history of the influence of movies, see Keith Reader, *Culture in Celluloid* (London: Quartet Books, 1981): 11–107; but also Martha Wolfenstein and Nathan Leites, *Movies: A Psychological Study* (New York: Atheneum, 1970). On the influence of television, see the collection of essays in Jennings Bryant and Dolf Zillmann (eds.), *Perspectives on Media Effects* (Hillsdale, N.J.: Lawrence Erlbaum Associates, 1986). I have examined the wider issue of how information technologies and other media were used in the United States in *Making the Information Society: Experience, Consequences, and Possibilities* (Upper Saddle River, N.J.: Prentice-Hall PTR, 2002): 136–248.

20. While there is no full-length study of the role of computers in the plots and scenes of American movies, there is a useful website listing movies by theme: "Hollywood and Computers," http://www.cabi.umn.edu/resources/hollywood.html (last accessed 6/16/2004).

21. "Welcome to Cybercinema," http://www2.english.uiuc.edu/cybercinema/,ain.htm (last accessed 5/18/2004). For a review of this site, see Hunter Crowther-Heyck, "Artificial Intelligence in Film: A Review of Cybercinema," *Iterations: An Interdisciplinary Journal of Software History* 3 (May 10, 2004): 1–2.

22. Mark Cotta Vaz and Patricia Rose Duignan, *Into the Digital Realm: Industrial Light & Magic* (New York: Ballantine Books, 1996): 286–294.

23. There were other sources of innovations in the 1950s and 1960s that eventually made it into films. Most notably, some scientists attempted to write animation using programming languages, such as Fortran and later Beflix, which contained instructions such as PAINT, ZOOM, and DISOLV to help scientists create primitive images. For details, see *The Focal Encyclopedia of Film & Television Techniques* (London: Focal Press, 1969): 206–207.

24. C. M. Theiss, "A Computer Aided Movie-Producing Program for the Simulation Program of Roadside Energy Conversion Systems" (Buffalo, N.Y.: Cornell Aeronautical Laboratory, Inc., of Cornell University, 1969), CBI 32, NBS Computer Literature Collection, Box 420, folder 12, Archives of the Charles Babbage Institute, University of Minnesota, Minneapolis.

25. Peggy Anne Talbot, "Animator: A System for Using the DEC-338 as an Input Terminal for Movie Making" (unpublished M.S. thesis, Moore School of Electrical Engineering, University of Pennsylvania, 1969), copy in CBI 32, NBS Computer Literature Collection, Box 460, folder 10, Archives of the Charles Babbage Institute, University of Minnesota, Minneapolis.

26. "System/7 Also Stars at Academy Awards," *Think Magazine* (April 1973): 9; and Jack M. Goetz, "Consolidated Film Industries and IBM Corporation Share Academy Award for Technical Achievement," press release, Consolidated Film Industries, March 27, 1973, both in IBM Archives, Somers, N.Y.

27. Patricia D. Netzley, *Encyclopedia of Movie Special Effects* (New York: Checkmark Books, 2001): 155; and for an early description of its use in *Star Wars*, Christopher Finch, *Special Effects: Creating, Movie Magic* (New York: Abbeville Press, 1989): 144–148.

28. Christopher W. Baker, *How Did They Do It? Computer Illusion in Film & TV* (Indianapolis, Ind.: Alpha Books, 1994): 7.

29. Ibid., 8.

30. Ibid., 162–166; for details on what was added to the DVD version, see Michael Rankins, "Snow White and the Seven Dwarfs," Case Number 1958, *DVD Verdict*, June 13, 2002, http://www.dvdverdict.com/reviews/snowwhite.php (last accessed 6/21/2004).

31. Baker, *How Did They Do It?* 166.

32. Ohanian, *Digital Nonlinear Editing*, 7–24.

33. Ibid., 25–29. Experimentation with video editing began in the mid-1970s, indeed further back in time than is normally acknowledged, providing additional evidence of how long it takes some computer-based technologies to develop before reaching a practical stage; in this case 20 years.

34. Ibid., 307.

35. Thompson and Bordwell, *Film History*, 522–524.

36. Ibid., 701–703.

37. Screen Digest, *Screen Digest Report on the Implications of Digital Technology for the Film Industry* (London: Screen Digest, September 2002): 13–15.

38. Ibid., 24.

39. Ibid., 24.

40. All quotes from IBM press release, "Film Placement by Computer," March 2, 1977, IBM Archives, Somers, N.Y.

41. Harrison Kinney, "Boffo: That's Hollywood for Big at the Box Office. And Now It's the Computer That's Boffo in Movieland," *Think Magazine* (May–June 1977): 6.

42. Ibid.

43. Ibid., 7.

44. Ibid.

45. For an example, see "Computer Provides 'Extra' Service for Universal Studios," IBM Press Release, January 23, 1981, IBM Archives, Somers, N.Y.

46. The Pew Research Center For The People & The Press, "Online News Audience Larger, More Diverse," news release, June 8, 2004, p. 4, http://www.people-press.org/ (last accessed 6/9/2004).

47. Data developed by Jessica Reif Cohen at Merrill Lynch, and reported in Dave McNary, "Writer-Producer Talks Could Spin Out Over DVDs," *Variety.com*, December 23, 2003, http://www.variety.com/index (last accessed 12/24/2003).

48. Screen Digest, *Screen Digest Report on the Implications of Digital Technology for the Film Industry*, 49–60.

49. See, for example, for *The Matrix* (1999–2003) series of movies, http://www.whatisthematrix.com and for *Memento* (2000), http://www.otnemem.com.

50. Screen Digest, *Screen Digest Report on the Implications of Digital Technology for the Film Industry*, 11.

51. Ibid., 73–75. The negative press coverage is ironic since all publishing industries faced similar problems, although to a far lesser degree. One can suppose that the opportunity to discuss a big industry picking on children and otherwise near helpless individuals was too tempting for even a fellow media industry to resist.

52. Ralf Ludemann, "Film Piracy Business Set to Increase," press release, July 12, 2004, http://www.screandaily.com (last accessed 7/12/2004).

53. Ibid., for example, 75.

54. Ibid., 78.

55. Quoted in "The Sorry State of Digital Hollywood," *Red Herring*, November 13, 2000.

56. Material for the previous two paragraphs and the subsequent one drawn from Kristin Thompson, "Fantasy, Franchises, and Frodo Baggins: *The Lord of the Rings* and Modern Hollywood," *The Velvet Light Trap* 52 (Fall 2003): 45–63.

57. John M. Moran, "Film Industry Caught in Speed Trap," *Wisconsin State Journal*, June 26, 2004, p. D–5; 840 gigabytes of data was transferred over the Internet in 27 minutes, making it possible to download a movie in just a few seconds.

58. For details on volumes of illegal DVDs and the cost implications for 2003–2004, see Leon Forde, "VoD Roll Out Threatened by DVD Boom," press release, July 12, 2004, http://www.screandaily.com (last accessed 7/12/2004).

59. This is the whole "creative destruction" issue; however, for a brief introduction and examples of how industries are facing and dealing with the problem, see two books by Clayton M. Christensen, *The Innovator's Dilemma: When New Technologies Cause Great Firms to Fail* (Boston, Mass.: Harvard University Press, 1997); and with Michael E. Raynor, *The Innovator's Solution: Creating and Sustaining Successful Growth* (Boston, Mass.: Harvard Business School Press, 2003).

60. Particularly the reliance on fantasy tales, described in a history of the games business, Brad King and John Borland, *Dungeons and Dreamers: The Rise of Computer Game Culture from Geek to Chic* (New York: McGraw-Hill/Osborne, 2003): 11–84.

61. Thompson, "Fantasy, Franchises, and Frodo Baggins," 53–58, quote, p. 57.

62. Ibid., 58.

63. Ibid., 59.

64. "Digital Cinema: Year 5 The Future Is Still on Hold," *Screen Digest* (May 2004): 138–140.

65. Douglas Gomery, "Radio Broadcasting and the Music Industry," in Compaine and Gomery, *Who Owns the Media?* 285–358; Robert Burnett, *The Global Jukebox: The International Music Industry* (London: Rutledge, 1996); Dick Weissman, *The Music Business: Career Opportunities and Self-Defense*, 2nd rev. ed. (New York: Three Rivers Press, 1997); Vogel, *Entertainment Industry Economics*, 148–172; M. William Krasilovsky and Sydney Shemel, *The Business of Music: The Definitive Guide to the Music Industry*, 9th ed. (New York: Billboard Books, 2003); Geoffrey P. Hull, *The Recording Industry* (Boston, Mass.: Allyn and Bacon, 1998).

66. U.S. Department of Labor, Bureau of Labor Statistics, *Monthly Labor Review* (November 1995); see also U.S. Department of Commerce, Bureau of the Census, annual reports in *Statistical Abstract of the United States*; Hull, *The Recording Industry*, 4.

67. Hull, *The Recording Industry*, 19–24.

68. Ibid., 29.

69. Ibid., 28–35; for the financial and economic elements driving industry behavior, see Robert G. Picard, *The Economics and Financing of Media Companies* (New York: Fordham University Press, 2002); Geoffrey P. Hull, "The Structure of the Recorded Music Industry," in Albert N. Greco (ed.), *The Media and Entertainment Industries* (Boston, Mass.: Allyn and Bacon, 2000): 76–98.

70. For a description of their work see Hull, *The Recording Industry*, 38–40.

71. Ibid., 47.

72. For example, in Krasilovsky and Shemel, *This Business of Music*, devoted over 400 pages out of 530 in the 9th edition (2003) to the subject. Hull, *The Recording Industry*, 201–255; hardly a month goes by without the industry's leading magazine, *Billboard*, reporting on the matter.

73. Hull, *The Recording Industry*, 96–106; Paul Hirsch, *The Structure of the Popular Music Industry* (Ann Arbor, Mich.: University of Michigan Institute for Social Research, 1969).

74. Recording Industry Association of America (RIAA), *1996 Statistical Survey*.

75. Krasilovsky and Shemel, *This Business of Music*, 4.

76. Ibid., 5; but see also RIAA for continuous supply of new data and comments.

77. Krasilovsky and Shemel, *This Business of Music*, 5– 6.

78. Ibid., 6.

79. Any of the industry surveys discuss the legal environment of the 1990s, but the most extensive coverage is by Krasilovsky and Shemel across many chapters in *The Business of Music*.

80. Without knowledge of the industry, computer scientists and engineers used computers to make musiclike sounds. In the construction of the UNIVAC II in the early 1950s, engineer David E. Lundstrom and colleagues attached a small amplifier and loudspeaker to the system to help identify technical problems, naming these added-on devices the Music Maker, David E. Lundstrom, *A Few Good Men from Univac* (Cambridge, Mass.: MIT Press, 1987): 34–35. Meanwhile, at the same time a programmer at the U.S. Air Force had written programs that played tunes and yet another programmer at another agency wrote "Anchors Away," Virginia C. Walker, "Meetings in Retrospect: Washington Computer Reminiscences," *Annals of the History of Computing* 1, no. 1 (July 1979): 65–66; many people who used the UNIVAC I computers often added games and music; and one computer operator at the Census Bureau's office in Philadelphia had composed music using a UNIVAC.

81. Jeff Goodell, "Steve Jobs: The *RollingStone* Interview," *RollingStone*, December 3, 2003, http://www.rollingstone.com/features/featuregen.asp?pid=2529 (last accessed 12/25/2003).

82. Marco Iansiti, "Managing Chaos," in David B. Yoffie (ed.), *Competing in the Age of Digital Convergence* (Boston, Mass.: Harvard Business School Press, 1997): 413–444; John V. Pavlik, *New Media Technology: Cultural and Commercial Perspectives* (Boston, Mass.: Allyn and Bacon, 1998): 133–164; Thompson and Bordwell, *Film History*, 720–722; John V. Pavlik and Shawn McIntosh, *Converging Media: An Introduction to Mass Communication* (Boston, Mass.: Pearson, 2004).

83. A synthesizer is an electronic musical instrument designed to artificially produce sound, using various techniques such as modeling synthesis or phase modulation, creating sounds by direct manipulation of electrical currents. A sampler is a variation of a synthesizer that takes an existing sound, such as a digital sound file, and replays samples of it in various ranges of pitches, often using software to alter the samples.

84. Hull, *The Recording Industry*, 17–18; for extensive coverage of various devices and technologies see, "Synthmuseum.com" at http://www.synthmuseum.com (last accessed 7/5/2004); "The History of Electronic Music," at http://www.phinnweb.com/history/ (last accessed 7/5/2004); Kristine H. Burns, "History of Electronic and Computer Music Includ-

ing Automatic Instruments and Composition Machines," http://music.dartmouth.edu/ADDUNLASYMBOLwowem/electronmedia/music/eamhistory.html (last accessed 7/5/2004).

85. Raymond Kurzweil has commented extensively on digital technologies. For the most useful source that includes discussion of the area of his machine development, see *The Age of Intelligent Machines* (Cambridge, Mass.: MIT Press, 1992); see also his company's website for details, http://www.kurzweilmusicsystems.com/about.html?Id=20 (last accessed 7/5/2004).

86. For modern technologies, Thomas E. Rudoph and Vicent A. Leonard, Jr., *Recording in the Digital World: Complete Guide to Studio Gear and Software* (Boston, Mass.: Berklee Press, 2001); but for historical perspective, Ben Kettlewell, *Electronic Music Pioneers* (Vallejo, Calif.: ArtistPro, 2001), and Iara Lee and Peter Shapiro (eds.), *Modulations: A History of Electronic Music* (London: Distributed Art Publishers, 2000).

87. For lists and discussions of such groups, begin with "The History of Electronic Music," http://www/phinnweb.com/history (last accessed 7/5/2004).

88. "Office Automation Applications Updating Supplement No. 8," Publications Department, Automation Consultants, Inc., May 1958, H4; "Music Logging Automated by BMI," CBI 55; "Market Reports," Box 70, folder 4, Archives of the Charles Babbage Institute, University of Minnesota, Minneapolis.

89. "Computer Helps Reap Royalties for Song Writers, Publishers," press release, February 1, 1978, IBM Corporation, Archives IBM Corporation, Somers, N.Y.

90. "Punched Card System Makes Good Music," Office Automation Applications Updating Service, July 1961, CBI "Marketing Report," Box 70, folder 7, Archives of the Charles Babbage Institute, University of Minnesota, Minneapolis.

91. "Computer Helps Reap Royalties for Song Writers, Publishers."

92. Ibid.

93. Hull, *The Recording Industry*, 142–143; Ed Christman, "EMI-Capitol Creates Marketing Arm," *Billboard*, April 13, 1996, p. 5; Ed Christman, "WEA Reduces Wholesale Prices on CDs," Ibid., March 13, 1993, p. 9.

94. Hull, *The Recording Industry*, 146–147.

95. Ibid., 147.

96. Ibid., 170–172.

97. Ibid.,172–173.

98. "Soundata Consumer Panel," *NARM Sounding Board*, March 1996, http://www.narm.com/publications/sb96/0396/035.htm (last accessed 7/7/2004).

99. Others have made the same point. For example, David Moschella wrote, "To listen to the brouhaha over Napster and subsequent forms of online music, one would think that the music industry has never had to accommodate a new technology before. . . . Yet somehow society managed to find a way to make these new technologies acceptable to artists and the business interests of the times," *Customer-Driven IT: How Users Are Shaping Technology Industry Growth* (Boston, Mass.: Harvard Business School Press, 2003): 75.

100. Krasilovsky and Shemel, *This Business of Music*, 399–400.

101. Ibid., 401.

102. Joseph Menn, *All the Rave: The Rise and Fall of Shawn Fanning's Napster* (New York: Crown Business, 2003): 223–307.

103. Krasilovsky and Shemel, *This Business of Music*, 402–403.

104. Steven M. Zeitchik, "The Digits On the Wall," *Publishers Weekly* 246, no. 34 (August 23, 1999): 25.

105. Andrew Ross Sorkin, "Zapping the Music Pirates," *International Herald Tribune*, May 5, 2003, p. 1.

106. *Time*, May 5, 2003, unpaginated.

107. Mary Madden and Amanda Lenhart, "Pew Internet Project Data Memo," July 2003, http://www.pewinternet.org (last accessed 7/8/04).

108. Mary Madden and Amanda Lenhart, "Pew Internet Project Data Memo," July 2003, http://www.pewinternet.org (last accessed 5/27/2003); also "Record Industry to Sue Downloaders," July 25, 2003, http://cnnmoney.printthis.clickability.com/pt/cpt?action-cpt&expire=&urlID=6726521&fb (last accessed 6/25/2003).

109. Ted Bridis, "Music-Sharing Subpoenas Target Parents," http://www.washingtonpost.com/wp-dyn/articles/A38406-2000July24.html (last accessed 7/24/2003); also Tim Ruzek, "Subpoenas Chill Music Downloads," *Wisconsin State Journal*, August 8, 2003, p. A-1, A-9.

110. "Music Subpoenas Illegal," *CNNMoney*, December 19, 2003, http://money.cnn.com/2003/12/19/technology/riaa_downloads.reut/index.htm (last accessed 12/19/2003).

111. Dawn C. Chmielewski, "Music Suits Aim at Schools," *Mercury News*, March 24, 2004, pp. 1C, 9C; "493 More U.S. Music Swappers Sued," *CNNMoney*, May 24, 2004, http://cnnmoney.printthis.clickability.com/pt/cpt?action=cpt&title=Record+industry+sues (last accessed 5/24/2004).

112. Recording Industry Association of America, "Anti-Piracy," undated, http://www.riaa.com/issues/piracy/default.asp (last accessed 12/14/2003).

113. "Downloads Did Not Cause The Music Slump, But They Can Cure It, Reports Forrester Research," press release, August 13, 2002, http://www.forrester.com/ER/Press/Release/0,1769,741,FF.html (last accessed 7/2/2004); for more details behind this study, Dan Bricklin, "The Recording Industry Is Trying to Kill the Goose That Lays the Golden Egg," undated [2004], http://www.bricklin.com/recordsales.htm (last accessed 7/2/2004).

114. Two European economists, Martin Peitz and Patrick Waelbroek, reported that downloads had hurt the industry in 2001, but only for less than 25 percent of the decline in CDs in the United States in 2002. Martin Peitz and Patrick Waelbroeck, "The Effect of Internet Piracy on CD Sales: Cross-Section Evidence," CESIifo Working Paper No. 1122, January 2004, http://www.CESifo.de (last accessed 7/3/2004).

115. Felix Oberholzer and Koleman Strumpf, "The Effect of File Sharing on Record Sales An Empirical Analysis," unpublished paper, March 2004, 3–4, http://www.unc.edu (last accessed 3/11/2005). This paper includes an account of prior economic studies of the problem by others and an excellent bibliography.

116. Ibid.

117. Ibid., 24.

118. "Did Big Music Really Sink the Pirates?" *BusinessWeek Online*, January 16, 2004, http://www.businessweek.com/technology/content/jan2004/tc20040116_9177_tc024.htm (last accessed 7/13/2004); "Crackdowns Don't Slow Internet Piracy," *CNNMoney*, July 13, 2004, http://money.cnn.com/2004/07/13/technology/internet_piracy.reut/index.htm (last accessed 7/13/2004).

119. Ron Harris, "Digital Music at Crossroads," *Palo Alto Daily News*, May 17, 2003, p. 14.

120. Jefferson Graham, "Net Music Services Finally Giving Users Something to Sing About," *USA Today*, February 26, 2003, http://www.usatoday.com/tech/webguide/internetlife/notablesites/2003-02-26-music-service (last accessed 2/27/2003).

121. Goodell, "Steve Jobs: The *Rolling Stone* Interview."

122. Ibid.

123. Ibid. The American press universally hailed the arrival of his offerings. For examples, see May Wong, "Apple's Online Music Store Is Quick Success," AP story, May 10, 2003, *Wisconsin State Journal*, p. C-8; Joseph P. Kahn, "The Dawn of Pocket Music Players Has Sparked an iRevolution," Ibid., April 12, 2004, pp. B-1, B-5; "iTunes for Windows Launched," *CNNMoney*, October 16, 2003, http://cnnmoney.printthis.clickability.com/pt/cpt?action=cpt&expire=&urlID=20455&fb= (last accessed 10/16/2003); "Apple May Soon Sell Lower-cost iPods," December 24, 2003, Ibid. (last accessed 12/24/2003).

124. Evidence began mounting in 2004, however, that the major media firms were quietly

tracking what was being downloaded not to sue people but as a source of marketing data on what music was popular and thus should be further promoted. For one of the first reliable accounts of this development, see Dawn C. Chmielewski, "Music Labels Use File-Sharing Data to Boost Sales," *The Mercury News*, March 31, 2004, http://www.mercurynews.com/mld/mercurynews/news/8318571.htm?1c (last accessed 7/13/2004).

125. Darryl C. Sterling, Brian P. Irwin, and Gary Rylander, *Uncharted Territory Ahead for Media and Entertainment Industry* (Somers, N.Y.: IBM Corporation, 2002): 3.

Chapter 12

1. Michael J. Wolf, *The Entertainment Economy: How Mega-Media Forces Are Transforming Our Lives* (New York: Times Books, 1999): xxi.

2. Melissa A. Schilling, "Technological Leapfrogging: Lessons from the U.S. Video Game Console Industry," *California Management Review* 45, no. 3 (Spring 2003): 6–32; Harold L. Vogel, *Entertainment Industry Economics: A Guide for Financial Analysis* (Cambridge: Cambridge University Press, 2001): 251–264; C. L. Hays, "The Road to Toyland Is Paved with Chips," *New York Times*, February 17, 2000. Even an industry association used various names. The Entertainment Software Association (ESA) used alternative names for the industry it served, such as "interactive entertainment industry," http://www.theesa.com/programs_main.html (last accessed 7/10/2004).

3. These same issues were evident in the formation of the Office Appliance Industry, a precursor to the Computer Industry. For details, see James W. Cortada, *Before the Computer: IBM, NCR, Burroughs & Remington Rand and the Industry They Created, 1865–1956* (Princeton, N.J.: Princeton University Press, 1993): 275–280.

4. Ernest Adams, *Break into the Game Industry: How to Get a Job Making Video Games* (New York: McGraw-Hill/Osborne, 2003): xvii.

5. Interactive Digital Software Association, *Economic Impacts of the Demand for Playing Interactive Entertainment Software* (Washington, D.C.: IDSA, 2001): 3; "Adults at Play," *Wisconsin State Journal*, February 20, 2005, p. G10.

6. In addition to Schilling, "Technological Leapfrogging," see Martin Campbell-Kelly, *From Airline Reservations to Sonic the Hedgehog: A History of the Software Industry* (Cambridge, Mass.: MIT Press, 2003): 269–288; J. C. Herz, *Joystick Nation: How Videogames Ate Our Quarters, Won Our Hearts, and Rewired Our Minds* (Boston, Mass.: Little, Brown and Company, 1997); David Sheff and Andy Eddy, *Game Over: Press Start to Continue* (New York: GamePress, 1999); Steven L. Kent, *The Ultimate History of Video Games: From Pong to Pokemon—The Story Behind the Craze That Touched Our Lives and Changed the World* (New York: Three Rivers Press, 2001); Arthur Asa Berger, *Video Games: A Popular Phenomenon* (New Brunswick, N.J.: Transaction Publishers, 2002); Van Burnham, *Supercode: A Visual History of the Videogame Age, 1971–1984* (Cambridge, Mass.: MIT Press, 2001); Mark J. P. Wolf, *Medium of the Video Game* (Austin, Tex.: University of Texas Press, 2001); Rusel DeMaria and Johnny L. Wilson, *High Score! The Illustrated History of Electronic Games*, 2nd ed. (New York: McGraw-Hill/Osborne, 2004); Leonard Herman, *Phoenix: The Rise and Fall of Videogames*, 2nd ed. (Union, N.J.: Rolenta Press, 1999).

7. For an account of many very early computer games, see Donald D. Spencer, *Game Playing with Computers* (Rochelle Park, N.J.: Hayden Book Company, 1968). The subject drew much attention in the 1950s and 1960s; perhaps the earliest publication on the subject is G.A.W. Boehm, *The New World of Math* (New York: Dial Press, 1959), but for an early overview there is J. O. Harrison, *Computer-Aided Information Systems for Gaming* (Washington, D.C.: U.S. Department of Commerce, September 1964).

8. On early chess playing, G. W. Baylor and Herbert A. Simon, "A Chess Mating Combinations Program," *Proceedings—Spring Joint Computer Conference, 1966* (Washington, D.C.: Spartan Books, 1966): 431–447; Claude E. Shannon, "A Chess-Playing Machine,"

Scientific America 182 (February 1950): 48–51; Shannon, *Programming a Computer for Playing Chess* (Murray Hill, N.J.: Bell Telephone Laboratories, October 8, 1948); David Levy, *1975 U.S. Computer Chess Championship* (Woodland Hills, Calif.: Computer Science Press, 1976); for an early brief history, Brad Leithauser, "The Space of One Breadth," *New Yorker* 63 (March 9, 1987): 41–73; "Runner-Up: Chess Playing Computer, 704," Ibid., 34 (November 29, 1958): 43–44.

9. Monty Newborn, *Kasparov Versus Deep Blue: Computer Chess Comes of Age* (New York: Springer-Verlag, 1997); Feng-Hsiung Hsu, *Behind Deep Blue: Building the Computer That Defeated the World Chess Champion* (Princeton, N.J.: Princeton University Press, 2002).

10. At the core of this issue is the work done on artificial intelligence (AI), of which there is a massive body of literature. I have been most influenced by Howard Gardner's thoughtful study, *The Mind's New Science: A History of the Cognitive Revolution* (New York: Basic Books, 1985); and even more so by Pamela McCorduck, *Machines Who Think* (San Francisco, Calif.: W. H. Freeman, 1979) for the early years; and for a more recent account, Rodney Allen Brooks, *Cambrian Intelligence: The Early History of the New AI* (Cambridge, Mass.: MIT Press, 1999). For bibliography on the early history of AI, see James W. Cortada, *A Bibliographic Guide to the History of Computing, Computers, and the Information Processing Industry* (Westport, Conn.: Greenwood Press, 1990): 437–446.

11. The emphasis of most histories has been on the games themselves, their development and use, and less about the emerging industry that spawned them. The two most frequently cited histories are emblematic of the more widely embraced approach, DeMaria and Wilson, *High Score*; and Kent, *The Ultimate History of Video Games*, both of which are well researched and detailed.

12. DeMaria and Wilson, *High Score*, 304–319.

13. Adams, *Break Into the Game Industry*, 8.

14. Philip E. Orbanes, *The Game Makers: The Story of Parker Brothers from Tiddledy Winks to Trivial Pursuit* (Boston, Mass.: Harvard Business School Press, 2004): 179. The story of this firm in the prior and subsequent paragraphs is drawn from this source, pp. 171–189.

15. Ibid., 199.

16. For that sense of young computer "nerds" and other gamers launching the process, see Brad King and John Borland, *Dungeons and Dreamers: The Rise of Computer Game Culture from Geek to Chic* (New York: McGraw-Hill/Osborne, 2003): 1–84.

17. Vogel, *Entertainment Industry Economics*, 259.

18. For the early history of the firm, see S. Cohen, *Zap: The Rise and Fall of Atari* (New York: McGraw-Hill, 1984); for the later period, see Herz, *Joystick Nation*, 14, 20, 35, 39.

19. On Nintendo, see David Sheff, *Game Over: How Nintendo Zapped an American Industry, Captured Your Dollars and Enslaved Your Children* (New York: Random House, 1993); Kent, *The Ultimate History of Video Games*, 281–285, 371–377, *passim*; and on Sega, Ibid., 447–451, 577–579.

20. Schilling, "Technological Leapfrogging," 6–30.

21. Campbell-Kelly, *From Airline Reservations to Sonic the Hedgehog*, 280–283.

22. Ibid., 283; Adams, *Break into the Game Industry*, 49–75; Alice LaPlante and Rich Seidner, *Playing for Profit: How Digital Entertainment Is Making Big Business Out of Child's Play* (New York: John Wiley & Sons, 1999).

23. Michael S. Malone, *Infinite Loop: How the World's Most Insanely Great Computer Company Went Insane* (New York: Doubleday, 1999): 23, 42–43, 47–49, 213, 227–228; Owen W. Linzmayer, *Apple Confidential: The Real Story of Apple Computer, Inc.* (San Francisco, Calif.: No Starch Press, 1999): 17–20, 64.

24. Schilling, "Technological Leapfrogging," 14–15.

25. Andrew Hargadon, *How Breakthroughs Happen: The Surprising Truth About How Companies Innovate* (Boston, Mass.: Harvard Business School Press, 2003): 193–196.

26. Ibid., 17.

27. James W. Cortada, *The Digital Hand: How Computers Changed the Work of American Manufacturing, Transportation, and Retail Industries* (New York: Oxford University Press, 2004): 216–224; Campbell-Kelly, *From Airline Reservations to Sonic the Hedgehog,* 89–162.

28. Campbell-Kelly, *From Airline Reservations to Sonic the Hedgehog*, 280; see also Rama Dev Jager and Rafael Ortiz, *In the Company of Giants: Candid Conversations with Visionaries of the Digital World* (New York: McGraw-Hill, 1997): 186.

29. Campbell-Kelly, *From Airline Reservations to Sonic the Hedgehog*, 285.

30. Adams, *Break Into the Game Industry*, 23–48; Schilling, "Technological Leapfrogging," 17–28.

31. Adams, *Break into the Game Industry*, 27.

32. "Face-Off: Battle of the Game Books," *Folio*, November 1, 2002, http://foliomag.com/magazinearticle.asp?magazinearticleid=160145&magazineid=125&sit (last accessed 2/21/2004); and for magazines in the gaming market, Adams, *Break into the Game Industry*, 48.

33. Diane Roback, "CD-ROM Games: Worth the Gamble?" *Publishers Weekly* 242, no. 45 (November 16, 1995): 41–43.

34. "Is It Live or Is It Silicon?" *CNNMoney*, November 19, 2002, http://money.cnn.com/2002/11/18/commentary/game_over/column_gaming/index.htm (last accessed 11/20/2002); "Sony Boosts Online Gaming," *CNNMoney*, February 27, 2003, http://money.com/2003/02/27/technology/sony_games.reut/index.htm (last accessed 2/27/2003).

35. Geoff Keighley, "Is Nintendo Playing the Wrong Game?" *Business 2.0* (August 2003), http://www.business2.com/subscribers/articles/mag/print/0,1643,50984,00.html (last accessed 8/4/2003).

36. For a brief description of the industry's antipiracy initiatives, see "Conducting a Worldwide Anti-Piracy Program," statement by the Entertainment Software Association, http://www.theesa.com/programs_main.html (last accessed 7/10/2004).

37. Quote and material for the paragraph, "Consortium to Develop Break-Through Solution for Digital Entertainment," IBM press release, July 14, 2004.

38. "Computer & Video Games Guide," Amazon.com, http://www.amazon.com/exec/obidos/tg/browse/-/748500/102-4951863-4033751 (last accessed 12/5/2003). This is an excellent source of information on all kinds of gaming products and technologies, quite thorough and clear.

39. "Sony Takes on GameBoy," *CNNMoney*, May 13, 2003, http://money.com/news/specials/e3/ (last accessed 5/13/2003).

40. King and Borland, *Dungeons and Dreamers*. This account of the origins of the industry is replete with such commentary,

41. Campbell-Kelly, *From Airline Reservations to Sonic the Hedgehog*, 269, 272.

42. Ibid., 276.

43. Vogel, *Entertainment Industry Economics*, 259. The digital hand appeared in various segments of the Toy Industry, for example with the popular Furby animal, but also in other toys. For an early account of this process, see Mark Pesce, *The Playful World: How Technology Is Transforming Our Imagination* (New York: Ballantine Books, 2000), especially 19–25, and about chess and computers, 43–53.

44. Schilling, "Technological Leapfrogging," 9–11.

45. King and Borland, *Dungeons and Dreamers*, 11–84; Kent, *The Ultimate History of Video Games*, 15–26.

46. Kent, *The Ultimate History of Video Games*, commented throughout his book about the broadening demographics; see also Herz, *Joystick Nation*, 171–182.

47. Data from last two paragraphs drawn from DFC Intelligence reports (February 2002), IDC; and HorizonWatch, "HorizonWatch On-Line Gaming," May 2003.

48. As in other media, there has been much discussion of gender issues in gaming. Useful

in understanding the issues are Jo Bryce and Jason Rutter, "Gender Dynamics and the Social and Spacial Organization of Computer Gaming," *Leisure Studies* 22 (June 2003): 1–15; and their essay, "The Gendering of Computer Gaming: Experience and Space," in S. Fleming and I. Jones (eds.), *Leisure Cultures: Investigations in Sport, Media and Technology* (Eastbourne, Eng.: Leisure Studies Association, 2003): 3–22.

49. For example, by Electronics Conservancy, Inc., at http://www.videotopia.com (last accessed 7/2/2004).

50. HorizonWatch, "HorizonWatch On-Line Gaming," May 2003.

51. Steve Jones, "Let the Games Begin: Gaming Technology and Entertainment Among College Students," Pew and Internet American Life Project, July 6, 2003, http://www.pewinternet.org (last accessed 6/7/2003).

52. "Video Game Sales Up 7%," *CNNMoney*, January 21, 2003, http://www.money.cnn.com/2003/01/21/technology/game_sales.reut/index.htm (last accessed 1/21/2003).

53. Adams, *Break into the Game Industry*, 43.

54. For a fascinating look at the issues and the effects of video games on learning, for example, or about violence and parents, see James Paul Gee, *What Video Games Have to Teach Us About Learning and Literacy* (New York: Macmillan, 2004), especially pp. 13–50.

55. Kristin Thompson, "Fantasy, Franchises, and Frodo Baggins: *The Lord of the Rings* and Modern Hollywood," *The Velvet Light Trap* 52 (Fall 2003): 59–61.

56. For a very early history of this class of games, see Herz, *Joystick Nation*, 197–213; and on the more recent period, Kent, *The Ultimate History of Video Games*, 549–553.

57. Douglas Collins, *The Story of Kodak* (New York: Harry N. Abrams, 1990): 343.

58. Photo Marketing Association International, *Photo Industry 2003: Review and Forecast* (Jackson, Mich: Photo Marketing Association, International, February 2003): 3, http://www.pmai.org/pdf/Photo_I.PDF (last accessed 12/29/2003).

59. Benjamin M. Compaine (ed.), *Understanding New Media: Trends and Issues in Electronic Distribution of Information* (Cambridge, Mass.: Ballinger Publishing Company, 1984): 99.

60. PMAI, *Photo Industry 2003*, p. 3.

61. U.S. Census Bureau, Table No. 1005, "Consumer Electronics and Electronic Components—Factory Sales by Product Category: 1990 to 1999," *Statistical Abstract of the United States: 2001* (Washington, D.C.: U.S. Census Bureau, 2001): 634.

62. In Kodak's case, see one well-done study of the firm that is highly critical of it, Alecia Swasy, *Changing Focus: Kodak and the Battle to Save a Great American Company* (New York: Crown, 1997): 184, 187.

63. Mary Bellis, "History of the Digital Camera," http://www.inventors.about.com/library/inventors (last accessed 4/14/2004).

64. Even with these products we can argue that they included electronics, using batteries to drive their mechanisms.

65. Leeridert Drukker, "Electronic Imaging," *Popular Photography* 92, no. 1 (January 1985): 55–56, 86–87.

66. "The Future," Ibid., 93, no. 7 (July 1986): 82–83.

67. "Elecronic Imaging Today," Ibid., 94, no. 9 (September 1987): 68–773.

68. "Kodak to Try Out New Machine," *Kodakery* 14, no. 13 (March 29, 1956): 4.

69. Automation Consultants, Inc., *Office Automation Applications* (New York: Automation Consultants, Inc., undated, [late 1950]): IV-E-19, CBI 55, Market Reports Box 70, folder 2, Archives of the Charles Babbage Institute, University of Minnesota, Minneapolis.

70. "Intricate Installation," *Kodakery* 15, no. 18 (May 9, 1957): 8.

71. "New 7080 Computer Double Memory Capacity," *Kodakery* 20, no. 20 (May 17, 1962): unpaginated. Other articles in the company magazine documented the story, see for example, "What Makes DACOM Tick?" Ibid., 18, no. 37 (September 15, 1960); "What's New in Kodak's Telecommunications," Ibid., 21, no. 15 (April 18, 1963); "Mathematics At

Work," Ibid., 22, no. 38 (September 17, 1964); "Computer Assists KP Engineering Division On Construction Projects," Ibid., 23, no. 10 (March 11, 1965).

72. "What's New in Kodak's Telecommunications," Ibid., 21, no. 15 (April 18, 1963).

73. "What Makes DACOM Tick?" Ibid., 18, no. 37 (September 15, 1960).

74. "Mathematics At Work," Ibid., 22, no. 38 (September 17, 1964).

75. Alecia Swasy, *Changing Focus: Kodak and the Battle to Save a Great American Company* (New York: Times Business, 1997): 29–40.

76. Ibid., 40–45.

77. Ibid., 122–23, 128–29.

78. Dan Richards, "Electronic Bridge to the Future?" Ibid., 97, no. 12 (December 1990): 17, 132, quote, p. 17.

79. Kodak, "History of Kodak: Milestones," http://www.kodak.com/US/en/corp/kodak History/1990_1999.shtml (Last accessed 4/13/2004).

80. "17 Top Electronic Cameras," *Popular Photography* 98, no. 12 (December 1991): 108.

81. Larry White and Michael J. McNamara, "Computer Imaging: A Photographer's Guide," Ibid., 57, no. 5 (September 1993): 38–43; Michael J. McNamara, "Digital Photography Comes of Age," Ibid., 58, no. 9 (September 1994): 33–40.

82. "News: Report Says Photographers Have Gone Digital," TrendWatch Graphics Arts, press release, November 19, 2002, refers to its report, *Digital Photography: How Creative Professionals Are Buying and Using Digital Cameras*, http://trendwatchgraphicarts.com (last accessed 12/5/2002).

83. Peter Kolonia and Aaron Schindler, "Photo Sites of Cyberspace," *Popular Photography* 59, no. 12 (December 1995): 80–84. This was the magazine's first article about the role of the Internet, and several times over the next nine years it would publish other pieces on cyberspace.

84. Michael J. McNamara, "What Price Digital Quality?" Ibid., 60, no. 4 (April 1996): 71.

85. Sony introduced a digital camera that worked with PCs in 1997 that received considerable attention; see John Nathan, *Sony: The Private Life* (Boston, Mass.: Houghton Mifflin, 1999): 292.

86. "Digital by Default," Ibid., 63, no. 15 (May 1999): 68–69, 74.

87. PMA Marketing Department, *Digital Makes Its Debut* (Jackson, Mich: Photo Marketing Association International, 2001): 1.

88. Michael McNamara, "Film versus Digital," *Popular Photography* 65, no. 3 (March 2001): 50–58; Douglas Carver, "Digital Still versus Digital Video," Ibid., 65, no. 10 (October 2001): 46–48, 182. Concerns about the supplies of the high-tech equipment have long been important in the information-processing industry as well. For example, in the 1920s and 1930s, over 4 percent of IBM's revenues came just from punched cards, and they generated far more than 4 percent of the corporation's profits.

89. Peter Kolonia, "Online Photo Processing," Ibid., 64, no. 1 (January 2000): 60–63, 66, 68; Mason Resnick, "Image Editing Software Round-Up," Ibid., 64, no. 2 (February 2000): 50–54; Peter Kolonia, "Smile: You're on Candid Cell Phone," Ibid., 67, no. 5 (May 2003): 70–72.

90. PMA Marketing Department, *Digital Makes Its Debut*, 1–3; Photo Marketing Association International, *Photo Industry 2004: Review and Forecast* (Jackson, Mich: Photo Marketing Association, International, February 2004): 5, 7, 9. These studies are available at http://www.pmai.org.

91. Photo Marketing Association International, *Photo Industry 2004*, 3.

92. Ibid., 7. Various press reports in late 2004 declared that this had just happened; more sober research will eventually suggest a date, although late 2004 is not an unreasonable one.

93. Jefferson Graham, "Picture This: Digital Eyes 2nd Banner Year," *USA Today*, February 11, 2004, p. 3B.

94. Pamela Paul, "The New Family Album," *Time* Bonus Section *Connections* (April 2004), unpaginated.

95. "Management Letter to Our Shareholders," *Eastman Kodak Company Annual Report 2003*, http://www.kodak.com/US/en/corp/annualReport03/letter/letter1.shtml (last accessed 8/5/2004).

96. Ibid.

97. Ibid.

98. "Kodak to Cut Up to 15,000 Jobs," *CNNMoney*, January 22, 2004, http://money.cnn.com/2004/01/22/news/companies/kodak.reut/index.htm (last accessed 1/22/2004).

99. "Fujifilm Facts" and "Fujifilm Firsts," both located at the company's website, http://www.fujifilm.com (last accessed 8/4/2004).

100. PMA, *PMA Industry Trends Report, Retail Markets* (Jackson, Mich.: PMA, 2001): 19. For the period from the mid-1980s forward, PMA published a similar report for each year and periodically would accumulate into tables ten or more years of data.

101. Chris Isidore, "AT&T, Kodak, IP Out of Dow," *CNNMoney*, April 1, 2004, http://money.cnn.com/2004/04/01/markets/dow/index.htm (last accessed 4/1/2004).

Chapter 13

1. Roger Alcaly, *The New Economy* (New York: Farrar, Straus and Giroux, 2003): 74–80.

2. Jeremy Rifkin, *End of Work: The Decline of the Global Labor Force and the Dawn of the Post-Market Era* (New York: G. P. Putnam's Sons 1995); (ed.), Peter Cappelli, Laurie Bassi, Harry Katz, David Knoke, Paul Osterman, and Michael Useem, *Change at Work: How American Industry and Workers Are Coping with Corporate Restructuring and What Workers Must Do to Take Charge of Their Own Careers* (New York: Oxford University Press, 1997); Simon Head, *The New Ruthless Economy: Work and Power in the Digital Age* (New York: Oxford University Press, 2003).

3. Major themes explored in chapter 9.

4. Now the subject of much study. For useful introductions, see Thomas W. Malone, Robert Laubacher, and Michael S. Scott Morton (eds.), *Inventing the Organizations of the 21st Century* (Cambridge, Mass.: MIT Press, 2003); Raghu Garud, Arun Kumaraswamy, and Richard N. Langlois (eds.), *Managing in the Modular Age: Architectures, Networks, and Organizations* (Oxford: Blackwell Publishing, 2003).

5. While the literature is vast, for current examples, see Frank Levy and Richard J. Murnane, *The New Division of Labor: How Computers Are Creating the Next Job Market* (Princeton, N.J.: Princeton University Press, 2004); Charles Granthan, *The Future of Work: The Promise of the New Digital Work Society* (New York: CommerceNet Press, 2000); Thomas W. Malone, *The Future of Work: How the New Order of Business Will Shape Your Organization, Your Management Style, and Your Life* (Boston, Mass.: Harvard Business School Press, 2004); and for some earlier ones, L. M. Applegate, James I. Cash, and D. Q. Mills, "Information Technology and Tomorrow's Managers," *Harvard Business Review* 66 (November–December 1988): 128–136; F. Becker and F. Steele, *Workplace by Design: Mapping the High-Performance Workplace* (San Francisco, Calif.: Jossey Bass, 1995); N. F. Crandall and M. J. Wallace, *Work and Rewards in the Virtual Workplace* (New York: American Management Association, 1998); M. C. Er, "A Critical Review of the Literature on the Organizational Impact of Information Technology," *IEEE Technology and Society Magazine* (June 1989): 17–23; Charles Handy, *The Age of Unreason* (Boston, Mass.: Harvard Business School Press, 1989).

6. Materials for the two previous paragraphs drawn from Richard N. Langlois, "The Vanishing Hand: The Changing Dynamics of Industrial Capitalism," *Industrial and Corporate*

Change 12, no. 2 (2003): 351–385; his bibliography is also quite comprehensive on the theme.

7. Ibid.

8. Carliss Y. Baldwin and Kim B. Clark, *Design Rules: The Power of Modularity* (Cambridge, Mass.: MIT Press, 2000), vol. 1; W. Michael Cox and Richard Alm, "The Right Stuff: America's Move to Mass Customization," in Federal Reserve Bank, *Federal Reserve Bank of Dallas Annual Report* (Dallas: Federal Reserve Bank, 1998); but see also B. Joseph Pine II, *Mass Customization: The New Frontier in Business Competition* (Boston, Mass.: Harvard Business School Press, 1993).

9. Key messages of Jeremy Rifkin, *The End of Work: The Decline of the Global Labor Force and the Dawn of the Post-Market Era* (New York: G. P. Putnam's Sons, 1995); Frank Levy and Richard J. Murnane, *The New Division of Labor: How Computers Are Creating the Next Job Market* (Princeton, N.J.; Princeton University Press, 2004); and Head, *The New Ruthless Economy*, 2003), especially pp. 6–9.

10. Richard R. Nelson, *The Sources of Economic Growth* (Cambridge, Mass.: Harvard University Press, 1996): 100–101.

11. Ibid., 113.

12. Michael Tushman and P. Anderson, "Technological Discontinuities and Organizational Environments," *Administrative Science Quarterly* (September 1986): 439–465.

13. Nelson, *The Sources of Economic Growth*, 118.

14. Ibid.

15. This activity continues to the present. For example, the major Hollywood studios agreed in September 2004 to extend the life of a group set up to develop technical standards for new digital movie projection systems, "Hollywood Extends Digital Film Group," *CNNMoney*, September 9, 2004, http://money.cnn.com/2004/09/09technology/digital_cinema.reut/index.htm (last accessed 9/9/2004).

16. In the beginning, some newspapers were even prepared to suffer losses on Internet versions of their papers, Pablo J. Boczkowski, *Digitizing the News: Innovation in Online Newspapers* (Cambridge, Mass.: MIT Press, 2004): 67.

17. Graham Y. Tanaka, *Digital Deflation: The Productivity Revolution and How It Will Ignite the Economy* (New York: McGraw-Hill, 2004): 25, his italics.

18. Ibid., see especially pp. 74–97.

19. Alcaly, *The New Economy*, 9. "Galbraithian" refers to the ideas presented by economist John Kenneth Galbraith in *The New Industrial State* (New York: New American Library, 1968), in which he predicted the economy would continue evolving into one with large integrated corporations along the lines described by Alfred Chandler Jr., and that Alcaly said evolved into different forms than predicted by Galbraith.

20. Alcaly, *The New Economy*, 102–103.

21. Ibid., 20.

22. David S. Landes, *The Unbound Prometheus: Technological Change and Industrial Development in Western Europe from 1750 to the Present* (Cambridge: Cambridge University Press, 1972): 1–6.

23. Alcaly, *The New Economy*, 25–42; Kevin J. Stiroh, "Information Technology and the U.S. Productivity Revival: What Do Industry Data Say?" *American Economic Review* 92, no. 5 (December 2002): 1559–1576; Brent R. Moulton, Eugene P. Seskin, and Daniel F. Sullivan, "Annual Revision of the National Income and Product Account," *Survey of Current Business* (August 2001): 10–14; Martin Neil Baily, "The New Economy: Post Mortem or Second Wind?" *Journal of Economic Perspectives* (Spring 2002): 3–22; Stephen D. Oliner and Daniel E. Sichel, "The Resurgence of Growth in the Late 1990s: Is Information Technology the Story?" Ibid. (Fall 2000): 3–22; Dale W. Jorgenson and Kevin J. Stiroh, "Raising the Speed Limit: U.S. Economic Growth in the Information Age," *Brookings Papers on*

Economic Activity 1 (2000): 125–211; Paul David, "The Modern Productivity Paradox in a Not-Too-Distant Mirror," in OECD, *Technology and Productivity* (Paris: OECD, 1991): 315–347.

24. Alcaly, *The New Economy*, 24; but see also the classic studies by Moses Abramovitz, "Resource and Output in the United States Since 1870," *American Economic Review* 46, no. 2 (May 1956): 5–23; Robert R. Solow, "Technical Change and the Aggregate Production Function," *Review of Economics and Statistics* 39, no. 3 (August 1957): 312–320, and his "Perspectives on Growth Theory," *Journal of Economic Perspectives* 8, no. 1 (Winter 1994): 45–54.

25. Betty W. Su, "The U.S. Economy to 2012: Signs of Growth," *Monthly Labor Review* (February 2004): 34, and also her footonote, number 17, p. 36.

26. Nelson, *The Sources of Economic Growth*, 100–119.

27. Alcaly, *The New Economy*, 90.

28. Jessica Litman, *Digital Copyright* (Amherst, N.Y.: Prometheus Books, 2001): 151–170; Jonathan Rosenoer, *CyberLaw: The Law of the Internet* (New York: Springer, 1997): 20–21.

29. For a useful discussion of the pros and cons of using the Internet in media and entertainment, see William W. Fisher III, *Promises to Keep: Technology, Law, and the Future of Entertainment* (Stanford, Calif.: Stanford University Press, 2004): 18–37.

30. Ed Treleven, "11 Sued in Madison for Music Piracy," *Wisconsin State Journal*, September 1, 2004, pp. A1, A8.

31. Fisher, *Promises to Keep*, 8–9.

32. Ibid., 9.

33. For an example of this sentiment expressed by a public official see the memoirs of an FCC chairman, Reed E. Hundt, *You Say You Want A Revolution: A Story of Information Age Politics* (New Haven, Conn.: Yale University Press, 2000).

34. Ibid., 199–258.

35. Ibid., 133.

36. Nelson, *The Sources of Economic Growth*, 53.

37. Nicholas G. Carr, *Does IT Matter? Information Technology and the Corrosion of Competitive Advantage* (Boston, Mass.: Harvard Business School Press, 2004): 46–48, 114.

38. Nelson, *The Sources of Economic Growth*, 56.

39. For examples, Nathan Rosenberg, *Perspectives on Technology* (Cambridge: Cambridge University Press, 1976); Everett M. Rogers, *Diffusion of Innovations*, 3d ed. (New York: Free Press, 1983); David C. Mowery and Richard R. Nelson (eds.), *Sources of Industrial Leadership: Studies of Seven Industries* (Cambridge: Cambridge University Press, 1999); Linsu Kim and Richard R. Nelson (eds.), *Technology, Learning, and Innovation: Experiences of Newly Industrializing Economies* (Cambridge: Cambridge University Press, 2000); and for a useful bibliography, Nelson, The *Sources of Economic Growth*, 308–328.

40. Nelson, *The Sources of Economic Growth*, 59.

41. Joseph A. Schumpeter, *Business Cycles: A Theoretical, Historical, and Statistical Analysis of the Capitalist Process* (New York: McGraw-Hill, 1939), vol. 1, pp. 87–192.

42. See, for example, Richard Foster and Sarah Kaplan, *Creative Destruction: Why Companies That Are Built to Last Underperform the Market—and How to Successfully Transform Them* (New York: Currency, 2001); Lee W. McKnight, Paul M. Vaaler, and Raul L. Katz (eds.), *Creative Destruction: Business Survival Strategies in the Global Internet Economy* (Cambridge, Mass.: MIT Press, 2001); Clayton M. Christensen and Michael E. Raynor, *The Innovator's Solution: Creating and Sustaining Successful Growth* (Boston, Mass.: Harvard Business School Press, 2003); Bhaskar Chakravorti, *The Slow Pace of Fast Change: Bringing Innovations to Market in a Connected World* (Boston, Mass.: Harvard Business School Press, 2003); Andrew Hargadon, *How Breakthroughs Happen: The Surprising Truth About How Companies Innovate* (Boston, Mass.: Harvard Business School Press, 2003); Henry Ches-

brough, *Open Innovation: The New Imperative for Creating and Profiting from Technology* (Boston, Mass.: Harvard Business School Press, 2003).

43. See, for example, the studies by Levy and Murnane, *The New Division of Labor*; Grantham, *The Future of Work*; Malone, *The Future of Work*; Head, *The New Ruthless Economy*.

44. Levy and Murnane, *The New Division of Labor*, 37.

45. Ibid., 41.

46. Ibid., 43.

47. Ibid., 44.

48. The reader should be aware that there are two authors who write about media with very similar names. There is Michael J. Wolf, whom I cite extensively in this book; he is a consultant working at Booz-Allen & Hamilton. Michael Wolff (note the different spelling of the last name) is a writer commenting on the media world. Both have written extensively on this sector.

49. Michael J. Wolf, *The Entertainment Economy: How Mega-Media Forces Are Transforming Our Lives* (New York: Times Books, 1999): 3–5, 28–29.

50. Michael McCarthy, "Media Giants Suit Up to Take on Video Games; TV Viewers Play Rather Than Watch," *USA Today*, August 27, 2004, http://global.factiva.com/en/arch/displayasp (last accessed 9/8/2004).

51. One of the earliest and still most thoughtful discussions of the economic behavior of content and the Internet, and that has enormously influenced my own research, is Carl Shapiro and Hal R. Varian, *Information Rules: A Strategic Guide to the Network Economy* (Boston, Mass.: Harvard Business School Press, 1999): 85–87, 101.

52. Grantham, *The Future of Work*, 25–29.

53. Head, *The New Ruthless Economy*, 6–7.

54. A personal example illustrates the point. At the time I wrote this chapter I had staff at IBM located in Japan, Netherlands, Belgium, and in the United Kingdom. We met every Thursday by way of a conference call to review our various projects; we conversed in English, notes were sent back and forth via e-mail. My management worked in Cambridge, Massachusetts, while I lived in Madison, Wisconsin. We had operated in such a global fashion for so long that it was not a problem to get our work done. The real challenge was finding a time to meet by phone when someone was not being forced to participate in the middle of the night. My staff spanned over 14 time zones.

55. Langlois, "The Vanishing Hand," 351–385.

56. Thomas J. Misa, *Leonardo to the Internet: Technology and Culture from the Renaissance to the Present* (Baltimore, Md.: Johns Hopkins University Press, 2004): 269.

57. There is a highly relevant academic study useful for managers, so much so that while out of print, is worth acquiring through secondhand book dealers and studying because the issues and recommendations reviewed in the study are remarkably relevant today: William G. Howard Jr. and Bruce R. Guile (eds.), *Profiting from Innovation: The Report of the Three-Year Study from the National Academy of Engineering* (New York; Free Press, 1992), especially pp. 39–132.

58. The classic work on the subject is George Stalk Jr. and Thomas H. Hout, *Competing Against Time* (New York: Free Press, 1990).

59. James W. Cortada and Heather Fraser, "Mapping the Future in Science-Intensive Industries: Lessons from the Pharmaceutical Industry," *IBM Systems Journal* 44, no. 1 (2005): 163–183; Cortada, James Spohre, Douglas McDavid, and Paul Maglio, "Convergence and Coevolution: Towards a Services Science," forthcoming; Clayton M. Christensen, Scott D. Anthony, and Erik A. Roth, *Seeing What's Next: Using the Theories of Innovation to Predict Industry Change* (Boston, Mass.: Harvard Business School Press, 2004).

60. Misa, *Leonardo to the Internet*, 268–276.

61. The subject of James W. Cortada, *Making the Information Society: Experience, Consequences, and Possibilities* (Upper Saddle River, N.J.: Financial Times/Prentice Hall, 2002); and James W. Cortada and Alfred D. Chandler, Jr. (eds.), *A Nation Transformed by Information: How Information Has Shaped the United States from Colonial Times to the Present* (New York: Oxford University Press, 2000). For a history of technology that virtually ignores the role of information in American society, see Ruth Schwartz Cowan's otherwise excellent history, *A Social History of American Technology* (New York: Oxford University Press, 1997).

62. Economists looking at a variety of technologies over three decades confirmed these criteria for the use of technology. An important study, which includes bibliography on this theme, is Diego Comin and Bart Hobijn, "Cross-Country Technology Adoption: Making the Theories Face the Facts," Staff Report no. 169, June 2003, a paper sponsored by the Federal Reserve Bank of New York.

63. "How Did Computing Go Global? The Need for an Answer and a Research Agenda," *IEEE Annals of the History of Computing* 26, no. 1 (January–March 2004): 53–58; James W. Cortada, *The Digital Hand: How Computers Changed the Work of American Manufacturing, Transportation, and Retail Industries* (New York: Oxford University Press, 2004): 385–387.

64. Misa, *Leonardo to the Internet*; Daniel R. Headrick, *The Tools of Empire: Technology and European Imperialism in the Nineteenth Century* (New York: Oxford University Press, 1981): 3–12; Peter J. Hugill, *Global Communications Since 1844: Geopolitics and Technology* (Baltimore, Md.: Johns Hopkins University Press, 1999): 1–23, 223–251.

65. For full details, see U.S. Department of Labor, "Table 1. Employment by Major Industry Division, 1992, 2002, and 2012," http://www.bls.gov/news.release/ecopro.t01.htm (last accessed 9/11/2004). For a detailed report on this topic, see Jay M. Berman, "Industry Output and Employment Projections to 2012," *Monthly Labor Review* (February 2004): 58–79.

Appendix A

1. Michael E. Porter, *Competitive Strategy: Techniques for Analyzing Industries and Competitors* (New York: Free Press, 1980), and his *Competitive Advantage: Creating and Sustaining Superior Performance* (New York: Free Press, 1985).

2. For a useful discussion of how historians view economic models, see Carlo M. Cipolla, *Between Two Cultures: An Introduction to Economic History* (New York: W. W. Norton, 1991): 69–70.

3. Fritz Machlup, a professor of economics at Princeton University, wrote many articles and books on this topic. But the first book that magisterially explained his ideas was *The Production and Distribution of Knowledge in the United States* (Princeton, N.J.: Princeton University Press, 1962); his other works expanded on his ideas all through the 1970s and early 1980s. For a brief discussion of his ideas, see James W. Cortada (ed.), *Rise of the Knowledge Worker* (Boston: Butterworth-Heinemann, 1998). Peter Drucker usually gets credit for being the first to call people's attention to the knowledge economy, perhaps because his books enjoyed such wide circulation. However, it was Machlup who did the serious, early empirical work that so many business professors, economists, and historians have relied upon over the past four decades.

4. Michael E. Porter, *The Competitive Advantage of Nations* (New York: Free Press, 1990).

5. Ibid., p. 240.

6. The observation was universally accepted by economists. For context, see Martin Feldstein (ed.), *The American Economy in Transition* (Chicago, Ill.: University of Chicago Press, 1980); Norman Frumkin, *Tracking America's Economy* (Armonk, N.Y.: M. E. Sharpe, 1998): 242–243; and a grand, earlier work by a distinguished economist, W. W. Rostow, *The World Economy: History and Prospect* (Austin, Tex.: University of Texas, 1978).

7. The literature is vast, but begin with Alvin Toffler, *Future Shock* (New York: Random

House, 1970), which probably triggered the whole notion for the public at large; the highly influential book by Daniel Bell, *The Coming of Post-Industrial Society: A Venture in Social Forecasting* (New York: Basic Books, 1973); a thoughtful study by Jorge Reina Schement and Terry Curtis, *Tendencies and Tensions of the Information Age: The Production and Distribution of Information in the United States* (New Brunswick, N.J.: Transaction Publishers, 1995); Alfred D. Chandler and James W. Cortada (eds.), *A Nation Transformed by Information: How Information Has Shaped the United States from Colonial Times to the Present* (New York: Oxford University Press, 2000); I have also commented on the subject in *Making the Information Society: Experiences, Consequences, and Possibilities* (Upper Saddle River, N.J.: Financial Times/Prentice-Hall, 2002).

8. Comment was made by Italian historian M. Salvati, quoted in Cipolla, *Between Two Cultures*, 69.

9. Recent examples include Margaret B. W. Graham and Alec T. Shuldiner, *Corning and the Craft of Innovation* (New York: Oxford University Press, 2001); Fred Dalzell and Rowena Olegario, *Lessons from 165 Years of Brandbuilding at Procter & Gamble* (Boston: Harvard Business School Press, 2003); Timothy Jacobson and George Smith, *Cotton's Renaissance: A Study in Market Innovation* (Cambridge: Cambridge University Press, 2001).

10. This is an ongoing process resulting in the redefinition of industry codes, dollar volumes, and so forth. This initiative is being driven by the U.S. Census Bureau and thus to keep current on this work, and for access to the "translation" from SIC to NAICS, visit this agency's website: www.census/gov.

Appendix B

1. A. Michael Noll, *Introduction to Telephones and Telephone Systems*, 3d ed. (Boston, Mass.: Artech House, 1998): 1–125; Stephen M. Walters, *The New Telephony* (Upper Saddle River, N.J.: Prentice-Hall PTR, 2002): 15–107; for pure analog systems, see David Talley, *Basic Telephone Switching Systems* (New York: Hayden, 1969). The body of knowledge about digital telephony was large even by the time of AT&T's breakup and had been codified and widely available in the industry; see, B. E. Keiser and F. Strange, *Digital Telephony and Network Integration* (New York: Van Nostrand Reinhold Company, 1985).

2. Because of the flexibility of digital technology, voice and data appeared the same on a line and thus even the FCC did not track the differences between one kind of traffic or another, per conversation between Jim Lande of the Industry Analysis Division of the FCC and the author, October 21, 2003. For the kind of tracking the FCC does, see Linda Blake and Jim Lande, *Trends in the International Telecommunications Industry* (Washington, D.C.: Federal Communications Commission, April 21, 2001). The only tracking was done mainly by various government agencies and ISPs interested in the amount of traffic going through the Internet.

3. Walters, *The New Telephony*, 43, 48, 50; Jane Laino, *The Telecom Handbook: Understanding Business Telecommunications Systems and Services*, 4th ed. (New York: CMP Books, 2002): 19–20.

4. Susan E. McMaster, *The Telecommunications Industry* (Westport, Conn.: Greenwood Press, 2002): 153–175.

BIBLIOGRAPHIC ESSAY

This brief bibliographic essay discusses some of the most obvious and useful sources for those interested in exploring in more detail the subject of this book. Citations in the endnotes point to sources and to additional materials used in highly specific ways, for example, Internet addresses; these are not repeated below. I emphasize books rather than articles because the former cover broader subjects more suitable for the purposes of this essay. One might also look at the bibliographic essay in the first volume of the *The Digital Hand*, which covers such topics as archival sources, computing applications, and economics—topics that I do not repeat below but upon which I relied in writing this second book.

Banking Industry

The literature on this industry is vast, particularly articles and white papers. The most important source on the role of IT in this industry, however, can be tracked for the entire period through the American Bankers Association's monthly magazine, *Banking* (later renamed *ABA Banking Journal*). It explained new technologies and applications, discussed their implications for banking, reviewed managerial concerns, and offered hundreds of case studies. The publication was available for the entire period of this study. For background on money itself, two recent studies are of use: Jack Weatherford, *The History of Money: From Sandstone to Cyberspace* (New York: Crown Publishers, 1997); and Elinor Harris Solomon, *Virtual Money: Understanding the Power and Risks of Money's High-Speed Journey into Electronic Space* (New York: Oxford University Press, 1997). For a modern analysis of the economy, and the role the digital is playing, a good introduction is a collection of essays in Erik Brynjolfsson and Brian Kahin (eds.), *Understanding the Digital Economy: Data, Tools, and Research* (Cambridge, Mass.: MIT Press, 2000); but see also Robert E. Litan and Alice M. Rivlin (eds.), *The Economic Payoff from the Internet Revolution* (Washington, D.C.: Brookings Institution Press, 2000).

On the industry itself, several comprehensive guides provide background on how the industry was structured and worked. Begin with the massive work of reference by Charles J. Woelfel, *Encyclopedia of Banking and Finance* (Chicago, Ill.: Probus Publishing Company, 1994). The standard overview on the industry, with discussion of IT and other issues, is by Allen N. Berger, Anil K. Kashyap, and Joseph M. Scalise, "The Transformation of the U.S. Banking Industry: What a Long, Strange Trip It's Been," *Brookings Papers on Economic Activity* 2 (1995): 55–201. For the industry midway through our story, see William H. Baughn and Charles E. Walker (eds.), *The Bankers' Handbook* (Homewood, Ill.: Dow Jones-Irwin, 1966, 1978). See also the nice introduction to the industry, circa 1960s and 1970s, by Arnold A. Heggestad and William G. Shepherd, "The Banking Industry," in Walter

Adams (ed.), *The Structure of American Industry*, 6th ed. (New York: Macmillan, 1982): 319–347. Because of its focus on both technology and business issues, see Patrick T. Harker and Larry W. Hunter, "Retail Banking," in David C. Mowery (ed.), *U.S. Industry in 2000: Studies in Competitive Performance* (Washington, D.C.: Academy Press, 1999): 179–214. For the period of the early 1990s, there is Dwight B. Crane and Zvi Bodie, "The Transformation of Banking," *Harvard Business Review* (March–April 1996): 109–117.

On the role of computing in the industry, consult ABA's *Banking*. However, several specialized studies are also useful. On credit cards, the major study is David Evans and Richard Schmalensee, *Paying with Plastic: The Digital Revolution in Buying and Borrowing* (Cambridge, Mass.: MIT Press, 1999); but do not overlook the earlier book by Matty Simmons, *The Credit Card Catastrophe: The 20th Century Phenomenon That Changed the World* (New York: Barricade Books, 1995). On EFTS, an early study is helpful, Mark J. Flannery and Dwight M. Jaffee, *The Economic Implications of an Electronic Monetary Transfer System* (Lexington, Mass.: Lexington Books, 1973). On the effects of technologies on the industry there is the excellent study by Thomas D. Steiner and Diogo B. Teixeira, *Technology in Banking: Creating Value and Destroying Profits* (Homewood, Ill.: Business One Irwin, 1990). A weaker study whose strength lies in the several chapters on the value and role of databases in banking, circa early 1990s, is by William S. Sachs and Frank Elston, *The Information Technology Revolution in Financial Services: Using IT to Create, Manage, and Market Financial Products* (Chicago, Ill.: Probus Publishing, 1994). On ERMA, there is James L. McKenney, with Duncan C. Copeland and Richard O. Mason, *Waves of Change: Business Evolution Through Information Technology* (Boston, Mass.: Harvard Business School Press, 1995). The earliest large study on computing in this industry is Robert S. Alsom et al., *Automation in Banking* (New Brunswick, N.J.: Rutgers University Press, 1962). For applications from the 1990s, see two studies that look at applications and their consequences: Mary J. Cronin (ed.), *Banking and Finance on the Internet* (New York: John Wiley & Sons, 1998); and the massive anthology by Jessica Keyes (ed.), *Financial Services Information Systems* (Boca Raton, Fla.: Auerbach, 2000). The endnotes in the first three chapters provide additional citations from government and industry studies which appeared throughout the period.

Insurance Industry

This industry is at least as complex to understand as the Banking Industry, and the volume of material available on it is massive. There are no formal book-length histories available of the industry of the post–World War II period, although a few articles exist, all cited in the endnotes to chapter 4. There are, however, many useful textbooks and other publications that describe the work of this industry. A good example of this kind of material is G. F. Michelbacher and Nestor R. Roos (eds.), *Multiple-Line Insurers: Their Nature and Operations* (New York: McGraw-Hill, 1970), which provides insight into the workings of this industry in the 1950s and 1960s. For a similar volume on the 1980s and 1990s, see J. D. Cummins and A. M. Santomero (eds.), *Changes in the Life Insurance Industry* (Norwell, Mass.: Kluwer Academic Publishers, 1999). However, because of the limited availability of that book, consult two other studies: Edward I. Altman and Irwin T. Vanderhoof (eds.), *The Strategic Dynamics of the Insurance Industry: Asset/Liability Management Issues* (Burr Ridge, Ill.: Irwin, 1996); and Robert E. Litan, *The Revolution in U.S. Finance* (Washington, D.C.: Brookings Institution, 1991). Because of the profound influence of regulators in this industry, to understand the issues, see Alan Gart, *Regulation, Deregulation, Reregulation: The Future of the Banking, Insurance, and Securities Industries* (New York: John Wiley & Sons, 1994), a book that remains remarkably current and is readily available. In addition, there are a number of magazines and journals published within the industry, the most useful of which are *Best's Review*, which through most of the period appeared in two editions, one for the life/health sector of the industry, the other for the property/casualty side of the business. Sometimes that second magazine was called the property/liability edition, and toward the end of the 1990s, the two were merged into one. Titles changed over time, but the volume numbers were kept sequential all through the period, so one should not be confused and think there were more than two, later one, *Best's Review*. These magazines were published monthly and covered all aspects of the industry's activities, making them the essential body of material with which to begin understanding both the history of the industry and how it approached IT.

While the endnotes contain many citations regarding the use of the digital in insurance, there are several major sources that help us piece together the story. First, many issues of *Best's Review* had case studies, product announcements, and other articles on the use of IT during the period—indeed hundreds of these. Second, LOMA published a series of conference proceedings, beginning in 1959 and extending to the end of the 1980s, that did the same. Third, textbooks aimed at the industry describe the technology, applications, and managerial considerations involved in the use of IT over the entire period. The most comprehensive review of IT is Charles H. Cissley's, *Systems and Data Processing in Insurance Companies* (Atlanta, Ga.: FLMI Insurance Education Program, LOMA, 1982), the standard work on the subject for the period of the 1960s through the late 1980s. For another LOMA text covering the period of the late 1980s and reflective of the situation through the majority of the 1990s, see Kenneth Huggins, Dani L. Long, and Vicki Mason Pfrimmer, *Information Management in Insurance Companies* (Atlanta, Ga.: FLMI Insurance Education Program and LOMA, 1995). But see also H. Thomas Krause, Douglas R. Bean, Robert L. C. Thorn, and Robert A. Oakley, *Insurance Information Systems*, 2 vols. (Malvern, Penna.: Insurance Institute of America, 1990). The American Institute for Property and Liability Underwriters published a great deal of material on the nuts-and-bolts of how to do the work of insurance firms, and did publish a study on computing, Michael D. Gannt, Alan J. Turner, and James Gatza, *Computers in Insurance* (Malvern, Penna.: American Institute for Property and Liability Underwriters, 1985). There are few good materials yet on the role of the Internet; trade journals remain the best source. However, there are a number of chapters on the subject in Jessica Keyes (ed.), *Financial Services Information Systems* (Boca Raton, Fla.: Auerbach, 2000); but see also Mary J. Cronin (ed.), *Banking and Finance on the Internet* (New York: John Wiley & Sons, 1998).

We actually have a good supply of data on deployment. Besides period surveys published by industry magazines (all cited in the endnotes in chapter 4), there are two important studies done by the U.S. Bureau of Labor Statistics, *Technology and Labor in Five Industries*, Bulletin 2033 (Washington, D.C.: U.S. Department of Labor, 1979), and *Technology and Its Impact on Employment in the Life and Health Insurance Industries*, Bulletin 2368 (Washington, D.C.: U.S. GPO, September 1990). But do not overlook the important survey done by Raymond Dash, *Data Processing Users Survey*, Special Release No. 8 (Atlanta, Ga.: LOMA, November 1978). We now have a major history of the role of information processing in this industry written by JoAnne Yates, *Structuring the Information Age: Life Insurance and Technology in the Twentieth Century* (Baltimore, Md.: Johns Hopkins University Press, 2005).

Brokerage Industry

There are a number of useful general histories of the industry that serve as good introductions to what it has done (and continues to do), particularly focused on Wall Street. These include Charles R. Geisst, *Wall Street: A History* (New York: Oxford University Press, 1997); Francis L. Eames, *The New York Stock Exchange* (New York: Thomas G. Hall, 1984); of all the books written by the late Robert Sobel, see *The Big Board: A History of the New York Stock Exchange, 1935–1975* (New York: Weybright and Talley, 1975); and finally, Walter Werner and Steven T. Smith, *Wall Street* (New York: Columbia University Press, 1991). Robert Sobel also wrote, *Amex: A History of the American Stock Exchange, 1921–1971* (New York: Weybright and Talley, 1972); but also useful is Stuart Weems Bruchey, *Modernization of the American Stock Exchange, 1971–1989* (New York: Garland, 1991). We have a history of the role of the Securities and Exchange Commission that is essential for understanding the role of the industry by Joel Seligman, *The Transformation of Wall Street: A History of the Securities and Exchange Commission and Modern Corporate Finance* (Boston: Houghton Mifflin, 1982). Finally, a memoir by an advertising agent working in the industry is very useful for the 1960s and 1970s, Alex Benn, *The Unseen Wall Street of 1969–1975: And Its Significance for Today* (Westport, Conn.: Quorum Books, 2000). For the period of the 1970s and 1980s, there is the very well-informed study by Marshall E. Blume, Jeremy J. Siegel, and Dan Rottenberg, *Revolution on Wall Street: The Rise and Decline of the New York Stock Exchange* (New York: W.W. Norton, 1993).

Literally hundreds of books have been published on the structure and operations of the industry. Reading these from each decade is essential to allow one to appreciate the daily goings on of the industry. Some of the most useful are those that went through multiple editions, demonstrating from one volume to another how things changed. See Charles J. Woelfel, *Encyclopedia of Banking and Finance*, particularly the 10th edition (Chicago, Ill.: Probus Publishing, 1994). For a very early description of the industry's operations, there is Charles Amos Dice and Wilford John Eitman, *The Stock Market* (New York: McGraw-Hill, 1952); and George L. Leffler, *The Stock Market*, in particular the 2nd and 3rd editions (New York: Ronald Press, 1957, 1963). One of the most useful references to the industry is Richard J. Teweles and Edward S. Bradley, *The Stock Market*, particularly the sixth edition (New York: John Wiley & Sons, 1998). For the start of the middle period, see also Frank G. Zarb and Gabriel T. Kerekes (eds.), *The Stock Market Handbook: Reference Manual for the Securities Industry* (Homewood, Ill.: Dow Jones-Irwin, 1970). On the paperwork crisis of the 1960s, see "Wall Street vs. Paper Work," *The Exchange* (September 1969): 11–35. Particularly useful in understanding the OTC phenomenon of the 1970s, which led to the broadening of the market over time, is Richard D. Kirshberg (ed.), *The Over-the-Counter Market Place: Myths, Mysteries and Mechanics* (New York: Practicing Law Institute, 1974); David M. Weiss, *After the Trade Is Made: Processing Securities Transactions* (New York: Institute of Finance, 1986); Stephen P. Rappaport, *Management on Wall Street: Making Securities Firms Work* (Homewood, Ill.: Dow Jones-Irwin, 1988); Charles P. Jones, *Investments: Analysis and Management*, especially the 2nd edition (New York: John Wiley & Sons, 1988); Charles P. Jones, *Investments and: Analysis and Management*, all three editions (New York: John Wiley & Sons, 1985, 1988, 1991).

The general surveys of the industry that cover the period from the 1980s forward contain commentaries on the role of computing and telecommunications. There are few book-length discussions on the role of IT in the Brokerage Industry, but the endnotes offer many citations of articles and various government reports on the topic. One early exception, aimed at users of online systems, is Jeremy C. Jenks and Robert W. Jenks, *Stock Selection: Buying and Selling Stocks Using the IBM PC* (New York: John Wiley & Sons, 1984). There are two excellent books on the history of Charles Schwab Corporation, an extensive user of IT. The first, *Clicks and Mortar: Passion Driven Growth in an Internet Driven World*, was written by two executives at the firm, David S. Pottruck (president and co-CEO) and Terry Pearce (an executive at the firm) (San Francisco, Calif.: Jossey-Bass, 2000); and John Kador, *Charles Schwab: How One Company Beat Wall Street and Reinvented the Brokerage Industry* (New York: John Wiley & Sons, 2002). For an excellent description of day trading by a day trader, see Gregory J. Millman, *The Day Traders: The Untold Story of the Extreme Investors and How They Changed Wall Street Forever* (New York: Times Business, 1999). The one economic monograph that is particularly relevant to the role of the Internet is Robert E. Litan and Alice M. Rivlin (eds.), *The Economic Payoff from the Internet Revolution* (Washington, D.C.: Brookings Institution Press, 2001).

Also useful for the Internet, see Mary J. Cronin (ed.), *Banking and Finance on the Internet* (New York: John Wiley & Sons, 1997).

Useful materials on the role of IT in the Brokerage Industry exist at the Charles Babbage Institute at the University of Minnesota in the Burroughs Papers, particularly for the 1950s and 1960s. The same is true for IBM's corporate archives, which cover the entire twentieth century. The New York Stock Exchange also has an archive which has preserved a small collection of materials concerning the exchange's use of IT and telecommunications. The Brokerage Industry press has periodically reviewed the role of IT, particularly when the exchanges introduced new applications or when brokerage firms introduced offerings that were dependent on computing.

Telecommunications Industry

For a broad history of all communications technologies going back into the nineteenth century, see George P. Oslin, *The Story of Telecommunications* (Macon, Ga.: Mercer University Press, 1992). For a short history, with an extensive emphasis on legal and regulatory issues, see Alan Stone, *How America Got On-Line: Politics, Markets, and the Revolution in Telecommunications* (Armonk, N.Y.: M. E. Sharpe, 1997). For a balanced yet brief account, skewed toward the post-1950 period, there is Susan E. McMaster, *The Telecommunications Industry* (Westport, Conn.: Greenwood Press, 2002). Finally, for a more technical history, and with European content, see John Bray, *The Communications Miracle: The Telecommunication Pioneers from Morse to the Information Superhighway* (New York: Plenum Press, 1995).

No history of the industry can be written or understood without examining the activities of AT&T. On the technology on which the entire Bell System relied, there is a four-volume history commissioned by Bell Labs that is detailed, useful, and indispensable: M. D. Fagen (ed.), *A History of Engineering and Science in the Bell System: National Service in War and Peace (1925–1975)* (Murray Hill, N.J.: Bell Telephone Laboratories, 1978); A. E. Joel, Jr. and staff, *Switching Technology (1925–1975)*, published in 1982; S. Millman (ed.), *Communications Sciences (1925–1980)*, published in 1984; F. M. Smits (ed.), *Electronics Technology (1925–1975)*, published in 1985. For a short version of the entire story of technology and Bell Labs, there is Prescott C. Mabon, *Mission Communications: The Story of Bell Laboratories* (Murray Hill, N.J.: Bell Telephone Laboratories, 1975). The most readable history, however, is by Jeremy Bernstein, *Three Degrees Above Zero: Bell Labs in the Information Age* (New York: Charles Scribner's Sons, 1984); the title refers to the average temperature above absolute zero of the universe. On Western Electric there is now an excellent business history by Stephen B. Adams and Orville R. Butler, *Manufacturing the Future: A History of Western Electric* (Cambridge: Cambridge University Press, 1999). The classic work analyzing the breakup of AT&T is also another fine business history by Peter Temin and Louis Galambos, *The Fall of the Bell System* (Cambridge: Cambridge University Press, 1987; published in paperback in various subsequent editions, first in 1989); if you could only read one book on AT&T, this would be the one. Two useful journalistic accounts of AT&T are by Sonny Kleinfield, *The Biggest Company on Earth: A Profile of AT&T* (New York: Holt, Rinehart and Winston, 1981); and Steve Coll, *The Breakup of AT&T* (New York: Atheneum, 1986).

A number of corporate histories fill in many details about the activities of AT&T's rivals. Philip L. Cantelon's massive history of MCI is essential reading, *The History of MCI: 1968–1988, The Early Years* (Dallas, Tex: Heritage Press, 1993); for a shorter study written by a journalist, see Larry Kahaner, *On the Line: The Men of MCI—Who Took on AT&T, Risked Everything, And Won* (New York: Warner Books, 1986). Most recent is Lorraine Spurge, *MCI: Failure Is Not an Option: How MCI Invented Competition in Telecommunications* (Encino, Calif.: Spurge.ink!, 1998). For a history of an independent phone company, published to celebrate an anniversary of the firm, there is K. C. August, *TDS: The First Twenty Years* (Chicago, Ill.: Telephone and Data Systems, Inc., 1989; rev. ed., 1993). On the

Canadian company Northern Telecom (later Nortel), there is a useful history by Larry Macdonald, *Nortel Networks: How Innovation and Vision Created a Network Giant* (Toronto: John Wiley & Sons, 2000). The most useful sources on Cisco are by David Bunnell, *Making the Cisco Connection: The Story Behind the Real Internet Superpower* (New York: John Wiley & Sons, 2000); and Jeffrey S. Young, *Cisco: Unauthorized: Inside the High-Stakes Race to Own the Future* (Roseville, Calif.: Forum, 2001). However, one should not ignore Robert Slater's *The Eye of the Storm: How John Chambers Steered Cisco Through the Technology Collapse* (New York: HarperBusiness, 2003). For the most current round of business activities of multiple firms, there is Om Malik, *Broadbandits: Inside the $750 Billion Telecom Heist* (New York: John Wiley & Sons, 2003). For introductions to the cell phone, two company histories, although of a European firm, provide much detail: Dan Steinbock, *The Nokia Revolution: The Story of an Extraordinary Company that Transformed an Industry* (New York: AMACOM, 2001); and Martti Haikio, *Nokia: The Inside Story* (London: Pearson, 2001).

All general surveys and analyses of the role of the industry and of its technologies in society begin with Ithiel de Sola Pool (ed.), *The Social Impact of the Telephone* (Cambridge, Mass.: MIT Press, 1977), and extend to another book he wrote, *Technologies Without Boundaries: On Telecommunications in a Global Age* (Cambridge, Mass.: Harvard University Press, 1990), published after his death but written a few years earlier. On social issues, I would suggest the outstanding history by Claude S. Fischer, *America Calling: A Social History of the Telephone to 1940* (Berkeley, Calif.: University of California Press, 1992; also available in paperback). A number of accounts on general activities in the industry in the 1980s and 1990s add context and detail. For an exuberant (and so far inaccurate) forecast for the future of broadband, see George Gilder, *Telecosm: How Infinite Bandwidth Will Revolutionize Our World* (New York: Free Press, 2000). A useful account of various corporate activities and players is provided by Kevin Maney, *Megamedia Shakeout: The Inside Story of the Leaders and Losers in the Exploding Communications Industry* (New York: John Wiley & Sons, 1995); on wireless, see Alex Lightman with William Rojas, *Brave New Unwired World: The Digital Big Bang and the Infinite Internet* (New York: John Wiley & Sons, 2002); and James B. Murray, Jr., *Wireless Nation: The Frenzied Launch of the Cellular Revolution in America* (Cambridge, Mass.: Perseus Publishing, 2001); but see also Louis Galambos and Eric John Abrahamson, *Anytime, Anywhere: Entrepreneurship and the Creation of a Wireless World* (Cambridge: Cambridge University Press, 2002). Since no discussion about telecommunications in the 1980s and 1990s would be complete without understanding the convergence of computing and communications, a useful collection of papers on the subject can be found in Stephen P. Bradley, Jerry A. Hausman, and Richard L. Nolan (eds.), *Globalization, Technology, and Competition: The Fusion of Computers and Telecommunications in the 1990s* (Boston, Mass.: Harvard Business School Press, 1993).

Because the literature on the history of the Internet is so massive, I would suggest looking at two books, both of which have excellent bibliographies: Arthur L. Norberg and Judy E. O'Neill, *Transforming Computer Technology: Information Processing for the Pentagon, 1962–1986* (Baltimore, Md.: Johns Hopkins University Press, 1996); and Jane Abbate, *Inventing the Internet* (Cambridge, Mass.: MIT Press, 1999). For a contemporary analysis of the role of the Internet and telecommunications, see Martin Fransman, *Telecoms in the Internet Age: From Boom to Bust to . . . ?* (Oxford: Oxford University Press, 2002).

Regulatory and policy activities proved so important in this industry that their role must be understood. While all histories of the industry address the topic, there are several other publications that are specific and informative. For an overview of issues, see Gerald W. Brock, *Telecommunications Policy for the Information Age* (Cambridge, Mass.: Harvard University Press, 1994); and Jean-Jacques Laffont and Jean Tirole, *Competition in Telecommunications* (Cambridge, Mass.: MIT Press, 2000); but then examine the memoirs of President Clinton's chairman of the FCC, Reed E. Hundt, *You Say You Want A Revolution: A Story of Information Age Politics* (New Haven, Conn.: Yale University Press, 2000); and the analysis of the U.S. economy in the 1990s by President Clinton's chairman of the Council of Economic Advisors, Joseph E. Stiglitz, *The Roaring Nineties* (New York: W.W. Norton, 2003). For a description and discussion of the 1996 law, there is Dale E. Lechman and Dennis L. Weisman, *The Telecommunications Act of 1996: The "Costs" of Managed Competition* (Boston, Mass.: Kluwer Academic Publishers, 2000); and Sam Peltzman and Clifford Winston (eds.), *Deregulation of Network Industries: What's Next?* (Washington, D.C.: AEI-Brookings Joint Center for Regulatory Studies, 2000) but see also Charles H. Ferguson, *The Broadband Problem: Anatomy of a Market Failure and a Policy Dilemma* (Washington, D.C.: Brookings Institution Press, 2004). Since economics are closely tied to regulatory issues, see Robert J. Saunders, Jeremy J. Warford, and Bjorn Wellenius, *Telecommunications and Economic Development* (Baltimore, Md.: World Bank and The Johns Hopkins University Press, 1994).

There are hundreds of guides to the technology of this industry. The standard reference used within the industry is by Harry Newton, *Newton's Telecom Dictionary* (New York: Flatiron Publishing, in nearly annual additions, beginning in the 1980s and extending to the present with nearly 20 editions so far); it is massive, quirky, yet clearly written. See also Richard Grigonis, *Computer Telephony Encyclopedia* (Gilroy, Calif.: CMP Books, 2000). The classic one-volume introduction to the whole subject, and with historical perspective, is A. Michael Noll, *Introduction to Telephones and Telephone Systems* (Boston, Mass.: Artech House, 3rd ed., 1998; 2nd ed. appeared in 1991; and the 1st in 1985). For a nice blend of technology and business issues for the late 1990s, see Stephen M. Walters, *The New Telephony: Technology Convergence, Industry Collision* (Upper Saddle River, N.J.: Prentice-Hall PTR, 2002). For a similar treatment covering uses of telephony in the 1970s and 1980s, there is Thomas J. Housel and William E. Darden III, *Introduction to Telecommunications: The Business Perspective* (Cincinnati, Oh.: South-Western Publishing Co., 1988). On satellites, consult David J. Whalen, *The Origins of Satellite Communications, 1945–1965* (Washington, D.C.: Smithsonian Institution Press, 2002).

There is a growing body of publications on wireless technologies. One of the earliest was by Raymond Steele, *Mobile Radio Communications* (London: Pentech Press, 1992). On the contemporary technologies and their uses, see Jouini Paavilainen, *Mobile Business Strategies: Understanding the Technologies and Opportunities* (London: Addison-Wesley and IT Press, 2001); and John P. Burnham, *The Essential Guide to the Business of U.S. Mobile Wireless Communications* (Upper Saddle River, N.J.: Prentice-Hall PTR, 2002). Finally, there is Paul Levinson, *Cellphone: The Story of the World's Most Mobile Medium and How It Has Transformed Everything!* (New York: Palgrave Macmillan, 2004). Because regulations influence all new technologies in this industry, consult Jonathan E. Nuechterlein and Philip J. Weiser, *Digital Crossroads: American Telecommunications Policy in the Internet Age* (Cambridge, Mass.: MIT Press, 2005).

The most important archive for the history of this industry is the AT&T Corporate Archives. It contains material on the entire Bell System dating back to the nineteenth century. The materials are very well-organized. No history of this industry can be written without using the records of this facility.

Media and Entertainment Industries

It would seem that no aspect of post-1990's U.S. society has been the subject of so much study as those industries comprising media and entertainment. However, it is also a set of industries about which little has been written concerning their use of technologies and the effects of these tools on their affairs prior to the 1980s. This is a set of industries that can best be understood in an ahistorical fashion by looking at its current situation and then going backwards. Begin with Michael J. Wolf, *The Entertainment Economy: How Mega-Media*

Forces Are Transforming Our Lives (New York: Times Books, 1999). Then consult three studies of the industries" underpinning media and entertainment, providing business and economic information: Benjamin M. Compaine and Douglas Gomery, *Who Owns the Media? Competition and Concentration in the Mass Media Industry*, published in three editions (Mahwah, N.J.: Lawrence Erlbaum Associates, 1979, 1982, 2000), with chapters on individual industries; Harold L. Vogel, *Entertainment Industries Economics: A Guide for Financial Analysis*, published in five editions (Cambridge: Cambridge University Press, 1986, 1990, 1994, 1998, 2001); and Robert G. Picard, *The Economics and Financing of Media Companies* (New York: Fordham University Press, 2002). Because these industries changed so quickly, consulting multiple editions of the same book is necessary as one works backwards through the history of most of these industries.

Because so many of the digital and telecommunications technologies were used in many of the media and entertainment industries, a number of useful surveys exist that provide a composite view of technology. A textbook now exists that serves as a detailed and excellent introduction to technologies of the 1990s and beyond: John V. Pavlik and Shawn McIntosh, *Converging Media: An Introduction to Mass Communication* (Boston, Mass.: Pearson Education, 2004). Also useful for the same period is Michael M. A. Mirabito and Barbara L. Morgenstern, *The New Communications Technologies: Applications, Policy, and Impact*, 4th ed. (Boston, Mass.: Focal Press, 2001). To begin providing historical perspective, consult an earlier study of converging technologies reflecting patterns of the late 1980s and early 1990s: John V. Pavlik and Everette E. Dennis, *Demystifying Media Technology* (Mountain View, Calif.: Mayfield Publishing Company, 1993).

Book Publishing Industry

The activities of this industry are extremely well-documented by industry insiders, information specialists (such as librarians), IT community, economists, historians, and writers. For each decade there are histories, accounts of how the industry functioned, memoirs, company histories, and descriptions of the technologies and processes of the time. There is a large collection of articles and books for the entire period dealing with the "future of the book" and its role in society as well. For a general introduction to the industry (although badly dated since it was written before the impact of the Internet was felt in the industry) is Albert N. Greco, *The Book Publishing Industry* (Boston, Mass.: Allyn and Bacon, 1997); there it has been recently reprinted, but sadly not updated, under the same title (Mahway, N.J.: Lawrence Erlbaum Associates, 2004). For an excellent history of the industry, see John Tebbel, *Between Covers: The Rise and Transformation of Book Publishing in America* (New York: Oxford University Press, 1987). On Amazon.com, there are 2 useful studies. The best is by Robert Spector, *amazon.com: Inside the Revolutionary Business Model that Changed the World* (New York: HarperBusiness, 2000); and a more narrowly focused management guide based on this firm's activities by Rebecca Saunders, *Business the amazon.com Way: Secrets of the World's Most Astonishing Web Business* (Dover, N.H.: Capstone, 1999). For a memoir by an employee from 1996 to 2001, see James Marcus, *Amazonia: Five Years at the Epicenter of the Dot.com Juggernaut* (New York: The New Press, 2004).

Accounts of how book publishing function in the period standardly include some discussion about the technologies used, including computing and advanced printing systems. Useful publications include Chandler B. Grannis (ed.), *What Happens in Book Publishing* (New York: Columbia University Press, 1957; 2nd ed., 1967); Clive Bingley, *Book Publishing Practice* (Hamden, Conn.: Archon Books, 1966); Kathryn Luther Henderson (ed.), *Trends in American Publishing* (Champaign, Ill.: University of Illinois Graduate School of Library Science, 1968); Herbert S. Bailey, Jr., *The Art and Science of Book Publishing* (New York: Harper & Row, 1970); John P. Dessauer, *Book Publishing: What It Is, What It Does* (New

York: R. R. Bowker, 1974); and Lewis A. Coser, Charles Kadushin, and Walter W. Powell, *Books: The Culture and Commerce of Publishing* (New York: Basic Books, 1982), which is a sociological look at the industry's activities.

There are some studies of the specific technologies of the industry. The U.S. Bureau of Labor Statistics published a useful survey of existing technologies and how they were used by printers in the 1960s and 1970s, *Outlook for Technology and Manpower in Printing and Publishing*, Bulletin 1774 (Washington, D.C.: U.S. Government Printing Office, 1973); also useful along the same lines is the *CIS Survey of Computerized Typesetters* (Los Angeles, Calif.: Composition Information Services, 1968). For a bibliography of early works on the general theme of technology in the industry, consult Lowell H. Hattery and George P. Bush (eds.), *Automation and Electronics in Publishing* (Washington, D.C.: Spartan Books, 1965). For the more modern period, a good introduction is provided by Oldrich Standera, *The Electronic Era of Publishing: An Overview of Concepts, Technologies and Methods* (New York: Elsevier, 1987); Don Evedon, *Introduction to Electronic Publishing* (Silver Springs, Md.: Association of Information & Image Management, 1992); but see also two other books, Michael M. A. Mirabito and Barbara L. Morgenstern, *The New Communications Technologies: Applications, Policy, and Impact* (Boston, Mass.: Focal Press, 2001), and a textbook full of current information by John V. Pavlik and Shawn McIntosh, *Converging Media: An Introduction to Mass Communications* (Boston, Mass.: Pearson, 2004). On the economics of the Internet and other recent technologies, see Brian Kahin and Hal R. Varian (eds.), *Internet Publishing and Beyond: The Economics of Digital Information and Intellectual Property* (Cambridge, Mass.: MIT Press, 2000). An early study on e-books is by Geoffrey Nunberg, "The Place of Books in the Age of Electronic Reproduction," *Representations* 42 (1993): 13–37.

Copyright issues and technology have long been the subject of intense interest in the industry. For the relevant literature for the earlier days, see George P. Bush (ed.), *Technology and Copyright: Annotated Bibliography and Source Material* (Mt. Airy, Md.: Lomond Systems, Inc., 1972); but see also a thorough examination of the issues up through the late 1960s by Stephen Breyer, "The Uneasy Case for Copyright: A Study of Copyright in Books, Photocopies, and Computer Programs," *Harvard Law Review* 84, no. 2 (December 1970). Two recent publications provide the necessary background on the issues and history of this subject: Jonathan Rosenoer, *CyberLaw: The Law of the Internet* (New York: Springer-Verlag, 1997); and Jessica Litman, *Digital Copyright* (Amherst, N.Y.: Prometheus Books, 2001).

In addition to this rich collection of books, there is a vast amount of literature in the form of articles; useful examples can be found in the endnotes to chapter 9. However, the most important source of information is *Publisher's Weekly*, the industry's key magazine which appeared throughout the entire period and carried hundreds of pages of discussion about the technology issues in the industry. One cannot study book publishing without this magazine. In addition, while many magazines and journals are published on the industry, from time to time *Publishing Research Quarterly* publishes in-depth studies of the role of technology in this industry.

Newspaper Industry

Robert G. Picard and Jeffrey H. Brody, *The Newspaper Publishing Industry* (Boston, Mass.: Allyn and Bacon, 1997) provides a useful introduction to the structure of the modern industry, but because it was published just before the Internet came into full flower, it missed that new phase in the life of the industry. For a more current study grounded in economics and business issues, and which compares the U.S. experience to that of other countries, consult Timothy Majoribanks, *News Corporation, Technology and the Workplace: Global Strategies, Local Change* (Cambridge: Cambridge University Press, 2000). Looking at descriptions of the industry in different periods sheds light not only on the nature of the industry but also on the timing and effects of various technologies on it. Alan Woods, *Modern Newspaper Production* (New York: Harper & Row, 1963), offers an excellent introduction to the pre-computer era in the industry, circa 1950s and 1960s. The changes become more evident when reading the various editions of Frank W. Rucker and Herbert Lee Williams, *Newspaper Organization and Management* (Ames, Ia.: Iowa State University Press, 1955, 1965, 1969, 1974). Along a similar vein there is Edmund C. Arnold, *Modern Newspaper Design* (New York: Harper & Row, 1969); Robert H. Giles, *Newsroom Management: A Guide to Theory and Practice* (Indianapolis, Ind.: R. J. Berg, 1987); Mario R. Garcia, *Contemporary Newspaper Design: A Structural Approach* (Englewood Cliffs, N.J.: Prentice-Hall, 1987).

For an introduction to offset presses, which began the modern revolution in printing for books, newspapers, and magazines, see David Genesove, *The Adoption of Offset Presses in the Daily Newspaper Industry in the United States* (Cambridge, Mass.: National Bureau of Economic Research, 1999). For a primer published after the arrival of computing into the industry, see Editors of the *Harvard Post, How to Produce a Small Newspaper* (Harvard, Mass.: Harvard Common Press and Port Washington, N.Y.: Independent Publishers Group, 1978); but also see, Dineh Moghdam, *Computers in Newspaper Publishing. User-Oriented Systems* (New York: Marcel Dekker, 1978). For a balanced review of business and technological issues of the 1960s and 1970s, see Anthony Smith, *Goodbye Gutenberg: The Newspaper Revolution of the 1980s* (New York: Oxford University Press, 1980). Martin L. Gibson squarely deals with technology in *Editing in the Electronic Era* (Ames, Ia.: Iowa State University Press, 1979, 1984). On the recent migration to online news services see Pablo J. Boczkowski, *Digitizing the News: Innovation in Online Newspapers* (Cambridge, Mass.: MIT Press, 2004). With chapters on the U.S. experience with graphics, such as at *USA Today*, see Rein Houkes (ed.), *Information Design and Infographics* (Rotterdam, The Netherlands: European Institute for Research and Development of Graphic Communication. While intended for electronic media journalism (such as, radio and TV), there is useful material on newspaper/journalistic practices in Carl Hausman, *Crafting the News for Electronic Media: Writing, Reporting and Producing* (Belmont, Calif.: Wadsworth, 1992).

For events in the industry and technologies in use in the late 1990s and early 2000s, all the various industry websites are useful. Citations in the endnotes for chapter 9 demonstrate the nature of those sources. But also consult the general literature on desktop publishing cited earlier. For a survey of the state of newsroom technologies as of the mid-1990s, see *ASNE/SND Technology Survey 1996* (New York: American Society of Newspaper Editors, 1996).

Magazine Industry

The best single volume history of the industry is John Tebbel and Mary Ellen Zuckerman, *The Magazine in America, 1741–1990* (New York: Oxford University Press, 1991). The only problem with it is that it only takes the story into the 1980s. However, this introduction to the industry's activities and processes partially fixes that problem: Charles P. Daly, Patrick Henry, and Ellen Ryder, *The Magazine Publishing Industry* (Boston, Mass.: Allyn and Bacon, 1997). An economic chapter-length perspective of the industry can be found in Benjamin M. Compaine and Douglas Gomery, *Who Owns the Media?* (White Plains, N.Y.: Knowledge Industry Publications, 1982). See also Albert N. Greco (ed.), *The Media and Entertainment Industries* (Boston, Mass.: Allyn and Bacon, 2000.)

There are no major studies on the role of computing in this industry; the story has to be teased out of various sources. The Printing Industry's various trade publications fill in details about the new printing equipment of the 1960s and 1970s. However, the best source for magazines is its own industry publication, *Folio*, which has periodically published on the topic over many years. An essential publication is a lengthy article by Lisa M. Guidone, "The Magazine at the Millennium: Integrating the Internet," *Publishing Research Quarterly*

16, no. 2 (Summer 2000): 14–33, which includes an extensive list of citations from other magazines and websites. Finally, on the role of electronic publishing for scholars, one anthology covers all the major issues, Richard Ekman and Richard E. Quandt (eds.), *Technology and Scholarly Communication* (Berkeley, Calif.: University of California Press, 1999).

Radio Industry

For a general introduction to the industry there is no better guide than Alan B. Albarran and Gregory G. Pitts, *The Radio Broadcasting Industry* (Boston, Mass.: Allyn and Bacon, 2001); it also includes additional bibliography on the industry. For a combination of history and contemporary radio issues, there is the large anthology of essays in Michele Hilmes and Jason Loviglio (eds.), *Radio Reader: Essays in the Cultural History of Radio* (New York: Routledge, 2002). For statistics and other contemporary issues related to the industry, consult the websites for both the U.S. Federal Communications Commission and that of the U.S. Bureau of the Census. Most of the publications cited above under media and entertainment also have chapters dealing with the business of the Radio Industry.

One of the most important sources on both radio and television in the United States is *Broadcasting*, an industry trade journal that covers the entire period to the present. While its coverage of IT and telecommunications is surprisingly limited, its coverage of regulatory and market issues is quite extensive. *Radio World* is another source, both in print and online; the online version is particularly useful for technological issues faced by the industry today because it covers the issues more extensively than *Broadcasting*. Mass trade magazines covering the Internet also discuss the role of radio in this new environment.

Television Industry

This industry is the perennial favorite of so many commentators on modern American media. Many histories have been written of it, along with economic studies. There are no histories of computing in the industry, although in recent years publications have appeared on the recent migration toward the digital. Furthermore, many of the publications listed above on media in general routinely include chapters on television. Starting points for studying this industry are two books that describe how the industry is set up, providing historical context, discussions of its technologies, and a rich body of bibliography: Patrick R. Parsons and Robert M. Frieden, *The Cable and Satellite Television Industries* (Boston, Mass.: Allyn and Bacon, 1998); and James Walker and Douglas Ferguson, *The Broadcast Television Industry* (Boston, Mass.: Allyn and Bacon, 1998). For an overview of Public Television, but with minimal discussion of technologies, see James Day, *The Vanishing Vision: The Inside Story of Public Television* (Berkeley, Calif.: University of California Press, 1995). For economic descriptions and analysis, see Harold L. Vogel, *Entertainment Industry Economics, A Guide for Financial Analysis* (Cambridge: Cambridge University Press, 2001) Benjamin M. Compaine and Douglas Gomery, *Who Owns the Media?* (White Plains, N.J.: Knowledge Industry Publication, 1982) and Robert G. Picard's, *The Economics and Financing of Media Companies* (New York: Fordham University Press, 2002). Albert N. Greco, *The Media and Entertainment Industries*, (Boston, Mass.: Allyn & Bacon, 1999) places TV into the broader context of all media industries.

On recent events in the industry, three studies focus on the quality of television programming, recent cable activities, and a broad analysis of media industries and the Internet; see William F. Baker and George Dessart, *Down the Tube: An Inside Account of the Failure of American Television* (New York: Basic Books, 1998); Mark Robichaux, *Cable Cowboy: John Malone and the Rise of the Modern Cable Business* (New York: John Wiley & Sons, 2002); and John Motavalli, *Bamboozled at the Revolution* (New York: Viking, 2002). Bruce M. Owen has written a useful overview of the industry with particular attention to its relationship with the Internet, *The Internet Challenge to Television* (Cambridge, Mass.: Harvard University Press, 1999); one should also consult his earlier book with Steven Wildman, *Video Economics* (Cambridge, Mass.: Harvard University Press, 1992); both contain excellent bibliographies. On regulatory matters, Reed E. Hundt, *You Say You Want a Revolution* (New Haven, Conn.: Yale University Press, 2000), provides extensive discussion about the FCC and television in the 1990s, and for a useful discussion of the FCC at the dawn of cable television, the standard work is by Don R. LeDuc, *Cable Television and the FCC: A Crisis in Media Control* (Philadelphia, Pa.: Temple University Press, 1973).

Numerous books cover the use of technology in the industry over the past decade. A useful starting point is the textbook by John Pavlik and Shawn McIntosh, *Converging Media* (Boston, Mass.: Allyn & Bacon, 2003); followed by Sydney W. Head, Christopher H. Sterling, Lemuel B. Schofield, Thomas Spann, and Michael A. McGregor, *Broadcasting in America: A Survey of Electronic Media*, 8th ed. (Boston, Mass.: Houghton Mifflin Co., 1998). Also useful and quite detailed is Michael M. A. Mirabito and Barbara L. Morgenstern, *The New Communications Technologies: Applications, Policy, and Impact*, 4th ed. (Boston, Mass.: Focal Press, 2001). The last, designed to be a textbook, is clearly written and contains valid reflections of existing technologies of the 1990s and beyond. For an analysis of interactive TV with some historical context, consult Phillip Swann, *TVdotCom: The Future of Interactive Television* (New York: TV Books, 2000).

Several technologies have been studied in some detail. There is a large body of articles and books about the VCR, famously about the format battles; however, a useful introduction to its early history is Gladys D. Ganley and Oswald H. Ganley, *Global Political Fallout: The VCR's First Decade, 1976–1985* (Cambridge, Mass.: Program on Information Resources Policy, Harvard University, 1987). An essential source on VCRs is Frederick Wasser, *Veni, Vidi, Video: The Hollywood Empire and the VCR* (Austin Tex.: University of Texas Press, 2001). For a description of video and how it was used in the last years of the twentieth century, there is Kathryn Shaw Whitver, *The Digital Videomaker's Guide* (Studio City, Calif.: Michael Wiese Productions, 1995). For a technical description of digital broadcasting in the late 1990s, consult Lars Tvede, Peter Pircher, and Jens Bodenkamp, *Data Broadcasting: The Technology and the Business* (New York: John Wiley & Sons, 1999). For a discussion of how one group of television viewers connect on the Internet, forming a virtual community interested in soap operas, see Nancy K. Baym, *Tune In, Log On: Soaps, Fandom, and Online Community* (Thousand Oaks, Calif.: Sage, 2000). For a description of how digital technologies have affected newsroom operations, the standard work now is by Stephen Quinn, *Knowledge Management in the Digital Newsroom* (Oxford: Focal Press, 2002). For a short but fact-filled introduction that includes discussion of HDTV, see Walter Oleksy, *Entertainment: The Information Revolution* (New York: Facts on File, 1996). And to keep up with ongoing developments, consult the industry's premier magazine, *Broadcasting*.

Movie Industry

There are a number of useful overviews of the industry's business operations. For introductions, see Benjamin M. Compaine and Douglas Gomery, *Who Owns the Media?*; and also, Harold L. Vogel, *Entertainment Industry Economics* (Cambridge: Cambridge University Press, 2001). Barry R. Litman has written many articles and book chapters describing the industry, but his most complete work is *The Motion Picture Mega-Industry* (Boston, Mass.: Allyn and Bacon, 1998), which also includes useful bibliographies. Also quite useful, if now dated, is D. Lees and S. Berkowitz, *The Movie Business* (New York: Random House, 1981). An early, now also dated, but well-done overview of the industry is a collection of contributed essays compiled by Tino Balio (ed.), *The American Film Industry* (Madison, Wis.: University of Wisconsin Press, 1985). The two most useful histories of the industry are Douglas Gomery, *Shared Pleasures: A History of Movie Presentation in the United States* (Madison, Wis.: University of Wisconsin Press, 1992); and the standard major work in the field, covering the industry all over the world, Kristin Thompson and David Bordwell, *Film History: An Introduction*, 2nd ed. (Boston, Mass.: McGraw-Hill, 2003). The latter is massive and is scheduled for yet a new edition to appear in 2007–2008.

It is important to understand film as a medium and art form in order to appreciate how important it was for technological innovations to serve the artistic nature of the industry. Standard works in the field include David Bordwell, *On the History of Film Style* (Cambridge, Mass.: Harvard University Press, 1997); Kristin Thompson, *Storytelling in the New Hollywood: Understanding Classical Narrative Technique* (Cambridge, Mass.: Harvard University Press, 1999); and David Bordwell and Kristin Thompson, *Film Art: An Introduction*, 6th ed. (New York: McGraw-Hill, 2001). For a cornucopia of insights on types of films, production, analysis of art forms, and research collections described in an extended bibliographic discussion, see Robert A. Armour, Film: A *Reference Guide* (Westport, Conn.: Greenwood Press, 1980). It desperately needs to be updated, however, to reflect developments of the past quarter-century.

On the use of digital technologies in animation and graphics in film, there is a rapidly growing body of literature that surprisingly provides developments in historical context written by practioners in the industry. Almost all these publications are richly illustrated to demonstrate the effects of digital technology. The key works in this new literature include the first of such, Christopher Finch, *Special Effects: Creating Movie Magic* (New York: Abbeville Press, 1984), providing solid evidence of early uses of computing; a "how to" guide for those wanting to use digital methods in filmmaking by Thomas A. Ohanian, *Digital Nonlinear Editing: New Approaches to Editing Film and Video* (Boston, Mass.: Focal Press, 1993), which is clearly written and rich in detail. For a short book rich with history and clear explanations of the use of technology from the 1960s to the early 1990s, almost as if describing how magicians did their magic, is Christopher W. Baker, *How Did They Do It: Computer Illusion in Film and TV* (Indianapolis, Ind.: Alpha Books, 1994). Mark Cotta Vaz and Patricia Rose Duignan tell the story of George Lucas and his company, ILM, in *Industrial Light & Magic: Into the Digital Realm* (New York: Ballantine Books, 1996); if you could only consult one book on the evolution and application of digital technology in the movies, this is it. For an overview of techniques, circa end of the century, consult Piers Bizony, *Digital Domain: The Leading Edge of Visual Effects* (New York: Billboard Books, 2001); but also see Zoran Perisic, *Visual Effects Cinematography* (Boston, Mass.: Focal Press, 2000); Patricia D. Netzley, *Encyclopedia of Movie Special Effects* (New York: Checkmark Books, 2001); and the earlier study by Thomas A. Ohanian and Michael E. Phillips, *Digital Filmmaking* (Boston, Mass.: Focal Press, 1996). For a contemporary study of the subject, which clearly states the case for why digital will dominate future developments in the industry, there is Matt Hanson, *The End of Celluloid: Film Futures in the Digital Age* (Mies, Switzerland: RotoVision S.A., 2004). There are literally dozens of current descriptions of how to use the technology, but for a simple introduction that covers most aspects of the subject consult Scott Billups, *Digital Moviemaking: The Filmmaker's Guide to the 21st Century* (Studio City, Calif.: Wiese Productions, 2000).

While there is a paucity of material on the role of computing and digital technologies in other aspects of the industry's activities, there are, nonetheless, a few resources. One excellent study on VCRs and Hollywood is the essential book by Frederick Wasser, *Veni, Vidi, Video: The Hollywood Empire and the VCR* (Austin, Tex.: University of Texas Press, 2001). Two other studies deal with technologies affecting both TV and movies: Eugene Marlow and Eugene Secunda, *Shifting Time and Space: The Story of Videotape* (New York: Praeger, 1991); and Aaron Foisi Nmungwun, *Video Recording Technology: Its Impact on Media and Home Entertainment* (Hillsdale, N.J.: Lawrence Erlbaum Associates, 1989). For the implications of the digital on entertainment and movies, consult Walter Oleksy, *Entertainment: The Information Revolution* (New York: Facts on File, 1996). On the current state of the industry and technology's influence on it, see Screen Digest, *Screen Digest Report on the Implications of Digital Technology for the Film Industry* (London: Screen Digest, September 2002); it is both excellent and well-written, a must-read. For a critique of the industry's response to the Internet and that also discusses the music and video game industries, see J. D. Lasica, *Darknet: Hollywood's War Against the Digital Generation* (New York: John Wiley & Sons, 2005).

Industry publications are usually valuable sources on the role of technology; however, that is not the case with the Movie Industry. *Variety* was surprisingly a thin source of information, as were both *Screen Digest* and *The Velvet Light Trap*. In fairness to the last two, however, when they focused on IT their coverage was excellent. The shortfall is made up, however, by articles that appeared in other publications, such as in IT technical journals. Note also that publishers of books on this industry tend to produce series on film studies, so their catalogues should be routinely consulted for additional material. In this category consider Focal Press, McGraw-Hill, University of Wisconsin Press, and University of Texas Press.

Music Recording Industry

The single most useful introduction to this industry was written by Geoffrey P. Hull, *The Recording Industry* (Boston, Mass.: Allyn and Bacon, 1998). Also helpful are Benjamin M. Compaine and Douglas Gomery, *Who Owns the Media?* (Mahwah, N.J.: Lawrence Erlbaum Associates, 2000) Harold L. Vogel, *Entertainment Industry Economics*, (Cambridge: Cambridge University Press, 2001) and John V. Pavlik, *New Media Technology* (Boston Mass.: Allyn & Bacon, 1997). The most complete discussion of business practices in the industry, with extensive coverage of copyright issues, can be found in M. William Krasilovsky and Syndney Shemel, *The Business of Music: The Definitive Guide to the Music Industry* 9th ed. (New York: Billboard Books, 2003), which is the most complete. But also consult Dick Weissman, *The Music Business: Career Opportunities and Self-Defense* (New York: Three Rivers Press, 1997), either edition, but the second is the most comprehensive and current. We now have a study of Napster by Joseph Menn, *All the Rave: The Rise and Fall of Shawn Fanning's Napster* (New York: Crown Books, 2003). A useful overview with historical perspective is Robert Burnett, *The Global Jukebox: The International Music Industry* (London: Routledge, 1996).

There is a large body of material on digital music; some of it is reflected in the endnotes. However, a major reference is Robert L. Wick, *Electronic and Computer Music: An Annotated Bibliography* (Westport, Conn.: Greenwood Press, 1997), reviewing some 250 books on all aspects of the subject. It would be difficult to ignore comments by Raymond Kurzweil, who has published a number of books on computing and society, but consult *The Age of Intelligent Machines* (Cambridge, Mass.: MIT Press, 1992). For a contemporary user's guide to digital music, see Thomas E. Rudoph and Vicent A. Leonard, Jr., *Recording in the Digital World: Complete Guide to Studio Gear and Software* (Boston, Mass.: Berklee Press, 2001); but also consult Paul Theberge, *Any Sound You Can Imagine: Making Music/Consuming Technology* (Hanover, N.H.: University Press of New England for Wesleyan University Press, 1999). There are three useful historical treatments: Ben Kettlewell, *Electronic Music Pioneers* (Vallejo, Calif.: ArtistPro, 2001); A. J. Millard, *America on Record: A History of Recorded Sound* (New York: Cambridge University Press, 1995); and Iara Lee and Peter Shapiro (eds.), *Modulations: A History of Electronic Music* (London: Distributed Art Publishers, 2000).

The complex issues concerning the Internet are best studied today by reading contemporary American newspapers and magazines, as it is a topic too early to have its own historian. As many of the citations above suggest, existing works on the industry were published just prior to the subject becoming critical to the industry. Each industry website carries statements about the industry's position, while leading entertainment magazines report on events, such as *RollingStone and Billboard.* Because of the highly public nature of the row, leading trade publications also discuss the issue from time to time, such as *Newsweek, Fortune*, and *Time*. For an overall discussion and set of recommendations about file-sharing and copyright issues in this and other media industries and the Internet, see William W. Fisher III, *Promises to Keep: Technology, Law, and the Future of Entertainment* (Stanford, Calif.: Stanford University Press, 2004). On Steve Jobs and the industry, consult Jeffrey S. Young and William L. Simon, *iCon: Steve Jobs, The Greatest Second Act in the History of Business* (New York: John Wiley & Sons, 2005).

Video Games Industry

For such a young industry, it is surprising how many important studies have been published about it. Most focus on the creation of games and descriptions of them and their developers, but some contain industry data, as well. Standard works, many of them beautifully illustrated, include Leonard Herman, *Phoenix: The Rise and Fall of Videogames*, 2d ed. (Union, N.J.: Rolenta Press, 1999); Steven L. Kent, *The Ultimate History of Video Games* (Roseville, Calif.: Prima, 2001); Arthur Asa Berger, *Video Games: A Popular Phenomenon* (New Brunswick, N.J.: Transaction Publishers, 2002); Van Burnham and Ralph H. *Baer, Supercode: A Visual History of the Videogame Age, 1971–1984* (Cambridge, Mass.: MIT Press, 2001); Mark J. P. Wolf, *Medium of the Video Game* (Austin, Tex.: University of Texas Press, 2001); and Rusel DeMaria and Johnny L. Wilson, *High Score! The Illustrated History of Electronic Games*, 2d ed. (New York: McGraw-Hill/Osbourne, 2004). For accounts of the industry with particular emphasis on the gamers themselves and their culture, there is J. C. Herz, *Joystick Nation: How Videogames Ate Our Quarters, Won Our Hearts, and Rewired Our Minds* (Boston, Mass.: Little, Brown and Company, 1997); and Brad King and John Borland, *Dungeons and Dreamers: The Rise of Computer Game Culture from Geek to Chic* (New York: McGraw-Hill/Osborne, 2003), which includes a great deal on Atari as well. Computing is also working its way into toys in general, and for that story see the highly readable book by Mark Pesce, *The Playful World: How Technology Is Transforming Our Imagination* (New York: Ballantine Books, 2000). For connections to other media, there is Geoff King and Tanya Kryzwinska (eds.), *ScreenPlay: Cinema/Videogames/Interfaces* (New York: Wallflower, 2002). On skills needed to play video games and also one of the earlier publications on video games in general, see David Sudnow, *Pilgrim in the Microworld* (New York: Warner Books, 1983).

Publications are beginning to appear on the economic and business aspects of the industry itself. Harold L. Vogel, *Entertainment Industry Economics* (Cambridge: Cambridge University Press, 2001), provides a brief introduction, while Martin Campbell-Kelly, *From Airline Reservations to Sonic the Hedgehog: A History of the Software Industry* (Cambridge, Mass.: MIT Press, 2003), devotes a sizable part of a chapter to it. The best introduction to business strategies and the role of technologies in defining the work of this industry is an article by Melissa A. Schilling, "Technological Leapfrogging: Lessons From the U.S. Video Game Console Industry," *California Management Review* 45, no. 3 (Spring 2003): 6–32. A book by a game producer describes work in the industry in the early 2000s and includes history and major practices: Ernest Adams, *Break into the Game Industry: How to Get a Job Making Video Games* (New York: McGraw-Hill/Osbourne, 2003); see also Alice LaPlante and Rich Seidner, *Playing for Profit: How Digital Entertainment Is Making Big Business Out of Child's Play* (New York: John Wiley & Sons, 1999). Histories of companies are becoming available. On Atari, there is S. Cohen, *Zap: The Rise and Fall of Atari* (New York: McGraw-Hill, 1984); but see also Brad King and John Borland, *Dungeons and Dreamers* (New York: McGraw-Hill, 2003). The major work on Nintendo is David Sheff, *Game Over: How Nintendo Zapped an American Industry, Captured Your Dollars, and Enslaved Your Children* (New York: Random House, 1993). Two histories of Sony Corporation include material on its videgame business: John Nathan, *Sony: The Private life* (New York: Mariner Books, 2001);

and the more detailed account, Shu Shin Luh, *Business the Sony Way: Secrets of the World's Most Innovative Electronics Giant* (New York: John Wiley & Sons, 2003).

There is a growing number of Internet sites for material on video games. The problem, of course, is that there is no guarantee that any of them will remain in existence or not change their site addresses. However, with that caveat, there is "Computer Gaming Bibliography," compiled by Matthew Southern et al. http://www.digiplay.org.uk/a-g.php; and "Timeline: Video Games," compiled by Amanda Kudler, http://www.infoplease.com/spot/gamestimeline1.html.

Photographic Industry

There are no book-length overviews of the industry at large, either as economic or historical studies. There are thousands of books on photography, however, and others on the history of cameras and photography. For our purposes, there are two business histories worth consulting. The first is a classic table-top, highly illustrated volume on the history of Kodak by Douglas Collins, *The Story of Kodak* (New York: Harry N. Abrams, 1990), which is long on pictures and short on scholarship but nonetheless a useful introduction to the firm in the years prior to the emergence of digital photography. A book highly critical of the company that was written by a journalist and covers the last two decades of the twentieth century in detail is Alecia Swasy, *Changing Focus: Kodak and the Battle to Save a Great American Company* (New York: Times Books, 1997). There are two studies of Sony that offer a great deal of information about its consumer electronics business, but each provides only limited accounts of digital photography. They are Shu Shin Luh, *Business the Sony Way: Secrets of the World's Most Innovative Electronics Giant* (Oxford: Capstone Publishing, 2003); and the better of the two, John Nathan, *Sony: The Private Life* (Boston, Mass.: Houghton Mifflin, 1999). The latter also has material on its games products. None of the media industry surveys cited earlier in this bibliography have chapters on the Photographic Industry.

There are literally hundreds of "how to" books on digital photography that provide descriptions of how digital cameras work, how best to take great photographs, and how to print them. They keep changing as the technology does, so citing them here would not add much value. Every large bookstore in industrialized economies seems to have many of them; few, however, have made their way into academic libraries. The best source of material on the emergence of digital photography and trends related to this technology is the industry's leading consumer publication, *Popular Photography*, which was published through the twentieth century on a monthly basis. There are both electronic and print versions. The Photo Marketing Association International (PMAI), the industry's leading trade association, publishes continuously on various trends in the industry, most notably about retail sales and consumer behavior. Its materials are crucial to any understanding of the industry, but most circulate among member companies. However, their website does make some data available to the public; citations can be found in the endnotes to chapter 12. Finally, the American business-trade press publishes a constant stream of news articles on the industry's key players and events that are worthwhile examining. Kodak does not maintain a corporate archive, although its media relations department does have access to recent information about the firm and the industry in which it operates.

Changing Nature of Work

General studies on work and industry transformations are useful additions to the literature focused on specific industries. The following were most influential in the preparation of the second volume of *The Digital Hand*. Because industries were transforming so rapidly in the late twentieth century, it should be no surprise that various books would begin appearing on the theme. Jeremy Rifkin published a widely circulated study that essentially argued the negative case for globalization and decline of standards of living and conditions for workers in the advanced economies, caused largely by technology in general and computers more specifically, in *The End of Work: The Decline of the Global Labor Force and the Dawn of the Post-Market Era* (New York: G. P. Putnam's Sons, 1995). A more rigorous academic study was prepared by Peter Cappelli and a team of researchers; it centered on the need for new types of skills across all industries: *Change at Work: How American Industry and Workers Are Coping with Corporate Restructuring and What Workers Must Do to Take Charge of Their Own Careers* (New York: Oxford University Press, 1997). Charles Grantham addressed issues workers faced due to changing technologies, not the least of which involved computers, in *The Future of Work: The Promise of the New Digital Work Society* (New York: McGraw-Hill, 2000). I focused one of my earlier books on the effects on management and what they needed to do in James W. Cortada, *21st Century Business: Managing and Working in the New Digital Economy* (Upper Saddle River, N.J.: Financial Times/Prentice-Hall, 2001). Simon Head prepared a study for the Century Foundation on workers and how computing was destroying their salary levels and how the nature of work was changing: *The New Ruthless Economy: Work and Power in the Digital Age* (New York: Oxford University Press, 2003). A broader view of the transformation of the U.S. economy in recent years is provided by Roger Alcaly, *The New Economy: What It Is, How It Happened, and Why It Is Likely to Last* (New York: Farrar, Straus and Giroux, 2003); but also examine a deeply thoughtful, very readable economic analysis that includes discussion of creative destruction theories of Joseph Schumpeter in modern times by economist Richard R. Nelson, *The Sources of Economic Growth* (Cambridge, Mass.: Harvard University Press, 1996). In a deeply innovative study of what tasks people do in work versus what computers can perform, we see new issues regarding the nature of work, Frank Levy and Richard J. Murnane, *The New Division of Labor: How Computers Are Creating the Next Job Market* (New York: Russell Sage Foundation, and Princeton, N.J.: Princeton University Press, 2004). Thomas W. Malone offers a management primer that reflects many of the changes that occurred in the workplace of the 1990s and early 2000s, in *The Future of Work: How the New Order of Business Will Shape Your Organization, Your Management Style, and Your Life* (Boston, Mass.: Harvard Business School Press, 2004). Because all important industries are global and increasingly integrated (thanks in part to the capabilities of computers and telecommunications), getting an introduction to the increasing role of multinational companies is important. In a novel approach, in which geographical graphics techniques are used to provide an atlas of the subject, we have a short book packed with statistical data on some of the industries described in this book: Medard Gabel and Henry Bruner, *GlobalInc.: An Atlas of the Multinational Corporation* (New York: New Press, 2003). A similar graphical representation of media industries is also available, prepared by Mark Balnaves, James Donald, and Stephanie Hemelryk Donald, *The Penguin Atlas of Media and Information: Key Issues and Global Trends* (New York: Penguin Putnam, 2001). For those used to reading books with text and few graphics, these two books can be described as "fun" to look at; they have almost no narrative text.

INDEX

ABC Radio, 276, 277, 338
ABC Television, 350, 366
Abramavitz, Moses, 457
Access Content System Licensing Administrator (AACSLA), 423
Accounting
  in Brokerage Industry, 14
  digital technologies in, 263
  mortgage, 56–57
  in Movie Industry, 385–386
  in Music Recording Industry, 402–403
  software for, 265
Accounting Receivable Conversion (ARC), 51
Actual radio airplay, 405
Adams, Ernest, 412, 413
Adding machines, 18–19, 45
Addison-Wesley, 305
Adobe Systems' *Illustrator* and *Photoshop*, 329
Advance Publications, 322
Adventure International, 420
Advertising agencies, 297
Aetna Life and Casualty, 126, 135
Agfa, 431
Agricultural sector, 480, 481
Alcaly, Roger, 456, 457
Alcatel, 210
Aldus Corporation, 307
Allstate Insurance, 130
  Sears' acquisition of, 42
Alternative Entertainment Network TV (AENTV), 365
A&M, 279
AMA computers, 235
Amazon.com, 105, 153, 287, 289, 296, 309–312, 324, 334, 406, 422
American Automobile Insurance Company, 130
American Banking Association (ABA)
  on ATMs, 69, 70
  on automation, 82
  on check processing, 49, 50
  on computer use, 84
  on consumer loans, 56
  on coordination of technology adoptions, 37–38
  on Internet banking, 51, 54, 102, 105
  leadership provided by, 86
  as lobbyist, 39
  on profitability, 109
  on standardization efforts, 46, 47
  survey of consumer habits, 72, 73
American Banking Association (ABA) Banking Journal, 102
American Bible Society, 299
American Century Companies, 185
American economy. *See* United States economy
American Express, 33, 64, 69, 470
  credit cards of, 61, 470
American industries. *See also specific*
  telecommunication applications across, 234
American Movie Classics, 277
American Society of Newspaper Editors, 324
American Stock Clearing Corporation, 166

American Stock Exchange (AMEX), 14, 154, 155, 162, 166, 168, 169, 173
America Online (AOL), 98, 269, 288, 290, 415
Ameritech, 202
Analog signals, 486
Analog video, 359
Animation
in Movie Industry, 372, 401
in Music Recording Industry, 401
Anytime-anyplace banking, 68
Apple Computer, 268, 269, 279, 280, 286, 329, 411, 415, 419, 450, 452, 460, 463, 467, 473
Arcade video games, 415
Archipelago, 186
Arch Wireless, 218
Area codes, increasing demand for, 488
Argonaut Underwriters, 130
Armor, Paul, 45
Armstrong, Michael, 223
Artificial intelligence (AI)
in Insurance Industry, 135
in Video Games Industry, 414
Asset overhang, 219
Associated Press (AP), 316
Atari, 279, 415, 419, 420
Video Game Computer System, 417, 425
AT&T, x, 77, 91, 100, 123, 192, 194, 205, 208, 217, 220, 229, 321, 447, 470, 486
acquisition of TCl, 369
Automatic Message Accounting computer at, 230
awareness of importance of telecommunications, 238
breakup of, 194, 197, 201, 203, 222, 224, 226, 227–228, 232, 237, 240, 261
computer use by, 234
credit cards of, 26
employment level of, 196
fiber-optics network at, 243
institutional rigidity of, 202
management resistance to Internet, 451
market share of, 203
modernization at, 241
research and development by, 207, 239–240
service level provided by, 261
settlement between Antitrust Division and, 200–202
transistor development by, 339
AT&T Universal Card, 64
AT&T Wireless, 211, 212, 219
AudioNet, 348
AudioVault drives, 349
Authorization services, credit cards and, 63
Automated attendant systems, 258
Automated Clearing House (ACH), 52, 75
growth of transactions of, 89–90
Automated newsroom systems, 317
Automated stock transfers, 14
Automated teller machines (ATMs), 4, 12, 470
in Banking Industry, ix, 39, 66–73, 88, 369, 444
in Brokerage Industry, 157–158
debit cards as cards for, 65–66
fraud losses through, 70
growth in numbers of, 88
impact of, 24
Automated Trading System (ATS) in Brokerage Industry, 172
Automatic call distributor (ACD), 240
Automatic Message Accounting computer at AT&T, 230
Automation, 19
in Banking Industry, 45, 47–48, 82–87
in Brokerage Industry, 14, 172, 187
in Print Media Industries, 317
in Telecommunications Industries, 258
Automation Consultants, Inc., 326
Automobile Club of Southern California, 135
Automobile insurance, 118
demand for, 117
Automobiles
consumer loans financing of, 55
installment purchase of, 33
Automotive Industry
advertisements of, 409
remote control devices and, 359

Baby Bells, 219, 276
Bache & Company, 163
Back-office infrastructure in Banking Industry, 65
Back-office productivity in Brokerage Industry, 184
Baker & Taylor, 310
BancOne, 72
Bandwidth, 489
auctioning of, to service providers, 253
BankAmerica, 61, 64, 72, 106
BankAmeriCard, 33, 62
Bank holding companies, 41
Banking and Securities Industry Committee (BASIC), 166, 187
Banking Industry, x, 6, 37–79, 113, 448, 483, 484
ACH transactions in, 89–90
artificial intelligence in, 135
assets in, 39, 40, 41, 116
ATMs in, ix, 39, 67–73, 88, 369, 444
automation in, 45, 47–48, 82–87
back-office operations in, 65, 94
balance sheets in, 93
bar code technology in, 50
base technologies in, 93
branch banking in, 72, 99–100
centralized facilities in, 115
check processing in, 44–52
competition in, 58, 64–65, 69, 71, 100, 107, 118–119, 120
competitive advantages in, 93
computers in, 43–79, 120–126, 442
consolidations in, 94, 346
consumer account management in, 102
consumer influence in, 97–98, 120
cost justifications in, 48, 69–71, 84
credit cards in, 45, 60–67, 72, 93, 110
cross selling in, 109
database management systems in, 124
data in, viii
data processing in, 85, 89, 113–114
data transfer in, 110
debit cards in, 60–67, 71, 72, 93, 100, 107, 110
decentralization in, 115
decline in number of companies, 38
demand deposit accounting in, 44–52
deregulation in, 92, 94
digital applications in, 80–112
distinct personality of, 80–81
Electronic funds transfer (EFT) system in, 39, 42, 44, 45, 49–50, 73–78, 89, 91, 94, 141
employee productivity in, 89
employment levels in, 9, 39, 41
evolution of, 37, 38–43, 116–120
expert systems in, 59, 134–135
export of data entry and file management in, 119
government regulation in, 15–16
gross domestic product of, 8–9
gross total assets in, 41
growth of, 9, 39
home banking in, 102–103
independent agents in, 141–142
information technology in, 95–96, 137–143
institutions in, 39–40
Insurance Industry and, 5
Internet in, 72–73, 96–109, 143–147
intra-industry cooperation in, 38
labor productivity in, 86
loan demand in, 41
loan processing in, 55–59, 93
machinery in, 82, 88
mail-based billing and payments in, 110–111
managerial maturity in, 100–101
mergers and acquisitions in, 41, 64, 90
movement of cash and credit across economy, 152
in networked world, 96–109
noncomputerized office equipment in, 45–46, 52
normalization of computerized banking in, 87–95
OCR technology in, 140
operational efficiencies in, 482
optical character recognition in, 135
output in, 86
paperwork in, 19
payroll processing and, 43
PC banking in, 91–92
physical security in, 100
precomputer technologies in, 19

Banking Industry (*continued*)
privacy issues in, 97, 106–107
punch-card tabulating equipment in, 121
record keeping by, 12
regulation of, 41, 113
retail banking in, 94
return on investment calculations for, 108
saving and investment applications, 52–55
service bureaus in, 83, 87
services offered in, 11–12, 30, 40, 106
structure of, 148
tasks of, 10–12
telecommunications applications in, 80–112, 115
trends and current issues in, 110–111
turnovers in, 157–158
Websites in, 145
word processing in, 127–128
Banking (magazine), 47
Bank of America, 46, 47–48, 48, 49, 61
universal credit cards of, 61
Bank of International Settlements (BIS), 98
Bank One Texas, 71
Bankruptcy protection, 218
Banks. *See also* Banking Industry
defined, 8
Bank Wire facility, 74
Bannan, Peter J., 49
Bantam, 305
Baran, Paul, 222, 488
Bar codes
in Banking Industry, 50
in Grocery Industry, 24, 482
in Print Media Industries, 308
Barnes & Noble, 272, 287, 296, 306, 311
Barton, Randolph, 416
Basic Books, 334
Batch computer systems, 35
switch from online systems to, 54
BBC, 352, 470
Bear, Stearns and Company, 163
Bedroom bands, 402
Bell, Daniel, 458
Bell Atlantic, 202, 205
Bell Heads, 489–490
Bell Laboratories, 191, 202, 231
automated equipment at, 237
computer use at, 230
employment level of, 196, 226
research and development by, 207, 208–209, 225
BellSouth, 202, 219, 321
Bell System Dataphones, 433
Bell Systems, 190, 200, 202, 240, 366
billing computer applications at, 235–237
breakup of, 196, 197, 201, 240, 257
computer use in, 230–233, 237
data processing in, 230, 234
employee level in, 226
internal publications of, 232
Benchmarking, 472–473
Benn, Alex, 161
Berger, Allen N., 94
Berninger, Daniel, 251
Bertelsmann, 272, 310, 394
Best, 119
Best's Review, 133–134, 136, 143, 146
Better Homes & Gardens, 327–328
Bezos, Jeff, 309–310, 311
Billboard, 405
Bill presentment in Internet banking, 106
Bitpop, 402
Blackboards in Brokerage Industry, 161, 167
Block Automation System in Brokerage Industry, 172
Blockbuster Video, 264
Blogs, 273
Bloomberg, 322
BMG, 279, 394
Board games, 415–417
Bodie, Zvi, 42, 44
Boettinger, Henry M., 229
Boezcowski, Pablo J., 334–335
Bond swaps, 14
Book Publishing, 264, 267, 271–272, 294–314, 334, 463, 467, 469
categories in, 417–418
challenges in, 312–314
computer-based skills of workers in, 297
firms in, 272
game sales by, 422

information technology in, 296, 298–309
Internet in, 272, 309–313
inventory control in, 266, 296, 302, 308
mergers and acquisitions in, 311
packagers in, 297
players in, 295
product evolution in, 297
production in, 295
sales in, 295, 310
self-publishing in, 474
subgroups in, 296–297
trends in, 297, 309–312
wholesalers in, 296
Boorstin, Daniel J., 33
Bootlegged DVDs, 286
Borders, 296
Boston Mutual Life, 123
Bowker, R. R., 299
BPI Communications, 405
Brainerd, Paul, 307
Branch banking in Banking Industry, 99–100
Branch offices, expansion of, 243
Branded credit cards, 61–62
Brandin, David H., 245
BREW, 212
Brick-and-mortar operations, 290
business models in, 468
Bridis, Ted, 407
Broadband, 488
costs of, 218
Broadband cell phone services, 255–256
Broadcast.com, 3
Broadcast Data Systems (BDS), 405
Broadcasting, 354
Broadcast Music, Inc. (BMI), 402–403, 404, 405
Broadcast television, 267, 276–277
Broderbund, 280, 420
Brokerage Industry, 151–188, 452
adding and calculating machines in, 161
automated teller machines in, 157–158
Automated Trading System (ATS) in, 172
automation in, 14, 172, 187
back-office productivity in, 184
blackboards in, 161, 167
ComEx in, 172
day traders in, 183
Designated Order Turnabout in, 172, 173
digital applications in, 160–180
efficiency of, 152
electronic communications networks in, 159
evolution of, 153–159
firms in, 165
floor traders in, 164
information technology in, 171, 266
initial public offerings in, 154, 157
institutional investors in, 156–157
Interactive Data Corporation (IDC) in, 175
Internet in, 180–185
Manning Rules in, 159
matching orders in, 167–168
mergers and acquisitions in, 158
national best bid and offer in, 159
national market system in, 169
online access to information in, 174–175
Opening Order Automated Report Service (OARS) in, 173–174
optical scanning equipment in, 170–171
Order Audit Trail System (OATS) in, 179
passing of Old Order, 151
personal computers in, 174, 175
R4 (Registered Representative Rapid Response) in, 176–177
record keeping in, 465
recovery issues related to 9/11, 186
reforms in, 161
regional exchanges in, 166
regulation of, 17–18, 151
responsibilities of, 152
Securities and Exchange Commission in, 151, 153, 154, 158, 159, 165–166, 168–169
self-policing in, 168
service bureaus in, 167–168
Small Order Execution System (SOES) in, 178
stock brokers in, 152

Brokerage Industry (*continued*)
SuperDOT in, 172
telegraphic systems in, 167
telephonic systems in, 167
teletypewriters in, 167
ticker tape system in, 161–162
transformation of, 23
trends in, 185–186
turnovers in, 157–158
Y2K in, 179–180, 185
Browsers, availability of, and Internet use, 245–246
Bubble economists, 225
Bulletin boards, 246
Bunches of mortgages, 42
Bunker-Ramo Corporation, 167
Burroughs, 52, 130, 353
Bushnell, Nolan, 417
Business, emergence of new style of, 22–30
Business applications
across industries, sectors and economic models, 479–484
in Brokerage Industry, 151–188
in Insurance Industry, 113–150
in Movie Industry, 373, 384–392
in Music Recording Industry, 402–405, 410
in Radio Industry, 338–339
in Telecommunication Industry, 221–224
in Television Industry, 351–352
Business historians, 479–480
Business justification, xii
Business models, 490
in Media and Entertainment Industry, 287–292
in Music Recording Industry, 410
for online content services, 287–288
Business-to-business publications, 273
Business-to-business transactions, electronic processing of, 74–75

Cable Act (1984), 357
Cable News Network (CNN), 466, 469
Cable television, 240, 244, 267, 276–277, 351–352, 460
FCC and, 357–358
Cablevision System, 277
Cahill, Thaddeus, 401
Calculators, 18–19
Caller ID, 199, 261
Call forwarding, 217, 235
Campbell-Kelly, Martin, 420–421, 425
Canon, 268, 431, 433, 435
Capitalist economy, features of, 464
Capitol, 279, 393
Card associations, emergence of, 61
Card-reading device, 60
Carlos, Walter, 401
Carlos, Wendy, 401
Carroll, Glenn R., 32
Carte Blanche, universal credit cards at, 61
Carterphone, 203
Cartoon Network, 277
CashBox, 405
Cashless societies, 34, 110
Casio, 212
Castells, Manuel, 189–190
Casualty Insurance, 12, 13
adoption of computing in, 129–137
employment in, 117
gross domestic product in, 117–118
independent operations in, 119
number of firms, 117
revenues in, 118
CBS, 279
CBS News, 366
CBS Radio, 276
CBS Television, 276, 277, 338, 346, 350, 352, 354, 362
CCS, 187
CDC, 353
CDnow.com, 406
CDs, 40, 285, 308, 313
early, 306
sales of, 279, 395
in Video Games Industry, 415
Cell phones, 210–211, 211, 450
broadband, 255–256
functions available on, 254, 255
photography and, 437–438
in Telecommunications Industry, 252–257, 261
in Telephone Industry, 199
Census Bureau, U.S.
publication of reports on media usage, 195
tracking of economic data by, 193

Center for Research and Security Prices at the University of Chicago, 175
Central Certificate Service (CCS), 166
Centralized facilities in Insurance Industry, 115
Centralized file sharing, 462
Century Foundation, 469
Cert, Vinton, 189, 191, 269, 488, 489
Chandler, Alfred D., Jr., 448, 478
Charge account banking, 62
Chase Manhattan Bank, universal credit cards at, 61
Chat rooms, 244
Checkless society, 12, 49
Checks
  cashing, 12
  clearing of, 84
  handling bounced, 10
  in Internet banking, 106
  processing, 50–52, 77
Check Truncation Act (2003), 110
Chemical Banking Corporation, 64
Chicago Board of Trade, 14, 177
Chief executive officers (CEOs), roles and missions of, 25
Chief financial officers (CFOs), roles and missions of, 25
Chip music, 402
Christmas Clubs, 52
Churn, opportunities created by, 473
Circuit breakers, 178
Circuit switching, 489
Cirrus, 71
  building of national networks by, 70
Cisco, 194, 210, 292, 445
Citibank, 106, 229
Citicorp, 7, 64, 93
Citigroup Inc., 146
ClariNet Communications, 245
Clear Channel Communications, 346
Clearing and settlement services, 14
Clearing House Interbank Payments Systems (CHIPS), 75
Clemons, Eric K., 184
Clinton, Bill, 204
  administration of, 97, 206, 462
CNN, 276, 277, 365, 366
COBOL, 125, 138, 179
Coca Cola, 281
Coleco, 420
Columbia, 279, 393
Comcast, 276, 277
ComEx, in Brokerage Industry, 172
Commerce, U.S. Department of, in driving Internet traffic, 246–247
Commerce Bank, 99–100
Commercial banks, 7, 39–40
Commercial loans, 55, 58
Commodore, 279, 415
Common Message Switch (CMS), 166
Communications applications in United States, 239–240
Communications Workers of America (CWA), 317–318
Compaine, Benjamin M., 267, 290
Compaq, 212, 253, 268
Competition
  in Banking Industry, 58, 64–65, 69, 71, 100, 107, 118–119, 120
  effects of regulatory practices on, 16
  between MasterCard and Visa, 67
  in Photographic Industry, 431
  in Telecommunication Industry, 203, 217, 231
Competitive Advantage of Nations, 480–481
Competitive advantages in Banking Industry, 93
Composition, 300–301, 302
Compton, 306
CompuBank, 106
CompuServe, 307, 415
Compustat Services, Inc., Standard & Poor's (S&P) establishment of, 175
Comp-U-Store, 181
Computer applications, patterns and practices in adoption of, 18–22
Computer-assisted journalism (CAJ), 324
Computer-based composition tools, 328
Computer-based games, 280
Computer-based retail store systems in Book Publishing Industry, 302
Computer-based skills of workers in Book Publishing Industry, 297
Computer chips, ix–x, 208
  in Video Games Industry, 414
Computer-enabled desegregation, 64
Computer games, film-related, 391–392

Computer graphics
  in Movie Industry, 401
  in Music Recording Industry, 401
Computer Industry, 413, 484
  patterns in, 225–226
Computer revolution, lack of, ix
Computer technologies, viii
Computer vendors, role of, 19–20
Computing
  emergence of books about, 305–306
  in Print Media Industry, 293–335
  in services sector, 4
Concentration in Radio Industry, 346
Conferencing, 217
Congress, U.S., regulatory role of, 17
Consoles in Video Game Industry, 418, 424, 425–426, 429
The Consortium, 423
Consumer Electronics Industry, 268, 420, 450
  products in, 268, 269
Consumer loans, 55–56, 93
Consumers
  in Banking Industry, 97–98
  changing patterns of, for financial services, 30–33
  financial assets of, 42–43
  in Music Recording Industry, 410–411
  tastes of, 454
  in Video Games Industry, 417
Continental, 280
Continuous Net Settlement System (CNS), 166
Convergence
  digital technologies in, 263–264
  in Television Industry, 361
  trends in, 264
Cooper, C. Russell, 83–84
Copy print machines, 431–432
Copyrights, 459, 460, 461–462
  laws on, 297, 313
  in Music Recording Industry, 278, 393, 394, 395–396, 402, 405
Corinthian Broadcasting Company, 342
County Trust Company (White Plains, NY), 37
Courseware, 304–305
Court TV, 277
Cox Enterprises, 275, 276, 277, 322
Craftlike manufacturing, 22
Crane, Dwight B., 42, 44
Creative destruction, 449, 463
Credit card associations, 62, 63
Credit cards
  in Banking Industry, 45, 60–67, 72, 93, 110
  branded, 61–62
  early, 7
Credit-scoring, 63
  applications, 94
  use of computers in, 57, 58
Credit unions, 39
Cronin, Mary J., 33–34
Cross-industry applications, 235
Cross-market restrictions, 18
Cross selling in Banking Industry, 109
Cross-state banking, 18
CRTs
  in Newspaper Industry, 317, 318, 319, 320
  in Print Media Industries, 300–301, 326–327
Cryptograpic security, increasing demand for, 241
Culberson, James M., Jr., 72
Curtis, Richard, 313
Customer ledger record, production of detailed, 53
Customer-support centers, computerized, 240
Cybernetics in Video Games Industry, 414
Cycle-time improvement, 473

D-1s, 231
D-50 portable audio product, 340
Dalton, 296
Database management system (DBMS) applications, in Insurance Industry, 124, 126
Data collection, digital technologies in, 263
Data Communications Corporation, 355
Data compression standards, 360–361
DATAmatic 1000 general purpose computer system, 53
Datamost, 420
Data network, 487

Data processing
in Banking Industry, 89, 113–114
at Bell Systems, 230, 234
Data streaming, 217
Data transfer in Banking Industry, 110
Data transmissions, 215, 488, 489
increase in demand for services, 237
Datek, 183
Davenport, Thomas H., 35
David, Paul, 457
Day traders in Brokerage Industry, 183
DC Comics, 264
Dean Witter, 158, 163, 186
Sears' acquisition of, 42
Debit cards in Banking Industry, 60–67, 71, 72, 93, 100, 107, 110
Decentralized computing in Banking Industry, 96
Decentralized file sharing, 462
Deintegration of services, 481
Dekom, Peter, 390
Dell, 268
Demand deposit accounting and check processing in Banking Industry, 44–52
Dennis, Everette E., 268
Department stores, credit card use by, 61
Depository Institutions Act (1982), 16
Depository Trust Company, 166
Deregulation
in Banking Industry, 92, 94
in Radio Industry, 345
in Telecommunication Industry, 201–202, 206
Deregulatory initiatives, 18
Desegregation
of banking, 64
computer-enabled, 64
Designated Order Turnabout (DOT) in Brokerage Industry, 172, 173
Desktop Publishing, 307, 326, 327, 333
technical components of, 328–329
De Sola Pool, Ithiel, 207–208
Detroit News, 316
Detroit Stock Exchange, 163
Dial-it services, 240
Dialog, 290, 296
Digital applications
in accounting, 263
in Banking Industry, 80–112
in Brokerage Industry, 160–180
in data collection, 263
in Electronic Media Industries, 336–370
in Entertainment Industry, 371–441
in Insurance Industry, 113–150
management attitude toward, 283
in Movie Industry, 282, 381
in Music Recording Industry, 282
in Photography Industry, 281, 431, 432, 434–439, 450
in Telecommunication Industry, 221–224, 227–262
in Telephone Industry, 228–240, 262
in Television Industry, 282
transformations made possible by, 263
uses of, ix–x
Digital audio broadcasting (DAB), 347, 358
Digital audio tape (DAT), 395
Digital broadcasting, 364
Digital Camera System (DCS), 435
Digital circuitry, 487
Digital Economy, existence of, 246
Digital Equipment Corporation (DEC), 209, 237, 253, 317
Digital exhibition, 392
Digital games, 283
Digital graphics, advances in, 267
Digital Insight, 109
Digital media, volume of, 284–285
Digital Millennium Copyright Act (1998) (DMCA), 348, 423
Digital PBX systems, 240
Digital photography, 281, 431, 432, 434–439, 450
Digital Portal business model, 288
Digital Print Station, 434
Digital signals, 486–487
Digital styles, 23, 25
Digital subscriber line (DSL) service, 249, 255
Digital switching, 208, 487
advances in, 235
Digital video recorder (DVR), 368
Diners Club, 61
Direct deposit of payroll checks, 110
Discover card, 64

Discovery Channel, 277
Displacement, 472
Disruptive technologies, 278, 396
Distribution, digital technologies in, 263
DLJdirect, 181
Document processing, 27
Dodge, W., 303
Dot-com shakeout, 331
Dow Jones, 274
Dowjones News, 307
Dun & Bradstreet, 303
DuPont & Company, 162
DVDs, 285, 286
  bootlegged, 286, 389
  in Movie Industry, 285–286, 386–388
  in Video Games Industry, 415

E. W. Scripps, 274–275
Eastman Kodak Company, 430–432
E-banking services, 106
eBay, 290, 469
Ebbers, Bernie, 203
E-books, 272, 308, 311–312, 313
E-commerce, 251
  gaming as form of, 423
  security advances in, 247–248
Economic activities, role and use of industries, sectors, and economic models as concepts to understand, 479–484
Economic influences
  in Radio Industry, 338–339
  in Television Industry, 351–352
Economic models, 479, 481–482
Economy. *See also* Global economy; New Economy; Old Economy; United States economy
  capitalist, 464
  Digital, 246
  entertainment, 368, 466–469
  Information, 267, 481
  information, viii, 481
  Internet, 288–289, 288–290
  manufacturing-based, 3
  new, viii
  old, viii
  physical, vii, viii
  post-industrial, viii
  Services, 3–4, 443, 481
EDI-based services for companies, 77
Editing systems in Newspaper Industry, 317
EDS, 100
Educational software, 304–305
800 services, 231
Eisner, Michael, 466
Electricity, introduction of, 22–23
Electronic Arts (EA), 391–392, 422
Electronic check sorters, 46
Electronic communication networks (ECNs), 186
  in Brokerage Industry, 159
Electronic Data Interchange (EDI), 74
  in Telecommunications Industry, 213–214, 246
Electronic data processing (EDP) in Banking Industry, 85
Electronic eavesdropping and larceny, 102
Electronic Funds Transfer Act (1979), 15–16
Electronic funds transfer (EFT) system
  in Banking Industry, 39, 42, 44, 45, 49–50, 73–78, 89, 91, 94, 141
  in Insurance Industry, 141
Electronic Gaming Monthly, 422
Electronic media, x
Electronic Media Industries. *See also* Radio Industry; Television Industry
  digital applications in, 336–370
Electronic music, 401
Electronic news services, 323
Electronic payments, 78, 110
Electronic prepress, 327
Electronic publishing, 306–309
Electronic Recording Machine Accounting (ERMA), 46, 48–49
Electronic Services, Inc., 123
Electronics in loan data management, 56
E-Legend, 268
E-mail, 211, 215, 244–246, 251
  availability of, over cell phones, 255
EMI, 279
E-money, 4, 12
Employers of Wausau, 126
Employment levels by sectors and industries, 477
Encyclopedia, electronic versions of, 306
ENIAC, 120

Entertainment economy, 368
United States as, 466–469
Entertainment Industry, 266. *See also* Movie Industry; Music Recording Industry; Photographic Industry; Video Games Industry
consolidation in, 444
digital applications in, 371–441, 412–441
Internet in, 287–292
mergers and acquisitions in, 264
in New Economy, 455–459
Entertainment sector. *See also* Game and Toy Industry; Music Recording Industry; Photographic Industry
in the American Economy, 277–281
patterns and practices in adoption of computer applications in, 281–287
Epyx, 280
Equitable Life Insurance Company of Iowa, 122, 123
Ericsson, 210, 211
ersonal Identification number (PIN), 66
ESS, 237
Essex Corporation, 146
E-Trade, 106, 183
Eurodollars, 40
European Union (EU), 211
E-wallets, 51
Executive support system (ESS), 234
Exodus, 218
Expert systems
in Banking Industry, 59
in Insurance Industry, 134–135
Extranets, 246
Exxon, merger of, 346

Facsimile transmissions, 258
Fanning, Shawn, 406
Farmers Insurance Group, 122
Fawcett Publications, 299
Faxing, availability of, over cell phones, 255
FDR, 64
Federal Communications Commission (FCC), 462
Cable television and, 357–358
cell phones and, 253, 261
competition in telephones and, 231
digital audio broadcasting and, 347
high definition radio and, 349
objectives of, 223–224
Radio Industry and, 275, 276, 338, 345
regulatory activities of, 193, 194, 195, 199–201, 203, 204–205, 214
telecommunications sector and, 193
telephone usage and, 251
Television Industry and, 277, 350, 351, 356, 357, 364
universal access and, 259
wireless communications and, 217
Federal Deposit Insurance Corporation (FDIC), 38
Federal Deposit Insurance Corporation Improvement Act (1991), 16
Federal Express (FedEx), 333
Federal Home Loan Mortgage Corporation (Freddie Mac), 42
Federal National Mortgage Association (Fannie Mae), 42
Federal Reserve Banks, 7, 38, 74, 107, 110, 462
charging for services, 50
electronics at, 74
telecommunications and, 86
Federal Reserve Wire System, 75
Federal Trade Commission (FTC), 16–17
Federated Mutual, 126
Federici, Joseph, 299
Fiber optics, 203, 257, 445, 489
advances in, 243
Fidelity, 181, 183
Filesharing applications, 410
File transfer protocols, 245
Film-based photography, reduced demand for, 280
Film-related companies, digital applications in, 282
Film-related computer games, 391–392
Films, hacking of, 389–390
Film versus digital debate in Photographic Industry, 436
Financial Industry, 480
computing in, 445
deregulation of, 4
in New Economy, 455–459
in U.S. economy, 3–36
Y2K problem and, 101

Financial investments, 13
Financial markets, non-financial institutions in, 6
Financial sector, x, 480
 in American economy, 6–15
 components of, 6
 data-handling requirements and applications, 20
 defined, 6
 gross domestic product of, 8
 industry committees and organizations in, 20
 major segments of, 7
 role of computing across, 5
Financial services
 consumer uses of, 30–33
 to Internet, 34
Financial Times, 322
Fire insurance, 13
First Data Corporation, 65
First Industrial Revolution, 455, 457
First Internet Bank of Indiana, 106
First National Bank of Boston, 53
First National Bank & Trust, 72
First Tennessee, 106
First Union National Bank (Charlotte, NC), 100, 106
Fischer, Claude S., 228
Fisher, William W., III, 461–463
Fisher Price, 268
Fitzpatrick, John D., 233
Flat screens, 361, 367
Floor traders in Brokerage Industry, 164
Folio, 329, 331
Food and Drug Administration (FDA), 462, 471
Forbes, Special Interest Publications, 334
Fordist style, 22
Ford Motor Company, Financial Services Group of, 64
Ford Motor Credit Company, 43
Forrester Research, 282, 408
401(k) retirement savings plans, 31
Four-layer model of Internet Economy, 289–290
Fowler, Jude W., 100
FOX, 350, 365, 366
Franchises, 390
Franklin, Benjamin, 333
Free Press, 264
Frei, Frances X., 91
Fuji, 435
Funston, Keith, 170

Galambos, Louis, 201, 222
Game and Toy Industry, 268, 279–280, 411, 413. *See also* Video Games Industry
 sales of, 425–426
GameBoy, 414, 428
GameCube, 423
Games, availability of, over cell phones, 255
Game theory, 414
Gannett Hearst, 274, 276, 277, 322
Gartner, 282
Gates, Bill, 290
Gateway Digital, 268
Gavin Report, 405
GEISCO, 77
General credit card business, 43
General Electric (GE), 48, 49, 120, 355
General Insurance Company, 130
General Motors Acceptance Corp. (GMAC) financing, 7, 9, 483
General Motors (GM), 7, 9, 100, 241, 445, 483
General Telephone of Indiana, 233
Generic aspects of new technology, 463–464
Genesis, 415
Genuity, 220
Georges, Joseph St., 342
Gilder, George, 252
Glass-Steagall Act (1933), 15
Global Crossing, 218, 220
Global economy
 Banking Industry in, 96–109
 Book Publishing Industry in, 310
 digital technology and, 264
 DVD sales and, 387
 implications of patterns in, 469–472
 Video Games Industry in, 414
Global industries, 479
Go-America, 212
Gold standard, 4
Goodbody & Company, 167
Google, 273

Government regulations, role of, 15–18
Gramm-Leach-Billey Act (1999), 18
Green field project, 242
Greenig, Robert A., 124
Greenspan, Alan, 225
Grid computing, 280, 422
Grocery Industry
bar code in, 24, 482
consolidations in, 346
Groliers, 306
Gross domestic product (GDP), 3
in Financial sector, 8–9
in Life/Casualty Industry, 117–118
Gross national product (GNP) in Telecommunication Industry, 189
GTE, 205
Gulf+Western, 385
Gutenberg, Johannes, 297, 334

Hallock, Lowell G., 69
Handspring, 211–212, 268
Hannan, Michael T., 32
Hard-core gamers, 428
Harker, Patrick T., 91
Harper, Marion, Jr., 354
HarperCollins, 308
Harper & Row, 305
Harte, Dick, 309
Harvard Business School Press, 473
Hasbro Games, 417
Haupt & Co., 162
Hayden, Edward P., 319
Head, Simon, 469
Health Industry, 117, 471
Health insurance, 12
independent operations in, 119
numbers of companies, 116
Hearst Newspapers, 275
Hewlett-Packard (HP), 212, 237, 253
Hickey, John P., 71–72
High definition radio, 349
High-Definition Television (HDTV), 360–361, 364
Hill, Vernon W., III, 99–100
Historical precedence, 80
History Channel, 352
Hitchcock Publishing Company, 299
Hitt, Lorin M., 184
Hogan, Donald, 299
Hollywood Model, 418, 444, 447–448, 468, 474
Hollywood System, 472
Holzer, Norman R., 238–239
Home banking, 23, 102–103
Home Box Office (HBO), 277, 352
Home game consoles, 415
Home offices, growth in, 243
Homeowners insurance, 118
Home video games, 267, 280
Horizontal integration, 201
Hot metal technology, 319
Houghton Mifflin, 305
House insurance, demand for, 117
Houston, Frank, 325
Hull, Geoffrey P., 404
Hundt, Reed E., 204, 206, 253, 369
Hunter, Larry W., 91
Hush-a-Phone, 203
W. E. Hutton & Company, 162
Hypertext, 273

IBM, 292
in Banking Industry, 49, 52, 92, 100
in Electronic Media Industries, 353
in Game and Toy Industries, 280, 281, 413, 423
in Insurance Industry, 123, 130
marketing strategies of, 480
in Print Media Industries, 317
products offered by, 29
in Telecommunications Industry, 237, 253
in Television and Radio, 341–342
use of training materials at, 286
IBM 47 Tape-to-Card Converter, 403
IBM 305, 53, 56
IBM 360, 48, 124–125, 233
arrival of, 125
in Brokerage Industry, 162
as general purpose type technology, 20
Hitchcock Publishing's upgrade to, 299
installation of, by Meredith Publishing, 327–328
operating back-office settlements systems on, 181
ordering of, by American Bible Society, 299

IBM 360 (*continued*)
for setting type and production tasks, 328
at Texas Bank & Trust Company, 62
IBM 370
BMI's installation of, 403
Illinois Bell's adoption of, 233
Universal Studios installation of, 385
Warner Brothers installation of, 386
IBM 650, 53
acquisition of, to brokerage firms, 162
arrival of, 125
availability of, 122
initiation into the Bell System, 231
installation of, at Backe & Company, 163
Nielsen's installation of, 353
review of installations of, 353
similarity to older calculators, 138, 139
IBM 702, 48
$64,000 *Question* TV show appearance of, 353
IBM 704 system, initiation into the Bell System, 231
IBM 705 computer system
for handling stock transactions, 163
Kodak's installation of, 433
IBM 1013 data transmission card terminals, data-transmission equipment using, 433
IBM 1440, 299
IBM 1460, installation of, by Fawcett Publications, 299
IBM 1620 system, 316
IBM 2740 communications terminals, 62
IBM 7080, Kodak's upgrade to, 433
IBM Card Programmed Calculator (CPC), 231
IBM Corporate Archives, 233
IBM Friden Flexowriter, 403
IBM System/3 minicomputer, 342
IBM System 3 Model 6, Meyer Broadcasting Company's installation of, 354
ICTV, 360
IDG, 306
IDVD, 268
ILife suite of products, 268
Illinois Bell Telephone Company, 211
computer use by, 233–234
IM, 251
Image-editing software, 436
IMovie, 268
Independent phone providers, 229
Independents, 196
Industrial Age, 193
Industries. *See also specific*
business applications across, 479–484
defining, 483
within sectors, 480
study of, 479
Industry-centric applications, xi
Infocom, 420
Information, xi
Information Age, viii, 22, 190, 292, 443
Information Economy, 267, 481
Information highway, 200
Information Processing Industry, 413
Information providers in Book Publishing Industry, 296
Information sector, 480
Information Society, viii
Information technology (IT), 22, 192
assets in, 116
in Banking Industry, 43
in Book Publishing Industry, 303–309
in Brokerage Industry, 171
influence on daily life, vii
in Insurance Industry, 137–143
in Magazine Industry, 326–330
in Music Recording Industry, 402–403
in Newspaper Industry, 315–320
patterns of adoption, 451–455
in Radio Industry, 341–345
in Service Industry, 447–451
in Television Industry, 352–361
Information Technology (IT) Industry, 413
Ingram Book Group, 272, 309, 310
Initial public offerings (IPOs), 13, 14
in Brokerage Industry, 154, 157
Innovation, 464–465
Institutional investors in Brokerage Industry, 156–157
Institutional rigidity of AT&T, 202

Insurance
computers in, 442
types of, 12, 13
Insurance Industry, x, 6, 462
artificial intelligence in, 135
banks and, 5
business patterns and digital applications in, 113–150
changes in operations in, 471
computing in, 5
consumer role in, 120
core competencies of, 12
defined, 8
electronic funds transfer in, 141
expert systems in, 134–135
fragmentation of, 5
government regulation of, 16–17
gross domestic product of, 8
image of, 5
independent agents in, 141–142
information technology in, 137–143
Internet in, 143–147, 149–150
IVANS in, 480
optical character recognition technology in, 135, 140
risk management in, 152–153
sectors of, 350
services offered by, 31
software applications in, 28
structure of, 148
tasks of, 12–13
Websites in, 145
word processing in, 127–128
Y2K in, 149
Insurance malls, 145
Insurance Value Added Network Services (IVANS), 127, 136
Integrated circuits, advances in, 240–241
Intel, 268, 423
Intellivision, 420
Interactive Data Corporation (IDC), in Brokerage Industry, 175
Interactive journalism, 324
Interactive television (ITV), 364–365
Interbank Card, 61
Intermarket Trading System, 169
Internal Revenue Service (IRS)
computer use by, 233
Information Return Form 1099, 52
International banking, 7
International Telephone and Telegraph Corporation (ITT), 192
International Typographical Union (ITU), 317–318, 319
Internet, 23, 73, 211, 267
in Banking Industry, 72–73, 96–109, 143–147
in Book Publishing Industry, 272, 309–313
in Brokerage Industry, 180–185
credit card use and, 61
daily uses in, 248
downloading of music off of, 247, 250, 366, 388, 389
in Entertainment Industry, 287–292
financial services to, 34
in Insurance Industry, 143–147, 149–150
in Magazine Industry, 330–331
in Media and Entertainment Industry, 287–292
in Movie Industry, 287, 386, 388–390
in Music Recording Industry, 283–284, 386, 405–409
in Newspaper Industry, 287, 321–322
photography and, 438
radio broadcasting and, 347–349
in Telecommunication Industry, 215–216, 244–252, 261
telephony and, 488
in Television Industry, 361–363, 366–367
video games and, 426–427
Internet access providers, 250
Internet banking, 12, 51, 54, 74, 96–109, 99–100
Internet-based publications, 326
Internet-based stock purchases, 444
Internet companies, 292
Internet Economy, 288–289
four-layer model of, 289–290
Internet Service Providers (ISPs), 407, 490
Internet VoD platform, efforts to establish, 390
Internet/Web Ad Push business model, 288
Intranets, 245–246

Intuit, 102
Inventory control
in Book Publishing Industry, 266, 296, 302, 308
in Movie Industry, 385–386
Investigatory journalism, 324
Investment-banking services, 14
Invisible hand, 448
Iowa-Illinois Telephone Company, computer use by, 233
IPhoto, 268
iPods, 279, 408, 450, 460, 467, 473
Irvine, M. M., 230
ITunes, 268, 269
iTunes Music Store, 408
IVANS in Insurance Industry, 480

J. P. Morgan, 7, 92
Jacobs, Donald P., 63
Java, 212
Jobs, Steve, 269, 400, 408–409, 411, 419, 450, 463
John Hancock, 126
John Wiley, 305, 306
Josephson, Stanley, 50
Journalism
computer-assisted, 324
interactive, 324
investigatory, 324
Jupiter Media Metrix, 406
Justice, U.S. Department of, Antitrust Division, 200
settlement between AT&T and, 200–202

K250, 401
KaZaA software, 250, 406
Kearsarge Telephone Company, 235
Kidder, Peabody & Co., 162
Kinko's Graphics Corporation, 333–334
Kinko's Impress, 334
Kley, John A., 37
KLIF (Dallas, TX), 348
Knight-Ridder, 274, 321, 322
Kodak, 281, 411, 415, 416, 433, 438–439
introduction of Digital Print Station by, 434
introduction of Photo CD by, 434
Konica, 435
Kurzweil, Raymond, 401

Landes, David S., 35, 457
Langlois, Richard N., 447–448, 448, 471–472
Laperouse, Stewart, 371
Laptops, 320
Law of 1934, 204
Lee, Charles R., 258
Lesser, David, 407
Lessig, Lawrence, 408
Levy, Frank, 465, 466
LexisNexis, 290, 296
Liberty Mutual, 126
Liberty National Bank and Trust Company of Oklahoma, 85–86
Libraries in Magazine Industry, 332
Life and Casualty Insurance Company, 130
Life Insurance Industry, 12, 13, 39, 118
adoption of computing by, 126–129
combined value of policies in, 116–117
companies in, 8, 116, 148–149
independent operations in, 119
Life Office Management Association (LOMA), 120
Linotype, 301
Linux, 212
Lipinski, Andrew J., 222
Liquid crystal displays (LCDs), 367
Liquidity Quote, 186
Litan, Robert E., 95
Literary agents, 296–297
Lithography, 326–327
Live entertainment, use of digital tools by, 266–267
LMDS, 256
Loan management, 84
Loan processing in Banking Industry, 55–59, 93
Local retirement funds, 39
Lockbox services, 52–53, 77–78
Long Distance Discount Service, 203
Longman, Pearson, 469
Lord of the Rings, 390, 391–392
Los Angeles Times, 318, 321
Lucent Technologies, 194, 203, 210, 211
Luddite behavior, 452

Lugables, 320
Lyman, Peter, 215
Lynch, Daniel C., 245
Lyons, William, 185

Ma Bell telephone system, 212
Machlup, Fritz, 480
Macmillan, 306
Magazine Industry, 267, 272–273, 325–332, 463
  computers in, 442
  information technology in, 326–330
  Internet in, 330–331
  libraries and, 332
  ownership in, 273
  subscription management in, 328
  trends in, 330–332
Magnetic Ink Character Recognition (MICR) process, 20, 46, 49, 84, 87, 482
Mail-based billings and payments, 110–111
Malik, Om, 220
Maloney Act (1936), 17
Management
  in age of New Economy, 472–475
  attitude toward digital applications, 283
Manning Rules in Brokerage Industry, 159
Manufacturers Hanover Trust, 64
Manufacturing, transformation from, 22
Manufacturing-based economy, 3
Manufacturing sector, 477, 480, 481
  workforce in, 477
Marine insurance, 13
Market crossovers, 264
Market Data Systems, 166, 171–172
Market reach, 338
Mark II Music Synthesizer, 401
Massachusetts Mutual Life Insurance, 123
Mass-production manufacturing, 22, 475
MasterCard, 33, 43, 60, 61
  acquisition of Cirrus by, 71
  as branded credit card, 108
  building of national networks by, 70
  competition between Visa and, 67
  displacement of retail store-brand cards by, 63
  employee's paycheck to account by, 110
  instant updating of files by, 66
  siphoning of business away from banks by, 64
  usablity of cards of, 470
MasterCard International, 63
Masterson, Norton E., 113
Matching orders in Brokerage Industry, 167–168
Matsushita, 211
Mattrick, Don, 391
May, Randolph J., 216
Mayday 1975, 168
Mayo, John S., 242–243
MCA, 279
McCarran-Ferguson Act (1945), 16
McGraw-Hill, 276, 301, 303, 305, 312
MCI (Microwave Communications, Inc.), 194, 202, 203, 205, 209, 220, 222, 231, 447
  fiber-optics network at, 243
Media, changing definition of, 270–271
Media Industry, x
  emerging definitions of, 266–271
  Internet in, 287–292
  mergers and acquisitions in, 264
  in New Economy, 455–459
  technological events affecting, 396–400
MediaNews Group, 274
Media sector. *See also* Book Publishing; Magazine Industry; Newspaper Industry; Radio Industry; Television Industry
  in the American Economy, 271–277
  computer applications in, 281–287
Megabanks, 41
Megaservice sector, 481
Melcher, Daniel, 299
Mellon National Bank and Trust Company, 19
Mercury, 279, 393
Meredith Publishing, 326, 327–328
Mergers and acquisitions, 469
  in Banking Industry, 41–42, 64, 90
  in Book Publishing Industry, 311
  in Brokerage Industry, 158
  in Magazine Industry, 273
  in Media and Entertainment Industry, 270
  in Music Recording Industry, 411
  in Petroleum Industry, 346

Mergers and acquisitions (*continued*)
  in Radio Industry, 346
  in service industries, 448
  in Telecommunication Industry, 205
Merrill-Lynch, 77, 158, 162, 184, 186
Mertz, Harry E., 3
Mesh wireless, 256
Metropolitan Life Insurance Company, 120, 122
Meyer, Philip, 335
Meyer Broadcasting Company (Bismarck, North Dakota), 354
MGM, 386
Miami *Herald*, 316
Microchips, embedding in smart cards, 65
Microsoft, 100, 253, 268, 280, 290, 363, 369, 419, 422, 423
  Online Games Industry and, 413
  publications of, 102
  system software and applications of, 304
  Web TV, 365
  Windows operating systems, 212
  Word, 452
  Xbox, 415, 419–420, 422, 423, 424
Microwave communications, 194, 195, 203, 212, 231
Midway Telephone Company, 235
Millstein, Lincoln, 323
Miniaturization of electronics, 337
Misa, Thomas J., 472
MIT Press, 411
MMDS, 256
Mobile, merger of, 346
Mobile client applications, 212
Mobile communications, availability of, over cell phones, 255
MODAC computers, 298
Modulation, 486
Money. 103–104
Money market accounts, 13–14, 39, 40
Money messages, 67
Moog, Robert, 401
Moog's device, 401
Moore Business Forms, 123
Moore's Law, 368, 458
  consumer embrace and application of, 454–455
Moran, John M., 391
Morgan Stanley Dean Witter, 184
Mortgage accounting, 56–57
Mortgage-backed securities markets, 42
Mortgage loans, 55, 58
Mosaic, 309
Moschella, David, 309
Mossberg, Walter, 255
Motion Picture Association, 391
Motorola, 194, 211–212, 212, 268, 436
Mt. Vernon Telephone Company, 235
Movie Industry, 372–392, 443, 444, 448, 449, 453, 458, 459, 461
  accounting in, 385–386
  animation in, 372, 401
  artificial intelligence in, 377–378
  computer-based graphics in, 401
  computerization of business operations, 373, 384–392
  computers as move stars in, 377–379
  computers in making movies in, 379–384
  data compression standards and, 360–361
  digital imaging in, 282, 381
  DVDs in, 285–286, 386–388
  employment level in, 375–376
  firms in, 272, 374
  growth of, 376
  Internet in, 287, 386, 388–390
  inventory control in, 385–386
  motion control in, 380
  non-linear editing in, 383–384
  as oligopoly, 373, 374–375, 376
  prerecorded videotapes and, 359
  radio and, 372
  regulations in, 373–374
  relationships between Television Industry and, 372, 374–375
  unions in, 376
  value chain in, 376
  video in, 359
  viewership, 372, 376
Movies, downloading of pirated copies, 389
Moving Picture Experts Group (MPEG), 406
MP3 files, 406
  downloading, 250
MSN, 288
MSNBC, 366, 369

Multimedia, concept of, 303
Multiple services sectors, 481
Murdoch, Rupert, 466
Murnane, Richard J., 465, 466
Music, electronic, 401
Musical Instrument Digital Interface (MIDI), 401
Music Recording Industry, 269, 278–279, 392–409, 443, 456, 459, 461–462, 467
- accounting applications in, 402–403
- business applications in, 402–405, 410
- companies in, 272, 310, 392–393
- computers in, 401, 442
- copyright issues in, 278, 303, 393, 394, 395–396, 402, 405
- digital technology in, 282
- distribution process in, 285–286
- downloading of music as issue in, 247, 250, 366, 388, 389
- economic role in, 393
- history of technology in, 396–402
- influence of consumers on, 410–411
- information technology in, 402–403, 458
- Internet in, 283–284, 287, 386
- look-and-feel elements of, 420
- mergers and acquisitions in, 411
- music videos in, 394–395
- as oligopoly, 393–394
- online subscription services in, 408–409
- piracy issue in, 396, 405–409
- punch-card software-based systems in, 401–402
- radio and, 372, 393, 394
- self-publishing in, 474
- shelf life in, 418
- sources of income in, 393–394
- television in, 393, 394
- Universal Product Codes (UPCs) in, 404

Music videos in Music Recording Industry, 394–395
Mutual Benefit Life Insurance Company, 122
Mutual funds, 14
Mutual of New York, 124, 135
Mutual savings banks, 39
MyMp3.com, 406
N2K's Music Boulevard, 406
Napster, 249–250, 269, 406, 410
NASCENT consumerism, 33
National Aeronautics and Space Administration (NASA), 432
National Association of Securities Dealers (NASD), 169
National Association of Security Dealers Automated Quotation (NASDAQ), 7, 154, 158, 169–170, 173, 177–178, 186
National best bid and offer in Brokerage Industry, 159
National Cash Register, 123
National credit cards, 61
National market system in Brokerage Industry, 169
National Public Radio (NPR) network, 345
National Securities Clearing Corporation, Automated Customer Account Transfer System (ACATS), 179
National Telecommunications Network, fiber-optics network at, 243
Nationwide Insurance, 122
Natural monopoly, 202
NBC, 276, 277, 338, 350, 354
NBC News, 366
NCR, 52, 130, 203
NEC, 426
Negotiable Order of Withdrawal (NOW), 16, 40
Nelson, Paula, 37
Nelson, Richard R., 449–450, 457, 463–464
NetB@nk, 106
Net Heads, 489–490
Netscape, 98, 309
Netscape Netcenter, 288
Network television, 267
New Century Network, 322
New Economy, 292, 443, 455
- features of, 456–457
- Financial Industry in, 455–459
- management's role in age of, 472–475
- Media and Entertainment Industry in, 455–459
- Telecommunication Industry in, 455–459

New Economy (*continued*)
wrestling with old problems and new issues in, 459–466
Newhouse Newspapers, 274
News Corporation, 276, 277, 310
Newspaper Association of America, 323
Newspaper Guild, 317–318
Newspaper Industry, 263, 274–275, 314–325, 334, 443
activity brands of, 314–315
computers in, 282
CRTs in, 318, 319, 320
growth of, 267–268
information technology deployment in, 315–320
Internet in, 287, 321–322
niche specialties in, 324–325
printing technologies in, 283
trends in, 320–325
unionized printers in, 301
word processing/editing systems in, 317
Newspapers, 267, 462–463
New York Life, 123
New York Stock Exchange (NYSE), 7, 152, 154, 155, 156, 158–159, 160–161, 162, 166, 168, 169, 170, 173, 177, 178, 185, 186, 187, 452
adoption of computing by, 188
Broker Volume services, 186
Fact Book, 175–176
public position of, 165
reforming basic processes at, 164
technology in, 163, 175–176
New York Times, 270, 274, 276, 316, 321, 322–323, 323
New York Times Digital, 323
New York Toy Fair (1976), 416
Niche markets
for magazines, 273
publications aimed at, 325
Niche specialties in Newspaper Industry, 324–325
A. C. Nielsen Company, 354
Nikon, 431, 435
Nintendo, 268, 280, 415, 417, 420, 421, 422–423, 424, 428
GameCube, 420
No. four ESS switching systems, advances in, 237
No.1 Electronic Switch Systems (ESS), 231
No Electronic Theft Act (1997), 395–396
Nokia, 211–212, 252–253, 268, 280, 436
Nondigital media, 285
Non-financial institutions in financial markets, 6
Norfolk and Dedham Mutual Fire Insurance Company, 130
Nortel, 210, 211
North American Free Trade Agreement (NAFTA), 193
North American Industry Classification System (NAICS)
Electrical Contractors, 194
Power and Communication Transmission Line Construction, 193–194
Telecommunications Resellers, 193
Wired Telecommunications Carriers, 193
North American Industry Classification System (NAICS) code, 266, 287
North American Industry Classification System (NAICS) taxonomy, 483
Northern Telecom, 208, 210, 237
Northwestern Mutual Life, 123, 126
Noyelle, Thierry, 95–96
Number of firms in Brokerage Industry, 165
NYNEX, 202, 205

Oberholzer, Felix, 408
October 19, 1987, stock market crash of, 178
Oettinger, Anthony G., 92
Offset printing, 316, 326–327
Ohio Bell Telephone Company, 233
Old Economy, 443, 445
Old Order, passing of, in Brokerage Industry, 151
Oligopoly
Movie Industry as, 373, 374–375, 376
Music Recording Industry as, 393–394
Olivier, Bertrand, 95–96
Olympus, 268, 435
OmniSky, 212

On-demand printing, 303, 307
One-year savings accounts, 52
Online access to information in Brokerage Industry, 174–175
Online banking, 51
Online bulletin board systems, 245
Online computer systems, switch from batch systems to, 54
Online content services, business models evident for, 287–288
Online Games Industry, 280, 413, 429
emergence of, 422
Online loan applications in Internet banking, 106
Online magazines, 272–273
Online news, demand for, 323
Online photo processing, 436
Online securities purchases in Internet banking, 106
Online subscription services in Music Recording Industry, 408–409
OpenBook, 186
Open-end investment companies, 39
Opening Order Automated Report Service (OARS), in Brokerage Industry, 173–174
Operations research, computer applications in, 57
Optical character recognition (OCR) technology, 316
in Insurance Industry, 135, 140
Optical fiber, 218, 488
Optical scanning equipment in brokerage Industry, 170–171
Orbanes, Philip E., 416–417
Order Audit Trail System (OATS) in Brokerage Industry, 179
Orfalea, Paul, 333
Organization for Economic Cooperation and Development (OECD), 476
Orr, Bill, 80
Outsourcing of IT, 101
Over-the counter (OTC) market, 14, 166
Overton, Thomas N., 62
Owen, Bruce M., 362, 363
Oxford University Press, 312, 471, 474

Pacific Stock Exchange, 172
Pacific Telsis, 202
Packet-switching, 208, 209, 237, 240, 283, 488–489
PacTel, 205
Pagers, 194, 195, 211, 212
Paid Internet business model, 288
PaineWebber, 186
Palm.net, 212
Palm Pilot, 211, 268
Pan-American Life Insurance Company, 122
Pan American Software, 192
Panasonic, 268, 423, 435
Paper Industry, 297
Paperwork in Banking Industry, 45
Paramount Communications, 264, 385
Parker Brothers, 415–416, 417
Party line, 229
Pavlik, John V., 268
Payment Services, 27
Payroll, processing of, in Banking Industry, 43
PC Magazine, 325
PDF, 331
Peachpit Press, 306
Pearson, 312, 313
Peer-to-peer (P2P) filesharing, 249–250
Personal computers
in Banking Industry, 12, 91–92
in Brokerage Industry, 174, 175
games for, 415
Personal digital assistants (PDAs), 195, 211, 212
availability of functions over cell phones, 254
Petersmeyer, C. Wrede, 342
Petroleum Industry, mergers and acquisitions in, 346
Pew Foundation, 407
Pew Internet and American Life Project, 98–99, 246–247, 429, 438
Pew Research Center for the People & Press, 386
Pharmaceutical Industry, 480
Philadelphia Savings Fund Society, 53
Philadelphia Stock Exchange, 172
Phister, 214
Photo CD system, 435
coming of age, 434–439
Photocomposition, 319, 326–327

Photocompression, 317
Photocopying services, 303, 333
Photographic Industry, 268, 280–281, 411, 430–439, 443, 456
  adapting to computing in, 415
  coming of age, 434–439
  competition in, 431
  digital technologies in, 432–434
  film versus digital debate, 436
  fragmentation of, 440
  types of cameras in, 431
Photo Marketing Association International, 437
Photo Marketing Association (PMA), 439
Photo-offset printing techniques, 301
Physical economy, vii, viii
Physical security in Banking Industry, 100
Picard, Robert G., 338
Picturephones, 240
Picture radio, 371
Piracy in Music Recording Industry, 396, 405–409
Pitt, Harvey L., 185
Plain old telephone service (POTS), 190, 210, 225, 488, 489
Plasma screens, 367
PlayStation, 415, 422
PLUS, 71
Pocket Books, 264
Pocket switching, 487
Point-of-sale (POS) terminals, 67, 73, 78, 296
Polaroid, 433
Political preconditions, 80
Political realities, 80
PolyGram, 279, 394
Pools of mortgages, 42
Popular Photography, 435–436, 438
Portables, 320
Portal and Personal Portal business model, 288
Porter, Michael E., 479, 480–481, 484
Portfolio management, 14
Postal Service, United States (USPS), 192
Post-industrial economy, viii
PostScript, 329
Powell, Michael, 216
Precomputer technologies in Banking Industry, 19
Prentice-Hall, 305, 306, 469
President's Task Force on Communications Policy, 355
Priceline, 290
Primary sources, 482
Printing Industry, 334
  capital expenditures in, 301
Print magazines, 272–273
Print Media Industry, x, 468. *See also* Book Publishing; Magazine Industry; Newspaper Industry
  regulation of, 15
  uses of computing in, 293–335
Privacy concerns in Banking Industry, 97
Private branch exchange (PBX), 447
  capabilities of, 241
Private pension funds, 39
Product evolution in Book Publishing Industry, 297
Production in Book Publishing Industry, 295
Productivity
  changes in American economy, 457–458
  competitive pressures to improve, 452
  measurement of, 457
  variations in, 464
Productivity Paradox, 25, 455
PROFS system, 214
Program trading, 173
Progressive Insurance, 145
Projection TVs, 367
Project on Technology and Workplace, 469
Proof machines, 45–46
Property Insurance, 8, 12, 148–149
  adoption of computing in, 129–137
  independent operations in, 119
  number of firms, 117
  revenues in, 118
Prudential Insurance, 7, 113, 186
Prudential Securities, 158
Prusak, Laurence, 35
PSINet, 218, 220
Public sector, 480
Publishers Weekly, 299, 304

Publishing Industry, impact of Internet on, 287
Punch-card technology
in Insurance Industry, 121
in Music Recording Industry, 401–402
Pushing the envelope, 464–465

Qualcomm, 211–212
QuarkXPress, 329
Quicken, 103–104
Quotron, 167, 181
QVC, 71, 277
Qwest, 205

R. R. Donnelly & Sons, 295, 300, 308
R4 (Registered Representative Rapid Response) in Brokerage Industry, 176–177
Rack jobbing, 395, 404
Radio airplay charts, 405
Radio Industry, 275–276, 337–350, 461, 462
business model in, 468
concentration in, 346
conomic influences on adoption of technology, 338–339
deregulation and, 345
information technology deployment in, 341–345
Internet in, 347–349
mergers and acquisitions in, 346
sections in, 337
special role of transistor radio in, 339–340
trends in, 345–350
Radios, 195, 335
Federal Communications Commission and, 338
in Music Recording Industry, 393, 394
transistor, 337, 339–340
Railroad Industry, 283
Raleigh, John N., 56
Random House, 272, 305, 306, 310
RCA 301, 354
RCA Victor, 279, 317, 355, 393
Reader's Digest, 298, 305
Real Estate Industry, 6
Recordable CDs, 462
Recorded music, 267. *See also* Music Recording Industry
Recording Industry. *See* Music Recording Industry
Recording Industry Association of America (RIAA), 406, 407
Recording studios, 404
Reel-to-reel magnetic tape, 337–338
Regency TR-1, 339, 340
Regional Bell operating companies, 209, 237
awareness of importance of telecommunications, 238
service level provided by, 261
Regional exchanges in Brokerage Industry, 166
Remaindering, 302
Remington Rand, 123
Remote control devices (RCDs), 358–359
Republic Bank, 50
Research in Motion, 212
Resellers, 203
Retail banking, 7, 94
Retail Industry, 484
Retail petroleum stores, credit card use by, 61
Return on investment (ROI) calculations for Banking Industry, 108
Revenue Accounting Offices (RAOs), 236–237
Revenue Act (1962), 52
Reverse engineering strategy, 416
Rhythm machines, 401
Richards, Dan, 435
Risks, managing, 13
Rockstar Games, 422
Rosenberg, Nathan, 457
Rosenbloom, Richard S., 343
Ross, Ian M., 231–233
R & R, 405
Ruckh, Patrick, 101

Salem Five Cents Savings, 106
Samsung, 211, 252–253, 268, 436
San Diego Trust & Savings Bank, 69
Sanyo, 436
Satellite communications, 231

Satellites
  in radio, 276, 337
  in Telecommunication Industry, 212–213
Satellite television, 276–277, 352
Saving and investment applications, 52–55
Savings and loans, 5, 39
SBC/Bell South, 211, 219
SBC Communications Inc. (SBC), 192, 205
Scantlin Electronic Corporation, 167
Schilling, Melisa A., 413, 425–426
Scholarly journals, evolution of, 331–332
Schumpeter, Joseph A., vii, ix, 278, 369, 442, 449, 457, 459, 463, 464, 473
Schwab, Charles, & Co., 7, 106, 153–154, 181, 182, 183, 450, 452, 473
Scribner, 264
Search engines, 245
Sears, Roebuck
  acquisition of Allstate and Dean Witter, 42
  financial counseling services of, 26
Second Industrial Revolution, 455
Sectors
  defined, 480, 483
  emerging view of, 480–481
  industries within, 480
Secure Digital Music Initiative (SDMI), 389
Securities Act (1933), 17
Securities Act Amendments (1975), 17–18
Securities and Exchange Commission (SEC), 16–17, 153, 154, 158
  Brokerage Industry and, 151, 153, 154, 158, 159, 165–166, 168–169
  Radio Industry and, 275
Securities Exchange Act (1934), 15, 17–18, 158
Securities Industry, services offered by, 31–32
Securities Industry Automation Corporation (SIAC), 166, 169, 187
Securities Investor Protection Act (1970), 17
Securities Investor Protection Corporation (SIPCO), 17
Securities markets, evolution of, 156–157
Securitization, 42, 58
Security brokers, 39
Security First Network Bank (SFNB), 81, 104, 106
Security Industry Software Applications, 29
Sega, 280, 417, 419, 420, 426, 428
Self-policing in Brokerage Industry, 168
Self-service banking, 67–68
Semiconductor Industry, 268
September 11, 2001, terrorist attacks, 100, 185, 186
  communication use following, 247
Service bureaus
  in Banking Industry, 83, 87
  in Brokerage Industry, 167–168
  television broadcasting and, 355
Service Industry
  adoption of computing by, 444
  effect of computing on, 442
  personnel and work practices of, 468–469
  role of information technology in, 447–451
Services
  defined, 481
  deintegration of, 481
Services Economy, 3–4, 443, 481
Services sector, 479, 480
  growth of, 477
  range of industries in, 443–444
  role of computing in, 4
Shapiro, Carl, 25, 366
Sharp, 212
Shearson Lehman, 158
Sherman, Cary, 407
Siemens, 210
Sierra Online, 280, 420
SilverDial, 268
Simon & Schuster, 264, 305, 308
Single-slot pay phones, 231
Sirius, 280
Small Order Execution System (SOES), in Brokerage Industry, 178
Smart cards, 51, 65, 100, 258
Smart networks, 487
Smith, Adam, 447, 448, 478
Smith Barney, 186

Snail mail, 111
Snyder, Richard, 308
Social safety net, 31
Social Security, 31
Society for Worldwide Interbank Financial Telecommunications (SWIFT), 75
Software
accounting, 265
educational, 304–305
image-editing, 436
in Insurance Industry, 28
in Music Recording Industry, 401–402
in typesetting, 300
in Video Games Industry, 414
Software Industry, 268, 304, 413, 421, 446, 466
licensing and outsourcing techniques in, 420
Solid-state devices, 208
Solomon, Elinor Harris, 67
Solomon, Ezra, 177, 178
Solow, Robert, 457
Somerset Telephone Company, 235
Sony, 211–212, 268, 276, 280, 281, 339, 394, 419, 420, 423, 424, 450
PlayStation, 419–420, 423, 426
Walkman, 340
Sony Ericsson, 436
Sony Music, 279, 394
SoundScan, 404–405
Sound storage technology, 337–338
The Source, 307
Southwestern Bell, 202, 233
Spam, problem of, 248–249
Speed calling, 235
Sperry Univac, 123, 317
Spinner.com, 348
Sports Report, 181
Sprint, 194, 202, 203, 205, 219, 438
digital advances at, 242
fiber-optics network at, 243
Sprint PCS, 211
S&S, 309
Standard Industrial Classification (SIC) code, 192, 194, 266, 287, 483
Standard & Poor's (S&P), 119, 303
establishment of Compustat Services, Inc., 175
Stanford Research Institute (SRI), 46
Stanzione, Daniel C., 254
Starbucks, wireless capability at, 256
State Farm, 7, 123, 130
State retirement funds, 39
State Street Bank & Trust Company, 42
Station WVCG (Coral Gables, FA), 341
Steel Industry, 283
Steiner, Thomas D., 92–93
Stiglitz, Joseph E., 206, 225
Stock brokerage, 6, 14. *See also* Brokerage Industry
computing in, 5
defined, 8
services of, 14
tasks of, 13–15
Stock brokers in Brokerage Industry, 152
Stock Clearing Corporation, 166
Stock market, growing participation of consumers in, 31
Stockmaster, 167
Stop payments in Internet banking, 106
Stored-program electronic control, 222
Strategic tools, use of information technology as, 44
Streaming video, 365
Strumpf, Koleman, 408
Subgroups in Book Publishing Industry, 296–297
Subscription management in Magazine Industry, 328
Sun, 329
SuperDOT in Brokerage Industry, 172
Surety Industry, 13
Switched-On Bach, 401
Switching, 486
advances in, 231
Synapse, 420
Synthesizer, 401, 410

T-1 rates, 488
Tabulators, 18–19
Tanaka, Graham Y., 455, 456–457
Tandy, 279
Tape-based systems, 326–327
Taylorism, 96
Taylorization work, 448
TCP/IP line, 245
TDS phone company, 235

TD Waterhouse, 183
Technology
  adoption of, 453
  declining costs of, 453
  impact on barriers, 458
  influence on daily work, 482–483
Teixeira, Diogo B., 92–93
Telcommunications sector, changing, in American economy, 195–206
Telecommunication Act (1996), 358, 364
Telecommunications
  applications across American industries, 234
  banking applications and, 73–74
  build-out of implementations, 228, 240–243
  deployment of applications in Banking Industry, 80–112
  in telephone industry, ix
Tele-Communications, Inc. (TCI), 205, 276, 277
  AT&T's acquisition of, 369
  National Digital Television Center, 360
Telecommunications Act (1996), 203, 204, 228, 261, 345
Telecommunications Industry, x, 187, 188, 442, 458, 480
  automation in, 258
  business patterns in, 221–224, 227–262
  business volumes in, 257–258
  capital efficiency of, 228
  cell phones in, 252–257, 261
  competition in, 203, 217, 231
  computer applications in, 213–216
  consolidation in, 219, 220–221
  defining, 192–195
  deregulation in, 201–202, 206
  digital applications in, 221–224, 227–262
  electronic data interchange in, 213–214, 246
  employee level, 465
  gross national product in, 189
  handset or terminal in, 485
  in Insurance Industry, 115
  Internet in, 189–190, 195, 215–216, 220, 244–252, 261
  investments in, 217–220
  job of, 206–213
  key competitive milestones in, 200
  mergers and acquisitions in, 192, 205
  in New Economy, 455–459
  origin of digital in, 228–240
  reasons for early disarray in, 190
  recent developments in, 216–221
  resistance in, 191
  satellites in, 212–213
  significance of, 190
  size of, 195
  threat to existing business models in, 283
  in United States economy, 189–226
Telecommunications wiring installation contractors, 193–194
Teleconferencing, 258
Telegraph, 196
Telegraphic systems in Brokerage Industry, 167
Telegraphy, decline in, 212
Teleharmonium, 401
Telematics, availability of, over cell phones, 255
Telemetry, availability of, over cell phones, 255
Telephone
  basic concept of, 486
  long-distance calls on, 229–230
  paying bills by, 69
  workings of, 486
Telephone-based services in banks, 12
Telephone data transfer, 270
Telephone Industry, x, 461
  cellular telephones in, 199
  digital technologies in, 228–240, 262
  elements of, 485–486
  employment in, 198
  Internet in, 198–199
  productivity in, 197–198
  revenues of, 198
  size of, 196–197
  telecommunications in, ix
  usage level of, 197
Telephone numbers, increasing demand for, 488
Telephone value chain, 210–211
Telephonic systems in Brokerage Industry, 167

Telephonic technologies, viii
Telephony
effects of Internet on, 488
technology underpinning, 485
Teleport Communications Group, 205
Telequote III, 167
Teleregister Corporation, 167
Teletypewriter in Brokerage Industry, 167
Television Industry, 195, 263, 276–277, 335, 350–368, 443, 449, 461
business and economic influences on adoption of technology, 351–352
convergence in, 361
digital technology in, 282
FCC and, 364
history of, 351
information technology deployment in, 352–361
Internet in, 361–363, 366–367
in Music Recording Industry, 393, 394
radio and, 372
relationship between Movie Industry and, 372, 374–375
trends in, 361–368
Telnet, 245
Temin, Peter, 201, 222
Terabeam, 256
Texas Bank & Trust Company, 62
Texas Instruments (TI), 279, 340, 432
Third Industrial Revolution, 22
Thompson, Kristin, 390, 391
Thompson Newspapers, 275
3-D graphics hardware, 415, 422
3G, 254
Three-digit area codes, 231
360Networks, 218
Thrift Supervision, U.S. Office of, 104
Ticker, 18–19
Ticker tape system in Brokerage Industry, 161–162
Times Mirror, 274, 322
Time Warner, 276, 277, 326, 364, 468
T-Is, 231
TiVo, 368
Toffler, Alvin, 458
Tokyo Stock Exchange, 178
Toshiba, 423
Total Systems, 64
Touchstone, 264
Touch-tone dialing, 231
Toy Industry. *See* Game and Toy Industry
Toyota approach, 22
TR-52, 339
Trade magazines, computers in publication of, 328
Transistor radios, 337, 339–340
Transistors, 208, 225
invention of, 190, 191, 339
Transmission speed, efforts to speed up, 488
Transmit data strobe (TDS), computer-based applications at, 241
Transportation Industry, 243
Travelers Group, 126, 146
Tribune, 275, 277
Tub files, 403
Turner, Ted, 277, 466
Turner Classic Movies, 277
Turnovers in Brokerage Industry, 157–158
Turnpiking effect, 49, 69
Twisted pair, 488–489
Tylecote, Andrew, 22
TYME, 77
Typesetting
software in, 300
use of CRTs in, 300–301
Typewriters, 282

Ultronic Systems, 167
Underwritings, 14
United Press International (UPI), 316
United Services Automobile Association (USAA), 135
United States economy
arrival of new, 475–478
computing's effects on, 45
Entertainment sector in the, 277–281
expansion of, 9, 266
financial industries in, 3–36
financial sector in, 6–15
Internet in, 108
media and entertainment industries in, 263–292
media sector in, 271–277
Telco sector in, 195–206
Video Games Industry in, 425–426
United States National Bank (Portland, OR), 56

U.S. West, 202, 205
Univac systems, 130, 237, 354
Universal credit cards, 61
Universal Product Codes (UPCs), in Music Recording Industry, 404
Universal service, 200
Universal Studios, 385
UNIX, 406
U.S. West Media Group, 277
USAA, 100
USA Network, 277
USA Today, 321, 322
Usenet, 245
UWB, 256

Vacuum tube radios, sales of, 340
Value-added networks (VANs), 194
Value-added services (VAS), 194
Value chain, 206–207, 481
  activities in, 484
  telephone, 210–211
Valueline, 303
Varian, Hal R., 25, 215, 366
Verizon, 211, 212, 219, 220, 258
Vertical disintegration, 448
Vertical integration, 201
VGA (video graphics array) tools, 415
Viacom, 264, 276, 277, 366, 468
Video arcade games, 415
Video cameras, 433
Videocassette recorders (VCRs), 358, 359, 386, 387
Video conferencing, 240
Video Games Industry, 413–430, 443, 446, 459–460, 466, 470
  artificial intelligence in, 414
  Beta standards for, 420
  computer chips in, 414
  consoles in, 418
  consumers in, 417
  cybernetics in, 414
  DVDs in, 415
  extent of deployment of, 424–430
  history of, 414–424
  software programming languages in, 414
  VHS standards for, 420
Video-on-demand (VoD), 391
Video poker games, 415
Video tape recorders (VTRs), 432
Video technology, 359–360
Videotext business model, 288
Viewdata Corporation of America, 321
Viewtron project, 321
Vinyl records, 285
Visa, 33, 43, 60, 61
  acquisition of PLUS by, 71
  as branded credit card, 108
  building of national networks by, 70
  competition between MasterCard and, 67
  displacement of retail store-brand cards by, 63
  employee's paycheck to account by, 110
  instant updating of files by, 66
  introduction of smart card, 65
  profits of, 468
  siphoning of business away from banks by, 64
  usablity of cards of, 470
Visa U.S.A., 63
Vivendi Universal Games, 391
VMX Incorporated, 240
Vogel, Harold L., 266–267, 336, 348, 413
Voice mail, 240
  availability of, over cell phones, 255
  options, 261
Voice network, 487
Voice over IP (VOIP), 249, 251
Voice signal, 486
VoiceStream, 211

Waldfogel, Joel, 291
Wall Street, 7
Wall Street Journal, 321, 322
Walt Disney Company, 270, 276, 277, 423, 466
WAP, 212
Warner Brothers, 386, 423
Warner Communications, 417
Warner Music Group, 279, 394
Washington Post, 321, 322
Wasser, Frederick, 359
Waunakee Telephone Company, 235
Wavelength Division Multiplexing (WDM), 489
WEA, 279, 394
Weaver, C. M., 82

Web, improved delivery of entertainment over, 460–461
Web aggregation, 106–107
Webcasting, 348, 462
Web-enabled magazine, 331
WebTV project, 363
Wells Fargo, 105, 106
West, 296
Western Electric, 202, 203, 230, 445
  automated equipment at, 237
  employment level of, 196
  manufacturing activities of, 209–210
Western Savings Fund Society of Philadelphia, 82
Western Union SICOM service, 168
Westlaw, 290
Wharton School Publishing, 313
Wheat, Willis J., 85
Wide-area telephone service (WATS) lines, 433
Wideband, 488
Wikler, Janet, 308
WiMax, 256
Winstar, 218
Wireless business
  expansion of, 204
  segmentation of equipment vendors, 253
Wireless telecommunications, 268
Wireless telephony, adoption of, 219
Wisconsin Telephone Company, 233
Wolf, Michael J., 412, 441, 466–467
Women
  in Banking Industry, 39
  use of digital cameras by, 436
  Video Gaming Industry and, 428
Wonder, Stevie, 401
Worcester County (MA) National Bank, 53
Word processing
  in Insurance Industry, 127–128
  in Newspaper Industry, 317
Workforce
  changing nature of skills in, 465
  in Manufacturing sector, 477
Workman's compensation, 118
Work practices, adaptation of, 453
WorldCom, 203, 205, 216, 218, 220
World Trade Center, 185
World Wide Web (WWW), 245–246, 364–365
Wozniak, Stephen, 419
Wriston, Walter, 229
WVCG Music Magazine, 341

Xerographic machines, 333
Xerox, 209

Y2K
  in Banking Industry, 101
  in Brokerage Industry, 179–180, 185
  in Insurance Industry, 149–150
Yahoo!, 105, 288, 290, 348, 429
Yamaha CS-80, 401
Yashica, 435
Yates, JoAnne, 138, 139, 148

Zenith, 358
Ziff-Davis, 306, 422
Zines, 326